# Asymptotic Theory of Supersonic Viscous Gas Flows

# Asymptotic Theory of Supersonic Viscous Gas Flows

**V. Ya. Neiland, V. V. Bogolepov, G. N. Dudin, and I. I. Lipatov**

ELSEVIER

AMSTERDAM • BOSTON • HEIDELBERG • LONDON • NEW YORK • OXFORD
PARIS • SAN DIEGO • SAN FRANCISCO • SINGAPORE • SYDNEY • TOKYO
Butterworth-Heinemann is an imprint of Elsevier

B H

Butterworth-Heinemann is an imprint of Elsevier
Linacre House, Jordan Hill, Oxford OX2 8DP, UK
30 Corporate Drive, Suite 400, Burlington, MA 01803, USA

First edition 2008



**British Library Cataloguing in Publication Data**
A catalogue record for this book is available from the British Library

**Library of Congress Cataloging-in-Publication Data**
A catalog record for this book is available from the Library of Congress

ISBN-13: 978-0-7506-8513-9

For information on all Butterworth-Heinemann publications
visit our web site at books.elsevier.com

Printed and bound in Hungary
08 09 10 10 9 8 7 6 5 4 3 2 1

Typeset by Charon Tec Ltd (A Macmillan Company), Chennai, India
www.charontec.com

*Dedicated to the 100th anniversary of the creation of theory of boundary layer by L. Prandtl and the 85th anniversary of the foundation of the Joukowski Central Aero-hydrodynamics Institute (TsAGI)*

# Contents

# Preface

An important role in fluid mechanics is played by flows with high Reynolds numbers. So far, the solution of the Navier–Stokes equations governing viscous gas flows presents considerable difficulties even when using modern computer technologies, though this line of inquiry has been characterized by certain successes. However, precisely high-Reynolds-number flows are most complicated and cumbersome for numerical solution. Moreover, the results of numerical studies are in a sense similar to experimental data: they require a theoretical analysis, the construction of models of phenomena and similarity laws, etc. For this reason, so far it is classical Prandtl's theory of boundary layer (Prandtl, 1904) that has been most commonly used. In this case, it is assumed that, since the Reynolds number (*Re*) is high, the viscous terms of the Navier–Stokes equations are unessential throughout almost the entire flowfield, except for small flow regions whose thicknesses reduce as *Re* increases. The outer inviscid flow is governed by the Euler equations. Their solution gives a part of boundary conditions for the boundary layer equations.

The classical theory of boundary layer has made a very strong impact on the development of fluid mechanics. In his article summing up the main achievements of the research in this field in the twentieth century Sir James Lighthill compared the importance of Prandtl's report (1904) with the influence of Einstein's work (1905) on the development of physics.

An essential assumption of boundary layer theory is the smallness of the longitudinal, or streamwise, gradients of flow parameters as compared with the transverse ones. For this reason, the Prandtl equations do not involve the higher-order derivatives with respect to the longitudinal variable, so that the equations are parabolic, which makes the solution of problems appreciably easier. Later Prandtl formulated the concept of successive refinement of the results equivalent to the theory of weak interaction between the outer inviscid flow and the boundary layer. From the solution of the Euler equations subject to the boundary condition of impermeability, the boundary conditions for the boundary layer equations can be obtained. Then the problem for the boundary layer is solved and corrections to the boundary conditions for the outer inviscid flow are determined, etc. It was assumed that this process of successive refinements of the solution can be convergent; later, a new term, namely, the boundary layer theory of second approximation, was introduced.

However, there exists a wide class of flows to which the classical boundary layer theory is inapplicable. These include, for example, the flows in the regions with a large local curvature of the body contour, vicinities of the boundary layer separation and reattachment points, points of shock incidence on boundary layers, separation zones of different kind, etc. The same category includes a wide class of problems which cannot be described within the framework of classical Prandtl's theory owing to the presence of other scales which are not inherent in boundary layer theory.

Problems associated with these flows can frequently acquire new physical and mathematical properties such as, for example, upstream disturbance propagation in spite of the fact that the outer flow is supersonic, the necessity of solving simultaneously equations for different regions interrelated by boundary conditions, large values of the longitudinal gradients of the flow parameters, etc. In these problems, the interaction between different flow regions is in no way weak either on the entire body surface (see, e.g., Chapter 4) or locally, near some singular places.

A particularly important case of strong local interaction is the flow near boundary layer separation points (see, e.g., Chapter 1).

Experimental investigation of the supersonic flows of this type got under way more than 60 years ago. Thus, as early as in 1940, Ferri (1940) noted that upon the interaction of a shock with the boundary layer on an airfoil in a supersonic flow, boundary layer separation is observable at a point, where, in accordance with the theory of inviscid flows, the pressure gradient should be negative. In the first systematic experimental studies of different separated flows (Leipman, 1946; Barry et al., 1951; Bogdonoff and Kepler, 1955; Gadd, 1957*a*; Greber et al., 1957; Chapman et al., 1958; Petrov and Bondarev, 1960; Bondarev and Petrov, 1960) the data on the general flow properties and the gas dynamic patterns, the nature of the pressure distribution, the critical pressure difference resulting in boundary layer separation, and the effect of the Mach and Reynolds numbers and the laminar–turbulent transition position on the separation zone dimensions were obtained for relatively simple flows.

One of the most important phenomena detected in the experimental studies was upstream propagation of pressure disturbances upon the incidence of a shock on a boundary layer. In first theoretical studies, an attempt was made to reduce the explanation of this phenomenon to a "purely" inviscid mechanism of disturbance propagation through the subsonic region of the boundary layer. With this purpose, Howarth (1948) considered the problem of two adjoining uniform semi-infinite flows, of which one was subsonic and the other supersonic, but both were inviscid. A small pressure disturbance was incident from the supersonic flow onto the interface and propagated in the subsonic flow thus exerting an inverse effect on the supersonic part of the flow. In the study of Tsien and Finston (1949), Chernyi G.G. (1952) the subsonic flow was not semi-infinite but bounded from below by a wall. In the same formulation of the problem, Lighthill (1950) took into consideration velocity nonuniformity in the subsonic inviscid flow. This approximate formulation of the problem did not include some essential features of the interaction process and its results were not in agreement with the experimental data for laminar boundary layers.

In the subsequent years, the studies devoted to separationless and, especially, separated flows of the type considered, both supersonic and hypersonic, grew sharply in number. These works are reviewed in the publications of Neiland and Kukanova (1965), Lapin et al. (1970), Golubinskii et al. (1973), Charwat (1970), Brown and Stewartson (1969), and Fletcher and Brigg (1970). Later there appeared the monograph of Chang (1970) which presents a vast survey of materials concerning different aspects of the separated flow problem.

So vast a material could hardly be reviewed in the preface; for this reason, here we will discuss the basic available methods of calculation and consider the results pertaining to the problems outlined in this book.

As noted above, there exist many studies containing different approximate methods for calculating separated and separationless supersonic flows with upstream disturbance propagation and interaction effects taken into account. However, for the most part they elaborate only a few lines of inquiry. One of these lines is associated with the use of integral boundary layer equations. The problem of the interaction, either separated or separationless, between the viscous flow region and the outer inviscid supersonic flow reduces to the integration of a system of nonlinear ordinary differential first-order equations. These equations can be derived by formal integration of the boundary layer equations over the transverse direction. They involve certain integral parameters of the boundary layer, such as the displacement, momentum, energy thicknesses, etc. Moreover, the system includes one more equation determining the relation between the pressure distribution in the inviscid supersonic flow and the viscous flow region displacement thickness. The information on the shapes of the velocity and enthalpy profiles across the boundary layer turns out to be lost and must be postulated in the form of certain families of curves dependent on some free parameters, the number of which is equal to the number of equations governing their distributions in the streamwise direction. The choice of a particular family of parameter profiles across the boundary layer is of great importance for obtaining satisfactory results. The unique criterion of the quality is the agreement with the experimental data.

The first, more or less successful application of this approach to the interaction and separation problems was the "mixing" theory suggested by Crocco and Lees (1952). Within the framework of this theory, it was proposed to consider a wide range of flows with interaction of laminar and turbulent boundary layers with shock waves, flows in overexpanded nozzles and in the base regions behind bodies, flows over different steps, etc. (Crocco and Lees, 1952; Tyler and Shapiro, 1953; Crocco, 1955; Bray, 1957; Hammit, 1958; Gadd and Holder, 1959; Bray et al., 1961; Glick, 1962; Rom, 1962; Vasiliu, 1962). The process controlling the interaction is assumed to be the mixing on the interfaces of different regions accompanied by momentum and mass transfer; hence the name of the theory.

In this theory, as distinct from the Pohlhausen method, the pressure gradient is not assumed to be uniquely related with the velocity profile. Therefore, it is necessary to specify an additional relation, which is preassigned in the form of a certain "universal" function of "mixing" determined on the basis of the experimental data (Crocco and Lees, 1952; Glick, 1962; Rom, 1962). Though the Crocco–Lees theory made it possible frequently to obtain qualitatively true description of phenomena, it turned out to be rather coarse and, which is most important, required the use of the experimental data.

In order to make the use of experimental data unnecessary, profiles in the form of polynomials (Pohlhausen method), families of self-similar solutions for the boundary layer including return flows (Stewartson, 1954), etc., can be taken for determining the relations between the integral parameters. This approach was used in several studies for analyzing both particular flow regions and the separation zone as a whole (Gadd, 1960; Curle 1961; Makofski, 1963; El'kin and Neiland, 1965; Bondarev, 1966; Zukoski, 1967). The comparison of the calculated results with the experimental data shows that in this case good agreement can be obtained for separationless flows and flows with small-sized separation zones over not-too-cooled surfaces. However, a rigid relationship between the velocity profile shape and the pressure gradient makes impossible the description of extended separation zones.

Both lines described above were naturally generalized by using additional higher-order moments of the boundary layer equations, which made it possible to avoid a rigid relationship between the velocity profiles and the local pressure gradient and the use of empirical functions (Abott et al., 1963; Lees and Reeves, 1964; Reeves and Lees, 1965; Webb et al., 1965; Grange et al., 1967). In very simple cases, the integral momentum equation and the first moment of the momentum equation are used. As before, the successful use of the method, which is often called the Lees–Reeves method, depends on the choice of one or another family of profiles. Thus, polynomials were used in the paper of Abott et al. (1963) for studying the shock/boundary layer interaction. However, this choice turned out to be unsuccessful. Lees and Reeves used the profiles of the families of the self-similar solutions of the boundary layer equations, including return flows, tabulated in the study of Cohen and Reshotko (1956). These profiles were considered as convenient functions for calculating the integral parameters rather than the solutions of the boundary layer equations; because of this, the relation between the local pressure gradient and the profile, which takes place in self-similar solutions, was dropped. In a very simple case, this approach resulted in three ordinary differential equations with three free parameters; the uniqueness of the solution was ensured by the condition that it passes through a singular saddle-type point located in the reattachment region.

One more group of studies (Nielsen, 1965; Holt, 1966; Crawford and Holt, 1968; Holt and Meng, 1968; Nielsen et al., 1968) used the boundary layer equations, together with the method of integral relations developed for separationless flows by Dorodnitsyn (1962). In this case, the boundary layer equations transformed to the Dorodnitsyn variables are multiplied by weight functions dependent only on the longitudinal velocity component $u$ and integrated across the boundary layer. After the passage to the independent variable $u$, the integrand is represented in the form of a function of $u$ and some free parameters, the number of which is equal to that of the equations obtained. It is interesting to note that in this method the separation zone is also terminated by a singular saddle-type point ensuring the choice of the solution (Holt and Meng, 1968). Since there is an infinite set of weight functions, for example, $(1-u)^n$, where $n = 1, 2, \ldots$, this approach apparently makes it possible to increase the accuracy of the solution for the separationless boundary layer by increasing $n$; in this sense, the approach belongs to the rational approximation methods, that is, it assumes the existence of a formal process for refining the results. However, for separated flows and flows with interaction the problem is not described by the Prandtl equations throughout the entire flow region. For this reason, the Nielsen–Holt method, as well as other methods considered above, does not ensure the proximity of the solution thus obtained to the solution of the Navier–Stokes equations in the $n \to \infty$ limit, that is, it is not a rational approximation to the problem solution in the meaning of Van Dyke's definition (1964).

One more line of inquiry of supersonic separated flows is associated with the separated flow model suggested by Korst and Chapman (Chapman, 1951; Korst et al., 1955; Korst, 1956; Chapman et al., 1958). The original basic form of the separated flow model is fairly simple and requires a comparatively small amount of calculations. It is assumed that the gas flow inside the separation zone can be neglected, while the flow in the mixing zone can be considered to be isobaric. The principal point of the theory is the assumption that at the mixing zone termination the stagnation pressure on the streamline arriving to the rear stagnation point is equal to the static pressure in the inviscid flow immediately behind the reattachment region. This approximation made it possible to obtain the results for base separation zones (Chapman, 1951; Bondarev and Yudelovich, 1960; Minyatov, 1961; Tagirov, 1961; El'kin et al., 1963; Baum et al., 1964; Neiland and Sokolov, 1964; Neiland, 1965), as well as separation zones ahead of flaps and forward separation zones (Brower, 1961; Neiland and Taganov, 1963*a* and *b*; Neiland, 1963; 1964; 1965*a*; Zavadskii and Taganov, 1968). Further investigations showed that the basic assumption leads to noticeable errors, particularly for turbulent flows. Attempts to make allowance for the effect of the initial thickness of the mixing zone and to introduce a correction to the Korst–Chapman criterion for the reattachment region were made in the studies of Kirk (1959), Nash (1963), Baum et al. (1964), Siriex et al. (1966), Tagirov (1966), and Neiland and Sokolov (1967). Obviously, even with account for the corrections, the nature of the theory is not in essence modified, though using semi-empirical methods of this type can be helpful for practical purposes and estimations of the flows with developed separation zones.

The use of approximate semi-empirical methods and different models based on considerably simplifying assumptions, such as, those described above, is undoubtedly justified and helpful in studying complicated flows which could hardly be described by the classical boundary layer theory. However, in principle, this approach cannot be recognized as satisfactory. These approximations do not infer any process of refining the results or a passage to a limit in which the solution tends to the exact solution. The relation between these results and the solution of the Navier–Stokes equations at high Reynolds numbers remains indefinite.

For laminar flows it is possible to use asymptotic methods departing from the Navier–Stokes equations. An increase in the Reynolds number leads ultimately to laminar–turbulent transition. However, the study of the limiting form of laminar flows and the corresponding solutions of the Navier–Stokes equations, as $Re \to \infty$, ensures the better understanding of the nature of these flows at high subcritical values of *Re*. Moreover, it makes possible the development of methods of calculation based on expansions in a small parameter, as $Re \to \infty$. The first and most important example of the application of the asymptotic approach is, in essence, furnished by classical Prandtl's theory itself, though it was primarily developed on the basis of physical considerations and an analysis of certain simple solutions (Prandtl, 1904). Later, using asymptotic methods turned out to be very helpful for obtaining higher-order approximations in boundary layer theory (Imai, 1957; Goldstein, 1960; Van Dyke, (1964); Brailovskaya and Chudov, 1967).

In the process of searching the scales for flow regions and functions some estimates obtained on the basis of "physical" considerations and the Navier–Stokes equations are systematically applied. This approach is quite similar to that of Prandtl in his fundamental work of 1904. We note that in further construction of the asymptotic expansions the correctness of the estimates obtained is to a certain degree checked when the asymptotic expansions are matched. As shown in what follows, this approach makes it frequently possible to construct the classification of possible flow patterns, which is particularly important in the cases in which the passage to the limit is carried out in several small parameters.

We note that the investigation of laminar flow regimes is not only of fundamental theoretical importance but can also give answers important for applications. Thus, for reentering orbiters of the type of *Shuttle* or *Buran* a heat flux peak is reached on the orbiter nose bluntness, where the boundary layer is laminar.

Several asymptotic methods are used in aerohydrodynamics. These are the matched asymptotic expansion method (or the method of internal and external expansions) developed by Friedrichs (1953; 1955), Kaplun (1954), Lagerstrom and Cole (1955), Lagerstrom (1957), and others; the method of boundary layer corrections proposed by Vishik and

Lusternik (1957; 1958), the Poincaré–Lighthill–Kuo method (Poincaré, 1892; Lighthill, 1949; Kuo, 1953), the method of many scales (Mahony, 1962), and some others; a brief description of these methods can be found, for example, in the books by Van Dyke (1964) and Cole (1968).

Typical for the problems, in which conventional methods of a small parameter cannot be applied, is the appearance of flowfield domains, where the ratio of the generalized scale lengths either vanishes or increases without bound as the smallness parameter tends to zero (e.g., the ratio of the characteristic boundary layer thickness to the longitudinal dimension of the body). This fact leads to different behavior of the solution in the regions, in which the flow is determined by the length scales of different orders with respect to the smallness parameter. For this reason, the conventional method of a small parameter does not provide the correct description for the entire flow region.

Asymptotic methods have found application in the majority of divisions of aerohydrodynamics. They were used for solving the problems of inviscid flows over the entire $M \in [0, \infty)$ range, in wing theory, in sonic boom theory, in viscous flow theory for $Re \to 0$ and $Re \to \infty$, and, finally, for the flows of radiating, relaxing, and, generally, real gases.

In this book, the matched asymptotic expansion method is used for solving different problems. In this case, additional asymptotic expansions are constructed in the flow regions with different length and flow parameter scales and then the matching principle is applied; different forms of this principle are discussed, for example, in the book of Van Dyke (1964). In all cases it is the boundary value problem for the Navier–Stokes equations with "natural" boundary conditions on the body and in the freestream that lies at the root. The emphasis is placed on an analysis of the limiting asymptotic flow structure as $Re \to \infty$, the derivation of the systems of equations and boundary conditions governing the flow in different characteristic regions, the solution of these problems, and the establishment of approximate limiting similarity laws, where it is possible.

In the first chapter we consider supersonic viscous high-Reynolds-number flows in the regions of strong local interaction between the supersonic flow and the boundary layer, where the pressure gradient induced by the variation of the boundary layer displacement thickness affects the flow in the viscous region even in the first approximation.

The problems of the viscous flow region and the outer inviscid supersonic flow cannot be separated and should be solved simultaneously. The most known flow of this type is the "free interaction" region located in the vicinity of the point of boundary layer separation from a smooth surface. (In what follows it is shown that some other flows or their regions pertain also to this type.)

Many properties of flows near separation points were established in experimental studies (Gadd, 1953; Gadd et al., 1954; Bogdonoff and Kepler, 1955; Chapman et al., 1958). It turned out that if the separation zone is developed, that is, contains a clearly defined region with a near-constant pressure, or a pressure "plateau", then in a certain vicinity of the separation point the flow is almost independent of the type of the separation-producing disturbance (a bend in the body contour, a step, a shock incident from outside, etc.). The boundary layer prehistory also influences the flow in the vicinity of the separation point only via the integral characteristics of the flow ahead of the beginning of the disturbed preseparation region. Naturally, this led to a hypothesis that the flow is chiefly determined by the local interaction of the boundary layer and the inviscid supersonic flow. In the paper of Chapman et al. (1958) the term "free interaction" was proposed. The mechanism of the upstream disturbance propagation received the following approximate explanation: let the pressure increase somewhat, then the boundary layer displacement thickness also increases. In turn, this results in a rise of the pressure. Thus, consistent disturbances of the pressure and the displacement thickness $\delta^*$ are attained.

An analysis of the flow near the separation point was drawn both separately for this region and simultaneously with the calculation of the entire separation zone using different approximate methods (Lees, 1949; Crocco and Lees, 1952; Crocco, 1955; Gadd, 1957*b*; Chapman et al., 1958; Hakkinen et al., 1959; Curle, 1960; Erdos and Pallone, 1962; Glick, 1962; El'kin and Neiland, 1965; Nielsen, 1965; Reeves and Lees, 1965). Emphasis was placed on the establishment of separation "criteria", that is, the relations between the pressure coefficients at the separation point and in the "plateau" region and the parameters of the undisturbed boundary layer and the inviscid flow. A fundamentally important step (apart from the demonstration of the fact that the flow is independent of distant regions) was made in the study of Chapman et al. (1958), in which an analysis was based on the estimation of the orders of local flow parameters.

In the work of Erdos and Pallone (1962) these ideas were used for deriving correlation formulas for the pressure distribution. Correlation formulas for different approximate schemes were also proposed in the study of Hakkinen et al. (1959) and some others. In the study of Lewis et al. (1968) the experimental data on the pressure distribution in the free interaction region were plotted in the variables proposed in the work of Curle (1960).

Lighthill (1953) considered the upstream disturbance propagation and, departing from the idea of the disturbance propagation through a subsonic sublayer developed in the earlier works of Howarth (1948), Tsien and Finston (1949), and Lighthill (1950), supplemented the model of the study of Lighthill (1950) by taking viscosity near the body surface into account. However, though in this case the solution was qualitatively more realistic (e.g., a correct estimate for the longitudinal scale of the interaction region was obtained), the linearized Lighthill solution remained inadequate near the separation point, since precisely near the body surface the undisturbed flow velocity is low and velocity disturbances are of the same order as the undisturbed velocity itself.

We will also dwell upon the general features of the work performed within the framework of the Crocco–Lees, Lees–Reeves, and Nielsen–Holt theories, as well as other theories using the integral boundary layer equations (Crocco and Lees, 1952; Tyler and Shapiro, 1953; Stewartson, 1954; Crocco, 1955; Bray, 1957; Hammit, 1958; Gadd and Holder, 1959; Gadd, 1960; Bray et al., 1961; Curle, 1961; Click, 1962; Rom, 1962; Vasiliu, 1962; Makofski, 1963; El'kin and Neiland, 1965; Bondarev, 1966; Zukoski, 1967; Abott et al., 1963; Lees and Reeves, 1964). In this case, the free interaction region is not considered separately, which makes difficult the establishment of similarity laws. The available calculated results for thermally insulated and not-too-cooled flows are in fairly good agreement with the experimental data. The ultimate judgment on the accuracy of such theories is made difficult by the fact that they involve rather arbitrary assumptions on the velocity profile shapes. However, even in this stage it can be noted that the integral representation of the profiles can lead to an inadequate description of flows, in which an important role is played by the narrow wall sublayer. Thus, in the study of Nielsen et al. (1968) it was shown that within the framework of the Nielsen–Holt method the free interaction mechanism no longer works for $T_w/T_0 < 0.133$. In the study of Ai (1967) it was shown within the framework of the Lees–Reeves method that, depending on the assumptions made on the profiles, the same flow can turn out to be both "subcritical" and "supercritical", according to Crocco's terminology (Crocco, 1955), that is, it either transmits disturbances upstream or not.

Other anomalies were investigated in the paper of Ai (1967). Using integral methods reduces in a more or less appropriate but always rather arbitrary way the description of the upstream transfer of disturbances to the properties of ordinary differential equations, which is not completely adequate to the physical problem under consideration even in the first approximation.

At the second half of the 1960s, a new line of inquiry began to be developed; it pertains to problems that cannot be described by the classical boundary layer theory but, like Prandtl, uses the asymptotic analysis of the solutions of the Navier–Stokes equations. This monograph is devoted to the results obtained within the framework of precisely this approach.

The first chapter of this book describes a class of flows characterized by the "free interaction" between a supersonic flow and a boundary layer. After the original work of Neiland (1968*a* and 1969*a*) had been published, in the paper of Stewartson and Williams (1969) a popular name *triple-deck flows* was proposed. We note that at our glance it is not quite successful, since it can pertain to flows of completely other types which also involve a three-layer structure (see Chapter 3 of this book and the paper of Neiland and Sychev, 1966).

Later, the same type of the solutions was used for describing laminar incompressible separated flows in the works of Sychev (1972) and Smith (1977).

In Sections 1.1 and 1.2 of this book, based on the results of the works of Neiland (1968; 1969*a*), the problem of the flow near a separation point is considered on the basis of an analysis of the limiting, as $Re \to \infty$, behavior of the solution of the Navier–Stokes equations. It is shown that in the vicinity of the separation point of length $O(Re^{-3/8})$ and with $\Delta p \sim Re^{-1/4}$ three layers of thicknesses of the orders $Re^{-3/8}$, $Re^{-1/2}$, and $Re^{-5/8}$ should be considered. In the first approximation, the outer region is described by the linear theory of supersonic flows. In the region of thickness $O(Re^{-1/2})$ the flow is locally inviscid, since the local pressure gradient there is greater in the order than the viscous terms. The most important role is played by the wall layer of thickness $O(Re^{-5/8})$, in which the disturbances of the velocity and the streamtube thickness are of the same order as their initial values at the boundary layer bottom in the undisturbed flow zone ahead of the interaction region. The problem is nonlinear. In the viscous sublayer, the flow is governed by the incompressible boundary layer equations. For this flow, the initial and boundary values are determined using the asymptotic expansion matching principle. Most important is that the pressure gradient distribution is not preassigned but should be determined in the process of the solution simultaneously with the solution of the wave equation governing the outer inviscid supersonic flow. This is equivalent to the substitution in the boundary layer momentum equation of the pressure gradient determined from the Ackeret formula which involves the slope formed by the viscous sublayer displacement thickness as a local body slope. (In the first approximation the variation of the streamtube thickness in the intermediate sublayer is out of order.)

Thus, the simultaneous solution of the Prandtl equation, which is parabolic for a given pressure distribution, and the hyperbolic wave equation interrelated via the boundary condition of interaction leads to the appearance in the boundary layer momentum equation of a term with the second derivative of the unknown function with respect to the longitudinal coordinate under the sign of integral with respect to a coordinate normal to the surface. This leads to the appearance of single-parameter families of solutions associated with compression and expansion flows. The solutions for compression flows pass through the separation point. Preassigning the separation point position makes it possible to single out the required solution. Since the separation point position is dependent on the flow regions located downstream of it, this corresponds to upstream transfer of disturbances. Expansion flows are considered later in connection with the problems of the expansion flow around a corner point on a body contour (Section 1.6 and the studies of Neiland (1969*b*; 1971*b*)).

The boundary value problem was solved numerically. The results obtained are in good agreement with the experimental data of other authors on both the pressure coefficients at the separation point and in the plateau region and the pressure distribution over the body. The local nature of the flow makes it possible to introduce similarity variables for all flow parameters. In these variables, the expression for the pressure coefficient coincides with that obtained previously in the studies of Chapman et al. (1958), Erdos and Pallone (1962), and others. For the longitudinal coordinate the form of the similarity variable is similar to those presented in the works of Chapman et al. (1958), Curle (1960), and Lewis et al. (1968).

In Sections 1.3 and 1.4 certain relatively simple steady flows with separation zones are considered. In the problem considered in Section 1.3, the separation point is located at a distance $\ell$ from the leading edge of a flat plate, while the separation zone is semi-infinite. The Reynolds number based on $\ell$ increases without bound, $Re \to \infty$. In this case, near the separation point there is a flow region with free interaction. In this region, $\Delta p \sim Re^{-1/4}$ and $\Delta x \sim Re^{-3/8}$. It is shown that downstream of the free interaction zone, as $\Delta x / Re^{-3/8} \to \infty$, the flow also breaks up into a mixing zone, an inviscid return flow region, and the boundary layer on the plate surface. For $Re^{-3/8} \ll \Delta x \ll 1$ the solutions are obtained in an analytical form. For $\Delta x \sim 1$ the solution is non-self-similar; however, using the results of the numerical solution obtained in the work of Chapman (1950) the solution for the inviscid return flow can be obtained in the quadrature form. As $x \to +\infty$, the solution approaches that determined previously for the separation zone beginning at the leading edge of the plate.

In Section 1.4 the solution is obtained for very simple semi-infinite separation zones which do not include reattachment regions (Neiland, 1971). The separation zone occurring near the leading edge of the plate is considered. In this case, the outer supersonic flow streams over a "liquid wedge" adjoined by a self-similar mixing zone rather than a "boundary layer".

The main part of the separation zone is filled with an inviscid return flow starting from the state of rest as $x \to +\infty$; this flow is driven by gas suction into the mixing zone. Since on the plate surface the no-slip conditions must be fulfilled, near the surface there is the boundary layer of the return flow which also starts as $x \to +\infty$. This solution of the boundary layer equations can exist in the case of an outer flow accelerating from the state of rest. The solutions for all three zones (mixing zone, inviscid return flow, and boundary layer on the plate) are self-similar. The solution derived in Section 1.4 resembles the self-similar solutions for turbulent flow zones obtained in the studies of Dem'yanov and Shmanenkov (1960) and Neiland and Taganov (1961) for the forward separation zones. However, there are considerable differences in these solutions. Thus, within the framework of the semi-empirical theories used, the return flow in turbulent separation zones turns out to be uniform, which makes impossible the satisfaction of the no-slip conditions if there is a body at the bottom of the inviscid return flow.

In Chapter 2 other types of flows described by free interaction theory are considered.

In Section 2.1 boundary layer flows, in which friction is small due to the action of the external pressure gradient, which is not connected with the free interaction process, are considered. In this case, the local free interaction zone is somewhat different from that considered above.

In Section 2.2 we consider the second family of solutions for the conventional free interaction but for expansion flows. This makes it possible to construct the expansion flow ahead of a corner point on a body in a supersonic stream for small angles of deflection or the far asymptotics of disturbance decay if the angle of turn is large. The same solutions describe the flows near the base section of a body for the corresponding values of the base pressure.

In Section 2.3 the solution of theory of flows with free interaction are used for describing a flow region, where a weak shock is incident on the boundary layer, or a flow around a corner close to $\pi$.

In Section 2.4 free interaction theory is used for describing how boundary layer separation could be prevented with the aid of slot suction of gas.

Chapter 3 is devoted to flows with strong local interaction, in which pressure gradients are finite, $O(1)$, rather than small, $O(Re^{-1/4})$, as in theory of free interaction.

Section 3.1 presents the results of the paper of Neiland and Sychev (1966) which was apparently the first work in which the structure of the asymptotic solution of the Navier–Stokes equations is of the three-layer type. In that work, an expansion flow over a region of the body contour turn by a finite angle $O(1)$ was considered, the body being in a supersonic stream. The radius of curvature of the contour is assumed to be of the same order as the boundary layer thickness. For this reason, the induced pressure gradient is high and the interaction is not free, since it is dependent on the body contour shape.

The three-layer (triple-deck) nature of the flow is due to the fact that in the main part of the boundary layer viscous stresses are of the same order as in the region of the boundary layer upstream of the turn region, whereas the orders of the pressure gradient and the inertial terms of the Navier–Stokes equations increase sharply. Thus, in the main part of the boundary layer the flow becomes locally inviscid. However, in the near vicinity of the wall a very thin viscous sublayer is developed owing to the fact that the no-slip conditions are imposed on the wall.

A similar flow structure is studied in Section 3.2 for expansion flows ahead of a base of a body, if the base pressure is by an order lower than that on the lateral surface of the body. However, in this case the pressure difference cannot be greater than a value corresponding to a certain "choking" condition considered in Section 3.2.

The nature of many separated flows depends in a considerable degree on how the flow reattaches the body or jets coalesce behind the base section. This determined a successful application of approximate semi-empirical theories based on the Chapman–Korst criterion. In applied problems, local peaks of heat fluxes related with reattachment regions are of great importance. Because of this, in Section 3.3 we draw a detailed asymptotic analysis of the Navier–Stokes equations. It is shown that the well-known Chapman–Korst criterion coincides with the first term of the expansion, while the corrections to the pressure difference are of the same order as the critical pressure difference in free interaction theory.

Sections 3.4 and 3.5 are devoted to the study of the asymptotic structure of the flows near the discontinuities in the boundary conditions.

In Chapter 4 we consider the flows in which the interaction of the outer supersonic flow with the boundary layer is non-weak along the entire length of the body. In Section 4.1 the estimates of free interaction theory are extended to the weak hypersonic interaction regime and in Section 4.2 it is shown that disturbances are transmitted along the entire body length when the interaction is not weak. In Section 4.3 the solutions of the problem are obtained for the strong hypersonic interaction regime and the well-known self-similar Lees–Stewartson solution is shown to be nonunique, since there exists a single-parameter family of solutions, which makes it possible to take account of the effect of the boundary conditions imposed on the trailing edge of a finite-length plate. The similarity laws established correlate well with the experimental data.

In Sections 4.4–4.7 supersonic flows with strong gas injection through the body surface are considered. The regions of boundary layer detachment from the body surface are considered, together with the nature of the upstream disturbance transfer.

In Chapter 5 we consider three-dimensional flows over wings of different planforms in the regime of non-weak hypersonic interaction. The problem of the nature of the upstream disturbance transfer from the plane of symmetry and the trailing edge of the wing are studied.

In this book particular attention is given to the regimes of two-dimensional (Chapter 6) and three-dimensional (Chapter 7) flows, in which the flow nature and the direction of the disturbance propagation in the interaction of the boundary layer with the supersonic inviscid flow can vary depending on the temperature factor. The influence of the temperature factor is exerted via the shape of the Mach number profile across the boundary layer.

It is shown that, depending on a certain mean-integral value of the Mach number, boundary layers can exhibit properties similar to those of subsonic, supersonic, and even transonic streamtubes. Following the terminology derived by Crocco from qualitative physical considerations in using the integral boundary layer equations, such terms as subcritical, transcritical, and supercritical boundary layers are used.

The investigation of this class of problems on the basis of the asymptotic solutions of the Navier–Stokes equations was initiated by the work of Neiland (1987).

In the last chapter we perform a systematic study of the flows over small barriers and depressions on the boundary layer bottom. The classification of possible types of the corresponding two-dimensional and three-dimensional flows is constructed, the corresponding boundary value problems are formulated, and the similarity parameters are derived. It is shown, in particular, that in certain cases the friction and heat fluxes can vary in the leading order.

Thus, using heuristic estimates and applying systematically asymptotic methods for solving the Navier–Stokes equations the book studies a wide range of flows which cannot be described by the classical boundary layer theory.

The authors wish to thank their colleagues I. V. Vinogradov, Yu. N. Ermak, A. Kh. Karabalaev, A. A. Kovalenko, R. V. Krechetnikov, L. A. Sokolov, and V. N. Trigub for participation in the studies performed.

The authors are grateful to the Russian Foundation for Basic Research for the support of the publication of this book in Russian under grant no. 03-01-14235d.

# 1 Flow in the Regions of Free Interaction Between a Supersonic Flow and a Boundary Layer

In this chapter we consider flow regions in which the pressure distribution over the body surface depends on the variation of the boundary layer displacement thickness even in the first approximation. In turn, the induced pressure gradient also affects the boundary layer flow in the first approximation. On the main part of the body surface, as $Re \to \infty$ or $\varepsilon = (Re)^{-1/2} \to 0$, where $Re = \rho_\infty u_\infty \ell / \mu_\infty$ is the Reynolds number based on a certain scalelength and the freestream density, velocity, and dynamic viscosity, the pressure gradient is determined in the leading term by the body curvature and is of the order $O(1)$. On a flat plate it is induced only by the boundary layer and is of the order $O(\varepsilon)$. All flow functions are assumed to be dimensionless and normalized by their freestream values, except for the pressure which is scaled on the double dynamic head. The coordinates are divided by the characteristic dimension of the body.

Thus, for the main part of the body surface the boundary layer induced pressure gradient can appear in the boundary layer equations only in the higher-order approximations.

However, in the vicinity of regions with a large local curvature of the body contour, separation and reattachment points, and regions of shock/boundary layer interaction the situation can be different. The interaction can have a noticeable effect on the flow. Thorough and detailed experimental studies performed in the vicinity of the separation point of a laminar-developed separation zone (Gadd, 1953; Bogdonoff and Kepler, 1955; Chapman et al., 1958) showed that the flow is locally independent of the way by which flow separation is produced and depends chiefly on the local parameters of the boundary layer and the supersonic flow ahead of the interaction region. This enabled Chapman et al. (1958) to suggest the term "free interaction." In this chapter, as in the papers of Neiland (1968; 1969*a*, *b*; 1971*a*), the application of the asymptotic approach makes it possible to derive the equations and boundary conditions governing different regions of the flow of this type, as well as to generalize, refine, and substantiate the similarity law obtained previously from physical considerations.

It turned out that within the framework of the same theory a fairly large group of flows with small disturbance amplitudes ($\Delta p \ll 1$) can be described; these are the flows near corner points on body contours, points of incidence of shocks on boundary layers, etc.

Numerical calculations were carried out for flows near boundary layer separation points, flat-plate flows ahead of the base section, compression and expansion flows near corner points on body contours, and flows in the regions of boundary layer/weak shock interaction.

In studying flows with finite disturbance amplitudes, $\Delta p \approx 1$, these solutions are necessarily used for certain important zones, for example, in the problems of flow attachment to the body surface or flows near corner points on body contours (Chapter 3).

## 1.1 Derivation of the equations and boundary conditions

### *1.1.1 Estimates of the scales and characteristic values of the functions in disturbed flow regions*

For the sake of simplicity, we will first consider the laminar boundary layer on a flat plate. Let a region with a pressure disturbance $\Delta p \ll 1$ occur at a distance $O(1)$ from the leading edge. The disturbance can be due, for example, to the turn of the body contour by a small angle or the incidence of a weak shock (Fig. 1.1). Since in the boundary layer there is always a subsonic flow region, the disturbance will travel upstream (as well as downstream) through a distance $\Delta x \sim \lambda$, thus bringing about a variation $\Delta\delta^*$ in the boundary layer displacement thickness. In accordance with the linear theory of supersonic flows we have

$$\Delta p \sim \frac{\Delta\delta^*}{\lambda} \tag{1.1}$$

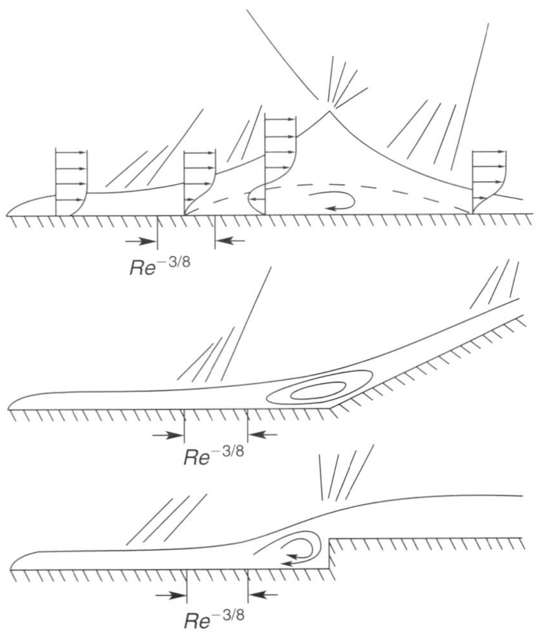


Fig. 1.1.

In the main part of the boundary layer of thickness $\delta \sim \varepsilon$, the velocity is $u \sim 1$ (region 2 in Fig. 1.2). For this reason, using the longitudinal momentum equation, together with the equations of continuity and state, the following estimates for the flow parameter disturbances along a fixed streamline can be easily obtained:

$$\Delta p \sim \Delta u^{(2)} \sim \Delta\rho^{(2)} \sim \frac{\Delta\delta^{*(2)}}{\delta}, \quad \delta \sim \varepsilon \tag{1.2}$$

Here, the superscript refers to the number of the flow region, in accordance with Fig. 1.2.

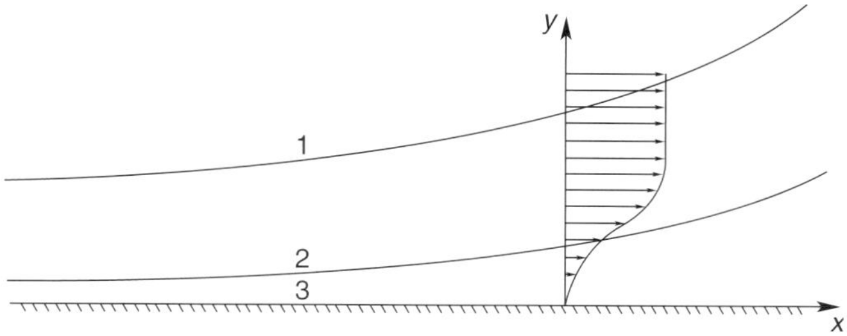


Fig. 1.2.

However, near the body surface, where the flow velocity vanishes for any $\Delta p$, there is always a region in which $\Delta u^{(3)} \sim u^{(3)}$. For this region, the longitudinal momentum equation gives a different estimate

$$u^{(3)} \sim \Delta p^{1/2} \sim \frac{\Delta \delta^{*(3)}}{\delta} \tag{1.3}$$

The latter estimate in Eq. (1.3) follows from the fact that in the undisturbed boundary layer, near the body surface we have $u \sim n/\delta$, where $n$ is the distance measured from the body surface. Comparing Eqs. (1.2) and (1.3) we can note that $\Delta\delta^{*(3)} \gg \Delta\delta^{*(2)}$. This means that in the disturbed flow region the total variation of the displacement thickness is $\Delta\delta^{*} \sim \Delta\delta^{*(3)}$. From the physical standpoint this can be easily understood if it is recalled that for $\Delta p \ll 1$ only slow near-wall streamtubes can vary strongly in thickness. Solving the equations for the scales (1.1)–(1.3) we arrive at the estimates

$$\Delta p \sim \Delta u^{(2)} \sim \Delta \rho^{(2)} \sim \left(\frac{\varepsilon}{\lambda}\right)^2, \quad \Delta\delta^{*(2)} \sim \frac{\varepsilon^3}{\lambda^2}, \quad u^{(3)} \sim \frac{\varepsilon}{\lambda} \tag{1.4}$$

$$\Delta\delta^{*} \sim \Delta\delta^{*(3)} \sim \frac{\varepsilon^2}{\lambda}, \quad \frac{\mathrm{d}p}{\mathrm{d}x} \sim \frac{\varepsilon^2}{\lambda^3}$$

It remains to note that these estimates no longer hold for $\lambda \leq \varepsilon$, since in this case the condition $\Delta p \ll 1$ is violated. Flows with $\Delta p \sim 1$ are considered in Chapter 3. Moreover, we will not consider flows in which the pressure gradient induced upon interaction is smaller in the order than that on the main part of the body surface, where it is of the order $O(1)$. Thus, we obtain

$$\varepsilon^{2/3} \gg \lambda \sim \Delta x \gg \varepsilon \tag{1.5}$$

The flows considered in this chapter are local ($\lambda \ll 1$) and the disturbance amplitudes are small ($\Delta p \ll 1$) but the gradients of the pressure and other functions must be large ($\partial p/\partial x \gg 1$).

### 1.1.2 Asymptotic representations, equations, and boundary conditions

In what follows, it is convenient to represent the Navier–Stokes equations in the von Mises variables

$$\rho u\frac{\partial u}{\partial x}+\frac{\partial p}{\partial x}-\rho v\frac{\partial p}{\partial \psi}=\varepsilon^2\left\{\rho u\frac{\partial}{\partial \psi}\left[\rho\mu u\frac{\partial u}{\partial \psi}+\mu\left(\frac{\partial v}{\partial x}-\rho v\frac{\partial v}{\partial \psi}\right)\right]\right.$$
$$\left.+\left(\frac{\partial}{\partial x}-\rho v\frac{\partial}{\partial \psi}\right)\left[\frac{4}{3}\mu\left(\frac{\partial u}{\partial x}-\rho v\frac{\partial u}{\partial \psi}\right)-\frac{2}{3}\rho\mu u\frac{\partial u}{\partial \psi}\right]\right\} \tag{1.6}$$

$$\rho u\frac{\partial v}{\partial x}+\rho u\frac{\partial p}{\partial \psi}=\varepsilon^2\left\{\rho u\frac{\partial}{\partial \psi}\left[\frac{4}{3}\rho\mu u\frac{\partial v}{\partial \psi}-\frac{2}{3}\mu\left(\frac{\partial u}{\partial x}-\rho v\frac{\partial u}{\partial \psi}\right)\right]\right.$$
$$\left.+\left(\frac{\partial}{\partial x}-\rho v\frac{\partial}{\partial \psi}\right)\left[\mu\left(\frac{\partial v}{\partial x}-\rho u\frac{\partial v}{\partial \psi}\right)+\rho\mu u\frac{\partial u}{\partial \psi}\right]\right\} \tag{1.7}$$

$$\frac{\partial n}{\partial x}=\frac{v}{u},\quad \frac{\partial n}{\partial \psi}=\frac{1}{\rho u} \tag{1.8}$$

$$\rho u\frac{\partial h}{\partial x}-u\frac{\partial p}{\partial x}=\frac{\varepsilon^2}{\sigma}\left[\rho u\frac{\partial}{\partial \psi}\left(\rho\mu u\frac{\partial h}{\partial \psi}\right)+\left(\frac{\partial}{\partial x}-\rho v\frac{\partial}{\partial \psi}\right)\left(\mu\frac{\partial h}{\partial x}-\rho\mu v\frac{\partial h}{\partial \psi}\right)\right]$$
$$+\varepsilon^2\mu\left[\left(\rho u\frac{\partial u}{\partial \psi}+\frac{\partial v}{\partial x}-\rho v\frac{\partial v}{\partial \psi}\right)^2+2\left(\frac{\partial u}{\partial x}-\rho v\frac{\partial u}{\partial \psi}\right)^2\right.$$
$$\left.+2\left(\rho u\frac{\partial v}{\partial \psi}\right)^2-\frac{2}{3}\left(\frac{\partial u}{\partial x}-\rho v\frac{\partial u}{\partial \psi}+\rho u\frac{\partial v}{\partial \psi}\right)^2\right] \tag{1.9}$$

Here, $\sigma$ is the Prandtl number and $h$ is the enthalpy divided by the square of the freestream velocity.

On the basis of the above estimates and the linear theory of supersonic flows, in region 1 we introduce the coordinates and flow functions of the order $O(1)$ (here and in what follows, subscripts refer to the variables of the order $O(1)$ within the corresponding flow region)

$$x_1=\frac{x-x_0}{\lambda},\quad \psi_1=\frac{\psi}{\lambda} \tag{1.10}$$

$$p\sim\frac{1}{\gamma M^2}+\left(\frac{\varepsilon}{\lambda}\right)^2 p_1+\cdots,\quad \rho\sim 1+\left(\frac{\varepsilon}{\lambda}\right)^2\rho_1+\cdots,\quad n\sim\lambda\psi_1+\frac{\varepsilon^2}{\lambda}n_1+\cdots$$

$$u\sim 1+\left(\frac{\varepsilon}{\lambda}\right)^2 u_1+\cdots,\quad v\sim\left(\frac{\varepsilon}{\lambda}\right)^2 v_1+\cdots,\quad h\sim\frac{1}{(\gamma-1)M^2}+\left(\frac{\varepsilon}{\lambda}\right)^2 h_1+\cdots$$

Here, $x_0$ is the coordinate of a point at which strong local interaction occurs (e.g., a separation point).

Substituting formulas (1.10) in the Navier–Stokes equations and passing to the limit $\varepsilon \to 0$ and $\lambda \to 0$ with $\varepsilon^{2/3} \gg \lambda \gg \varepsilon$ we obtain

$$\frac{\partial u_1}{\partial x_1} + \frac{\partial p_1}{\partial x_1} = 0, \quad \frac{\partial v_1}{\partial x_1} + \frac{\partial p_1}{\partial \psi_1} = 0, \quad \frac{\partial n_1}{\partial x_1} = v_1 \tag{1.11}$$

$$\frac{\partial n_1}{\partial \psi_1} = -\rho_1 - u_1, \quad \frac{\partial h_1}{\partial x_1} - \frac{\partial p_1}{\partial x_1} = 0$$

The equation of state gives the relation

$$h_1 = \frac{\gamma}{\gamma - 1} p_1 - \frac{1}{(\gamma - 1)M^2} \rho_1 \tag{1.12}$$

Equations (1.11) and (1.12) reduce to the wave equation

$$\frac{\partial^2 p_1}{\partial \psi_1^2} - (M^2 - 1)\frac{\partial^2 p_1}{\partial x_1^2} = 0, \quad \frac{\partial^2 v_1}{\partial \psi_1^2} - (M^2 - 1)\frac{\partial^2 v_1}{\partial x_1^2} = 0 \tag{1.13}$$

whose solution is well known.

We will write down relations valid at $\psi_1 = 0$ which will be necessary in what follows

$$\sqrt{M^2 - 1}\, p_{1w}(x_1) = v_{1w}(x_1) \tag{1.14}$$

Here, the subscript $w$ refers to the values of the variables at the body surface in the undisturbed boundary layer, directly ahead of the interaction region.

In deriving Eq. (1.14) it was taken into account that all disturbances decay as $x_1 \to -\infty$.

The solution should be modified somewhat in the case in which a pressure discontinuity of amplitude $[p_1]$ propagates from the outside along the $x_1 + \sqrt{M^2 - 1}\psi_1 = 0$ characteristic and is reflected at point $x_1 = 0$ from the line $\psi_1 = 0$ without a change of the pressure on it (as from a free surface). Then for the region to the right of point $x_1 = 0$, instead of relation (1.14) we obtain

$$\sqrt{M^2 - 1}\, p_{1w}(x_1) = v_{1w}(x_1) + 2\sqrt{M^2 - 1}\, |p_1|, \quad x_1 > 0 \tag{1.15}$$

For region 2 the asymptotic representations and the scales take the form:

$$x_2 = x_1 = \frac{x - x_0}{\lambda}, \quad \psi_2 = \frac{\psi}{\varepsilon} \tag{1.16}$$

$$u \sim u_{2,0}(\psi_2) + \left(\frac{\varepsilon}{\lambda}\right)^2 u_2 + \cdots, \quad v \sim \left(\frac{\varepsilon}{\lambda}\right)^2 v_2 + \cdots$$

$$n \sim \varepsilon n_{2,0}(\psi_2) + \frac{\varepsilon^2}{\lambda} n_{2,1} + \frac{\varepsilon^3}{\lambda^2} n_{2,2} + \cdots, \qquad h \sim h_{2,0}(\psi_2) + \left(\frac{\varepsilon}{\lambda}\right)^2 h_2 + \cdots$$

$$p \sim \frac{1}{\gamma M^2} + \left(\frac{\varepsilon}{\lambda}\right)^2 p_2 + \cdots, \qquad \rho \sim \rho_{2,0}(\psi_2) + \left(\frac{\varepsilon}{\lambda}\right)^2 \rho_{2,1} + \cdots$$

The terms $u_{2,0}(\psi_2)$, $n_{2,0}(\psi_2)$, and $h_{2,0}(\psi_2)$ are determined by matching with the solution in the undisturbed boundary layer ahead of the interaction region. If ahead of the interaction region the body represents a flat plate, then it is simply the Blasius solution at a distance $O(1)$ from the plate edge. We recall that the second term in the expansion for $n$ is caused by the region 2 displacement due to the variation in the streamtube displacement thickness in region 3. Substituting formulas (1.16) in the equation of state and the Navier–Stokes equations and passing to the limit $\varepsilon \to 0$ and $\lambda \to 0$ we arrive at the system of equations

$$\rho_{2,0} u_{2,0} \frac{\partial u_2}{\partial x_2} + \frac{\partial p_2}{\partial x_2} = \frac{\varepsilon}{\lambda} \rho_{2,0} v_2 \frac{\partial p_2}{\partial \psi_2} + \frac{\lambda^3}{\varepsilon^2} \frac{d}{d\psi_2} \left( \rho_{2,0} u_{2,0} \mu_{2,0} \frac{du_{2,0}}{d\psi_2} \right) + \cdots \tag{1.17}$$

$$\frac{\partial p_2}{\partial \psi_2} = -\frac{\varepsilon}{\lambda} \frac{\partial v_2}{\partial x_2} + \cdots \tag{1.18}$$

$$\frac{\partial n_{2,1}}{\partial x_2} = \frac{v_2}{u_{2,0}}, \qquad \frac{\partial n_{2,1}}{\partial \psi_2} = 0 \tag{1.19}$$

$$p_2 = \frac{1}{M^2} \frac{\rho_2}{\rho_{2,0}} \tag{1.20}$$

In view of restriction (1.5), the right-hand sides of Eqs. (1.17) and (1.18) can be taken to be equal to zero. Equations (1.19) show that $n_{2,1}$, that is, the main variable part of the displacement thickness, is dependent only on $x_2$ and, therefore, is to be determined in matching the solutions in regions 2 and 3. Taking into account that the disturbances vanish, as $x_2 \to -\infty$, and integrating Eq. (1.17) we obtain

$$p_2 + \rho_{2,0} u_{2,0} u_2 = 0 \tag{1.21}$$

For region 3 the scales and the asymptotic representations take the form:

$$x_3 = x_1 = \frac{x - x_0}{\lambda}, \qquad \psi_3 = \frac{\psi \lambda^2}{\varepsilon^3} \tag{1.22}$$

$$u \sim \frac{\varepsilon}{\lambda} u_3 + \cdots, \qquad v \sim \left(\frac{\varepsilon}{\lambda}\right)^3 v_3 + \cdots, \qquad p \sim \frac{1}{\gamma M^2} + \left(\frac{\varepsilon}{\lambda}\right)^2 p_3 + \cdots$$

$$h \sim h_w + \frac{\varepsilon}{\lambda} h_3 + \cdots, \qquad \rho \sim \rho_w + \cdots, \qquad \mu \sim \mu_w + \cdots, \qquad n \sim \frac{\varepsilon^2}{\lambda} n_3 + \cdots$$

Substituting Eq. (1.22) in the Navier–Stokes equations and passing to the limit $\varepsilon \to 0$ and $\lambda \to 0$ results in the system of equations for the incompressible boundary layer

$$\rho_w u_3 \frac{\partial u_3}{\partial x_3} + \frac{dp_3}{dx_3} = \frac{\lambda^4}{\varepsilon^3} \rho_w u_3 \frac{\partial}{\partial \psi_3}\left(\rho_w \mu_w u_3 \frac{\partial u_3}{\partial \psi_3}\right) \tag{1.23}$$

$$\frac{\partial p_3}{\partial \psi_3} = 0, \quad \frac{\partial n_3}{\partial x_3} = \frac{v_3}{u_3}, \quad \frac{\partial n_3}{\partial \psi_3} = \frac{1}{\rho_w u_3}$$

$$\frac{\partial h_3}{\partial x_3} = \frac{\lambda^4}{\varepsilon^3} \frac{\rho_w \mu_w}{\sigma} \frac{\partial}{\partial \psi_3}\left(u_3 \frac{\partial h_3}{\partial \psi_3}\right)$$

The most general case is associated with $\lambda = \varepsilon^{3/4}$; let us consider it. However, it should be noted that in the cases in which the boundary conditions correspond to $\varepsilon^{2/3} \gg \lambda \gg \varepsilon^{3/4}$ or $\varepsilon \ll \lambda \ll \varepsilon^{3/4}$ the solution can be found by the corresponding passage to limit starting from the solution obtained for $\lambda \sim \varepsilon^{3/4}$. Both cases admit appreciable simplifications and are used below in studying certain problems.

On the body surface conventional boundary conditions are imposed

$$u_{3w} = v_{3w} = h_{3w} = 0 \tag{1.24}$$

For determining the initial and boundary conditions on the outer boundary of region 3, the asymptotic expansion matching principle should be applied to the solutions in regions 2 and 3 and the undisturbed boundary layer. We note that $p_2 = p_3$, since both $p_2$ and $p_3$ are independent of $\psi_2$ and $\psi_3$.

In matching the flow functions in regions 2 and 3 the passage to the limit $\varepsilon \to 0$ for the functions of region 2 should be performed at fixed $x$ and $\psi_3$; this corresponds to the expansion of the functions as $\psi_2 \to 0$. The function $u_{2,0}(\psi_2) \to 0$ is as follows:

$$u_{2,0} \sim \sqrt{\frac{2a}{\rho_w}} \psi_2^{1/2} + \cdots \tag{1.25}$$

where $a = (\partial u / \partial n_2)_w$ is the dimensionless viscous stress ahead of the interaction region in the undisturbed boundary layer.

Using Eqs. (1.21) and (1.25) we obtain the required expansions for the solution in region 2 in terms of the variables of region 3

$$u \sim \varepsilon^{1/4}\left(\sqrt{\frac{2a}{\rho_w}} \psi_3^{1/2} - \frac{p_3}{\sqrt{2a\rho_w}\, \psi_3^{1/2}}\right) + \cdots \tag{1.26}$$

Relation (1.26) determines the behavior of $u_3$ when the passage to the limit $\varepsilon \to 0$ is performed at a fixed $\psi_2$, that is, when $\psi_3 \to \infty$. Thus, as $\psi_3 \to \infty$, the function $u_3$ must increase as $\psi_3^{1/2}$, that is, linearly in $n_3$ with a slope determined by the friction in the undisturbed boundary layer

$$u_3 \to \sqrt{\frac{2a}{\rho_w}}\psi_3^{1/2} \quad \text{as } \psi_3 \to \infty \tag{1.27}$$

The condition for the energy equation is derived analogously

$$h_3 \to b\sqrt{\frac{2a}{\rho_w}}\psi_3^{1/2} \quad \text{as } \psi_3 \to \infty \tag{1.28a}$$

where $b = (\partial h/\partial n_2)_w$ in the undisturbed boundary layer.

Let us perform the matching procedure for $n$. In region 2, as $\psi_2 \to 0$, we obtain

$$n \sim \varepsilon\sqrt{\frac{2\psi_2}{\rho_w a}} + \varepsilon^{5/4} n_{2,1}(x) + \cdots \tag{1.28b}$$

For region 3 we can write

$$n \sim \varepsilon^{5/4}\int_0^{\psi_3}\frac{d\psi_3}{\rho_w u_3} = \varepsilon^{5/4}\left[\sqrt{\frac{2\psi_3}{\rho_w a}} + \int_0^{\psi_3}\left(\frac{1}{u_3} - \sqrt{\frac{\rho_w}{2a\psi_3}}\right)\frac{d\psi_3}{\rho_w}\right]$$

Hence, in view of Eq. (1.19), we obtain

$$n_{2,1}(x) = \lim_{\psi_3\to\infty}\left(\int_0^{\psi_3}\frac{d\psi_3}{\rho_w u_3} - \sqrt{\frac{2\psi_3}{\rho_w a}}\right) \tag{1.29}$$

This relation makes it possible to determine the value of $v_2(x_2, \psi_2 \to \infty)$ using the first equation (1.19)

$$v_2(x_2, \psi_2 \to \infty) \to \frac{d}{dx_2}\left[\lim_{\psi_3\to\infty}\left(\int_0^{\psi_3}\frac{d\psi_3}{\rho_w u_3} - \sqrt{\frac{2\psi_3}{\rho_w a}}\right)\right] \tag{1.30}$$

Taking into account that from the matching principle for regions 1 and 2 it follows that

$$v_1(x_1, 0) = v_2(x_2, \infty) \quad \text{and} \quad p_1|_{\psi_1=0} = p_2 = p_3 \tag{1.31a}$$

we obtain an additional boundary condition for region 3, which makes it possible to close the problem using formulas (1.14) or (1.15)

$$\sqrt{M^2-1}\,p_3 = \frac{d\Delta_3^*}{dx_3}, \quad \Delta_3^* = \lim_{\psi_3\to\infty}\left(\int_0^{\psi_3}\frac{d\psi_3}{\rho_w u_3} - \sqrt{\frac{2\psi_3}{\rho_w a}}\right) \tag{1.31b}$$

$$\sqrt{M^2-1}\,p_3 = \frac{d\Delta_3^*}{dx_3} + 2\sqrt{M_1^2-1}\,|p_1|, \quad x_1 > 0 \tag{1.31c}$$

It can be easily seen that $\Delta_3^*$ is the dimensionless displacement thickness of region 3.

The initial conditions for region 3 are obtained from the condition of matching with the "bottom" of the undisturbed boundary layer

$$u_3(-\infty, \psi_3) = \sqrt{\frac{2a}{\rho_w}}\psi_3^{1/2}, \quad p_3(-\infty) = 0, \quad h_3(-\infty, \psi_3) = b\sqrt{\frac{2a}{\rho_w}}\,\psi_3^{1/2} \tag{1.31d}$$

Thus, the flow in region 3 is determined by Eqs. (1.23) with the initial conditions (1.31d), the boundary conditions on the body (1.24) and the outer boundary (1.27), and the conditions of the interaction with the outer flow (1.31b).

After the problem for region 3 has been solved, the flow in region 2 with a known $p_2(x_2)$ dependence can be easily determined from formulas (1.18)–(1.21).

## 1.2 Flow near the separation point of the laminar boundary layer in a supersonic flow

### *1.2.1 Formulation of the problem and similarity law*

We will consider the flow near the separation point of the laminar boundary layer in a supersonic flat-plate flow. It is known that the boundary layer can be separated from the smooth surface of a body of small curvature only in the presence of a positive (adverse) pressure gradient. On a flat plate in a uniform supersonic flow aligned with the plate in the undisturbed region, the pressure gradient ahead of an obstacle or the point of incidence of a shock (Fig. 1.1) can be caused only by a variation in the boundary layer displacement thickness. Since this induced pressure gradient affects the boundary layer even in the first approximation, we arrive at the problem of interaction of the same type as that considered in Section 1.1.

The flow in the vicinity of the separation point must include a separationless region at $x < 0$ (origin is placed at the separation point) and a return flow region at $x > 0$. The first equation (1.23) indicates that the case $\Delta p \gg \varepsilon^{1/2}$, or $\lambda \ll \varepsilon^{3/4}$, is impossible. If this were the case, viscous terms would disappear from Eq. (1.23), so that separationless solutions could not exist near the body surface for $\Delta p > 0$ (within the limits of this order). Therefore, below we assume that $\lambda = \varepsilon^{3/4}$ and $\Delta p \sim \varepsilon^{1/2}$. An analysis of the asymptotics of this solution as $x_3 \to -\infty$, equivalent to the study of flows with $\Delta p \ll \varepsilon^{1/2}$, makes it possible to conclude that these flows do not include a separation point. Thus, in order to obtain a solution in the entire vicinity of the separation point, it is necessary to let $\lambda = \varepsilon^{3/4}$ and solve the problem for complete nondegenerate equations (1.23). Since in this section for calculating the main flow parameters it is sufficient to solve the problem formulated in Section 1.1 for region 3, below, for the sake of brevity, we will omit the subscript referring to the region.

For solving numerically the problem it is convenient to introduce variables in which the beginning of integration is mapped from $x = -\infty$ to the finite point $\xi = 0$

$$\xi = \left[\frac{2(M^2-1)^{1/2}}{a\mu_w}\right]^{1/2} p, \quad \eta = \frac{a\rho_w^{1/2}}{p^{1/2}} n = \left[\frac{2(M^2-1)^{1/2}\rho_w^2 a^3}{\mu_w}\right]^{1/4} n\xi^{-1/2} \tag{1.32}$$

$$X = [2^3(M^2-1)^{3/2}\rho_w^2\mu_w a^5]^{1/4} x$$

$$\psi = \left[\frac{\mu_w}{2a(M^2-1)^{1/2}}\right]^{1/2} \xi f(\xi,\eta), \quad g = \frac{a}{b}\left(\frac{\rho_w}{p}\right)^{1/2} h$$

In these variables, the equations and the boundary conditions take the form:

$$f''' = \beta[1 + 0.5f'^2 - ff'' + \xi(f'\dot{f}' - \dot{f}f'')] \tag{1.33}$$

$$\frac{g''}{\sigma} = \beta[0.5f'g - fg' + \xi(f'\dot{g} - \dot{f}g')]$$

$$\beta = \xi^{1/2}\frac{\mathrm{d}\xi}{\mathrm{d}x} = \frac{\xi^{3/2}}{\dfrac{\mathrm{d}}{\mathrm{d}\xi}\left\{2\xi^{1/2}\lim\limits_{\eta\to\infty}[\eta - (2f)^{1/2}]\right\}}$$

$$f(\xi,0) = f'(\xi,0) = g(\xi,0) = 0, \quad f''(\xi,\infty) = g'(\xi,\infty) = 1$$

$$f(0,\eta) = 0.5\eta^2, \quad g(0,\eta) = \eta$$

Here, primes refer to the differentiation with respect to $\eta$ and dots to $\xi$. The boundary value problem (1.33) does not contain the parameters $M$, $Re$, and $\gamma$ and gives a universal solution. Thus, formulas (1.32) determines the similarity law for the flow region with free interaction.

In accordance with Eq. (1.32), the pressure coefficient $c_\mathrm{p}$ is determined by the formula

$$c_\mathrm{p} = \frac{\xi c_{\mathrm{f}0}^{1/2}}{(M^2-1)^{1/4}} \tag{1.34}$$

where $c_{\mathrm{f}0}$ is the conventional friction drag coefficient for the undisturbed boundary layer ahead of the disturbed region.

We will also note that $f''_w$ and $g'_w$ are equal to the ratios of the local viscous stress and heat transfer rate to their values in the undisturbed boundary layer ahead of the separation zone.

Of course, the above consideration is valid for not only a flat plate but also a body with a curved generator, if only in the free interaction region its curvature is less in the order than that induced by the boundary formed by the boundary layer displacement thickness (e.g., for a body with the curvature of the order of unity). Then the corresponding local values of $c_{\mathrm{f}0}$, $M$, and the temperature factor should be substituted in formulas (1.32) and (1.34). The results are also applicable for bodies of revolution in a flow aligned with the axis of symmetry, if only the longitudinal dimension of the interaction region $\varepsilon^{3/4}$ is considerably larger than the distance from the body surface to the axis of symmetry.

### *1.2.2 Asymptotic behavior of the solution at minus infinity and results of the numerical solution of the problem*

The integration of the boundary value problem (1.33) begins at $\xi = 0$ and $X \to -\infty$. We will construct a series expansion of the solution in the vicinity of this point by introducing the following variables and expansions for the flow functions

$$(\xi, \eta) \to (\xi, N), \quad N = \eta \xi^{1/2} \tag{1.35}$$

$$f \sim \frac{N^2}{2\xi} + \Phi(N) + \cdots, \quad g \sim \frac{N}{\xi^{1/2}} + \xi^{1/2} G(N) + \cdots, \quad \xi \to 0$$

Substituting Eqs. (1.35) in Eqs. (1.33) leads to the following result:

$$\beta \sim -\frac{\xi^{3/2}}{2\Phi'(\infty)} + \cdots, \quad -2\Phi'(\infty)\Phi''' = 1 + N\Phi' - \Phi \tag{1.36}$$

$$\Phi(0) = \Phi'(0) = \Phi''(\infty) = 0$$

A simple change of variables makes it possible to reduce Eq. (1.36) to the standard form:

$$\Phi = \overline{\Phi}, \quad N = [-2\Phi'(\infty)]^{1/3}\overline{N}, \quad \overline{\Phi}''' = 1 + \overline{N}\overline{\Phi}' - \overline{\Phi} \tag{1.37}$$

The numerical solution gives the value $\overline{\Phi}'(\infty) = -1.45$.
Then the asymptotic form of the pressure distribution law as $x \to -\infty$ takes the form:

$$\xi \sim \exp[+0.49(X - X_0)], \quad X \to -\infty. \tag{1.38}$$

The solution is determined correct to an arbitrary constant $X_0$. The behavior of the solution as $X \to +\infty$ is discussed in Section 1.4 in considering developed separation zones.

### *1.2.3 Results of calculations and comparison with experimental data*

Problem (1.33) was numerically solved using two different approaches described briefly in Section 2.3.4. The distributions of the pressure $\xi$, friction $f''_w$, and heat flux $g'_w$ in the free interaction region are presented in Fig. 1.3, while in Fig. 1.4 we have plotted the distribution of the parameter $\beta$ characterizing the pressure gradient. Integration was carried out through the separation point with no evidence of instability. Certain difficulties arose first near the plateau region, that is, as $X \to +\infty$, where the characteristic values of the pressure gradient and parameter $\beta$ became small. In this region, the variation of the integration step $\Delta\xi$ from 0.2 to 0.05 led to the change in the final value of $\xi$ from 0.35 to 2.16. The final results

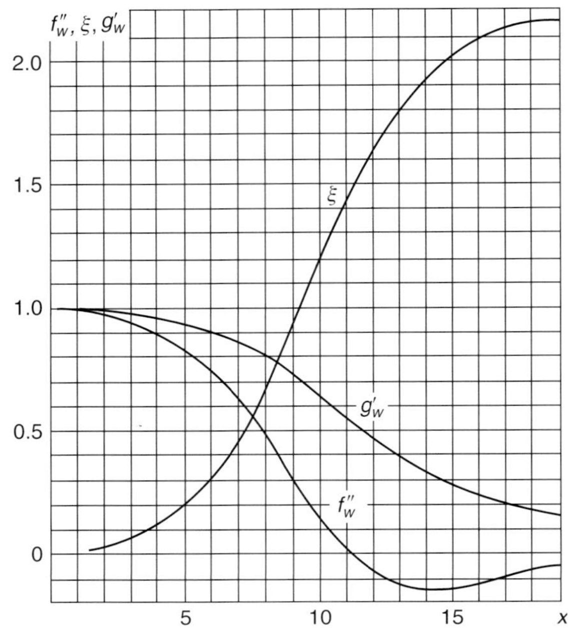


Fig. 1.3.

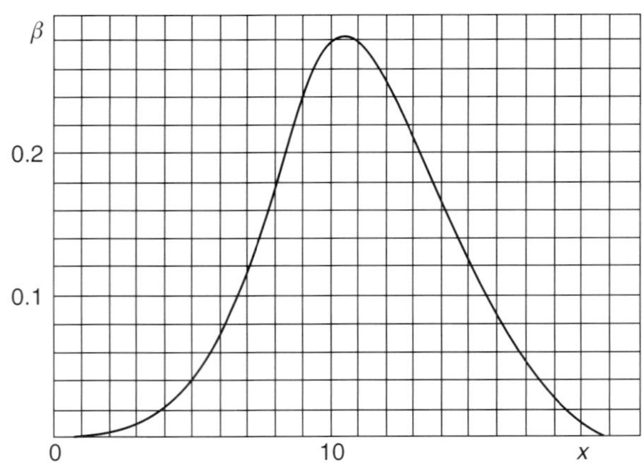


Fig. 1.4.

were obtained for $\Delta\xi = 0.02$ with account for the asymptotic behavior of the functions as $X \to +\infty$ studied in Section 1.4 in connection with the problem of the flow in the developed separation zone. An analysis of the solution as $X \to +\infty$ shows that, in accordance with the tendencies presented in Fig. 1.3, the pressure approaches a finite limit. We note that the calculation of the pressure in the plateau region on the basis of the integral boundary layer equations does not lead to satisfactory results and depends appreciably on the chosen form of velocity profiles. Friction on the body surface passes through zero at the separation point, reaches a minimum, and then starts to increase (decrease in absolute magnitude). Within the limits of the separation point vicinity under consideration, the heat flux to the body decreases monotonically.

In Figs. 1.5 and 1.6 we have plotted parameter profiles across region 3 (see Fig. 1.2) of the boundary layer. In the main part of the layer the profiles of the velocity and other parameters are of little interest, since in region 2 with thickness of the order of $\varepsilon$ any variations are small. It can be seen that the return flow velocities are relatively low.

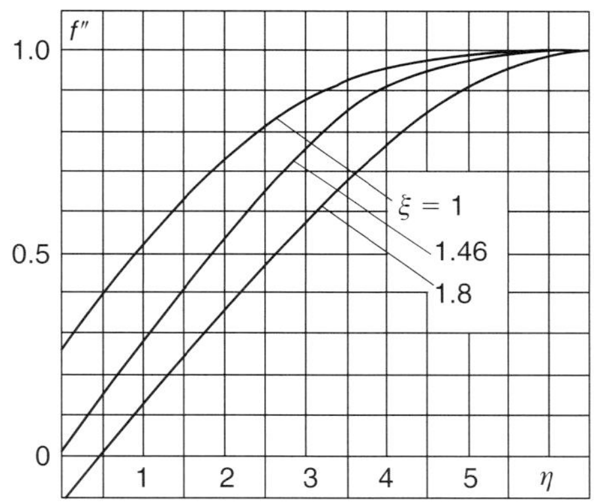


Fig. 1.5.

In estimating the accuracy of the presented results (not only calculated but also more general, e.g., following from the similarity law) it should be remembered that they were obtained in the first approximation and their relative error is $O(Re^{-1/8})$. This estimate follows from taking account of the region 2 displacement thickness, the Busemann term in the condition on the outer boundary, etc. In fact, a direct comparison of the calculated results with the experimental data on the free interaction region dimensions shows that the calculated lengths of the interaction regions are about 15% larger than the experimental ones (Fig. 1.7).

The comparison of the calculated and experimental results concerning the pressure coefficient in the plateau region of the separation zone drawn in Fig. 1.8 turns out to be more favorable. We note the important role played by this parameter in various approximate and engineering methods for calculating separation zones. The experimental data presented in Fig. 1.8 are taken from the work of Erdos and Pallone (1962), they pertain to the pressure measurements in the plateau regions of developed separation zones occurring ahead of a step, a deflected flap, and near the point of shock incidence on the boundary layer. The theoretical curve plotted in accordance with Eq. (1.34) and the calculated value $\xi = 2.16$ (Fig. 1.3) is in good agreement with the experimental data. The lower group of points which relates to the pressure coefficient at the separation point, lies about 15% below the calculated curve. This discrepancy is not beyond the accuracy of the first-approximation theory. However, it

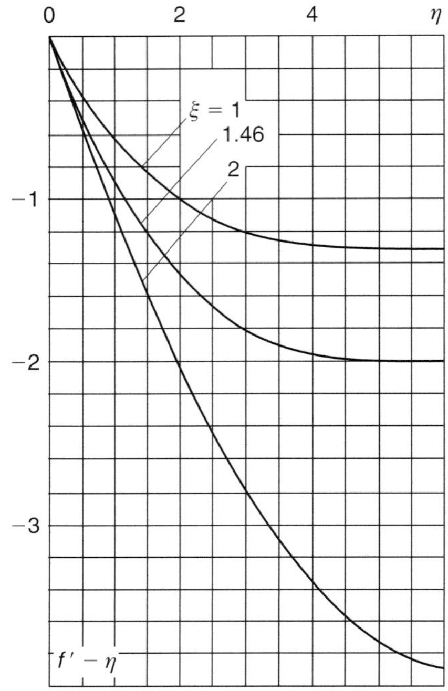


Fig. 1.6.

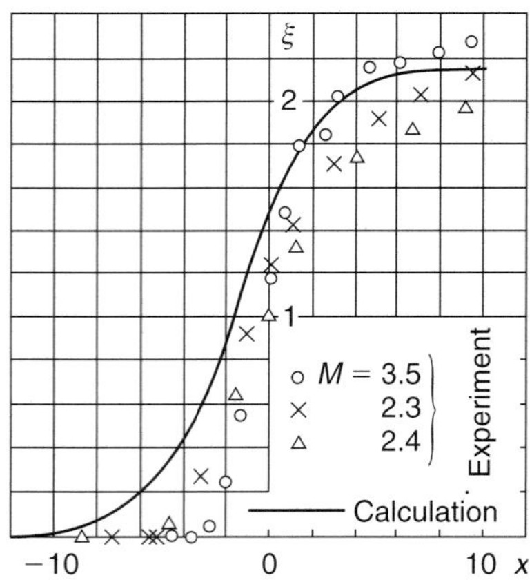


Fig. 1.7.

should be noted that the good agreement of the data for the plateau region sets one thinking on possible systematic errors in interpreting the experimental data on the separation point position. The pressure is measured fairly accurately. However, in all three works, the results of which are gathered in the paper of Erdos and Pallone (1962) and presented in Fig. 1.8, the position of the separation point was defined as the site of accumulation of oil deposited

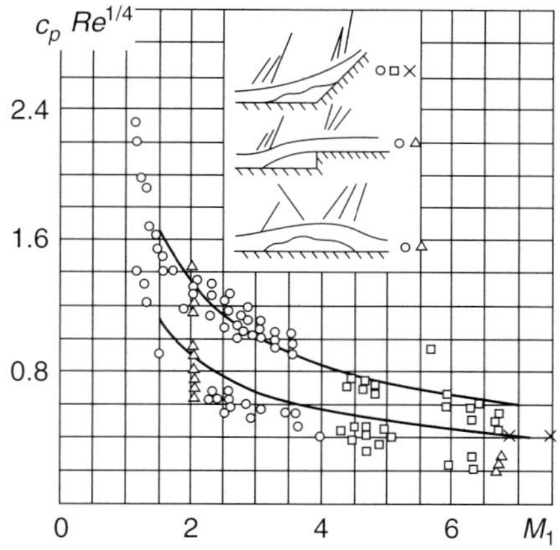


Fig. 1.8.

in the form of a film onto the model surface before the experiments had been started. It should be noted that at the separation point the friction is zero, while the oil was subjected to an upstream-directed pressure gradient. In this connection it may be suggested that the oil accumulation line is actually displaced somewhat upstream, where the pressure gradient is counterbalanced by a small positive friction.

### *1.2.4 Note on the nature of upstream disturbance propagation in the interaction between the boundary layer and the outer flow*

The boundary value problem specified by the equations and boundary conditions (1.33) provides a unique solution for the flow in the vicinity of the separation point only if an additional condition is preassigned in the form of the separation point coordinate

$$f_w''|_{x=0} = 0 \tag{1.39}$$

In the absence of condition (1.39) there exists a single-parameter family of solutions which in essence makes it possible to take account of the boundary conditions imposed downstream by varying, for example, the separation point position. As another example, we will consider the flow presented in Fig. 1.9a. Let the flap deflection angle be $O(Re^{-1/4})$. Ahead of point $O$ the solution must also be described by pieces of the universal solution presented above; with increase in the deflection angle the interval of the integral curve corresponding to this flow region with the characteristic length $O(Re^{-3/8})$ increases and the region of primarily separationless flow with interaction expands upstream. An analogous situation takes place for the flow on the upper side of the body presented in Fig. 1.9c. For $p_\infty - p_b \sim O(Re^{-1/4})$, ahead of point $O$ there appears a flow region with free interaction, such that compression disturbances propagate upstream for $p_b - p_\infty > 0$, while for $p_b - p_\infty < 0$ expansion flows are formed (solutions for expansion flows are presented in Section 2.2). Thus, for uniquely

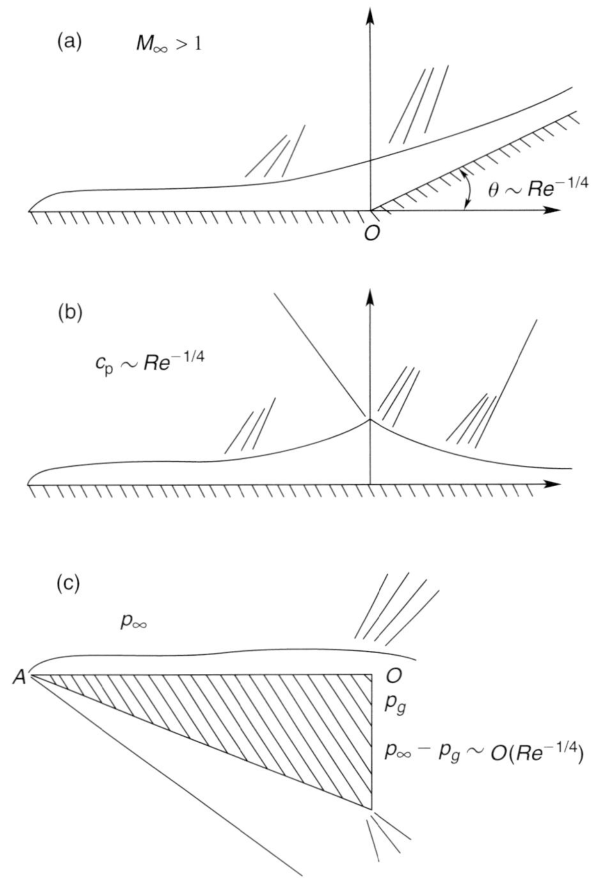


Fig. 1.9.

determining the solution it is necessary to preassign one more constant, namely, the pressure disturbance value at the end of the body.

That the upstream disturbance propagation effect is present in the interaction between the supersonic flow and the boundary layer is unexpected from neither a physical nor a mathematical standpoint. In the boundary layer near a body surface there is always a flow region with subsonic velocities. Moreover, in the exact formulation the problem is governed by the complete elliptic Navier–Stokes equations.

However, an analysis of the situation which takes place when using the asymptotic first-approximation theory requires certain explanations. In this case, the flow outside the boundary layer is governed by the Euler equations for an inviscid supersonic flow. It is well known that in this case disturbances can propagate only downstream, within Mach cones. The boundary layer equations are parabolic. When conventional initial and boundary conditions and a pressure distribution are preassigned, they also describe only downstream transport of disturbances and do not admit a possibility to satisfy any boundary conditions preassigned downstream of the region under consideration if separation and return flows are absent.

The problems for flows with interaction presented above, which possess the property of upstream disturbance transfer, differ from the conventional problems in that in their formulation the boundary layer pressure distribution is not given but is determined in the joint integration of the inviscid flow and boundary layer equations. Thus, it is precisely the interaction mechanism and the corresponding boundary conditions that ensure the possibility of partial upstream transfer of the information.

An additional boundary condition relates the pressure disturbance with the first derivative of the boundary layer displacement thickness $\delta^*$ with respect to $x$. However, $\delta^*$ is always the integral of a function dependent on the longitudinal velocity component $u$ and, possibly, some other flow functions (e.g., for hypersonic flows considered in Chapter 4). Therefore, substituting the expression for $\mathrm{d}p/\mathrm{d}x$ in the longitudinal momentum equation for the boundary layer leads to the appearance in this equation of a term containing $\partial^2 u/\partial x^2$ under the sign of integral with respect to $y$. If the derivative $\partial^2 u/\partial x^2$ did not appear under the integral sign, this would lead to the necessity of preassigning an arbitrary function of $y$ at the body end. However, since $\partial^2 u/\partial x^2$ is under the sign of integral with respect to $y$, the problem admits preassigning a single constant at the end of the body; in this case, it takes the form of an arbitrary additive constant at $x$.

The second note should be made in connection with the results of the numerical integration of the equations for the flow with "free interaction" behind the separation point. It is known that at a given pressure distribution the boundary layer equations possess a singularity which makes impossible the numerical solution of the problem. A detailed survey of the analytical and numerical results concerning this issue can be found in the paper of Brown and Stewartson (1969). The analytical nature of the singularity at the separation point was studied in the works of Goldstein (1948), Stewartson (1958), Terril (1960), and Catherall, et al. (1965). However, in the paper of Catherall and Mangler (1966) it was noted that for the pressure distributions of particular form the increase without bound of the derivatives $\mathrm{d}\delta^*/\mathrm{d}x$ and $\mathrm{d}\tau_w/\mathrm{d}x$ can be removed. For this purpose, in the paper of Catherall and Mangler (1966) a special approach was employed. First, the incompressible boundary layer equations were numerically integrated in a usual fashion, that is, at a given pressure distribution. At a small distance ahead of the separation point the function $\delta^*(x)$ in the form of a second- or third-order polynomial was preassigned rather than the pressure $p(x)$, while the pressure was determined. In this case, it became possible to pass through the separation point and, for small-sized zones, even through the reattachment point. In this study, as in the work mentioned in the survey of Stewartson and Williams (1969), such an approach was not needed. The possibility of passing through the separation point was ensured by the condition $p \sim \mathrm{d}\delta^*/\mathrm{d}x$. We note that the question concerning the determination of a unique solution in the region downstream of the separation point cannot be considered as completely answered. However, as shown in Sections 1.3 and 1.4, the solution obtained here corresponds to the passage into a semi-infinite separation zone (or the stagnation point of the return flow in a closed separation zone). In these sections, the analytical form of the solution far away from the separation point, as $\Delta x\, Re^{3/8} \to \infty$, is obtained under assumption that at these lengths the body temperature is constant in the first approximation.

## 1.3 Separation far from the leading edge

We will now consider the flow in which the separation point is at a distance of the order of unity from the plate nose (Fig. 1.10). The Reynolds number *Re* is based on the parameters in the undisturbed part of the incident supersonic flow 1 and a certain length $\ell$. In accordance with the results of Section 1.1, $p_0 - p_1 \sim O(\varepsilon^{1/2})$. In the immediate vicinity of the separation point there is a free interaction region of length $x \sim \varepsilon^{3/4}$ which contains a viscous sublayer 3 of thickness $\varepsilon^{5/4}$ and region 2 of locally inviscid, slightly disturbed flow.

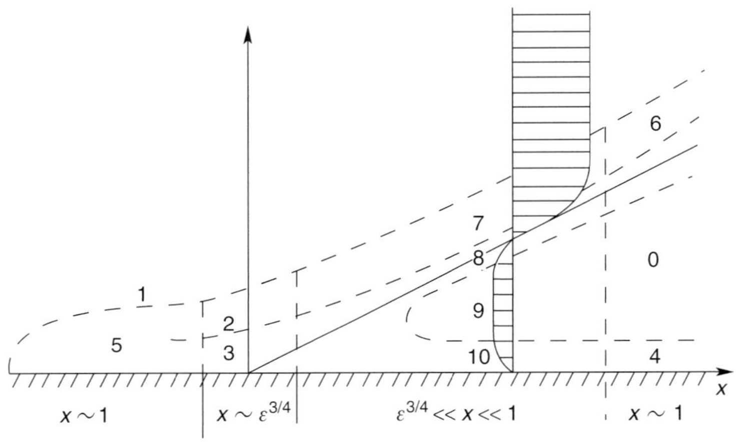


Fig. 1.10.

In region 3 the coordinate scales and the asymptotic expansions for the flow functions take the form:

$$x_3 = \frac{x}{\varepsilon^{3/4}}, \quad \psi_3 = \frac{\psi}{\varepsilon^{3/2}}, \quad n \sim \varepsilon^{5/4} n_3 + \cdots \tag{1.40}$$

$$u \sim \varepsilon^{1/4} u_3(x_3, \psi_3) + \cdots, \quad v \sim \varepsilon^{3/4} v_3(x_3, \psi_3) + \cdots, \quad \mu \sim \mu_w + \cdots$$

$$p \sim \frac{1}{\gamma M_1^2} + \varepsilon^{1/2} p_3(x_3) + \cdots, \quad h \sim h_w + \varepsilon^{1/4} h_3(x_3, \psi_3) + \cdots, \quad \rho \sim \rho_w + \cdots$$

In the first approximation, the problem is governed by the equations and boundary conditions (1.32) and (1.33). We will study the behavior of problem (1.33) as $x_3 \to +\infty$. For this purpose, it is convenient to introduce the variable

$$x_i = x\varepsilon^{-\alpha} = x_3 \varepsilon^b, \quad i = 7 \text{ to } 10, \ \ 0 < \alpha < 3/4, \ \ b = 3/4 - \alpha \tag{1.41}$$

Regions 7–10 will be introduced below. For a while, we only note that, as $\varepsilon \to 0$ and for $x_i \sim O(1)$, we obtain $x_3 \to +\infty$.

We will first present some physical considerations. The transverse dimension of the region below the $\psi = 0$ streamline increases, $y \sim \varepsilon^{1/2}x$, since $p_0 - p_1 \sim \varepsilon^{1/2}$. In region 7 of thickness $\sim\varepsilon$, as in the undisturbed boundary layer, ahead of the separation zone $u \sim 1$ and $\Delta u \sim \varepsilon^{1/2} \ll 1$. Therefore, near the streamline $\psi = 0$ the friction is of the same order $\varepsilon$ as in the original boundary layer.

Below there is region 8, in which at distances $x_8 \sim O(1)$ gas particles continue to move downstream driven only by friction forces. Because of this, in region 8 the leading viscous and inertial terms are of the same order. This condition leads to the following estimate for the region 8 thickness:

$$y \sim x\varepsilon^{1-2\alpha/3} \ll \varepsilon^{1/2}x \quad \text{for } \alpha < 3/4 \tag{1.42}$$

Therefore, for $x_8 = x_9 \sim O(1)$ an inviscid return flow region 9 with a transverse size $y \sim \varepsilon^{1/2}x$ must arise. Similar to the previous problem, in region 9 all streamlines begin from the rest region corresponding to $x_9 \to +\infty$ and inflow to the mixing zone 8. The flow is driven by the longitudinal pressure gradient induced by the interaction between the region 9 displacement thickness and the outer supersonic flow. Finally, the boundary layer 10 must be introduced in order to satisfy the no-slip condition on the body surface.

We will now formally obtain the equations and boundary conditions for regions 8–10 by passing to the limit $x_3 \to +\infty$ or $\varepsilon \to 0$ for $x_i \sim O(1)$, where $i = 8, 9, 10$, in the boundary value problem (1.33). For region 8, taking into account the above considerations, we introduce the variables

$$\eta_8 = [\eta_3 - r(x)]^{-1/3}x_3, \quad f_3(\xi_3, \eta_3) = x_3^{2/3} f_8(\xi_8, \eta_8) \tag{1.43}$$

Here, the subscripts refer to the numbers of the corresponding regions, while $r(x)$ is the value of $\eta_3$ for the $\psi = 0$ streamline.

We let all the variables in the original formulas (1.33) be referred by the subscript 3. Substituting Eq. (1.43) in Eq. (1.33) and passing to the limit $\varepsilon \to 0$ for $x_8 \sim O(1)$ leads to the following result:

$$f_8''' = \xi_{30}^{3/2}\left[\frac{1}{3}f_8'^2 - \frac{2}{3}f_8 f_8'' + x_8(f_8' f_8'^{*} - f_8^{*} f_8'')\right] \tag{1.44}$$

$$f_8'' \to 1 \quad \text{for } \eta_8 \to +\infty$$

The conditions on the body surface are outside of region 8. However, in view of Eq. (1.43), $\psi = 0$ for $\eta_8 = 0$. Moreover, as shown below, the longitudinal velocities of region 9 are out of order in region 8. For this reason, two more conditions can be added

$$f_8(x_8, 0) = 0, \quad f_8'(x_8, -\infty) = 0 \tag{1.45}$$

In Eq. (1.44) the constant $\xi_{30}$ is equal to $\xi_3(+\infty)$.

Problems (1.44) and (1.45) admit the self-similar solution

$$f_8(x_8, \eta_8) = \left(\frac{3}{2}\right)^{2/3} \xi_{30}\varphi_8(\lambda_8), \quad \eta_8 = \left(\frac{3}{2}\right)^{1/3} \xi_{30}^{1/2}\lambda_8 \tag{1.46}$$

$$\varphi_8''' + \varphi_8\varphi_8'' - \frac{1}{2}\varphi_8'^{\,2} = 0, \quad \varphi_8''(+\infty) = 1, \quad \varphi_8(0) = \varphi_8'(-\infty) = 0$$

The physical meaning of the result obtained is that, as $x_3 \to +\infty$, the outer part of region 3 becomes the zone 8 of mixing between the flow with a linear velocity profile (the lower part of the profile of the separated boundary layer, or region 7) and a semi-infinite (on the scale of region 8) region 9 of slow return flows. In what follows, we need the quantity $\varphi_8(-\infty) = -0.9569$.

Let us consider region 9, whose boundary, as $x_3 \to +\infty$, in accordance with Eq. (1.32), takes the form:

$$y \to \varepsilon^{1/2} x\xi_{30}[0.5a\mu_w(M_1^2 - 1)^{1/2}]^{1/2} \tag{1.47}$$

Using expression (1.32) for $\psi$, the matching principle in regions 8 and 9, and solutions (1.43)–(1.46) we obtain the value of $\varphi$ on the boundary

$$\psi = \varepsilon^{1+2\alpha/3} x_9\varphi_8(-\infty)\left(\frac{3}{2}\right)^{2/3}\left[\frac{\mu_w}{2a(M_1^2 - 1)^{1/2}}\right]^{1/2} \tag{1.48a}$$

Expressions (1.47) and (1.48a) make it possible to determine the scales for the coordinates and flow functions in region 9. Substituting them in Eq. (1.33) and passing to the limit $\varepsilon \to 0$ for $x_9 \sim O(1)$ leads to the Euler equations but with the momentum equation of the same form as in the Prandtl equation. Omitting simple algebra, we will present the final results

$$n_9 = \varepsilon^{-1/2-\alpha} n, \quad u \sim \varepsilon^{1/2-\alpha/3} u_9 + \cdots \tag{1.48b}$$

$$p \sim p(+\infty) + \varepsilon^{1-2\alpha/3} p_9 + \cdots$$

$$u_9(x_9) = -Kx_9^{-1/3}, \quad v_9 = -\frac{1}{3}Ky_9 x_9^{-4/3}, \quad \rho \sim \rho_w + \cdots$$

$$p_9 = -0.5\rho_w K^2 x_9^{-2/3}, \quad K = -\frac{\left(\frac{3}{2}\right)^{2/3}\varphi_8(-\infty)}{\xi_{30}a(M_1^2 - 1)^{1/2}}$$

Let us, finally, consider region 10, that is, the boundary layer of the inviscid return flow in region 9. Matching with the solution for region 9 gives the scale and the outer boundary condition for $u$ and $p$ in region 10. Using also the condition of the equality of the orders of the leading viscous and inertial terms of the equations we obtain the following scales for the variables

$$x_{10} = x\varepsilon^{-\alpha}, \quad n_{10} = \varepsilon^{-3/4-2\alpha/3} n \tag{1.49}$$

$$u \sim \varepsilon^{1/2-\alpha/3} u_{10}(x_{10}, n_{10}) + \cdots, \quad v \sim \varepsilon^{-1/4} v_{10}(x_{10}, n_{10}) + \cdots$$

$$p_{10} = p_9, \quad \mu_{10} = \mu_w, \quad \rho_{10} = \rho_w$$

Substituting Eq. (1.49) in Eq. (1.33) and passing to the limit $\varepsilon \to 0$ for $x_{10} \sim O(1)$ leads to the conventional equations for the incompressible boundary layer. The boundary conditions take the form:

$$u_{10}(x_{10}, 0) = v_{10}(x_{10}, 0) = 0, \quad u_{10}(x_{10}, +\infty) = -Kx_{10}^{-1/3} \tag{1.50}$$

The latter condition is obtained in accordance with the velocity distribution in region 9 and formula (1.48). In the region $0 < x < +\infty$ the problem for the boundary layer with boundary conditions (1.50) is self-similar

$$u_{10} = -Kx_{10}^{-1/3}\varphi'_{10}(\eta_{10}), \quad \eta_{10} = \left(\frac{K}{3\nu_w}\right)^{1/2} y_{10}x_{10}^{-2/3} \tag{1.51}$$

$$\varphi'''_{10} - \varphi_{10}\varphi''_{10} + (1 - \varphi'^2_{10}) = 0, \quad \varphi_{10}(0) = \varphi'_{10}(0) = 0, \quad \varphi'_{10}(\infty) = 1$$

The results of the numerical integration of Eq. (1.51) are presented in Fig. 1.12 as dashed curves.

Thus, far from the separation point in the region $\varepsilon^{3/4} \ll x \ll 1$ $(x_3 \to +\infty)$ the friction on the body and the pressure disturbances decay following the laws

$$\varepsilon^2\left(\frac{\partial u}{\partial y}\right)_w \sim -\varepsilon^{7/4}x^{-1}\varphi''_{10}(0)\left(\frac{K^3}{6\nu_w}\right)^{1/2} \tag{1.52}$$

$$p - p(+\infty) \sim -\frac{\varepsilon\rho_w K^2}{2x^{2/3}}$$

For finite-sized separation zones the solution obtained for $\varepsilon^{3/4} \ll x \ll 1$ is needed for describing transition from the free interaction region to the vicinity of the stagnation point in the return flow. Further, it is invalid. For a semi-infinite separation zone the solution must be also considered for $x \geq O(1)$.

At distances $x \sim 1$ the thickness of the mixing zone 6, in which the leading viscous and inertial terms of the Navier-Stokes equations are of the same order, is equal to $O(\varepsilon)$. The distance from the $\psi = 0$ line to the body is of the order $\varepsilon^{1/2}$. Thus, a region $O$ must exist, in which the flow is inviscid in the first approximation. In fact, on the scale of region $O$, region 6 is simply a line into which the gas with $v \sim \varepsilon$ is sucked. In region $O$ we have $y \sim \varepsilon^{1/2}x$; therefore, the continuity equation gives $u \sim \varepsilon^{1/2}$. Then in the momentum equation for $x \sim 1$ the inertial terms are $\rho_w u u_x \sim \varepsilon$, while the viscous terms are $\mu u_{yy} \sim \varepsilon^{3/2}$. Thus, in region $O$ the flow is inviscid, as in region 9. The solutions are different only owing to the boundary condition $\Delta p \sim \varepsilon$ imposed on the outer boundary.

In region 6 the Prandtl equations must be solved. The problem is non-self-similar, since matching with the flows in regions 7, 2, and 5 shows that the initial profile 6 coincides with the final profile of the region 5 of the undisturbed boundary layer. Regions 2 and 7 contain a locally inviscid flow in which $\Delta p \sim \varepsilon^{1/2}$. The solution for the zone of isobaric mixing with an initial velocity profile was obtained numerically in the work of Denison and

Baum [1961] in connection with the problem of the calculation of the base pressure behind a wedge. Here, the results of that work can be used for determining the boundary condition on the outer boundary of region $O$

$$\psi_6(x, -\infty) = \psi_0(x, y_{0e})$$

Obviously, the problem is non-self-similar but the solution for region $O$ at a given $\psi_6(x, -\infty)$ can be easily determined in quadratures. We also note that in the $x \to +\infty$ limit the solution goes over to the solution of the problem with separation near the leading edge of a plate considered in Section 1.4.

## 1.4 Separation from a leading edge

We will consider a very simple case of the separation zone beginning in a small vicinity of the leading edge of a thin flat plate (Fig. 1.11). A supersonic flow streams over a semi-infinite stagnation region, where the flow is driven only by viscosity forces on the boundary $OB$. The pressure difference is $p_0 - p_2 \sim O(1)$. Thus, boundary layer separation can begin only from a small vicinity of point $O$. If at the separation point $Re_x \gg 1$, then, in accordance with the results of this chapter, it should be $p_0 - p_2 \sim Re^{-1/4} \ll 1$.

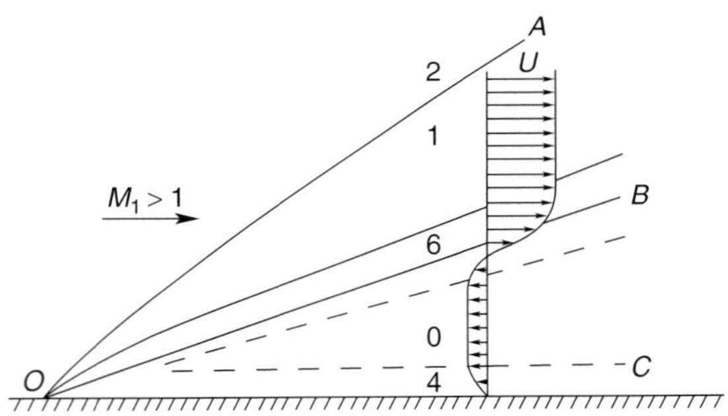


Fig. 1.11.

If the gas were inviscid, the flow of this type would consist of the uniform freestream 2, the uniform supersonic flow 1 behind the shock $OA$, and the dead air region 0. For the given freestream 2 parameters and the pressure $p_0$ this flow is completely determined by the Rankine–Hugoniot relations for an oblique shock. However, an inviscid gas flow cannot provide the limiting solution of the Navier–Stokes equations as $Re \to \infty$ throughout the entire flow region (here, $\ell$ is a certain fictitious length introduced only for the sake of convenience; it is absent from the final results, since, by virtue of dimensionality theory, at distances considerably larger than the "viscous length" $\mu/\rho u$ the solution must be self-similar). The point is that in the "inviscid" solution the velocity undergoes a discontinuity across the line $OB$. In accordance with the classical boundary layer theory, a mixing zone 6 of thickness $\delta \sim \varepsilon = Re^{-1/2}$ must be considered near $OB$. The solution of this problem is self-similar; it was obtained, for example, in the paper of Chapman (1950). In accordance with this solution, gas continuously arrives at region 6 due to the action of viscosity forces.

In region 6 the velocity component normal to $OB$ is of the order of $\varepsilon$. Therefore, the gas starts to flow also in region 0. In accordance with the asymptotic expansion matching principle, in region 0 the transverse velocity component is also of the order of $\varepsilon$ ($v \sim \varepsilon$; this is also obvious from physical considerations). Since $BOC \sim O(1)$, from the continuity equation it follows that in region 0 we have $u \sim v \sim \varepsilon$. Then the momentum equation indicates that the variable part of the pressure $\Delta p \sim O(\varepsilon^2)$. In this case, in the first approximation the boundary $OB$ of region 0 must remain rectilinear. Small perturbation theory applied to the supersonic flow 1 shows that the deviation of the slope $OB$ from a straight line is $O(\varepsilon^2)$. In order for a steady solution could be obtained, in region 0 the gas temperature $T_0$ must be equal in the first approximation to the wall temperature $T_w$. Then in the first approximation the gas density $\rho_0$ is constant and corresponds to the values $p = p_0$ and $T = T_0$. Substituting these estimates in the Navier–Stokes equations and passing to the limit $\varepsilon \to 0$ shows that in region 0 the flow is described by the complete Euler equations for an inviscid incompressible fluid. The flow remains to be irrotational, since all streamtubes begin from the state of rest, as $x_0 \to +\infty$, and then inflow to the mixing zone. For the stream function the Laplace equation can be written as

$$\Delta_0 \psi_0 = 0, \quad \psi_0(x_0, 0) = 0, \quad \psi_0(x_0, n_0 = Kx_0) = -Nx_0^{1/2} \tag{1.53}$$

The last boundary condition can be easily obtained by matching the solutions for regions 6 and 0. It expresses the equality of the gas flow rates for the gas sucked to region 6 and the gas carried away from region 0 through the boundary $OB$; $K$ is determined from the oblique shock relations. Here, it is worth to recall once more that in region 0 the pressure in the leading term (rather than the flow-induced addition $\Delta p \sim \varepsilon^2$) represents the boundary condition of the problem. Here, $N$ is a constant determining the gas flow rate to the self-similar mixing zone 6, for example, in accordance with the data of Chapman (1950).

The solution of Eq. (1.53) takes a simple form:

$$\psi_0 = -Nx_0^{1/2} \sinh\left(\frac{K\xi_0}{2}\right) \cosh\left(\frac{1}{2} \operatorname{arsinh} K\right), \quad \frac{n_0}{x_0} = \sinh(K\xi_0) \tag{1.54}$$

We will now note that in the return flow determined by Eq. (1.54) the no-slip condition on the body surface is violated

$$u_{0w} = u_0(x_0, 0) = -\frac{N}{E}, \quad E = 2x_0^{1/2} \sinh\left(\frac{1}{2} \operatorname{arcosh} K\right) \tag{1.55}$$

This means that the boundary layer 4 on the return flow bottom must be considered. The equations and boundary conditions for region 4 are as follows:

$$u_4 \frac{\partial u_4}{\partial x_4} + v_4 \frac{\partial u_4}{\partial n_4} = u_{0w} \frac{\mathrm{d}u_{0w}}{\mathrm{d}x_0} + \nu_w \frac{\partial^2 u_4}{\partial n_4^2}, \quad \frac{\partial u_4}{\partial x_4} + \frac{\partial v_4}{\partial n_4} = 0 \tag{1.56}$$

$$u_4(x_4, 0) = v_4(x_4, 0) = 0, \quad u_4(x_4, +\infty) = u_0(x_4, 0), \quad \frac{\partial p_4}{\partial n_4} = 0$$

The solution of problem (1.56) is as follows:

$$\xi_4 = x_4^{-1/2}, \quad \eta_4 = \left(\frac{4\xi_4^3}{4\nu_w}\right)^{1/2}, \quad u_0 = -L\xi_4 \tag{1.57}$$

$$L = \frac{1}{2}N\cosh\left(\frac{1}{2}\operatorname{arsinh} K\right), \quad u_4(\xi_4, \eta_4) = u_o(\xi_4)f_4'(\eta_4)$$

$$f_4''' - f_4 f_4'' + 2(1 - f_4'^2) = 0, \quad f_4(0) = f_4'(0) = 0, \quad f_4'(\infty) = 1$$

For the problem of the boundary layer in a flow accelerating from an infinitely distant point with the velocity increasing proportional to $x^{-n}$, where $0 < n < 1$, the equation of the type of Falkner–Skan with a negative coefficient of the second term of Eq. (1.57) and a positive coefficient of the third term is obtained (Neiland and Sychev, 1966). If $n \leq 0$, the solution does not exist, while for $n = 1$ the second term of Eq. (1.57) vanishes and for $n > 1$ it is positive and the equation takes the conventional form. In Fig. 1.12 solution (1.57) is presented as solid curves.

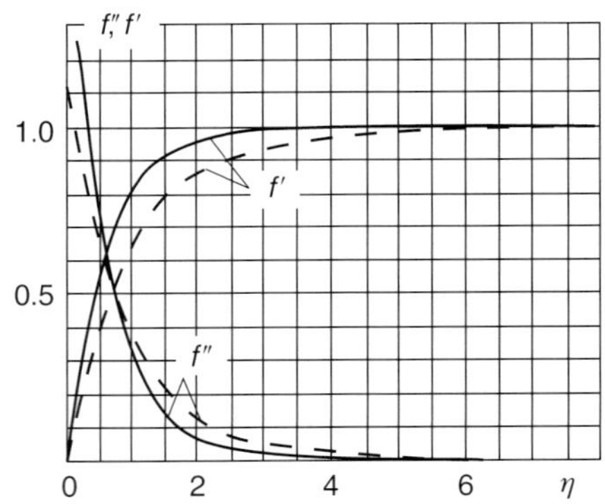


Fig. 1.12.

# 2 Other Types of Flows Described by Free Interaction Theory

In Section 2.1 we present the results of the investigation of preseparated flows under conditions of small skin friction in the boundary layer. Section 2.2 is devoted to the consideration of expansion flows described by free interaction theory. In Section 2.3 we present the results of an analysis of flows with free interaction which manifests itself upon the incidence of a weak shock on the boundary layer and in the flow past a body with a curvilinear contour.

Finally, Section 2.4 presents the results of the investigation of the boundary layer flow subject to slot suction ensuring the conservation of separationless flow in the boundary layer.

## 2.1 Laminar boundary layer separation in a supersonic flow under conditions of low skin friction

Investigations of laminar boundary layer separation due to an adverse pressure gradient showed (Landau and Lifshitz, 1944; Goldstein, 1948) that the solution of the system of boundary layer equations includes a singularity at a point, where the surface friction vanishes and cannot be continued through the separation point. In these and further studies of boundary layer separation, the survey of which can be found in the paper by Brown and Stewartson (1969), it was concluded that boundary layer theory is inapplicable for describing the flow in the vicinity of a separation point.

An alternative theory was proposed in the works of Neiland (1968*a*; 1969*a*) and Stewartson and Williams (1969); it is outlined in the previous sections.

As shown above, an important effect accompanying separation is the interaction between the boundary layer flow and the outer supersonic flow.

It can be suggested that taking the interaction into account makes it possible to obtain a nonsingular solution in the vicinity of the point of separation caused by an adverse pressure gradient in the outer flow. An attempt to take account for the interaction of the laminar boundary layer flow with the outer flow was made in the paper of Stewartson (1970*b*); however, it turned out that the solution of the system of equations governing the flow in the vicinity of the separation point contains a singularity.

### *2.1.1 Formulation of the problem. Estimation of the scales and characteristic values of the flow functions in the wall region*

We will consider a laminar viscous gas flow over a flat plate. Let at a distance $\ell$ from the leading edge of the plate the boundary layer separates due to an adverse pressure gradient. An orthogonal coordinate system is so chosen that its origin coincides with the leading edge, while the longitudinal coordinate is measured along the surface. It is assumed that in the outer supersonic flow the pressure increases over a distance comparable with the distance $\ell$.

The pressure increase can be induced by a certain superstructure on the surface of the body in the flow. Other situation can also take place, in which the unfavorable pressure distribution is generated by the shape of the original surface. Generally, this can lead to the formation of a curvilinear shock; for further analysis, the assumption of isentropicity of a local flow region located above the interaction region is important. It is assumed that the Reynolds number $Re = \rho_\infty u_\infty \ell / \mu_\infty$ is high but not higher than the critical value, so that the flow regime is still laminar.

In accordance with classical Prandtl's scheme, as $Re \to \infty$ an inviscid flow region and a thin, as compared with the longitudinal body dimension, boundary layer can be singled out. The solution describing the boundary layer flow near the point, where the surface friction vanishes, ceases to be uniformly accurate which leads to the necessity of introducing a wall viscous flow region and an inviscid flow region (Landau and Lifshitz, 1944; Goldstein, 1948). As the zero friction point is approached, the boundary layer displacement thickness grows, which leads to the appearance of an induced pressure gradient. From an analysis of the estimates made below, it follows that for the flow regime under consideration in the vicinity of the zero friction point the boundary layer induced pressure gradient is of the same order as the given pressure gradient $\mathrm{d}p/\mathrm{d}x = K = O(1)$. The limiting case, in which the induced-to-given pressure gradient ratio is low, leads to a scheme, in which the flow is described in the leading term by a solution which disregards the interaction and, thus, is incapable of eliminating the singularity and continuing the solution through the zero friction point (Stewartson, 1970*b*).

In other limiting case, in which the induced-to-given pressure gradient ratio is high, the "free interaction" regime governing another flow is realized.

We will consider the flow in the wall region containing a zero friction point, whereby assuming the inertia and viscosity forces are of the same order and the pressure gradient causes nonlinear variations of the longitudinal velocity. It is assumed that ahead of the interaction region the velocity profile is as follows:

$$u = \left(\frac{y}{\varepsilon}\right)^2 + \frac{a(x)y}{\varepsilon} + \cdots, \qquad \varepsilon = Re^{-1/2} \tag{2.1}$$

where the function $a(x)$ is proportional to the surface friction and is determined from the solution describing the boundary layer flow upstream of the interaction region. Using the equation for the variation of the longitudinal momentum we obtain the following relation between the thickness and the length of the nonlinear variation region: $\Delta y \sim \varepsilon(\Delta x)^{1/4}$. The condition of the equality of the orders of the given and induced pressure gradients allows us to determine the interaction region length: $\Delta x \sim \varepsilon^{4/7}$.

Naturally, other flow regimes can also exist, such that the orders of the given and induced pressure gradients are different. The realization of one or another flow pattern depends on the conditions imposed downstream, for example, on the value of the base pressure at the rear of the body. The given pressure distribution is, as it were, fitted to the surface, whereas the induced pressure distribution can change with the variation of the conditions imposed downstream. The choice of the flow pattern under consideration is due to the fact that it is the most general case from the mathematical standpoint.

In deriving the last relation it was assumed that the interaction region length $\Delta x$ is greater in order than the boundary layer thickness ahead of the interaction region $\delta_0^* \sim \varepsilon$. It can be easily seen that the estimate obtained satisfies this condition. It was also assumed that the variation of the thickness of the wall region, where $\Delta y$ varies nonlinearly, is greater than the variation of the thickness of the main part of the boundary layer $\Delta\delta_0^*$. The estimates obtained maintain this assumption

$$\Delta y \sim \varepsilon^{8/7}, \quad \Delta\delta_0^* \sim \varepsilon^{11/7}, \quad \Delta y \gg \Delta\delta_0^*$$

The estimates for the disturbances of the longitudinal velocity, transverse velocity, and pressure take the form:

$$u \sim \Delta u \sim \varepsilon^{2/7}, \quad v \sim \varepsilon^{6/7}, \quad \Delta p \sim \varepsilon^{4/7}$$

The estimates obtained indicate that ahead of the interaction region on the plate surface the friction is $O(\varepsilon^{-6/7})$. Thus, as the boundary layer friction decreases to $O(\varepsilon^{-6/7})$, the role played by the processes of the boundary layer/outer supersonic flow interaction becomes important. As a result, in the interaction region the variation of the surface friction is more rapid than that in the preceding region. In accordance with the results obtained in the works of Landau and Lifshitz (1944) and Goldstein (1948) the surface friction is proportional to $(-x_0)^{1/2}$, where $x_0$ is the distance measured from the zero friction point. The above estimates show that the interaction results in upstream displacement of the separation point through a distance equal to $O(\varepsilon^{2/7})$.

### 2.1.2 *Equations and boundary conditions*

In the flow near the separation point, apart from the wall region two more regions must be singled out for constructing a uniformly accurate solution (Fig. 2.1). In region I, whose transverse and longitudinal dimensions are equal and of the same order as the interaction region

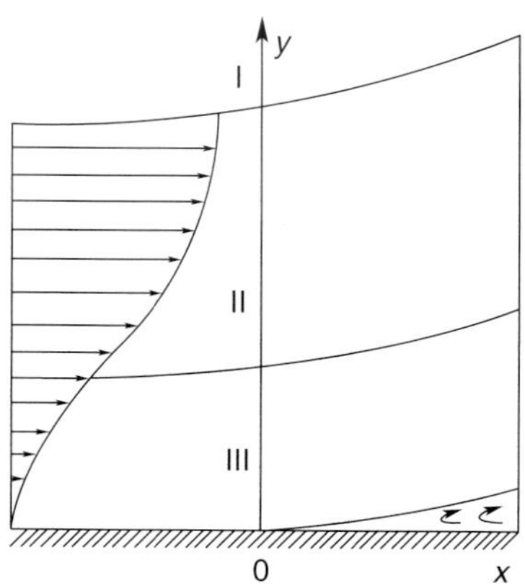


Fig. 2.1.

length $\Delta x$, the flow functions, in accordance with the above estimates, can be represented in the form of the following asymptotic expansions in new coordinates $x_1$ and $y_1$:

$$\begin{aligned} x &= 1+\varepsilon^{4/7}x_1, \quad y=\varepsilon^{4/7}y_1 \\ u &= 1+\varepsilon^{4/7}(u_0(x_1,y_1)+u_1(x_1,y_1))+\cdots \\ v &= \varepsilon^{4/7}(v_0(x_1,y_1)+v_1(x_1,y_1))+\cdots \\ p &= \varepsilon^{4/7}(p_0(x_1,y_1)+p_1(x_1,y_1))+\cdots \\ \rho &= 1+\varepsilon^{4/7}(\rho_0(x_1,y_1)+\rho_1(x_1,y_1))+\cdots \end{aligned} \tag{2.2}$$

where the terms with the subscript 0 relate to the flow function variations associated with the given pressure disturbance, while those with the subscript 1 to the variations caused by the boundary layer displacement thickness. By virtue of the linearity, the solution of the problem can be sought in the form of superposition of the given and induced disturbances.

Substituting Eq. (2.2) in the system of Navier–Stokes equations and passing to the limit $Re\to\infty$ leads to the linearized system of Euler equations, whose solution, which will be needed for further analysis, is as follows:

$$\beta p_1(x_1,0)=v_1(x_1,0), \quad \beta=(M_\infty^2-1)^{1/2} \tag{2.3}$$

In region II, whose thickness is equal to that of the boundary layer upstream of the separation point and length is equal to the interaction region length, we introduce the following representations of the coordinates and flow functions:

$$y=\varepsilon y_2, \quad x=1+\varepsilon^{4/7}x_1 \tag{2.4}$$

$$u=u_0(y_2)+\varepsilon^{2/7}u_2(x_1,y_2)+\cdots, \qquad v=\varepsilon^{4/7}v_2(x_1,y_2)+\cdots$$

$$p=\varepsilon^{4/7}p_2(x_1)+\cdots, \qquad \rho=\rho_0(y_2)+\varepsilon^{4/7}\rho_2(x_1,y_2)+\cdots$$

Substituting Eq. (2.4) in the system of Navier–Stokes equations and passing to the limit $Re\to\infty$ leads after some transformations to the following system of equations:

$$u_0\frac{\partial u_2}{\partial x_1}+v_2\frac{\mathrm{d}u_0}{\mathrm{d}y_2}=0, \quad \frac{\partial u_2}{\partial x_1}+\frac{\partial v_2}{\partial y_2}=0 \tag{2.5}$$

$$u_2=-A(x_1)\frac{\mathrm{d}u_0}{\mathrm{d}y_2}, \quad v_2=\frac{\mathrm{d}A}{\mathrm{d}x_1}u_0(y_2)$$

where $A_1(x)$ is an indefinite function.

In region III we make the change of variables and introduce the following asymptotic representations of the functions

$$y=\varepsilon^{8/7}y_3, \quad x=1+\varepsilon^{4/7}x_1 \tag{2.6}$$

$$u = \varepsilon^{2/7} u_3(x_1, y_3) + \cdots, \qquad v = \varepsilon^{6/7} v_3(x_1, y_3) + \cdots$$

$$p = \varepsilon^{4/7} p_3(x_1) + \cdots, \qquad \rho = \rho_w + \varepsilon^{4/7} \rho_3(x_1, y_3) + \cdots$$

Substituting Eq. (2.6) in the system of Navier–Stokes equations and passing to the limit $Re \to \infty$ yields the system of equations

$$\rho_w u_3 \frac{\partial u_3}{\partial x_1} + \rho_w v_3 \frac{\partial u_3}{\partial y_3} + K + \frac{\partial p_3}{\partial x_1} = \mu_w \frac{\partial^2 u_3}{\partial y_3^2} \tag{2.7}$$

$$\frac{\partial u_3}{\partial x_1} + \frac{\partial v_3}{\partial y_3} = 0, \qquad \frac{\partial p_3}{\partial y_3} = 0$$

The initial conditions are determined by matching the solution for region III with that for the region located upstream, that is, for $x_1 \to -\infty$. Using the representation of the velocity profile $u_0(y_2)$ as $y_2 \to 0$ we obtain

$$u_3(x_1 \to -\infty, y_3) = \frac{K y_3^2}{2\mu_w} + B y_3 \tag{2.8}$$

where $B$ is constant. Matching the solutions for regions III and II makes it possible to determine the lacking boundary conditions

$$u_3(x_1, y_3 \to \infty) = \frac{K y_3^2}{2\mu_w} + B y_3 - A(x)\left(\frac{K y_3}{\mu_w} + B\right) + O(1) \tag{2.9}$$

Matching the solutions for regions I and II makes it possible to determine the relation between the functions $A$ and $p_3$

$$p_3(x_1) = \beta^{-1} \frac{\mathrm{d}A}{\mathrm{d}x_1} \tag{2.10}$$

Using a change of variables we can bring the boundary value problem (2.7)–(2.10) into the form:

$$\begin{aligned} x_3 &= \mu_w^{2/7} \rho_w^{-1/7} \beta^{-4/7} K^{-5/7} X, \quad y_3 = \mu_w^{4/7} \rho_w^{-1/7} \beta^{-1/7} K^{-3/7} Y \\ u_3 &= \mu_w^{1/7} \rho_w^{-4/7} \beta^{-2/7} K^{1/7} U, \quad v_3 = \mu_w^{3/7} \rho_w^{-5/7} \beta^{1/7} K^{3/7} V, \\ p_3 &= \mu_w^{2/7} \rho_w^{-1/7} \beta^{-4/7} K^{2/7} P, \quad B = \mu_w^{-3/7} \rho_w^{-3/7} \beta^{-1/7} K^{4/7} B_0 \\ A &= \mu_w^{4/7} \rho_w^{-2/7} K^{-3/7} A_0 \end{aligned} \tag{2.11}$$

$$U \frac{\partial U}{\partial X} + V \frac{\partial U}{\partial Y} + 1 + \frac{\partial P}{\partial X} = \frac{\partial^2 U}{\partial Y^2}$$

$$\frac{\partial U}{\partial X} + \frac{\partial V}{\partial Y} = 0, \quad \frac{\partial P}{\partial Y} = 0, \quad P = -\frac{dA_0}{dX}$$

$$U(X \to -\infty, Y) = \frac{Y^2}{2} + B_0 Y, \quad U(X, 0) = V(X, 0) = 0$$

$$U(X, Y \to \infty) = \frac{Y^2}{2} + B_0 Y - A_0(X)(Y + B_0) + O(1)$$

### 2.1.3 Solution of the linear boundary value problem

As $X \to -\infty$, the solution of the boundary value problem (2.11) can be represented in the following form:

$$U = \frac{Y^2}{2} + B_0 Y + C_0 U_0(Y) \exp(\alpha X) \tag{2.12}$$

$$V = C_0 \left( V_0 + \frac{Y^2}{2} + B_0 Y \right) \exp(\alpha X), \quad P = C_0 \exp(\alpha X)$$

Substituting Eq. (2.12) in Eq. (2.11) brings the latter boundary value problem into the form:

$$V_0''' - \alpha V_0' \left( \frac{Y^2}{2} + B_0 Y \right) + V_0 \alpha (Y + B_0) + \alpha^2 = 0 \tag{2.13}$$

$$V_0'(0) = -B_0, V_0'(\infty) = 0, \quad V_0''(\infty) = 0, \quad V_0(0) = 0$$

For large values of the parameter $B_0$ the change of variables (2.14) makes it possible to obtain the following boundary value problem:

$$\alpha = B_0^{5/4} \alpha_1, \quad Y = B_0^{3/4} Y_1, \quad V_0 = B_0^{1/4} V_1 \tag{2.14}$$

$$V_1''' - \alpha_1 V_1' Y + V_1 \alpha_1 + \alpha_1^2 = 0 \tag{2.15}$$

$$V_1'(0) = -1, \quad V_1'(\infty) = 0, \quad V_1''(\infty) = 0, \quad V_1(0) = 0$$

This problem describes the linear "free interaction" regime; Lighthill (1953) was the first who obtained its solution. In particular, the expression for the parameter $\alpha_1$ is as follows:

$$\alpha_1 = [-3\text{Ai}'(0)]^{3/4}$$

where $\text{Ai}(Y_1)$ is the Airy function (see Abramowitz and Stegun, 1964).

For small values of the parameter $B_0$ the change of variables (2.16) brings the boundary value problem (2.13) into the form:

$$\alpha = B_0^{2/3} \alpha_2, \quad Y = B_0^{1/6} Y_2, \quad V_0 = B_0^{5/6} V_2 \tag{2.16}$$

$$V_2''' - \frac{\alpha_2 V_2' Y^2}{2} + V_2 \alpha_2 Y_2 + \alpha_2^2 = 0 \tag{2.17}$$

$$V_2'(0) = -1, \quad V_2'(\infty) = 0, \quad V_2''(\infty) = 0, \quad V_2(0) = 0$$

One more change of variables transforms the differential equation for the function $V_2$ to the second-order differential equation for the function $f$

$$V_2' = \frac{zY_2^2}{2}, \quad z' = t^{-5/4} f, \quad t = 2^{3/2} \alpha_2^{-1/2} t_1, \quad Y_2 = t^{1/2}$$

$$t_1^2 f'' + t_1 f' - \left(t_1^2 + \frac{1}{16}\right) f = -\alpha_2^{-13/6} 2^{7/8} t_1^{5/4} \tag{2.18}$$

The general solution of Eq. (2.18) can be represented in the form:

$$f = c_1 I_{1/4}(t_1) + c_2 K_{1/4}(t_1) + L_{1/4}(t_1)$$

where $I_{1/4}$ and $K_{1/4}$ are the modified Bessel functions and $L_{1/4}$ is the modified Struve function. Using the boundary conditions we can derive the following expression for the parameter $\alpha_2$:

$$\alpha_2 = 2\pi^{-1/3} \left[ \frac{\Gamma\left(\frac{5}{4}\right)}{\Gamma\left(\frac{3}{4}\right)} \right]^{2/3}$$

where $\Gamma$ is the gamma function. An analogous result can be obtained using the results of the works of Stewartson (1970*b*) and Ruban (1982).

For the values of the parameter $B_0 = O(1)$ the solution of the boundary value problem (2.13) was obtained by numerical integration. The $\alpha(B_0)$ dependence is plotted in Fig. 2.2.

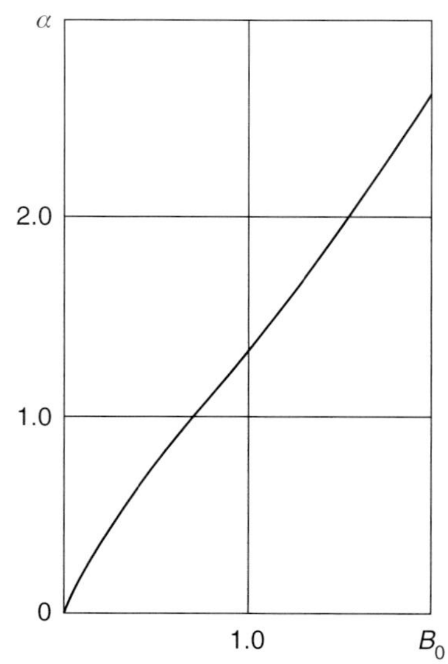


Fig. 2.2.

The results of the solution of the linear boundary value problem (2.13) were taken for the initial conditions corresponding to $X \to -\infty$ in the numerical integration of the nonlinear boundary value problem (2.11). The system of equations was approximated by second-order finite differences in the variable $Y$ and first-order differences in the variable $X$.

At $X = \text{const}$ the sweeping technique was used for integrating the system of equations. The interaction condition (2.10) was satisfied in the course of an iteration procedure. The results of the solution are presented in Figs. 2.3 and 2.4 for three values of the parameter $B_0$. In Fig. 2.3 we have plotted the distribution of the pressure disturbance $P$ against $X$ in the interaction region ahead of the separation point. Clearly, an increase in the parameter $B_0$ leads to the shortening of the free interaction region which is in agreement with the results obtained in analyzing the linear boundary value problem. In Fig. 2.4 the distribution of the

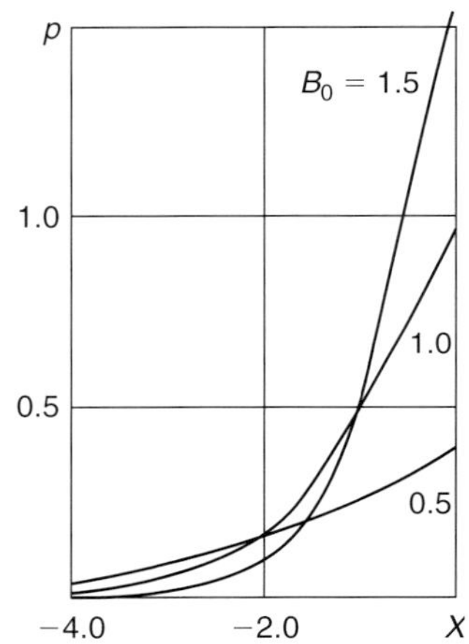


Fig. 2.3.

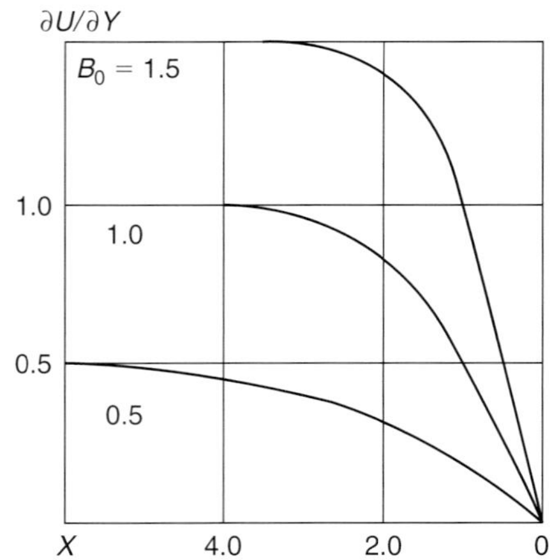


Fig. 2.4.

viscous stress over the plate surface is plotted; the stress is normalized on the friction in the boundary layer ahead of the interaction region.

The boundary value problem (2.11) incorporates the similarity parameter $B_0$ proportional to the ratio of the surface friction ahead of the interaction region to the given pressure gradient. For determining the value of $B_0$ the separation point position is required to be known; the latter depends on the conditions imposed downstream, in particular, at the reattachment point.

The boundary value problem (2.11) describes not only the flow in the vicinity of the separation point but also some other flows in which interaction effects are important. These include, for example, the supersonic flow over a flat plate in the presence of an adverse pressure gradient if at the point, where the surface friction is equal in order to $O(\varepsilon^{-6/7})$, a pressure differing from the pressure in the inviscid flow at the same point by $O(\varepsilon^{4/7})$ is preassigned. The pressure difference can be both positive and negative; the choice of the corresponding solution is determined by the sign of the constant $C_0$ in solution (2.12). In the interaction region the flow parameters depend on the similarity parameter $B_0$. For large values of $B_0$ the flow is described by "free interaction" theory (Neiland, 1968; 1969*a*; Stewartson and Williams, 1969). For small values of $B_0$ the problem reduces to that studied in the paper of Stewartson (1970*b*), the results of which are applicable in the region, where the surface friction is positive and does not vanish.

## 2.2 Expansion flow

The free interaction equations (1.23) for $\lambda = \varepsilon^{3/4}$, subject to the boundary conditions (1.24) and (1.31) and the initial conditions (1.31d), have also solutions corresponding to expansion flows in which $p_3 < 0$. These solutions are associated, for example, with the flows presented in Fig. 1.9c for $p_b - p_\infty < 0$, the flow ahead of the corner point for $\theta < 0$ in Fig. 1.9a, and the flow far away from the corner point for $|\theta| = O(1)$ (the last problem is considered in Chapter 3).

For studying the expansion flows we will introduce variables analogous to (1.32) with the only difference that in the first formula (1.32) the sign ahead of $p$ is reversed in order for $\xi$ could remain positive. Then, instead of problem (1.33), we obtain the following problem:

$$f''' = \beta[-1 + 0.5f'^2 - ff'' + \xi(f'\dot{f}' - \dot{f}f'')] \quad (2.19)$$

$$\frac{g''}{\sigma} = \beta[0.5f'g - fg' + \xi(f'\dot{g} - \dot{f}g')]$$

$$\beta = \frac{-\xi^{3/2}}{\dfrac{d}{d\xi}\left\{2\xi^{1/2}\lim_{\eta\to\infty}[\eta - (2f)^{1/2}]\right\}}$$

$$f(\xi, 0) = f'(\xi, 0) = g(\xi, 0) = 0, \quad f''(\xi, \infty) = g'(\xi, \infty) = 1$$

$$f(0, \eta) = 0.5\eta^2, \quad g(0, \eta) = \eta$$

### 2.2.1 Asymptotic behavior of the solution, as $\xi \to 0$ and $\xi \to \infty$

We will first consider the behavior of the solution as $\xi \to 0$ or $X \to -\infty$.

We introduce new variables and expansions for the functions

$$(\xi, \eta) \to (\xi, N), \quad N = \eta\xi^{1/2} \tag{2.20}$$

$$f \sim \frac{N^2}{2\xi} - \Phi(N) + \cdots, \quad g \sim \frac{N}{\xi^{1/2}} - \xi^{1/2}G(N) + \cdots, \quad \xi \to 0$$

Substituting Eq. (2.20) in Eq. (2.19) and passing to the limit $\xi \to 0$ leads to Eqs. (1.36)–(1.38). Thus, for both expansion and compression flows the flow function disturbances decay following the same law as $X \to -\infty$ (of course, the only difference is in the signs of $\Delta p$, $\Delta f''_w$, $\Delta g'_w$, etc.).

The behavior of the solution as $\xi \to \infty$ is much more complicated. In what follows it is shown that, as $\xi \to \infty$, $X$ approaches a certain finite limit. This limiting value can be conveniently taken for the origin. In the main region, where $\eta \sim O(1)$, the solution can be easily found as $\xi \to \infty$, since outside a narrow wall layer the viscous terms turn out to be inessential. Omitting elementary algebra, we will present the final solution

$$f(+\infty, \eta) \sim \frac{(\eta + \sqrt{2})^2}{2} - 1 + \cdots, \quad g \sim \sqrt{2f} + \cdots \tag{2.21}$$

$$\beta \sim \frac{\xi^2}{\sqrt{2}} + \cdots, \quad (-X) \sim \frac{2^{3/2}}{\xi^{1/2}} + \cdots, \quad \xi \sim \frac{8}{(-X)^2} + \cdots$$

However, this limiting solution does not satisfy the conditions imposed on the body surface. Therefore, it is necessary to consider separately the thinner wall layers, for which viscosity and heat conduction effects in the momentum and energy equations remain to be essential as $\xi \to \infty$ and $\eta \to 0$.

Formulas (2.1) indicates that, as $\eta \to 0$, we have

$$f \sim \sqrt{2}\eta + \cdots, \quad g \sim 2^{3/4}\eta^{1/2} + \cdots \quad \eta \to 0 \tag{2.22}$$

In accordance with the asymptotic expansion matching principle, in the region $\eta \sim 1$ and in the region $\eta \to 0$, whose scale is as yet unknown, expressions (2.22) must be used in determining the outer boundary conditions for the viscous flow region. This helps us to guess the required form of the variables in the region $\eta \to 0$:

$$\bar{\eta} = \frac{8\eta}{2^{1/4}\bar{X}^2}, \quad \bar{X} = -X, \quad f = \frac{2^{1/4}}{8}\bar{X}^2\bar{f}, \quad g = \frac{\bar{X}}{2^{5/8}} \frac{\bar{g}}{\left(\ln \frac{1}{\bar{X}}\right)^{1/4}} \tag{2.23}$$

$$\bar{f}''' = -1 + \frac{1}{2}\bar{f}'^2 + O(\bar{X}), \quad \bar{f}_w = \bar{f}'_w = 0, \quad \bar{f}'_e = \sqrt{2} \tag{2.24}$$

$$\bar{g}'' = 0 + O(\bar{X}), \quad \bar{g}(0) = 0 \tag{2.25}$$

The form of Eq. (2.25) itself shows that, as regards the energy equation, the $\eta \sim O(1)$ and $\overline{\eta} \sim O(1)$ regions do not overlap, since, as $\overline{X} \to 0$, the degenerate equation (2.25) has no solution which would behave as $\overline{\eta}^{1/2}$ when $\overline{\eta} \to \infty$ (in accordance with Eq. (2.22), this should be necessary for matching the solutions).

Thus, we are obliged to introduce an additional intermediate region in which the variables and functions take the form:

$$\overline{\eta}_1 = \frac{4\eta}{2^{1/4}\overline{X}^2\left(\ln \frac{1}{\overline{X}}\right)^{1/2}}, \quad \overline{X} = -X, \quad f = \frac{2^{1/4}\overline{X}^2}{4}\left(\ln \frac{1}{\overline{X}}\right)^{1/2}\overline{f}_1 \tag{2.26}$$

$$g = \frac{\overline{X}}{2^{1/8}}\left(\ln \frac{1}{\overline{X}}\right)^{1/4}\overline{g}_1, \quad \overline{f}_1 = \sqrt{2}\overline{\eta}_1 + \cdots$$

$$\frac{1}{\sigma}\overline{g}_1''' = \sqrt{2}\left(\frac{1}{2}\overline{g}_1 - \overline{\eta}_1\overline{g}_1'\right), \quad \overline{g}_1(0) = 0, \quad \overline{g}_1 \to \overline{\eta}^{1/2} \quad \overline{\eta}_1 \to \infty$$

Matching the solutions for $\overline{g}_1$ and $\overline{g}$ makes it possible to determine the boundary condition for $\overline{g}$ and to find the solution for the asymptotic heat flux

$$\overline{g}_e' = \frac{1}{2}\overline{g}_{1w}', \quad \overline{g}_{1w}' \approx 0.938\sigma^{1/4} \tag{2.27}$$

Thus, at the acceleration of the viscous flow region accompanied by an increase in the pressure gradient, the viscosity and heat conduction effects vanish from the first approximation in an increasingly larger part of the boundary layer. However, near the wall there are always regions in which these effects are essential. The necessity of introducing the $\overline{\eta}_1 \sim O(1)$ region shows that the heat conduction effects decay somewhat more slowly, though the difference between the viscous and heat-conducting layer scales is small. Their ratio is of the order $(\ln(1/\overline{X}))^{1/2}$. As $X \to 0$, the friction and heat flux increase as

$$f_w'' \sim \frac{8\overline{f}_w''}{2^{1/4}(-X)^2} + \cdots, \quad g_w' \sim \frac{2^{13/8}0.938\sigma^{1/4}}{(-X)\left[\ln \frac{1}{(-X)}\right]^{1/4}} + \cdots$$

### 2.2.2 *Results of calculations*

From the physical standpoint, the expansion flows with free interaction between the supersonic flow and the boundary layer can be qualitatively described as follows. Let small rarefactions penetrate upstream through the subsonic part of the boundary layer. Thus, in Fig. 1.9a the flap is deflected downward by an angle $O(Re)^{-1/4}$ or in Fig. 1.9c $p_b - p_\infty = O(Re)^{-1/4} < 0$. Then a decrease in the thickness of accelerated subsonic streamtubes results in the downward deflection of the outer supersonic flow. In turn, this leads to further decrease in the thickness of near-wall streamtubes, etc., until a new pressure distribution is attained. The solution of Eq. (2.19) governs such a flow. Naturally, too large disturbance amplitudes, $\Delta p \sim O(1)$, cannot be described by free interaction theory. On the $O(Re^{-1/4})$ scale, these $\Delta p$ are infinitely large in magnitude. For this reason, the study of expansion flows with $\Delta p \sim O(1)$ requires also the consideration of other flow regions (see Chapter 3), while transition to these regions within the framework of free interaction theory corresponds to $\xi \to \infty$.

Equations (2.19) was solved numerically using the same methods as for compression flows. The results for the distributions of the pressure $\xi$, friction $f''_w$, and heat flux $g'_w$ coefficients are presented in Figs. 2.5–2.10. It should be noted that in these flows the friction increases much more rapidly than the heat flux.

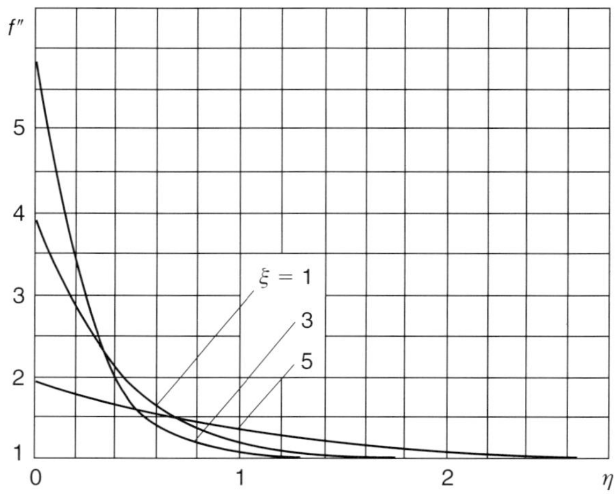


Fig. 2.5.

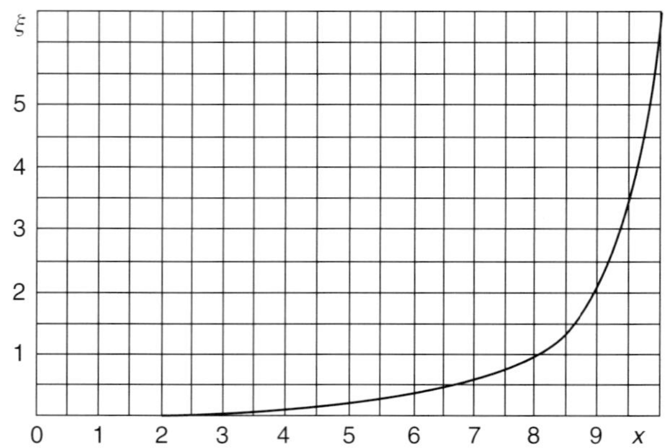


Fig. 2.6.

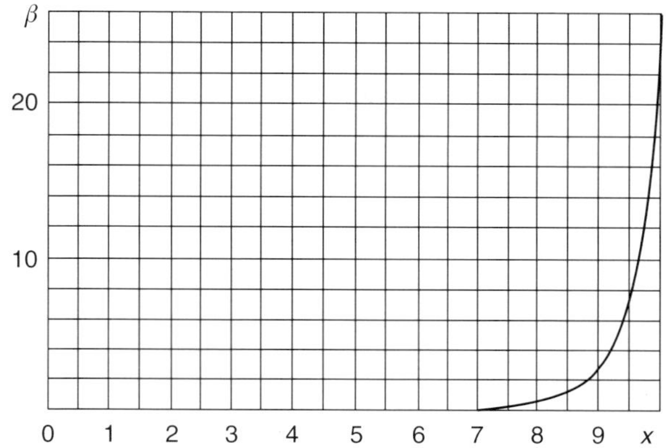


Fig. 2.7.

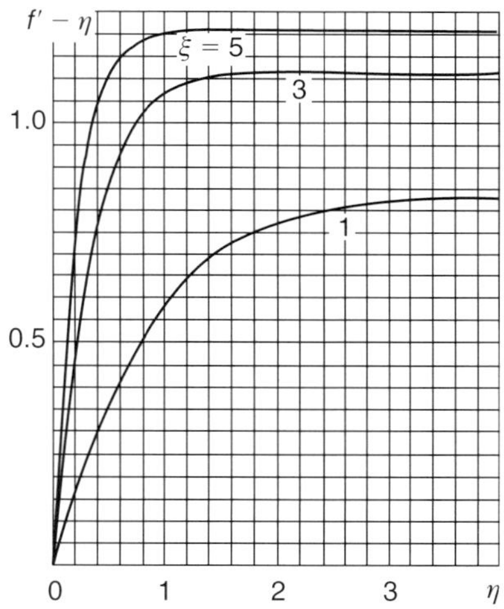


Fig. 2.8.

We will note once more that the dimensionless function distributions presented above give a universal solution for the corresponding problems (see, e.g., Fig. 1.9) in the similarity variables related with the physical variables by formulas (1.32). Thus, the local flow is only slightly dependent on the boundary layer prehistory.

## 2.3 Other types of flows described by free interaction equations

The problems considered in Sections 1.2 and 2.2 pertain to flat-plate compression and expansion flows. However, the fairly general and simple form of the similarity laws for flows with

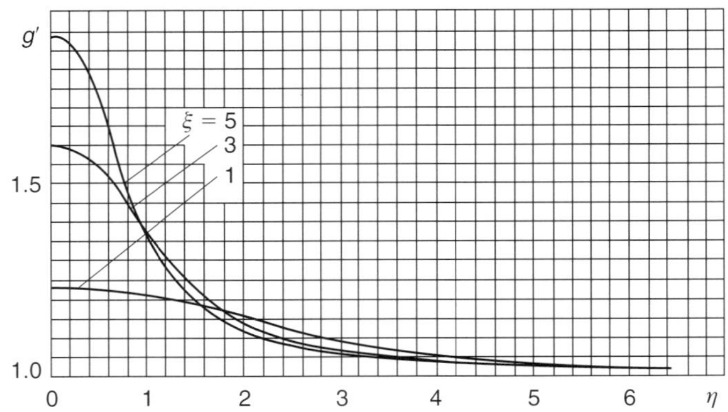


Fig. 2.9.

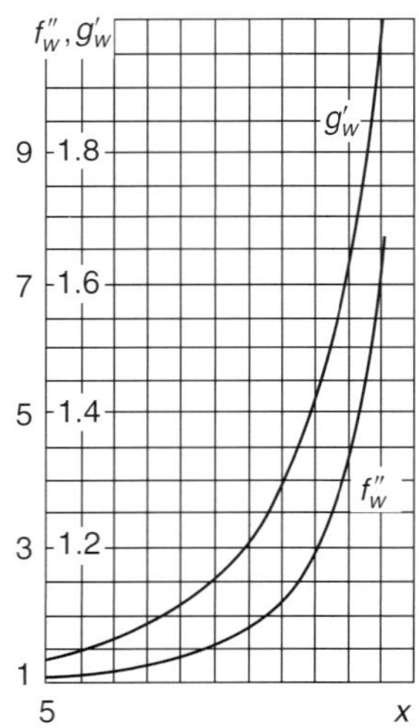


Fig. 2.10.

free interaction, the relatively simple form of the equations and boundary conditions, and, finally, the fact that even in the first approximation the results obtained have a satisfactory accuracy for not-too-large disturbance amplitudes, are exact in the limit, lead to a clear idea on the contributions of different physical factors, and stimulate the development of theory applications to a wider class of flows. For certain of these flows, such as the flow inside a corner somewhat smaller than $\pi$ and the flow in the region of interaction between a shock and a boundary layer, numerical solutions are obtained. For other flows we present only the formulation of the problems, the equations and the boundary conditions, and some ideas concerning the flow nature.

### *2.3.1 Equations and boundary conditions for the case of a curvilinear body contour*

We will first consider the expansion (and compression) flow presented in Fig. 2.11. We will assume that the region, where the flow is deflected by an angle $\theta_w \approx O(\varepsilon^{1/2})$, connects two rectilinear regions of the body contour and its length is of the order $O(\varepsilon^{3/4})$. Along with the Cartesian coordinate system with the $x$ axis parallel to the oncoming supersonic flow and the body contour ahead of the turn region, we introduce a curvilinear orthogonal coordinate system $(s, n)$ fitted to the body contour and normals to it (Fig. 2.11).

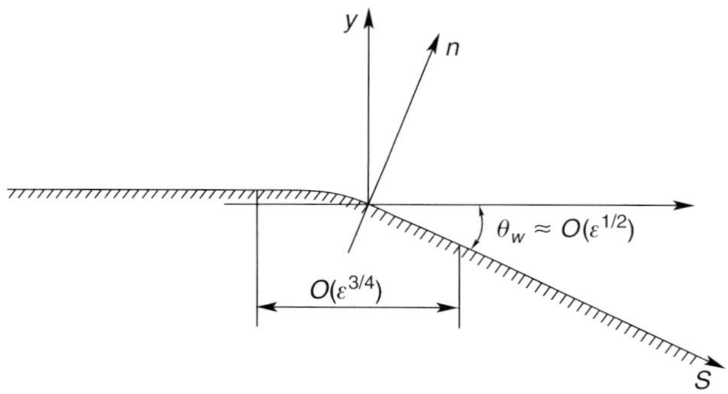


Fig. 2.11.

The system of free interaction equations (in the first approximation in $\varepsilon$) is quite analogous to system (1.23)

$$\rho_w u_3 \frac{\partial u_3}{\partial s_3} + \rho_w v_3 \frac{\partial u_3}{\partial n_3} + \frac{\mathrm{d}p_3}{\mathrm{d}s_3} = \mu_w \frac{\partial^2 u_3}{\partial n_3^2}, \qquad \frac{\partial p_3}{\partial n_3} = 0 \tag{2.28}$$

$$\rho_w u_3 \frac{\partial h_3}{\partial s_3} + \rho_w v_3 \frac{\partial h_3}{\partial n_3} = \frac{\mu_w}{\sigma} \frac{\partial^2 h_3}{\partial n_3^2}, \qquad \frac{\partial u_3}{\partial s_3} + \frac{\partial v_3}{\partial n_3} = 0$$

In the first approximation, the longitudinal curvature effect is absent from Eq. (2.28). Let the body shape be given by a function $y_{3w}(x_3)$

$$y_w = \varepsilon^{5/4} y_{3w}(x_3), \qquad \frac{\mathrm{d}y_w}{\mathrm{d}x_3} = \varepsilon^{1/2} \theta_{3w} \tag{2.29}$$

The difference from the problem for the flat plate is in the boundary condition replacing condition (1.31); on its right-hand side it involves not only the variation of the boundary layer displacement thickness but also the body contour slope.

We introduce the similarity law and integration variables as follows:

$$\xi = |P| = |p_3|\left[\frac{2(M^2-1)}{\mu_w a}\right]^{1/2}, \quad \xi f = \Psi = \psi_3\left[\frac{2a(M^2-1)^{1/2}}{\mu_w}\right]^{1/2} \tag{2.30}$$

$$X = S = [8(M^2-1)^{3/2}\rho_w^2\mu_w a^5]^{1/4}x_3$$

$$\xi^{1/2}\eta = N = n_3\left[\frac{2(M^2-1)^{1/2}\rho_w^2 a^3}{\mu_w}\right]^{1/4}, \quad Y_w = y_{3w}\left[\frac{32(M^2-1)\rho_w^2 a^3}{\mu_w}\right]^{1/4}$$

$$\theta_w = \theta_{3w}\left[\frac{2}{\mu_w a(M^2-1)^{1/2}}\right]^{1/2}$$

where $P$, $S$, $X$, $N$, $Y_w$, $\theta_w$, and $\Psi$ are the similarity law variables. For them the equations of type (2.28) or (1.23) do not incorporate the constants $\rho_w$, $a$, and $\mu_w$ characterizing a particular flow.

For the case of a curvilinear body contour the equations and boundary conditions in the integration variables (1.38) or (2.19) remain the same, except for the outer boundary condition

$$\beta = \xi^{1/2}\frac{d\xi}{dS} = \frac{\pm\xi^{3/2}(\xi \mp \theta_w)}{\dfrac{d}{d\xi}\left\{2\xi^{1/2}\lim\limits_{\eta\to\infty}[\eta-(2f)^{1/2}]\right\}} \tag{2.31}$$

Here, the upper and lower signs correspond to the compression and expansion flows, respectively.

### 2.3.2 *Flow inside a corner somewhat smaller than $\pi$ and region of weak shock incidence on a boundary layer*

Let us consider the flow near the corner point on the contour of a body consisting of two flat plates. The corner point is at a distance $l$ from the nose of the leading plate set parallel to the oncoming flow. The deflection angle is $\theta_w^0 \approx O(\varepsilon^{1/2})$ or $O(Re^{-1/4})$. The equations and boundary conditions governing the flow in the vicinity with the longitudinal dimension $O(Re^{-3/8}) = O(\varepsilon^{3/4})$ are presented above. If the origin is placed at the corner point, the boundary condition (2.31) for this flow takes the form:

$$\theta_w = \begin{cases} 0 & \text{for } S < 0 \\ \theta_w^0 = \text{const} & \text{for } S > 0 \end{cases} \tag{2.32}$$

In the case $\theta_w^0 > 0$ we have a compression flow. We will first show that in the free interaction region with the longitudinal dimension $\sim\varepsilon^{3/4}$ this flow in the approximation under consideration is quite equivalent to a flat-plate flow, at the point $x = 0$ of which a weak shock with an appropriately chosen pressure difference is incident. For the flow with a shock the outer boundary condition was derived in Chapter 1 (second formula (1.31)). It can be easily seen that the boundary condition (2.31) for the problems with shock incidence on the boundary layer (1.31*b*) and the flow around the corner point on the body contour (2.32) are the same provided

$$2(M_1^2 - 1)[p_3] = \theta_{3w}^0 \tag{2.33}$$

The total pressure increase turns out to be the same in both cases.

A considerable difference in the flows takes place in the immediate vicinity of the shock incidence point and the corner point on the body contour. Let us consider this flow region in more detail. Its transverse dimension is determined by the boundary layer thickness and is of the order $\varepsilon$. Since in the supersonic flow the shock inclination angle is $O(1)$, the longitudinal dimension of the region, within which the shock is reflected from the vortex part of the flow (boundary layer), must be of the same order as the transverse dimension ($\sim\varepsilon$). The shock amplitude is $\Delta p \approx O(\varepsilon^{1/2})$. Then in the small region under consideration with the scales $\varepsilon \times \varepsilon$ the longitudinal and transverse derivatives of the pressure are of the same order $\varepsilon^{-1/2}$. It can be easily seen that the viscous terms of the Navier–Stokes equations are small everywhere in this region, except for a thin wall sublayer of thickness $\sim\varepsilon^{3/2}$. These locally inviscid flows are studied in detail in Chapter 3. For studying the flows with free interaction it is only important that within the region under consideration of length $\sim\varepsilon$ the pressure on the body surface remains constant in the leading order: $\Delta p \approx O(\varepsilon^{1/2})$. Precisely this condition was preassigned in deriving the boundary condition (1.31c).

Let us first assume that this condition is violated and within the region of length $\sim\varepsilon$ the pressure $\Delta p \approx O(\varepsilon^{1/2})$. Then in a sublayer of thickness $\sim\varepsilon^{5/4}$, in which the velocities are of the order $\varepsilon^{1/4}$, the velocities and, hence, the streamtube thicknesses change by their leading order. However, in this case the complete boundary layer displacement thickness, the variation of which in the leading order is determined, as shown above, by the variation of the displacement thickness of the sublayer under consideration, changes by $\sim\varepsilon^{5/4}$ over a length $\sim\varepsilon$. However, by virtue of the outer boundary condition (Ackeret formula), we have $\Delta p \approx \varepsilon^{1/4}$; thus, we arrive at a contradiction. Thus, within the region of length and thickness $\varepsilon \times \varepsilon$ the flow is locally inviscid, that is, the Bernoulli equation holds, while at both ends of the region, where the transverse pressure gradient $\partial p/\partial y$ vanishes, the pressure is the same in the leading term $\sim\varepsilon^{1/2}$, which makes it possible to disregard this short zone in solving the free interaction problem and only to take account of the discontinuity in the constants in the boundary conditions (2.31) and (2.32).

The problems for compression and expansion flows were solved numerically by integrating Eqs. (1.33) and (2.31) subject to the boundary condition (2.32). Since each value of $\theta_w^0$ is associated with its own value of the pressure coefficient $\xi$ at point $S = 0$, integration was performed using an iteration procedure for determining $\theta_w^0$ at a fixed value of $\xi(0)$.

Numerical integration for $S < 0$ at $\theta_w = 0$ in formula (2.32) was also carried out using the method described below in Section 2.3.4. Then at a certain value $\xi$, for which it was taken $S = 0$, the search for the value of $\theta_w^0$ was started. The branch of the solution associated with $S > 0$ and $\theta_w = \theta_w^0$ was integrated to $\xi = \theta_w^0$ or the band, where condition (2.31) could not be satisfied. The required solution ensuring the attainment of the undisturbed solution with $f''_w = 1$ for $\xi = \theta^0_w$ separated the two above-mentioned solution families obtained in the course of the iteration procedure.

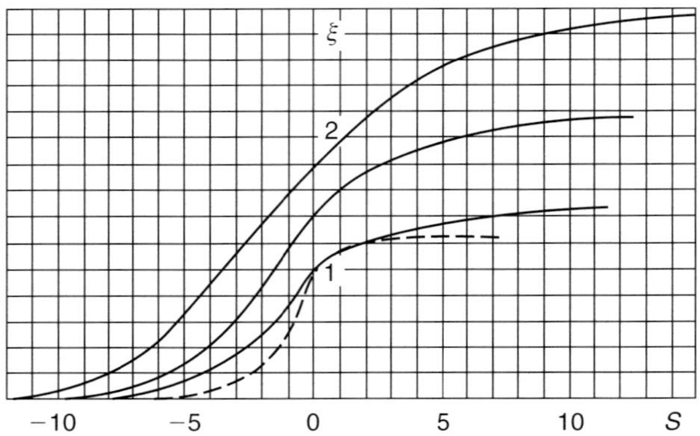


Fig. 2.12.

In Fig. 2.12 we have plotted the calculated pressure distributions for the flows over bodies with different deflection angles at point $S = 0$ or with shock incidence at this point. Solid curves present the results for compression flows. At fairly large disturbance amplitudes (values of $\theta^0_w$) the flow contains separation regions. The separation and reattachment point positions are marked by points at the curves. It was interesting to calculate the disturbance amplitude corresponding to incipient separation. Calculations show that the complete pressure coefficient for the disturbance producing separation of "zero" length is as follows:

$$c_\text{p} \sim 2.3 \left[ \frac{c_{f0}}{(M^2 - 1)^{1/2}} \right]^{1/2} \tag{2.34}$$

Accordingly, for the incident shock $c_\text{p}$ is equal to half this value. Of course, the experimental data on the pressure coefficient for the disturbance producing incipient separation are always somewhat greater than the calculated value, since in most cases for flows with small-sized separation zones the typical inflection on the pressure curves is hardly noticeable. Visually, only a fairly large separation zone is observable.

### *2.3.3 Formulation of other problems for flows with free interaction*

In this section we will discuss some problems whose formulation can take advantage of the above theory for flows with free interaction, though with small modifications. The examples of such flows are presented in Fig. 2.13. It should be remembered that the field of application is not exhausted, even in the main, by these examples.

The first example (Fig. 2.13a) represents the flow downstream of the trailing edge of a finite-length plate. In the region $x > 0$ (the origin is at the edge) the no-slip boundary condition ($u_w = 0$), which take place for $x < 0$, is replaced by the symmetry condition $\partial u/\partial y = 0$. Then the viscosity-driven gas acceleration is accompanied by a decrease in

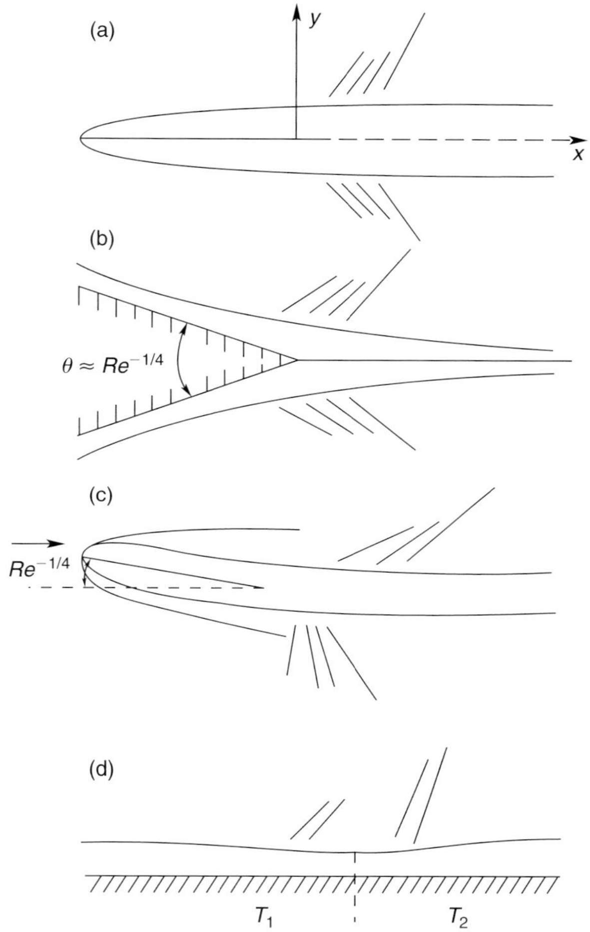


Fig. 2.13.

the boundary layer displacement thickness. Estimates similar to those made in Section 1.1 show that at distances $\sim Re^{-3/8}$ from the edge the induced pressure variation is $\Delta p \approx Re^{-1/4}$. The flow in the layer of thickness of the order $\sim Re^{-5/8}$ governing the pressure variation is described by equations of type (2.31). The results for the flat-plate flows with free interaction presented in Sections 1.2 and 2.2 make it possible to conclude that in the $x < 0$ region the pressure decreases monotonically; since somewhere downstream it must reach the original value (in the leading order $\Delta p \approx \varepsilon^{1/2} = Re^{-1/4}$), a region of pressure recovery must be located at $x > 0$.

The flow presented in Fig. 2.13b is somewhat more complicated. In this case, at point $x = 0$, apart from the change in the boundary condition at the "wall", the "outer" boundary condition (1.31) also changes in the same fashion as in the problem of the flow around a corner point on the body contour. It can be easily understood that, depending on the wedge vertex angle (within the limits of the order $Re^{-1/4}$), on the body there can occur an expansion flow (at small vertex angles), a compression flow, and even flow separation. An example of a nonsymmetric flow is shown in Fig. 2.13c in which the flow over the rear part of a plate set at an angle of attack $\alpha \approx Re^{-1/4}$ in a supersonic stream is considered. At a fairly large value of the ratio $\alpha/Re^{-1/4}$ a separation zone starts to develop on the upper side of the plate. In this case, the algorithm of the numerical solution must be more complicated than that for symmetric problems.

Finally, it should be noted that the free interaction theory equations do not necessarily correspond to an incompressible boundary layer. In Fig. 2.13d we have presented an example of the flow over a plate on which in a region of length $\sim Re^{-3/8}$, or shorter, the wall temperature varies by the leading order.

Another example of the flow, in which the density is variable and the momentum and energy equations cannot be separated, is considered in Chapter 3 in connection with the problem of jet attachment to a body surface. There the equations and boundary conditions are presented. Here, we will note that at finite distances $\Delta x \approx 1$ from the region of rapid variation in the wall temperature, the boundary layer gas temperature and density and hence the displacement thickness change in the leading order. This produces a pressure gradient $\sim Re^{-1/2}$ in the outer supersonic flow, which affects the boundary layer flow only in the second approximation. However, at distances $\varepsilon^{3/4} \approx Re^{-3/8}$ and on "heating" a sublayer of thickness $\sim Re^{-5/8} = \varepsilon^{5/4}$, the variation of the displacement thickness of this sublayer produces the pressure variation $\Delta p \approx \varepsilon^{1/2} = Re^{-1/4}$ which affects the sublayer flow already in the first approximation. Again we deal with a free interaction problem in which the pressure first decreases (on cooling) and then is restored (in the leading order, where $\Delta p \approx \varepsilon^{1/2}$) to the original value.

### *2.3.4 Integration of the equations*

While in integrating Eq. (1.33) for increasing the accuracy near the outer boundary in $\eta$ we introduce the variables

$$\tilde{f} = f - \frac{1}{2}\eta^2, \quad \tilde{g} = g - \eta$$

The equations and boundary conditions take the form:

$$\tilde{f}''' = \beta\left(1 + \frac{1}{2}\tilde{f}'^2 - \tilde{f}\tilde{f}'' + \eta\tilde{f}' - \frac{1}{2}\eta^2\tilde{f}'' - \tilde{f}\right) + \xi\beta(\tilde{f}'\tilde{f}'^* - \tilde{f}^*\tilde{f}'' + \eta\tilde{f}'^* - \tilde{f}^*)$$

$$\frac{\tilde{g}''}{\sigma} = \beta\left[\frac{1}{2}\tilde{f}'\tilde{g} - \tilde{f}\tilde{g}' + \frac{1}{2}\eta(\tilde{f}' + \tilde{g}) - \frac{1}{2}\eta^2\tilde{g}' - \tilde{f}\right] + \xi\beta(\tilde{f}'\tilde{g}^* - \tilde{f}^*\tilde{g}' + \eta\tilde{g}^* - \tilde{f}^*)$$

$$\tilde{f}(\xi, 0) = \tilde{f}'(\xi, 0) = \tilde{f}(0, \eta) = \tilde{g}(0, \eta) = 0$$

$$\tilde{f}''(\xi, \infty) = \tilde{g}'(\xi, \infty) \to 0$$

The integration was performed using two methods. The first method was developed by Petukhov (1964) for integrating the boundary layer equations at a given pressure distribution.

In TsAGI there is a semi-standard program developed by Seliverstov on the basis of the method proposed by Petukhov (1964). Within the framework of this method, the equations are replaced by finite differences with errors $O(\Delta\xi^2)$ and $O(\Delta\eta^4)$. The solution is sought on $\xi = \text{const}$ lines using an iteration procedure and the sweeping technique. Since in the problem under consideration the pressure is not given, we were led to preassign $\beta(\xi)$ at each characteristic, to determine the solution using the semi-standard program, and then to check the fulfillment of the boundary condition for $\beta(\xi)$; thus, $\beta(\xi)$ was obtained by iteration. The latter method was different from the former only in that the derivatives with respect to $\eta$ were not replaced by finite differences and it was ordinary differential equations that were integrated at the characteristics, with an additional iteration procedure aimed at the fulfillment of the outer boundary condition $\widetilde{f}''(\infty) = 0$ (or $\widetilde{g}'(\infty) = 0$). In the latter case, the required computer time is greater, while the results are almost the same in using both methods.

## 2.4 Elimination of boundary layer separation by means of slot suction

In many cases boundary layer separation is undesirable; for this reason, different ways of controlling the boundary layer are used for eliminating separation. One of the most effective techniques of controlling the boundary layer is gas suction. The first investigation of the suction effect on the boundary layer was performed by Prandtl in 1904. There is much work, both experimental and theoretical, devoted to the study of flows with boundary layer suction; their survey can be found in the books by Lachmann (1961) and Schlichting (1968).

The suction effect on laminar boundary layer separation can be investigated only provided the flow in the vicinity of the boundary layer separation point is comprehensively described. In the paper of Neiland (1971*b*) the problem of the flow over a flat plate with a deflected flap was considered. On the basis of an asymptotic analysis of the behavior of the solutions of the Navier–Stokes equations it was shown that the pressure coefficient, or the corresponding flap deflection angle, at which in the laminar boundary layer there appears a zero-length separation zone, is determined by the following formula:

$$C_{\text{pi}} = 2.23 C_{f0}^{1/2} (M_\infty^2 - 1)^{-1/4} \tag{2.35}$$

where $M_\infty$ is the Mach number in the inviscid flow above the boundary layer and $C_{f0}$ is the friction drag coefficient in the undisturbed boundary layer ahead of the flap. From relation (2.35) it follows that the permissible pressure difference, at which the flow past the deflected flap is still separationless, increases with the friction drag coefficient. In this connection, in the paper of Lipatov and Neiland (1974) it was shown that on a flat plate with uniform gas suction at dimensionless suction velocities $Re^{-1/2} \ll v_w \ll 1$ the flow remains separationless, if the coefficient of the pressure disturbance induced, for example, by shock incidence or control deflection is not greater than the maximum value

$$C_{\text{pi}} = 2.23 (\rho_w v_w)^{1/2} (M_\infty^2 - 1)^{-1/4}$$

Distributed suction from the plate, starting from its leading edge, forms a boundary layer with an "asymptotic" velocity profile in which the dimensionless friction is of the order $O(v_w)$. In flows, in which the shock position is given, there is no need in gas suction along the entire body length.

In order to form a boundary layer with friction $O(v_w)$ it is sufficient to suck off gas with the flow rate $\sim O(Re^{-1/2})$ over a length $\Delta x \sim O(Re^{1/2} v_w^{-1})$ ahead of the shock incidence point or a bend in the body contour. From physical considerations it can be supposed that the narrower the region in which suction is concentrated the more effective the suction, since the length of the zone, in which the sucked-off boundary layer is subjected to the decelerating action of friction on the body, decreases with the extent of the above-mentioned region. For this reason, in this study we consider the slot suction of gas, which is, moreover, easier to implement.

In this section, we consider the asymptotic, as $Re \to \infty$, behavior of the solution of the Navier–Stokes equations governing the laminar flow over a flat plate in the presence of slot suction.

### 2.4.1 Formulation of the problem

We will consider the uniform supersonic viscous flow over a flat plate. At a distance $\ell$ from the leading edge the plate contour is deflected by an angle $\theta > 0$ (Fig. 2.14).

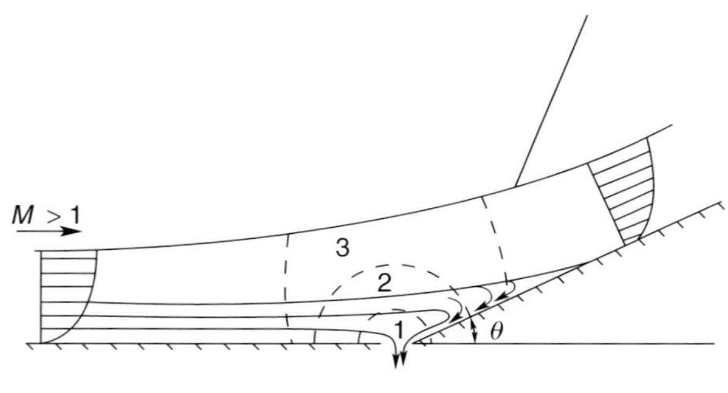


Fig. 2.14.

It is assumed that the characteristic Reynolds number $Re = \rho_\infty u_\infty \ell / \mu_\infty$ is high but subcritical, so that the boundary layer is still laminar.

When the wall is deflected by an angle $\theta$, in the inviscid supersonic flow the pressure must increase by $\Delta p \sim \theta$. Therefore, for fairly large values of $\theta$ boundary layer separation might be expected. The theory describing separation in this flow was developed in the works of Neiland (1969; 1971). It was shown that separation begins at $\theta \sim \varepsilon^{-1/2}$, $\varepsilon \sim Re^{-1/2}$. The pressure disturbances resulting in separation are transmitted both upstream and downstream through a distance $\Delta x \sim \varepsilon^{3/4}$. Since $\Delta p \sim \theta \ll 1$, the gas velocity changes by its leading order only in a narrow wall layer of thickness $\sim \varepsilon^{5/4}$, in which the velocities of streamtubes are $u \sim \Delta p^{1/2} \sim \varepsilon^{1/4}$. These streamtubes overcome the pressure increase driven only by viscosity forces.

The thickness of the layer, in which the dynamic head varies in the leading term, increases with $\theta$ and $\Delta p$. In fact, in the undisturbed boundary layer near the wall we have

$$u \sim \frac{y}{\varepsilon} \tag{2.36}$$

while, in accordance with the momentum equation, the region with nonlinear velocity variations is determined by the relation

$$\Delta p \sim u^2 \tag{2.37}$$

Then from Eqs. (2.36) and (2.37) there follows the estimate for the thickness of the zone with nonlinear velocity variations

$$\delta \sim \varepsilon \Delta p^{1/2} \tag{2.38}$$

We also note, following the work of Matveeva and Neiland (1967), that the thickness of the outer part of the boundary layer varies only by $\Delta\delta \sim \varepsilon\Delta p$. For this reason, upstream of the corner pressure disturbances are induced in the flow over the boundary layer with a thickness changed by $\delta$. Then, in accordance with the linear theory of supersonic flows

$$\Delta p \sim \frac{\delta}{\Delta x} \tag{2.39}$$

where $\Delta x$ is the disturbed flow zone length. From Eqs. (2.39) and (2.38), taking into account that $\Delta p \sim \theta$, we can obtain the following estimates:

$$\Delta x \sim \frac{\varepsilon}{\theta^{1/2}} \sim \frac{\varepsilon}{\Delta p^{1/2}} \gg \varepsilon \quad \text{for } \Delta p \ll 1 \tag{2.40}$$

Let us now compare the orders of the leading viscous term and the pressure gradient in the equations for the sublayer with nonlinear viscosity disturbances

$$\varepsilon^2 \left( \frac{\partial^2 u}{\partial y^2} : \frac{\partial p}{\partial x} \right) \sim \frac{\varepsilon}{\theta^2} \tag{2.41}$$

In accordance with Eq. (2.41), for $\theta \sim \varepsilon^{1/2}$ this ratio is of the order of unity; therefore, a separationless flow is possible. However, for $\theta \gg \varepsilon^{1/2}$ this ratio is small and the inviscid flow thus obtained is not able to overcome the increase in the pressure.

From this consideration there follows the estimate for the dimensionless gas flow rate $Q$ which must be sucked off through a slot near the corner point in order to ensure separationless flow past a flap with $\theta \gg \varepsilon^{1/2}$. This is the flow rate in the sublayer with nonlinear velocity disturbances, which, in accordance with Eqs. (2.36)–(2.41), for a given $\theta$ is as follows:

$$Q \sim \varepsilon \Delta p \sim \varepsilon\theta \tag{2.42}$$

Let a $b$-long slot be located near a bend in the body contour. We will assume that the pressure difference across the slot produced by the boundary layer suction system is fairly high and a critical "sonic" outflow regime has been attained. Then, according to Eq. (2.42), we obtain the estimate for the slot width

$$b \sim Q \sim \varepsilon\theta \tag{2.43}$$

From the above estimates it follows that in order for the flow to be separationless the gas suction for a given $\theta$ must be such that the pressure disturbance ahead of the slot is negative, $\Delta p < 0$. If the pressure disturbance is positive, $\Delta p > 0$, it cannot be greater than $\varepsilon^{1/2}$. This means that for $\theta \gg \varepsilon^{1/2}$ the pressure ahead of the slot can only decrease in the leading order. In the separationless flow the wall pressure increase may occur only near the slot, $\Delta x \ll \varepsilon$, or downstream of it.

Let us now consider the local flow pattern in a small vicinity of the slot. The characteristic dimensions of region 1 (Fig. 2.14) are determined by the slot dimensions $\Delta x \sim \Delta y \sim b$, while the velocities are $u \sim v \sim 1$, since $v_w \sim 1$. At these distances from the slot, the flow velocity in the undisturbed boundary layer is small. According to Eqs. (2.36)–(2.42), it is of the order $b/\varepsilon \ll 1$. At distances $\Delta x \gg b$ the decay of the suction-induced velocity can be estimated using the continuity equation from the formula

$$u \sim \frac{b}{\Delta x} \tag{2.44}$$

From Eqs. (2.44) and (2.36) it can be seen that in this case at distances $\Delta x \sim \Delta y \sim \delta$ the suction-induced velocity must become of the order of the velocities in the undisturbed boundary layer at distances from the wall

$$u \sim \frac{\delta}{\varepsilon} \tag{2.45}$$

For this reason, it is necessary to introduce also region 2 (Fig. 2.14) with the scales $\Delta x \sim \Delta y \sim \varepsilon\theta^{1/2}$. From Eq. (2.41) it follows that for $\Delta x \ll \varepsilon^{3/4}$ the flow regions with nonlinear velocity disturbances become inviscid. Then viscosity turns out to be essential only in thinner layers and has no effect on the pressure distribution in the leading term. The estimates for the layer thicknesses are derived below.

We will now note that the pressure distribution over the surface is determined by the body shape and the boundary layer thickness distribution for distances greater than the boundary layer thickness and the form of the solution for regions 1 and 2 for $\Delta x \ll \varepsilon$. Naturally, in deriving the global solution a region with the scales $\Delta x \sim \Delta y \sim \varepsilon$ should be considered. However, for further analysis of the problem the possible local solutions for regions 1 and 2 and the nonlinear layer must be preliminarily derived for $\Delta x \gg \varepsilon$.

### *2.4.2 Derivation of the equations and boundary conditions for regions 1 and 2*

In accordance with the estimates obtained above, we introduce the following coordinates and flow functions which conserve the values $O(1)$ in region 1 when passing to the limit in the Navier–Stokes equations and the boundary conditions

$$\varepsilon \to 0, \quad \theta \to 0, \quad \varepsilon^{3/4} \le \theta \ll 1 \tag{2.46}$$

$$x = bx_1, \quad y = by_1$$
$$u = u_1(x_1, y_1) + \cdots, \quad v = v_1(x_1, y_1) + \cdots \tag{2.47}$$
$$p = p_1(x_1, y_1) + \cdots, \quad \rho = \rho_1(x_1, y_1) + \cdots$$

Substituting Eq. (2.47) in the system of the Navier–Stokes equations and passing to limit (2.46) leads to the complete system of Euler equations for a compressible inviscid gas

$$\rho_1 u_1 \frac{\partial u_1}{\partial x_1} + \rho_1 v_1 \frac{\partial u_1}{\partial y_1} + \frac{\partial p_1}{\partial x_1} = 0 \tag{2.48}$$

$$\rho_1 u_1 \frac{\partial v_1}{\partial x_1} + \rho_1 v_1 \frac{\partial v_1}{\partial y_1} + \frac{\partial p_1}{\partial y_1} = 0$$

$$\rho_1 u_1 = \frac{\partial \psi_1}{\partial y_1}, \quad \rho_1 v_1 = -\frac{\partial \psi_1}{\partial x_1}$$

with the boundary conditions

$$u_1 = v_1 = 0 \quad \text{for } \sqrt{x_1^2 + y_1^2} \to \infty$$

$$v_1\left(|x_1| > \frac{1}{2}, 0\right) = 0 \tag{2.49}$$

It is assumed that downstream of the slot the pressure is fairly low, so that the sonic velocity is reached at the edges $x_1 = \pm 1/2$, while the sonic line shape is determined from the solution of the problem. The first boundary condition (2.49) can be derived from the gas flow rate balance. There exists an exact solution of the system of equations (2.48) subject to the boundary conditions (2.49) obtained by Frankl (1947). We will present the expression for the gas flow rate $Q$ which will be necessary in what follows:

$$Q = -0.85 \left(\frac{\gamma+1}{2}\right)^{-(\gamma+1)/[2(\gamma-1)]} \left[\left(1 - \frac{\gamma-1}{2} M_\infty^2\right) \frac{T_w}{T_0}\right]^{-1/2} \frac{b}{M_\infty} \tag{2.50}$$

In order to impose the boundary condition in region 2 we will present the form of the solution for region 1, as $x_1^2 + y_1^2 \to \infty$

$$\psi_1 = Q \arctan\left(\frac{y_1}{x_1}\right) + Q\pi \tag{2.51}$$

which corresponds to the solution for a point sink with the flow rate $Q$ located at the origin in an incompressible fluid.

In region 2 we introduce the following representations for the flow functions and coordinates:

$$x = \varepsilon\theta^{1/2} x_2, \quad y = \varepsilon\theta^{1/2} y_2$$

$$u = \theta^{1/2} u_2(x_2, y_2) + \cdots, \qquad v = \theta^{1/2} v_2(x_2, y_2) + \cdots \tag{2.52}$$

$$\rho = \rho_w + \cdots, \qquad p = \frac{1}{(\gamma M_\infty^2)} + \theta p_2(x_2, y_2) + \cdots$$

Substituting expansions (2.52) in the complete system of Navier–Stokes equations and passing to limit (2.46) leads to the following system of equations:

$$\rho_w u_2 \frac{\partial u_2}{\partial x_2} + \rho_w v_2 \frac{\partial u_2}{\partial y_2} + \frac{\partial p_2}{\partial x_2} = 0 \tag{2.53}$$

$$\rho_w u_2 \frac{\partial v_2}{\partial x_2} + \rho_w v_2 \frac{\partial v_2}{\partial y_2} + \frac{\partial p_2}{\partial y_2} = 0$$

$$\frac{\partial u_2}{\partial x_2} + \frac{\partial v_2}{\partial y_2} = 0, \qquad \rho_w u_2 = \frac{\partial \psi_2}{\partial y_2}, \qquad \rho_w v_2 = -\frac{\partial \psi_2}{\partial x_2}$$

with the boundary conditions

$$v_2(x_2, 0) = 0; \quad p_2(x_2 \to -\infty) = p_{20} \leq 0 \tag{2.54}$$

Moreover, at $x_2 = y_2 = 0$ there is the sink determined by Eq. (2.51). For $\theta \gg \varepsilon^{1/2}$ in region 2 the gas flow rate is much larger than that in the preceding viscous disturbed flow zones. For this reason, the vorticity and total pressure distributions in $\psi_2$ are determined from the matching with the lower part of the profile of the undisturbed boundary layer ahead of the interaction region.

Therefore, we have

$$\omega_2 = -a, \quad p_2 + \rho_w \frac{u_2^2 + v_2^2}{2} = a\psi_2 \tag{2.55}$$

The system of equations (2.53) can be reduced to the Poisson equation

$$\Delta \psi_2 = -a\rho_w \tag{2.56}$$

The solution of the problems (2.53)–(2.56) takes the form:

$$\psi_2 = \frac{\rho_w a y_2^2}{2} + \left(\frac{Q_2}{\pi}\right)\left[\arctan\left(\frac{y_2}{x_2}\right) - \pi\right] + y_2\sqrt{-2p_{20}\rho_w} \tag{2.57}$$

where $Q_2 = Q/(\varepsilon\theta)$.

Clearly, Eq. (2.57) represents the superposition of three terms corresponding to a shear flow, a sink flow of intensity $Q_2$, and a uniform flow with a velocity corresponding to a possible flow acceleration over a length $\Delta x \gg \varepsilon$ ahead of region 2. Generally, this acceleration can be due to upstream propagation of the suction-induced pressure disturbance $p_{20} \leq 0$. Thus, the solution for region 2 depends on two parameters, namely, $Q_2$ and $p_{20}$. The flow rate $Q_2$ is determined from the solution for region 1. The value of $p_{20}$ should be determined from the global solution of the problem and for a time has not been found.

For the purposes of further analysis we will determine the total pressure difference on the body within region 2 using solution (2.57) and the Bernoulli integral (2.55). It is as follows:

$$\Delta p_{2w} = p_2(x_2 \to \infty) - p_{20} \tag{2.58}$$

$$\Delta p_{2w} = -aQ_2 > 0$$

Thus, the total pressure difference on the body in region 2 is independent of $p_{20}$ and completely determined by the sink intensity. However, the type of the flow is different for $p_{20} = 0$ and $p_{20} \neq 0$. In Figs. 2.15 and 2.16 solid curves present the pressure distribution over the body surface and the streamline pattern for the former case and the dashed curves for the latter. As can be seen from the further consideration, the former case describes the flow

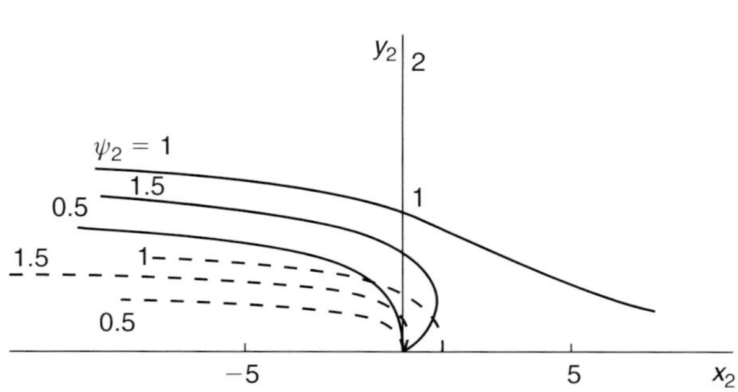


Fig. 2.15.

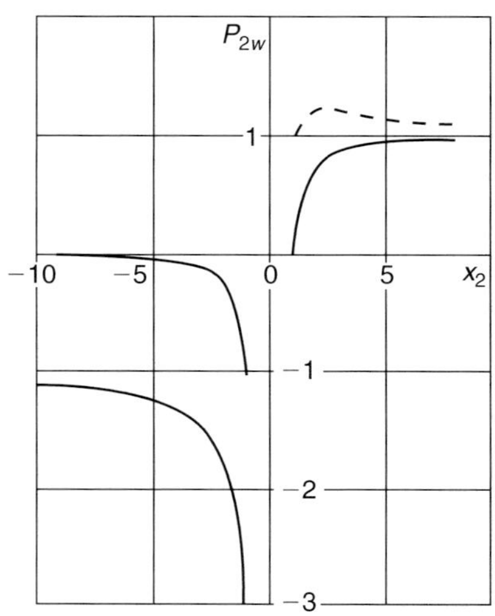


Fig. 2.16.

with the least possible value of $Q_2$ ensuring a separationless flow past the flap with $\theta \gg \varepsilon^{1/2}$. It is characterized by the absence of a stagnation point in the locally inviscid flow region. Dashed curves in Figs. 2.15 and 2.16 pertain to the flow with gas suction rate $Q_2$ greater than the minimum gas flow rate. These flows contain a stagnation point $x_2^*$ and a pressure maximum $p_2^*$

$$x_2^* = \frac{-Q_2}{\sqrt{-2p_{20}\rho_w}}, \quad p_{20}^* = -aQ_2 \tag{2.59}$$

### 2.4.3 *Solutions for nonlinear inviscid flow regions for* $\Delta x \gg \varepsilon$

Let us consider flow regions in which the pressure (its total difference) varies in the leading order ($1 \gg \Delta p \sim \theta \gg \varepsilon^{1/2}$) over lengths $\Delta x \gg \varepsilon$. In accordance with the work of Matveeva and Neiland (1967) and the estimates obtained above, the pressure variation is due to the body slope and the variation of the displacement thickness of a narrow wall layer with nonlinear disturbances of the flow functions. For this layer we will introduce the following variables and asymptotic expansions:

$$x = \varepsilon\theta^{-1,2}x_3, \quad \psi = \varepsilon\theta\psi_3$$

$$u = \theta^{1,2}u_3(x_3, \psi_3) + \cdots, \quad v = \theta^{3,2}v_3(x_3, \psi_3) + \cdots \tag{2.60}$$

$$p = \frac{1}{\gamma M_\infty^2} + \theta p_3(x_3, \psi_3) + \cdots, \quad \rho = \rho_w + \cdots, \quad y = \theta^{1/2}y_3(x_3, \psi_3) + \cdots$$

Substituting Eq. (2.60) in the Navier–Stokes equations written in the von Mises variables and passing to limit (2.46) yields the following system of equations:

$$\rho_w u_3 \frac{\partial u_3}{\partial x_3} + \frac{\partial p_3}{\partial x_3} = 0, \quad \frac{\partial p_3}{\partial \psi_3} = 0 \tag{2.61}$$

$$\frac{\partial y_3}{\partial \psi_3} = \frac{1}{\rho_w u_3}, \quad \frac{\partial y_3}{\partial x_3} = \frac{v_3}{u_3}$$

The first equation (2.61) admits the Bernoulli integral

$$\frac{\rho_w u_3^2}{2} + p_3 = \frac{\rho_w u_3^2(-\infty, \psi_3)}{2} \tag{2.62}$$

The pressure distribution $p_3(x_3)$ can be obtained using the Ackeret formula of the linear theory of supersonic flows

$$(M_\infty^2 - 1)^{1/2}p_3 = \frac{\mathrm{d}}{\mathrm{d}x_3}(\Delta\delta_3^* + \delta_{3w}) \tag{2.63}$$

Here, $\delta_{3w}$ is the local body thickness and $\Delta\delta_3^*$ is the variable part of the displacement thickness of region 3

$$\delta_{3w}(x_3 < 0) = 0, \quad \delta_{3w}(x_3 > 0) = x_3$$

$$\frac{d\delta_3^*}{dx_3} = \frac{d}{dx_3} \int\limits_{\psi_{3w}}^{\infty} \left[ \frac{1}{\rho_w u_3(x_3, \psi_3)} - \frac{1}{\rho_w u_3(-\infty, \psi_3)} \right] d\psi_3 \tag{2.64}$$

where

$$\psi_{3w}(x_3 < 0) = 0, \quad \psi_{3w}(x_3 > 0) = -Q_2, \quad p_3(-\infty) = 0$$

Using Eq. (2.62) and integrating Eq. (2.64) we bring Eq. (2.63) into the form:

$$(M_\infty^2 - 1)^{1/2} p_3 = H(x_3) - \frac{d}{dx_3} \sqrt{\frac{2\psi_{3w}}{\rho_w a} - \frac{2p_3}{\rho_w a^2}} \tag{2.65}$$

where $H(x_3)$ is the Heaviside function.

For the region $x_3 < 0$ we obtain the solution of Eq. (2.64) in the form:

$$p_3 = -\left[ \frac{1}{\sqrt{-p_{20}}} - \frac{\rho_w^{1/2} a x_3 \sqrt{M_\infty^2 - 1}}{2^{1/2}} \right]^{-2} \tag{2.66}$$

where $p_{20}$ is the pressure disturbance at the beginning of region 2, that is, for $x_3 = -0$, which is for a while unknown.

In the region $x_3 > 0$ Eq. (2.65) can be brought into the form:

$$\frac{dp_3}{dx_3} = (\beta p_3 - 1) 2\rho_w a^2 \left( -\frac{2p_3}{\rho_w a^2} - \frac{2Q_2}{\rho_w a} \right)^{1/2} \tag{2.67}$$

where $\beta = \sqrt{M_\infty^2 - 1}$. Clearly, for $p_3(0) \neq 1/\beta$ solution (2.67) does not contain the branches satisfying the boundary condition at the end of the interaction region, where, in accordance with Eq. (2.63), $p_3(\infty) \to 1/\beta$. Therefore, throughout the entire region 3 the pressure $p_3(0) = 1/\beta$, if only the flap is fairly long, and downstream of the interaction region the pressure must amount to the value following from inviscid flow theory, namely, $1/\beta$. Taking into account the results obtained above for region 2 (Eq. (2.59)) we obtain

$$p_3(+0) = \frac{1}{\beta} = p_{20} - aQ_2, \quad p_{20} = \frac{1}{\beta} + aQ_2 \tag{2.68}$$

Since for separationless flows the pressure disturbance $p_{20}$ ahead of the suction region cannot be positive in the leading term, the least possible value of the suction rate ensuring a separationless flow is determined by the relation

$$Q_{2\,\mathrm{min}} = -\frac{1}{a\beta} \tag{2.69}$$

In Fig. 2.17 we have plotted the pressure distribution in the separationless flow past the flap for different values $Q_2 \geq Q_{2\,\mathrm{min}}$. It can be easily seen that for $Q_2 = Q_{2\,\mathrm{min}}$ on the scale of region 3 the pressure is constant everywhere, except for the origin, where its jumplike variation is equal to $aQ_{2\,\mathrm{min}}$. All pressure variations in the leading term $\Delta p \sim \theta \gg \varepsilon^{1/2}$ are concentrated over smaller lengths. If $Q_2 > Q_{2\,\mathrm{min}}$, then ahead of the flap an expansion zone is formed, in which the minimum pressure $p_{20}$ is determined by formula (2.68).

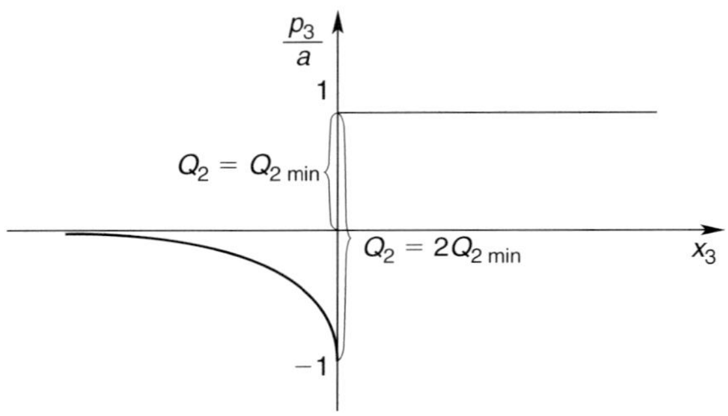


Fig. 2.17.

From formula (2.69) there follows an inference important for practice. The minimum gas flow rate through the slot ensuring a separationless flow past the flap with $\theta \gg \varepsilon^{1/2}$ can be determined by analogy with the known criterion of developed separation zone reattachment proposed on the basis of an analysis of the experimental data in the works of Chapman et al. (1958) for laminar flows and Korst et al. (1956) for turbulent flows. A rigorous derivation of this criterion and corrections to it was carried out on the basis of an analysis of the asymptotic solutions of the Navier–Stokes equations (Neiland, 1970*a*; it is also presented in Section 3.3). Formula (2.75) indicates that for a given value of $\theta$ it is necessary to suck off all the streamlines of the original undisturbed boundary layer at which the total pressure is less than the static pressure in the inviscid supersonic flow at the flap at the end of the interaction region. The final formula for the dimensional gas flow rate for $1 \gg \theta \gg \varepsilon^{1/2}$ for a laminar boundary layer can be presented in the following form:

$$Q_\Phi = -2\rho_\infty u_\infty \ell \theta [Re\, C_{f0}(M_\infty^2 - 1)]^{-1} \left(\frac{T_w}{T_\infty}\right)^{\omega} \tag{2.70}$$

Solution (2.26) of systems (2.61) and (2.63) does not satisfy the no-slip condition at the wall. Therefore, a region 4, in which the viscosity effect is essential, should be introduced.

Using Eqs. (2.36)–(2.41) and equating the orders of the leading viscous term and the pressure gradient in the momentum equation, we can determine the scales and orders of the flow functions in region 4

$$x_4 \sim x_3 \sim \varepsilon \Delta p_3^{-1/2}, \quad y_4 \sim \varepsilon^{3/2} \Delta p_3^{-1/2}$$

$$u_4 \sim \Delta p_3^{1/2}, \quad \Delta u_4 \sim \Delta p_3^{1/2}, \quad v_4 \sim \varepsilon^{1/2} \Delta p_3^{1/2} \tag{2.71}$$

Substituting Eqs. (2.71) in the Navier–Stokes equations and passing to limit (2.46) leads to the system of boundary layer equations with a velocity distribution given on the outer boundary. The solution of the system of equations thus obtained makes it possible to satisfy the no-slip condition at the wall.

### *2.4.4 Solutions for finite-length flaps and bodies with a bend in the contour for $1 \gg \theta \gg \varepsilon^{1/2}$*

We will continue the analysis of possible solutions of the problem for region 3 at $x_3 > 0$ for bodies with different shapes of the contour behind the slot. As noted above, for $p_3(0) \neq 1/\beta$ Eq. (2.67) has two more families of solutions. If $p_3(0+) > 1/\beta$, then $p_3$ increases but cannot become greater than the maximum value

$$p_{3\,\max} = -Q_2 a \tag{2.72}$$

The integral curves, at which $p_3(0+) < 1/\beta$, correspond to unbounded decrease of $p_3(x_3)$. Thus, solutions for flaps with length $\Delta x \sim \varepsilon\theta^{-1/2}$ can be obtained. On the end of such a flap the pressure must be preassigned within the limits $-\infty < p_3(x_{3k}) < -Q_2 a$. The left limit corresponds to finite pressure differences $\Delta p < 0$, while the right limit indicates that for a given flap length and a flow rate $Q_2$ in the separationless flow the base pressure must not be greater than the maximum value (2.72).

If $p_3(x_{3k})$ is given, then the pressure distribution over the entire region 3 is described by the formula

$$p_3 = -Q_2 a - \left(-Q_2 a - \frac{1}{\beta}\right) \frac{1 + c\exp\left(x_3 2^{3/2} \rho_w^{1/2} a\beta \sqrt{-Q_2 a - \frac{1}{\beta}}\right)}{1 - c\exp\left(x_3 2^{3/2} \rho_w^{1/2} a\beta \sqrt{-Q_2 a - \frac{1}{\beta}}\right)} \tag{2.73}$$

where

$$c = \frac{1}{2\sqrt{-Q_2 a - \frac{1}{\beta}}} \ln \frac{\sqrt{-p_3(x_{3k}) - Q_2 a} - \sqrt{-Q_2 a - \frac{1}{\beta}}}{\sqrt{-p_3(x_{3k}) - Q_2 a} + \sqrt{-Q_2 a - \frac{1}{\beta}}} - 2^{1/2} \rho_w^{1/2} a\beta x_{3k}$$

Solution (2.73) describes the flow past a finite-length flap with a given base pressure and the flow past a body with a bend in the generator at $x_3 = x_{3k}$.

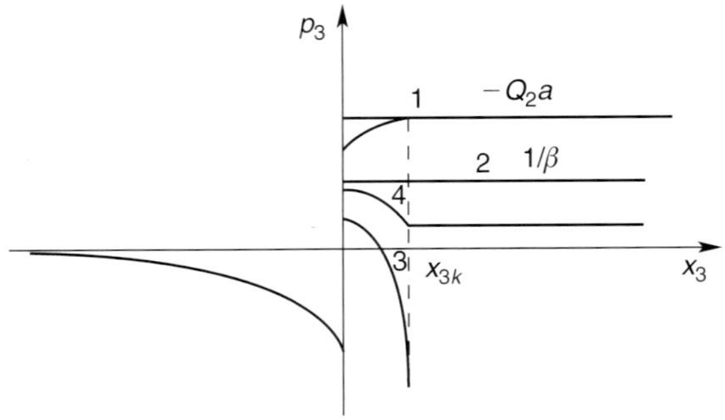


Fig. 2.18.

In Fig. 2.18 we have sketched the solutions described by formula (2.73). Curve 1 relates to the solution with a given pressure in the flow past a finite-length flap and the greatest permissible (for given $Q$) base pressure. Curve 2 describes the solution for a semi-infinite flap $p_3 = 1/\beta$. As follows from Eq. (2.67), preassigning the base pressure $p_3(x_{3k}) < 1/\beta$ results in the appearance of a negative pressure gradient over the entire finite-length flap. The branch of the solution of this type, marked by number 3 in the figure, ends at a singular point, where $p_3 \to \infty$; therefore, any value $p_3(x_{3k}) < 1/\beta$ can be preassigned.

Solution (2.73) also describes the flow past a body with a bend in its contour. In Section 2.4.3 it is shown that solutions of Eq. (2.73) do not involve branches satisfying the boundary condition $p_3 \to \theta_1/\beta$, as $x_3 \to \infty$, where $\theta_1$ is the contour deflection angle behind the bend point $x_3 = x_{3k}$. Then for $x_3 > x_{3k}$ we have $p_3 = \theta_1/\beta$, while for $x_3 < 0 < x_{3k}$ the solution is described by Eq. (2.73). In Fig. 2.18 this solution is presented by curve 4. Thus, the pressure distribution over bodies of complicated shape can be nonmonotonic and have local maxima.

### 2.4.5 *Flow past a flap deflected by an angle* $\theta \sim \varepsilon^{1/2}$

From estimates (2.60) and (2.71) it follows that for $\theta \sim \varepsilon^{1/2}$ the orders of the transverse dimensions of regions 3 and 4 are the same. Therefore, the pressure variation in the leading order is induced by the interaction between the displacement thickness of region 3 and the outer supersonic flow. As shown in the previous chapter, the solution of the system of equations governing the flow in region 3 is uniquely determined by imposing an additional boundary condition, for example, the pressure disturbance at a certain downstream point. In the problem under consideration, solutions in regions 2 and 3 downstream of the slot must be found for determining the value of $p_{20}$. The solution for region 3 at $x_3 < 0$ determines the vorticity distribution $\omega(\psi)$ in region 2. In the more general case $\omega(\psi) \neq \text{const}$ the solution of systems (2.54)–(2.56) cannot be written in an explicit form; therefore, it is necessary to obtain a numerical solution on the basis of which the dependence of the pressure difference $\Delta p_{2w}$ in region 2 on the gas flow rate could be established. On the scale of region 3 at $x_3 = 0$ there occurs the transformation of the velocity profile determined by preassigning $\Delta p_{2w}$. Since the transformation takes place over lengths shorter than the longitudinal dimension of region 3, on these short distances the flow is, as shown above, locally inviscid and, therefore, the velocity variation along a streamtube can be determined from the Bernoulli equation. The velocity profile thus obtained determines the initial conditions for the system of equations governing the flow in region 3 for $x_3 > 0$.

It should be taken into account that downstream of the slot the pressure variation in the leading order is formed due to not only the variation of the displacement thickness of region 3, but also the body contour deflection $\Delta\delta_3^* \sim \delta_w \sim \varepsilon^{5/4}$.

For given values of the gas flow rate $Q$ and the flap deflection angle $\theta$ the process of the solution of the complete problem can be presented as follows. First, a value $p_{20}$, which allows us to single out the required branch of the solution in region 3 for $x_3 < 0$, is preassigned. The solution of systems (2.54) and (2.55) with account for the dependence $\omega(\psi)$ obtained makes it possible to determine $\Delta p_{2w}$ and hence the velocity profile in region 3 for $x_3 \geq 0$.

As $x_3 \to \infty$, the solution in region 3 must satisfy the condition $\mathrm{d}p_3/\mathrm{d}x_3 \to 0$; thus, if the solution obtained does not satisfy this condition, the procedure is repeated for other value of $p_{20}$ until the required solution branch is singled out.

Generally speaking, it is easier to solve a problem in which the flap deflection angle $\theta$ is not given beforehand. Then for fixed values of $p_{20}$ and $Q$ the value of $\theta$ is determined ensuring the fulfillment of the condition $\mathrm{d}p_3/\mathrm{d}x_3 \to 0$, as $x_3 \to \infty$.

We can also formulate a problem of determining the least possible gas flow rate $Q$ for which the flow is still separationless. In the preceding section the maximum value of the pressure disturbance $p_{20}$ associated with zero friction at $x_3 = -0$ was presented. In that case it was $\theta$ and $p_{20}$ that were fixed and the parameter $Q$ was determined in the course of an iteration procedure. Here, such calculations are not carried out. However, from the above reasoning it follows that for $\theta = Re^{-1/4}$ the gas flow rate ensuring separationless flow is $Q = Re^{-3/4}$. As $Q\,Re^{3/4} \to 0$, the permissible value of $\theta_{\max}$ approaches the value determined by formula (2.35). Thus, the approximate range of applicability of relation (2.70) is established for small values of $\theta$.

### 2.4.6 *Flow patterns in the laminar boundary layer for finite flap deflection angles*

From estimates (2.36)–(2.45) it follows that for $\theta \sim 1$ the dimensions of regions 1 and 2 are equal in the order to the boundary layer thickness

$$\Delta x \sim \Delta y \sim \varepsilon \tag{2.74}$$

It can be shown that in this case the flow in a region with the scales $x \sim y \sim \varepsilon$ can be described by the system of Euler equations for an inviscid compressible gas. Without solving the complete system of equations, an approximate method for calculating the minimum flow rate ensuring the separationless flow past a flap with an angle $\theta$ can be proposed. As shown above, for deflection angles $\theta \gg \varepsilon^{1/2}$ the $Q(\theta)$ dependence is determined in the leading order on the basis of the solutions describing an inviscid flow. Viscosity has an effect only on higher-order corrections. The physical meaning of Eq. (2.72) is that it is that wall region of the boundary layer or those streamtubes, for which the total pressure is less than the static pressure on the flap, that are needed to be sucked off for eliminating separation. For a given flap deflection angle $\theta$ and the Mach number $M_\infty$ of the undisturbed oncoming flow the ratio $p_1/p_\infty$ can be determined, where $p_1$ is the static pressure on the flap.

It is necessary to suck off the streamtubes in the boundary layer between the body surface and the dividing streamline, on which the dynamic head is determined from the isentropic deceleration condition by the formula

$$\frac{\rho u^2}{2} = \frac{1}{(\gamma - 1)M_\infty^2}\left[\left(\frac{p_1}{p_\infty}\right)^{(\gamma-1)/\gamma} - 1\right] \tag{2.75}$$

The equation relating the dynamic head and the gas flow rate on the dividing streamline can be derived if the dependence of the velocity and the enthalpy on the flow rate in the undisturbed boundary layer is known

$$\frac{\rho u^2}{2} = \frac{2}{(\gamma - 1)M_\infty^2}\frac{u(\psi)}{g - u^2(\psi)}, \quad g = \frac{2c_p T}{u_\infty^2} + u^2 \tag{2.76}$$

In the particular case in which the Prandtl number $\sigma = 1$ and $n = 1$, where $n$ is the exponent in the viscosity–temperature dependence, Eq. (2.76) takes the form:

$$\rho u^2 = \frac{2}{(\gamma - 1)M_\infty^2}\frac{f'^2}{(g_e - g_w)f' + g_w - f'^2} \tag{2.77}$$

where $g_e = 1 + [2/(\gamma - 1)M_\infty^2]$ and $f' = u$ is the velocity profile in the Blasius solution

$$f = \frac{\psi}{\rho_\infty u_\infty \ell}\left(\frac{Re\, \rho_\infty \mu_\infty}{2\rho_w \mu_w}\right)^{1/2}$$

The $\psi = \psi(\rho u^2)$ dependence is presented in Fig. 2.19. In the book of Lachmann (1961) the results of an experimental investigation of the slot suction effect on laminar boundary layer

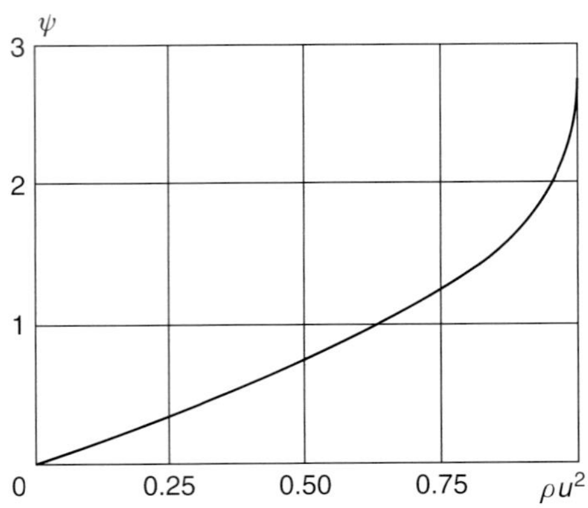


Fig. 2.19.

separation in a supersonic flow are presented. Since in the work of Ball and Korkegi (1968) the dependence of the minimum slot width, at which the boundary layer is still unseparated, on the flap deflection angle was derived, the relation between the gas flow rate and the slot width must be known for determining the minimum flow rate. Obviously, relation (2.50) yields only an approximate dependence $\psi(\theta)$ for $\theta \ll 1$, since for finite flap deflection angles the conditions far away from the slot are changed; moreover, the slot section configuration can be different from the slot geometry for which Eq. (2.50) was derived. The paper of Ball and Korkegi (1968) does not contain more detailed data.

For approximately calculating the $\psi(\theta)$ dependence theoretical results for a plane compressible sink can be used. In Fig. 2.20 we have plotted the $\psi(\theta)$ curves from the data of the work of Ball and Korkegi (1968). For determining the gas flow rate from those data we used both formula (2.50) (curve 1) and the formula for the sink in a compressible fluid (curve 4). Clearly, the absence of the data on the shape of the channel leads to a noticeable undefiniteness in the interpretation of the experimental data concerning the gas flow rate.

For theoretical approximate determination of the gas flow rate on the basis of Eq. (2.77) the static pressure on the flap near the bend in the body contour should be preassigned. At small distances downstream of the slot a shock can have no time to be formed; then the static pressure on the flap is determined from the condition of isentropic compression of a supersonic flow deflected by the angle $\theta$.

In Fig. 2.20 we have plotted the dependence $\psi(\theta)$ obtained on the basis of Eq. (2.77); here, the static pressure on the flap was determined from the conditions behind the shock (curve 2) and from the isentropic compression conditions (curve 3).

Curve 3 can serve as the boundary of the minimum gas flow rate ensuring separationless flow past the flap, since in isentropic compression the pressure variation is always greater

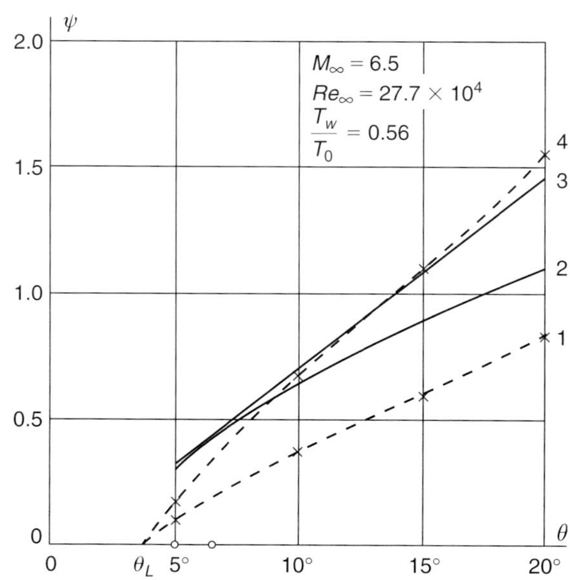


Fig. 2.20.

than that across a shock. The qualitative agreement of the behavior of solutions (1, 4) and (2, 3) can also be noted.

From the comparison of curves 1 and 2 it follows that the calculated values of the gas flow rate are greater than the experimental data, which is attributable to the viscosity effect disregarded in the calculation formulas. From the calculations it would follow that $\psi = 0$ for $\theta = 0$, while the experiment gives the value $\psi = 0$ for $\theta_i = 3°45'$.

Therefore, the calculation formulas can be employed on the $\theta > \theta_i$ range. The comparison of the experimental value $\theta_{max}$ in the absence of suction with the theoretical value obtained with account for viscosity in the work of Neiland (1971) indicates good agreement of the results. At the same time, the fact that the theoretical value of $\theta_{max}$ is somewhat in excess indicates that in the experiment of Ball and Korkegi (1968) the values of $\theta_{max}$ are somewhat underestimated, since the comparison with the results of other experimental studies conducted with the special purpose of determining $\theta_{max}$ in the absence of suction always gave a certain excess over the theoretical value. This can be attributed to different criteria used for determining $\theta_{max}$ in theory and experiment. In the work of Neiland (1971) it was zero friction at one or another point on the body surface that was associated with $\theta_{max}$, whereas in the experimental studies a later stage of the separation development was recorded.

# 3 Viscous Gas Flows in Regions with Developed Locally Inviscid Zones and High Local Pressure Gradients

The main assumption of Prandtl's classical boundary layer theory (Prandtl, 1904) is the smallness of the longitudinal gradients of the flow functions (velocity, temperature) in the boundary layer, as compared with the transverse gradients. However, there are many problems of viscous high-Reynolds-number gas dynamics for which this assumption is not fulfilled. These include, in particular, the problems with different local singularities, such as corner points on the body contour, separation zone reattachment points, etc. In this chapter we study the flows, in which the pressure in the supersonic flow near the body surface varies by its leading order over short distances, for example, of the order of the boundary layer thickness. For this purpose, we study the asymptotic behavior of the solutions of the Navier–Stokes equations in typical flow regions thus formed using the well-known principle of matching asymptotic expansions representing the solutions for different regions.

## 3.1 Formulation of the problem of the expansion flow near a corner point on a body in supersonic flow

We will consider certain general properties of the asymptotic solutions of the Navier–Stokes equations when the characteristic Reynolds number increases without bound. For the sake of definiteness, we will consider the problem of the supersonic viscous flow past a body with a characteristic dimension $\ell$. It can easily be established that, as $Re \to \infty$, in the most part of the flow the viscosity effect vanishes and the Navier–Stokes equations go over to the Euler equations. In the limit, near the body surface a contact discontinuity is formed, thanks to which the no-slip condition can be fulfilled; under certain conditions, it can separate from the body surface. If the longitudinal gradients of the flow parameters are fairly small along this surface, then, as is well known, in the first approximation its structure is described by equations of the type of the Prandtl boundary layer equations.

Let us now assume that the main assumption of boundary layer theory, that is, the smallness of the longitudinal gradients, is fulfilled everywhere, except for a region, whose length is equal in the order to the boundary layer thickness. Let along this length the unknown flow functions (e.g., the pressure) vary by a finite value, that is, a value which is always equal in the order to the characteristic value in the flow under consideration. In the $Re \to \infty$ limit the length of this region vanishes, while the function undergoes a discontinuity across a certain point on the contact surface under consideration.

We will choose a curvilinear coordinate system $(s, n)$ with the origin at that point on the contact surface, where in the $Re \to \infty$ limit the parameter discontinuity is localized. Let $s$ be measured along the contact surface (e.g., the body surface) and $n$ normal to it (Fig. 3.1). For

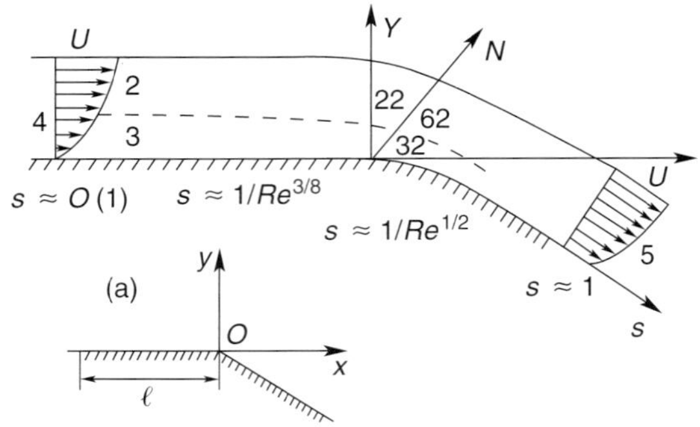


Fig. 3.1.

the scale length $\ell$ we take the distance from the beginning of the contact surface (boundary layer) to the origin. As before, the main smallness parameter is as follows:

$$\varepsilon = Re^{-1/2} \tag{3.1}$$

For a two-dimensional flow the Navier–Stokes equations can be written in the form:

$$(\rho u)_s + [(1+kn)\rho v]_n = 0 \tag{3.2}$$

$$\begin{aligned}
&\varepsilon^{-2}\left[\rho\left(\frac{uu_s}{1+kn} + vu_n + \frac{k}{1+kn}uv\right) + \frac{p_s}{1+kn}\right] \\
&= \left[\mu\left(u_n + \frac{v_s - ku}{1+kn}\right)\right]_n + \frac{2}{1+kn}\left(\mu\frac{u_s - kv}{1+kn}\right)_s \\
&\quad + \frac{2\mu k}{1+kn}\left(u_n + \frac{v_s - ku}{1+kn}\right) + \frac{1}{1+kn}\widetilde{\mu}\left(\frac{u_s - kv}{1+kn} + v_n\right)_s
\end{aligned} \tag{3.3}$$

$$\begin{aligned}
&\varepsilon^{-2}\left[\rho\left(\frac{uv_s}{1+kn} + vv_n - \frac{ku^2}{1+kn}\right) + p_n\right] \\
&= 2(\mu v_n)_n + \frac{1}{1+kn}\left[\mu\left(u_n + \frac{v_s - ku}{1+kn}\right)\right]_s + \frac{2\mu k}{1+kn}v_n \\
&\quad - \frac{2\mu k}{1+kn}\frac{u_s - kv}{1+kn} + \widetilde{\mu}\left(\frac{u_s + kv}{1+kn} + v_n\right)_n
\end{aligned} \tag{3.4}$$

$$\begin{aligned}
&\varepsilon^{-2}\left[\rho\left(\frac{uT_s}{1+kn} + vT_n\right) - \frac{up_s}{1+kn} - vp_n\right] \\
&= \frac{1}{1+kn}\left[\frac{\mu T_s}{\sigma(1+kn)}\right]_s + \left(\frac{\mu}{\sigma}T_n\right)_n + \frac{k\mu T_n}{\sigma(1+kn)} \\
&\quad + \mu\left[2\left(\frac{u_s + kv}{1+kn}\right)^2 + 2v_n^2 + \left(u_n + \frac{v_s - ku}{1+kn}\right)^2\right] + \widetilde{\mu}\left(\frac{u_s + kv}{1+kn} + v_n\right)^2
\end{aligned} \tag{3.5}$$

Here, $\widetilde{\mu}$ is the bulk viscosity, $k$ is the curvature of the $n=0$ surface, and $u$ and $v$ are the tangential and normal velocity components.

As before, the dimensionless quantities are obtained by scaling on the values of the functions in the undisturbed inviscid flow.

We will perform our investigation with reference to the problem of the supersonic flow past a body of a comparatively simple shape, as shown in Fig. 3.1. The forebody and the afterbody are flat plates. The angle they form is $O(1)$ and in the deflection region the characteristic value of the contour curvature is of the order $k \sim O(\varepsilon^{-1})$. Below it is shown that the main results can also be used for other flows: the flow ahead a base section (Section 3.2) and the region of supersonic flow attachment to the smooth surface of a body (Section 3.3). In Fig. 3.1a we have presented the limiting form of the flow as $Re \to \infty$ on the scale $(x, y) \sim O(1)$.

### *3.1.1 Asymptotic expansions*

Following the classical boundary layer theory, for the outer main part of the flow (region 1 in Fig. 3.1) we can introduce the following scales for the independent variables and asymptotic representations for the functions

$$
\begin{aligned}
&s_1 = s, \quad n_1 = n \\
&u(s, n; \varepsilon) \sim u_1(s_1, n_1) + \cdots, \qquad v(s, n, \varepsilon) \sim v_1(s_1, n_1) + \cdots \\
&p(s, n; \varepsilon) \sim p_1(s_1, n_1) + \cdots, \qquad \rho(s, n, \varepsilon) \sim \rho_1(s_1, n_1) + \cdots \\
&h(s, n; \varepsilon) \sim h_1(s_1, n_1) + \cdots, \qquad H = h + \frac{1}{2}(u^2 + v^2)
\end{aligned} \tag{3.6}
$$

Substituting Eq. (3.6) in the Navier–Stokes equations (3.2)–(3.5) and passing to the limit $\varepsilon \to 0$ leads, obviously, to the complete Euler equations. They are associated with the conventional boundary conditions for inviscid flow problems including the equality to zero of the normal velocity component on the body surface and the compatibility conditions on shocks and contact surfaces if there are any in the flow.

In order to satisfy the no-slip condition on the body surface, it is necessary to introduce the conventional boundary layer variables in the regions outside a vicinity of the corner point (regions 4 and 5 in Fig. 3.1)

$$
\begin{aligned}
&s_i = s, \quad n_i = \frac{n}{\varepsilon} \\
&u(s, n; \varepsilon) \sim u_i(s_i, n_i) + \cdots, \qquad v(s, n, \varepsilon) \sim \varepsilon v_i(s_i, n_i) + \cdots \\
&p(s, n; \varepsilon) \sim p_i(s_i, n_i) + \cdots, \qquad \rho(s, n, \varepsilon) \sim \rho_i(s_i, n_i) + \cdots \\
&h(s, n; \varepsilon) \sim h_1(s_i, n_i) + \cdots, \qquad i = 4, 5
\end{aligned} \tag{3.7}
$$

Substituting Eq. (3.7) in the Navier–Stokes equations and passing to the limit $\varepsilon \to 0$ leads, as is well known, to the boundary layer equations; here, for the brevity sake, they are omitted.

In the vicinity of the origin of length $\sim\varepsilon$ (region 22 in Fig. 3.1) for the streamtubes passing through the boundary layer (region 4) the independent variables and the asymptotic expansions take the form:

$$
\begin{aligned}
&s_{22} = \frac{s}{\varepsilon}, \quad n_{22} = \frac{n}{\varepsilon} \\
&u(s,n;\varepsilon) \sim u_{22}(s_{22},n_{22}) + \cdots, \quad v(s,n;\varepsilon) \sim v_{22}(s_{22},n_{22}) + \cdots \\
&p(s,n;\varepsilon) \sim p_{22}(s_{22},n_{22}) + \cdots, \quad \rho(s,n,\varepsilon) \sim \rho_{22}(s_{22},n_{22}) + \cdots \\
&h(s,n;\varepsilon) \sim h_{22}(s_{22},n_{22}) + \cdots, \quad k(s) \sim \frac{1}{\varepsilon}K(s_{22})
\end{aligned}
\tag{3.8}
$$

Substituting Eq. (3.8) in the Navier–Stokes equations and passing to the limit $\varepsilon \to 0$ leads again to the Euler equations

$$
\begin{aligned}
&\rho_{22}\left(\frac{u_{22}}{1+Kn_{22}}\frac{\partial u_{22}}{\partial s_{22}} + v_{22}\frac{\partial u_{22}}{\partial n_{22}} + \frac{K}{1+Kn_{22}}u_{22}v_{22}\right) + (1+Kn_{22})\frac{\partial p_{22}}{\partial s_{22}} = 0 \\
&\rho_{22}\left(\frac{u_{22}}{1+Kn_{22}}\frac{\partial v_{22}}{\partial s_{22}} + v_{22}\frac{\partial v_{22}}{\partial n_{22}} - \frac{K}{1+Kn_{22}}u_{22}^2\right) + \frac{\partial p_{22}}{\partial n_{22}} = 0 \\
&\frac{\partial(\rho_{22}u_{22})}{\partial s_{22}} + \frac{\partial}{\partial n_{22}}[(1+Kn_{22})\rho_{22}v_{22}] = 0 \\
&u_{22}\frac{\partial}{\partial s_{22}}\left(h_{22} + \frac{u_{22}^2+v_{22}^2}{2}\right) + v_{22}\frac{\partial}{\partial n_{22}}\left(h_{22} + \frac{u_{22}^2+v_{22}^2}{2}\right) = 0
\end{aligned}
\tag{3.9}
$$

The solution of system (3.9) must satisfy the condition $v_{22}(s_{22},0)=0$ imposed on the body surface. For obtaining “initial conditions” it is necessary to perform the matching with the solution for region 4 (in what follows, we will introduce region 2 between regions 4 and 22; however, since in this region $\Delta u \to 0$, as $\varepsilon \to 0$, its presence does not change the results)

$$
s_{22} = \frac{s_4}{\varepsilon}, \quad u_4(0,n_4) = u_{22}(-\infty,n_{22}) \tag{3.10}
$$

On the scale of region 1, region 22 degenerates in limit to a point; therefore, the matching leads to the formulas of the following form:

$$
\begin{aligned}
&u_1(s_1 \to 0,\ n_1 \to 0) \to u_{22}(s_{22} \to \infty,\ n_{22} \to \infty) \\
&p_1(s_1 \to 0,\ n_1 \to 0) \to p_{22}(s_{22} \to \infty,\ n_{22} \to \infty) \\
&v_1(s_1 \to 0,\ n_1 \to 0) \to v_{22}(s_{22} \to \infty,\ n_{22} \to \infty), \quad s_{22} = \frac{s_1}{\varepsilon}, \quad n_{22} = \frac{n_1}{\varepsilon} \\
&\frac{n_1}{s_1} = \frac{n_{22}}{s_{22}} = \text{const}, \quad (n_1 \to 0,\ s_1 \to 0,\ s_{22} \to \infty,\ n_{22} \to \infty)
\end{aligned}
\tag{3.11}
$$

Expressions of type (3.11) lead to a physically obvious result: with distance from the body surface, on the scale of region 22 the flow must be described by the Prandtl–Meyer formulas which represent the solution for the outer inviscid supersonic flow past the body formed by the displacement thickness of region 22. For further analysis it is convenient to write down the relation between the angle $\theta$ of the velocity vector inclination to the body surface, the contour inclination $\theta_{10}$ to the $x$ axis, and the Prandtl function $\nu_e$

$$\theta_e + \theta_w + \nu_e = \nu_{e4}$$

$$\nu_e = \sqrt{\frac{\gamma+1}{\gamma-1}}\arctan\sqrt{\frac{\gamma-1}{\gamma+1}(M_e^2-1)} - \arctan\sqrt{M_e^2-1} \qquad (3.12)$$

Here, $\nu_{4e}$ is the value of $\nu$ in the undisturbed region ahead of the turn region, where $\theta_e = \theta_w = 0$. The values of $\theta_e$ and $M_e$ correspond to the outer boundary of region 22.

Finally, matching the solutions for regions 22 and 5 yields the initial parameter distributions in region 5 for $s_5 = 0$. This follows from the matching principle and the following relations:

$$s_5 = \varepsilon s_{22}, \quad n_{22} = n_5, \quad \lim_{s_{22}\to+\infty} u_{22}(s_{22}, n_{22}) = u_5(0, n_5)$$

$$\lim_{s_{22}\to+\infty} h_{22}(s_{22}, n_{22}) = h_5(0, n_5), \quad \lim_{s_{22}\to+\infty} p_{22}(s_{22}) = p_5(0)$$

It should be noted that for determining the initial conditions in region 5 there is no need to solve the problem for region 22. Equations (3.9) are the complete Euler equations; therefore, in region 22 the entropy and the stagnation enthalpy are conserved along streamlines. Because of this, the initial conditions can directly be calculated given the boundary layer profiles at the end of region 4 ($s_4 = 0$) and the pressure immediately behind the turn ($s_5 = 0$). Of course, this is true only for separationless flows and the flows in which overexpansion in region 22 is not followed by shock formation.

The solution of Eq. (3.9) cannot satisfy the no-slip conditions. Therefore, near the body surface there is a flow region in which, as in the boundary layer, the leading viscous terms become essential (region 32 in Fig. 3.1). The corresponding coordinate scales and asymptotic expansions are as follows:

$$s_{32} = \frac{s}{\varepsilon}, \quad n_{32} = \frac{n}{\varepsilon^{3/2}}$$

$$u(s,n;\varepsilon) \sim u_{32}(s_{32}, n_{32}) + \cdots, \quad v(s,n,\varepsilon) \sim \varepsilon^{1/2} v_{32}(s_{32}, n_{32}) + \cdots \qquad (3.13)$$

$$p(s,n;\varepsilon) \sim p_{32}(s_{32}, n_{32}) + \cdots, \quad \rho(s,n;\varepsilon) \sim \rho_{32}(s_{32}, n_{32}) + \cdots$$

$$H(s,n;\varepsilon) \sim H_w + e(\varepsilon) H_{32}(s_{32}, n_{32}) + \cdots, \quad k(\varepsilon) \sim \frac{1}{\varepsilon} K(s_{32})$$

The scale $e(\varepsilon)$ is not known beforehand and should be determined in matching with the solutions for other regions. Substituting Eq. (3.13) in the Navier–Stokes equations and passing to the limit $\varepsilon \to 0$ leads to the equations

$$\frac{\partial}{\partial s_{32}}(\rho_{32} u_{32}) + \frac{\partial}{\partial n_{32}}(\rho_{32} v_{32}) = 0, \quad \frac{\partial p_{32}}{\partial n_{32}} = 0$$

$$\rho_{32}\left(u_{32}\frac{\partial u_{32}}{\partial s_{32}} + v_{32}\frac{\partial u_{32}}{\partial n_{32}}\right) + \frac{\partial p_{32}}{\partial n_{32}} = \frac{\partial}{\partial n_{32}}\left(\mu\frac{\partial u_{32}}{\partial n_{32}}\right) \qquad (3.14)$$

$$\rho_{32}\left(u_{32}\frac{\partial H_{32}}{\partial s_{32}} + v_{32}\frac{\partial H_{32}}{\partial n_{32}}\right) = \frac{\partial}{\partial n_{32}}\left(\frac{\mu}{\sigma}\frac{\partial H_{32}}{\partial n_{32}}\right) + \frac{\partial}{\partial n_{32}}\left[\frac{(\sigma-1)}{\sigma e(\varepsilon)}\mu u_{32}\frac{\partial u_{32}}{\partial n_{32}}\right]$$

Here, it is proposed to consider the following passage to the limit: $\varepsilon \to 0$ and $A = (\sigma - 1)/\sigma e(\varepsilon) \to O(1)$. Of course, the solutions corresponding to $\varepsilon \to 0$ and $\sigma - 1 = O(1)$ ($A \to \infty$) could also be considered. However, on the Reynolds and Prandtl number ranges, interesting for practice ($10^4 \leq Re \leq 10^6$, $0.5 \leq \sigma \leq 1$), we have $A \sim O(1)$, so that the passage to the limit $\varepsilon \to 0$, $\sigma - 1 = O(1)$ will not be considered in this study as overestimating the role of the dissipative terms in the energy equation.[1]

Certain boundary conditions for Eq. (3.14) are obvious:

$$u_{32}(s_{32}, 0) = v_{32}(s_{32}, 0) = H_{32}(s_{32}, 0) = 0$$

Since it is assumed that $e(\varepsilon) \to 0$ as $\varepsilon \to 0$, from the equation of state it follows that

$$H_w - \frac{1}{2}u_{32}^2 = \frac{\gamma}{\gamma - 1}\frac{p_{32}}{\rho_{32}}$$

Matching for the longitudinal velocity component yields

$$u_{22}(s_{22}, 0) = u_{32}(s_{32}, \infty)$$

Other boundary and initial conditions call for further investigation; they are obtained below.

### *3.1.2 Upstream disturbance decay*

The initial condition (3.10) indicates that the original velocity profile of the locally inviscid flow 22 includes a region of subsonic velocities. Therefore, disturbances must propagate upstream. We will study the disturbance decay in region 22 as $s_{22} \to -\infty$. Let us consider a class of bodies for which the body surface $y_{22w}$ approaches the $x$ axis following the law

$$y_{22w} \approx \frac{-A}{(-s_{22})^{\alpha}}, \quad A > 0, \ \alpha > 0, \ s_{22} \to -\infty \tag{3.15}$$

Then for $\theta_w$ and the curvature $K$ we obtain

$$\theta_w \sim -A\alpha(-s_{22})^{-(\alpha+1)}, \quad K \sim -A\alpha(\alpha + 1)(-s_{22})^{-(\alpha+2)} \tag{3.16}$$

As $s_{22} \to -\infty$, we obtain $\theta_w \to 0$ and the pressure disturbance $\Delta p_{22} \to 0$, while, in accordance with the initial condition (3.10), the velocity profile is progressively less different from that in the undisturbed boundary layer ahead of the corner point. In the case of small disturbances, the boundary condition (3.12) can, in accordance with the linear theory of supersonic flows, be transformed to the Ackeret formula

$$(M_{e4}^2 - 1)^{1/2}\Delta p_{22} \approx \theta_w + \frac{\mathrm{d}\Delta\delta_{22}^*}{\mathrm{d}s_{22}} \tag{3.17}$$

where $\Delta\delta_{22}^*$ is the variation of the displacement thickness of region 22 due to the appearance of pressure disturbances $\Delta p_{22} \ll 1$.

[1] It was studied in the paper of Matveeva and Neiland (1967).

In the main part of region 22 the velocity $u \sim O(1)$ and $\Delta u \sim \Delta p_{22}$; therefore, here, as in the case of flows with small pressure disturbances considered in Chapter 1, $\Delta\delta^*_{22}$ is formed by a relatively thin wall layer in which $u \sim \Delta u \sim \sqrt{\Delta p}$, while its thickness varies by its leading order. In accordance with Eq. (3.10), for $\Delta p_{22} = 0$ in this layer we have

$$u_{22} \to a y_4 + \cdots = \left(\frac{2\psi_{22} a}{\rho_w}\right)^{1/2} + \cdots, \quad s_{22} \to -\infty, \quad \psi_{22} \to 0 \tag{3.18}$$

while in the region $0 < |\Delta p_{22}| \ll 1$, by virtue of the Bernoulli equation, we have

$$u_{22} \sim \left(\frac{2\psi_{22} a}{\rho_w} - \frac{2}{\rho_w}\Delta p_{22}\right)^{1/2} \quad \text{for } \psi_{22} \to 0 \tag{3.19}$$

where $a$ is the constant dimensionless friction at the end of region 4.

Formula (3.19) makes it possible to determine the velocity distribution over the bottom of region 22, that is, precisely in the wall layer determining the value of $\Delta\delta^*_{22}$; therefore, we have

$$\Delta\delta^*_{22} \approx \lim_{\psi_{22}\to\infty}\left(\int_0^{\psi_{22}} \frac{d\psi_{22}}{\rho_w\sqrt{\frac{2\psi_{22}a}{\rho_w} - \frac{2}{\rho_w}\Delta p_{22}}} - \int_0^{\psi_{22}} \frac{d\psi_{22}}{\rho_w\sqrt{\frac{2\psi_{22}a}{\rho_w}}}\right) = -\sqrt{\frac{-2\Delta p_{22}}{\rho_w a^2}}$$

From Eq. (3.17) we obtain the following equation

$$(M_{e4}^2 - 1)^{1/2}\Delta p_{22} \approx \theta_w - \frac{d}{ds_{22}}\left(-\frac{2\Delta p_{22}}{\rho_w a^2}\right)^{1/2} \tag{3.20}$$

We should seek such a solution of Eq. (3.20) in which $\Delta p_{22} \to 0$ as $s_{22} \to -\infty$. We note that for $\theta_w < 0$ we have $\Delta p_{22} < 0$. The first term on the right side of Eq. (3.20) takes account for the body surface slope, while the second term accounts for the slope produced by the variation of the displacement thickness $\Delta\delta^*_{22}$ or, to be more precise, the narrow wall layer. An elementary analysis shows that if in Eq. (3.16) $\alpha < 1$, then the effect of the second term in Eq. (3.20) is small and it is the wall slope effect that is predominant when $s_{22} \to -\infty$. For $\alpha = 1$ both terms are of the same order. In the most interesting case $\alpha > 1$ it is the disturbance propagation through the sublayer that is predominant when $s_{22} \to -\infty$.

When $s_{22} \to -\infty$, the asymptotic solutions (3.20) take the form:

$$\Delta p_{22} \sim \frac{-A\alpha}{(M_{e4}^2 - 1)^{1/2}(-s_{22})^{\alpha+1}} \quad \text{for } \alpha < 1 \tag{3.21}$$

$$\Delta p_{22} \sim -\frac{B^2}{(-s)^2}, \quad B = \frac{1 + \sqrt{1 + 2\rho_w a^2 (M_{e4}^2 - 1)^{1/2} A}}{\sqrt{2\rho_w a^2 (M_{e4}^2 - 1)}} \quad \text{for } \alpha = 1$$

$$(s_{22} \to -\infty)$$

$$B = \sqrt{\frac{2}{2\rho_w a^2 (M_{e4}^2 - 1)}} \quad \text{for } \alpha > 1$$

For $\alpha = 1$ the solution incorporates the effects of both the wall layer and the body shape. Clearly, as $A \to \infty$, it must go over to the solution for $\alpha < 1$ when $\alpha \to 1$, while for $A \to 0$, when the role played by the body slope decreases, to the solution for $\alpha > 1$.

For the bodies corresponding to $\alpha \geq 1$, when the wall layer displacement thickness effect on the disturbance decay asymptotics is important, the range of applicability of solutions (3.21) is restricted. In fact, in deriving Eq. (3.20) the Bernoulli equation (3.19) was used for calculating $\Delta\delta^*_{22}$. However, there exist distances $|s_{22}| \gg 1$ on which the leading viscous terms become equal in the order to the inertial terms near the body surface, namely, in the wall layer producing the main part of the region 22 displacement thickness (region 3 in Fig. 3.1), though in the outer part (region 2 in Fig. 3.1) the viscosity effect is still small. From the equality of the leading viscous and inertial terms in the wall layer producing the main part of the variation of the displacement thickness $\Delta\delta^*_2$, in which $\Delta u \sim u \sim \sqrt{\Delta p}$, we obtain $s \sim \varepsilon^{3/4} = Re^{-3/8}$, that is, $s \sim \varepsilon^{3/4}$. It can easily be seen that regions 2 and 3 of length $s \sim \varepsilon^{3/4} = Re^{-3/8}$ are fully equivalent to regions 2 and 3 in Fig. 1.2 for the flows with free interaction studied in detail in Chapter 1. In the first approximation, the flow in region 2 is inviscid, with $\Delta p \sim Re^{-1/4}$. This confirms the correctness of the derived condition (3.10). In Section 1.6 for the flat-plate flow, which corresponds to $\alpha > 1$ in formula (3.21), the limiting solution for region 3 was derived in an explicit form. It can easily be seen that the asymptotic forms of the pressure distributions (1.78) and (3.21) coincide. This ensures the matching of the profiles of the velocity and other parameters in the overlapping region ($s_3 \to 0$, $s_{22} \to \infty$), since in the inviscid part the Bernoulli equation holds.

### *3.1.3 Boundary conditions for the viscous sublayer 32*

The viscous sublayer 32 is described by the Prandtl equations (3.14). For determining the initial conditions as $s_{32} \to -\infty$ and the boundary conditions for $H_{32}$ as $n_{32} \to +\infty$ it is necessary to match the solutions for regions 3 and 32. Let us determine the limiting solution for region 3 as $s_3 \to 0$. Let $\alpha \geq 1$ (the case $\alpha < 1$ is less interesting and will not be considered). The matching for the corresponding inviscid regions made it possible to obtain the pressure distribution (3.21). The case $\alpha > 1$ is considered in Section 3.2. The more general case $\alpha = 1$ differs only by the value of the constant $B$ in formula (3.21) for the pressure distribution. Therefore, it might be expected that the complicated limiting structure of the solution with a viscous sublayer lying on the bottom of a somewhat thicker "heat-conducting" layer discovered in Section 3.2, will retain in the general case $\alpha = 1$.

In what follows, it will be useful to derive the solution in the von Mises variables. Bearing in mind Eq. (3.21) we will seek for the solution for the viscous sublayer in the form:

$$u_3 = \left(\frac{2B^2}{\rho_w}\right)^{1/2} \frac{\psi(\overline{\varphi}_3, s_3)}{-s_3}, \quad \overline{\psi}_3 = \frac{\psi_3}{\left(\mu_w\sqrt{2B^2\rho_w}\right)^{1/2}}, \quad \Delta p = -\frac{B^2}{(-s_3)^2} \tag{3.22}$$

Substituting Eq. (3.22) in the momentum equation

$$\rho_w u_3 \frac{\partial u_3}{\partial s_3} + \frac{\partial \Delta p}{\partial s_3} = \rho_w u_3 \frac{\partial u_3}{\partial \psi_3}\left(\rho_w \mu_w u_3 \frac{\partial u_3}{\partial \psi_3}\right) \tag{3.23}$$

yields the following equation

$$\Phi(\Phi\Phi')' = \Phi^2 - 1 + (-s_3)\Phi\Phi \tag{3.24}$$

As $s_3 \to 0$, the solution of Eq. (3.24) subject to the boundary conditions $\Phi(0,0)=0$ and $\Phi(\infty,0)=1$ takes the form:

$$\overline{\psi}_3 = \int_0^{\phi} \frac{\Phi\, d\Phi}{\left(\frac{2}{3}\Phi^3 - 2\Phi + \frac{4}{3}\right)^{1/2}} \tag{3.25}$$

The limiting solution obtained indicates that in regions 3 and 32 the orders of $\Phi$ are the same, though at flow acceleration ($\Delta p \sim \varepsilon^{1/2}$ in region 3 and $\Delta p \sim 1$ in region 32) the transverse dimension reduces by a factor of $\varepsilon^{1/4}$ ($\sim\sqrt{\Delta p}$). Formula (3.25) gives the initial conditions for region 32 as $s_{32} \to -\infty$.

The limiting solution for the energy equation

$$\frac{\partial H_3}{\partial s_3} = \frac{\rho_w \mu_w}{\sigma} \frac{\partial}{\partial \psi_3}\left(u_3 \frac{\partial H_3}{\partial \psi_3}\right) \tag{3.26}$$

is more complicated

$$H_3 = b\left(\frac{2}{\rho_w a}\right)^{1/2}\left[-4\rho_w\mu_w\left(\frac{2B^2}{\rho_w}\right)^{1/2}\ln\frac{1}{(-s_3)}\right]^{1/4} Q\left(s_3, \overline{\overline{\psi}}_3\right) \tag{3.27}$$

$$\overline{\overline{\psi}} = \frac{\psi_3}{\left[4\rho_w\mu_w\sqrt{\frac{2B^2}{\rho_w}}\ln\frac{1}{(-s_3)}\right]^{1/2}}$$

Substituting Eq. (3.27) in Eq. (3.25) and taking in account that in the $\overline{\overline{\psi}}_3 \sim 0(1)$ region $\overline{\psi}_3 \to \infty$ when $(-s_3) \to 0$ and hence $\Phi = 1$ leads to the equation

$$\frac{Q''}{\sigma} = Q - 2\overline{\overline{\psi}}_3 Q' - 4(-s_3)\ln(-s_3)Q \tag{3.28}$$

Using the Bernoulli equation and the form of transformations (3.27) we obtain the following boundary conditions for Eq. (3.28):

$$Q(s_3, 0) = 0, \quad Q(s_3, \overline{\overline{\psi}}_3 \to 0) \to \overline{\overline{\psi}}^{1/2} \tag{3.29}$$

Equation (3.28) subject to the boundary conditions (3.29) was solved numerically for $s_3 = 0$. As a result, it was established that as $\overline{\overline{\psi}}_3 \to 0$, the function $Q(0, \overline{\overline{\psi}}_3)$ takes the form:

$$Q(0, \overline{\overline{\psi}}_3) \sim 1.022\sigma^{1/4}\overline{\overline{\psi}}_3 + \cdots \tag{3.30}$$

For $s_3 \sim \varepsilon^{1/4}$ the $\overline{\overline{\psi}}_3 \sim O(1)$ region is in the order thicker than the region $\psi_3 \sim O(1)$ or, what is the same, thicker than the $\psi_{32} \sim O(1)$ region. This means that for investigating the stagnation enthalpy $H$ distribution in the maximum disturbance zone $s_{32} \sim O(1)$ it is necessary to take account for the existence of the "nonheat-conducting" zone 62 (Fig. 3.1) in which, due to the heat conduction effect in zone 3, the stagnation enthalpy gradient behaves on its top ($\psi_{62} = \overline{\overline{\psi}}_3 \to +\infty$) as $\overline{\overline{\psi}}_3^{\,1/2}$, while on its bottom ($\psi_{62} = \overline{\overline{\psi}}_3 \to 0$ or $\overline{\overline{\psi}}_3 \to \infty$) we have $H - H_w \sim \psi_{62}$.

The distribution of $\Delta H = H - H_w$ over this region is determined by the solution of Eq. (3.28) subject to the boundary conditions (3.29). The numerically determined asymptotic representation (3.30) must be used for small $\psi_{62}$ for matching with the solution of the energy equation in region 32. This allows us to determine the as yet indefinite scale $e(\varepsilon)$ in Eq. (3.14) and the outer boundary condition for the increment of the stagnation enthalpy $H_{32}$ in region 32. Bearing in mind the relations

$$s_3 = \varepsilon^{1/4} s_{32}, \quad \overline{\overline{\psi}}_3 = \psi_{62}, \quad \psi_{32} = \psi_3, \quad H \sim H_w + \varepsilon^{1/4} H_3 + \cdots$$

as well as Eqs. (3.27)–(3.30) and (3.13), and requiring that, in accordance with the matching principle, the external limit of the internal expansion (solution for region 32 as $\psi_{32} \to \infty$) be equal to the internal limit of the external expansion (solution for region 62 as $\psi_{62} \to 0$) we obtain

$$e(\varepsilon) = \left( \frac{\varepsilon}{\ln \frac{1}{\varepsilon^{1/4}}} \right) \tag{3.31}$$

$$H_{32}(s_{32}, \psi_{32} \to \infty) \to b \left( \frac{2}{\rho_{3w} a} \right)^{1/2} \frac{1.022\sigma^{1/4} \psi_{32}}{\left( 4(\rho_3 \mu_3)_w \sqrt{\frac{2B^2}{\rho_{3w}}} \right)^{1/4}}$$

In accordance with Eq. (3.13), the characteristic dimension of the viscous sublayer 32 is $n \sim \varepsilon^{3/2}$, while the velocity and enthalpy differences are $\Delta u \sim 1$ and $\Delta H \sim \left(\varepsilon / \ln \frac{1}{\varepsilon^{1/4}}\right)$. For this reason, in the deflection region there occurs a sharp increase in the viscous stress and a somewhat smaller increase in the heat fluxes, so that we obtain

$$\tau \sim \mu \frac{\partial u}{\partial y} \sim \varepsilon^{1/2}, \quad q \sim \lambda \frac{\partial T}{\partial y} \sim \frac{\varepsilon^{3/4}}{\left(\ln \frac{1}{\varepsilon^{1/4}}\right)}$$

instead of the orders characteristic of the boundary layer, where, as is well known, $\tau \sim \varepsilon$ and $q \sim \varepsilon$.

### *3.1.4 Bringing the equations for region 33 into the standard form*

The methods for solving the boundary layer equations for upstream and downstream unbounded (on their scales) flows were discussed in the paper of Neiland (1966). In these cases the important role is played by the behavior of the velocity and pressure disturbances, as $s_{32} \to -\infty$, where $u_{32e} \to 0$ and $\Delta p_{32} \to 0$. Following the paper of Neiland (1966) we introduce variables analogous to those of Lees and Dorordnitsyn

$$\zeta = C \int_{-\infty}^{s_{33}} (\rho\mu)_{33w} u_{33e} \, ds_{33}, \quad \lambda = m u_{33e} \int_{0}^{n_{33}} \rho_{33} \, dn_{33} \tag{3.32}$$

$$u = u_e f(\zeta,\lambda), \quad g = \frac{m}{M} H_{32}(\zeta,\lambda)$$

$$C = m^2 = \frac{a(M_{e4}^2 - 1)^{1/2}}{2\mu_w}, \quad M^* = \frac{1.022\sigma^{1/4} b\left(\frac{2}{\rho_w a}\right)^{1/2}}{(4\mu_w\sqrt{2p_w B^2})^{1/4}}$$

The equation for region 33 takes the form:

$$(Nf'')' + \frac{\mathrm{d}u_{33e}}{\mathrm{d}\zeta}\left(\frac{\rho_{33e}}{\rho_{33}} - f'^2\right) = u_{33e}(f'\dot{f}' - \dot{f}f'')$$

$$\left(\frac{N}{\sigma}g'\right)' + \frac{\Delta m}{M^*}u_{33e}^2(Nf'f'')' = u_{33e}(f'\dot{g} - \dot{f}g') \tag{3.33}$$

$$\lambda = \frac{\sigma - 1}{\sigma}\left(\frac{\ln\frac{1}{\varepsilon^{1/4}}}{\varepsilon}\right)^{1/4}, \quad N = \frac{\rho_{33}\mu_{33}}{\rho_{33w}\mu_{33w}}$$

The boundary conditions are as follows:

$$f(\zeta,0) = f'(\zeta,0) = g(\zeta,0) = 0, \quad f'(\zeta,\phi) = g'(\zeta,\infty) = 1 \tag{3.34}$$

Thus, the integration starts from $\zeta = O(s_{33} \to -\infty)$. The $u_{33e}$ distribution must be determined in solving the problem for the inviscid flow in region 22. However, the asymptotic solution (3.21) for $s_{32} = s_{33} \to -\infty$ obtained above makes it possible to check the fulfillment of the initial conditions, that is, the matching with solutions (3.24) to (3.28). Using formula (3.21) we obtain $u_{33e} \sim \zeta + \cdots$, $N \to 1$, and $\rho_{33}/\rho_{33e} \to 1$. Then the momentum equation (3.33) takes the form:

$$f''' + 1 - f'^2 = 0, \quad f(0) = f'(0) = 0, \quad f'(\infty) = 1 \tag{3.35}$$

For $\zeta \to 0$ and $s_3 \to 0$ we perform the substitutions

$$\Phi = f', \quad \overline{\psi} = f, \quad \frac{\mathrm{d}f}{\mathrm{d}\lambda} = \Phi, \quad \frac{\mathrm{d}^2 f}{\mathrm{d}\lambda^2} = \Phi\frac{\mathrm{d}\Phi}{\mathrm{d}\overline{\psi}}, \quad \frac{\mathrm{d}^3 f}{\mathrm{d}\lambda^3} = \Phi\frac{\mathrm{d}}{\mathrm{d}\overline{\psi}}\left(\Phi\frac{\mathrm{d}\Phi}{\mathrm{d}\overline{\psi}}\right) \tag{3.36}$$

It can easily be seen that substituting formulas (3.36) in Eq. (3.35) reduces the latter to Eq. (3.24) for $s_3 = 0$. Thus, the matching of the solutions in regions 3 and 33 is ensured.

Therefore, the problem for the viscous sublayer is reduced to the standard form (3.33), (3.34) and can be integrated numerically using any standard method for the boundary layer equations. However, the velocity distribution over the outer boundary should be preliminarily determined by solving the problem for region 22.

### *3.1.5 Solution of the problem in the region of locally inviscid flow 22*

The flow in region 22 governed by the complete Euler equations (3.9) was calculated by the method of integral relations of Dorodnitsyn (1958). In the first, one-strip approximation, after simple but cumbersome algebra the Euler equations written as conservation laws can be brought into the following form:

$$
\begin{aligned}
&\frac{d\delta}{ds} = (1 + K\delta)\tan\theta_e \\
&\left[\frac{1}{2}\sin(2\theta_e)\frac{d\rho_e V_e^2}{dM_e} - \rho_e V_e^2 \frac{dv_e}{dM_e}\cos(2\theta_e)\right]\frac{dM_e}{ds} \\
&\quad = -K\rho_e V_e^2 \cos(2\theta_e) + \frac{2}{\delta}\left[p_w - (1 - \delta K)p_e + \frac{1}{2}\delta K(p_e - p_w + \rho_e V_e^2\cos^2\theta_e\right. \\
&\qquad\qquad \left. + \rho_w V_w^2) - \frac{1}{4}(1 + K\delta)\rho_e V_e^2 \sin(2\theta_e)\tan\theta_e\right] \\
&\frac{dT_w}{ds} = -\left[\left(\frac{d\rho_e V_e}{dM_e}\cos\theta_e + p_e V_e \frac{dv_e}{dM_e}\sin\theta_e\right)\frac{dM_e}{ds} - K\rho_e V_e \sin\theta_e\right. \\
&\qquad\qquad \left. + \frac{1}{\delta}(1 + K\delta)(\rho_e V_e \cos\theta_e + T_w)\tan\theta_e\right] \\
&\frac{d\Pi_w}{ds} = u_w \frac{dT_w}{ds}
\end{aligned}
\tag{3.37}
$$

For the sake of brevity, here we omitted the subscripts referring to the numbers of regions. We also introduced the variables facilitating the passage through a singular point occurring at $M_w = 1$ proposed in the paper of Lun'kin et al. (1956)

$$
\Pi = p + pu^2, \quad T = pu, \quad V = \sqrt{u^2 + v^2} \tag{3.38}
$$

In the above equations $\delta$ is the thickness of region 22 which coincides, as $s \to -\infty$, with the outer edge of the boundary layer. Clearly, preassigning $\delta$ when using a nonasymptotic one-strip method is to a certain degree arbitrary. In the calculations it was assumed that $n = \delta$ as $s \to -\infty$, when $\left|\frac{u - u_e}{u_e}\right| = 0.01$.

The quantities $\rho_e V_e$, $d\rho_e V_e/dM_e$, $dv_e/dM_e$, $\rho_e V_e^2$, $d\rho_e V_e^2/dM_e$, and $p_e$ entering into Eq. (3.37) can be expressed in terms of $M_e(s)$ and $M_e(-\infty)$ using isentropic relations.

On a near-wall streamline we have

$$
\begin{aligned}
p_w &= \Pi_w - T_w u_w \\
u_w &= \frac{\gamma}{\gamma + 1}\frac{\Pi_w}{T_w} \times \left[1 \mp \sqrt{1 - \frac{(\gamma - 1)(1 + \gamma)}{\gamma^2}\frac{H_w}{H_e}\frac{1 + ((\gamma - 1)/2)M_e^2(-\infty)}{((\gamma - 1)/2)M_e^2(-\infty)}\frac{T_w^2}{\Pi_w^2}}\right]
\end{aligned}
\tag{3.39}
$$

Here, minus and plus signs relate to the subsonic and supersonic velocities near the body.

Thus, the system of equations (3.37) incorporates four equations for determining $T_w$, $\Pi_w$, $M_e$, and $\delta$.

The initial conditions are as follows:

$$\delta \to \delta(-\infty), \quad \Pi_w \to p(-\infty), \quad T_w \to 0, \quad M_e \to M_e(-\infty) \quad \text{for } s \to (-\infty) \tag{3.40}$$

In passing through the "sonic" point $M_w = 1$ the derivatives of $T_w$ must, by virtue of the known properties of gas dynamic functions, turn to zero

$$\frac{dT_w}{ds} = 0 \quad \text{for } M_w = 1 \tag{3.41}$$

This fact is used in solving the problem. System (3.37) is integrated numerically by the Runge–Kutta method. The values of $T_w$ and $\Pi_w$ can be found at point $M_w = 1$ for given initial conditions (3.40). The parameter $\delta(M_w = 1)$ turns out to be inessential, since in what follows the difference between $\delta(M_w = 1)$ and $\delta(-\infty)$ can be eliminated by transformation of all linear scales of problem (3.37). This is possible, since system (3.37) is invariant relative to the transformation in which all linear dimensions are multiplied by the same constant. The values of $s$ and $M_e$ remain unknown at point $M_w = 1$. If $M_e$ is arbitrarily preassigned, then $s$ is determined from condition (3.41). Then the fulfillment of conditions (3.40), except for the first condition for $\delta$, is checked by numerical integration toward $s \to -\infty$.

After $M_e$ has been determined by an iteration procedure at $M_w = 1$, system (3.37) is integrated toward $s \to +\infty$. No iteration procedure is needed, since in this direction $M > 1$ everywhere.

By way of illustration, we present in Fig. 3.3 the calculated results for a body with the shape $x_{32w} = A_0 y_{22w} - A/(-y_{22w})^{\alpha}$. The calculations were carried out for $M_e(-\infty) = 3$, $H_w/H_e = 0.5$, $A = 1$, $\alpha = 9$, and $A_0 = -3.7726$. In Figs. 3.2–3.5 an appreciable pressure

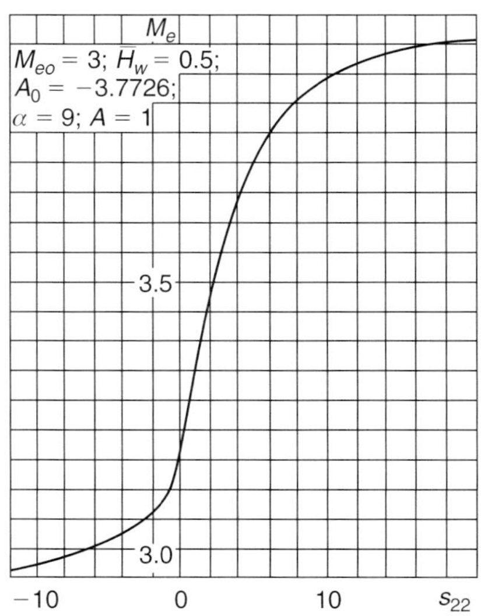


Fig. 3.2.

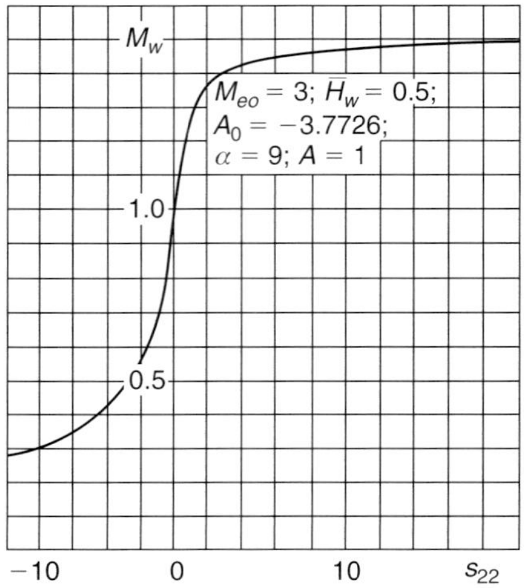


Fig. 3.3.

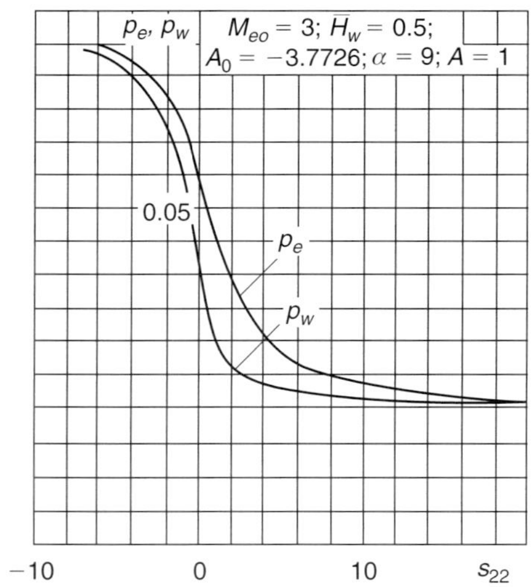


Fig. 3.4.

difference in the disturbed region is clearly visible. Due to the action of centrifugal forces, the pressure on the body is lower than that on the outer boundary. At large deflection angles this can lead to a decrease of the pressure $p_{w22}$ lower than the value $p_{22}(+\infty)$, that is, to overexpansion. Then on the body there must appear a region with a pressure increase, where separation of the viscous sublayer 22 is possible.

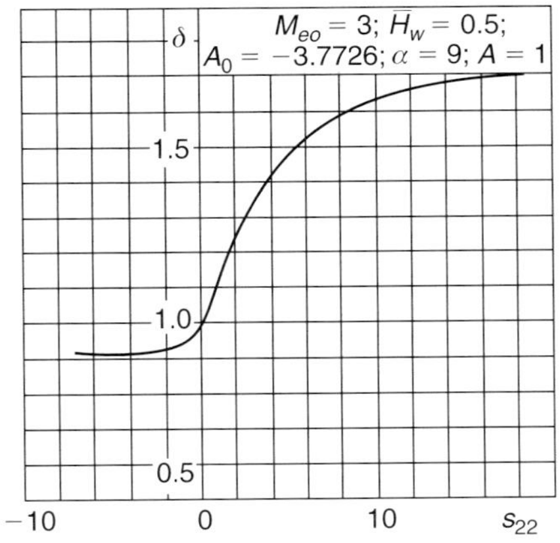


Fig. 3.5.

## 3.2 Flow ahead of the base section of a body

We will consider the flow near the base section of a body in a supersonic gas stream. If the base pressure is different from that on the lateral surface of the body, then it should be expected that disturbances will propagate upstream owing to the existence of a subsonic flow region in the boundary layer. This can have a considerable effect on the local flow characteristics (pressure, friction, and heat transfer distributions) and lead to a certain displacement of the center of pressure of the body. Moreover, the study of the flow near the base section is needed for the investigation of the near wake behind the body.

### *3.2.1 Formulation of the problem and characteristic flow regions*

Let for the sake of simplicity the forebody is a flat plate or a wedge in an uniform supersonic flow. For the axisymmetric flow past a circular cone it should be assumed that the base radius is considerably greater than the characteristic size of the disturbed region and the Stepanov–Mangler transformation is applicable. We will first assume that the base pressure is less than that on the lateral surface of the body by $O(1)$. The rarefaction propagates upstream along subsonic streamtubes whose cross-sections decrease in the process of gas acceleration. This leads to deflection of the supersonic part of the flow toward the wall and the formation of expansion waves. In accordance with theory of inviscid supersonic flows, for $\Delta p \sim O(1)$ in the outer supersonic flow (region 1 in Fig. 3.6) we have $v/u \sim O(1)$. The boundary layer thickness is of the order $O(\varepsilon)$, where $\varepsilon = Re^{-1/2}$. However, the outer supersonic flow streams past a "body" formed by the displacement thickness of the vortical part of the flow. Then in this region we have also $v \sim u \sim O(1)$ and, by virtue of the continuity equation, we must consider region 22 (see Fig. 3.6) in which $v \sim u \sim 1$ and $s \sim n \sim \varepsilon$. However, then the asymptotic representations and scales for region 22 are no longer different from those given by formulas (3.8). The equations for region 22 are obtained from Eqs. (3.9), if $K = 0$ is let in the latter equations. The boundary conditions at the body $v_{22w} = 0$ remain unaltered, while in the condition at the outer edge (4.12) we should let $\theta_w = 0$. An important difference

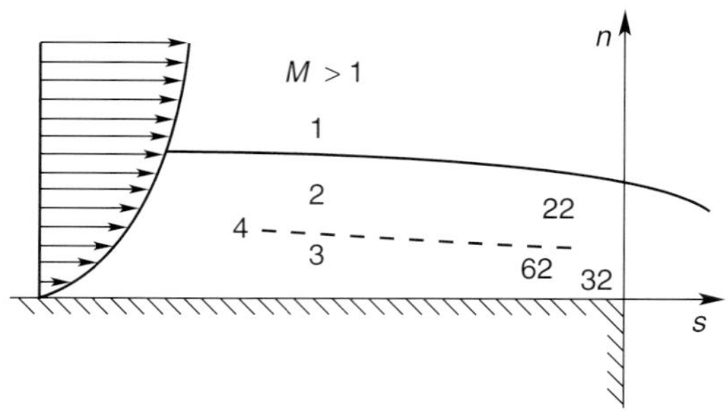


Fig. 3.6.

consists in the nature of the solution near the sonic point on the body $M_w = 1$. However, this is considered below and for a while we note that the asymptotics of the pressure disturbance decay, as $s_{22} \to -\infty$, studied in Section 3.1 can also be used for the flow ahead of the base section if it is assumed that $\alpha > 1$ in formula (3.21). In fact, it was established above that for bodies with $\alpha > 1$ the term $\theta_w$ vanishes from Eq. (3.20) as $s_{22} \to -\infty$, since it is the boundary layer displacement thickness that is predominant, while the body shape turns out to be unimportant.

Then, following Section 3.1, we must introduce a free interaction region of length $O(\varepsilon^{3/4})$ which would contain the viscous sublayer 3 and the inviscid, weakly perturbed vortical flow 2 (see Fig. 3.6); the solutions for the flat-plate flows with free interaction were studied in Section 1.6.

In the main disturbed flow region with $s \sim O(\varepsilon)$ solutions of the Euler equations for region 22 cannot satisfy the no-slip conditions at the body surface. As in Section 3.1, we must introduce a viscous boundary sublayer 32 (Fig. 3.6) with the coordinate and function scales (3.13) and Eq. (3.14). Then matching the solutions of the energy equations for regions 3 and 32 leads to the appearance of region 62 in which the stagnation enthalpy distribution is described by Eqs. (3.27)–(3.30).

Thus, the flow ahead of the base section is to a considerable degree a particular case of the theory developed for the flow near the "rounded" corner point considered in Section 3.1.

### 3.2.2 Solution of the problem and comparison with experimental data

For calculating the flow in the locally inviscid region 22 we used the first (single-strip) approximation of Dorodnitsyn's method of integral relations. The approximating system of equations is obtained by letting $K = \theta_w = 0$ in Eqs. (3.37).

It should be noted that, due to the absence of $\theta_w$ and $K$, the solution governing the flow ahead of the base section always involves a singular point at $M_w = 1$ (since the derivative $\mathrm{d}M_w/\mathrm{d}s = \infty$). This becomes obvious if we recall that the left side of the third equation (3.37) can be written in the form:

$$\frac{\mathrm{d}T_w}{\mathrm{d}s} = \frac{\mathrm{d}(\rho u)_w}{\mathrm{d}M_w}\frac{\mathrm{d}M_w}{\mathrm{d}s}$$

For $M_w = 1$ the specific flow rate $(\rho u)_w$ reaches a maximum and the coefficient of $\mathrm{d}M_w/\mathrm{d}s$ vanishes, whereas the right side of Eq. (3.37) does not contain the geometric parameters $K(s)$ and $\theta_w$ and is generally nonzero. The necessity of precisely this behavior becomes obvious from physical considerations. In fact, with decrease in the pressure the subsonic streamtubes near the body surface become narrower. This makes possible the enlargement of supersonic streamtubes. However, when the value $M_w \approx 1$ is reached, further flow expansion along the body surface becomes impossible. Further flow expansion takes place in the expansion wave centered at the corner point at which $M_w = 1$. Upon transition to the base region, the Mach number on the streamline lying on the body varies jumpwise, which gives $\mathrm{d}M_w/\mathrm{d}s = \infty$. Of course, in order for a sonic point to appear, the pressure ratio in the base region and on the lateral surface ahead of the disturbed region must be less than the known critical value $(2/\gamma + 1)^{\gamma/\gamma - 1}$; for example, for $\gamma = 1.4$ this value is about 0.528. Further decrease of the base pressure has no effect on the flow ahead of the base section. There occurs the phenomenon of "choking" quite analogous to that which takes place in nozzles. If the pressure difference is smaller than the critical value, then at the corner point the pressure is equal to the base pressure and $M_w < 1$. In the first approximation the approximate solutions for small pressure differences are described by the third formula (3.21).

The theory developed was extended to the limiting case of the hypersonic flow in the weak interaction regime $[M_e(-\infty) \gg 1]$ by Olson and Messiter (1969). Here, we will compare the results presented in the study of Matveeva and Neiland (1967) with the experimental data of Hama (1966) and the calculation performed in the first approximation. The results of the paper of Matveeva and Neiland (1967) are in good agreement with the experimental data, while the results Ollson and Messiter (1969) obtained in the hypersonic approximation lie somewhat lower. The latter fact can be understood taking into account that in the experiment the Mach number varied from 2.35 to 4.02. The comparison is presented in Fig. 3.7. As in the study of Olson and Messiter, along the abscissa axis the ratio of the distance from the corner point to the undisturbed boundary layer displacement thickness $\delta_0^*$ calculated in the hypersonic approximation is measured. Clearly, the length of the region with $\Delta p \sim 1$ amounts to about five boundary layer thicknesses.

For calculating the friction and heat transfer distributions we used the system of equations (3.32) and (3.33) for region 32 and the solutions obtained in Chapter 2 for region 3. In Fig. 3.8 we have plotted the friction and heat transfer distributions for $M_e(-\infty) = 3$, the Prandtl number $\sigma = 1$, and the temperature factor $H_w/H_e = 0.515$. Clearly, the friction increases considerably as the flow accelerates; the heat flux also increases but somewhat weaker than the friction. We note that possible scorching of the surface near the base section at distances $\sim(3 \text{ to } 5)\delta_0^*$ can result in the loss of stability of hypersonic flight vehicles.

## 3.3 Reattachment of a supersonic flow to the body surface

The problem of the supersonic flow reattachment to the body surface or the coalescence of two parts of a supersonic flow is of its own interest. Moreover, it is very important in theory of separated flows.

The results obtained above for developed locally inviscid flows with free interaction make it possible to study the asymptotic structure of the flow in the vicinity of viscous supersonic stream attachment as the Reynolds number increases without bounds. In this section we will

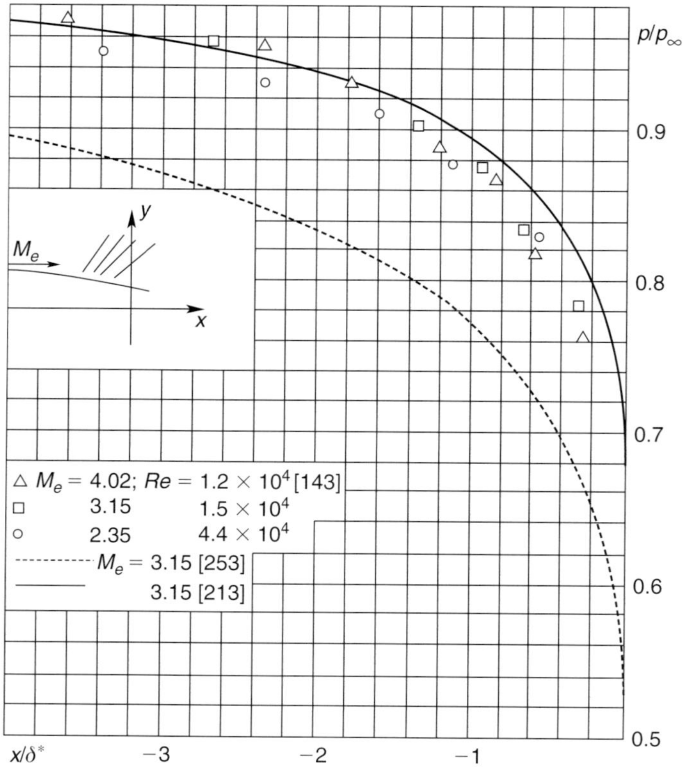

y
$M_e$
x
$p/p_\infty$
0.9
0.8
0.7
0.6
0.5
△ $M_e = 4.02$; $Re = 1.2 \times 10^4$ [143]
□ 3.15 $1.5 \times 10^4$
○ 2.35 $4.4 \times 10^4$
$M_e = 3.15$ [253]
3.15 [213]
$x/\delta^*$
−3
−2
−1


Fig. 3.7.

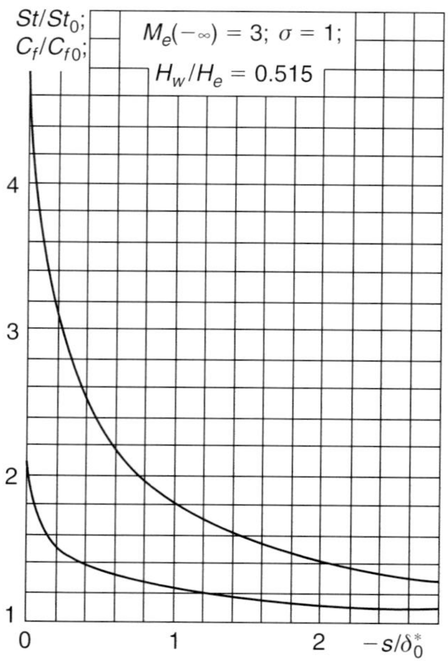

$St/St_0$;
$C_f/C_{f0}$;
$M_e(-\infty) = 3$; $\sigma = 1$;
$H_w/H_e = 0.515$
4
3
2
1
0
1
2
$-s/\delta_0^*$


Fig. 3.8.

consider a very simple case of the incidence of a plane semi-infinite supersonic jet onto an infinite plane at angles which would correspond to flow deflection in an attached oblique shock, if the flow were inviscid and have not include a mixing zone. In the next chapter the relation between the solution obtained and that of the problem of a developed laminar separation zone in a supersonic flow is established.

### *3.3.1 Formulation of the problem and main flow regions*

The diagram of the flow under consideration is presented in Fig. 3.9. Above and to the left of point $A$ there is a uniform supersonic flow in which the velocity is directed by an angle $\alpha$ to an infinite plane aligned with the $s$ axis. An infinite region 7 is that of dead gas. It is assumed that mixing starts from point $A$. The distance from the mixing zone beginning to the origin along the line $AO$ parallel to the undisturbed supersonic flow velocity is equal to $\ell$ and used as the main scale length on which all linear dimensions of the problem are normalized. For the velocity, density, viscosity, etc., scales we take their values at the outer edge of the boundary layer, far away to the right of the attachment region. Point $O$ does not generally coincides with the stagnation point in the flow.

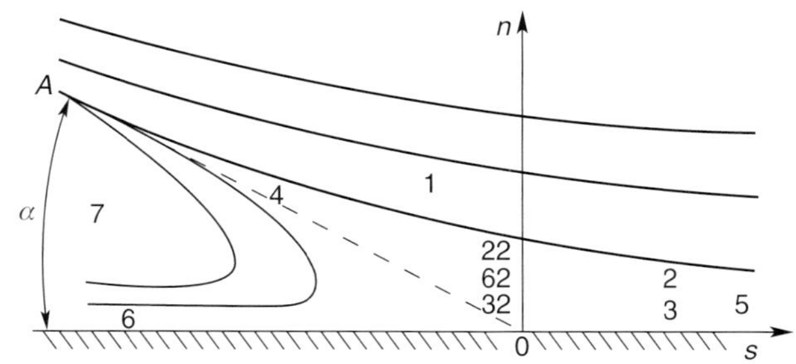


Fig. 3.9.

The Reynolds number $Re$ is based on the parameters of the undisturbed supersonic flow far away from the attachment region, while the scale length is $\ell$. Passing to the limit $Re \to \infty$ at $(s, n) = O(1)$ we obtain the limiting flow pattern of the form presented in Fig. 3.10. The

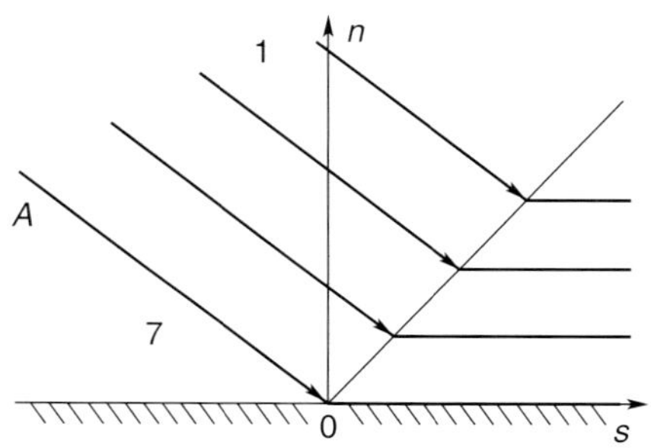


Fig. 3.10.

outer supersonic flow 1 passes across the attached shock proceeding from point $O$. At point $O$ the pressure varies jumpwise from the value $p_7$ in the stagnation region 7 (or in the undisturbed supersonic flow) to the value $p_5$ which can easily be determined from the Rankine–Hugoniot relations, given the angle $\alpha$ and the undisturbed flow parameters.

We will now note that this solution does not represent a uniformly accurate first approximation to the solutions of the Navier–Stokes equations. Firstly, along the line $AO$, in accordance with the classical boundary layer theory, a viscous mixing zone (region 4 in Fig. 3.9) with thickness $O(\varepsilon)$, where $\varepsilon = Re^{-1/2}$, should be introduced instead of the discontinuity in the tangential (relative to $AO$) velocity component. Obviously, the flow in the mixing zone is governed by the Prandtl equations. In region 4 the velocity component tangential to $AO$ is of the order $O(1)$; it varies across the mixing zone from $u_{4e} \sim O(1)$, on transition to region 1, to zero, on transition to region 7. In a certain vicinity of point $O$ the outer supersonic flow is deflected by an angle $\alpha \sim O(1)$. In this region we also have $\Delta p \sim O(1)$. Because of this, a part of streamtubes with a low dynamic head cannot penetrate to the elevated-pressure region. These streamtubes turn to the left, thus marking the beginning of the boundary layer 6. The outer part of the mixing zone 4 goes to the right and forms the beginning of the boundary layer 5. Let us now consider directly the deflection zone; here, $\alpha \sim O(1)$, $\Delta p \sim O(1)$, and $v/u \sim 1$. The characteristic dimension of the vortical part of the flow, whose streamtubes proceed from the mixing zone 4, is $\varepsilon$, so that, in view of the continuity equation, we have $s/n \sim v/u \sim 1$ and $s \sim n \sim \varepsilon$. Thus, it is necessity to introduce the scales and asymptotic expansions analogous to those used for region 22 in Sections 3.1 and 3.2 (see Eq. (3.8)). As before, substituting Eq. (3.8) in the Navier–Stokes equations and passing to the limit $\varepsilon \to 0$ leads in the first approximation to the Euler equations (3.9) for region 22. The general structure of the solution in region 22 determines the nature of the solutions in all other flow regions; for this reason, we will consider it in more detail.

### 3.3.2 Nature of the locally inviscid flow in region 22

As shown above, in the first approximation the flow is described by the Euler equations. Let us derive the boundary conditions. At the body surface we have

$$v_{22}(s_{22}, 0) = 0 \tag{3.42}$$

For matching the solutions in regions 4 and 22 it is convenient to go over to the von Mises variables. This gives the parameter distributions in the "oncoming flow" for region 22:

$$\begin{aligned} &s_4 = \varepsilon s_{22}, \quad \Psi_4 = \Psi_{22} \\ &u_4(0, \Psi_4) = u_{22}(-\infty, \Psi_{22}), \quad v_4(0, \Psi_4) = v_{22}(-\infty, \Psi_{22}) \\ &\quad p_4 = p_{22}(-\infty, \Psi_4), \quad \rho_4(0, \Psi_4) = \rho_{22}(-\infty, \Psi_{22}) \end{aligned} \tag{3.43}$$

The compatibility conditions for regions 1 and 22 can be determined in the same fashion as above for expansion flows (see Eq. (3.11)). The locally outer supersonic flow 1 streams past a smooth surface formed by the displacement thickness of the vortical, locally inviscid flow 22. If the effect of the submerged shock formed in a small vicinity of point $O$ can be neglected (e.g., $\alpha$ is not too large and then the difference between an isentrope and a shock polar, which is of order $O(\alpha^3)$, can be neglected), then the compatibility condition is simplified and can be presented in the form:

$$\theta_e + \nu_e = -\alpha + \nu_{e4} \tag{3.44}$$

Here, $\nu_{e4}$ is the value of $\nu$ in the undisturbed flow 1 on the outer boundary of region 4 and $\theta_e$ and $\nu_e$ correspond to the outer boundary of the vortical inviscid flow 22.

Let us denote the pressure far away from the turn region by $p_{22\infty}$. In the streamtube which arrives at the body surface, the stagnation pressure $P_{22}$ in the leading term of expansion $[\Delta p \sim O(1)]$ cannot, in accordance with the Bernoulli equation, be less than $p_{22\infty}$. If $P_{22w} > p_{22\infty}$, then in the flow on the body surface there exists a stagnation point at which the $\Psi_{22} = 0$ streamline branches. An alternative flow pattern is that in which $P_{22} = p_{22\infty}$. The streamline patterns for the cases $P_{22w} > p_{22\infty}$ and $P_{22} = p_{22\infty}$ are plotted in Figs. 3.11 and 3.12, respectively, on the scale of region 22 as $\varepsilon \to 0$. In the latter case the $\Psi_{22} = 0$ streamline approaches the body surface asymptotically, as $s_{22} \to +\infty$, and the velocity $u_{22}(s_{22} \to +\infty, \Psi_{22} = 0) \to 0$. In any finite part of the flow in region 22 there is no stagnation point. This fact (the absence of a stagnation point at the equality of the total pressure on the zero streamtube and the static pressure on the right as $s \to +\infty$) was demonstrated in connection with the problem of incompressible flow (Taganov, 1968) on the basis of the monotonicity theorems proved by A.A. Nikol'skii. For choosing the solution of the type shown in Fig. 3.12 in an incompressible flow Taganov in his study of (1968) made a hypothesis on the instability of flows of the type presented in Fig. 3.11. The supersonic flow considered in this study is subject to essentially different boundary conditions (other important difference consists in the fact that in the work of Taganov (1968) the inviscid flow was introduced on the scale of the order of the body size, whereas in this study it appears only in a local region of size $O(\varepsilon)$ as a result of the passage to the limit $\varepsilon \to 0$). In certain important cases, the necessity of the fulfillment of the condition $P_{22}(0) = p_{22\infty}$ and the realization of a flow with no stagnation point in a locally inviscid flow region can be proved. (As shown in what follows, this means that the stagnation point is located in the free interaction region and in its vicinity the important role is played by viscosity forces.)

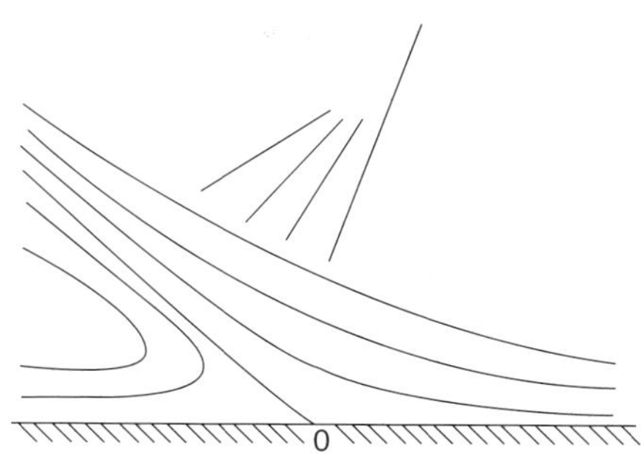


Fig. 3.11.

We will now demonstrate that the stagnation point cannot exist in a finite region of the locally inviscid flow 22 if under the action of small pressure disturbances $\Delta p_{22}$ the vortical

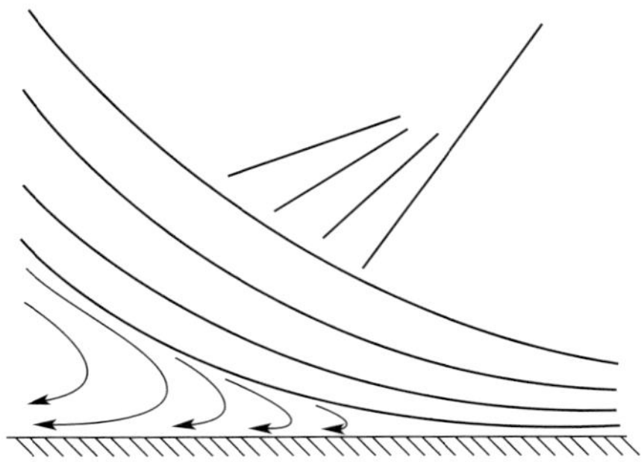

Fig. 3.12.

part of the jet changes the displacement thickness $\delta^*_{22}$ as a subsonic streamtube. Let us assume that a stagnation point exists. Then to the right of it there is no return flow and the flow becomes plane-parallel as $s_{22} \to +\infty$, while $d\delta^*/ds_{22} \to 0$. The pressure can approach a limiting value either increasing or decreasing. If the pressure approaches the limiting value decreasing, $(dp_{22}/ds_{22}) < 0$, then, as for the subsonic streamtube, we have $d\delta^*/ds_{22} < 0$. However, in this case $d^2\delta^*/ds^2_{22} > 0$ and the pressure in the outer supersonic flow past the concave body formed by the boundary $\delta^*_{22}$ must increase, so that $dp_{22}/ds_{22} > 0$, which is in contradiction with the original assumption.

We will now consider the second possibility. Let us assume that $dp_{22}/ds_{22} > 0$. Then $d\delta^*/ds_{22} > 0$ and $d^2\delta^*/ds^2_{22} < 0$. In this case, the effective body is convex and has disturbances with $dp_{22}/ds_{22} < 0$ that must be induced in the supersonic flow. Again we have arrived at a contradiction. These contradictions do not arise in the flow presented in Fig. 3.12 in which streamlines can turn for all $s_{22}$. The solutions are derived below. The original assumption that for $s_{22} \to +\infty$ the vortical part of the flow changes the displacement thickness as a subsonic streamtube is obligatorily fulfilled at least in two important cases: (1) when the Mach number is not too high and (2) when the jet incidence angle $\alpha$ is small. In the latter case, in view of the Bernoulli equation, on the streamline $\Psi_{22} = 0$ we have $\Delta p_{22} \sim \alpha$ and the velocity $u_{22} \sim \Delta u_{22} \sim \alpha^{1/2}$. In these flows $d\delta^*_{22}/ds_{22}$ is determined in the leading term by the variation of the wall layer displacement thickness, which is $\sim\alpha^{1/2}$, while the thickness of the main part of the vortex layer 22 changes by $O(\alpha)$. But in the wall sublayer we have $M \sim O(\alpha^{1/2}) < 1$.

### *3.3.3 Solution for the problem of the locally inviscid flow in region 22*

We will first consider the asymptotic behavior of the solution for region 22 as $s_{22} \to +\infty$. The static pressure $p_{22}$ approaches the limiting value $p_5$ behind the deflection region, while on the body the velocity $u_{22w} \to 0$. For a while, we denote $\Delta p = p_{22} - P_w = O(\alpha^*)$, where $\alpha^* \ll 1$ is a smallness parameter. In the main part of region 22 $[n_{22} \sim O(1)$, where $u_{22} \sim O(1)]$ the velocity variation is $\Delta u_{22} \sim O(\alpha^*)$ and that of the displacement thickness is $O(\alpha^*)$. This follows from the Bernoulli and continuity equations. Near the body surface there is always a layer with $\Delta u_{22} \sim u_{22} \sim O(\alpha^{*1/2})$, since $u_{22w} \to 0$, as $\Delta p \to 0$, by virtue of the same Bernoulli equation. In this layer, the streamtube thickness varies by the leading order, which is obvious from the continuity equation. Let us estimate the thickness of this layer and its contribution to the variation of the complete displacement thickness $\Delta\delta_{22}$ of region 22. We note that, as $s_{22} \to +\infty$, the velocity profile subject to the condition $u_{22w}(s \to \infty) = 0$ can easily be obtained, given the parameter profiles at the end of the mixing zone 4, using the conservation of the stagnation enthalpy and the entropy along streamlines in the locally inviscid region 22. For a while, we only note that $\partial u_{22}/\partial n_{22} = O(1)$ as $s_{22} \to +\infty$. Therefore, the thickness of the sublayer, in which $\Delta u_{22} \sim u_{22} \sim \alpha^{*1/2}$, is of the order $u_{22} \sim O(\alpha^{*1/2})$. However, since it varies in the leading order, precisely it determines the total variation of the thickness of the entire region 22: $\Delta\delta^*_{22} \sim n_{22} \sim \alpha^{*1/2}$. For $\Delta p \sim \alpha^*$ on the outer boundary of region 22 the streamline slope is, in accordance with the boundary condition (3.44), also of the order $\alpha^*$, while the complete variation of the region thickness is $\sim\alpha^{*-1/2}$. Therefore, its length is $\Delta s_{22} \sim O(\alpha^{*-1/2})$.

For deriving the asymptotic form of the solution for region 22 as $s_{22} \to +\infty$ on the basis of the above estimates, for $s_{22} \sim \alpha^{*-1/2}$ we will consider two layers, namely, the main layer with $n_{22} \sim O(1)$ and a sublayer with $n_{22} \sim O(\alpha^{*1/2})$. In the main layer the following asymptotic expansions and scales are introduced

$$
\begin{aligned}
&S_{22} = \alpha^{*1/2} s_{22}, \quad \Psi_{22} \qquad\qquad (3.45)\\
&u_{22} \sim u_{22}(+\infty, \Psi_{22}) + \alpha^* u^{(1)}_{22}(S_{22}, \Psi_{22}) + \cdots, \quad p_{22} \sim P_5 + \alpha^* p^{(1)}_{22}(S_{22}, \Psi_{22})\\
&\rho_{22} \sim \rho_{22}(+\infty, \Psi_{22}) + \alpha^* \rho^{(1)}_{22}(S_{22}, \Psi_{22}) + \cdots, \quad v_{22} \sim \alpha^* v^{(1)}_{22}(S_{22}, \Psi_{22}) + \cdots\\
&n_{22} \approx n_{22}(+\infty, \Psi_{22}) + \alpha^{*1/2} n^{(0)}_{22}(S_{22}, \Psi_{22}) + \alpha^* n^{(1)}_{22}(S_{22}, \Psi_{22}) + \cdots
\end{aligned}
$$

Substituting Eq. (3.45) in the Euler equations (3.9) and passing to the limit $\alpha^* \to 0$ leads to the equations

$$
\begin{aligned}
&\frac{\partial p^{(1)}_{22}}{\partial S_{22}} + \rho_{22}(\infty, \psi_{22}) u_{22}(+\infty, \psi_{22}) \frac{\partial u^{(1)}_{22}}{\partial \psi_{22}} = 0, \quad \frac{\partial n^{(0)}_{22}}{\partial \psi_{22}} = 0 \qquad (3.46)\\
&\frac{\partial p^{(1)}_{22}}{\partial \psi_{22}} = 0, \quad \frac{\partial n^{(0)}_{22}}{\partial S_{22}} = \frac{v^{(1)}_{22}}{u_{22}}(+\infty, \psi_{22}) + \cdots
\end{aligned}
$$

The first equation is the linearized form of the Bernoulli equation. From the other results it is the absence of the pressure difference $p^{(1)}_{22}$ and the conservation of the quantity $n^{(0)}_{22}$ across region 22 that are important.

The corresponding scales and asymptotic expansions for the sublayer, in which the velocity disturbances are of the order of the undisturbed velocity, are as follows:

$$
\begin{aligned}
&S_{22} = \alpha^{*1/2} s_{22}, \quad \Psi_{22} = \frac{\psi_{22} - \psi_{22w}}{\alpha^*}\\
&u_{22} \sim \alpha^{*1/2} U_{22}(S_{22}, \Psi_{22}) + \cdots, \quad v_{22} \sim \alpha^{*3/2} V_{22}(S_{22}, \Psi_{22}) + \cdots \qquad (3.47)\\
&\rho_{22} \sim \rho_{22w} + \cdots, \quad n_{22} \sim \alpha^{*1/2} N_{22}(S_{22}, \Psi_{22}) + \cdots
\end{aligned}
$$

The expansion for $p_{22}$ coincides with Eq. (3.45). Substituting Eq. (3.47) in the Euler equations (3.9) and passing to the limit $\alpha \to 0$ yields

$$\frac{\partial N_{22}}{\partial S_{22}} = \frac{V_{22}}{U_{22}}, \quad \frac{\partial N_{22}}{\partial \Psi_{22}} = \frac{1}{\rho_{22w} U_{22}} \tag{3.48}$$

$$p_{22}^{(1)}(S_{22}) + \frac{\rho_{22w} U_{22}^2(S_{22}, \Psi_{22})}{2} = \frac{\rho_{22w}}{2} U_{22}^2(+\infty, \Psi_{22})$$

In the third equation (3.48) having the form of the Bernoulli equation for an incompressible fluid we used the conditions $p_{22} \to p_5$ and $p_{22}^1 \to 0$ for $S_{22} \to \infty$; $U_{22}(+\infty, \Psi_{22})$ is the velocity profile near the body in region 22 after the completion of the flow turn.

The streamline at which the flow turns and $\Psi_{22}$ reaches a minimum is determined from the condition

$$p_{22}^{(1)} = \frac{\rho_{22w}}{2} U_{22}^2(+\infty, \Psi_{22\,\mathrm{min}}) \tag{3.49}$$

We will integrate the second equation (3.48)

$$N_{22}(S_{22}, \Psi_{22}) = 2 \int\limits_{\Psi_{22\,\mathrm{min}}(S_{22})}^{0} \frac{\mathrm{d}\Psi_{22}}{\rho_{22w}\sqrt{U_{22}^2(+\infty, \Psi_{22}) - \dfrac{2p_{22}^{(1)}}{\rho_{22w}}}} + \int\limits_{0}^{\Psi_{22}} \frac{\mathrm{d}\Psi_{22}}{\rho_{22w}\sqrt{U_{22}^2(+\infty, \Psi_{22}) - \dfrac{2p_{22}^{(1)}}{\rho_{22w}}}} \tag{3.50}$$

Applying the matching principle to the function $n_{22}$ in the outer and inner regions for determining $n_{22}^{(0)}$ and taking into account that for $\alpha^* \ll 1$ the outer boundary condition for region 22 leads, obviously, to the Ackeret formula, we can write

$$[M_{22e}^2(+\infty) - 1]^{1/2} p_{22}^{(1)} = \frac{\mathrm{d}}{\mathrm{d}S_{22}} \left\{ \lim_{\Psi_{22} \to +\infty} [N_{22}(S_{22}, \Psi_{22} - N_{22}(+\infty, \Psi_{22})] \right\} \tag{3.51}$$

Equation (3.51) means that the outer, weakly perturbed supersonic flow streams past the effective body formed by the displacement thickness of region 22.

Equations (3.50) and (3.51) can be solved, thus determining the pressure disturbance $p_{22}^{(1)}(S_{22})$ decay law, if the velocity profile $U_{22}(\infty, \overline{\Psi}_{22})$ is known; the latter is determined by the velocity profile in the mixing zone 4 ahead of the turn; actually, only the information on its behavior near $\Psi_{22} = \psi_{22}$ is needed, since in the sublayer $\Psi_{22} \sim O(1)$, in accordance with Eq. (3.47), there are only streamtubes for which $\psi_{22} = \psi_{22w} + O(\alpha^*)$ as $\alpha^* \to 0$ (or $s_{22} \to +\infty$ for $S_{22} \sim O(1)$). Since generally it is not the streamtube that passes through point $A$ (see Fig. 3.9) that arrives at the wall in region 22, we assume that $\psi_{22w} \neq 0$.

In accordance with Eq. (3.50) we have

$$N_{22}(+\infty, \Psi_{22}) = \int_0^{\Psi_{22}} \frac{d\Psi_{22}}{\rho_{22w} U_{22}(+\infty, \Psi_{22})} \tag{3.52}$$

since $p_{22}^{(1)}(+\infty) = 0$. However, the velocity profile in region 22 after the turn ($S_{22} \to 0$) takes the form:

$$U_{22}(+\infty, \Psi_{22}) \sim aN_{22}(+\infty, \Psi_{22}) + \cdots \tag{3.53}$$

where the constant $a$ will be determined below from the data for the profile in region 4 using the Bernoulli and isentrope equations.

From Eqs. (3.52) and (3.53) there follows

$$\begin{aligned} U_{22}(+\infty, \Psi_{22}) &\sim \sqrt{\frac{2a}{\rho_{22w}} \Psi_{22}^{1/2}} + \cdots, \\ U_{22}(S_{22}, \Psi_{22}) &= \pm\sqrt{\frac{2a}{\rho_{22w}} \Psi_{22} - \frac{2}{\rho_{22w}} p_{22}^{(1)}(S_{22})} \end{aligned} \tag{3.54}$$

In the second equation (3.54) plus and minus signs relate to the flows above and below the streamline $\Psi_{22\,\min}$, respectively. Thus, $\Psi_{22\,\min}(S_{22}) = (1/a)p_{22}^{(1)}(S_{22})$. Substituting Eq. (3.54) in Eqs. (3.50) and (3.51) makes it possible to obtain the solution

$$p_{22}^{(1)} = -\frac{2}{\rho_{22w} a^2 [M_{22e}^2(\infty) - 1] S_{22}^2} \tag{3.55}$$

The artificial smallness parameter $\alpha^*$ drops out from the solution

$$p_{22} \sim p_5 - \frac{2}{\rho_{22w} a^2 [M_{22}^2(\infty) - 1] s_{22}^2} + \cdots \quad \text{for } s_{22} \to +\infty \tag{3.56}$$

The expression for the velocity in region 22 near the body surface $[(\psi_{22} - \psi_{22w} \to \infty]$ takes the form:

$$u_{22}(s_{22}, \psi_{22}) \sim \pm\sqrt{\frac{2a}{\rho_{22w}}(\psi_{22} - \psi_{22w}) + \frac{4}{\rho_{22w}^2 a^2 [M_{22e}^2(\infty) - 1] s_{22}^2}} + \cdots \tag{3.57}$$

The $\psi_{22\,\min}(s_{22})$ dependence can be derived from Eq. (3.57) by equating $u_{22}(s_{22}, \psi_{22\,\min})$ to zero.

The stagnation point of the flow is at infinity ($s_{22} \to +\infty$). We note that for the flow with a stagnation point $\Psi_{22\,\min}$ should be equated to zero in Eq. (3.50) in the region to the right of

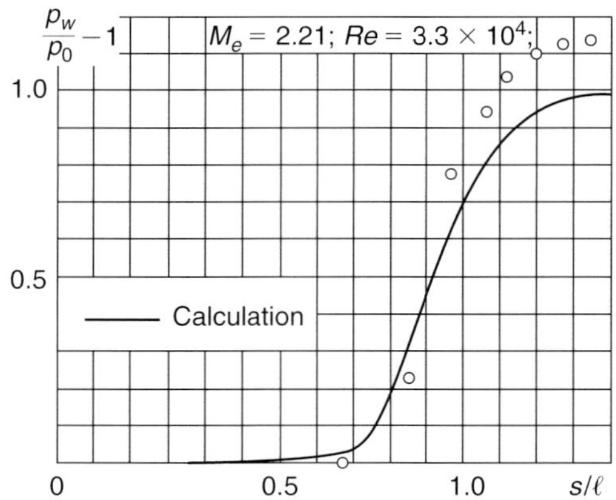


Fig. 3.13.

this point. Elementary calculations show that, as it might be expected, the solution of such a problem does not exist. Numerical calculations for the entire region 22 were performed using the modified method of integral relations by Dorodnitsyn. We note that the two-strip approximation was used with the strip boundaries coinciding with the "outer" streamline of region 22 and the upper branch of the $\psi_{22w}$ streamline. In view of the physical features of the problem, a refined approximation of the velocity and density profiles was used on the lower strip. For this purpose, the information on the stagnation enthalpy and entropy distributions in streamlines was used. In Fig. 3.13 the calculated pressure distribution in the flow reattachment region is compared with the experimental data of Chapman et al. (1958). Since in the experiment flow separation began almost from the leading edge, the parameter distributions at the end of zone 4 required for the calculation were taken in accordance with the well-known self-similar solution (see, e.g., Chapman, 1950). The agreement of the calculated and measured results in the region, in which their gradients are large, is satisfactory. This is important for the further consideration of friction and heat transfer. However, we note that at the end of region 22 the calculated pressure is somewhat lower than the experimental value. The reason is the viscosity effect in the vicinity of the stagnation point as $s_{22} \to \infty$; below this question is considered in detail.

### *3.3.4 Viscous flow regions*

The solution of the problem for the inviscid region 22 cannot provide a uniformly accurate first approximation for the problem solution as $Re \to \infty$. First, the locally inviscid flow does not satisfy the no-slip boundary condition on the body. This requires introducing the viscous sublayer 32 (see Fig. 3.9), on the scale of which the leading viscous terms are of the same order as the inertial terms. The layer 32 is considered below. Secondly, from the asymptotic formulas (3.56) and (3.57) it follows that for $s_{22} \sim \varepsilon^{-1/4}$ in the lower part of region 22, which, as shown above, "controls" the pressure distribution as $s_{22} \to +\infty$, and for which at these distances $\psi_{22} - \psi_{22w} \sim \varepsilon^{1/2}$, the leading viscous terms also become of the same order as the inertial terms (the situation is similar to that considered in Section 3.2

for expansion flows). Thus, it is necessary to consider regions 2 and 3 with the longitudinal scale $s \sim \varepsilon^{3/4}$ (since $s_{22} = s/\varepsilon$) and the pressure disturbances $\Delta p \sim \varepsilon^{1/2}$, as in all previously considered flows with free interaction (Chapter 1 and Sections 3.1 and 3.2). The scales of the independent variables and the asymptotic expansions for the functions can be obtained from Eq. (3.47) for $\alpha^* = \varepsilon^{1/2}$ with account of Eq. (3.8)

$$\begin{aligned}
s_3 &= \frac{s}{\varepsilon^{3/4}}, \quad \psi_3 = \Psi - \frac{\psi_w}{\varepsilon^{3/2}}, \quad n_3 = \frac{n}{\varepsilon^{5/4}} \\
p &\sim p_5 + \varepsilon^{1/2} p_3(s_3, n_3) + \cdots, \quad \rho \sim \rho_3(s_3, n_3) + \cdots \\
H &\sim H_3(s_3, n_3) + \cdots, \quad u \sim \varepsilon^{1/4} u_3(s_3, n_3) + \cdots \\
v &\sim \varepsilon^{3/4} v_3(s_3, n_3) + \cdots
\end{aligned} \tag{3.58}$$

Substituting Eq. (3.58) in the Navier–Stokes equations and passing to the limit $\varepsilon \to 0$ leads to the conventional equations for a compressible boundary layer

$$\begin{aligned}
&\rho_3 \left( u_3 \frac{\partial u_3}{\partial s_3} + v_3 \frac{\partial u_3}{\partial n_3} \right) = -\frac{\partial p_3}{\partial s_3} + \frac{\partial}{\partial n_3} \left( \mu \frac{\partial u_3}{\partial n_3} \right) \\
&\frac{\partial p_3}{\partial n_3} = 0, \quad \frac{\partial \rho_3 u_3}{\partial s_3} + \frac{\partial \rho_3 v_3}{\partial n_3} = 0 \\
&\rho_3 \left( u_3 \frac{\partial H_3}{\partial s_3} + v_3 \frac{\partial H_3}{\partial n_3} \right) = \frac{\partial}{\partial n_3} \left( \frac{\mu_3}{\sigma} \frac{\partial H_3}{\partial n_3} \right)
\end{aligned} \tag{3.59}$$

At these distances, the results for the main part of the vortex flow (region 2 in Fig. 3.9) are obtained by substituting $\alpha^* = \varepsilon^{1/2}$ in formulas (3.46). Since this does not lead to appearance of any additional terms in Eqs. (3.47), Eqs. (3.46) and (3.47) can be used for region 2 letting $\alpha^* = \varepsilon^{1/2}$ and replacing the subscript 22 by 2. The parameter distributions in region 2 can easily be obtained in the quadrature form if the pressure distribution $p_2^{(1)} = p_3$ is known.

Let us specify the boundary conditions for region 3. At the body surface we have

$$u_3(s_3, 0) = v_3(s_3, 0) = 0, \quad H_3(s_3, 0) = H_{3w} \tag{3.60}$$

Matching the solutions for regions 2 and 3 yields

$$H_3(s_3, +\infty) = H_2(s_2, \psi_{2w}) \tag{3.61}$$

We note that for $\psi_2 = \psi_{2w}$ we have $H_2 = H_4(0, \psi_{2w})$ if $\psi_{2w}$ denotes the value at the streamline arriving at the body surface. Generally, $H_{3w} \neq H_{3e}$. Because of this, in region 3 $\rho_3$ and $\mu_3$ are not constant, as it was the case for the flows with free interaction considered in Chapter 1 and Sections 3.1 and 3.2. However, the other outer boundary conditions obtained by matching the solutions in regions 2, 3, and 1 are derived in the same fashion as before

$$u_3(s_3, \psi_3) \to \sqrt{\frac{2a}{\rho_{3e}}} \psi_3^{1/2} \quad \text{for } \psi_3 \to \infty \tag{3.62}$$

$$[M_2^2(+\infty)-1]^{1/2}p_3=\frac{\mathrm{d}}{\mathrm{d}s_3}\left\{\lim_{\psi_3\to+\infty}\left[\int_0^{\psi_3}\frac{\mathrm{d}\psi_3}{\rho_3 u_3}-\left(\frac{2\psi_3}{\rho_{3e}a}\right)^{1/2}\right]\right\} \tag{3.63}$$

Here, as before, the constant $a$ is the velocity gradient at the wall in region 2 as $\Delta p\to 0$:

$$a=\left(\frac{\partial u_2}{\partial n_2}\right)_w \quad \text{for } s_2\to+\infty \tag{3.64}$$

Let us investigate the behavior of solution (3.59)–(3.64) as $s_3\to 0$ and perform the matching with the corresponding solutions in regions with the longitudinal scale $s\sim\varepsilon$, for example, region 22 as $s_{22}\to+\infty$.

We will first make the following change of variables which reduces the boundary value problem to the form with a smaller number of parameters

$$\begin{aligned}
\xi &= \left\{\frac{2[M_{22e}^2(+\infty)-1]^{1/2}}{a\mu_{3e}}\right\}(-p_3)\\
\eta &= \left\{\frac{2[M_{22e}^2(+\infty)-1]^{1/2}a^3}{\rho_{3e}^2\mu_{3e}}\right\}\frac{\int_0^n \rho_3\,\mathrm{d}n_3}{\xi^{1/2}}\\
X &= \left\{8[M_{22e}^2(+\infty)-1]^{3/2}\rho_{3e}^2\mu_{3e}a^5\right\}^{1/4} s_3\\
\psi_3 &= \left\{\frac{\mu_{3e}}{2a[M_{22e}^2(+\infty)-1]^{1/2}}\right\}^{1/2}\xi f(\xi,\eta)
\end{aligned} \tag{3.65}$$

Substituting Eq. (3.65) in Eqs. (3.59)–(3.63) yields

$$\begin{aligned}
&(Nf'')'=\xi^{1/2}\frac{\mathrm{d}\xi}{\mathrm{d}X}\left(-g+\frac{1}{2}f'^2-ff''\right)+\xi^{3/2}(f'\dot{f}'-f''\dot{f})\\
&\left(\frac{N}{\sigma}g'\right)'=-\xi^{1/2}\frac{\mathrm{d}\xi}{\mathrm{d}X}fg'+\xi^{3/2}(f'\dot{g}-\dot{f}g')\\
&f(X,0)=f'(X,0)=0,\quad g(X,0)=g_w\\
&f''(X,\infty)=g(X,0)=1,\quad N=\frac{\rho_3\mu_3}{(\rho_3\mu_3)_e}\\
&\xi=-\frac{\mathrm{d}}{\mathrm{d}X}\left[2\sqrt{\xi}\lim_{\eta\to\infty}\left(\int_0^{\eta}g\,\mathrm{d}\eta-\sqrt{2f}\right)\right]
\end{aligned} \tag{3.66}$$

Since $H_{3e} - H_{3w} \sim O(1)$ and the velocities are of the order $\varepsilon^{1/4}$, in the first approximation the static and total enthalpies coincide and the energy equation does not include a dissipative term. We also have

$$\frac{\rho_e}{\rho} = g, \quad N = g^{\omega - 1}, \quad \text{if } \frac{\mu_3}{\mu_{3e}} = \left(\frac{h}{h_{3e}}\right)^{\omega} \tag{3.67}$$

For matching with region 22 the passage to the limit $X \to 0$ must be performed for $\eta \sim O(1)$

$$f(X, \eta) \sim f_0(\eta) + O(X^2), \quad g \sim 1 + O(X^2), \quad \xi = \frac{8}{X^2} + O(1) \tag{3.68}$$

Expansion (3.68) can be substituted in Eq. (3.66) with the passage to the limit $X \to 0$

$$f_0(\eta) = \frac{1}{2}(\eta - \sqrt{2})^2 - 1 \tag{3.69}$$

Expressions (3.68) and (3.69) coincide with solutions (3.56) and (3.57) for the lower part of region 22 as $s_{22} \to +\infty$, that is, the initial profile in the main part of region 3 for $s_3 = 0$ is determined by the lower parts of the profiles in region 22.

However, the solution is obtained not completely, since expressions (3.68) and (3.69) do not satisfy the conditions at the body surface. This might also be expected from the physical considerations, since as $s_3 \to 0$ the pressure gradients increase and viscous effects vanish from the main part of region 3 but persist near the body.

We will seek the limiting, as $X \to 0$, solutions for small $\eta$, that is, for the layers in which the viscous terms of Eqs. (3.66) are conserved. Since on the outer boundaries of these wall layers the velocity remains, in accordance with Eq. (3.69), of the order $O(1)$, we introduce the variables

$$g(X, \eta) = g(X, \overline{\eta}), \quad f(X, \eta) = A(X)\Phi(X, \overline{\eta}), \quad \overline{\eta} = \frac{\eta}{A(X)} \tag{3.70}$$

Substituting Eq. (3.70) in the energy equation (3.66) yields

$$\left(\frac{N}{\sigma} g'\right) = \left(\frac{32\sqrt{2}}{X^4} A^2 - \frac{16\sqrt{2}}{X^3} A\dot{A}\right) \Phi g' + \frac{16\sqrt{2} A^3}{X^3} (\Phi' \dot{g} - \dot{\Phi} g') \tag{3.71}$$

In order to satisfy both boundary conditions for $g$ in Eq. (3.66) it is necessary to require that both the convective term and the term with the higher-order derivative with respect to $g$ be retained as $X \to 0$. For this purpose, it is sufficient that the following equalities hold

$$\frac{32\sqrt{2}}{X^4} A^2 - \frac{16\sqrt{2}}{X^3} A\dot{A} = \frac{1}{\sigma}, \quad A^2 = \frac{1}{8\sqrt{2}\sigma} X^4 \ln \frac{1}{X} \tag{3.72}$$

Substituting Eq. (3.72) with $X \to 0$ in the energy equation yields

$$(g^{\omega - 1} g')' = \Phi g', \quad \Phi' = -\sqrt{2g}, \quad g(0) = g_w, \quad g(\infty) = 1 \tag{3.73}$$

Thus, in the limit in the "thermal" layer the viscous terms drop out from the momentum equation. Therefore, in order to satisfy the no-slip boundary condition $f'(X, 0) = 0$ it is necessary to additionally introduce a "viscous" layer at the bottom of the "thermal" layer as $X \to 0$. At the outer boundary of this layer, which represents the "thermal" layer bottom, the velocity $f' = \Phi' = -\sqrt{2g_w}$ and the stagnation enthalpy $g_w$ is the same as at the wall. In order to retain the "viscous" term in the momentum equation, thus ensuring the fulfillment of the condition $f'_w = 0$ and the matching of the solutions for the "viscous" and "thermal" layers, it is necessary to introduce the following variables as $X \to 0$

$$f(X,\eta) = X^2\Phi^*(X,\eta^*), \quad g(X,\eta) = g(X,\eta^*), \quad \eta^* = \frac{\eta}{X^2} \tag{3.74}$$

The choice (3.74) can be substantiated in the same fashion as it was done above for the thermal layer. Substituting Eq. (3.74) in Eq. (3.66) and passing to the limit $X \to 0$ yields

$$(g_w^{\omega-1}\Phi''^*)' = 32\sqrt{2}\left(g_w - \frac{1}{2}\Phi'^{*2}\right), \quad \left(\frac{g^{\omega-1}}{\sigma}g'\right)' = 0 \tag{3.75}$$

The boundary conditions are as follows:

$$\eta = 0\colon \Phi^* = \Phi'^* = 0, \quad g = g_w$$

$$\eta \to \infty\colon \Phi'^* \to -\sqrt{2g_w}, \quad g \to g_w$$

The energy equation has the obvious solution

$$g = g_w$$

The physical meaning of the results obtained is that on transition to the region with large pressure gradients, $s \sim \varepsilon$, the "thermal" and "viscous" layers start to develop at the return flow bottom, the viscosity effects being concentrated on the bottom of a narrow zone with a near-constant stagnation enthalpy.

Let us return to the investigation of the viscous and heat-conducting flow region with $s \sim \varepsilon$, that is, with $s_{22} \sim O(1)$. The limiting solutions of Eqs. (3.70)–(3.75) obtained above make it possible to determine the scales of the coordinates and functions in these regions, together with the initial conditions.

We will first consider region 62 (Fig. 3.9) in which the stagnation enthalpy varies from the value corresponding to the $\psi_{22} = 0$ streamline of the inviscid flow 22 to the value corresponding to the body surface temperature. The corresponding streamtubes flow out of the return flow region at the lower part of region 3; they are described by solution (3.70)–(3.73) obtained above. We will use these results, together with the matching principle, for determining the transverse scale of the "thermal" layer 62

$$s_{62} = \frac{s}{\varepsilon}, \quad \psi_{62} = \frac{\psi}{\varepsilon^{3/2}}\left(\ln\frac{1}{\varepsilon}\right)^{1/2}$$

$$u \sim u_{62}(s_{62},\psi_{62}) + \cdots, \quad p \sim p_{62}(s_{62},\psi_{62}) + \cdots \tag{3.76}$$

$$\rho \sim \rho_{62}(s_{62},\psi_{62}) + \cdots, \quad v \sim \varepsilon^{1/2}\left(\ln\frac{1}{\varepsilon}\right)^{1/2} v_{62}(s_{62},\psi_{62}) + \cdots$$

$$H \sim H_{62}(s_{62},\psi_{62}) + \cdots, \quad n \sim \varepsilon^{3/2}\left(\ln\frac{1}{\varepsilon}\right)^{1/2} n_{62}(s_{62},\psi_{62}) + \cdots$$

Substituting Eq. (3.76) in the Navier–Stokes equations and passing to the limit $\varepsilon \to 0$ in the first approximation yields

$$\rho_{62}u_{62}\frac{\partial u_{62}}{\partial s_{62}} + \frac{\partial p_{62}}{\partial s_{62}} = 0, \quad \frac{\partial p_{62}}{\partial \psi_{62}} = 0, \quad \frac{\partial n_{62}}{\partial s_{62}} = \frac{v_{62}}{u_{62}}, \quad \frac{\partial n_{62}}{\partial \psi_{62}} = \frac{1}{\rho_{62}u_{62}}$$

$$\frac{\partial H_{62}}{\partial s_{62}} = 0, \quad H_{62} = H_{62}(\psi_{62}), \quad p_{62} = p_{62}(s_{62}) \tag{3.77}$$

The solution is completely obtained if $H_{62}(\psi_{62})$ and $p_{62}(s_{62})$ are determined. The latter function is determined by matching with region 22 as $\psi_{22} \to 0$

$$p_{62}(s_{62}) = p_{22}(s_{22}, \psi_{22} = 0) \tag{3.78}$$

The function $H_{62}(\psi_{62})$ is determined by matching with the solution in region 3 as $s_3 \to 0$ and $\overline{\eta} \sim O(1)$. Formally, for this purpose it is sufficient to express $H_{62}$ and $\psi_{62}$ in terms of the variables of region 3 and then substitute $\varepsilon^{1/4}s_{62}$ for $s_3$ and pass to the limit $\varepsilon \to 0$ for $s_{62} \sim O(1)$. Then the following formulas are obtained:

$$H_{62} = g(0, \overline{\eta})H_{3e}, \quad \psi_{62} = 4a^{1/3}\mu_{3e}^{2/3}\Phi(0, \overline{\eta}) \tag{3.79}$$

The functions $g(0, \overline{\eta})$ and $\Phi(0, \overline{\eta})$ are determined by formulas (3.73). The variable $\overline{\eta}$ can be eliminated by the following changes

$$g(0, \overline{\eta}) = g(\Phi), \quad \frac{d}{d\overline{\eta}} = -\frac{1}{\sqrt{2g}}\frac{d}{d\Phi}, \quad W^* = \frac{1}{2^{1/4}\Phi}$$

$$\frac{d}{dW^*}\left(g^{\omega-1/2}\frac{dg}{dW^*}\right) = -\frac{W^* dg}{dW^*}, \quad g(\Phi = 0) = g_w \tag{3.80}$$

$$g(W^* \to -\infty) = 1$$

Thus, in region 62 in the first approximation the flow is inviscid but the stagnation parameter distributions in $\psi_{62}$ are determined by heat conduction effects in region 3.

From Eqs. (3.77)–(3.80) we obtain

$$u_{62}(s_{62}, 0) = -\sqrt{2H_{2w}\left[1 - \left(\frac{p_{62}}{p_{62}(+\infty)}\right)^{(\gamma-1)/\gamma}\right]} \tag{3.81}$$

$$H_{62}(s_{62}, \psi_{62} \to 0) \sim H_{62w} + H_{62e}g'(0)\left\{\frac{a[M_{22e}^2(\infty) - 1]^{1/2}\sigma}{\mu_{3e}}\right\}\psi_{62} + \cdots$$

The quantity $g'(0)$ is determined by numerically solving Eq. (3.80). For $\omega = 1/2$ Eq. (3.80) can easily be integrated in the quadrature form. Formulas (3.81) indicate that $u_{62}(s_{62}, 0) \neq 0$ and in order for the no-slip boundary conditions to be fulfilled for $s \sim \varepsilon$, it is necessary to consider a wall region 32 in which the leading viscous terms of the Navier–Stokes equations are of the same order as the inertial terms. Since, in accordance with the asymptotic expansion matching principle for regions 62 and 32, formulas (3.81) must ensure the outer boundary conditions for region 32, the scales of the independent variables and functions must take the form:

$$s_{32} = s_{62} = \frac{s}{\varepsilon}, \quad \psi_{32} = \frac{\psi}{\varepsilon^{3/2}}, \quad n_{32} = \frac{n}{\varepsilon^{3/2}}$$

$$u \sim u_{32}(s_{32}, n_{32}) + \cdots, \quad v \sim \varepsilon^{1/2} v_{32}(s_{32}, n_{32}) + \cdots \tag{3.82}$$

$$p \sim p_{32}(s_{32}, n_{32}) + \cdots, \quad H \sim H_w + \left(\ln\frac{1}{\varepsilon}\right)^{-1/2} H_{32}(s_{32}, n_{32}) + \cdots$$

Substituting Eq. (3.82) and passing to the limit $\varepsilon \to 0$ yields the system of equations for region 22:

$$\rho_{32}\left(u_{32}\frac{\partial u_{32}}{\partial s_{32}} + v_{32}\frac{\partial u_{32}}{\partial n_{32}}\right) + \frac{dp_{32}}{ds_{32}}$$

$$= \frac{\partial}{\partial n_{32}}\left(\mu_{32}\frac{\partial u_{32}}{\partial n_{32}}\right) \frac{\partial(\rho_{32}u_{32})}{\partial s_{32}} + \frac{\partial(\rho_{32}v_{32})}{\partial n_{32}} = 0, \quad \frac{\partial p_{32}}{\partial n_{32}} = 0$$

$$\rho_{32}\left(u_{32}\frac{\partial H_{32}}{\partial s_{32}} + v_{32}\frac{\partial H_{32}}{\partial n_{32}}\right) = \frac{\partial}{\partial n_{32}}\left(\frac{\mu_{32}}{\sigma}\frac{\partial H_{32}}{\partial n_{32}}\right) + \frac{\left(\ln\frac{1}{\varepsilon}\right)^{1/2}(\sigma - 1)}{\sigma}$$

$$\times \frac{\partial}{\partial n_{32}}\left(\mu_{32}u_{32}\frac{\partial u_{32}}{\partial n_{32}}\right) \tag{3.83}$$

It is assumed that $(\sigma - 1) \to 0$ as $\varepsilon \to 0$, but in such a way that $(\ln 1/\varepsilon)^{1/2}(\sigma - 1) = O(1)$. This passage to the limit results in minimal degeneration of the equation of energy and for this reason is most interesting. Practically, for $Re \sim 10^4$ to $10^6$ and $\sigma \sim 0.7$, the quantity $(\ln 1/\varepsilon)^{1/2}(\sigma - 1)$ is, in fact, close to unity.

The boundary conditions at the body are obvious:

$$u_{32}(s_{32}, 0) = v_{32}(s_{22}, 0) = H_{32}(s_{32}, 0) = 0 \tag{3.84}$$

The outer boundary conditions are obtained by matching the solutions for regions 32 and 62

$$p_{32}(s_{62}) = p_{62}(s_{32}) = p_{22}(s_{22}, 0), \quad u_{32}(s_{32}, +\infty) = u_{62}(s_{62}, 0)$$

$$H_{32}(s_{32}, n_{32}) \to H_{62e}g'(0)\left\{\frac{a[M_{22e}^2(\infty) - 1]^{1/2}\sigma}{\mu_{3e}}\right\}\psi_{32} + \cdots \tag{3.85}$$

In region 32, at the return flow bottom ($u_{62w} < 0$) the boundary layer develops from the region $s_{32} \to +\infty$, where the velocity at its outer edge vanishes. The initial conditions for region 32 coincide with the solution for region 3 as $s_3 \to 0$ and $\eta^* \sim O(1)$ and are determined by formulas (3.74) and (3.75).

On the basis of the above estimates it can be seen that in regions 3, 5, and 6 the friction and heat flux at the body surface are of the same order (with respect to $\varepsilon$ or $Re$) as in the undisturbed boundary layer and only in region 32 are larger than this quantity

$$q \sim \left(\frac{\varepsilon}{\ln\frac{1}{\varepsilon}}\right)^{1/2} \gg \varepsilon, \quad \tau \sim \varepsilon^{1/2} \gg \varepsilon \tag{3.86}$$

On the basis of inequalities (3.86) and the behavior of the solution in the zone of "overlapping" of regions 32 and 3, as well as from qualitative physical considerations, it is clear that for determining the maximum heat flux values in the reattachment region it is necessary to solve numerically the boundary value problem for region 32. This will be done in the next section.

Let us now consider the boundary layers of the direct and return flows, that is, regions 5 and 6 in Fig. 3.9. The scales of the longitudinal and transverse coordinates for these regions are conventional:

$$u_5(s_5, n_5 \to +\infty) = u_{1w}, \quad u_6(s_6, n_6 \to -\infty) = 0$$

The initial conditions should be obtained by matching the corresponding profiles of $u_i(\psi_i)$ and $H_i(\psi_i)$ $(i = 5, 6)$ with the profiles of the inviscid regions 22 and 2, respectively. Without writing down these conditions we only note that they are easily obtained from the profiles of the mixing zone 4, since along the streamlines of regions 2 and 22 the stagnation enthalpy and the entropy are conserved.

### *3.3.5 Solution for the region with maximum friction and heat flux values*

The friction and the heat flux reach maximum values in region 32. For solving problem (3.83)–(3.86) it is necessary to integrate first the equations of the inviscid flow in region 22. The method for solving the problem for region 22 was outlined above.

We will bring Eqs. (3.83) to the standard form which admits the application of well-developed methods for solving the boundary layer equations. For this purpose we make the change of variables

$$\zeta = m^2 \int_{+s_{32}}^{\infty} \rho_{32w}\mu_{32w}u_{32w}^2 \, \mathrm{d}s_{32}, \quad \lambda = -mu_{32e} \int_0^{n_{32}} \rho_{32} \, \mathrm{d}n_{32}$$

$$u_{32} = u_{32e}f'(\zeta, \lambda), \quad H_{32} = \frac{A_0}{m}g_{32}, \quad N = \frac{(\rho\mu)_{32}}{(\rho\mu)_{32w}} \tag{3.87}$$

$$(Nf'')' - \left(\frac{\mathrm{d}u_{32e}}{\mathrm{d}\zeta}\right)\left(\frac{\rho_{32e}}{\rho_{32}} - f'^2\right) = -u_{32e}(f'\dot{f}' - f''\dot{f}\,)$$

$$\left(\frac{N}{\sigma}g_{32}'\right)' + Ku_{32e}^2(Nf'f'')' = -u_{3e}(f'\dot{g}_{32} - \dot{f}g_{32}')$$

It should be remembered that

$$u_{32} < 0, \quad 0 \le \zeta < \infty, \quad 0 \le \lambda < \infty$$

$$f(\zeta,0) = f'(\zeta,0) = g_{32}(\zeta,0) = 0, \quad f'(\zeta,\infty) = g'_{32}(\zeta,\infty) = 1 \tag{3.88}$$

$$m^2 = \frac{a[\rho_{3e}\rho_{3w}(M^2_{22e}(\infty) - 1)]^{1/2}}{2\rho_{3w}\mu_{3w}}$$

where $g'(0)$ is obtained by solving Eq. (3.80).

The computer solution of Eqs. (3.87) subject to the boundary conditions (3.88) was performed using the semi-standard program of Seliverstov. The numerical study made it possible to determine the dependence of the maximum Stanton number $St_{\max}$ for different regimes of the jet incidence on the plane. A parameter $\Pi$ dependent on the Reynolds number and the jet incidence angle turned out to be convenient for presenting the results. The introduction of this parameter can be validated on the basis of the following, purely qualitative reasoning.

Taking into account that the longitudinal dimension of the reattachment region is $\sim\delta/\sin\alpha$, where $\delta$ is the mixing zone thickness and $\alpha$ is the jet incidence angle, and using the Dorodnitsyn–Lees variables we obtain the following estimates for the maximum heat flux: $g_{\max} \sim \mu_w H_e g'_w (\rho_w u_e \sin\alpha/\mu_e\delta)^{1/2}$.

Then it might be expected that the following parameters can be helpful for correlating the results

$$St_{\max} \sim \frac{q_{\max}}{\rho_w u_e H_e} \sim \left(\frac{\mu_w \sin\alpha}{\rho_w u_e \delta}\right)^{1/2} = \Pi$$

The calculated dependence $St_{\max}(\Pi)$ is plotted in Fig. 3.14 for the case in which the Mach number of flow 1 ahead of the incidence is $M_1 = 3$ and for a wide range of the Reynolds number, the temperature factors $g_w = H_w/H_{1e}$ and $g^*_w = H_7/H_{1e}$, and the value of the streamfunction $\psi_{4w}$ arriving at the body surface which corresponds to different jet

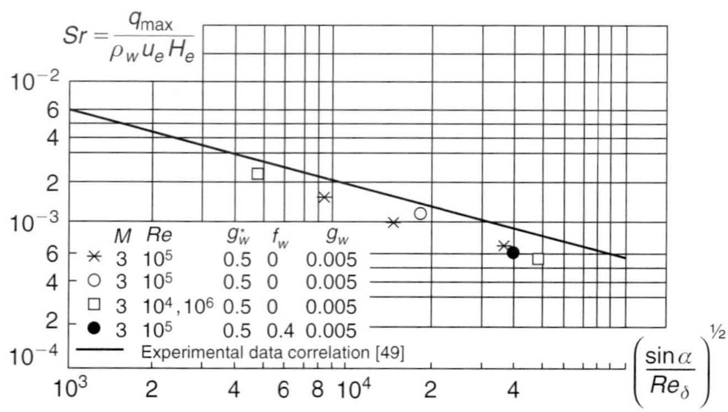


Fig. 3.14.

incidence angles (in Fig. 3.14 the dimensionless value of the streamfunction of the mixing zone 4 is presented: $f_w = \psi^*_{4w}/\sqrt{2\rho_{e4}\mu_{e4}u_{e4}\ell}$). On the log scale all the calculated points lie on the same straight line. For the sake of comparison in the same figure we have presented the correlation line plotted in accordance with the experimental data of different authors (Bushnell and Weinstein, 1968). All the experimental data were obtained for Mach numbers ranging from 10 to 20 on bodies with laminar separation zones. In processing these data, in the study of Bushnell and Weinstein (1968) it was assumed that ahead of the reattachment region the mixing layer thickness was equal to the sum of the thicknesses of the separated boundary layer and the self-similar mixing zone beginning at the separation point.

The calculated results are in good agreement with the correlated experimental data. In Figs. 3.15–3.29 we have plotted the distributions of the pressure, the friction drag coefficient $c_f$, and the heat transfer coefficient $St$ over the body surface.

### 3.3.6 *Discussion of the Chapman–Korst criterion*

The important role played by reattachment in the formation of separation zones in supersonic flows was noted by Chapman et al. (1958) and Korst. In developing a simple model of the

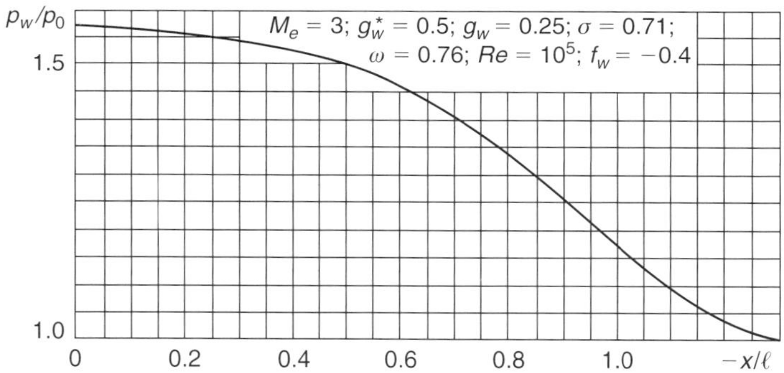


Fig. 3.15.

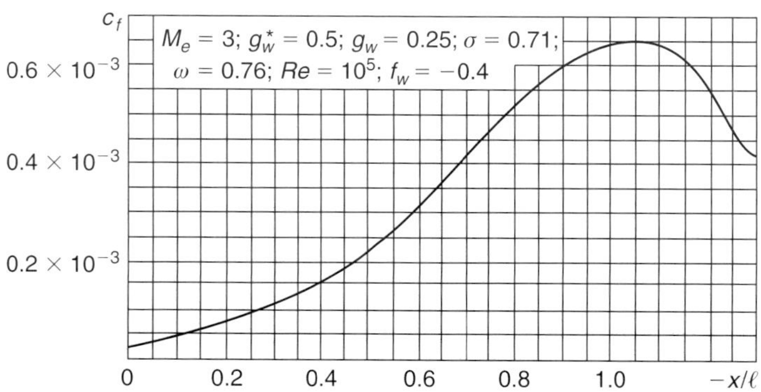


Fig. 3.16.

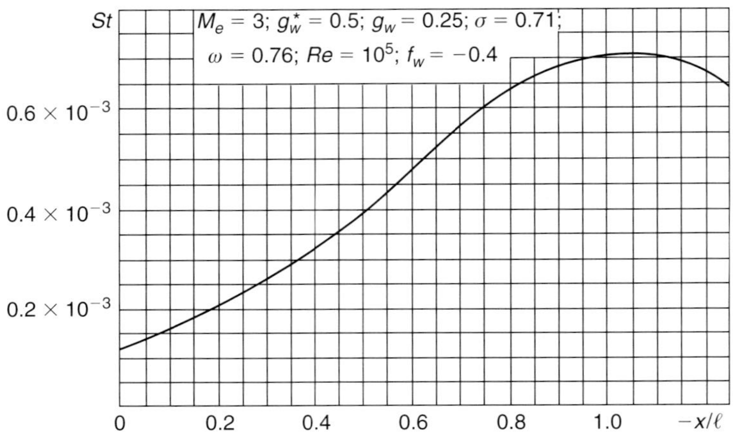


Fig. 3.17.

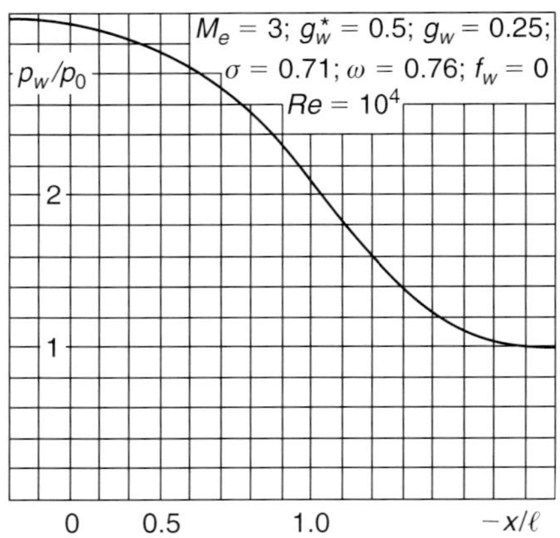


Fig. 3.18.

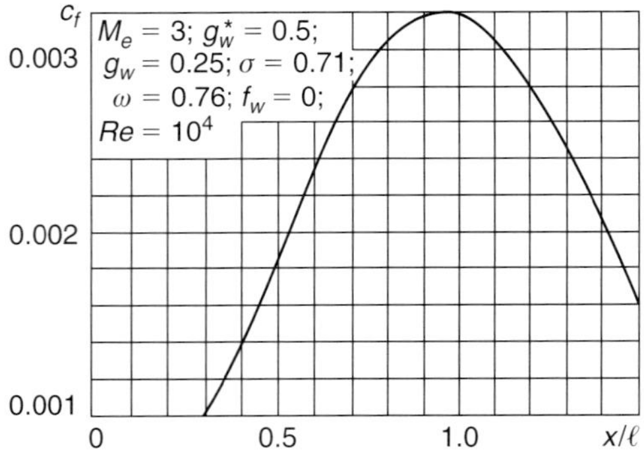


Fig. 3.19.

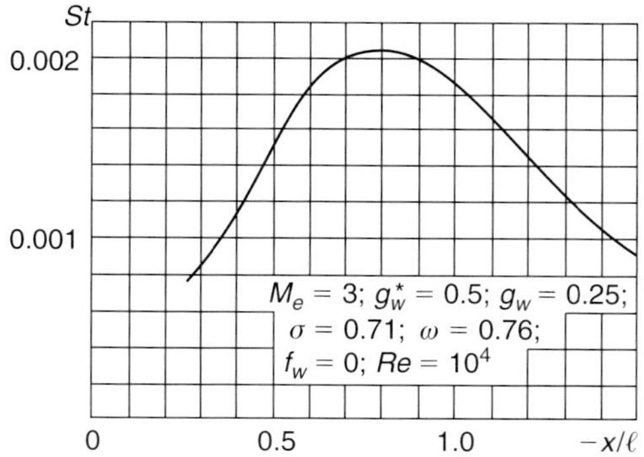


Fig. 3.20.

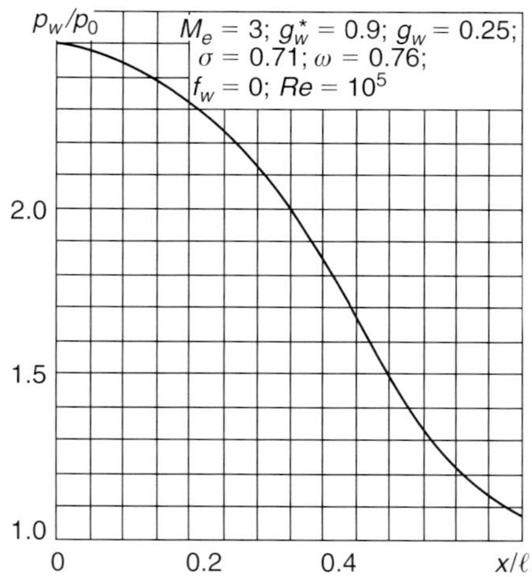


Fig. 3.21.

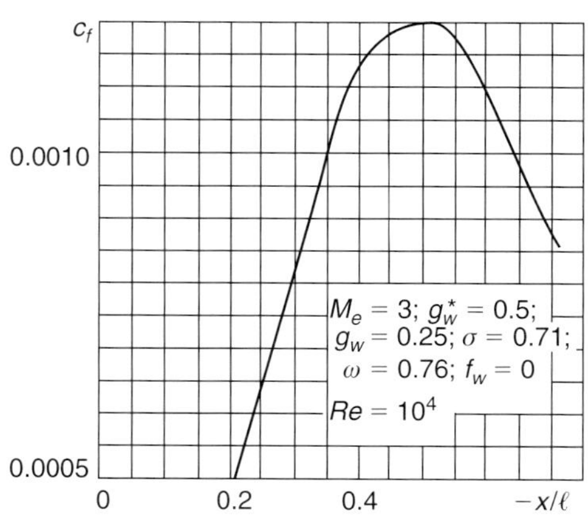


Fig. 3.22.

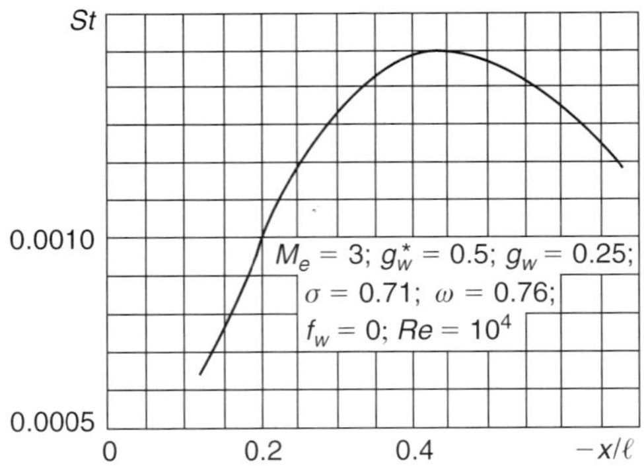


Fig. 3.23.

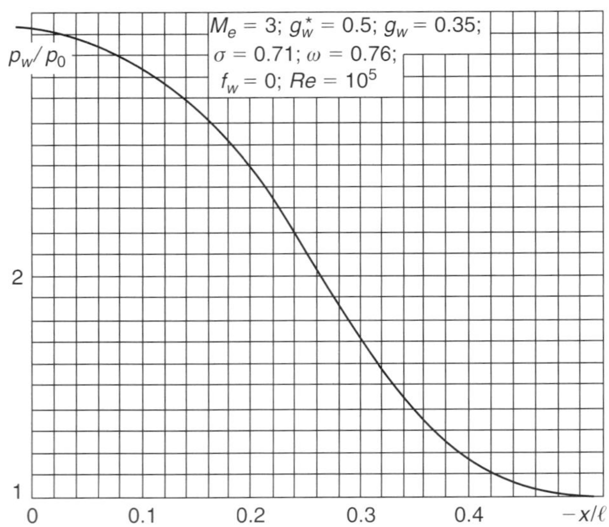


Fig. 3.24.

developed separation zone they took for the condition determining the flow in the reattachment region the assumption that the total pressure at the dividing streamline of the mixing zone (here, $P_{22w}$) must be equal to that in the flow downstream of the reattachment zone ($p_5$). Later, in the works of Kirk (1959), Nash (1963), Neiland and Sokolov (1967), Siriex et al. (1966), and Tagirov (1966) it was noted that the Chapman–Korst criterion ignores, in essence, the role of the viscosity and it was proposed to introduce certain corrections based either on the experimental data (Nash, 1963; Siriex et al., 1966) or approximate methods of calculation (Tagirov, 1966; Neiland and Sokolov, 1967).

The above results show that the Chapman–Korst criterion is actually accurate in the $Re \to \infty$ limit, since region 22 contains no stagnation point and $p_{22}(s_{22} \to +\infty) = P_{22w}$. However, there is a stagnation point in the free interaction region 3, where the pressure

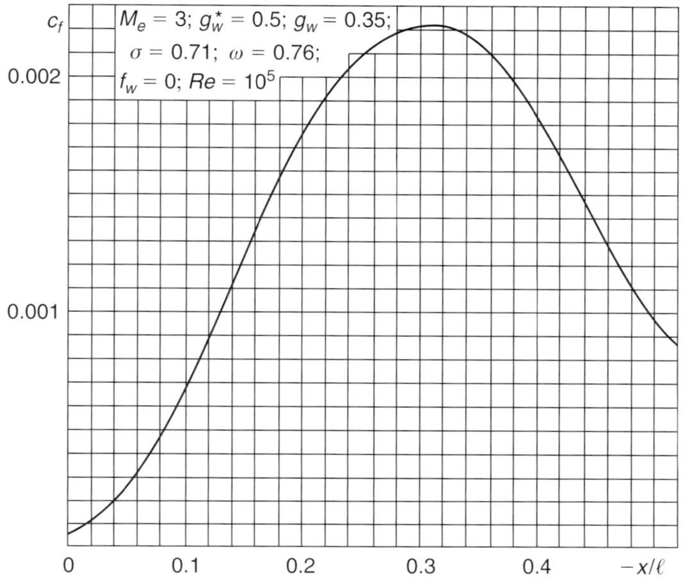


Fig. 3.25.

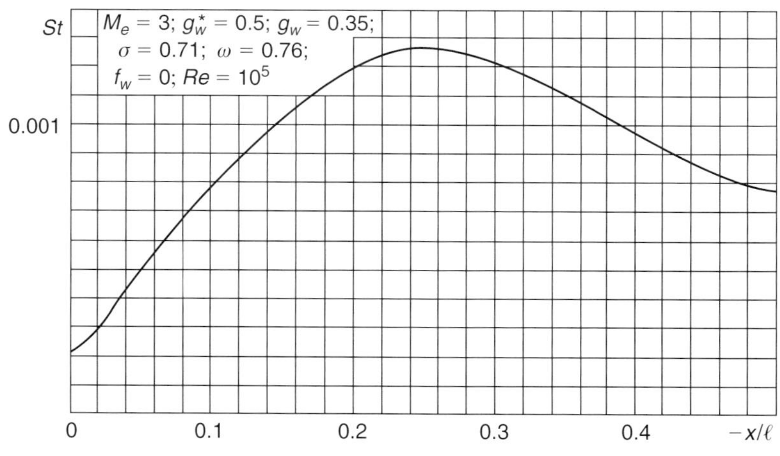


Fig. 3.26.

difference is $O(\varepsilon^{1/2})$. This makes it possible to determine the correction to the Chapman–Korst criterion. Using Eqs. (3.65) and (3.68) we obtain

$$p_5 = P_{22w} + \varepsilon^{1/2} \left[ \frac{a\mu_{3e}}{2(M_{5e}^2 - 1)^{1/2}} \right]^{1/2} C(\sigma, g_w) + \cdots \tag{3.89}$$

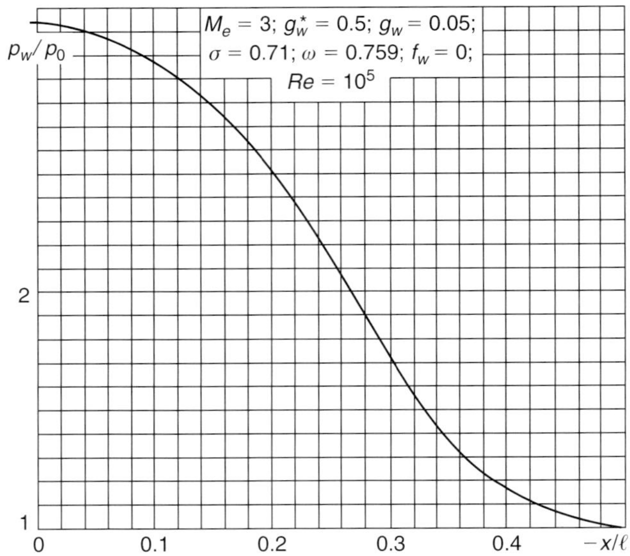


Fig. 3.27.

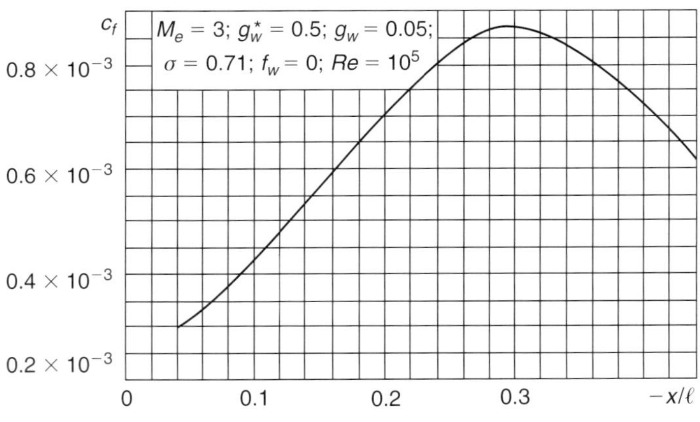


Fig. 3.28.

The constant $C(\sigma, g_w)$ can be determined by solving the boundary value problem (3.66). For a thermally insulated flow ($g_w = 1$) $C$ is simply a number.

Formula (3.89) is very similar to expression (1.32) and (1.34) for the flow near the separation point. In formula (3.89), as $s_{22} \to +\infty$, in the locally inviscid region 22 the parameter $a = (\partial u/\partial n)_{22w}$ is expressed in terms of the velocity and friction on the "dividing" streamline of the mixing zone 4 by means of the equations of conservation of the entropy and the stagnation enthalpy along the streamlines. Formally, $a$ is analogous to the parameter in the formula for flows near the separation point, where it is determined by the local value of

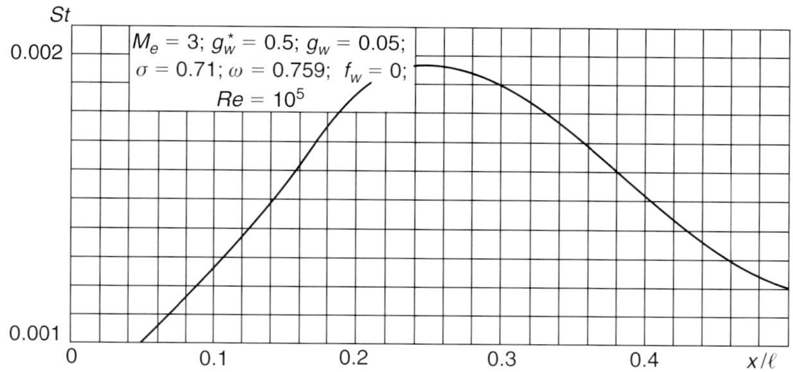


Fig. 3.29.

$c_f$ in the undisturbed boundary layer. Thus, the correction to the Chapman–Korst criterion is fairly large and is of the order $\varepsilon^{1/2} \sim Re^{-1/4}$.

## 3.4 Problems with discontinuous boundary conditions describing laminar high-Reynolds-number flows

In boundary layer theory discontinuous boundary conditions can be encountered in problems describing the flow near the edges of airfoils or wings, near the bends in surfaces, etc. The problems of this kind are usually related with the control of the boundary layer flow. The control can be performed using different techniques, such as tangential injection, suction, setting the surface into motion, etc. The effect of the motion of the surface on the flow is studied comparatively poorly, though the efficiency of this technique of controlling the boundary layer flow was demonstrated by Prandtl in the experiment with a rotating cylinder as long ago as in 1904. The book by Schlichting (1968) contains the description of experiments carried out by Favre in which it was established that the flow over an airfoil with a movable upper side remains separationless up to angles of attack as high as 55°. However, this technique of flow control is not easy to implement and for this reason it has not found wide application in practice. It should be noted that preassigning a nonzero velocity of the surface can describe not only a flow over a movable surface but some other flows as well.

In the works of Zhuk and Ryzhov (1979), Zhuk (1980), and Sokolov (1980) the interaction between a shock moving at a constant velocity and a laminar boundary layer was studied and it was shown that in certain cases the resulting flow can be described by the system of equations for the steady-state "free interaction" regime at nonzero surface velocity. Preassigning a nonzero surface velocity also turns out to be necessary for describing certain regimes of interaction between an outer supersonic flow and a boundary layer in which a gas jet is injected along the surface for ensuring separationless flow pattern or reducing the heat flux to the surface. When the surface is set suddenly into motion or when the motion suddenly stops, the discontinuity in the boundary conditions brings a perturbation in the original boundary layer flow. The classical boundary layer theory can be inapplicable for describing such flows. For this reason, the problems related with the effect of the surface motion onset or cessation require a special analysis.

An analysis of the problems with suddenly changing boundary conditions can be based upon the matched asymptotic expansion method, which was in essence first applied by Prandtl in formulating boundary layer theory, and the coordinate expansion method used by Goldstein (1931) in studying the wake flow behind a flat plate. The change in the boundary conditions which takes place in the boundary layer does not necessarily lead to the violation of the boundary layer theory assumptions; in this case the effect of the variation in the boundary conditions is described by the problem for regular disturbances. In this formulation, the effect of the variation in the boundary conditions was studied in the works of Shmanenkov (1966), Dem'yanov (1969), and Eliseev (1972). In the last of the cited papers it was assumed that the variation in the boundary conditions has an effect on the flow even in the first approximation and the resulting flow is described by boundary layer theory. In essence, it was an inverse problem that was solved in which the body contour was chosen in a special way for eliminating the boundary layer effect on the outer flow.

Transition from the boundary layer flow on a zero-thickness flat plate set at zero incidence to the wake flow is related with a discontinuity of the normal derivative of the longitudinal velocity which is proportional to the friction on the plate surface and is zero at the wake axis.

The investigation of the flows of this type on the basis of the boundary layer equations showed that on the outer boundary of the wake the vertical velocity increases without bound as the trailing edge of the plate is approached. Thus, the assumptions of boundary layer theory are violated.

On the basis of an asymptotical analysis of the Navier–Stokes equations (Messiter, 1970; Stewartson, 1970) it was shown that on the outer boundary of the viscous flow the vertical velocity due to the variation of the wake displacement thickness is bounded and cannot be greater than the values at which the pressure disturbance induced in the outer inviscid flow starts to have an effect on the variation of the displacement thickness. Further analysis (Veldman, 1976) showed that in the vicinity of the plate's trailing edge there occurs a complicated flow pattern including a series of embedded regions in which the flow is governed by the complete system of Navier–Stokes equations, the system of the boundary layer equations with an induced pressure gradient, etc.

Other types of discontinuities, namely, those in the surface velocity (Lipatov and Neiland, 1982), the derivative of the stream function (distributed suction) (Lipatov, 1976), the surface temperature (Sokolov, 1975; Bogolepov et al., 1990) are also characterized by complicated structures of the disturbed flow regions.

Time-dependent disturbances generated by discontinuous boundary conditions are studied considerably worse, though the study of such processes is necessary for determining the characteristics of nonlinear stability of laminar flows at high Reynolds numbers.

### *3.4.1 Structure of disturbed flow regions*

As an example, we will consider the problem of the flow near the velocity discontinuity region on a flat plate having a region moving at a velocity $u_w$ at a distance $\ell$ from the leading edge (Fig. 3.30). The same designations as in the preceding sections are taken for

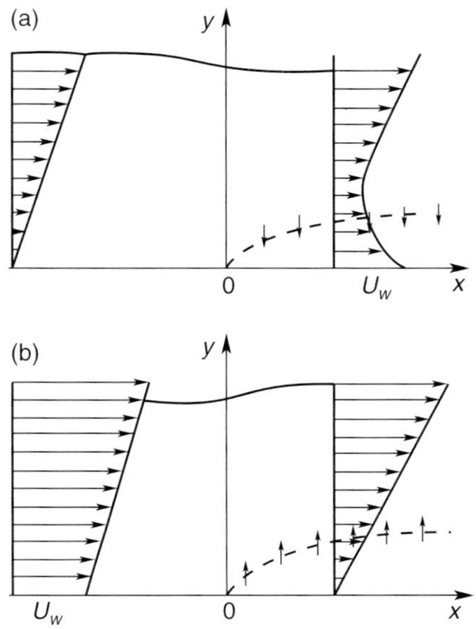


Fig. 3.30.

the Cartesian coordinates measured along the surface and normal to it: time, the velocity components, the density, the viscosity, and the total enthalpy.

Let us consider the structure of the disturbed steady flow for which $u(x<1)=0$ and $u(x>0)=u_w>0$. The difference in the velocities of streamtubes flowing near the surface at a velocity $u_w$ for $x>1$ and those with near-zero velocities for $x<1$ can lead to the formation of a new boundary layer downstream of the point of discontinuity in the boundary condition.

The estimate for the thickness of the newborn boundary layer follows from the condition of the equality of the orders of the terms that describe the effects of the inertia and viscosity forces in the longitudinal momentum equation

$$\varepsilon = Re^{-1/2}, \quad Re = \frac{\rho_\infty u_\infty \ell}{\mu_\infty} \tag{3.90}$$

$$y \sim \varepsilon u_w^{-1/2} x_0^{1/2}, \quad x_0 = x - 1 \tag{3.91}$$

At a fixed surface velocity and a variable thickness of the newborn boundary layer the friction in the latter decreases monotonically with increasing longitudinal coordinate. Using estimate (3.91) we can determine the distance $x_1$ at which the friction in the new boundary layer becomes comparable with that in the main boundary layer

$$\frac{u_w}{\varepsilon} \sim \frac{1}{\varepsilon}, \quad x_1 \sim u_w^3 \tag{3.92}$$

This formula can be conveniently presented in the form of the dependence $\ln x_i/\ln \varepsilon = f(\ln u_w/\ln \varepsilon)$; then Eq. (3.92) is presented as line *OB* in Fig. 3.31.

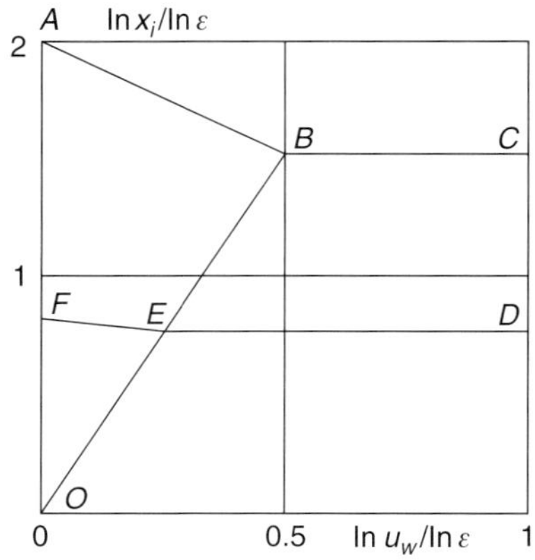


Fig. 3.31.

From estimate (3.91) we also can determine the distance $x_2$ from the point of the discontinuity in the boundary condition, at which the nonlinear disturbance region thickness and length become of the same order and where, in essence, the assumptions of boundary layer theory are violated

$$x_2 \sim \frac{\varepsilon^2}{u_w} \tag{3.93}$$

Equation (3.93) is presented by line *AB* in Fig. 3.31.

Then the coincidence of the longitudinal and transverse scales leads to the equality of the orders of the disturbed longitudinal and transverse velocities. Since Eq. (3.91) was obtained under assumption of the equality of the orders of the inertia and viscosity forces, it can be shown that the flow in a region with the scales $x_2 \sim y_2 \sim \varepsilon^2/u_w$ is described by the complete system of incompressible Navier–Stokes equations. An analogous region appears also in considering the flow in the vicinity of the leading edge of a zero-thickness flat plate.

Let us estimate the effect of the boundary layer formed as a result of the discontinuity in the boundary conditions on the flow in the main boundary layer formed near the fixed plate (since at the bottom of the main boundary layer a new boundary layer is formed and these layers need to be distinguished, in what follows we will call them the main and newborn boundary layers). Physically, this effect manifests itself as gas absorption from the main boundary layer. The estimate for the vertical velocity in the newborn boundary layer follows from Eq. (3.91) and the continuity equation and takes the form:

$$v \sim \varepsilon u_w^{1/2} x^{-1/2}, \quad \psi \sim vx \sim \varepsilon u_w^{1/2} x^{1/2}$$

Absorption of this flow rate from the original boundary layer over the length $x$ leads to a variation of its thickness. For determining this variation we will use the representation $u \sim y/\varepsilon$ of the velocity profile in the main boundary layer at small distances as compared with the boundary layer thickness. Accordingly, at the distance $y$ from the surface the gas flow rate through the main boundary layer can be estimated as follows: $\psi \sim y^2/\varepsilon$. Therefore, the estimate of the variation of the boundary layer displacement thickness takes the form:

$$\Delta\delta \sim \varepsilon^{1/2}\psi^{1/2} \sim \varepsilon u_w^{1/4} x^{1/4}$$

This variation of the displacement thickness induces the corresponding pressure variation in the outer inviscid flow

$$\Delta p \sim \frac{\Delta\delta}{x} \sim \varepsilon u_w^{1/4} x^{-3/4}$$

The latter estimate follows from the linear theory of inviscid (both subsonic and supersonic) flows. Using this theory is justified if the distance $x_3$, over which the above-mentioned effects are important, is greater in the order than the main boundary layer thickness $\delta \sim \varepsilon$. The fulfillment of the condition $x_3 > O(\varepsilon)$ can be verified if the estimate of the distance over which the interaction effects manifest themselves is obtained.

The estimate for the pressure disturbance makes it possible to determine the distance $x_3$ over which the induced pressure gradient has a nonlinear effect on the wall region of the main boundary layer. For further analysis it is important to note that a gas flow rate is absorbed from precisely this region and the variation of the thickness of precisely this region determines the total variation of the boundary layer displacement thickness

$$\Delta p \sim u_3^2, \quad x_3 \sim \varepsilon^{4/5} u_w^{-1/5} \tag{3.94}$$

The second relation (3.94) is presented by line *EF* in Fig. 3.31. Using estimate (3.94) we can write the condition at which the interaction region length is greater than the boundary layer thickness

$$u_w < \frac{1}{\varepsilon}$$

This inequality is assumed to be necessarily fulfilled.

The characteristic points *B* and *E* are at intersections of line *OB* with lines *AB* and *FE*. Typical for flow regimes corresponding to points *B* and *E* is the coincidence of the orders of the friction in the disturbed zone and the main boundary layer. Then point *B* is associated with the flow described by the system of Navier–Stokes equations and point *E* with the flow described by the system of equations of free interaction theory.

Estimate (3.93) is invalid for the region of variation of the parameter $u_w$ located to the right of point *B* and in the cases in which the linear disturbance regime is realized due to a greater relative viscosity effect. The equality of the orders of the terms describing the effects of the viscosity and inertia forces leads to the estimate

$$x_4 \sim \varepsilon^{3/2} \tag{3.95}$$

Relation (3.95) is presented by line *BC* in Fig. 3.31. Similar considerations can be used for determining the distance $x_5$ over which linear processes of viscous–inviscid interaction occur. Relation (3.94) is invalid in the region of variation of the parameter $x_5$ located to the right of point *E*. For this distance the following estimate holds

$$x_5 \sim \varepsilon^{3/4} \tag{3.96}$$

which is associated with line *ED*.

The diagram of the disturbed flow regions plotted in Fig. 3.31 makes it possible to determine the dimensions of these regions and the nature of the corresponding flows for a given amplitude of the parameter $u_w$. Thus, the effect of the disturbance with an amplitude $O(\varepsilon^{1/4}) \leq u_w \leq O(1)$ consists in the appearance near the discontinuity of a region with dimensions determined by line *AR*, where the flow is described by the system of incompressible Navier–Stokes equations. Next in extent is the region whose longitudinal dimension is determined by line *EF*, where the flow in the first approximation is described by the Burgers equation. At intermediate distances, with variation of the parameters in the region between lines *AB* and *EF* the viscosity effect is inessential in the flow in the nonlinear disturbance region and the compensation interaction is realized (see Bogolepov and Neiland, 1976 and the corresponding section in Chapter 8). The absence of viscous terms from the equations governing the disturbed flow requires introducing a subdomain in which the viscosity and inertia forces are of the same order. At the same time, there exists a region with a length determined by line *OB* in which the viscosity effect is essential and the surface friction is of the same order as that in the original boundary layer. As noted above, point *E* corresponds to the general case in which nonlinear processes of equalization of the friction of interaction with the outer flow occur in the same region, namely, the free interaction region (Neiland, 1969*a*; Stewartson and Williams, 1969).

When the parameter $u_w$ varies on the range $O(\varepsilon^{1/4}) \leq u_w \leq O(1)$, apart from the above-mentioned region, in which the flow is described by the system of Navier–Stokes equations, there appears one more region whose length is determined by line *BE*. Here, the flow is described by the system of boundary layer equations with a compensation condition of interaction. In this region, the surface friction is equalized. Finally, for $u_w \sim \varepsilon^{1/2}$ friction is equalized directly in the region in which the flow is described by the system of Navier–Stokes equations.

Thus, near the discontinuity point (line) there occurs a system of embedded regions with different longitudinal scales. For given $u_w$ the dimension of each of these regions can be determined using the diagram in Fig. 3.31. It should also be borne in mind that each region, whose longitudinal dimension is greater in the order than the boundary layer thickness, consists of subdomains with different transverse dimensions.

Using the diagram in Fig. 3.31 and the above estimates we can determine the conditions at which time-dependent effects can manifest themselves in the disturbed flow regions considered. For this purpose we will determine the characteristic temporal scales equal to the ratios of the region lengths to the corresponding characteristic velocities. Thus, for the region under consideration, consisting of a system of embedded subdomains, the greatest characteristic time is associated with the subdomain with the least characteristic longitudinal velocity, while time-dependent processes in the subdomain with the greatest scale time are associated with quasi-stationary processes in the other subdomains. As follows from the estimates presented above, the least longitudinal velocity is characteristic of the region in which nonlinear variations take place.

It should also be taken into account that the regions with different longitudinal dimensions are associated with different characteristic times. Thus, for the regions corresponding to the lines presented in Fig. 3.31 we have the following estimates: $t_2 \sim \varepsilon^2 u_w^{-2}$ at *AB*, $t_3 \sim \varepsilon^{3/5} u_w^{-2/5}$ at *EF*, and $t_1 \sim u_w^2$ at *OB*.

For further analysis we note that for $u_w = O(\varepsilon^{1/2})$ the least time is characteristic of the flow region corresponding to line $AB$, then for $u_w > O(\varepsilon^{1/2})$ the next in duration is time for region $EF$, and the greatest characteristic time is that for the region corresponding to line $OE$.

### 3.4.2 *Analysis of the regimes described by free interaction theory*

For $u_w \sim \varepsilon^{1/4}$ the flow in the region with nonlinear variations of the flow functions is described by the system of equations for the free interaction regime. This system of equations is as follows:

$$\begin{aligned}
&\Psi'\dot{\Psi}' - \dot{\Psi}\Psi'' + \dot{P} = \Psi''' \\
&\dot{\Psi}'(X,\infty) = -P, \quad \Psi(X,0) = 0, \quad \Psi(-\infty, Y) = \frac{Y^2}{2} \\
&\Psi'(X<0,0) = 0, \quad \Psi'(X>0,0) = U_w \\
&()' = \frac{\partial}{\partial Y}, \quad \dot{()} = \frac{\partial}{\partial X}
\end{aligned} \tag{3.97}$$

Here, the following similarity variables are introduced

$$\begin{aligned}
X &= (a^5\beta^3\rho_w^2\mu_w\varepsilon^{-3})^{1/4}x, \qquad Y = (a^3\beta\rho_w^2\mu_w^{-1}\varepsilon^{-5})^{1/4}y \\
\Psi &= (a\beta\rho_w^2\mu_w^{-1}\varepsilon^{-3})^{1/2}\psi, \qquad P = (a^{-1}\beta\mu_w^{-1}\varepsilon^{1})^{1/2}p \\
U_w &= (a^{-1}\beta\rho_w^2\mu_w^{-1}\varepsilon^{-1})^{1/4}u_w, \quad \beta = (M_\infty^2 - 1)^{1/2} \\
a &= \varepsilon\frac{\partial u}{\partial y}(X \to -\infty, 0)
\end{aligned} \tag{3.98}$$

For $U_w \ll 1$ the solution of the boundary value problem (3.97) can be represented in the form:

$$\begin{aligned}
&\Psi = \frac{Y^2}{2} + f(X,Y)U_w, \quad P = U_w\overline{P} \\
&Y\dot{f}' - \dot{f} + \dot{\overline{P}} = f''' \\
&f'(X,\infty) = -\overline{P}, \quad f'(X<0,0) = 0, \quad f'(X>0,0) = 1 \\
&f(X,0) = 0, \quad f(-\infty, Y) = 0
\end{aligned} \tag{3.99}$$

Using the Fourier transformation the solution of the boundary value problem (3.99) can be obtained in the form:

$$P(X<0) = \frac{-3U_w\theta\exp(\theta X)}{4}, \quad \theta = [-3\mathrm{Ai}'(0)]^{3/4} \tag{3.100}$$

$$P(X>0) = -\frac{3^{1/2}U_w\theta}{2\pi}\int_0^\infty \frac{s^{4/3}\exp(-\theta s X)\,\mathrm{d}s}{1+s^{4/3}+s^{8/3}}$$

The disturbed flow near the point, at which the motion of the surface stops, can be similarly described. It is governed by the system of equations (3.97) in which the boundary conditions for the function $\Psi'(X,0)$ take the form:

$$\Psi'(X<0,0)=U_w,\quad \Psi'(X>0,0)=0$$

The solution of the corresponding linear boundary value problem for $U_w \ll 1$ is determined as follows:

$$P(X<0)=\frac{3U_w\theta\exp(\theta X)}{4} \tag{3.101}$$

$$P(X>0)=\frac{3^{1/2}U_w\theta}{2\pi}\int\limits_0^\infty \frac{s^{4/3}\exp(-\theta sX)\,\mathrm{d}s}{1+s^{4/3}+s^{8/3}}$$

It should be noted that for $U_w \ll 1$ the solutions governing the flows near the points of the beginning and cessation of the motion of the surface directed counter to the undisturbed oncoming flow differ from solutions (3.100) and (3.101) only in sign.

In the vicinity of point $X=+0$ for $Y=O(1)$ the flow functions can be represented in the form of the following coordinate expansions

$$\Psi=\Psi_0(X=-0,Y)+X^{1/2}\Psi_1(Y)+\cdots,\qquad P(X)=P_0(X=-0)+X^{1/2}P_{10} \tag{3.102}$$

The functional form of the expansions is determined from the conditions of the matching with the solution in the newborn boundary layer, that is, for $Y=O(1)$, where the flow functions can be represented in the form:

$$\Psi=2^{1/2}X^{1/2}U_w^{1/2}f(\eta)+\cdots,\qquad \eta=2^{-1/2}YU_w^{1/2}X^{-1/2} \tag{3.103}$$

Substituting Eq. (3.102) in the system of equations (3.97) leads to the following equation for the function $\Psi_1(Y)$:

$$\Psi_0'\Psi_1'-\Psi_1\Psi_0''+P_1=0$$

An analysis of expressions (3.102) and (3.103) and the interaction conditions shows that for nonzero values of the function $\Psi_1(Y\to\infty)$ nonzero disturbances imposed on the outer boundary of the region with nonlinear flow function variation lead to infinitely large negative values of the induced pressure; therefore, condition $\Psi_1(Y\to\infty)=0$ must be fulfilled. Thus, a rapid decrease or increase in the displacement thickness at the cost of the newborn boundary layer must be accompanied by the appearance of a large pressure gradient ensuring zero, in the leading term, total variation of the displacement thickness. The solution for the function $\Psi_1(Y)$ takes the form:

$$\Psi_1=-\Psi_0'P_{10}\int\limits_Y^\infty\frac{\mathrm{d}Y}{\Psi_0'^{\,2}}$$

Matching with the solution for the newborn boundary layer makes it possible to determine the parameter $P_{10}$; in the case of the beginning of the motion of the surface it is as follows:

$$P_{10} = 2^{1/2} C_0 a U_w^{1/2}, \quad C_0 \approx 1.229$$

and for the case of the cessation of the motion

$$P_{10} = -2^{1/2} C_1 a U_w^{-1/2} J^{-1}, \quad C_1 \approx 1.217, \quad J = \int_0^\infty \frac{\mathrm{d}Y}{\Psi_0'^2}$$

### 3.4.3 *Boundary value problem for the case* $\varepsilon^{1/4} \ll u_w \ll 1$ *in the vicinity of the point of the beginning of the motion of the surface (steady case)*

For $\varepsilon^{1/4} \ll u_w \ll 1$ in the vicinity of the point of the beginning of the motion of the surface in the region with nonlinear variations of the flow parameters the functions and the coordinates can be represented in the form:

$$\begin{aligned} x &= \varepsilon^{4/5} u_w^{-1/5} X_1, \quad y = \varepsilon^{6/5} u_w^{1/5} Y_1 \\ u &= \varepsilon^{1/5} u_w^{1/5} U_1(X_1, Y_1) + \cdots, \quad v = \varepsilon^{3/5} u_w^{3/5} V_1(X_1, Y_1) + \cdots \\ \psi &= \varepsilon^{7/5} u_w^{2/5} \Psi_1(X_1, Y_1) + \cdots, \quad p = \varepsilon^{2/5} u_w^{2/5} P_1(X_1, Y_1) + \cdots \end{aligned} \tag{3.104}$$

For further analysis it is convenient to go over to the von Mises variables. In these variables, the system of Navier–Stokes equations in the limit $Re \to \infty$, $u_w \to 0$, and $u_w Re^{-1/8} \to 0$ gives the following system of equations for the first terms of expansions (3.104)

$$\rho_w U_1 \frac{\partial U_1}{\partial X_1} + \frac{\partial P_1}{\partial X_1} = 0, \quad \frac{\partial P_1}{\partial \Psi_1} = 0 \tag{3.105}$$

$$\frac{\partial Y_1}{\partial X_1} = \frac{V_1}{U_1}, \quad \frac{\partial Y_1}{\partial \Psi_1} = \frac{1}{\rho_w U_1}$$

$$V_1(X_1 < 0, 0) = 0, \quad V_1(X_1 > 0, \Psi_w) = -2^{-1/2} \rho_w^{-1/2} \mu_w^{1/2} C_0 X_1^{-1/2}$$

$$\Psi_w = 2^{1/2} \rho_w^{1/2} \mu_w^{1/2} C_0 X_1^{1/2}$$

Upstream of the point of the beginning of the motion of the surface in the wall region of the undisturbed boundary layer the velocity profile is as follows:

$$V_1(-\infty, Y_1) = a Y_1, \quad U_1(-\infty, \Psi_1) = 2^{1/2} \rho_w^{-1/2} \Psi_1^{1/2} \tag{3.106}$$

Matching the unknown solution with that for the outer inviscid flow leads to the following relation

$$\beta P_1 = \frac{\mathrm{d}Y_1(X_1, \infty)}{\mathrm{d}X_1} \tag{3.107}$$

After the change of variables

$$X_1 = (2^3 a^6 \beta^4 \rho_w^3 \mu_w C_0^2)^{-1/5} \widetilde{X}_1, \quad P_1 = (2a^2 \beta^{-1} \rho_w \mu_w^2 C_0^4)^{1/5} \widetilde{P}_1$$

the system of equations (3.105) and boundary conditions (3.106) and (3.107) for the function $\widetilde{P}_1(\widetilde{X}_1)$ can be reduced to the following boundary value problem for the Riccati equation

$$\frac{d\widetilde{P}_1}{d\widetilde{X}_1} = \frac{1}{2\widetilde{X}_1^{1/2}} + \widetilde{P}_1(\widetilde{X}_1^{1/2} - \widetilde{P}_1)^{3/2}, \quad \widetilde{X}_1 > 0 \tag{3.108}$$

$$\frac{d\widetilde{P}_1}{d\widetilde{X}_1} = \widetilde{P}_1(-\widetilde{P}_1)^{3/2}, \quad \widetilde{X}_1 < 0$$

$$\widetilde{P}_1(\widetilde{X}_1 \to \pm\infty) = 0$$

The solution of the boundary value problem (3.108) is as follows:

$$\widetilde{P}_1 = \widetilde{X}_1^{1/2} - \left\{ \frac{\frac{d}{d\widetilde{X}_1}\left[\widetilde{X}_1^{1/2} K_{2/5}\left(\frac{4}{5\widetilde{X}_1}\right)^{5/4}\right]}{\widetilde{X}_1^{1/2} K_{2/5}\left(\frac{4}{5\widetilde{X}_1}\right)^{5/4}} \right\}, \quad \widetilde{X}_1 \geq 0$$

where $K_{2/5}$ is the modified Bessel function (Abramowitz and Stegun, 1964).

The solution of the boundary value problem (3.108) makes it possible to determine the value $P(X=0) = 2^{1/5} C_0^{4/5} C_2 U_w^{2/5}$, where $C_2 \approx 0.858$.

The flow in the vicinity of the point of the cessation of the motion of the surface for $u_w \sim \varepsilon^{1/4}$ can be considered in a similar way. It is described by the system of equations (3.97) with modified boundary conditions. With increase in the surface velocity $u_w$ the flow in the region, where the variation of the displacement thickness is formed, is in the first approximation inviscid.

Based on the equations of motion we can obtain that for $\varepsilon^{1/4} \ll u_w \ll 1$ a positive pressure disturbance conserves the same order as in the case $u_w \sim \varepsilon^{1/4}$, while the disturbed region length tends to zero.

It should be noted that the cessation of the motion of the surface does not lead to separation, since for $\varepsilon^{1/4} \ll u_w \ll 1$ the flow is governed by a linear system of equations. The functions and coordinates of the disturbed flow region can be represented in the form:

$$x = \varepsilon u_w^{-1} x_1, \quad \psi = \varepsilon u_w^2 \psi_1 \tag{3.109}$$

$$u = u_w u_0(\psi_1) + \varepsilon^{1/2} u_w^{-1} u_1(x_1, \psi_1) + \cdots, \quad v = \varepsilon^{1/2} u_w^2 v_1(x_1, \psi_1) + \cdots$$

$$y = \varepsilon u_w y_0(\psi_1) + \varepsilon^{3/2} u_w^{-1} y_1(x_1, \psi_1) + \cdots$$

$$p = \varepsilon^{1/2} p_1(x_1) + \cdots, \quad \rho = \rho_w + \cdots$$

Substituting expansions (3.109) in the system of Navier–Stokes equations and passing to the limit $u_w \to 0$, $u_w Re^{-1/8} \to 0$ leads to the following system of equations

$$\rho_w u_1 u_0 + p_1 = 0, \quad \frac{\partial p_1}{\partial \psi_1} = 0 \tag{3.110}$$

$$\frac{\partial y_1}{\partial x_1} = \frac{v_1}{u_0}, \quad \frac{\partial y_0}{\partial \psi_1} = \frac{1}{\rho_w u_0}, \quad \frac{\partial y_1}{\partial \psi_1} = -\frac{u_1}{\rho_w u_0^2}$$

$$\beta p_1(x_1) = v_1(x_1, \psi_1 \to \infty), \quad u_0(\psi_1) = \left(1 + \frac{2a\psi_1}{\rho_w}\right)^{1/2}$$

$$v_1(x_1 < 0, 0) = 0, \quad v_1(x_1 > 0, 0) = v_{1w}(x_1)$$

The system of equations (3.110) can be transformed to an ordinary differential equation

$$\beta p_1 = \frac{\mathrm{d}p_1}{\mathrm{d}x_1}(a\rho_w)^{-1} + v_{1w}(x_1) \tag{3.111}$$

On the surface the vertical velocity $v_{1w}(x_1)$ is determined by matching with the solution for the boundary layer formed downstream of the point of the variation of the boundary condition. In accordance with the work of Van Dyke (1964), the newborn boundary layer induces a vertical velocity $v_{1w}(x_1) = 2^{-1/2}\rho_w^{-1/2}\mu_w^{1/2}C_1 x_1^{-1/2}$. The solution of Eq. (3.111) satisfying the condition of the disturbance decay far downstream from the point of the variation of the boundary condition is as follows:

$$p_1(x_1) = \frac{C_1\mu_w^{1/2}\pi^{1/2}a^{1/2}\exp(\beta a\rho_w x_1)}{2^{1/2}\beta^{1/2}}\left[1 - \frac{\beta^{1/2}a^{1/2}\rho_w^{1/2}}{\pi^{1/2}}J_0(x_1)\right], \quad x_1 > 0$$

$$J_0 = \int_0^{x_1} \xi^{-1/2}\exp(-\beta a\rho_w \xi)\,\mathrm{d}\xi$$

$$p_1 = 2^{-1/2}\beta^{-1/2}C_1\mu_w^{1/2}\pi^{1/2}a^{1/2}\exp(\beta a\rho_w x_1), \quad x_1 < 0$$

From the solution obtained it follows that, as $u_w \to \infty$, the value $p(0)$ approaches the limit equal to $2^{-1/2}C_1\pi^{1/2}$. As follows from Eq. (3.109), with increase in $u_w$ the disturbed region length monotonically vanishes.

### *3.4.4 Numerical solution of the problem*

For solving numerically the boundary value problem (3.97) for $0 < X \leq 1$ the integration region was subdivided to two subdomains. In the first subdomain the equations were written for the function $\Psi(\xi, Y)$, where $\xi = X^{1/2}$, and in the second for the function $\varphi(\xi, \eta_1)$, where $\varphi = \Psi/\xi$ and $\eta_1 = Y/\xi$. The calculations were carried out at each $X = \mathrm{const}$ using the sweeping technique simultaneously in the two regions. The following difference grids were chosen: $\Delta Y = 0.1$, $\Delta X = 0.1$, $\Delta\xi = 0.01$, and $\Delta\eta = 0.01$, while the maximum values of the coordinates were $Y_e = 15$ and $\eta_{1e} = 10$. The computational procedure is described in detail in the paper of Lipatov (1976).

In Fig. 3.32 we have plotted the pressure disturbance distributions obtained by numerical solution of the boundary value problem (3.97). In this and following figures of this section numbers I and II indicate the flows near the points of the onset and cessation of the motion of the surface, respectively. The solution of the system of equations (3.97) was obtained for the parameter $U_w = 0.4$. We note that the singular behavior of the pressure gradient as $X \to +0$ is in agreement with formulas (3.102). Curve III describes the pressure disturbance distribution $\widetilde{P}_1(\widetilde{X}_1)$ in the vicinity of the point of the beginning of the motion of the surface for $\varepsilon^{1/4} \ll u_w \ll 1$ obtained by numerical integration of Eq. (3.97). We recall that in variables (3.98) the disturbed zone length in the vicinity of the point of the beginning of the motion of the surface is zero for $\varepsilon^{1/4} \ll u_w \ll 1$, while the pressure disturbance is infinitely large.

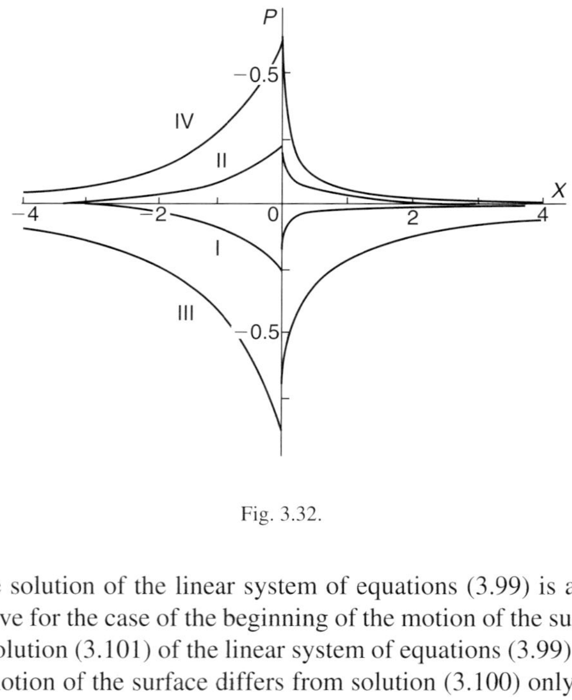


Fig. 3.32.

In Fig. 3.32 the solution of the linear system of equations (3.99) is also presented; the corresponding curve for the case of the beginning of the motion of the surface is marked by number IV. The solution (3.101) of the linear system of equations (3.99) for the case of the cessation of the motion of the surface differs from solution (3.100) only in sign and is not presented in Fig. 3.32.

In Fig. 3.33 we have plotted the viscous stress distributions over the surface for the case $U_w = 0.4$. The singularity in the friction distribution as $X \to +0$ is due to the formation of the boundary layer downstream of the point of the variation of the boundary condition. In the newborn boundary layer the longitudinal velocity is equal in the order to the surface velocity, while the newborn boundary layer thickness vanishes as $X \to +0$. Thus, as $X \to +0$, the sudden onset of the motion of the surface leads – in variables (3.98) – to infinitely large negative friction, whereas the sudden cessation of the motion of the surface leads to infinitely large positive friction.

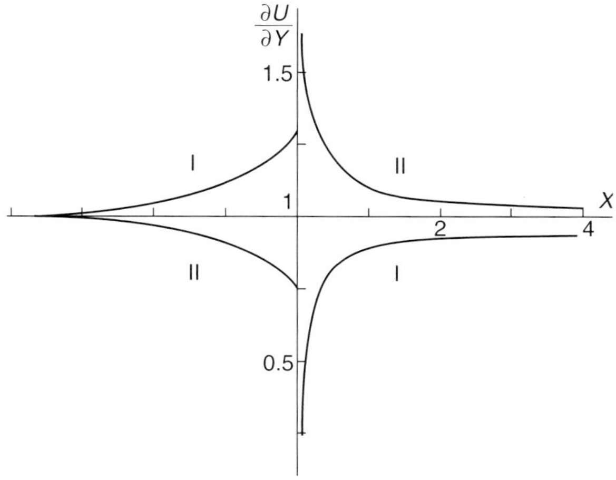


Fig. 3.33.

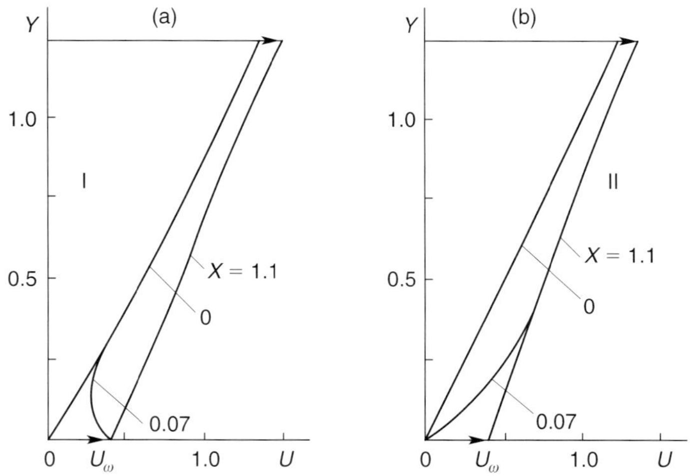


Fig. 3.34.

The flow velocity profiles $U(Y)$ directly ahead of the point of the variation of the boundary conditions and downstream of it are plotted in Fig. 3.34. We note that far downstream from the origin the velocity profiles acquire the shape typical of the shear flow above a fixed or moving surface.

### *3.4.5 Analysis of nonlinear time-dependent flow patterns*

For further analysis it is important that in the main approximation the disturbed flow in the region characterized by the smaller longitudinal scale has no effect on the flow in the region characterized by the greater longitudinal scale. A similar example is furnished by the flow near the leading edge of a zero-thickness flat plate set at zero incidence in the oncoming flow.

We will restrict ourselves to the consideration of the flow patterns corresponding to the region *EF* in Fig. 3.31. As shown above, the flow in region 3 turns out to be inviscid, while the discontinuity in the boundary conditions leads to the appearance of nonlinear disturbances. Apart from the above-mentioned hierarchy of the flow regimes in the longitudinal direction, each region, whose length is greater than the boundary layer thickness, is associated with subdomains, whose transverse dimensions are determined by the thickness of the newborn boundary layer, the thickness of the nonlinear variation region, the thickness of the main boundary layer, and the longitudinal dimension of the disturbed flow region near the discontinuity.

We will consider the flow in the region 3 of nonlinear variation of the longitudinal velocity characterized by the following representations of the flow functions and the coordinates

$$x = 1 + x_3 \varepsilon^{4/5} u_w^{-1/5}, \quad y = y_3 \varepsilon^{6/5} u_w^{1/5}, \quad t = t_3 \varepsilon^{3/5} u_w^{-2/5} \tag{3.112}$$

$$(u, v) = (\varepsilon^{1/5} u_w^{1/5} u_3, \varepsilon^{3/5} u_w^{3/5} v_3) + \cdots$$

$$(p, \rho) = \left( \frac{1}{\gamma M_\infty^2} + \varepsilon^{2/5} u_w^{2/5} p_3, \rho_w \right) + \cdots$$

Substituting expansions (3.112) in the system of Navier–Stokes equations and passing to the limit

$$\varepsilon \to 0, \quad u_w \to 0, \quad \frac{\varepsilon^{1/4}}{u_w} \to 0 \tag{3.113}$$

yields the following system

$$\frac{\partial u_3}{\partial t_3} + u_3 \frac{\partial u_3}{\partial x_3} + v_3 \frac{\partial u_3}{\partial y_3} + \frac{1}{\rho_w} \frac{\partial p_3}{\partial x_3} = 0 \tag{3.114}$$

$$\frac{\partial u_3}{\partial x_3} + \frac{\partial v_3}{\partial y_3} = 0, \quad \frac{\partial p_3}{\partial y_3} = 0$$

As $x_3 \to -\infty$, the boundary conditions are determined by the solution for the wall region in the main boundary layer

$$u_3 = a y_3, \quad v_3 = 0, \quad p_3 = 0 \tag{3.115}$$

For region 2 located above region 3 the following asymptotic expansions are valid

$$x = 1 + \varepsilon^{4/5} u_w^{-1/5} x_2, \quad y = \varepsilon y_2, \quad t = \varepsilon^{3/5} u_w^{-2/5} t_2 \tag{3.116}$$

$$u(x, y, t, \varepsilon, u_w) = u_0(y_2) + \varepsilon^{1/5} u_w^{1/5} u_2(x_2, y_2, t_2) + \cdots \tag{3.117}$$

$$v(x, y, t, \varepsilon, u_w) = \varepsilon^{2/5} u_w^{2/5} v_2(x_2, y_2, t_2) + \cdots$$

$$p(x, y, t, \varepsilon, u_w) = \frac{1}{\gamma M_\infty^2} + \varepsilon^{2/5} u_w^{2/5} p_2(x_2, y_2, t_2) + \cdots$$

$$\rho(x, y, t, \varepsilon, u_w) = \rho_0(y_2) + \varepsilon^{2/5} u_w^{2/5} \rho_2(x_2, y_2, t_2) + \cdots$$

Substituting expansions (3.117) in the system of Navier–Stokes equations and passing to the limit (3.113) leads to the following system

$$u_0 \frac{\partial u_2}{\partial x_2} + v_2 \frac{\partial u_2}{\partial y_2} = 0, \quad \frac{\partial u_2}{\partial x_2} + \frac{\partial v_2}{\partial y_2} = 0 \tag{3.118}$$

The solution of system (3.118) is as follows:

$$u_2 = A(x_2, t_2) \frac{\mathrm{d}u_0}{\mathrm{d}y_2}, \quad v_2 = -u_0 \frac{\partial A}{\partial x_2} \tag{3.119}$$

In region 1 which contains the streamtubes of the outer inviscid flow, we can introduce the following representations of the functions and coordinates

$$x = 1 + \varepsilon^{4/5} u_w^{-1/5} x_1, \quad y = \varepsilon^{4/5} u_w^{-1/5} y_1, \quad t = \varepsilon^{3/5} u_w^{-2/5} t_1$$

$$u(x, y, t, \varepsilon, u_w) = 1 + \varepsilon^{2/5} u_w^{2/5} u_1(x_1, y_1, t_1) + \cdots \tag{3.120}$$

$$\rho(x, y, t, \varepsilon, u_w) = 1 + \varepsilon^{2/5} u_w^{2/5} \rho_1(x_1, y_1, t_1) + \cdots$$

The representations for the other functions coincide – correct to the subscript – with Eq. (3.117).

Substituting expansions (3.120) in the system of Navier–Stokes equations and passing to the limit (3.113) leads to the following result

$$\frac{\partial u_1}{\partial x_1} + \frac{\partial p_1}{\partial x_1} = 0, \quad \frac{\partial v_1}{\partial x_1} + \frac{\partial p_1}{\partial y_1} = 0, \quad \frac{\partial \rho_1}{\partial x_1} + \frac{\partial u_1}{\partial x_1} + \frac{\partial v_1}{\partial y_1} = 0, \quad p_1 = \frac{\rho_1}{M_\infty^2} \tag{3.121}$$

Hence we can derive the well-known wave equation (for $M_\infty > 1$) of the linear theory of supersonic flows, whose solution is the Ackeret formula (Ferri, 1940)

$$\sqrt{M_\infty^2 - 1}\, p_1(x_1, 0, t_1) = v_1(x_1, 0, t_1) \tag{3.122}$$

For subsonic flows the solution of system (3.121) takes the form:

$$\frac{\partial p_1(x_1, 0, t_1)}{\partial x_1} = \frac{1}{\pi} \int\limits_{-\infty}^{\infty} \frac{1}{(s - x_1)} \frac{\partial v_1(s, 0, t_1)}{\partial s} \mathrm{d}s \tag{3.123}$$

Matching the solutions in regions 1 and 2 we obtain

$$p_1(x_1, 0, t_1) = \begin{cases} -\frac{1}{\sqrt{M_\infty^2 - 1}} \frac{\partial A}{\partial x_1}, & M_\infty > 1, \\ -\frac{1}{\pi} \int\limits_{-\infty}^{\infty} \frac{1}{s - x_1} \frac{\partial A}{\partial s} \, ds, & M_\infty < 1 \end{cases} \tag{3.124}$$

Matching the solutions for regions 2 and 3 yields

$$aA = u_3(x_3, y_3, t_3) - ay_3, \quad y_3 \to \infty \tag{3.125}$$

The solution for region 3 can be sought in the form:

$$u_3(x_3, y_3, t_3) = ay_3 + aA(x_3, t_3)$$

In this case the system of equations (3.114) takes the form:

$$a\frac{\partial A}{\partial t_3} + a^2 A \frac{\partial A}{\partial x_3} + \frac{1}{\rho_w} \frac{\partial p_3}{\partial x_3} + av_{3w} = 0 \tag{3.126}$$

the pressure disturbance being determined by formulas (3.124), since

$$p_3(x_3, t_3) = p_1(x_1, 0, t_1)$$

Thus, in the nonlinear disturbance region the flow in the vicinity of the discontinuity line is governed by the Benjamin–Ono equations for the subsonic outer flow or the Burgers equation for the supersonic outer flow.

It should also be noted that in the case of the flow around a slender cylinder the corresponding flow near the velocity discontinuity point is described by the Korteweg-de Vries equation (Karabalaev and Lipatov, 1996).

### *3.4.6 Examples of numerical solutions of nonlinear time-dependent problems*

In the examples presented below the disturbance introduced by the discontinuity in the boundary conditions has an effect on the flow in region 3 via the vertical velocity $v_w$. This suction (or injection) velocity is determined by the solution for the newborn boundary layer. The boundary conditions for the boundary layer equations are determined by the type of the discontinuity. It is important to note that all the discontinuities in the boundary conditions considered below are associated with the same formulation of problem (3.126), correct to algebraic change of variables and different velocity distributions. It should also be noted that, except for the first example considered, the flow in the newborn boundary layer is characterized by asymptotically large longitudinal velocities and, correspondingly, smaller values of the characteristic time than in region 3. Thus, assuming the time-dependent nature of the flow in region 3 we arrive at quasi-stationary processes in the newborn boundary layer.

1. *Suction at a velocity* $v_w$ *starting at the moment* $t_3 = 0$ *in the porous-surface region* $X_3 > 0$. As shown in the paper of Lipatov (1976), in region 3 suction generates a flow described by the system of equations (3.113) in the case $O(\varepsilon^{3/4}) < v_w < O(1)$; the stationary solution of the problem was also obtained. Below we study transition to this stationary solution. In this case, characteristic for region 3 are other asymptotic expressions

$$x = 1 + (\varepsilon^{-3}a^3\beta\rho_w v_w)^{-1/3}X, \quad t = (\varepsilon^{-3}a^3\beta\rho_w v_w^2)^{-1/3}T \tag{3.127}$$

$$p = \frac{1}{\gamma M_\infty^2} + (\beta^{-2}\rho_w v_w^2)^{1/3}P, \quad P = -\frac{\partial A}{\partial X}$$

Substituting expressions (3.127) in the system of equations (3.126) we obtain

$$\frac{\partial A}{\partial T} + A\frac{\partial A}{\partial X} - \frac{\partial^2 A}{\partial X^2} + F(X,T) = 0 \tag{3.128}$$

$$A(X,0) = 0, \quad A(-\infty,T) = 0, \quad \frac{\partial A(\infty,T)}{\partial X} = 0 \tag{3.129}$$

$$F(X,T) = \begin{cases} 0, & X < 0, \ T \geq 0, \\ 1, & X \geq 0, \ T > 0 \end{cases}$$

It should be noted that the Burgers equation can also describe other boundary layer flow regimes, not necessarily generated by discontinuities in boundary conditions. Examples of these flows were previously studied in the works of Zhuk and Ryzhov (1982) and Lipatov and Neiland (1987).

The solution of problem (3.128), (3.129) was obtained numerically, using the finite difference method. In Fig. 3.35 we have plotted the function $A(X,T)$ for different moments. For large times the numerical solution of the time-dependent problem coincides with the solution of the stationary problem presented above.

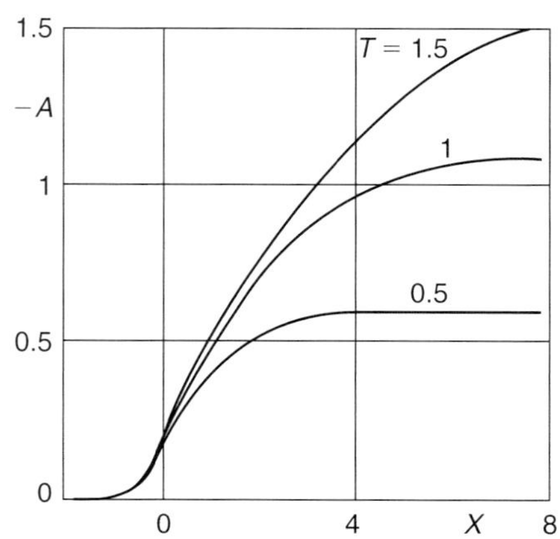


Fig. 3.35.

2. *The motion of the surface at a velocity* $u_w$ *starting at the moment* $t_3 = 0$ *in the region* $x_3 > 0$.
   The discontinuity of this type has an effect on the flow in region 3 via the occurrence of the suction velocity whose value is determined by ejection properties of the newborn boundary layer. The boundary layer flow is governed by the Blasius equation

$$f''' + ff'' = 0$$

$$\psi = (2\varepsilon^2 u_w \rho_w^{-1} \mu_w x)^{1/2} f(\eta), \quad \eta = (2\varepsilon^{-2} u_w \rho_w \mu_w^{-1} x^{-1})^{1/2} y$$

   with the boundary conditions

$$f(0) = 0, \quad f'(0) = 1, \quad f'(\infty) = 0$$

   corresponding to the problem under consideration.
   The solution required for further analysis takes the form:

$$f(\infty) = C_0, \quad C_0 \approx 1.23$$

   Then we have

$$v_{3w} = -(2^{-1} \rho_w^{-1} \mu_w C_0^2 x_3^{-1})^{1/2}$$

   The change of variables

$$x = 1 + (2\varepsilon^4 u_w^{-1} a^{-6} \beta^{-4} \rho_w^{-3} \mu_w^{-1} C_0^{-2})^{1/5} X$$

$$t = (4\varepsilon^3 u_w^{-2} a^{-7} \beta^{-3} \rho_w^{-1} \mu_w^{-2} C_0^{-4})^{1/5} T$$

$$p = (\gamma M_\infty^2)^{-1} + (4^{-1} \varepsilon^2 u_w^2 a^2 \beta^{-2} \rho_w \mu_w^2 C_0^{-4})^{1/5} P, \quad P = -\frac{\partial A}{\partial X}$$

   leads the equation to the form (3.2), where

$$F(X,T) = \begin{cases} 0, & X < 0, \ T \geq 0, \\ -X^{-1/2}, & X \geq 0, \ T > 0 \end{cases} \tag{3.130}$$

   with the boundary conditions

$$A(X,0) = 0, \quad A(-\infty,T) = 0, \quad \frac{\partial A(\infty,T)}{\partial X} = 0 \tag{3.131}$$

   The results of the numerical solution of problem (3.128), (3.130), and (3.131) are presented in Fig. 3.36 in which the function $A(X,T)$ is plotted for different moments. With time the solution describing the disturbed flow upstream of the discontinuity rapidly approaches the solution of the stationary problem. The solution describing the disturbed flow downstream of the discontinuity is characterized by the motion of a solitary wave. It is more clearly visible in Fig. 3.37 in which the pressure disturbance distributions are plotted for different moments.
3. *Tangential injection starting at the moment* $t_3 = 0$ *through a slot located at the surface at* $x_3 = 0$.
   It is assumed that near the body surface there occurs the flow described by the known self-similar solution (Akatnov, 1953) for the near-wall jet injected into the ambient space.

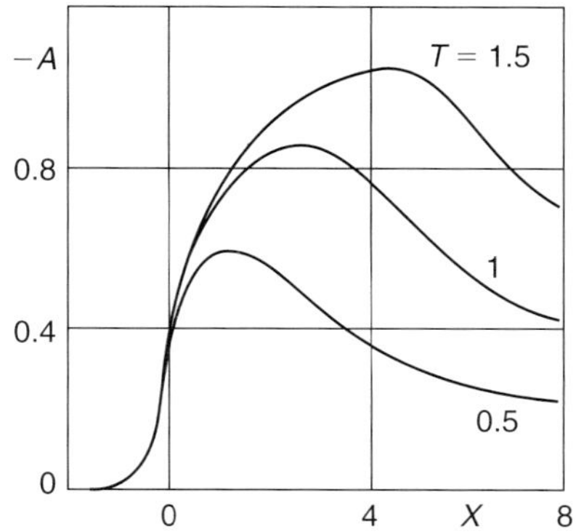


Fig. 3.36.

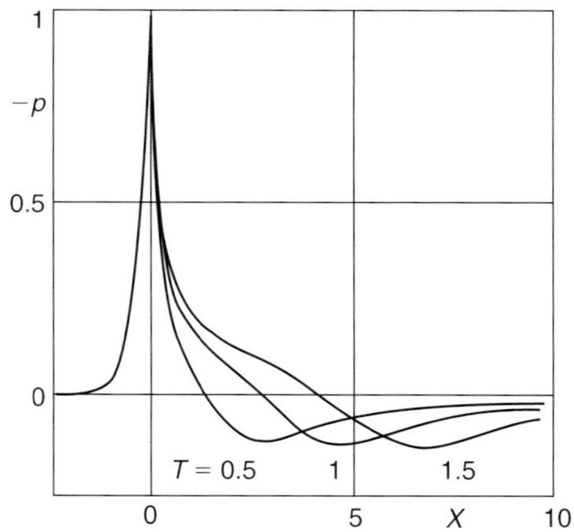


Fig. 3.37.

The stationary solution of this problem was studied by Lipatov (1983) who showed that when the invariant

$$I = \int_0^\infty u^2 \, \mathrm{d}\lambda \int_0^\lambda u \, \mathrm{d}y$$

ranges from $O(\varepsilon^{13/4})$ to $O(1)$, in region 3 there occurs a flow described by the Burgers equation (3.128). As in the previous case, the effect of the wall jet on the flow in region 3 is realized via the injection velocity on the outer boundary of the wall jet. For the problem under consideration the following expansions are valid

$$x = 1 + (\varepsilon^{10} a^{-12} \beta^{-8} \rho_W^{-7} \mu_w^{-1} I^{-1} C_1^{-4})^{1/9} X$$

$$t = (\varepsilon^2 a^{-15} \beta^{-7} \rho_w^{-5} I^{-2} C_1^{-8})^{1/9} T$$

$$p = (\gamma M_\infty^2)^{-1} + (\varepsilon^{-2} a^6 \beta^{-2} \rho_w^5 \mu_w^2 I^2 C_1^8)^{1/9} P$$

$$v_w = 4^{-1}(\varepsilon^2 \rho_w^{-1} \mu_w I C_1^4 x^{-3})^{1/4}, \quad C_1 \approx 2.515$$

The equation for the function $A(X, T)$ takes the form (3.128), where

$$F(X,T) = \begin{cases} 0, & X < 0,\ T \geq 0, \\ -X^{-3/4}, & X \geq 0,\ T > 0 \end{cases} \tag{3.132}$$

The results of the numerical solution of Eq. (3.128) subject to conditions (3.132) are presented in Fig. 3.38. As in the previous case of the motion of the surface, jet injection leads to the formation of a downstream-traveling solitary wave.

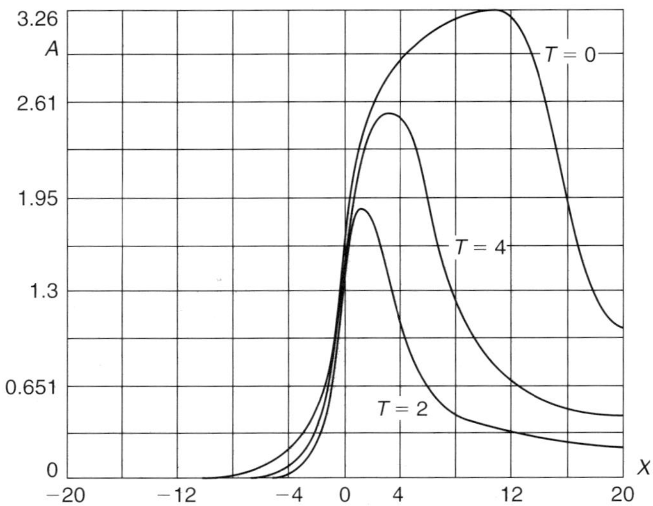


Fig. 3.38.

All the examples considered are characterized by the formation of an elevated-pressure region moving downstream. However, this does not lead to separation of the wall boundary layer, since, by assumption, the flow in this layer is in the first approximation independent of the flow in the nonlinear disturbance region.

## 3.5 Structure of chemically nonequilibrium flows at jumpwise variation of the temperature and catalytic properties of the surface

The problem of chemically nonequilibrium flow in the vicinity of the point of jumpwise variation of the temperature or catalytic properties of the surface is of undoubted importance, both theoretical and practical. Thus, in the studies of Chung et al. (1963), Springer and Pedley (1973), Stewart et al. (1981), and Gershbein et al. (1985) the effect of the discontinuity in the catalytic properties of the surface on the flow past a body was studied within the framework of the laminar boundary layer or hypersonic viscous shock layer models. In the studies of

Bespalov and Voronkov (1980), Gershbein et al. (1986), and Gershbein and Kazakov (1988) this problem was studied in the same formulation though with the introduction of a fictitious inner boundary layer immediately behind the discontinuity point; under some simplifying assumptions it became possible to derive analytical solutions for the flow functions in the vicinity of the point of discontinuity in the catalytic properties of the surface.

For describing the propagation of disturbances upstream of the discontinuity point, which does not take place in the boundary value problems for parabolic equations (Chung and Mirels, 1963; Springer and Pedley, 1973; Bespalov and Voronkov, 1980; Stewart et al., 1981; Gershbein et al., 1985; Gershbein et al., 1986; and Gershbein and Kazakov, 1988) the longitudinal diffusion effect was taken into account in a certain vicinity of the discontinuity point (Popov, 1975; Gershbein and Krupa, 1986; Brykina, 1988); for weak discontinuities this model was substantiated in the paper of Krupa and Tirskii (1988) on the basis of the matched asymptotic expansion method (Van Dyke, 1964).

In analyzing the flow in the vicinity of the point of discontinuity in the catalytic properties of the surface it should be borne in mind that on transition from a noncatalytic to an ideally catalytic surface the gas density increases by its characteristic value, that is, the streamlines are displaced toward the body surface which corresponds to the flow over a cavity in the body surface. For the flows of this type one of the assumptions of Prandtl's boundary layer theory of the smallness of the longitudinal gradients of the flow functions as compared with the transverse ones can be violated which makes necessary the use of the complete Navier–Stokes equations.

The physical processes described above have much in common with the processes characteristic of the flow near the point of discontinuity in the surface temperature, the solution of which is presented in the paper of Sokolov (1975).

In this section we present the results of the study of the flow in the vicinity of the point of discontinuity in the catalytic properties and the temperature of a surface (Bogolepov et al., 1990). It is assumed that a cold flat plate is in a laminar high-Reynolds-number supersonic chemically nonequilibrium flow of a binary mixture. It is shown that in the vicinity of the discontinuity point chemical reactions can proceed only on the catalytic surface. The main similarity parameters are determined, the distributions of the induced pressure disturbances, viscous stress, and the normal gradients of the enthalpy and mass concentration of atoms over the plate are presented, and the asymptotic laws of the variation of these parameters are derived.

### *3.5.1 Formulation of the problem*

We will consider the uniform supersonic viscous chemically nonequilibrium flow past a semi-infinite flat plate at high but subcritical Reynolds numbers $Re$. It is assumed that the gas is a binary mixture of atoms and diatomic molecules consisting of the same atoms, while the surface temperature is not higher than the level on which molecules begin to dissociate at the local pressure. We will study the effect of the jumpwise variation of the temperature and catalytic properties of the surface at a certain distance $\ell$ from the leading edge on the flow past the plate and its heating. We will construct the solution of the Navier–Stokes equations, together with the equation of conservation of the mass concentration of atoms, when

$Re = \rho_\infty u_\infty \ell / \mu_\infty = \varepsilon^{-2} \to \infty$. Below in this section we use the same dimensionless variables as in the previous sections; the temperature is referred to $mu_\infty^2/R$, where $m$ is the molecular weight of the molecular component of the gas and $R$ is the gas constant, the heat flux to $\rho_\infty u_\infty^3$, the specific heats to $R/m$, and the other flow functions to their freestream values.

Let at jumpwise variation of the temperature and catalytic properties there be $\Delta T \sim T \sim O(1)$ and $\Delta c \sim c \sim O(1)$, that is, on the plate surface the temperature $T$ and the mass concentration $c$ vary by their leading orders. Following the matched asymptotic expansion method (Van Dyke, 1964) we will first consider the region with the same longitudinal and transverse dimensions comparable with the body length, $x \sim y \sim O(1)$. At high Reynolds numbers the flow in this region is described by the Euler equations; in the case of the flow past a zero-thickness flat plate set at zero incidence it is the oncoming flow that is the solution of these equations. For satisfying the no-slip boundary condition on the plate surface it is necessary to introduce the region with characteristic dimensions $x \sim O(1)$ and $y \sim O(\varepsilon)$, that is, the Prandtl boundary layer. Solutions for this flow region in the presence of the discontinuity in the catalytic properties of the surface were obtained in the works of Chung et al. (1963), Springer and Pedley (1973), Stewart et al. (1981), and Gershbein et al. (1985); however, they do not describe a small vicinity of the discontinuity point.

A jumpwise variation of the temperature and catalytic properties of the surface can lead to the variation in the gas density $\Delta\rho \sim \rho \sim O(1)$ in the wall layer. Taking the Prandtl and Schmidt numbers $\sigma \sim Sc \sim O(1)$ we assume that in the most general case the thickness of this wall layer and those of the viscous, heat-conducting, and diffusion layers are of the same order. Then, using the estimate for the longitudinal velocity $u \sim O(\varepsilon)$ (in the wall layer at small distances from the surface the longitudinal velocity varies in proportion to this distance) and equating the orders of the convective and dissipative terms in the longitudinal momentum equation, we obtain the estimate for the disturbed wall layer thickness $\Delta y$ as a function of its length $\Delta x \le O(1)$

$$\Delta y \sim O(\varepsilon \Delta x^{1/3}) \le O(\varepsilon) \tag{3.133}$$

Variation of the wall layer thickness (3.133) leads to a proportional displacement of the outer edge of the boundary layer (Neiland, 1974) and induces, due to the interaction with the oncoming uniform supersonic flow, a pressure disturbance $\Delta p \sim O(\Delta y/\Delta x)$. Assuming that generally the pressure disturbance leads to nonlinear disturbances of the velocity $u$ in the wall layer, $u \sim \Delta u \sim \Delta p^{1/2}$, we obtain that the estimate for the pressure disturbance $\Delta p$ is in agreement with estimate (3.133) only provided

$$\Delta x \sim O(\varepsilon^{3/4}), \quad \Delta y \sim O(\varepsilon^{5/4}), \quad \Delta p \sim O(\varepsilon^{1/2}) \tag{3.134}$$

These estimates determine the disturbed flow region extent in the vicinity of the point of the discontinuity in the temperature and catalytic properties of the surface where theories of laminar boundary layer and hypersonic viscous shock layer no longer hold and the induced pressure disturbance must be taken into account. It should also be noted that the disturbed wall layer is located to both sides of the point of discontinuity in the boundary conditions rather than only downstream of it, as it was assumed in the studies of Bespalov and Voronkov (1980), Gershbein et al. (1986), and Gershbein and Kazakov (1988).

If the flow past the plate is in chemical equilibrium, then for the surface temperatures lower than the value, at which dissociation reactions proceed at the local pressure, the atomic component is absent from the wall layer (3.133): $c \approx \partial c/\partial y \approx 0$. Then only chemically nonreacting flows can be studied. In the case of the nonequilibrium flow in the equations of the mass concentration conservation for all components of the gas mixture all terms, including the term describing the rate of formation of an individual component, must be of the same order $O(1)$ (Gladkov et al., 1972). Estimates (3.134) indicate that in the vicinity under consideration large gradients of the flow functions are induced (e.g., $\partial p/\partial x \sim O(\varepsilon^{-1/4}) \gg 1$). Therefore, in this vicinity the term responsible for the rate of formation of an individual component of the mixture is unessential and the flow can be considered as chemically "frozen" with recombination reactions proceeding on the plate surface (Bogolepov, 1984). Since in the region under consideration the order of the pressure disturbance $\Delta p \sim O(\varepsilon^{1/2})$ is less than those of the temperature and mass atom concentration, $\Delta T \sim T \sim O(1)$ and $\Delta c \sim c \sim O(1)$, the barodiffusion effect is asymptotically small. The order of the term governing thermal diffusion is equal to those of the convective terms of the equation of the conservation of mass concentration of atoms but in this section it is neglected in view of its smallness (Gladkov et al., 1972).

### *3.5.2 Parameter scales, equations, and boundary conditions*

In accordance with the technique outlined in Chapter 1, we will first consider the disturbed region I representing the uniform supersonic flow over the interaction region with the following characteristic dimensions: $\delta \ll x \sim y \sim O(\varepsilon^{1/4}) \ll 1$ ($\delta \sim O(\varepsilon)$ is the boundary layer thickness), in which there hold the following independent variables and the asymptotic expansions of the functions

$$x = 1 + \varepsilon^{3/4} x_1, \quad y = \varepsilon^{3/4} y_1 \tag{3.135}$$

$$u = 1 + \varepsilon^{1/2} u_1 + \cdots, \quad v = \varepsilon^{1/2} v_1 + \cdots, \quad \rho = 1 + \varepsilon^{1/2} \rho_1 + \cdots$$

$$p = p_\infty + \varepsilon^{1/2} p_1 + \cdots, \quad h = h_\infty + \varepsilon^{1/2} h_1 + \cdots$$

Substituting expansions (3.135) in the Navier–Stokes equations and passing to the limit $\varepsilon \to 0$ shows that in the first approximation the flow in region I is described by the linearized Euler equations

$$\frac{\partial \rho_1}{\partial x_1} + \frac{\partial u_1}{\partial x_1} + \frac{\partial v_1}{\partial y_1} = 0, \quad \frac{\partial u_1}{\partial x_1} + \frac{\partial p_1}{\partial x_1} = 0, \quad \frac{\partial h_1}{\partial x_1} - \frac{\partial p_1}{\partial x_1} = 0 \tag{3.136}$$

The system of equations (3.136) is supplemented by the formula for calculating the Mach number for the frozen speed of sound and the relation for the enthalpy increment at frozen chemical composition of the mixture

$$M^2 = \frac{1 - \frac{\partial h}{\partial p}}{\frac{\partial h}{\partial \rho}}, \quad h_1 = \frac{\partial h}{\partial p} p_1 + \frac{\partial h}{\partial \rho} \rho_1 \tag{3.137}$$

Equations (3.136) and (3.137) reduce to the wave equation whose solution is known (d'Alembert solution for hyperbolic equations), so that for $y_1 \to 0$ we have

$$\sqrt{M^2 - 1}\, p_1(x_1, 0) = v_1(x_1, 0) \tag{3.138}$$

Since in the vicinity of the point of discontinuity in the surface properties the flow is chemically frozen, in region I the mass concentrations of the mixture components are conserved along streamlines.

Below we will consider the disturbed flow region II located within the boundary layer and having the following characteristic dimensions: $\delta \ll x \sim O(\varepsilon^{3/4} \ll 1$ and $y \sim \delta \sim O(\varepsilon)$. In this region the following new independent variables and asymptotic expansions of the flow functions are introduced

$$x = 1 + \varepsilon^{3/4} x_2, \quad y = \varepsilon y_2 \tag{3.139}$$

$$u = u_{20}(y_2) + \varepsilon^{1/4} u_{21} + \cdots, \qquad v = \varepsilon^{1/2} v_{21} + \cdots, \qquad \rho = \rho_{20}(y_2) + \varepsilon^{1/4} \rho_{21} + \cdots$$

$$p = p_\infty + \varepsilon^{1/2} p_2 + \cdots, \quad T = T_{20}(y_2) + \varepsilon^{1/4} T_{21} + \cdots, \quad c = c_{20}(y_2) + \varepsilon^{1/4} c_{21} + \cdots$$

where $u_{20}(y_2)$, $\rho_{20}(y_2)$, $T_{20}(y_2)$, and $c_{20}(y_2)$ are the profiles of the longitudinal velocity, density, temperature, and mass concentration of atoms in the undisturbed boundary layer ahead of the vicinity of the point of discontinuity in the boundary conditions. Substituting expansions (3.139) in the Navier–Stokes equations and the equation of conservation of the mass concentration of atoms (see, e.g., Gladkov et al., 1972) and passing to the limit $\varepsilon \to 0$ indicates that in region II the flow is described by the system of equations

$$\rho_{20} \frac{\partial u_{21}}{\partial x_2} + u_{20} \frac{\partial \rho_{21}}{\partial x_2} + \rho_{20} \frac{\partial v_{21}}{\partial y_2} + v_{21} \frac{\mathrm{d}\rho_{20}}{\mathrm{d}y_2} = 0, \quad u_{20} \frac{\partial u_{21}}{\partial x_2} + v_{21} \frac{\mathrm{d}u_{20}}{\mathrm{d}y_2} = 0$$

$$\frac{\partial p_2}{\partial y_2} = 0, \quad u_{20} \frac{\partial T_{21}}{\partial x_2} + v_{21} \frac{\mathrm{d}T_{20}}{\mathrm{d}y_2} = 0, \quad u_{20} \frac{\partial c_{21}}{\partial x_2} + v_{21} \frac{\mathrm{d}c_{20}}{\mathrm{d}y_2} = 0$$

which admits partial integration

$$u_{21} = D \frac{\mathrm{d}u_{20}}{\mathrm{d}y_2}, \quad v_{21} = -u_{20} \frac{\mathrm{d}D}{\mathrm{d}x_2}, \quad \rho_{21} = D \frac{\mathrm{d}\rho_{20}}{\mathrm{d}y_2} \tag{3.140}$$

$$T_{21} = D \frac{\mathrm{d}T_{20}}{\mathrm{d}y_2}, \quad c_{21} = D \frac{\mathrm{d}c_{20}}{\mathrm{d}y_2}, \quad p_2 = p_2(x_2)$$

where $D = D(x_2)$ is an arbitrary function.

In the viscous heat-conducting wall layer III with the characteristic dimensions $\delta \ll x \sim O(\varepsilon^{3/4}) \ll 1$ and $y \sim O(\varepsilon^{5/4}) \ll \delta$ we introduce the following independent variables and asymptotic expansions of the flow functions

$$x = 1 + \varepsilon^{3/4} x_3, \quad y = \varepsilon^{5/4} y_3 \tag{3.141}$$

$$u = \varepsilon^{1/4} u_3 + \cdots, \qquad v = \varepsilon^{3/4} v_3 + \cdots, \qquad \rho = \rho_3 + \cdots$$

$$p = p_\infty + \varepsilon^{1/2} p_3 + \cdots, \qquad T = T_{30} + \varepsilon^{1/4} T_{31} + \cdots, \qquad c = c_{30} + \varepsilon^{1/4} c_{31}$$

Substituting expansions (3.141) in the Navier–Stokes equations and the equation of conservation of the mass concentration of atoms and passing to the limit $\varepsilon \to 0$ shows that in the first approximation the flow in the wall layer III is described by the equations for the compressible boundary layer

$$\frac{\partial(\rho_3 u_3)}{\partial x_3} + \frac{\partial(\rho_3 v_3)}{\partial y_3} = 0, \quad \rho_3 \left( u_3 \frac{\partial u_3}{\partial x_3} + v_3 \frac{\partial u_3}{\partial y_3} \right) + \frac{\partial p_3}{\partial x_3} = \frac{\partial}{\partial y_3} \left( \mu \frac{\partial u_3}{\partial y_3} \right) \tag{3.142}$$

$$\frac{\partial p_3}{\partial y_3} = 0, \quad \rho_3 \left( u_3 \frac{\partial c_{30}}{\partial x_3} + v_3 \frac{\partial c_{30}}{\partial y_3} \right) = \frac{\partial}{\partial y_3} \left( \frac{\mu}{Sc} \frac{\partial c_{30}}{\partial y_3} \right), \quad p_0 = \rho_3 T_{30}(1 + c_{30})$$

$$\rho_3 c_p \left( u_3 \frac{\partial T_{30}}{\partial x_3} + v_3 \frac{\partial T_{30}}{\partial y_3} \right) + \frac{\partial p_3}{\partial x_3} = \frac{\partial}{\partial y_3} \left( \frac{\mu c_p}{Sc} \frac{\partial T_{30}}{\partial y_3} \right) + \frac{\mu}{Sc}(c_{p1} - c_{p2}) \frac{\partial T_{30}}{\partial y_3} \frac{\partial c_{30}}{\partial y_3}$$

where $c_p$, $c_{p1}$, and $c_{p2}$ are the specific heats at constant pressure of the atomic and molecular components and their mixture, respectively. From Eq. (3.141) it can be seen that Eq. (3.142) must not incorporate the terms describing barodiffusion $\varepsilon^2 \partial^2 p/\partial x^2 \sim \varepsilon \ll \varepsilon^2 \partial^2 c/\partial y^2 \sim \varepsilon^{-1/2}$ and the longitudinal diffusion

$$\varepsilon^2 \frac{\partial^2 c}{\partial x^2} \sim \varepsilon^{1/2} \ll \varepsilon^2 \frac{\partial^2 c}{\partial y^2} \sim \varepsilon^{-1/2} \tag{3.143}$$

The outer boundary conditions are obtained by matching the asymptotic expansions for the pressure (3.135), (3.139), and (3.141) in regions I, II, and III with account for Eqs. (3.138) and (3.140)

$$p_3(x_3) = p_2(x_2) = p_1(x_1, 0) = \frac{v_1(x_1, 0)}{\sqrt{M^2 - 1}} = \frac{v_{21}(x_2, \infty)}{\sqrt{M^2 - 1}} = -\frac{\frac{\mathrm{d}D}{\mathrm{d}x_2}}{\sqrt{M^2 - 1}} \tag{3.144}$$

The initial boundary conditions are derived by matching with the wall region of the undisturbed flat-plate boundary layer

$$u_3 \to A y_3, \quad T_{30} \to T_{20}(0), \quad c_{30} \to c_{20}(0), \quad p_3, D \to 0 \ \ x_3 \to -\infty \tag{3.145}$$

On the plate surface the no-slip and impermeability conditions must be fulfilled

$$u_3 = v_3 = 0 \quad (y_3 = 0) \tag{3.146}$$

If the origin of region III is located at the point of the jumpwise variation of the temperature and catalytic properties of the surface, then the conditions for the temperature at the plate surface can be written as follows:

$$T_{30} = T_{20}(0) \quad (x_3 < 0), \qquad T_{30} = (1 + \alpha) T_{20}(0) \quad x_3 \geq 0, \ \ y_3 = 0 \tag{3.147}$$

Here, $\alpha > -1$ is a parameter characterizing the temperature jump.

For the plate flow under consideration the condition for the mass concentration of atoms at a catalytic surface can be written as follows (Gladkov et al., 1972):

$$\frac{\partial c}{\partial y} = \frac{Sc\, k\rho c}{\varepsilon^2 \mu} \tag{3.148}$$

where the catalycity coefficient $k = \varepsilon^{\beta} K$, $K \sim O(1)$. Since the flow in the region under consideration is chemically frozen, it makes sense to dwell upon only such cases in which ahead of the point of discontinuity in the boundary conditions the plate surface is not ideally catalytic, that is, $c_{20}(0) > 0$ and $\beta = 1$. Otherwise, the atomic component would be absent from region III. Then in the variables of region III the boundary condition for $x_3 < 0$ takes the form:

$$\frac{\partial c_{30}(0)}{\partial y_3} = 0 \quad x_3 < 0, \;\; y_3 = 0 \tag{3.149}$$

From Eqs. (3.142) and boundary conditions (3.143), (3.145), (3.147), and (3.149) there follows:

$$T_{30}(x_3, y_3) = T_{20}(0), \quad c_{30}(x_3, y_3) = c_{20}(0) \;\; x_3 < 0 \tag{3.150}$$

Obviously, a nontrivial solution for the mass concentration of atoms can be determined only for the values of the parameter $\beta \leq 3/4$. Then in the variables of region III the boundary condition (3.148) takes the form:

$$\frac{\partial c_{30}}{\partial y_3} = \frac{Sc\, K\rho_3 c_{30}}{\mu} \quad \text{or} \quad c_{30} = 0 \qquad x_3 \geq 0, \;\; y_3 = 0 \tag{3.151}$$

For $x_3 < 0$ and when the condition (3.150) is fulfilled, it makes sense to consider the next terms of expansions (3.144) for the temperature and mass concentration of atoms. For these terms the following equations are obtained

$$\rho_3 \left( u_3 \frac{\partial c_{31}}{\partial x_3} + v_3 \frac{\partial c_{31}}{\partial y_3} \right) = \frac{\partial}{\partial y_3} \left( \frac{\mu}{Sc} \frac{\partial c_{31}}{\partial y_3} \right) \tag{3.152}$$

$$\rho_3 c_p \left( u_3 \frac{\partial T_{31}}{\partial x_3} + v_3 \frac{\partial T_{31}}{\partial y_3} \right) = \frac{\partial}{\partial y_3} \left( \frac{\mu}{\sigma} \frac{\partial T_{31}}{\partial y_3} \right)$$

Obviously, also in this approximation, barodiffusion and longitudinal diffusion are unessential.

The outer boundary conditions for the functions $T_{31}$ and $c_{31}$ are obtained by matching the asymptotic expansions for the flow functions (3.139) and (3.141) in regions II and III with account for Eq. (3.140)

$$T_{31} \to B(y_3 + D), \quad c_{31} \to C(y_3 + D) \;\; y_3 \to \infty \tag{3.153}$$

where $B = \mathrm{d}T_{20}/\mathrm{d}y_2$ and $C = \mathrm{d}c_{20}/\mathrm{d}y_2$ for $y_2 = 0$. The initial boundary conditions are determined by matching with the solution for the wall region of the undisturbed flat-plate boundary layer

$$T_{31} \to By_3, \quad c_{31} \to Cy_3 \;\; x_3 \to -\infty \tag{3.154}$$

At the plate surface the functions $T_{31}$ and $c_{31}$ must satisfy the conditions

$$T_{31} = 0, \quad \frac{\partial c_{31}}{\partial y_3} = C \quad x_3 < 0, \ y_3 = 0 \tag{3.155}$$

In variables (3.141) the viscous stress $\tau$ and the heat flux $q$ are expressed by the formulas

$$\tau = \varepsilon\mu\frac{\partial u_3}{\partial y_3} + \cdots, \qquad -q = \varepsilon^{3/4}q_1 + \varepsilon q_2 + \cdots \tag{3.156}$$

$$q_1 = \frac{\mu}{\sigma}\left[c_p\frac{\partial T_{30}}{\partial y_3} + Le\frac{\partial c_{30}}{\partial y_3}\left(\int_0^{T_{30}}(c_{p1} - c_{p2})\,\mathrm{d}T + h^0\right)\right]$$

$$q_2 = \frac{\mu}{\sigma}\left[c_p\frac{\partial T_{31}}{\partial y_3} + Le\frac{\partial c_{31}}{\partial y_3}\left(\int_0^{T_{30}}(c_{p1} - c_{p2})\,\mathrm{d}T + h^0\right)\right]$$

Here, $Le$ is the Lewis number and $h^0$ is the dimensionless heat of atom formation. Formulas (3.156) indicate that for $x_3 < 0$ the heat flux distribution $q$ is described by the second term (see Eq. (3.150)) which is equal in the order to the heat flux in the undisturbed flat-plate boundary layer. The jumpwise variation of the temperature and catalytic properties of the surface leads to the variation of the temperature in the order, so that for $x_3 \geq 0$ its distribution is chiefly determined by the first term.

In order to reduce the boundary value problem (3.142)–(3.147), (3.149), (3.151)–(3.155) to the form convenient for numerical integration, the following change of variables is carried out

$$x = \frac{x_3}{x^*}, \quad \eta = \frac{1}{\eta^*}\int_0^{y_3}\rho_3\,\mathrm{d}y_3, \quad \psi = \frac{\psi_3}{\psi^*}, \quad \frac{\partial\psi_3}{\partial y_3} = \rho_3 u_3, \quad \frac{\partial\psi_3}{\partial x_3} = -\rho_3 v_3 \tag{3.157}$$

$$p = \frac{p_3}{p^*}, \quad c_1 = c_{30}, \quad T_1 = \frac{T_{30}}{T_{20}(0)[1 + c_{20}(0)]}, \quad c_2 = \frac{c_{31}\rho^*}{C\eta^*}, \quad T_2 = \frac{T_{31}\rho^*}{B\eta^*}$$

$$d = \frac{D\rho^*}{\eta^*}, \quad \rho^* = \frac{p_0}{T_{20}(0)[1 + c_{20}(0)]}, \quad x^* = (\rho^*)^{-1/4}(M^2 - 1)^{-3/8}A^{-5/4}$$

$$\eta^* = (\rho^*)^{1/4}(M^2 - 1)^{-1/8}A^{-3/4}, \quad \psi^* = (\rho^*)^{-1/2}(M^2 - 1)^{-1/4}A^{-1/2}$$

$$p^* = (\rho^*)^{-1/2}(M^2 - 1)^{-1/4}A^{1/2}$$

The following simplifying assumptions (Gladkov et al., 1972) are also adopted

$$\rho_3\mu = 1, \sigma, Sc, c_p = \text{const}, \qquad \int_0^{T_{30}}(c_{p1} - c_{p2})\,\mathrm{d}T \ll h^0 \tag{3.158}$$

Then in variables (3.157) and under assumptions (3.158) the boundary value problem can be written in the standard form:

$$\psi''' = -\ddot{\mathrm{d}}(1+c_1)T_1 + \psi'\dot{\psi}' - \dot{\psi}\psi'' \tag{3.159}$$

$$\frac{c_i''}{Sc} = \psi'\dot{c}_i - \dot{\psi}c_i', \quad \frac{T_i''}{\sigma} = \psi' Tc_i - \dot{\psi} Tc_i, \quad i = 1, 2$$

$$\psi(x,0) = 0, \quad \psi'(x,0) = 0, \quad c_2'(x,0) = 1, \quad T_2(x,0) = 0$$

$$c_1'(x,0) = F\frac{c_1(x,0)}{1+c_1(x,0)}, \quad T_1(x,0) = \frac{1+\alpha}{1+c_{20}(0)} \quad x \geq 0$$

$$\psi'(x,\infty) \to \int_0^\infty T_1(1+c_1)\,\mathrm{d}\eta + \mathrm{d}, \quad c_1(x,\infty) \to c_{20}(0), \quad T_1(x,\infty) \to \frac{1}{1+c_{20}(0)}$$

$$T_2(-\infty,\eta) \to \eta, \quad \mathrm{d}(-\infty) \to 0, \quad c_1(-\infty,\eta) \to c_{20}(0)$$

$$T_1(x,\eta) = \frac{1}{(1+c_{20}(0))} \quad x < 0$$

$$F = \frac{Sc\,Kp_0^{5/4}}{(M^2-1)^{1/8}A^{3/4}T_{20}^{5/4}(0)(1+\alpha)[1+c_{20}(0)]^{1/4}}$$

Here, $(\,)' = \partial/\partial\eta$ and $(\dot{\,}) = \partial/\partial x$; $F$ is the local Damkeller number, while formulas (3.156) for the viscous stress and heat flux are brought into the form:

$$\tau = \psi''' + \cdots, \quad q_1 = T_1' + Le\,E_1 c_1', \quad q_2 = T_2' + Le\,E_2 c_2' \tag{3.160}$$

$$E_1 = \frac{h^0}{c_p T_{20}(0)[1+c_{20}(0)]}, \quad E_2 = \frac{Ch^0}{Bc_p}$$

where $\tau$, $q_1$, and $q_2$ are additionally divided by the quantities $\varepsilon A/\rho^*$, $c_p T_{20}(0)[1+c_{20}(0)]/\sigma\eta^*$, and $c_p B/\sigma\rho^*$, respectively. The similarity parameters $E_1$ and $E_2$ characterize the ratio of the heat fluxes due to the transport of energy by the diffusive gas mixture components and heat conduction.

Obviously, the amplification of the catalytic effect of the plate surface for $x_3 \geq 0$ must lead to a decrease in the mass concentration of atoms near the surface, that is, $c_1(x,0) \to 0$ as $x \to \infty$. Then using the variables

$$n = \frac{\eta}{(3x)^{1/3}}, \quad \psi = (3x)^{2/3}(\varphi + J_2 + \gamma_1 n), \quad d = (3x)^{1/3}\gamma_1 \tag{3.161}$$

makes it possible to reduce the boundary value problem (3.170) to the self-similar problem (for $x \to \infty$)

$$\frac{d^3\varphi}{dn^3} + 2(\varphi + \gamma_1 n + J_2)\frac{d^2\varphi}{dn^2} - \left(\frac{d\varphi}{dn} + 2\gamma_1 + 2J\right)\frac{d\varphi}{dn} + 2T_1(1 + c_1)\varphi$$

$$= (\gamma_1 + J)^2 - \frac{d}{dn}[T_1(1 + c_1)] - 2T_1(1 + c_1)(c_1 n + J_2) \qquad (3.162)$$

$$\frac{1}{Sc}\frac{d^2c_1}{dn^2} + 2(\varphi + \gamma_1 n + J_2)\frac{dc_1}{dn} = 0, \quad \frac{1}{\sigma}\frac{d^2T_1}{dn^2} + 2(\varphi + \gamma_1 n + J_2)\frac{dT_1}{dn} = 0$$

$$\varphi = 0, \quad \frac{d\varphi}{dn} = -\gamma_1, \quad c_1 = 0, \quad T_1 = \frac{(1+\alpha)}{[1 + c_{20}(0)]} \quad n = 0$$

$$\frac{d\varphi}{dn} = 0, \quad \frac{d^2\varphi}{dn^2} = 0, \quad c_1 = c_{20}(0), \quad T_1 = \frac{1}{[1 + c_{20}(0)]} \quad n \to \infty$$

$$J = \int_0^n T_1(1 + c_1)\,dn, \quad J_2 = \int_0^n dn \int_0^n T_1(1 + c_1)\,dn$$

The distinctive feature of the boundary value problem (3.162) is the absence of a term with the longitudinal pressure gradient. The problem involves an additional unknown parameter $\gamma_1$ obtained from the solution itself of the problem. The form (3.161) of the variables indicates that as $x \to \infty$ the viscous stress $\tau$ (see Eq. (3.160)) approaches its value ahead of the point of discontinuity in the boundary conditions in the undisturbed boundary layer, while the heat flux $q_1$ decreases in accordance with the following law

$$\tau = \psi''' = \frac{d\varphi^2}{dn^2} + T_1(1 + c_1), \quad q_1 = \frac{1}{(3x)^{1/3}}\left(\frac{dT_1}{dn} + LeE_1\frac{dc_1}{dn}\right) \qquad (3.163)$$

where *Le* is the Lewis number.

The asymptotics of the solutions of the boundary value problems of type (3.159), as $x \to 0$, were studied in the paper of Neiland (1974). Formulas (3.161) are functionally similar to the expansions constructed in the studies of Gershbein et al. (1986) and Gershbein and Kazakov (1988) for the inner boundary layer immediately behind the point of discontinuity in the plate surface properties. However, in those works the effect of the variation of the surface properties on the pressure, viscous stress, and heat flux disturbances was not taken into account, that is, only the cases of weak discontinuities were considered.

### *3.5.3 Analysis of the flow in region IV near the point of jumpwise variation of the temperature and catalytic properties of the surface*

For analyzing the solution of the boundary value problem (3.159) as $x \to 0$ we must additionally study region IV near the point of the jumpwise variation of the temperature and catalytic properties of the surface with a length smaller than the longitudinal dimension of the interaction region ($\varepsilon^{3/2} < x < \varepsilon^{3/4}$). If the variation of the plate surface characteristics does not result in boundary layer separation, then for the nonlinear, viscous, heat-conducting, and diffusive region IV the same relations, as for region III, are valid, while region IV lies totally inside region III. Then for flow in region IV the outer flow is that in the near-wall part of region III as $x \to -0$

$$u_3 \to A_w y_3, \quad v_3 = 0, \quad T_{30} \to T_{20}(0), \quad c_{30} \to c_{20}(0) \tag{3.164}$$
$$p_3 = p_w, \quad \rho_3 = \frac{p_0}{T_{20}(0)[1 + c_{20}(0)]}$$

Here $A_w = (\partial u_3/\partial y_3)_w$, while the subscript $w$ refers to $x \to -0$. In region IV the following independent variables and asymptotic expansions of the flow functions are valid

$$x = 1 + \Delta x\, x_4, \quad y = \varepsilon \Delta x^{1/3} y_4 \tag{3.165}$$
$$u = \Delta x^{1/3} u_4 + \cdots, \quad v = \left(\frac{\varepsilon}{\Delta x^{1/3}}\right) v_4 + \cdots, \quad \rho = \rho_4 + \cdots$$
$$p = p_0 + \varepsilon^{1/2} p_w + \cdots, \quad T = T_{40} + \cdots, \quad c = c_{40} + \cdots$$

Substituting Eq. (3.165) in the Navier–Stokes equations and the equation of conservation of the mass concentration of atoms and passing to the limit $\varepsilon \to 0$, $\varepsilon^{3/2} < \Delta x < \varepsilon^{3/4}$, shows that in the first approximation the flow in region IV is described by equations of type (3.142) with the subscript 4 substituted for 3. The outer and initial boundary conditions are obtained by matching expansions (3.165) with the solution for the near-wall part of region III as $x \to -0$ and have the form (3.143), (3.145), in which the subscript 4 should be substituted for 3, $D \equiv 0$, and $A$ should be replaced by $A_w$.

In order for the equation of conservation of the mass concentration of atoms to have a nontrivial solution for $x_4 > 0$, it is necessary that $\varepsilon^{\beta} \geq \varepsilon/\Delta x^{1/3}$, since for smaller values of the catalycity coefficient the mass concentration of atoms has no time to change in the leading order over the length $\varepsilon^{3/2} < \Delta x < \varepsilon^{3/4}$. Then the boundary conditions at the plate surface take the form (3.146), (3.147), (3.149), and (3.151) (with the corresponding replacement of the subscript 3 by 4).

In the case under consideration the interaction with the oncoming uniform supersonic flow does not take place, since otherwise an excessively large pressure disturbance $\Delta p \sim \Delta y/\Delta x \sim \varepsilon/\Delta x^{2/3} > \Delta x^{2/3}$ would be induced at $\Delta x < \varepsilon^{3/4}$. Therefore, in region IV there is a realized compensation flow pattern (Bogolepov and Neiland, 1976) in which the variation of the region IV thickness in compensated – in the first approximation – by the variation of the thickness of the near-wall part of the flow in region III directly ahead of the point of the variation of the plate surface properties. In the study of Bogolepov and Neiland (1976) it was obtained that in this case the induced pressure disturbance is determined from the relation

$$\rho_{20}(0) A_w v_4 + \frac{dp_4}{dx_4} \to 0 \quad y_4 \to \infty \tag{3.166}$$

The distinctive feature of the boundary value problem (3.142), (3.143), (3.145), (3.147), (3.149), and (3.151) is the compensation condition of interaction (3.166) which, however, does not increase the orders of the derivatives with respect to the longitudinal coordinate entering into this boundary value problem and does not induce disturbances ahead of the point of discontinuity in the boundary conditions (the problem remains parabolic). For this reason, here the initial boundary conditions can be preassigned at $x_4 = 0$, correct to certain additive constants which have no effect on the solution in region IV though they are important for region III. In variables (3.165) expressions (3.156) for the viscous stress $\tau$ and the heat flux $q$ take the form:

$$\tau = \varepsilon\mu \frac{\partial u_4}{\partial y_4} + \cdots \tag{3.167}$$

$$-q = \frac{\varepsilon\mu}{\Delta x^{1/3}\sigma}\left[c_p \frac{\partial T_{40}}{\partial y_4} + Le \frac{\partial c_{40}}{\partial y_4}\left(\int_0^{T_{40}} (c_{p1} - c_{p2})\,\mathrm{d}T + h^0\right)\right]$$

Introducing new variables

$$x = \frac{x_4}{x^*}, \quad n = \frac{1}{\eta(3x)^{1/3}} \int_0^{y_3} \rho_4\,\mathrm{d}y_4, \quad \psi_4 = \psi^*(3x)^{2/3}(\varphi + J_2) \tag{3.168}$$

$$\frac{\partial \psi_4}{\partial y_4} = \rho_4 u_4, \quad \frac{\partial \psi_4}{\partial x_4} = -\rho_4 v_4$$

$$p_4 = p^*(3x)^{2/3}\gamma_2, \quad c_1 = c_{40}, \quad T_1 = \frac{T_{40}}{T_{20}(0)[1 + c_{20}(0)]}$$

$$\eta^* = (x^*)^{1/3}\rho_{20}^{1/3}(0)A_w^{-1/3}(M^2 - 1)^{-1/8}A^{-3/4}, \quad \psi^* = (x^*)^{2/3}\rho_{20}^{-1/3}(0)A_w^{1/3}$$

$$p^* = (x^*)^{2/3}\rho_{20}^{-1/3}(0)A_w^{4/3}$$

and using the simplifying assumptions (3.158) makes it possible to reduce the boundary value problem (3.142), (3.143), (3.145)–(3.147), (3.151) (with the corresponding changes for region IV), and (3.166) for the partial differential equations to the self-similar problem

$$\frac{\mathrm{d}^3\varphi}{\mathrm{d}n^3} + 2(\varphi + J_2)\frac{\mathrm{d}^2\varphi}{\mathrm{d}n^2} - \left(\frac{\mathrm{d}\varphi}{\mathrm{d}n} + 2J\right)\frac{\mathrm{d}\varphi}{\mathrm{d}n} + 2T_1(1 + c_1)\varphi$$

$$= J^2 - \frac{\mathrm{d}}{\mathrm{d}n}[T_1(1 + c_1)] - 2T_1(1 + c_1)(\gamma_2 + J_2) \tag{3.169}$$

$$\frac{1}{Sc}\frac{\mathrm{d}^2 c_1}{\mathrm{d}n^2} + 2(\varphi + J_2)\frac{\mathrm{d}c_1}{\mathrm{d}n} = 0, \quad \frac{1}{\sigma}\frac{\mathrm{d}^2 T_1}{\mathrm{d}n^2} + 2(\varphi + J_2)\frac{\mathrm{d}T_1}{\mathrm{d}n} = 0$$

$$\varphi = 0, \quad \frac{\mathrm{d}\varphi}{\mathrm{d}n} = 0, \quad c_1 = 0, \quad T_1 = \frac{(1 + \alpha)}{[1 + c_{20}(0)]} \quad n = 0$$

$$\varphi = \gamma_2, \quad \frac{d^2\varphi}{dn^2} = 0, \quad c_1 = c_{20}(0), \quad T_1 = \frac{1}{[1 + c_{20}(0)]} \quad n \to \infty$$

where $\gamma_2$ is an additional unknown parameter determined from the solution itself of the problem. Obviously, for $\varepsilon^{\beta} \sim \varepsilon\Delta x^{1/3}$, $\varepsilon^{3/2} < \Delta x < \varepsilon^{3/4}$ the solution of the boundary value problem (3.169) describes the asymptotic behavior of the solution of the original non-self-similar boundary value problem in region IV for $c_{40}(x_4, 0) \to 0$, that is, $x_4 \to \infty$. For $\varepsilon^{\beta} > \varepsilon\Delta x^{1/3}$, $\varepsilon^{3/2} < \Delta x < \varepsilon^{3/4}$, that is, for $c_{40}(x_4, 0) \equiv 0$, these boundary value problems are identical. In variables (3.168) expressions (3.167) for $\tau$ and $q$ take the form:

$$\tau = \frac{d^2\varphi}{dn^2} + T_1(1 + c_1) + \cdots, \quad -q = \frac{1}{(3x)^{1/3}}\left(\frac{dT_1}{dn} + LeE_1\frac{dc_1}{dn}\right) + \cdots \tag{3.170}$$

Here, $\tau$ and $q$ are additionally divided by $\varepsilon A_w/\rho_{20}(0)$ and $\varepsilon/(\Delta x)^{1/3} c_p T_{20}(0)[1 + c_{20}(0)]/\sigma\eta^*$, respectively. Formulas (3.170) determine the behavior of the solution of the boundary value problem (3.168) as $x \to +0$. We note the functional similarity of Eq. (3.168) with the expansions constructed in the studies of Gershbein et al. (1986) and Gershbein and Kazakov (1988) for a small vicinity behind the point of discontinuity in the catalytic properties of the surface. As noted above, this flow representation is valid only for a weak discontinuity in the properties of the surface. Generally, it is necessary to solve the self-similar problem (3.169) determining the initial conditions for the inner layer behind the point of discontinuity in the boundary conditions.

The boundary value problem (3.169) describes the jumpwise variation of the flow functions in passing through the point of discontinuity in the boundary conditions on the plate surface. For studying a vicinity of the discontinuity point smaller than regions III and IV, in accordance with the matched asymptotic expansion method (Van Dyke, 1964) it is necessary to consider a region whose characteristic length and thickness are of the same order, $\Delta x \sim \Delta y$. Estimates (3.133) show that in this case $\Delta x \sim \Delta y \sim O(\varepsilon^{3/2})$, $u \sim v \sim O(\varepsilon^{1/2})$, and $\Delta p \sim O(\varepsilon)$ and the flow is governed by the complete system of Navier–Stokes equations and the equation of conservation of the mass concentration of atoms at a variable density. Only in this region the longitudinal diffusion is important, that is, the equations incorporate the terms of the form $\sim\partial^2 c/\partial x^2$. However, in this region the no-slip conditions are no longer fulfilled on the plate surface, since, owing to a finite disturbance of the temperature or the mass concentration of atoms $\Delta T \sim T \sim \Delta c \sim O(1)$ there appears the slip velocity $U \sim \varepsilon^2 \partial T/\partial x \sim O(\varepsilon^{1/2})$. Moreover, other effects of molecular gas dynamics can also be important (Kogan, et al., 1976). Therefore, taking account of the longitudinal diffusion, as it was done in the works of Popov (1975), Gershbein and Krupa (1986), Brykina (1988), and Krupa and Tirskii (1988), is justified only for weak discontinuities in the plate surface properties.

### 3.5.4 Results of numerical calculations

In integrating numerically the boundary value problem (3.159), in view of the interaction between the flows in regions III and I, the induced pressure must be determined in the process of computation. Taking the interaction into account imparts the property of weak ellipticity to the problem (Neiland, 1974), so that an additional boundary condition should be preassigned as $x \to \infty$ for obtaining a unique solution (an analysis of expressions (3.161) shows that $p \to 0$ as $x \to \infty$). Moreover, due to the discontinuity in the boundary conditions at $x = 0$ in the computation domain $x > 0$ a subdomain near the plate surface must be introduced and a special procedure of calculations in two regions must be constructed (Daniels, 1974; Lipatov and Neiland, 1982).

Numerical calculations of the boundary value problem (3.159) were carried out for the following controlling similarity parameters:

1. $F = 1/\sqrt{2}$, $\alpha = 0$, and $c_{20} = 1$;
2. $F \to \infty$, $\alpha = 0$, and $c_{20} = 1$; and
3. $F = 0$, $\alpha = 1$, and $c_{20} = 1$.

At a constant surface temperature (cases 1 and 2) a decrease in the atom concentration near the catalytic surface leads to an increase in the density and streamline displacement toward the plate surface, that is, an effective cavity is formed and an expansion flow is realized. Contrariwise, an increase in the temperature leads to a decrease in the gas density and streamline displacement away (upward) from the plate (case 3). In this case, a compression flow past an effective bump takes place.

In Figs. 3.39–3.42 curves 1–3 correspond to the numbers of the cases. The behavior of the solution of the boundary value problem (3.159) for $x < 0$ was studied in the paper of Neiland (1974); for $x \to -0$ all flow functions vary in a nonsingular way and take finite values.

In Fig. 3.39 we have plotted the distributions of the induced pressure $p$. The behavior of the pressure as $x \to +0$ is determined by Eq. (3.168); the numerical solution of the boundary value problem (3.169) makes it possible to determine the values of the unknown parameter $\gamma_2 = 0.22912$ and $0.46593$, for cases 2 and 3, respectively. As $x \to \infty$, the pressure disturbance distribution is described by Eq. (3.161) ($p = -\dot{d}$); the numerical solution of the boundary value problem (3.162) makes it possible to determine the values of the unknown parameter $\gamma_1 = 0.46094$ and $0.81717$ for cases 2 and 3, respectively.

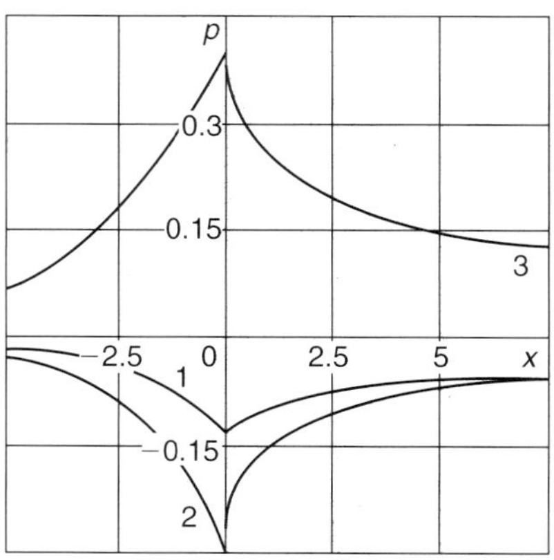


Fig. 3.39.

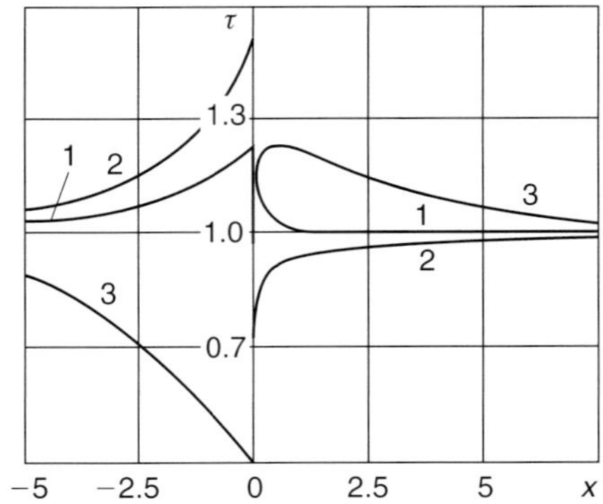


Fig. 3.40.

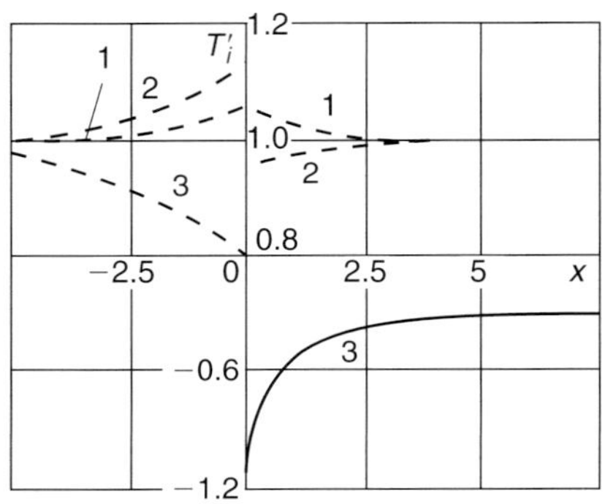


Fig. 3.41.

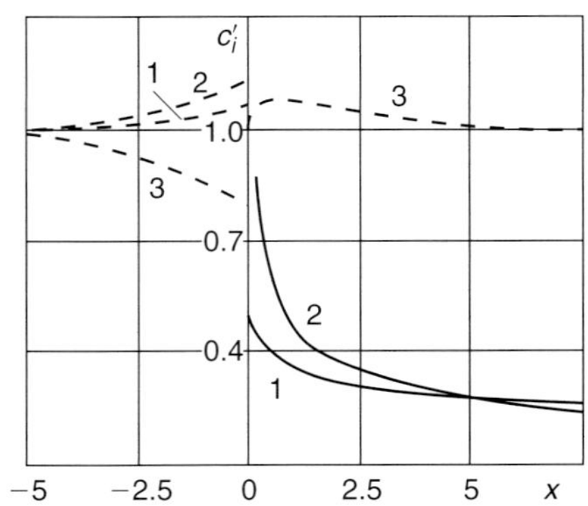


Fig. 3.42.

The viscous stress $\tau$ distributions over the plate surface are presented in Fig. 3.40. For the expansion flow (curves 1 and 2) the friction increases ahead of the point of discontinuity in the boundary conditions, whereas for the compression flow (curve 3) it decreases. The solution of the local boundary value problem (3.78) for $x \to +0$ describes the subsonic variable-density shear flow over an effective cavity (cases 1 and 2) or bump (case 3). Therefore, $\tau$ decreases sharply as $x \to +0$ in the former case and increases in the latter case. For case 1 the friction distribution is continuous, while for cases 2 and 3 the solution of the boundary value problem (3.78) for $x \to +0$ gives the following values of the viscous stress: $\tau = 0.720$ and 0.964, respectively. These values are in good agreement with the results of the integration of the boundary value problem (3.159). As $x \to \infty$, the viscous stress approaches its value in the undisturbed flat-plate boundary layer ahead of the vicinity of the point of discontinuity in the boundary conditions ($\tau \to 1$).

An analysis of the behavior of curve 2 shows that the greatest possible jumpwise increase in the catalytic activity of the surface is associated with an about 1.5-fold increase of the viscous friction as $x \to -0$. The twofold jumpwise increase in the surface temperature reduces the viscous stress ahead of the discontinuity point to 0.4 (curve 3), while the increase in the temperature by a factor of about 3.87 leads to incipient boundary layer separation (Sokolov, 1975).

In Figs. 3.41 and 3.42 the distributions of the normal gradients of the temperature $T_i'$ and mass concentration of atoms $c_i'$ over the plate surface are presented (solid curves relate to $i = 1$ and dashed curves to $i = 2$). The twofold jumpwise temperature increase reduces $T_2'$ by about 20% as $x \to -0$ (dashed curve 3). For $x > 0$ a strongly heated surface is cooled by the oncoming flow; in this case the normal temperature gradient varies in the order and $T_1' < 0$ (solid curve 3). As $x \to \infty$, the disturbances produced by the jumpwise variation in the surface properties decay, so that $T_1' \to 0$ and $T_2' \to 1$.

The distributions of $c_2'$ at $x \leq 0$ are similar with the $\tau$ and $T_2'$ distributions for the corresponding cases (dashed curves 1–3 in Fig. 3.42). The jumpwise increase of the catalytic activity of the surface at $x > 0$ leads to an increase in the normal gradient of the mass concentration of atoms by an order of the magnitude (solid curves 1 and 2 and Eq. (3.156)). The twofold jumpwise increase of the surface temperature has an only slight effect on the distribution of the parameter $c_2'$ (curve 3) as $x \to \infty$, $c_1' \to 0$, and $c_2' \to 1$.

A qualitative similarity of the present calculated results with those presented, for example, in the works of Gershbein et al. (1986) and Gershbein and Krupa (1986) stems from the functional similarity of the representation of the solutions. As for quantitative comparison of the solutions, it turned out to be impossible in view of the fact that different initial conditions were used.

# 4 Flows Under Conditions of the Interaction Between the Boundary Layer and the Outer Flow Along the Entire Body Length

In Section 4.1 it is shown that for weak interaction between a hypersonic flow and a boundary layer, free interaction regions (which can occur, e.g., near boundary layer separation points and within which disturbances can propagate upstream) are local, as at supersonic velocities of the outer inviscid flow. For this reason, the theory of these flows can be reduced, with slight modification of the form of the parameters, to the theory developed for supersonic flow regimes in Chapter 1.

Sections 4.2 and 4.3 of this chapter are devoted to the study of the moderate and strong interaction regimes; for these cases it is shown that the region of upstream propagation of disturbances remains finite, that is, the ratio of its length to the body length does not vanish when the Reynolds (and Mach) numbers increase. For a very simple problem of the flat-plate boundary layer a single-parameter family of non-self-similar solutions is found (it exists apart from the well-known self-similar Lees–Stewartson solution). The properties of these solutions, which make it possible to satisfy an additional boundary condition at the body end, are studied. They are obtained both in the form of series expansions in the vicinity of the leading edge and numerically. The similarity law obtained is used for correlating the experimental data.

As another example of the flow, in which the boundary layer/outer flow interaction leads to disturbance propagation through distances of the order of the body dimensions, we consider the supersonic flow past a finite-length wedge through the surface of which intense gas injection is realized (Sections 4.4–4.7).

## 4.1 Regime of weak interaction with the outer flow

We will consider the hypersonic flow of a viscous heat-conducting gas past a thin body. In this case, it is convenient to introduce the Reynolds number $Re_0$ based on the body length, the freestream velocity and density, and the viscosity calculated at the stagnation temperature (moreover, everywhere in this chapter the viscosity is divided by its value at the stagnation temperature). We will assume that the disturbed flow region thickness $\tau \approx \delta \ll 1$, where $\delta$ is the boundary layer thickness. This means that the body thickness $\delta_w \ll O(\delta)$.

We will consider the solutions of the Navier–Stokes equations in the first approximation at the double passage to the limit $M \to \infty$ and $Re_0 \to \infty$. In accordance with hypersonic small perturbation theory (Guiraud, 1966; Chernyi, 1966; Hayes and Probstein, 1966), for inviscid flows with $M\tau \ll 1$ we have

$$p \approx 1 + O(M\tau) \tag{4.1}$$

Then the boundary layer thickness $\delta \approx M/Re_0^{1/2} \approx \tau$.

Thus, on the main part of the body the interaction of the hypersonic flow with the boundary layer induces the pressure difference $\Delta p$

$$\Delta p \approx \chi \tag{4.2}$$

where $\chi = M\tau = M^2/Re_0^{1/2}$.

According to the standard terminology, the weak interaction regime corresponds to the passage to the limit

$$M \to \infty, \quad Re_0 \to \infty, \quad \chi \to 0 \tag{4.3}$$

In this case, $\chi$ is a small parameter of the problem. The pressure gradient $\Delta p \approx O(\chi)$ induced on the main part of the body appears in the boundary layer equations only in the second approximation (Hayes and Probstein, 1966).

Let us now consider for the weak interaction regime such flow regions in which the boundary layer induced pressure gradient affects the boundary layer flow even in the first approximation. By analogy with the problems considered in Chapter 1, this situation might be expected, for example, near the separation point (Fig. 1.1).

If a pressure difference $\Delta p \ll 1$ is applied to the undisturbed (in the first approximation) boundary layer, then in the main part of the boundary layer, where $u \approx 1$ and $\rho \approx M^{-2}$, in view of the momentum and continuity equations, the orders of the velocity, density, and streamtube thickness disturbances are as follows:

$$\Delta u^{(2)} \sim \frac{\Delta \rho^{(2)}}{\rho^{(2)}} \sim \frac{\Delta \delta^{(2)}}{\delta^{(2)}} \sim \frac{\Delta p^{(2)}}{p^{(2)}}, \quad u \sim 1, \quad \rho^{(2)} \sim M^{-2}, \quad \delta \sim M\, Re_0^{-1/2} \tag{4.4}$$

Here, the subscript 2 refers to the disturbed flow region of thickness $\sim\delta$ (see Fig. 1.2). However, near the body surface there is a layer in which $\Delta u \sim u$, since at the body surface the velocity is equal to zero. By virtue of the same equations, together with the velocity profile in the undisturbed boundary layer, in this layer the estimates are different

$$u^{(3)} \sim \Delta u^{(3)} \sim \left(\frac{\Delta p}{p}\right)^{1/2}, \quad u^{(3)} \sim \frac{\Delta \delta^{(3)}}{\delta} \tag{4.5}$$

Here, the subscript 3 refers to the variables in the wall layer, where $u \sim \Delta u$.

From Eqs. (4.4) and (4.5) there follows

$$\Delta \delta \sim \Delta \delta_3 \sim M\, Re_0^{-1/2} \left(\frac{\Delta p}{p}\right)^{1/2} \gg \Delta \delta_2 \tag{4.6a}$$

Thus, in the first approximation the disturbance of the total boundary layer displacement thickness $\Delta\delta$ is due to the variation of the "slow" streamtube thicknesses $\Delta\delta_3$.

In the vicinity of the separation point (as in the case of the supersonic flow considered in Chapter 1), the equation for region 3 must retain the leading viscous terms of the momentum equation

$$\frac{\rho_w u^{(3)2}}{\Delta x} \sim \frac{\mu_w u^{(3)}}{(\Delta\delta^{(3)})^2} \tag{4.6b}$$

Thus, Eqs.(4.1) and (4.6), together with two relations (4.5), give four equations for four unknown scales $\Delta p$, $\Delta x$, $u^{(3)}$, and $\Delta\delta^{(3)}$, so that we have

$$\Delta x \sim \chi^{3/4}, \quad u^{(3)} \sim \chi^{1/4}, \quad \Delta p \sim \frac{\chi^{1/2}}{M^2}, \quad \delta^{(3)} \sim \frac{\chi^{5/4}}{M} \tag{4.7}$$

Given estimates (4.7), the estimates for other variables can easily be obtained (like it was done in Chapter 1) in all three regions shown in Fig. 1.2.

For the inviscid outer flow region 1 the scales of the variables and the asymptotic representations take the form:

$$\begin{aligned}
&x = \chi^{3/4} x_1, \quad \psi = \frac{\chi^{3/4}}{M}\psi_1\\
&p \sim \frac{1}{M^2}\left(\frac{1}{\gamma} + \chi^{1/2} p_1 + \cdots\right), \quad \rho \sim 1 + \chi^{1/2}\rho_1 + \cdots\\
&n \sim \frac{\chi^{3/4}}{M}\psi_1 + \frac{\chi^{5/4}}{M} n_1 + \cdots, \quad h \sim \frac{1}{M^2}\left(\frac{1}{\gamma-1} + \chi^{1/2} h_1 + \cdots\right)\\
&u \sim 1 + \frac{\chi^{1/2}}{M^2} u_1 + \cdots, \quad v \sim \frac{\chi^{1/2}}{M^2} v_1 + \cdots
\end{aligned} \tag{4.8}$$

We note that the transverse dimension of the outer region is determined by the relation $\Delta x \sim \chi^{3/4}$, while the slope of the characteristics is of the order $1/M$. It should be noted that from the estimates for $\Delta u$ and $v$ it follows that the plane section law holds true in the linear theory for $M \gg 1$.

Substituting Eq. (4.8) in the Navier–Stokes equations (1.6)–(1.9), in which, in accordance with the new normalization of $\mu$, $Re_0^{-1}$ is substituted for $\varepsilon^2$, and passing to the limit $Re_0 \to \infty$, $M \to \infty$, and $\chi \to 0$ leads to the following equations:

$$\frac{\partial u_1}{\partial x_1} + \frac{\partial p_1}{\partial x_1} = 0, \quad \frac{\partial p_1}{\partial \psi_1} + \frac{\partial v_1}{\partial x_1} = 0, \quad \frac{\partial n_1}{\partial x_1} = v_1, \quad \frac{\partial n_1}{\partial \psi_1} = -\rho_1, \quad \frac{\partial h_1}{\partial x_1} - \frac{\partial p_1}{\partial x_1} = 0 \tag{4.9}$$

From the equation of state there follows

$$h_1 = \frac{\gamma}{\gamma-1} p_1 - \frac{\gamma}{\gamma-1}\rho_1 \tag{4.10}$$

The solution of Eqs. (4.9) and (4.10) is reduced to that of the wave equation

$$\frac{\partial^2 n_1}{\partial x_1^2} - \frac{\partial^2 n_1}{\partial \psi_1^2} = 0, \quad n_1 = A(x_1 - \psi_1) + B(x_1 + \psi_1)$$
$$p_1 = A'(x_1 - \psi_1) + B'(x_1 + \psi_1) \tag{4.11}$$

If there is no disturbances coming from outside, then $B \equiv 0$.

The value of the arbitrary function $A$ is determined by matching with the solutions for regions 2 and 3.

In region 2 the corresponding independent variables and functions take the form:

$$x = \chi^{3/4} x_2, \quad \psi = \frac{\chi^{3/4}}{M^3}\psi_2$$
$$p \sim \frac{1}{M^2}\left(\frac{1}{\gamma} + \chi^{1/2} p_2 + \cdots\right), \quad \rho \sim \frac{1}{M^2}\left(\rho_{2,0} + \chi^{1/2}\rho_2 + \cdots\right)$$
$$n \sim \frac{\chi}{M}\left(n_{2,0} + \chi^{1/4} n_{2,1} + \chi^{1/2} n_{2,2} + \cdots\right), \quad h \sim h_{2,0} + \chi^{1/2} h_2 + \cdots$$
$$u \sim u_{2,0} + \chi^{1/2} u_2 + \cdots, \quad v \sim \frac{\chi^{1/2}}{M} v_2 + \cdots \tag{4.12}$$

Substituting Eq. (4.12) in the Navier–Stokes equations and passing to the limit yields

$$\rho_2^{(0)} u_2^{(0)} \frac{\partial u_2}{\partial x_2} + \frac{\partial p_2}{\partial x_2} = 0, \quad \frac{\partial p_2}{\partial \psi_2} = 0, \quad \frac{\partial n_2}{\partial x_2} = \frac{v_2}{u_2^{(0)}}$$
$$\frac{\partial n_{2,0}}{\partial \psi_2} = \frac{1}{\rho_{2,0} u_{2,0}}, \quad \frac{\partial n_{2,1}}{\partial \psi_2} = 0, \quad \frac{\partial n_{2,2}}{\partial \psi_2} = -\frac{1}{\rho_{2,0} u_{2,0}}\left(\frac{\rho_2}{\rho_{2,0}} + \frac{u_2}{u_{2,0}}\right) \tag{4.13}$$

It is important to note that $n_{2,1}$ and $p_2$ depend only on $x$ and do not change across region 2.

In region 3 the scales and asymptotic representations take the form:

$$x = \chi^{3/4} x_3, \quad \psi = \frac{\chi^{3/2}}{M^3}\psi_3$$
$$p \sim \frac{1}{M^2}\left(\frac{1}{\gamma} + \chi^{1/2} p_3 + \cdots\right), \quad \rho \sim \frac{1}{M^2}(\rho_w + \cdots)$$
$$n \sim \frac{\chi^{5/4}}{M}(n_3 + \cdots), \quad h \sim h_w + \chi^{1/4} h_3 + \cdots$$
$$u \sim \chi^{1/4} u_3 + \cdots, \quad v \sim \frac{\chi^{3/4}}{M} v_3 + \cdots, \quad \mu \sim \mu_w + \cdots \tag{4.14}$$

Substituting Eq. (4.14) in the Navier–Stokes equations and passing to the limit $M \to \infty$, $Re_0 \to \infty$, and $\chi \to 0$ yields

$$\rho_w u_3 \frac{\partial u_3}{\partial x_3} + \frac{\partial p_3}{\partial x_3} = \rho_w^2 \mu_w u_3 \frac{\partial}{\partial \psi_3}\left(u_3 \frac{\partial u_3}{\partial \psi_3}\right), \quad \frac{\partial p_3}{\partial \psi_3} = 0$$
$$\frac{\partial n_3}{\partial x_3} = \frac{v_3}{u_3}, \quad \frac{\partial n_3}{\partial \psi_3} = \frac{1}{\rho_w u_3}$$
$$\frac{\partial h_3}{\partial x_3} = \rho_w \mu_w \frac{\partial}{\partial \psi_3}\left(\frac{u_3}{\sigma}\frac{\partial u_3}{\partial \psi_3}\right) \tag{4.15a}$$

The boundary conditions at the body are as follows:

$$u_{3w} = v_{3w} = h_{3w} = 0 \tag{4.15b}$$

Matching with the solution for region 2 (as in Chapter 1 for the supersonic flow) yields

$$u_3 \to \left(\frac{2a\psi_3}{\rho_w}\right)^{1/2}, \quad h_3 \to b\left(\frac{2\psi_3}{\rho_w a}\right)^{1/2} \quad \text{for } \psi_3 \to +\infty \tag{4.15c}$$

Matching with the solution for the undisturbed boundary layer yields

$$u_3 \to \left(\frac{2a\psi_3}{\rho_w}\right)^{1/2}, \quad h_3 \to b\left(\frac{2\psi_3}{\rho_3 a}\right)^{1/2} \quad \text{for } x_3 \to -\infty \tag{4.16}$$

Finally, $p_3$ should be related with the variation of the boundary layer displacement thickness. The absence of the pressure difference from regions 2 and 3 yields

$$p_1(x_3, 0) = p_2(x_3) = p_3(x_3) \tag{4.17}$$

The fifth equation (4.13) indicates that

$$n_1(x_3, 0) = n_2(x_3) \tag{4.18}$$

Now it remains to perform the matching for $n(x, \psi, \chi)$ in regions 2 and 3

$$n_1(x_1, 0) = n_2(x_3) = \int_0^\infty \frac{1}{\rho_w}\left(\frac{1}{u_3} - \sqrt{\frac{\rho_w}{2a\psi_3}}\right) \mathrm{d}\psi_3 \tag{4.19}$$

Thus, for the flow in the vicinity of the separation point, in which $B \equiv 0$ in Eq. (4.11), we obtain

$$p_3(x) = \frac{\mathrm{d}}{\mathrm{d}x_3} \int_0^\infty \frac{1}{\rho_w}\left(\frac{1}{u_3} - \sqrt{\frac{\rho_w}{2a\psi_3}}\right) \mathrm{d}\psi_3 \tag{4.20}$$

Equations (4.15a) and conditions (4.15b) coincide with (1.23) and (1.24) for $\lambda = \varepsilon^{3/4}$, conditions (4.15c) with (1.25), and (4.20), correct to a constant which can easily be eliminated by the change of variables of type (1.23), with (1.31b).

Thus, the free interaction theory for the weak interaction regime in a hypersonic flow represents the limiting case of the theory developed in Chapter 1, if the parameter $\chi = M^2/Re_0^{1/2}$ is introduced instead of $\varepsilon = Re^{-1/2}$.

## 4.2 Moderate and strong interactions in a hypersonic flow

On moderate and strong interactions between a hypersonic flow and a boundary layer the pressure gradient induced in the inviscid flow by the displacement thickness has an effect on the boundary layer flow even in the first approximation. It is well known (Hayes and Probstein, 1966) that the corresponding passages to the limit take the form:

$$M \to \infty, \quad Re_0 \to \infty$$

with $\chi = O(1)$ associated with the moderate and $\chi \to \infty$ with the strong interaction.

At weak interaction, near corner points, separation points, etc., there appear regions, in which, as shown above, the pressure gradients are in the order considerably higher than those on the rest of the body.

The results obtained above suggest that for $\chi \geq O(1)$ the solution of the problem can be nonunique over the whole length of the body.

### *4.2.1 Flow nature in the locations of rapid variation of the boundary conditions*

As a characteristic example, we will consider the flow over a flat plate terminated by a base section. Let the base pressure be varied arbitrarily, for example, by varying the flow conditions in other flow regions or at the cost of gas injection to or suction from the base region. In accordance with the conventional boundary layer theory, the flow over the plate consisting of an inviscid hypersonic shock layer and a viscous boundary layer must be independent of the variation of the base pressure.

Assuming that this is the case, we obtain that the pressure determined from the solution for the main part of the body in the region directly ahead of the base region must generally differ in the leading order from the base pressure. In accordance with hypersonic small perturbation theory, for $\chi \geq O(1)$ the pressure and the pressure gradient on the body are of the order $\tau^2$. The boundary layer has always a subsonic region, through which pressure disturbances can propagate upstream. In this case, two situations are possible: either the pressure disturbances modify the flow at distances $\Delta x \ll 1$ and produce high local pressure gradients in this region or the flow restructures itself as a whole for $\Delta x \sim 1$. In the latter case the problem of the strong or moderate interaction between the boundary layer and the inviscid flow subject to conventional initial and boundary conditions must admit a whole family of solutions, that is, the solution is nonunique. Then it becomes possible to satisfy an additional condition at the body end.

We will first assume that near the base section there appears a zone, in which the pressure varies in the leading order at small distances $\Delta x \ll 1$. Then in this region at distances of the order $\delta$ ($\delta$ is the thickness of the undisturbed boundary layer ahead of the beginning of the disturbed flow region) the pressure gradients are considerably greater than the viscous stresses in the entire boundary layer, except for a vanishingly thin wall layer. Obviously, this flow must behave as inviscid. In more detail, the locally inviscid flows were considered in Chapter 3.

As shown below, an important parameter of this inviscid boundary layer is the derivative $\mathrm{d}\delta/\mathrm{d}p$ calculated for given original parameter profiles (in accordance with the solution for the original undisturbed boundary layer) under the assumption that $\partial p/\partial y = 0$ and the flow is inviscid. It can be calculated that

$$\frac{\mathrm{d}\delta}{\mathrm{d}p} \sim \int_0^{\delta} \left( \frac{1}{M^2} - 1 \right) \mathrm{d}y \tag{4.21}$$

If $\mathrm{d}\delta/\mathrm{d}p < 0$, then with an increase in the pressure the narrowing of the supersonic streamtubes of the former boundary layer overcomes the broadening of the subsonic part of the flow, so that $\delta$ as a whole decreases. Thus, if $\mathrm{d}\delta/\mathrm{d}p < 0$ and the base pressure is greater than $p$ determined from the solution for the main part of the body for $x = 1$, then it is necessary to assume that in the locally inviscid region $\partial p/\partial y \sim \partial p/\partial x$. Otherwise, the original pressure distribution would induce rarefaction near the outer edge of the boundary layer. The presence of $\partial p/\partial y \sim \partial p/\partial x$ removes the paradox, since in this case the $\Delta p$ disturbances in the outer part of the boundary layer are smaller than those at the wall, so that for $\Delta p > 0$, $\Delta\delta$ remains positive as a whole. However, from the condition of the equality of the longitudinal and transverse pressure gradients it follows that the disturbed region length is of the same order as its thickness. Thus, from $\mathrm{d}\delta/\mathrm{d}p < 0$ it follows that $\Delta x \sim \delta$ and $\partial p/\partial x \sim \partial p/\partial y$. From the physical standpoint, this result is understandable, since the original disturbances are transferred upstream through the subsonic part of the boundary layer and then are carried away along characteristics in the supersonic part of the flow. Since for $M \approx 1$ the characteristic slope is also of the order of unity, it is necessary that $\Delta x \sim \Delta y$ in order for the transverse and longitudinal pressure gradients could be of the same order and the gradients at the top be smaller in absolute magnitude than those at the bottom.

If $\mathrm{d}\delta/\mathrm{d}p > 0$, then a decrease in the base pressure leads to a decrease in $\delta$. We will demonstrate that in this case disturbances cannot be local. In fact, over the lengths $\Delta x \ll 1$ a disturbance $\Delta p \sim \tau^2$ produces

$$\frac{\Delta p}{\Delta x} \gg O(\tau^2) \tag{4.22}$$

that is, the flow is locally inviscid. Since for any $\Delta x$ near the body $|\Delta p|$ is not smaller than that in the outer part of the flow, in accordance with the momentum, energy, and continuity equations, we obtain the following estimates for the disturbances

$$\Delta u \sim u, \quad \Delta\rho \sim \rho, \quad \Delta\delta \sim \delta \sim \tau \tag{4.23}$$

These induce the slope of the outer boundary of the layer

$$\frac{\Delta\delta}{\Delta x} \gg \tau \tag{4.24}$$

and, in accordance with the hypersonic small perturbation theory, the pressure disturbance at the outer boundary is as follows:

$$\Delta p \sim \left(\frac{\Delta\delta}{\Delta x}\right)^2 \gg \tau^2 \sim p \tag{4.25}$$

Thus, we have arrived at a contradiction. Therefore, the original hypothesis on the disturbance localization for $\mathrm{d}\delta/\mathrm{d}p > 0$ turned out to be untrue. Then it remains only to assume that in this case the disturbance propagates through the entire body length.

According to Eq. (4.21), the sign of $\mathrm{d}\delta/\mathrm{d}p$ depends on the Mach number distribution across the boundary layer. Integral (4.21) is improper and has a singularity at $y=0$, where $M=0$. For the conventional solution of the problem of the boundary layer on a thermally insulated body we have

$$g_w \sim O(1), \quad u \sim y(y \to 0) \tag{4.26}$$

Then it can easily be shown that

$$M(y) \sim y(y \to 0) \tag{4.27}$$

and integral (4.21) is divergent. This means that all flows of this type do not admit disturbance localization and in these flows $\mathrm{d}\delta/\mathrm{d}p > 0$ (see Eq. (4.22)). However, as shown below, a gradual decrease in the base pressure leads to the restructuring of the solution on the main part of the body in such a way that for a fairly low base pressure a singular point appears at which

$$u \sim y^{1/n}, \quad \text{where } n > 2 \tag{4.28}$$

Behind this singular point $\mathrm{d}\delta/\mathrm{d}p < 0$ and a local region is formed, in which $\Delta p \sim \tau^2$, while at the corner point of the body $M_w = 1$ (in the locally inviscid flow).

Another possible way of the formation of flows in which disturbances are not transferred over $\Delta x \sim 1$ is cooling and suction, since for $g_w \sim 0$ the following solutions can be obtained:

$$u \sim y, \quad M \sim y^{1/n} \quad \text{for } y \to 0, \ n > 2 \tag{4.29}$$

### *4.2.2 Equations and boundary conditions for the flat-plate flows in the presence of moderate and strong interactions*

We will consider the flow of a viscous, heat-conducting gas over a flat plate. The strong interaction regime corresponds to the passage to the limit

$$M \to \infty, \quad Re_0 \to \infty, \quad \chi \to \infty \tag{4.30}$$

In accordance with the conventional theory of strong interaction we will choose the following independent variables and asymptotic representations for the flow functions in the boundary layer:

$$x = x_2, \quad y = y_2\tau, \quad \tau = Re_0^{-1/4}, \quad \chi = M\tau$$

$$u(x, y, M, Re_0) \sim u_2(x_2, y_2) + \cdots, \qquad v(x, y, M, Re_0) \sim \tau v_2(x_2, y_2) + \cdots \tag{4.31}$$

$$p(x, y, M, Re_0) \sim \tau^2 p_2(x_2, y_2) + \cdots, \quad \rho(x, y, M, Re_0) \sim \tau^2 \rho_2(x_2, y_2) + \cdots$$

$$H(x, y, M, Re_0) \sim H_2(x_2, y_2) + \cdots, \qquad \mu(x, y, M, Re_0) \sim \mu_2(x_2, y_2) + \cdots$$

Here the enthalpy $H$ is divided by the square of the undisturbed flow velocity and the subscript 2 refers to the deformed variables which conserve the values $O(1)$ inside the boundary layer.

Substituting Eq. (4.31) in the Navier–Stokes equations and passing to the limit (4.30) we obtain the conventional boundary layer equations

$$\rho_2\left(u_2\frac{\partial u_2}{\partial x_2} + v_2\frac{\partial u_2}{\partial y_2}\right) = -\frac{\partial p_2}{\partial x_2} + \frac{\partial}{\partial y_2}\left(\mu_2\frac{\partial u_2}{\partial y_2}\right), \quad \frac{\partial p_2}{\partial y_2} = 0$$

$$\frac{\partial(\rho_2 u_2)}{\partial x_2} + \frac{\partial(\rho_2 v_2)}{\partial y_2} = 0, \quad H_2 = h_2 + \frac{1}{2}u_2^2 \tag{4.32}$$

$$\rho_2\left(u_2\frac{\partial H_2}{\partial x_2} + v_2\frac{\partial H_2}{\partial y_2}\right) = \frac{\partial}{\partial y_2}\left(\frac{\mu_2}{\sigma_2}\frac{\partial H_2}{\partial y_2}\right) + \frac{\partial}{\partial y_2}\left[\mu_2\left(1 - \frac{1}{\sigma}\right)u_2\frac{\partial u_2}{\partial y_2}\right]$$

The equation of state and the viscosity–enthalpy dependence are as follows:

$$p_2 = \frac{\gamma - 1}{\gamma}\left(H_2 - \frac{u_2^2}{2}\right)\rho_2, \quad \mu_2 = \left[2\left(H_2 - \frac{u_2^2}{2}\right)\right]^{\omega} \tag{4.33}$$

The boundary conditions take the form:

$$u_2(x_2, 0) = v_2(x_2, 0) = 0, \quad H_2(x_2, 0) = H_{2w} \tag{4.34}$$

$$u_2(x_2, \delta_2) = 1, \quad v_2(x_2, \delta_2) = \frac{d\delta_2}{dx_2}, \quad H_2(x_2, \delta_2) = \frac{1}{2}$$

Here $\delta_2$ is the boundary layer thickness and $\sigma$ is the Prandtl number.

In the first approximation, the outer edge of a hypersonic boundary layer is exactly determined on the scale of the thickness layer, since at the outer edge the temperature vanishes and the density increases without bound (of course, there is a third transitional layer but it has an effect on the main flow only in the higher approximations (Bush, 1966;

Lee and Cheng, 1969). In the boundary layer the gas flow rate is negligibly small as compared with the flow rate in the inviscid region of the disturbed flow, whereas the thicknesses may be of the same order. Precisely the difference in the orders of the gas flow rates makes it possible to develop a correct theory of the boundary layer.

The following variables can conveniently be introduced for solving the problem:

$$\xi = \int_0^x \rho_{2w}\mu_{2w}\,\mathrm{d}x_2, \qquad \eta = (2\xi)^{-1/2}\int_0^{y^2} \rho_2\,\mathrm{d}y_2 \tag{4.35}$$

$$u_2 = f'(\xi,\eta), \quad g = 2H_2, \quad \psi_2(x_2,y_2) = (2\xi)^{1/2} f(\xi,\eta)$$

Then the equations take the form:

$$(Nf'')' + ff'' - \beta(\xi)(g - f'^2) = 2\xi(f'\dot{f}' - f''\dot{f}) \tag{4.36}$$

$$\left(\frac{N}{\sigma}g'\right) + fg' + \frac{1}{2}\left[N\left(1 - \frac{1}{\sigma}\right)f'f''\right]' = 2\xi(f'\dot{g} - \dot{f}g')$$

Here, as usual, primes refer to the differentiation with respect to $\eta$ and dots to $\xi$.

The boundary conditions are as follows:

$$f(\xi,0) = f_w(\xi), \quad f'(\xi,0) = 0, \quad g(\xi,0) = g_w, \quad g(\xi,\infty) = 1, \quad f'(\xi,\infty) = 1$$

Other necessary relations are as follows:

$$N = \frac{\rho_2\mu_2}{(\rho_2\mu_2)_w}, \qquad \beta = \frac{\gamma - 1}{\gamma}\frac{\mathrm{d}\ln p_2}{\mathrm{d}\ln\xi} \tag{4.37}$$

$$\rho_2 = \frac{2\gamma}{\gamma - 1}\frac{p_2}{g - f'^2}, \qquad \mu_2 = \left(\frac{g - f'^2}{g_w}\right)^{\omega}$$

The expression for the boundary layer thickness $\delta_2$ takes the form:

$$\delta_2 = \sqrt{2\xi}\int_0^\infty \frac{\mathrm{d}\eta}{\rho_2} = \frac{\gamma - 1}{2\gamma}\frac{(2\xi)^{1/2}}{p_2}\int_0^\infty (g - f'^2)\,\mathrm{d}\eta \tag{4.38}$$

The inviscid part of the disturbed flow region must be considered on the basis of the hypersonic small perturbation theory (Guiraud, 1965; Chernyi, 1966; Hayes and Probstein, 1966). For studying the hypersonic flow past a thin body formed by the boundary layer displacement thickness we introduce the following coordinates and functions:

$$x = x_1; \quad y = \tau y_1$$

$$u \sim 1 + \tau^2 u_1 + \cdots, \quad v \sim \tau v_1 + \cdots \tag{4.39}$$

$$p \sim \tau^2 p_1 + \cdots, \quad \rho \sim \rho_1 + \cdots$$

In accordance with the books of, for example, Guiraud (1965) and Chernyi (1966) the system of equations and boundary conditions on the "body" $y_1 = \delta_2(x)$ and the shock $y_1 = G(x_1)$ take the form:

$$\frac{\partial \rho_1}{\partial x_1} + \frac{\partial \rho_1 v_1}{\partial y_1} = 0, \quad \rho_1 \left( \frac{\partial v_1}{\partial x_1} + v_1 \frac{\partial v_1}{\partial y_1} \right) + \frac{\partial p_1}{\partial y_1} = 0$$

$$\left( \frac{\partial}{\partial x_1} + v_1 \frac{\partial}{\partial y_1} \right) \frac{p_1}{\rho_1^\gamma} = 0, \quad v_1[x_1, \delta_2(x_1)] = \frac{d\delta_2}{dx_1}$$

$$\gamma p_1[x_1 G(x_1)] = \frac{2\gamma}{\gamma + 1} \left( \frac{dG}{dx_1} \right)^2 - \frac{\gamma - 1}{\gamma + 1} \frac{1}{\chi^2} \tag{4.40}$$

$$\rho_1[x_1, G(x_1)] = \left[ \frac{\gamma - 1}{\gamma + 1} + \frac{2}{\gamma + 1} \frac{1}{\chi^2} \left( \frac{dG}{dx_1} \right)^{-2} \right]^{-1}$$

$$v_1[x_1, G(x_1)] = \frac{2}{\gamma + 1} \frac{dG}{dx_1} \left[ 1 - \frac{1}{x^2} \left( \frac{dG}{dx_1} \right)^{-2} \right], \quad v_1[x_1, \delta_2(x_1)] = \frac{d\delta_2}{dx_1}$$

For practical purposes, fairly accurate results are furnished by the tangent wedge method for which we have

$$p = \frac{1}{\gamma \chi^2} + \frac{\gamma + 1}{4} \left( \frac{d\delta}{dx} \right)^2 + \frac{d\delta}{dx} \left[ \frac{1}{\chi^2} + \left( \frac{\gamma + 1}{4} \frac{d\delta}{dx} \right)^2 \right]^{1/2} \tag{4.41}$$

Here, the subscripts of $p$, $x$, and $\delta$ are omitted, since for these variables the normalization is the same in the inviscid shock layer (region 1) and in the boundary layer (region 2). It is important to note that as $\chi \to \infty$, $\chi$ is absent from all the relations and the formulation of the problem corresponds to the limiting flow with strong interaction.

### *4.2.3 Study of the nature of the nonuniqueness of the boundary value problem*

As $\chi \to \infty$, the boundary value problem (4.36) has the well-known self-similar solution obtained using the tangent wedge method (4.41) by Lees (1953) and, in a more accurate formulation, for example (4.40), by Stewartson (1955*b*). However, the principle formulated and proved at the beginning of this section, in accordance with which in a flow with $\chi \geq O(1)$ regions with high local pressure gradients cannot frequently occur, suggests that, strictly speaking, this solution can describe only the flow over a semi-infinite plate.

For other flows, other solutions, also satisfying Eqs. (4.36) and (4.40) or (4.41), must exist.

Another indicator of nonuniqueness is the boundary condition relating the pressure with the boundary layer thickness distribution. This fact was discussed in Chapter 1 in more detail.

We will now try to determine these solutions. For this purpose, we will first note that the boundary conditions are invariant with respect to the following group of transformations:

$$\xi = \lambda\overline{\xi}, \quad \eta = \overline{\eta}, \quad p = \frac{1}{\lambda}\overline{p}, \quad \rho_2 = \frac{1}{\lambda}\overline{\rho}_2, \quad u_2 = \overline{u}_2, \quad h_2 = \overline{h}_2$$

$$v = \frac{1}{\sqrt{\lambda}}\overline{v}, \quad y = \lambda^{3/2}\overline{y}, \quad x = \lambda^2\overline{x}, \quad \chi = \sqrt{\lambda}\overline{\chi}, \quad f = \overline{f}, \quad g = \overline{g} \tag{4.42}$$

$$\rho_1 = \overline{\rho}_1, \quad G = \lambda^{3/2}\overline{G}$$

where $\lambda$ is an arbitrary constant.

For the sake of simplicity we will study problems (4.36) and (4.41) letting $\chi = \infty$. We will first seek the solution in the vicinity of the point $\xi = 0$ in the form of series expansions with the preliminary change of variables

$$\zeta = \xi^b, \quad \eta = \eta \tag{4.43}$$

$$f(\xi,\eta) = f(\zeta,\eta), \quad g(\xi,\eta) = g(\zeta,\eta), \quad p(\xi) = \zeta^{-1/b}P(\zeta)$$

$$v_e(\xi) = \frac{\mathrm{d}\delta}{\mathrm{d}x} = \zeta^{-1/3b}V(\zeta), \quad \beta(\xi) = \beta(\zeta), \quad F = \int_0^\infty (g - f'^2)\,\mathrm{d}\eta$$

The equations and boundary conditions take the form:

$$(Nf'')' + ff'' - \beta(g - f'^2) = 2\zeta b\left(f'\frac{\partial f'}{\partial \zeta} - f''\frac{\partial f}{\partial \zeta}\right)$$

$$\left(\frac{N}{\sigma}g'\right)' + fg' + \frac{1}{2}\left[N\left(1 - \frac{1}{\sigma}\right)f'f''\right]' = 2\zeta b\left(f'\frac{\partial g}{\partial \zeta} - g'\frac{\partial f}{\partial \zeta}\right)$$

$$P = \frac{\gamma+1}{2}V^2, \quad \beta = \frac{\gamma-1}{\gamma}\left(b\frac{\mathrm{d}\ln P}{\mathrm{d}\ln\zeta} - 1\right) \tag{4.44}$$

$$V = \sqrt{2}\,b\zeta\frac{\mathrm{d}F}{\mathrm{d}\zeta} - \sqrt{2}\,b\frac{\mathrm{d}\ln P}{\mathrm{d}\ln\zeta} + \frac{3}{\sqrt{2}}F, \quad N = \left(\frac{g - f'^2}{g_w}\right)^{\omega-1}$$

$$f(\zeta,0) = f_w, \quad f'(\zeta,0) = 0, \quad g(\zeta,0) = g_w, \quad f'(\zeta,0) = g(\zeta,\infty) = 1$$

In these variables, the expansions turn out to be regular

$$f(\zeta,\eta)\sim\sum_{n=0}^{\infty}f_n(\eta)\zeta^n,\quad g(\zeta,\eta)\sim\sum_{n=0}^{\infty}g_n(\eta)\zeta^n,\quad N(\zeta,\eta)\sim\sum_{n=0}^{\infty}N_n(\eta)\zeta^n \tag{4.45}$$

$$P(\zeta)\sim\sum_{n=0}^{\infty}P_n\zeta^n,\quad V(\zeta)\sim\sum_{n=0}^{\infty}V_n\zeta^n,\quad F(\zeta)\sim\sum_{n=0}^{\infty}F_n\zeta^n$$

Substituting Eq. (4.45) in Eq. (4.44) and equating the third power terms we obtain the following recurrent systems:

$$(N_0f_0'')'+f_0f_0'''-\beta_0(g_0-f_0'^2)=0$$

$$\left(\frac{N_0}{\sigma}g_0'\right)'+f_0g_0'+\frac{1}{2}\left[N_0\left(1-\frac{1}{\sigma}\right)f_0'f_0''\right]=0 \tag{4.46}$$

$$N_0=\left(\frac{g_0-f_0'^2}{g_w}\right)^{\omega-1},\quad \beta_0=\frac{\gamma-1}{\gamma},\quad F_0=\int_0^{\infty}(g_0-f_0'^2)\,\mathrm{d}\eta$$

$$V_0=\frac{3}{2}F_0,\quad P=\frac{\gamma+1}{2}V_0^2$$

The boundary conditions are as follows:

$$f_0(0)=f_w,\quad f_0'(0)=0,\quad g_0(0)=g_w,\quad f_0'(\infty)=g_0(\infty)=1 \tag{4.47}$$

The system of equations (4.46) subject to the boundary conditions (4.47) corresponds to the known self-similar ("three-fourth") solution for the flat plate

$$(N_0f_1''+N_1f_0'')'+f_0f_1''+f_1f_0''''-\beta_1(g_0-f_0'^2)-\beta_0(g_1-2f_0'f_1')=2b(f_0'f_1'-f_1f_0'')$$

$$\left(\frac{N_0}{\sigma}g_1'+\frac{N_1}{\sigma}g_0'\right)'+f_0g_1'+f_1g_0'+\frac{1}{2}\left[\left(1-\frac{1}{\sigma}\right)(N_0f_0'f_1''+N_0f_1'f_0''+N_1f_0'f_0'')\right]'$$
$$=2b(f_0'g_1-f_1g_0')$$

$$N_1=(\omega-1)N_0\frac{g_1-2f_0'f_1'}{g_0-f_0'^2},\quad \beta_0P_1+\beta_1P_0=(b-1)\frac{\gamma-1}{\gamma}P_1 \tag{4.48}$$

$$P_1=(\gamma+1)V_0V_1,\quad F_1=\int_0^{\infty}(g_1-2f_0'f_1')\,\mathrm{d}\eta$$

$$V_1 = \frac{1}{g_w}\left[F_1\left(\frac{3}{\sqrt{2}} + b\sqrt{2}\right) - \sqrt{2}bF_0\frac{P_1}{P_0}\right]$$

The boundary conditions are as follows:

$$f_1(0) = f_0'(0) = f_1'(\infty) = g_1(0) = g_1(\infty) = 0 \tag{4.49}$$

Problem (4.48) for the variables $f_1$, $g_1$, $N_1$, $\beta_1$, $P_1$, $V_1$, and $F_1$ is the linear homogeneous problem with zero boundary conditions (4.49). It has a non-trivial solution only for a quite definite value of $b$ which can be called the eigenvalue of the problem. For certain particular problems this value was determined by numerical integration. The solution is definite, correct to one arbitrary constant.

The following systems of equations are linear but inhomogeneous. For example, the system for $f_2$, $g_2$, ... takes the form:

$$(N_0f_2'' + N_1f_1'' + N_2f_0'')' + f_0f_2'' + f_1f_1'' + f_2f_0''\beta_2(g_0 - f_0'^2)$$
$$- \beta_1(g_1 - 2f_0'f_1') - \beta_0(g_2 - 2f_0'f_2' - f_1'^2) = 2b(2f_0'f_2' + f_1'^2 - 2f_0'''f_2 - f_1''f_1)$$

$$\left[\frac{1}{\sigma}(N_0g_2' + N_1g_1' + N_2g_0')\right]' + f_0g_2' + f_1g_1' + f_2g_0'$$
$$+ \frac{1}{2}\left[\left(1 - \frac{1}{\sigma}\right)(N_2f_0'f_0'' + N_0f_2'f_0'' + N_0f_0'f_2'' + N_1f_1'f_0'' + N_0f_1f_1'' + N_1f_0'f_1''\right]'$$
$$= 2b(2f_0'g_2 + g_1f_1' - 2f_2g_0' - f_1g_1')$$

$$N_2 = (\omega - 1)N_0\left[\frac{g_2 - 2f_0'f_2' - f_1'^2}{g_0 - f_0'^2} + \frac{\omega - 2}{2}\left(\frac{g_1 - 2f_0'f_1'}{g_0 - f_0'^2}\right)^2\right] \tag{4.50}$$

$$\beta_0P_2 + \beta_1P_1 + \beta_2P_0 = (2b - 1)\frac{\gamma - 1}{\gamma}P_2, \quad P_2 = \frac{\gamma + 1}{2}(V_1^2 + 2V_0V_2)$$

$$\frac{1}{\sqrt{2}}(P_0V_2 + P_1V_1 + P_2V_0) = 2bPF_2 + bP_1F_1 - 2bF_0P_0 - bF_1P_1$$
$$+ \frac{3}{2}P_0F_2 + \frac{3}{2}P_1F_1 + \frac{3}{2}P_2F_0$$

$$F_2 = \int_0^\infty (g_2 - 2f_0'f_2' - f_1'^2)\,d\eta$$

The boundary conditions are trivial. System (4.50) is linear but inhomogeneous, so that for a fixed value of the arbitrary constant in the nontrivial solution of system (4.48) the solution of system (4.50) is quite definite. Thus, at least in the vicinity of the point $\xi = 0$, the strong interaction problem has a single-parameter family of non-self-similar solutions. Below these solutions are determined by numerical integration.

For the strong interaction problem ($\chi = \infty$) the existence of the group of transformations makes it possible to establish certain important properties of the non-self-similar solutions. First, the non-self-similar solutions obtained cannot exist over the entire range of $\xi$ from zero to infinity with conservation of the given boundary conditions. In fact, if such a solution would exist, then, in the presence of the group of transformations (4.42), it should be reduced to the uniquely determined self-similar solution. From the preceding it follows that the domain of existence of any non-self-similar solution is terminated by a singular point or a point $\xi_0$ at which an additional condition determining the value of a free parameter, say $P_1$, is preassigned. Precisely the appearance of the new characteristic dimension is responsible for the non-self-similarity of solution. Due to the existence of the group of transformations, the whole variety of non-self-similar solutions of problem (4.44) can be reduced to two "standard" solutions corresponding to $P_1 < 0$ and $P_1 > 0$. As for solutions of physical problems, these are described by different pieces of the two standard curves.

To make the situation clearer we will consider the flow over a flat plate with a base section under the assumption that the base pressure can be controlled in one or another way (injection, suction, etc.).

The self-similar solution of the strong interaction problem assumes a quite definite value of the pressure on the plate ahead of the base section ($x = 1$). However, if the base pressure is different from this value, then the pressure at the plate end must change, since, according to the assertion proved above, in a small vicinity of the base section there cannot appear a region, in which the pressure gradient is in the order greater than that on the rest of the plate, at least, up to the singular point $f''_w = \infty$.

Let us assume that by a proper choice of the parameter $\lambda$ (4.42) all non-self-similar solutions are reduced to two solutions corresponding to $P_1 = \pm 1$. The $P_1 = -1$ solution is terminated by a singular point corresponding to $f''_w = \infty$. For $P_1 = +1$ the solution includes a separation point.

Now, if the base pressure differs from that determined by the self-similar solution, then the two relations, $x = \lambda^2 \overline{x}$ and $p = (1/\lambda)\overline{p}$, together with the boundary condition $p(1) = p_b$ make it possible to choose the required piece of the standard non-self-similar solution $\overline{p}(\overline{x})$ governing the flow. In fact, we obtain an additional relation for determining the coordinates of the point on the "standard" curve $\overline{p}(\overline{x})$ corresponding to the base section

$$\overline{p}_b = \frac{p_b}{\sqrt{\overline{x}_b}}$$

*Thus, the non-similar solutions derived completely determine the plate flow if one more additional condition at the plate end is preassigned.* It can be the value of the pressure, the position of the singular point $f''_w = \infty$, and, generally, the condition of the matching with some other flow region located downstream.

### 4.2.4 Results of calculations and comparison of the similarity law with the experimental data

We will first note that for flows slightly differing from the self-similar solution the approximate series-form solution (4.45) was derived above; up to the third term, the coefficients of these series are determined by systems of equations (4.46)–(4.50). The calculations were carried out for some particular cases. It was assumed that the Prandtl number is $\sigma = \omega = g_w = 1$. For the flows with no injection ($f_w = 0$) the results were obtained for two values of $\gamma$ equal to 7/5 and 5/3

$$p \sim \frac{A}{\zeta^{1/b}}[1 \pm (\zeta + p_{11}\zeta^2 + \cdots)]$$

$$\beta \sim \frac{\gamma - 1}{2\gamma}\frac{1}{A}\zeta^{2/b}\left[\frac{1}{2} \mp \frac{\zeta}{2+b} + \frac{\zeta^2(1 \pm p_{11})}{2(1+b)} + \cdots\right] \tag{4.51}$$

| $\gamma$ | $b$ | $A$ | $p_{11}$ | $\beta_{11}$ |
|---|---|---|---|---|
| 7/5 | 46.82 | 8.708 | $\mp 6.221$ | $\mp 3.840$ |
| 5/3 | 24.76 | 8.92 | $\mp 4.335$ | $\mp 3.863$ |

In the table plus and minus signs relate to the "compression" and "expansion" flows corresponding to $P_1 > 0$ and $P_1 < 0$. However, the values of the coefficient $P_1$ are taken to be equal to $\pm A$ rather than $\pm 1$.

Formulas (4.51) make it possible to draw certain conclusions on the behavior of the solution. First, for large values of the exponent $b$ the distributions of the pressure and other flow functions over the forward part of the body are only slightly different from those determined by the self-similar solution but further they change very rapidly. This provides the explanation to the fact that in many cases using approximate methods based on the application of integral equations of boundary layer generates a need of introducing the notion of the "subcritical" and "supercritical" behavior of the boundary layer. These notions were first introduced in the paper of Crocco (1955). Now it is clear that in the integral description of the parameter profiles in the boundary layer the role of the subsonic wall layer was considered imperfectly, though in many cases this approach could lead to satisfactory results. It should be noted that the exponent $b$ is not always large and transition from the strong to slight influence region is not always rapid. Thus, the calculations of the flows with injection ($f_w < 0$) showed that with intensification of the injection the exponent $b$ decreases ($b = 1.16$ for $f_w = -10$). In the paper of Kozlova and Mikhailov (1970) it was shown that for the flow past a yawed flat plate $b$ decreases rapidly with increase in the yaw angle. Another example of the flows with small eigenvalues is considered in Section 4.4.

For obtaining the results outside a small vicinity of the leading edge of the plate a numerical method for solving the boundary value problem (4.36) was developed. The solutions were obtained for the compression and expansion flows at the following values of the relevant parameters: $\sigma = g_w = \omega = 1$, $f_w = 0$, and $P_1 = \pm A$ (we note that the terminology used – compression and expansion flows – is to a certain degree conditional, since even in the compression flows the pressure decreases near the plate nose). Apart from their convenience, the introduction of such terms can be justified by the fact that the expansion flows can be obtained from the self-similar solution by deflecting the rear part of the plate downward and the compression flows by deflecting it upward.

Some calculated results are presented in Figs. 4.1–4.5. We have plotted the distributions of the pressure and the ratio of the friction drag coefficient $c_f$ to its value $c_{f,self}$ in the self-similar flow. The distributions are plotted against both the natural integration variable $\zeta$ and the physical variable $x$. It should be noted that any point $x = x^*$ can be considered as the end of the body thus obtaining the flow parameter distributions over the body from the results

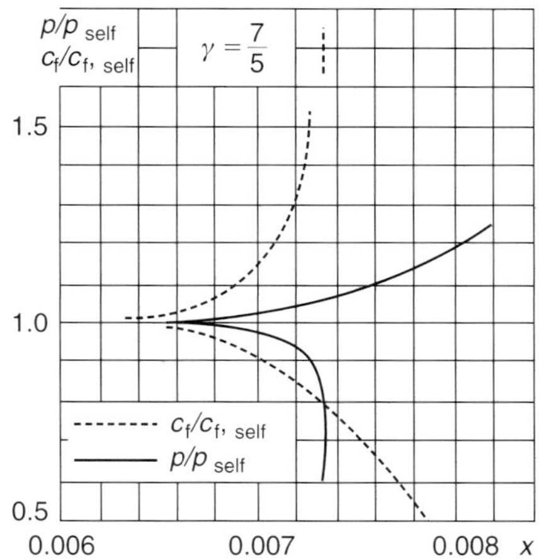


Fig. 4.1.

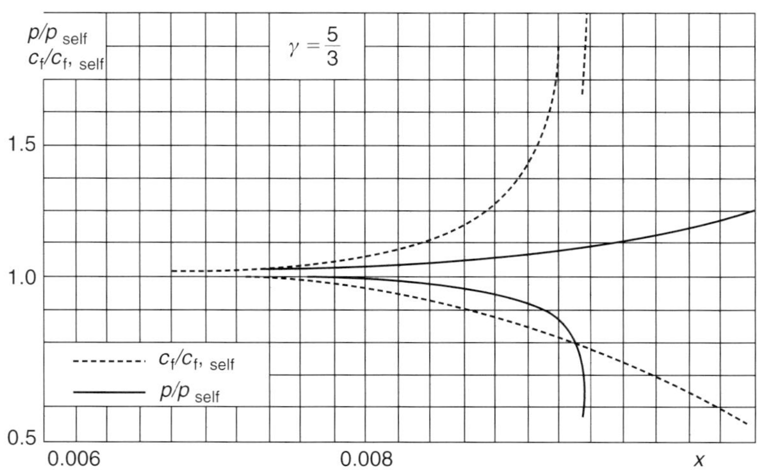


Fig. 4.2.

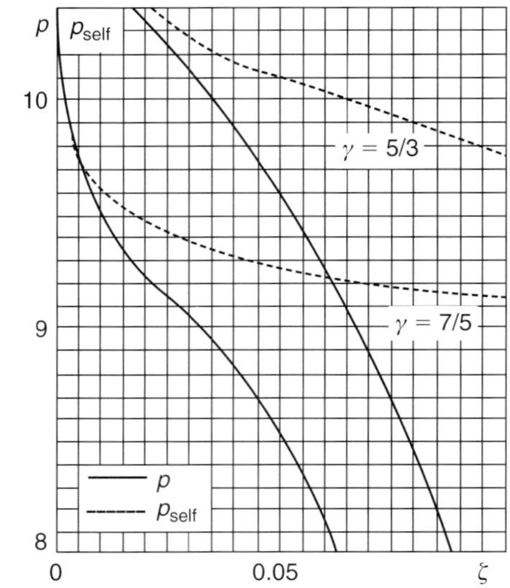


Fig. 4.3.

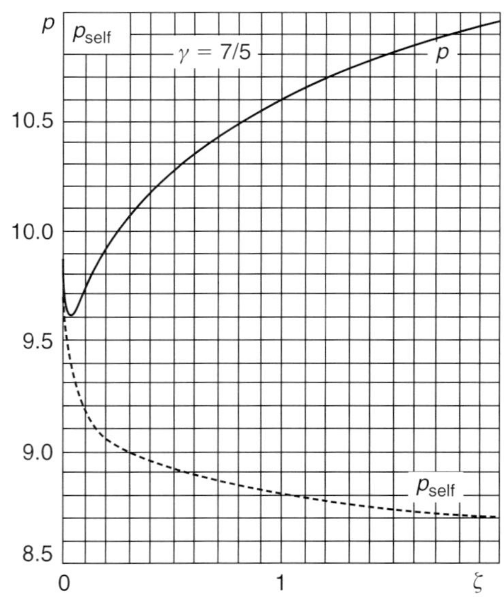


Fig. 4.4.

presented above with the use of the group of transformations (4.42). For this purpose, it is necessary that $\overline{x}^* = 1$ in the transformed variables; hence $\lambda = \sqrt{x^*}$. The results plotted versus the physical variable $x$ (Figs. 4.1 and 4.2) allow us to make certain practically interesting conclusions. Firstly, a comparison of the results obtained for $\gamma = 7/5$ and $5/3$ shows that the length of the region, where the pressure and friction distributions are appreciably different

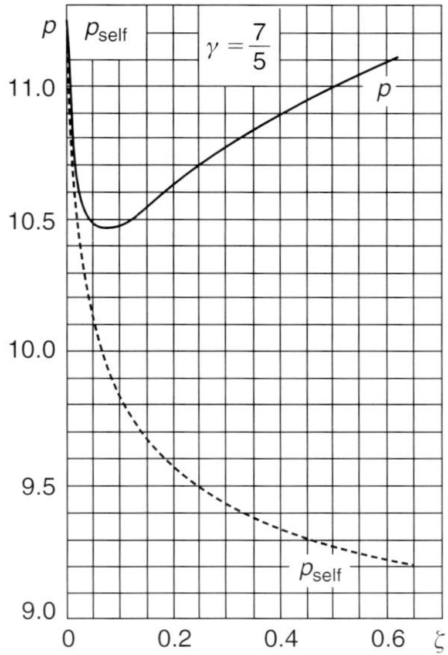


Fig. 4.5.

from those in the self-similar solution, increases almost twofold with $\gamma$. This can easily be understood by bringing Eq. (4.48) for $\omega = \sigma = g_w = 1$ into a simpler form. In this case, an "inhomogeneous" term proportional to $(\gamma - 1)/\gamma$ appears in the equation for $f_1$.

For compression flows ($P_1 > 0$) the disturbance propagation region turns out to be more extended (though the pieces of the integral curves presented in the figures are not plotted to the end). Because of this, a considerable effect on the total aerodynamic characteristics may be expected.

The group of transformations (4.42) represents a peculiar kind of a similarity law for plate flows with moderate and strong interactions. (It should be noted in passing that the presence of the effect of upstream disturbance propagation for $\chi \geq 1$ indicates that the classical similarity law for viscous hypersonic flows first formulated in the papers of Hayes and Probstein (1959) and Lunev (1959) is not complete for these flow regimes). Formulas (4.42) make it possible to obtain, by eliminating the free parameter $\lambda$, variables in which the dependences become universal. In Fig. 4.6 we have presented the experimental data obtained in a helium wind tunnel by Gorislavskii and Stepchenkova (1971). Figure 4.6a presents the pressure distributions over a plate for different deflection angles $\alpha$ of a flap mounted on the plate. The plate is thermally insulated, $Re_0 = 1.9 \times 10^4$, and $M = 23.3$. Figure 4.6b presents the same data in the similarity variables; the correlation is quite satisfactory. For large flap deflection angles a boundary layer separation zone arises ahead of the flap.

We will now consider the flow patterns in which the boundary layer separation point is at finite distances from the plate edges (for $Re \to \infty$ and $M \to \infty$). Since in the separation zones of this type the pressure and the longitudinal-to-transverse characteristic dimension ratio is determined by the flow in the vicinity of the separation point, in these zones $p \sim \tau^2$ and $y/x \sim \tau$. However, for these scales of the flow functions ($x \sim 1$, $u \sim 1$, and $\chi \geq O(1)$)

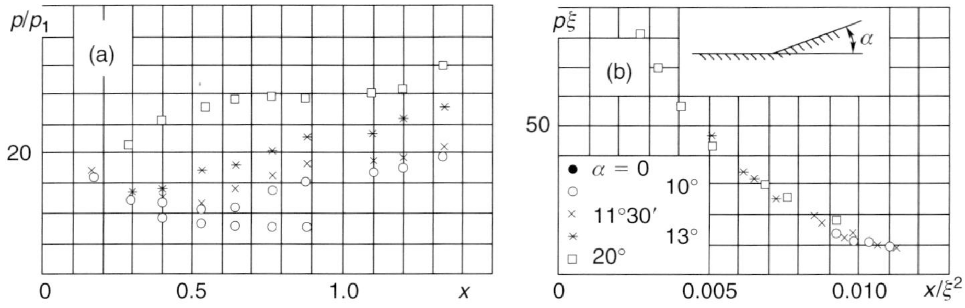


Fig. 4.6.

the flow in the first approximation is described by the boundary layer equations (though in the region containing $u < 0$ the formulation of the boundary value problem and the nature of disturbance transfer call for additional studies). However, the demonstration of the fact that this flow cannot contain regions with the pressure gradient on the body greater in the order than $O(\tau^2)$ remains valid. Hence it follows that for flap deflection angles greater in the order than $O(\tau^2)$ the separation point must be displaced upstream, into a small vicinity of the leading edge.

In conclusion, we will consider the distinctive features of the flow near the base section of a flat plate which can arise at fairly low base pressures $p_b$. In fact, for the family of the solutions of the flat plate flow problem obtained above, through each point of the plane of variables $(p, x)$ there passes either one or none integral curves. The latter case is realized if this point is located below the curve passing through the ends of the integral curves describing expansions flows, where these solutions have the singular points $f_w'' = \infty$ and $\mathrm{d}p/\mathrm{d}x = -\infty$ (for $\gamma = 1.4$ this curve is of the form $p^* = 0.636\, Re_{0x}^{-1/2}$).

Let $p_b$ be preassigned at the body end, that is, for $x = 1$. If an integral curve passes through the point $(x = 1, p = p_b)$ of the solution plane and the point is nonsingular $(p > p^*)$, then precisely this curve describes the solution on the entire body. At its end $p = p_b$, since everywhere on it $f_w'' = O(1)$, that is, $\mathrm{d}\delta/\mathrm{d}p > 0$, and, as shown above, in a small vicinity $\Delta x \ll 1$ of the base section a region with the pressure variation $\Delta p \sim O(1)$ cannot arise. If $p_b < p^*$, then further decrease of the pressure on the main part of the body for $x \approx 1$ cannot occur, since the solution cannot be continued through the singular point $p = p^*$ at which $\mathrm{d}\delta/\mathrm{d}p = 0$. From the outer boundary condition it can also be seen that at this point the pressure gradient increases without bound:

$$\frac{\mathrm{d}p}{\mathrm{d}x} = \frac{\sqrt{\dfrac{2}{(\gamma+1)p}}}{\dfrac{\mathrm{d}\delta}{\mathrm{d}p}}$$

This means that further flow expansion takes place in a short region with high local pressure gradients. In fact, as shown above, a region with $\Delta x \sim \Delta y \sim \delta$, where $\Delta p \sim p \sim \tau^2$ and $\partial p/\partial x \sim \partial p/\partial y$, appears if $\mathrm{d}\delta/\mathrm{d}p \le 0$ at high local pressure gradients, when viscous stresses in the main part of the boundary layer streamtubes become small as compared with the increasing gradients $\partial p/\partial x$. The general theory of these flows was considered in detail in Chapter 3 of this book. For a while, we only note that the flow in this region is governed by the complete Euler equations, since on the basis of the above-mentioned properties the following estimates are valid there:

$$\Delta u \sim u \sim 1, \quad v \sim u \sim 1, \quad \Delta p \sim p \sim \tau^2, \quad \Delta x \sim \Delta y \sim \delta \sim \tau$$

The boundary conditions are as follows: $v_w = 0$ and $v_e = 0$. The former condition is due to a narrowness of the region in which viscosity forces are essential over these lengths and the latter is caused by the fact that the variation of the outer boundary slope is of the order $\tau$, thus being out of order, since inside this region $v \sim 1$. Finally, for $(x/\delta) \to -\infty$ the boundary layer parameter profiles at the singular point $p = p^*$ for $x = 1$ should be taken as the initial conditions for the flow through this narrow channel.

Flow expansion in the region $x \sim \delta$ can continue only until the value $M_w = 1$ is reached at the edge of the base section. In fact, until this flow pattern is attained, supersonic streamtubes can expand only at the cost of the narrowing of the subsonic streamtubes. However, when the subsonic part vanishes, further acceleration can be realized only at the flow turn into the base region.

Thus, there exist three flow patterns. The first pattern is that with $p_b > p^*$. In this case, the variation of the base pressure changes in principle the whole flow up to the leading edge. For $[2/(\gamma+1)]^{\gamma/(\gamma-1)} p^* < p_b < p^*$ on the main part of the body the flow is described by the integral curve $p(x=1) = p^*$, while the variation of $p_b$ changes the flow only over a length $\Delta x \sim \delta$; in this case $M_w < 1$ at the body edge. Finally, for $p_b \leq (2/(\gamma+1))^{\gamma/(\gamma-1)} p^*$ we obtain $M_w = 1$ above the base section and further variation has no effect on the upstream flow, since disturbances are "choked."

## 4.3 Theory of hypersonic flow/boundary layer interaction for two-dimensional separated flows

In this section we will consider the formulation of the boundary value problem for two-dimensional separated flows in the regimes of strong or moderate interaction with the outer hypersonic flow which are almost everywhere governed by the boundary layer equations.

### *4.3.1 Formulation of the problem, equations, and boundary conditions*

Let us consider the flow over a flat plate. Let at a distance $\ell$ from the leading edge there be an obstacle leading to boundary layer separation. We will consider the flows in which the separation point is located at a finite distance from both the plate nose and the obstacle.

In accordance with the conventional estimates for the boundary layer in a hypersonic flow (Hayes and Probstein, 1966) we introduce the designations $\ell x$ and $y\tau\ell$ for the coordinates measured along the plate and normal to it, where $\tau = Re_0^{1/4}$, the Reynolds number $Re_0 = \rho_\infty u_\infty \ell/\mu_0$, $\rho_\infty$ and $u_\infty$ are the freestream density and velocity, and the viscosity

$\mu_0$ is calculated at the stagnation temperature of the freestream. We denote the velocity components, density, stagnation enthalpy, and viscosity in the boundary layer by $u_\infty u$, $\tau u_\infty v$, $\tau^2 \rho_\infty \rho$, $\tau^2 u_\infty^2 p$, $(u_\infty^2/2)g$, and $\mu_0 \mu$. Substituting these variables in the Navier–Stokes equations and passing to the limit

$$M_\infty \to \infty, \quad Re_0 \to \infty, \quad \chi = M_\infty^2 \tau^2 \geq O(1) \tag{4.52}$$

leads to the boundary layer equations

$$\rho\left(u\frac{\partial u}{\partial x} + v\frac{\partial u}{\partial y}\right) = -\frac{dp}{dx} + \frac{\partial}{\partial y}\left(\mu\frac{\partial u}{\partial y}\right)$$

$$\frac{\partial \rho u}{\partial x} + \frac{\partial \rho v}{\partial y} = 0, \quad p = \frac{\gamma - 1}{2\gamma}\rho(g - u^2) \tag{4.53}$$

$$\rho\left(u\frac{\partial g}{\partial x} + v\frac{\partial g}{\partial y}\right) = \frac{\partial}{\partial y}\left(\frac{\mu}{\sigma}\frac{\partial g}{\partial y}\right) + \frac{\partial}{\partial y}\left[\mu\left(1 - \frac{1}{\sigma}\right)\frac{\partial u^2}{\partial y}\right]$$

where $\sigma$ is the Prandtl number. The boundary conditions are as follows:

$$u(x,0) = v(x,0) = 0, \quad g(x,0) = g_w(x) \tag{4.54}$$

$$u(x,\delta) = 1, \quad v(x,\delta) = \frac{d\delta}{dx}, \quad g(x,\delta) = 1$$

In the hypersonic flow the outer edge $y$ of the boundary layer is determined exactly, since $\rho(x,\delta) = \infty$, while in the shock layer the gas density is greater in the order by a factor of $\tau^2$ than that in the boundary layer. The pressure distribution is not preassigned and must be determined in the process of the solution of the boundary value problems (4.53) and (4.54), together with the equations for the outer flow. Here, for the sake of simplicity the approximate formula of the tangent wedge method is used

$$p = \frac{1}{\chi\gamma} + \frac{\gamma+1}{4}\left(\frac{d\delta}{dx}\right)^2 + \frac{d\delta}{dx}\left\{\frac{1}{\chi} + \left[\frac{\gamma+1}{4}\frac{d\delta}{dx}\right]^2\right\}^{1/2} \tag{4.55}$$

In the paper of Neiland (1970*b*) the main results for separationless flows were rigorously obtained on the basis of the hypersonic small perturbation theory.

For studying the separated flow shown in Fig. 4.7 certain results of the work of Neiland (1970*b*) are important. Firstly, near the leading edge of the body and up to the separation point there exists a single-parameter family of the solutions of problems (4.53)–(4.55). The choice of a unique solution is determined by the separation point position which depends on the shape and dimensions of the obstacle. Secondly, throughout the entire separation region of length $\Delta x \sim O(1)$, except for a small vicinity of the obstacle, the flow is also described by Eqs. (4.53).

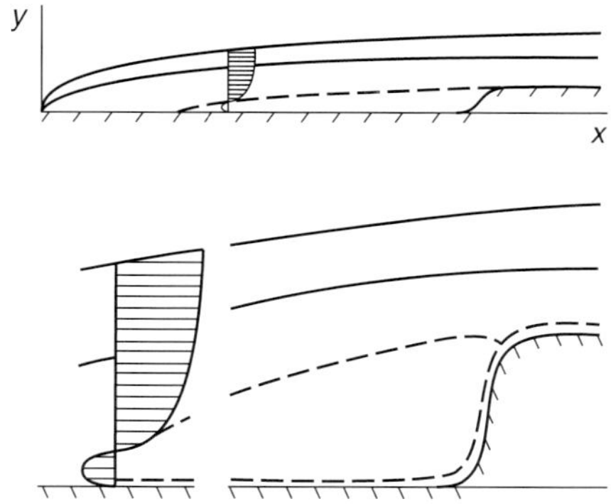


Fig. 4.7.

This can easily be understood from physical considerations. In fact, the boundary layer equations are derived from the Navier–Stokes equations by passing to limit (4.52) under certain assumptions on the scales of the longitudinal velocity, pressure, and streamtube slope. However, these assumptions must be fulfilled also behind the separation point. If the pressure in that region is greater in the order than $\tau^2$, the boundary layer streamlines cannot penetrate in the region behind the separation point, since on these lines $M = O(1)$ and then the total pressure is of the order $\tau^2$. The assumption on the pressure order smaller than $\tau^2$ and the order of the viscous region thickness smaller than $\tau$ does not permit us to conserve the gas flow rate order $\tau^3$.

Thus, the solutions of the boundary layer equations must be continued through the separation point up to a small ($\Delta x \ll 1$) vicinity of the obstacle. The integration of the boundary layer equations through the separation point in a supersonic flow was carried out for free interaction regions in the works of Stewartson and Williams (1969) and Neiland (1971*a*). However, in the work of Stewartson and Williams (1969) the integration was not completed, while in the paper of Neiland (1971*a*) it is terminated by a semi-infinite separation zone.

Let us consider the formulation of the boundary value problem between the separation point and the obstacle. For the sake of definiteness, we will assume that the obstacle is step-shaped, of height $O(\tau)$, and has rounded edges (Fig. 4.7). Let us draw a line $y_0(x) > 0$ on which $u_0(x) = 0$; above this line $u > 0$. For this part of the flow Eqs. (1.2) must be integrated in the positive direction along the $x$ axis and below it, in the $u \leq 0$ region, in the negative direction of the axis. If the separation point position is preassigned (for a while, it is arbitrary), that is, the free parameter in the series expansions of the boundary value problems (4.53)–(4.55) as $x \to 0$ is specified, then behind the separation point $x = x_0$ in the $y \geq y_0(x)$ region the boundary conditions (4.54) at $y = \delta$ remain unaltered. Moreover, it is necessary to specify three arbitrary functions $y_0(x)$, $v_a(x)$, and $g_0(x)$. Then Eqs. (4.53) can be integrated up to the point at which the obstacle is placed. In order to carry out the integration in the region $x_0 < x < x_1$, $0 \leq y < y_0(x)$, where $u \leq 0$, the profiles $u(x_1, y)$ and $g(x_1, y)$ must be known. These profiles should be obtained by considering the flow in a small vicinity of the obstacle, where, owing to the large slopes of the body contour, the Prandtl equations are inapplicable. For a while, let us assume that the above-mentioned profiles are known. Then

Eqs. (4.53) are integrated from $x_1$ to $x$ (since $u < 0$) subject to the boundary conditions

$$u[x, y_0(x)] = 0, \quad g[x, y_0(x)] = g_0(x), \quad v[x, y_0(x)] = v_0(x) \tag{4.56}$$
$$u(x, 0) = 0, \quad g(x, 0) = g_w$$

If the arbitrary functions $y_0(x)$, $v_a(x)$, and $g_0(x)$ can be so chosen that the following conditions

$$v(x, 0) = 0, \quad \frac{\partial u}{\partial y}[x, y_0^-(x)] = \frac{\partial u}{\partial y}[x, y_0^+(x)], \quad \frac{\partial g}{\partial y}[x, y_0^-] = \frac{\partial g}{\partial y}[x, y_0^+] \tag{4.57}$$

are satisfied, then we obtain a solution satisfying the boundary conditions (4.54) and Eqs. (4.53) throughout the entire boundary layer. In fact, in view of Eq. (4.53), the continuity of a function and its first derivatives on the curve $y_0(x)$ ensures also the continuity of the second derivatives of the functions $u(x, y)$, $g(x, y)$, etc., across $y_0(x)$. Of course, the above reasoning is rather heuristic than strict.

Now we must return to the question of the nature of the flow near the obstacle, the formulation of the initial conditions for the region $0 \le y \le y_0(x)$, and the choice of the separation point position, that is, the choice of a unique solution on the forward part of the body, ahead of the separation point.

If a solution of the boundary layer equations does not contain sections in which $u(x^\circ, y) = 0$ (except for the case $x^\circ \to \infty$), then ahead of the obstacle the velocity $u \sim O(1)$. The transverse dimension of the local region must be $O(\tau)$ both in the boundary layer and near the obstacle. In accordance with the boundary conditions on the obstacle surface, in this local region the streamline slope must be of the order $O(1)$. Since $u \sim O(1)$, this means that $v \sim O(1)$. Then from the continuity equation it follows that the longitudinal dimension of the turn region is of the order $O(\tau)$. Since in the boundary layer the gas density is $O(\tau^2)$, the pressure is also $O(\tau^2)$. Using the estimates obtained we introduce the local coordinates and asymptotic representations for a vicinity of the turn region

$$x - 1 = \tau X, \quad y = Y$$

$$u(x, y, \tau) \approx U(X, Y) + \cdots, \quad v(x, y, \tau) \approx \tau^{-1} V(X, Y) + \cdots \tag{4.58}$$

$$\rho(x, y, \tau) \approx R(X, Y) + \cdots, \quad p(x, y, \tau) \approx P(X, Y) + \cdots$$

$$g(x, y, \tau) \approx G(X, Y) + \cdots, \quad \delta(x, \tau) \approx \Delta \cdots$$

Substituting Eqs. (4.58) in the Navier–Stokes equations and passing to the limit (4.52) leads to the system of Euler equations

$$R\left(U\frac{\partial U}{\partial X} + V\frac{\partial U}{\partial Y}\right) = -\frac{\partial P}{\partial X}, \quad R\left(U\frac{\partial V}{\partial X} + V\frac{\partial V}{\partial Y}\right) = -\frac{\partial P}{\partial Y} \tag{4.59}$$

$$RU = \frac{\partial \psi}{\partial Y}, \quad RV = -\frac{\partial \psi}{\partial X}, \quad G = G(\psi) = \frac{\gamma}{\gamma - 1}\frac{P}{R} + \frac{U^2 + V^2}{2}$$

The limiting flow pattern near the obstacle is shown in Fig. 4.7 on the $X, Y$ scale. For the equations of motion for the inviscid fluid the impermeability conditions must be fulfilled on the body surface. On the outer streamline the condition

$$V(X, Y = \Delta) = 0 \tag{4.60}$$

must be fulfilled. In fact, the outer boundary slope is $O(\tau)$, since the pressure is $O(\tau^2)$. Therefore, for the transverse velocity component of the order $O(1)$ we obtain condition (4.60). Other conditions are obtained by matching the solutions corresponding to the boundary layer and the locally inviscid region. In accordance with the asymptotic expansion matching principle (Van Dyke, 1964), we obtain

$$u(x_1, y) = U(-\infty, Y), \quad g(x_1, y) = G(-\infty, Y) \tag{4.61}$$

$$p(x_1) = P(-\infty, Y), \quad \delta(x_1) = \Delta$$

The region thickness $\Delta$ does not change in the leading order, since its variation at distances $X \sim 1$ would mean that the outer boundary slope is $O(1)$, which would lead to the pressure $\Delta p \sim 1$, that is, much greater than $\tau^2$.

We will now return to the question of the initial conditions necessary for integrating the boundary layer equations in the return flow region. From Eqs. (4.59) and (4.61) it follows that the profiles $u(x_1, y)$, $g(x_1, y)$, and $\rho(x_1, y)$ for the streamlines lying below the dividing streamline passing through the separation point, obtained by integration of the upper part of the flow ($u > 0$), must be taken as the profiles for the return flow region if the losses in shocks can be neglected. In the locally inviscid region we have $G = G(\psi)$ and the entropy is conserved along streamlines, while in the matching region the pressures are the same. Therefore, after the turn the flow functions in all streamtubes take the values they had before the turn. If for a fairly high step streamtubes with fairly large supersonic velocities turn backward, then taking account for the variation of the entropy across the shocks is also required. In these cases, both problems must be solved consistently. Finally, in order to satisfy the no-slip boundary condition in the region $x - 1 \sim O(\tau)$ we can, following the general method developed in the study of Neiland and Sychev (1966), introduce a viscous sublayer near the body surface. It begins at the stagnation point of the locally inviscid flow (Fig. 4.7, bottom). Since its upper boundary coincides with the dividing streamline, the longitudinal velocity in the sublayer is of the order $O(1)$ and the longitudinal coordinate is $O(\tau)$. The condition of the equality of the leading viscous and inertial terms of the momentum equation makes it possible to determine the order of the viscous sublayer thickness; it is equal to $\tau^{3/2}$. The formulation of the boundary conditions and the form of the equations for the sublayer are conventional.

Conditions (4.61) close the solution of the problem for the boundary layer at a given separation point position. The position itself should be determined from the boundary conditions imposed downstream. We will present an example of these conditions. Let behind the obstacle, at a certain distance from it, there be a base section and the base pressure be low. Then the solution for the separationless boundary layer behind the reattachment region must contain a singular point, at which $dp/dx \to \infty$ and $p \to O(1)$. This type of the boundary condition which confines the region of upstream disturbance propagation, is considered in the paper of Neiland (1972). If the slope of the separation-producing obstacle is everywhere of the order $\tau$, then there is no need to consider the locally inviscid region. In this case, the problem represents a particular case of the problem considered above.

### *4.3.2 Similarity criteria*

The similarity of two-dimensional separationless flows in non-weak interaction of the boundary layer with a hypersonic flow with account for upstream disturbance propagation was first considered in the paper of Neiland (1970*b*). In the study of Kozlova and Mikhailov (1971) the similarity law was proposed in the form in which the disturbance transfer was taken into account by introducing an additional parameter equal to the dimensionless pressure at a characteristic point on the body surface. However, it is useful to introduce similarity parameters independent of the solution of the problem and valid for separated flows.

Let us consider the schematics of the introduction of such similarity criteria with reference to some particular flows with account for the results of the preceding section and the note in the paper of Neiland (1970*b*), in accordance with which a major part of the flow region including the separation zone is described by the boundary layer equations. Obviously, in this case the boundary conditions must be imposed in the entire region, where disturbances propagate upstream. For an arbitrary thin airfoil at a small angle of attack $\alpha \sim \tau$ the disturbance transfer region is closed at a certain point of the wake, where the disturbances are blocked owing to gas acceleration. The conditions in the blocking section were obtained in the paper of Neiland (1974*b*) for the general case of a three-dimensional flow.

Another possible case of the closure of the domain of influence corresponds to a sharp transition to the supercritical flow regime at the trailing edge of a wing via an expansion flow analogous to that considered in the work of Neiland (1972). However, from the fact itself of the existence of the blocking section it follows that the coincidence of the body shapes in the $x, y$ coordinates is necessary for the similarity (we recall that in introducing the dimensionless variables we used the body length $\ell$, together with the parameter $\tau = Re_0^{-1/4}$ for the coordinate $y$). Moreover, the parameters $\alpha/\tau$, $g_w$, $\gamma$, and $M_\infty \tau$ must coincide. In the flow past steps, both upstream- and downstream-facing, the parameter $\alpha/\tau$ is replaced by $h/\tau$ if the step is followed by an infinite horizontal plane or a base section. If the length of the horizontal plane region behind the step is finite, then the parameter $L/\ell$ is added to the similarity criteria, where $L$ is the rear region length. In the flow past a concave corner with an infinite backward side the parameter $\theta/\tau$ should be used instead of $\alpha/\tau$, where $\theta$ is the value of the angle. If the backward wall terminates and is followed by a base section with a very low base pressure, then the parameter $L/\ell$ should additionally be introduced, where $L$ is the length of the backward side of the corner.

Thus, in each case the form of the similarity parameters depends on the nature of the closure of the region, where disturbances can be transferred upstream.

## 4.4 Propagation of disturbances at strong distributed gas injection through the body surface to a supersonic flow

Below we will briefly consider one more example of the flow in which the interaction of the boundary layer with the outer supersonic flow leads to upstream disturbance propagation.

### *4.4.1 Formulation of the problem and derivation of the equations*

For the sake of simplicity we will consider the flow presented in Fig. 4.8. The upper side of the wedge in a supersonic flow is isolated from the lower side and the surface of the former is aligned with the freestream velocity. We introduce a Cartesian coordinate system fitted to the wedge surface. All flow functions are normalized on their freestream values and linear dimensions are scaled on the upper generator length $\ell$.

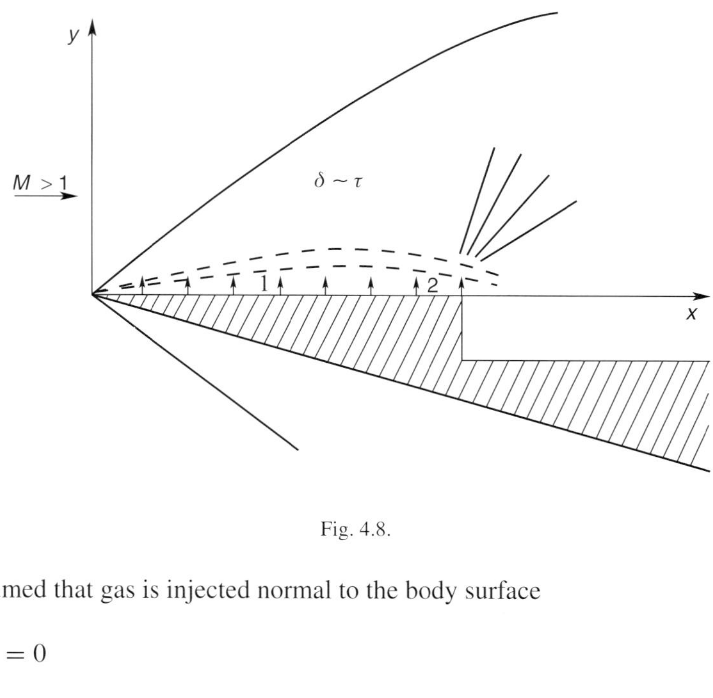


Fig. 4.8.

It is assumed that gas is injected normal to the body surface

$$u(x, 0) = 0 \tag{4.62}$$

The tangential momentum can appear in the injected gas only due to the action of a longitudinal pressure gradient or viscosity forces. Let the normal component of the injection velocity $v(x, 0)$ be such that the injected gas layer displacement thickness is of the order $\tau$. For $\tau \ll 1$ a pressure gradient is induced; in accordance with the linear theory of supersonic flows, it is also of the order $\tau \ll 1$.

The longitudinal velocity component $u$ cannot be greater than $O(1)$, as in the viscous boundary layer, and smaller than $O(\tau^{1/2})$, even if it is produced only by the induced pressure distribution. The second estimate follows from the continuity equation. Therefore, the ratio of the leading viscous and inertial terms of the Navier–Stokes equations cannot be greater than $(Re\,\tau^3)^{-1}$. For $Re \gg \tau^{-3}$ the viscosity forces are small and the injected gas velocity is determined by the longitudinal pressure gradient. Then from the longitudinal momentum and continuity equations there follow the estimates

$$u \sim \tau^{1/2}, \quad v \sim \tau^{3/2} \tag{4.63}$$

Precisely, this case will be considered below. Apparently, this formulation of the problem was first proposed in the paper of Cole and Aroesty (1968), in which solutions of certain inverse problems were presented (the injection distribution law was determined for a given pressure distribution).

From the above estimates it follows that for the main part of the flow region occupied by the injected gas (outside a narrow mixing zone of thickness $\sim Re^{-1/2}$ located between the injected gas and the outer shock layer) the following asymptotic representations for the coordinates and the flow functions must be introduced:

$$x = x_1, \quad y = \tau y_1, \quad \psi = \psi_0 \psi_1(x_1, y_1), \quad \psi_0 = \tau^{3/2}$$

$$u(x, y; \tau, Re) \sim \tau^{1/2} u_1(x_1, y_1) + \cdots, \quad v(x, y; \tau, Re) \sim \tau^{3/2} v_1(x_1, y_1) + \cdots \tag{4.64}$$

$$p(x, y; \tau, Re) \sim \frac{1}{\gamma M^2} + \tau p_1(x_1, y_1) + \cdots, \quad \rho(x, y; \tau, Re) \sim \rho_1 + \cdots$$

In the first approximation the density is constant, since the pressure difference is small, the wall is assumed to be isothermal, the velocities are low, and the injected gas temperature is equal to the surface temperature. The subscript 1 shows that the variables introduced are of the order $O(1)$ in region 1 (Fig. 4.8).

Substituting Eq. (4.64) in the system of Navier–Stokes equations and passing to the limit $\tau \to 0$, $Re \to \infty$, and $\tau^{5/2} Re \to \infty$ in the first approximation we obtain

$$u_1 \frac{\partial u_1}{\partial x_1} + v_1 \frac{\partial u_1}{\partial y_1} + \frac{1}{\rho_1} \frac{\partial p_1}{\partial x_1} = 0, \quad \frac{\partial p_1}{\partial y_1} = 0, \quad \frac{\partial u_1}{\partial x_1} + \frac{\partial v_1}{\partial y_1} = 0$$

or

$$u_1 = \frac{\partial \psi_1}{\partial y_1}, \quad v_1 = -\frac{\partial \psi_1}{\partial x_1} \tag{4.65}$$

By virtue of the assumptions made above, the boundary conditions takethe form:

$$u_1(x_1, 0) = 0, \quad \psi_1(x_1, \delta_1) = 0, \quad \psi_1(x_1, 0) = \psi_w(x_1) \tag{4.66}$$

where $\delta$ is the injected gas region thickness and $\psi_w(x_1)$ is a given function determined by the injection distribution over the body surface.

The pressure distribution is not preassigned. Instead, a condition following from the matching of the solutions in region 1 and the outer flow must be derived. The outer supersonic flow streams over a thin body formed by the thickness of region 1. Small disturbances are described by the wave equation. In accordance with the linear theory of supersonic flows we obtain

$$(M^2 - 1)^{1/2} p_1 = \mathrm{d}\delta_1/\mathrm{d}x_1 \tag{4.67}$$

We will show that the solution of the system of equations of the "inviscid boundary layer" (4.65) subject to the boundary conditions (4.66) and (4.67) is not uniquely defined. Like in the problem of the strong interaction of the hypersonic flow with the boundary layer considered in Section 4.2, here one more additional condition should be imposed at the end of region 1. We will restrict ourselves to power-law injection distributions

$$\psi_w(x_1) = -x_1^m, \quad 0 < m < 1 \tag{4.68}$$

### 4.4.2 *Analysis of the solutions for region 1*

We will first note that problems (4.65)–(4.67) are invariant relative to the following group of transformations:

$$\begin{aligned} &x_1 \to \lambda\widetilde{x}_1, \quad y_1 \to \widetilde{y}_1\lambda^{(2m+1)/3}, \quad u_1 \to \widetilde{u}_1\lambda^{(m-1)/3} \\ &v_1 \to \widetilde{v}\lambda^{m-1}, \quad \psi_1 \to \widetilde{\psi}_1\lambda^m \\ &\rho_1 \to \widetilde{\rho}_1, \quad p_1 \to \widetilde{p}_1\lambda^{2(m-1)/3} \end{aligned} \tag{4.69}$$

This group of transformations is associated with a self-similar solution of the problem which is valid for the semi-infinite body. In what follows, it is shown that, apart from the self-similar solution, there exists a single-parameter family of non-self-similar solutions.

Let us introduce the variables

$$\eta = \frac{y_1}{\delta_1}, \quad f(x_1, \eta) = \frac{\psi_1(x_1, y_1)}{\psi_w(x_1)}, \quad \delta_1 = [(M^2 - 1)^{1/2}\rho]^{1/3}\delta \tag{4.70}$$

Then formulas (4.65) and (4.66) take the form:

$$\frac{\psi_w}{\delta^2}\frac{\mathrm{d}\psi_w}{\mathrm{d}x_1} ff'' - \frac{\psi_w}{\delta}\frac{\mathrm{d}}{\mathrm{d}x_1}\left(\frac{\psi_w}{\delta}\right) f'^2 - \frac{\mathrm{d}^2\delta}{\mathrm{d}x_1^2} = \frac{\psi_w^2}{\delta^2}(f'\dot{f}' - \dot{f}f'') \tag{4.71}$$

$$f(x_1, 0) = 1, \quad f'(x_1, 0) = 0, \quad f(x_1, 1) = 0 \tag{4.72}$$

Here, primes refer to the differentiation with respect to $\eta$ and dots to $x_1$.

Equation (4.71) is of the second order; for it three boundary conditions (4.72) are formulated. One of these conditions must be used for determining $\delta(x)$ with $\delta(0) = 0$.

We will seek the solution of problems (4.71) and (4.72) in the form:

$$f(x_1, \eta) \sim f_0(\eta) + x^a f_1(\eta) + \cdots, \quad \delta(x_1) \sim Cx_1^k + C_1 x_1^{k_1} + \cdots \tag{4.73}$$

where $a > 0$ and $k_1 > k > 0$.

Substituting Eq. (4.73) in Eq. (4.71) and boundary conditions (4.72) yields

$$f_0 f'' - a f_0'^2 + K^2 = 0, \quad f_0(0) = 1, \quad f_0'(0) = f_0(1) = 0$$

$$\eta = \frac{1}{K} \int_{f_0}^{1} \left( \frac{a}{1 - \xi^{2a}} \right)^{1/2} d\xi, \quad K = \int_0^1 \left( \frac{a}{1 - \xi^{2a}} \right)^{1/2} d\xi \tag{4.74}$$

$$k = \frac{2m+1}{3}, \quad \alpha = \frac{m-1}{3m}, \quad C = \left[ \frac{9mK^2}{2(2m+1)(1-m)} \right]^{1/3}$$

Formulas (4.74) determine uniquely the self-similar solution of the problem corresponding to the group of transformations (4.69) and the flow past a semi-infinite body, for which next terms of expansions (4.73) must be assumed to be identically equal to zero. However, generally for the second term of the expansion we obtain

$$f_0 \Phi_1'' + \Phi_1 f_0'' - 2 f_0 f_0'' - D f_0'^2 - 2a \Phi_1' f_0' + E = \frac{a}{m} (f_0' \Phi_1' - f_0' \Phi_1) \tag{4.75}$$

$$f_1 = \frac{C_1}{C} \Phi_1, \quad \Phi_1(0) = \Phi_1' = \Phi_1(1) = 0,$$

$$k_1 = k + a, \quad D = \frac{\alpha[a + 2(m-k)]}{m-k}, \quad E = \frac{K^2(2m+3a+1)(2m+3a-2)}{2(2m+1)(m-1)}$$

Equation (4.75) contains one unknown parameter; it is of the second order but is subject to three boundary conditions. In solving numerically Eq. (4.75) a value of $a$ belonging to the $+\infty > a > 0$ domain was determined, such that it provides a nontrivial solution for $\Phi_1$. The calculated results are presented in Figs. 4.9–4.11. The values of $a$ are small, which indicates a considerable effect of the upstream disturbance propagation over the entire body surface. The constant $C_1$ is indefinite. The existence of the nontrivial solution for $\Phi_1$ indicates that for the original problem formulated in the functions $f(x_1, \eta)$ and $\delta$, the self-similar solution is not a unique solution of the problem. After the integration of Eq. (4.75) the solution near the nose is determined correct to the arbitrary constant $C_1$. We note that the next terms of the expansion of solution (4.71) near the nose are uniquely determined for $a$ obtained and $C_1$ given.

As in Section 4.2, here the first two terms of expansion (4.73) of the solution of problems (4.71) and (4.72) can be used for beginning the numerical integration of Eq. (4.71) in the downstream region at a certain value of $C_1$. Other non-self-similar solutions corresponding to other values of $C_1$ could be determined using the group of transformations (4.69).

The choice of the needed non-self-similar solution can be carried out by choosing the position of the base section and matching with the solution in a local vicinity of the base section. The latter solution is studied below. We also note that in this problem permissible solutions

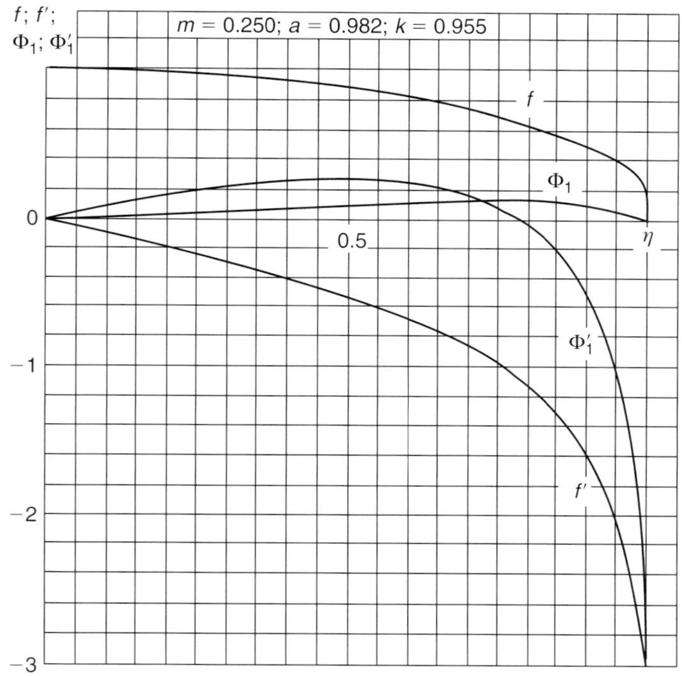


Fig. 4.9.

are only those in which the pressure decreases. This becomes obvious if we recall that in region 1 the Bernoulli equation $p_1 + (\rho_1 u_1^2/2) = P(\psi_1)$ must be fulfilled, together with the equation $\partial p_1/\partial y_1 = 0$ and the boundary condition (4.62). A streamtube flowing out from the body cannot move rightward if the pressure is non-decreasing.

### *4.4.3 Flow near the base section*

If the base pressure behind a step is smaller than the pressure in region 1 by $O(\tau)$, then preassigning the value $p_1(x_1 = 1)$ makes it possible to separate out the required branch of the non-self-similar solution in region 1 by determining $C_1$.

However, generally the base pressure can be smaller than the pressure in region 1 by $O(1)$, that is, correspond to the value $p_1(x_1 \to 1) \to -\infty$. We recall that in region 1 the pressure difference is $O(\tau)$, where $\tau$ is a small parameter. Let us consider the flow in region 2 (Fig. 4.8) lying near the end of the plate. Near the base section, the outer supersonic flow must be deflected by $O(1)$. However, in this case in region 2 $x \sim y$ and, in accordance with the continuity equation, $u \sim v$. Since in this region $\Delta p \sim O(1)$, we have $u \sim O(1)$. In region 2 the gas flow rate is the same as in region 1, that is, $O(\tau^{3/2})$; then the region 2 thickness $y \sim O(\tau^{3/2})$.

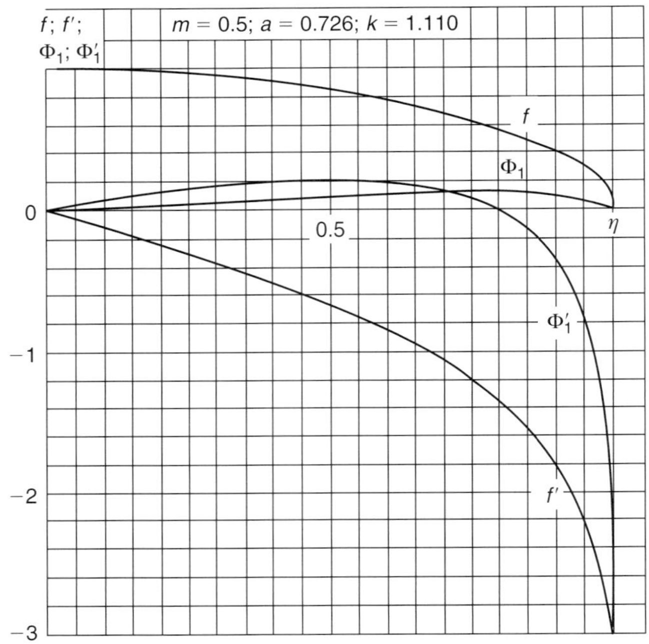


Fig. 4.10.

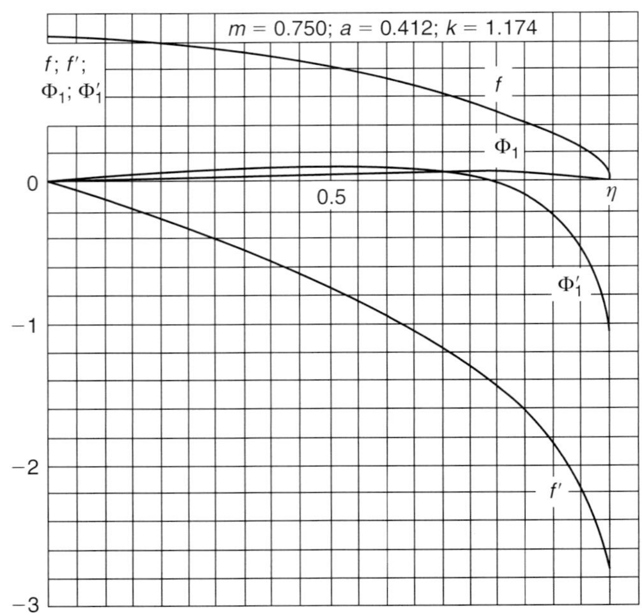


Fig. 4.11.

The estimates obtained make it possible to introduce the following asymptotic representations:

$$x - 1 = \tau^{3/2} x_2, \quad y = \tau^{3/2} y_2, \quad u \sim u_2(x_2, y_2) + \cdots, \tag{4.76}$$

$$v \sim v_2(x_2, y_2) + \cdots, \quad p \sim p_2(x_2, y_2) + \cdots$$

Since it was assumed that in region 1 the temperatures of the surface and the injected gas are equal in the first approximation, in region 2 the stagnation enthalpy is constant in the same approximation. Substituting Eq. (4.76) in the Navier–Stokes equations and passing to the limit $\tau \to 0$ and $Re \to \infty$ leads to the complete compressible Euler equations in region 2. Matching with the outer supersonic flow gives a relation between the pressure and the boundary slope in the form of the Prandtl–Meyer equations as the outer boundary condition. A similar problem was solved in Section 3.2 by the integral relation method. Here, we do not pursue the goal to obtain the numerical solution of the problem. Because of this, we will restrict ourselves to an analysis of the behavior of the solution in region 2 as $x_2 \to -\infty$, which corresponds to $x_1 \to 1$. Precisely this is necessary for establishing the additional boundary condition separating out the required branch of the solution for region 1 on the basis of the solution matching principle. Let us introduce a region 5, transitional from region 2 to 1, and introduce a new smallness parameter $g$

$$\tau \ll g \ll 1$$

Let in region 5 $\Delta p \sim g$. In view of the momentum equation, $u \sim g^{1/2}$. Since, in accordance with the outer boundary condition, $y/x \sim g$, the continuity equation gives $v \sim g^{3/2} \gg \tau^{3/2}$. (Therefore, in the first approximation injection can be neglected in this region, as in region 2.) Since the gas flow rate $\psi \sim \tau^{3/2}$, we have $y \sim \tau^{3/2}/g^{1/2}$; then $x \sim (\tau/g)^{3/2}$.

Thus, in region 5 the following variables can be introduced:

$$x - 1 = \left(\frac{\tau}{g}\right)^{3/2} x_5, \quad y = \left(\frac{\tau^{3/2}}{g^{1/2}}\right) y_5$$

$$p \sim \frac{1}{\gamma M^2} + g p_5(x_5, y_5) + \cdots, \quad \rho \sim \rho_5 + \cdots \tag{4.77}$$

$$u \sim g^{1/2} u_5(x_5, y_5) + \cdots, \quad v = g^{3/2} v_5(x_5, y_5) + \cdots$$

The equations and the boundary conditions for the first approximation take the form:

$$\sqrt{M^2 - 1} p_5 = \frac{d\delta_5}{dx_5}, \quad u_5 \delta_5 = \psi_w(x_1 = 1), \quad p_5 + \frac{\rho_5 u_2^5}{2} = 0 \tag{4.78}$$

The first equation (4.78) represents the outer boundary condition, the second is the mass conservation law, and the third is the Bernoulli integral. The first condition is linear, since $g \ll 1$.

For all streamtubes in region 5 the Bernoulli constant is equal to zero, since all the streamtubes flow out of region 1, where $u \sim \tau^{1/2} \ll g^{1/2}$ and $\partial p_5/\partial y_5 = 0$, since in region 5, $y \ll x$. Finally, the second equation states that the whole flow rate proceeding from region 1 passes through region 5.

The solutions of Eq. (4.16) take the following simple form:

$$u_5 = \left[\frac{2}{3\rho_5\sqrt{M^2-1}}\frac{1}{(-x_5)}\right]^{1/3}, \quad p_5 = \frac{-\rho_5}{2}\left[\frac{2}{3\rho_5\sqrt{M^2-1}}\frac{1}{(-x_5)}\right]^{2/3} \tag{4.79}$$

The passage to region 2 corresponds to $x_5 \to 0$ and to region 1 to $x_5 \to -\infty$. Formulas (4.79) indicate that the non-self-similar solution for region 1 must include a singular point at which $p_1(x_1 \to 1) \to -\infty$, since $x_5 \sim (x_1 - 1)$ (see Eq. (4.77)). The unique solution in region 1 must be chosen using the additional condition $p_1(x_1 \to 1) \to -\infty$.

### *4.4.4 Concluding remarks*

The results presented in Section 4.4 were published in the paper of Matveeva and Neiland (1970); in that study the nature of the flow near the plate nose and the outer boundary of region 1, where additional flow regions must be considered, was also studied. Here, for the sake of brevity, the consideration of these questions is omitted, since the purpose of Section 4.4 is only the investigation of the disturbance propagation upstream from the base section in the presence of interaction between the supersonic flow and a thin layer, within which the injected gas flows. The calculated eigenvalues presented in Figs. 4.9–4.11 are small. This means that the effect of the boundary condition imposed on the base is not concentrated near the base but is important over the entire body surface.

It is interesting to note that for the flows of the type under consideration transition from the conventional boundary layer theory to the strong injection regime accompanied by upstream disturbance propagation, takes place already at flow rates $v_w \sim Re^{-1/2}$.

In order to demonstrate this fact, we will consider a simple plate flow, for which a self-similar solution exists within the framework of boundary layer theory. We let $\psi_0 \sim Re^{-1/2}$ in Eq. (4.64) and $m = 0.5$ in Eq. (4.68). If there is the inviscid flow region 1, then, in accordance with Eq. (4.64), we have

$$\tau \sim y \sim Re^{-1/3}, \quad u \sim Re^{-1/6}, \quad v \sim Re^{-1/2}, \quad \psi \sim Re^{-1/2}, \quad \frac{\mu u_{yy}}{\rho u u_x} \sim Re^{-1/6} \tag{4.80}$$

However, between the outer flow and the region of a slow inviscid (in the first approximation) gas there is a mixing zone in which

$$y \sim Re^{-1/2}, \quad u \sim 1, \quad v \sim Re^{-1/2}, \quad \psi \sim Re^{-1/2}, \quad \frac{\mu u_{yy}}{\rho u u_x} \sim 1 \tag{4.81}$$

Clearly, in both zones the orders of the gas flow rates are the same. Then matching the solutions in the mixing zone and region 1 gives, instead of the condition $\psi_1(x_1, \delta_1) = 0$, the following condition:

$$\psi_1(x_1, \delta_1) = \psi^*(x_1, y^* \to -\infty) < 0 \tag{4.82}$$

where asterisks refer to the variables retaining the order $O(1)$ for the mixing zone and $y^*$ is measured from the streamline $\psi = 0$.

In the mixing zone the flow is isobaric, since the induced pressure gradient $\sim Re^{-1/3}$, while the velocities $u \sim O(1)$. Using the well-known self-similar solution for the mixing zone, together with Eqs. (4.82) and (3.60), we obtain

$$f(x_1, 1) = \frac{\psi_1(x_1, \delta_1)}{\psi_w(x)} = \frac{0.87}{E} = h \tag{4.83}$$

where $\psi_0 = E\,Re^{-1/2}$ and $E$ is a number characterizing the gas flow rate.

Then in the solution of the self-similar problem (4.74) for region 1 with the boundary condition (4.82), instead $f = 1$ (for $\eta = 1$) and $m = 0.5$, we have

$$K = \int_h^1 \sqrt{\frac{3}{\xi^{-2/3} - 1}}\,\mathrm{d}\xi, \quad C = \left(\frac{3}{2K}\right)^{2/3}$$

With increasing injection $E \to \infty$ and $h \to 0$ and we obtain solution (4.74) for $\psi_0 \gg Re^{-1/2}$.

If $h = 1$, region 1 vanishes and the entire injected gas flows in the viscous boundary layer. When the gas flow rate through the body surface is greater than the amount which can be absorbed by the mixing zone but is $O(Re^{-1/2})$, then from Eq. (4.83) it follows that $0 < h < 1$ and there appears region 1 in which viscosity forces are small and disturbances can be transferred upstream.

We also recall that the self-similar solution for the laminar flat-plate boundary layer at "self-similar" injection distribution cannot exist for all gas flow rates. It ceases to exist precisely for the value of gas flow rate corresponding to the flow rate of the gas sucked off to the isobaric mixing zone (Jaffe, 1970). The above solution demonstrates the variation of the flow pattern which occurs for large flow rates (of course, in the presence of a supersonic outer flow). It becomes clear why the flow restructuring results in the vanishing of friction in the solution for the boundary layer. In fact, for small gas flow rates boundary layer theory gives $\tau_w \sim Re^{-1/2}$, while for the flow with zone 1 estimates (3.69) lead to $\tau_w \sim Re^{-5/6} \ll Re^{-1/2}$.

The case of the uniform distribution of strong injection with the inviscid zone 1 was also studied in the paper of Matveeva and Neiland (1970); for the sake of brevity, here it is omitted. However, it is useful to present two results. On a semi-infinite body the "self-similar" solution for the uniform injection is degenerate. Since in this case the injection velocity vanishes, the injection region transforms into a wedge-shaped stagnation region. This solution has no physical meaning, since the omitted viscous terms of the Navier–Stokes equations become leading. For a finite-length body non-self-similar solutions exist, since the presence of base expansion induces the required pressure gradient. The contact surface shape is close to the straight line

$$\delta_1 \sim Cx_1 \left( \ln \frac{1}{x_1^3} \right)^{1/3} + \cdots$$

### 4.4.5 Integration of Eqs. (4.36)

Equations (4.36) were integrated subject to the additional condition (4.41). An analytical study of the problem shows that near the origin there is a singular point. An additional study performed in determining the eigenvalue $b$ showed that the values of $b$ are large. These two considerations show that for solving numerically the non-self-similar problem it is convenient to make the change of variables

$$f(\zeta,\eta) = f_0(\eta) + e\zeta f_1(\zeta,\eta), \quad p = \frac{A}{\xi}[1 + e\zeta p_1(\zeta)]$$

$$v_e(\zeta) = \frac{B}{\xi^{1/2}}\left[1 + \frac{e}{2}\zeta v_{e1}(\zeta)\right], \quad \beta(\zeta) = \beta_0 + e\zeta\beta_1(\zeta) \tag{4.84}$$

$$F(\zeta) = F_0 + e\zeta F_1 + \zeta^2 F_2(\zeta), \quad \zeta = \xi^b$$

where $e = \pm 1$ are the branches of the solution with the pressure greater and smaller than that in the self-similar solution. Thus, two branches are separated out from the non-self-similar solution families.

The formulas are written only for the case $\sigma = \omega = 1$, $g = g_w = 1$, and $\chi = \infty$.

For variables (4.84) we obtain the equations

$$f_0''' + f_0 f_0'' - \beta_0(1 - f_0'^2) = 0$$

$$\begin{aligned} f_1''' &= f_0 f_1'' - f_1 f_0'' + 2\frac{\gamma - 1}{\gamma} f_0' f_1' + \beta(1 - f_0'^2) + 2b(f_0' f_1' - f_1 f_0'' + 2b\zeta(f_0' \dot{f}_1' - f_0'' \dot{f}_1) \\ &\quad - e\zeta\left[f_1 f_1'' - \frac{\gamma - 1}{\gamma} f_1'^2 + 2\beta_1 f_0' f_1' - 2b(f_1'^2 - f_1 f_1'') - 2b\zeta(f_1' \dot{f}_1' - \dot{f}_1 f_1'')\right] \\ &\quad - \zeta^2 \beta_1 f_1'^2 \end{aligned} \tag{4.85}$$

$$\beta_0 = -\frac{\gamma}{\gamma - 1}, \quad B = \frac{3}{\sqrt{2}} F_0, \quad F_0 = \int_0^\infty (1 - f_1'^2)\, d\eta, \quad A = \frac{\gamma + 1}{2} B^2$$

$$\beta_1 = \frac{\gamma - 1}{\gamma} \frac{bp_1 + b\zeta \times p_1}{1 + e\zeta \times p_1}, \quad F_1 = -2\int_0^\infty f_0' f_1'\, d\eta, \quad F_2 = -\int_0^\infty f_1'^2 d\eta$$

$$p_1 - v_1 = \frac{1}{4e}\zeta v_1^2 \tag{4.86}$$

$$\frac{3}{2}F_0 v_1 = \left(3F_1 + 2bF_1 - \frac{2\gamma}{\gamma - 1}\beta_1 F_0 + 2b\zeta\dot{F}_1\right)$$
$$+ e\zeta\left(3F_2 + 4bF_2 - \frac{2\gamma}{\gamma - 1}\beta_1 F_1 + 2b\zeta\dot{F}_2\right) - \frac{2\gamma}{\gamma - 1}\zeta^2\beta_1 F_2$$

System of equations (4.85) and (4.86) is subject to the following boundary conditions:

$$f_0(0) = f_1(\zeta, 0) = f_0'(0) = f_1'(\zeta, 0) = 0, \quad f_0'(\infty) = 1 \quad f_1'(\zeta, \infty) = 0,$$
$$p_1(0) = v_1(0) = 1$$

First, a system including the equations for $f_0(\eta)$, $\beta_0$, $A$, $B$, and $F_0$ is solved for $\zeta = 0$; five relations for their determination are written in the first rows of Eqs. (4.85) and (4.86). Then for $\zeta = 0$ and known "zero" variables, with account for the relation $p_1(0) = v_1(0) = 1$, we obtain five equations for determining $f_1(0, \eta)$, $\beta(0)$, $F_1(0)$, $F_2(0)$, and $b$. We note that $e = \pm 1$ is absent from these variables. After $b$ has been determined and the complete solution of the problem has been obtained for $\zeta = 0$, we have six equations for further integration of the system in the region $\zeta > 0$ and the determination of $f_1(\zeta, \eta)$, $F_1(\zeta)$, $F_2(\zeta)$, $p_1(\zeta)$, $v_1(\zeta)$, and $\beta_1(\zeta)$; these are the equation for $f_1(\zeta, \eta)$, relations for $\beta_1$, $p_1$, $F_1$, and $F_2$, and, finally, the last equation (4.86).

The problem was solved by a finite-difference method using the semi-standard program of Seliverstov.

At each band the value of $v_1$ was determined by means of an iteration procedure. All the equations, except for the last equation (4.86), were used for determining the value of $\beta_1$ entering in the equation for $f_1(\zeta, \eta)$. The equation was integrated using the semi-standard program. Then the values of $F_1$ and $F_2$ were calculated and the fulfillment of the last equation (4.86) was checked. After this condition has been satisfied, the solution was sought on the next characteristic band $\zeta = \text{const}$.

## 4.5 Detachment of a laminar boundary-layer

In accordance with Prandtl's theory, at high Reynolds numbers the viscosity effect is concentrated in a thin boundary layer near the surface in a flow. This flow structure is due to the processes of vorticity diffusion and convection away from the body surface. For a low viscosity ($Re \gg 1$) the distance normal to the body surface through which vorticity diffuses is considerably smaller than the distance through which it is transferred along the body surface due to convection for the same time interval. At boundary layer separation the viscosity effect is no longer localized in a thin wall layer and can propagate to regions with greater scales. Boundary layer separation due to shock incidence, an adverse pressure gradient, etc., is accompanied by the appearance of a positive velocity component in the boundary layer; therefore, the vorticity is transferred normal to the surface due to not only diffusion but convection as well. This convection can arise as a result of not only the outer flow deceleration but also a distributed surface injection. Experimental and theoretical studies indicate that also in this case the boundary layer flow can restructure itself in such a way that the boundary layer becomes detached. For a power-law injection velocity distribution over the surface ($v_w = C_0 x^{-1/2}$) the solution of the Blasius equation exists only on a restricted range of the parameter $C_0$. As this parameter tends to a certain limiting value ($C_0 \to C_0^*$), the surface friction vanishes (Emmons and Leigh, 1954; Chernyi, 1955). For $C_0 = C_0^*$ the solution describes the mixing layer between the undisturbed flow and a semi-infinite stagnation region.

For uniformly distributed injection the solution of the system of boundary layer equations exists in a bounded region along the streamwise coordinate terminated by a point at which the surface friction vanishes (Catherall et al., 1965). The appearance of the zero friction, or boundary layer detachment, point can qualitatively be explained by the fact that the velocity at which the vorticity diffuses away from a certain (zero) streamline to the body surface is offset by the oppositely directed convection velocity. Downstream of the detachment point, the convection velocity is greater than the diffusion velocity; therefore, the longitudinal momentum is transferred from the outer flow only to a certain part of the gas injected normal to the body surface rather than to its whole mass. For imparting a longitudinal momentum to the remaining mass of the injected gas, other mechanisms are needed, for example, the action of a favorable pressure gradient.

In the works of Matveeva and Neiland (1970), Kassoy (1973) and Levin (1973), flows with intense injection leading to the formation of an inviscid wall flow over the entire body (plate) surface were studied. The pressure decrease along the surface leading to the injected gas acceleration was ensured by the shape of the contact surface which induced an accelerating pressure gradient in the subsonic (Kassroy, 1973) and supersonic (Matveeva and Neiland, 1970) outer flows at a particular choice of the injection velocity distribution. The negative pressure gradient can also be induced in the case of uniformly distributed injection due to the base pressure difference (Matveeva and Neiland, 1970). In all these cases an important role is played by the interaction between the inviscid wall flow and the outer flow.

The interaction effects were taken into account in the work of Lipatov (1977) in which the solution of a composite system of boundary layer equations governing the supersonic flow over a flat porous surface was obtained.

In this section, we present the results of the study of the flow pattern at uniformly distributed injection from a flat permeable surface under conditions ensuring transition from the boundary layer flow to a detached flow including a mixing layer and a region of inviscid wall flow.[1]

### *4.5.1 Formulation of the problem, equations, and boundary conditions*

We will consider the supersonic viscous flow past a wedge through the surface of which aligned with the oncoming flow a gas is injected at a uniformly distributed velocity. The

[1] Flow detachment at a power-law injection distribution was studied in the work of Matveeva and Neiland (1967), its results were outlined above, in Section 4.4.

Cartesian coordinate system is so chosen that its origin is at the leading edge, the *OX* axis is parallel to the porous surface, and the *OY* axis is normal to it (Fig. 4.12). The Reynolds number $Re = \rho_\infty u_\infty \ell/\mu_\infty$, where $\rho_\infty$, $u_\infty$, and $\mu_\infty$ are the freestream density, velocity, and dynamic viscosity, and $\ell$ is the wedge generator length, is assumed to be high but not higher than the critical value at which transitional or turbulent flow regions appear near the surface. It is also assumed that the system of Navier–Stokes equations governing the wedge flow has a stationary solution. The following designations are accepted for the coordinates measured along the *OX* and *OY* axes, the corresponding velocity components, the density, the pressure, the enthalpy, and the dynamic viscosity: $\ell x$, $\ell y$, $u_\infty u$, $u_\infty v$, $\rho_\infty \rho$, $\rho_\infty u_\infty^2 p$, $u_\infty^2 H/2$, and $\mu_\infty \mu$.

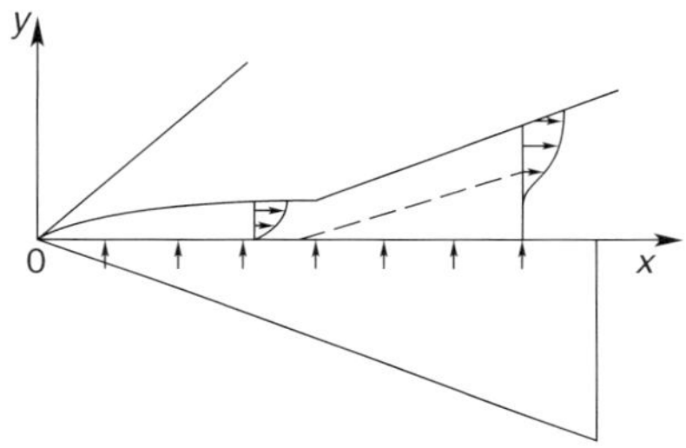


Fig. 4.12.

At high Reynolds numbers the system of Euler equations, to which the system of Navier–Stokes equations is reduced in the first approximation, has the following solution for $x \sim y \sim O(1)$: $u = 1$, $v = 0$. In order to satisfy the no-slip condition, a boundary layer should be introduced in a region with the scales $y = \varepsilon Y$, $x = X$, $\varepsilon = Re^{-1/2}$ near the wedge surface. There the flow functions take the form:

$$u = u_2(X, Y) + \cdots, \quad v = v_2(X, Y) + \cdots \tag{4.87}$$

$$p = \frac{1}{\gamma M_\infty^2} + \varepsilon p_2(X, Y) + \cdots, \quad g = g_2(X, Y) + \cdots$$

Using certain transformations brings the equations governing the boundary layer flow into the form:

$$\frac{\partial^2 \psi_2}{\partial x_2 \partial y_2}\frac{\partial \psi_2}{\partial y_2} - \frac{\partial^2 \psi_2}{\partial y_2^2}\frac{\partial \psi_2}{\partial x_2} = \frac{\partial^3 \psi_2}{\partial y_2^3} \tag{4.88}$$

$$\psi_2(x_2, 0) = -x_2, \quad \frac{\partial \psi_2}{\partial y_2}(x_2, 0) = 0, \quad \frac{\partial \psi_2}{\partial y_2}(x_2, y_2 \to \infty) = 1$$

$$u_2 = \frac{\partial \psi_2}{\partial y_2}, \quad y_2 = A_0 \int_0^Y \rho_2 \, \mathrm{d}Y$$

$$A_0 = v_{2w}\mu_w^{-1}, \quad x_2 = \rho_w v_{2w} A_0 X$$

It is assumed that the compositions of the injected and freestream gases are the same, the surface temperature is constant, and the viscosity–temperature dependence is linear. Numerical integration of problem (4.88) performed in the work of Catherall et al., (1965) showed that the surface friction, proportional to $(\partial^2\psi_2/\partial y_2^2)_w$, decreases monotonically with increasing coordinate $x_2$. The calculations were carried out up to the singular point $x_2 = x_2^*$ ($x_2^* \approx 0.7456$) at which the surface friction vanishes. For $x_2 = x_2^*$ solution (4.88) cannot be represented in the form of regular expansion in powers of $y_2$ as $y_2 \to 0$, since all derivatives $(\partial^n\psi_2/\partial y_2^n)_w$ $(n = 1, 2, \dots)$ turn to zero. This fact, together with the results of the numerical analysis, indicates the boundary layer flow transition to a flow with other transverse scale, greater than that suggested in Prandtl's boundary layer theory.

For further analysis it is important to assume that downstream of the detachment point $x_2 = x_2^*$ a region of inviscid slow flow is formed near the wedge surface and the thickness of this region is greater than that of the mixing layer located above. The estimates of the characteristic dimensions of this region and the flow functions within it are given below; it is important to note that in the first approximation the flow in the wall region has no effect on the mixing layer flow. It is useful to pass to the Crocco variables $x_2, u_2$ in Eq. (4.88)

$$\tau^2\frac{\partial^2\tau}{\partial u_2^2} = u_2\frac{\partial\tau}{\partial x_2}, \quad \tau = \frac{\partial u_2}{\partial y_2}, \quad \frac{\partial\tau}{\partial u_2}(x_2, 0) = 1, \quad \tau(x_2, 1) = 0 \tag{4.89}$$

For determining the solution of the boundary value problem (4.89) for $x_2 > x_2^*$ it should be taken into account that the boundary layer is detached from the surface and transforms to the mixing layer between the oncoming flow and the inviscid flow wall region. The ejection properties of the mixing layer, or the flow rate of the gas sucked off from the stagnation (in the first approximation) region, are not given beforehand and must be determined from the solution. Thus, for $x_2 > x_2^*$ the first boundary condition (4.89) must be omitted and replaced by the boundary condition $\tau(x_2, 0) = 0$.

The boundary value problem (4.89) with the modified boundary condition $\tau(x_2 > x_2^*, 0) = 0$ was numerically integrated. As a result, the distributions of the functions $\tau_w(x_2)$ for $x_2 < x_2^*$ and $v_{2w}(x_2) = (\partial\tau/\partial u_2)_w$ for $x_2 > x_2^*$ were obtained; they are plotted in Fig. 4.13. In the same figure the $\tau_w(x_2)$ distribution obtained in the work of Catherall

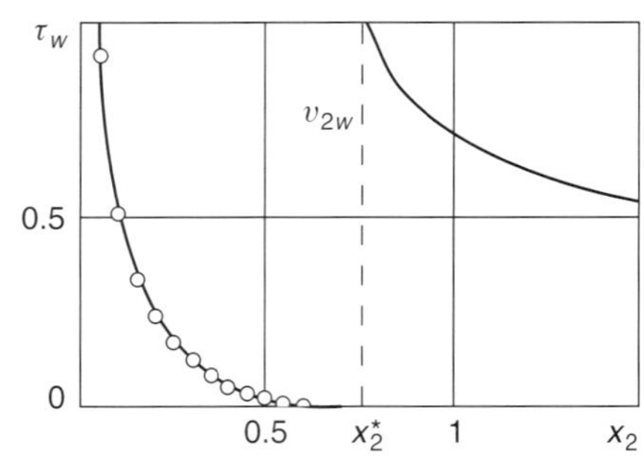


Fig. 4.13.

et al. (1965) is presented. It should be noted that the two $\tau_w(x_2)$ distributions are in agreement. For large values of $x_2$ the solution of the boundary value problem (4.89) is described in the first approximation by the self-similar solution governing the flow in the mixing layer between the uniform flow and the stagnation region. The solution of the self-similar problem leads to the following result for the function $(\partial\tau/\partial u_2)_w$ as $x_2 \to \infty$:

$$\left(\frac{\partial\tau}{\partial u_2}\right)_w = v_{1w} = C_1 x_2^{-1/2} \tag{4.90}$$

### 4.5.2 *Results of the solution*

The solution of the boundary value problem (4.89) makes it possible to obtain the velocity distribution $v_{2w}$ of gas suction from the inviscid flow wall region into the mixing layer. This velocity is less than the injection velocity; because of this, the unabsorbed part of the gas acquires a longitudinal momentum due to the pressure disturbance $\Delta p$ induced as a result of the interaction between the wall region flow and the outer supersonic flow. Generally, the length of the porous region of the wedge surface from the detachment point to the base section is finite. Since in the wall region the longitudinal velocity varies nonlinearly, $u \sim \Delta u \sim (\Delta p)^{1/2}$, for a given injection velocity $v_w \sim \varepsilon$ from the continuity equation it follows that $(\Delta p)^{1/2}\delta_3 \sim \varepsilon$, where $\delta_3$ is the inviscid flow wall region thickness.

Using the Ackeret formula for estimating the pressure disturbance we can obtain that $\Delta p \sim \varepsilon^{2/3}$, $\delta_3 \sim \varepsilon^{2/3}$, and $u \sim \varepsilon^{1/3}$. These estimates show that the inviscid flow wall region thickness is greater than that of the original boundary layer, while the longitudinal velocity is small as compared with that in the mixing layer, which confirms the above assumptions. In accordance with the estimates for this region, in which $x_3 = x_2 - x_2^*$ and $y = \varepsilon^{2/3} y_3$, the flow functions can be represented in the form of the following asymptotic expansions:

$$u = \varepsilon^{1/3} u_3(x_3, y_3) + \cdots, \qquad v = \varepsilon v_3(x_3, y_3) + \cdots \tag{4.91}$$

$$p = \frac{1}{\gamma M_\infty^2} + \varepsilon^{2/3} p_3(x_3, y_3) + \cdots, \qquad \rho = \rho_w + \cdots$$

Substituting Eqs. (4.91) in the system of Navier–Stokes equations and passing to the limit $Re \to \infty$ leads to the following system of equations:

$$\rho_w u_3 \frac{\partial u_3}{\partial x_3} + \rho_w v_3 \frac{\partial u_3}{\partial y_3} + \frac{\partial p_3}{\partial x_3} = 0 \tag{4.92}$$

$$\frac{\partial u_3}{\partial x_3} + \frac{\partial v_3}{\partial y_3} = 0, \qquad \frac{\partial p_3}{\partial y_3} = 0$$

$$u_3(x_3, 0) = 0, \quad v_3(x_3, 0) = v_{2w}, \quad v_3(x_3, \delta) = v_{1w}, \quad \delta = \delta_3 \varepsilon^{-2/3}$$

Due to the smallness of the disturbances introduced by injection, the interaction condition represents the Ackeret relation $(M_\infty^2 - 1)p_3 = \mathrm{d}\delta/\mathrm{d}x_3$. Introducing the following similarity variables reduces the boundary value problem (4.92) to the form:

$$\delta = [\rho_w^{-5} v_{2w}^{-4} (M_\infty^2 - 1)^{1/2}]^{1/2} \Delta, \quad p_3 = \rho_w^{1/2} v_{2w} (M_\infty^2 - 1)^{-1/2} P \tag{4.93}$$

$$\eta = \frac{y_3}{\Delta}, \quad \psi_3 = X_2 \rho_w v_{2w} f, \quad X_2 = x_3$$

$$ff'' - Af'^2 - B = X_2(f'\dot{f}' - \dot{f}f''), \quad A = \Delta \frac{\mathrm{d}}{\mathrm{d}X_2}\left(\frac{X_2}{\Delta}\right), \quad B = \frac{\Delta^2}{X_2}\frac{\mathrm{d}^2\Delta}{\mathrm{d}X_2^2}$$

$$f(X_2, 0) = 1, \quad f(X_2, 1) = \int_0^{X_2} v_{2w}(X_2)\,\mathrm{d}X_2$$

The boundary value problem analogous to (4.92) was formulated in thin layer theory (Matveeva and Neiland, 1970; Levin, 1973) describing the supersonic flow over porous flat surfaces for the injection velocities $O(\varepsilon) < v_w < O(1)$ at which detachment takes place in a small vicinity of the leading edge. For this regime the last boundary condition (4.93) takes the form $f(X_2, 1) = 0$, since the mixing layer absorbs zero (in first approximation) gas flow rate. Another difference from the flow pattern studied before is that in the case under consideration $P(0) = \text{const}$ (in thin layer theory at a constant injection velocity the pressure disturbance has a logarithmic singularity near the leading edge). The pressure disturbance $P(0)$ is not given beforehand and depends on the base pressure difference. The boundary value problem (4.93) describes the process of upstream disturbance transfer from the base section to the detachment point. The results of numerical integration of Eq. (4.93) are presented in Fig. 4.14 in which the $P(X_2)$ solutions are plotted for $P(0) = 1.01$ and 1.03. It should be noted that $P(0) > 0$; otherwise, the inviscid flow region could not exist.

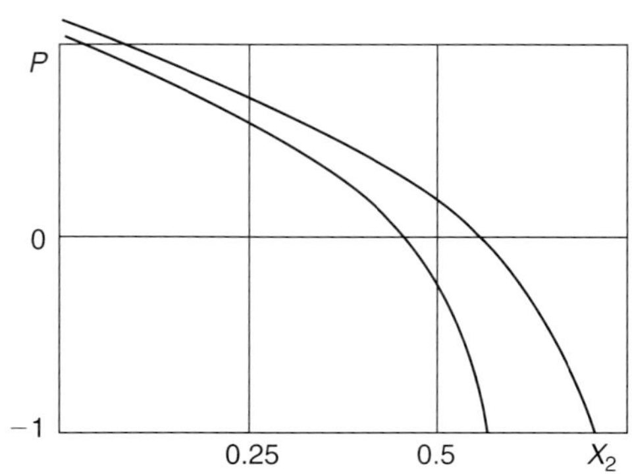


Fig. 4.14.

A pressure increase from the undisturbed value to $\varepsilon^{2/3}P(0)$ takes place in a local region located upstream of the detachment point.

## 4.6 Gas injection into a hypersonic flow

In this section we consider the flow arising at intense gas injection through the body surface and strong interaction of the hypersonic flow with the boundary layer. Using the Navier–Stokes equations and the asymptotic expansion matching principle makes it possible to establish the similarity parameters and the main flow patterns, to formulate boundary value problems, and to obtain certain solutions. Emphasis is placed on the effect of upstream disturbance transfer and the flow pattern near the base section both in the presence of injection and on an impermeable surface.

### *4.6.1 Formulation of the problem*

An interest to the flows occurring at intense gas injection through the surface of a flight vehicle into the flow is due to the fact that the use of this technique can reduce heat fluxes and friction and produce a required pressure distribution. Moreover, the intense injection regime can sometimes arise in a "natural" way upon ablation of thermal protection. Here, we will study the flows near thin bodies, whose thickness is not greater in the order than that of the region occupied by the injected gas, and bodies with small longitudinal curvatures of the surface (wedge, flat plate, etc.). In these cases, the pressure distribution is determined by the distribution of the displacement thickness of the injected gas layer, which, in turn, depends on the pressure distribution. The extensive literature on this subject is reviewed in the papers of Inger and Gaitatzes (1971) and Lees and Chapkis (1971). In the work of Matveeva and Neiland (1970) the effects that take place at gas injection into a supersonic flow were studied using the Navier–Stokes equations and the asymptotic expansion matching principle and transition from the flow pattern described by the classical boundary layer theory to the intense injection regime, in which a region of an inviscid (in the first approximation) flow of the injected gas is formed near the body surface, was considered. The flow is governed by the Prandtl equations with no viscous terms. The pressure distribution over the body surface is not known beforehand and is determined during the joint solution of the problem in the layer and the outer flow. A similar model was considered in the work of Cole and Aroesty (1968) in which certain inverse problems were solved. An analysis of the formulation of the boundary value problem drawn in the work of Matveeva and Neiland (1970) showed that for the flows of this type an important role is played by upstream disturbance propagation and a boundary condition should be imposed on the end of the body.

### *4.6.2 Equations and boundary conditions*

We will consider the flow near the upper side of a wedge, through the surface of which a distributed gas injection is realized normal to the surface (Fig. 4.15). The oncoming hypersonic flow is parallel to this side of the wedge. As shown below, the lower part of the

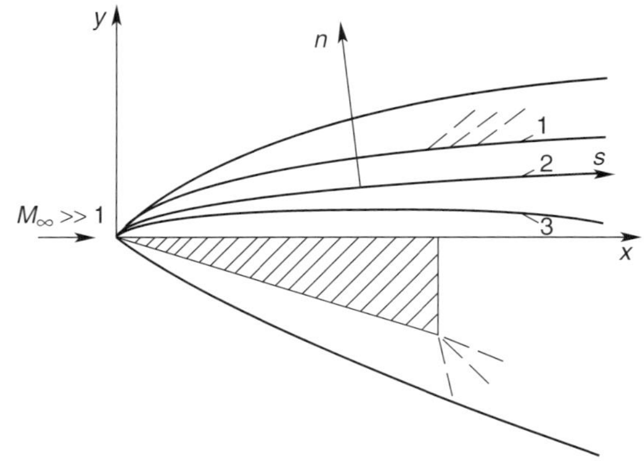


Fig. 4.15.

flow also has an effect – within certain limits – on the upper side, via the base pressure. However, here the base pressure is assumed to be preassigned (it can be controlled, e.g., changing the wedge thickness or acting on the base region flow).

Near the upper side of the wedge, the hypersonic flow streams over an "effective body" formed with account for viscosity and injection. We will consider only such flow regimes in which the characteristic slope of the effective body $\tau \ll 1$. Then the inviscid shock layer 1 (Fig. 4.15) is described – in the first approximation – by hypersonic small perturbation theory outlined, for example, in the book of Hayes and Probstein (1966). It can be shown that the weak interaction regime $M\tau \ll 1$, for which the induced pressure difference is small as compared with the freestream pressure, is in essence described by the theory developed in the paper of Matveeva and Neiland (1970) for moderate supersonic velocities. As shown in the paper of Neiland (1970*b*), the same situation takes place for flows with no injection. For this reason, here we consider the flows with $M\tau \gg 1$. Following hypersonic small perturbation theory we introduce the coordinates and functions for region 1

$$x = \ell x_1, \quad y = \tau \ell y_1, \quad \chi = M_\infty \tau \gg 1 \tag{4.94}$$

$$u = u_\infty[1 + \tau^2 u_1(x_1, y_1, \chi) + \cdots], \quad \rho = \rho_\infty[\rho_1(x_1, y_1, \chi) + \cdots]$$

$$p = \rho_\infty u_\infty^2 \tau^2[p_1(x_1, y_1, \chi) + \cdots], \quad v = u_\infty \tau[v_1(x_1, y_1, \chi) + \cdots]$$

where $\ell$ is the wedge length, $u$, $v$, $p$, $\rho$, and $\gamma$ are the longitudinal and transverse velocity components, the pressure, the density, and the specific heat ratio, the subscript $\infty$ refers to the freestream parameters, and the subscripts in the form of numbers relate to the dimensionless variables of the order $O(1)$ in the corresponding region.

Substituting Eq. (4.94) in the Navier–Stokes equations and passing to the limit $M_\infty \to \infty$, $Re \to \infty$, and $\chi \to \infty$ gives the conventional equations of the hypersonic small perturbation theory for an inviscid hypersonic flow (Hayes and Probstein, 1966).

In what follows we will use the approximate formula of the tangent wedge method for $\chi \to \infty$

$$p_1[x_1, \delta_1(x)] = \frac{\gamma + 1}{2}\left(\frac{d\delta_1}{dx_1}\right)^2 \tag{4.95}$$

where $\delta_1$ is the displacement thickness produced by injection and viscosity effects.

The conventional estimates of hypersonic boundary layer theory make it possible to introduce the following coordinates and functions in region 2 (Fig. 4.15) in which the leading viscous and inertial terms of the Navier–Stokes equations are of the same order:

$$s = s_2\ell, \quad n = \tau_2 \ell n_2, \quad Re_0 = \frac{\rho_\infty u_\infty \ell}{\mu_0}, \quad \tau_2 = \frac{Re_0^{-1/2}}{\tau}$$

$$p = \rho_\infty u_\infty^2 \tau^2 [p_2(s_2, n_2, \chi) + \cdots], \quad \rho = \rho_\infty \tau^2 [\rho_2(s_2, n_2, \chi) + \cdots]$$

$$u = u_\infty [u_2(s_2, n_2, \chi) + \cdots], \qquad v = u_\infty \tau_2 [v_2(s_2, n_2, \chi) + \cdots] \tag{4.96}$$

$$\mu = \mu_0 [\mu_2(s_2, n_2, \chi) + \cdots], \qquad H = \frac{u_\infty^2}{2}[H_2(s_2, n_2, \chi) + \cdots]$$

Here, $\mu$ and $H$ are the dynamic viscosity and the stagnation enthalpy, while the $s, n$ coordinates are fitted to the zero streamline (or the body surface in the absence of injection). The subscript 0 refers to the stagnation parameters in the freestream. Substituting Eqs. (4.96) in the Navier–Stokes equations and passing to the limit $M_\infty \to \infty$, $Re_0 \to \infty$, and $\chi \to \infty$ leads to the conventional Prandtl equations (see, e.g., Neiland, 1970*b*).

As shown below, with increase in the injection intensity layer 2 is detached from the body surface with the formation of region 3 of an inviscid (in the first approximation) flow (Fig. 4.15) in which the tangential momentum is produced only due to the induced accelerating pressure gradient. If region 3 has appeared, the coordinates and flow functions take the form:

$$x = x_3\ell, \quad y = \tau_3 \ell y_3, \quad H = H_w H_3$$

$$p = \rho_\infty u_\infty^2 \tau^2 (p_3 + \cdots), \quad \rho = \frac{\rho_\infty u_\infty^2}{2H_w}\tau_3^2(\rho_3 + \cdots) \tag{4.97}$$

$$u = (2H_w)^{1/2}(u_3 + \cdots), \quad v = (2H_w)^{1/2}(v_3 + \cdots)$$

Here, the subscript $w$ refers to the values of the parameters near the wall. Given the order of the pressure (4.88), these estimates can be obtained by assuming the stagnation enthalpy to be equal to $H_w$ and using the equation of motion for an inviscid gas, together with the equations of continuity and state.

Substituting Eq. (4.97) in the Navier–Stokes equations and passing to the limit $\tau_3 \to 0$ leads to the equations of the form:

$$\rho_3\left(u_3\frac{\partial u_3}{\partial x_3}+v_3\frac{\partial u_3}{\partial y_3}\right)+\frac{\partial p_3}{\partial x_3}=0,\qquad \frac{\partial p_3}{\partial y_3}=0 \tag{4.98}$$

$$\frac{\partial(\rho_3 u_3)}{\partial x_3}+\frac{\partial(\rho_3 v_3)}{\partial y_3}=0,\qquad H_3=1$$

Thus, as distinct from the $M_\infty \sim O(1)$ flow (Matveeva and Neiland, 1970), in region 3 compressibility is now important, that is, $M_3 \sim O(1)$.

The dimensionless thickness $\tau_3$ will be obtained more accurately in what follows; for a while, we will give its approximate estimate following from Eq. (4.97):

$$\tau_3 \approx \frac{\sigma_w(2H_w)^{1/2}}{\rho_\infty u_\infty^2 \ell \tau^2} \tag{4.99}$$

where, $\sigma_w$ is the gas flow rate through the body surface.

### 4.6.3 Self-similar solutions

Let us trace the sequence of the flow patterns giving way one another with increase in the injection intensity. We will begin with the weak injection regime for which formulas (4.96) are valid, if we assume $\tau = \tau_2 = Re_0^{-1/4}$ (Hayes and Probstein, 1966; Neiland, 1970*b*). It is convenient to go over from variables (4.89) to the Dorodnitsyn–Lees variables

$$\xi = \frac{2\gamma}{\gamma-1}\int_0^x p\,\mathrm{d}s,\qquad g=H,\qquad u=f'(\xi,\eta)$$

$$\eta = \frac{2\gamma p}{(\gamma-1)2\xi_1^{1/2}}\int_0^n \frac{\mathrm{d}y}{g-f'^2} \tag{4.100}$$

For the original regime of weak injection into the boundary layer (the subscript 2 is omitted, since for a while, region 3 is absent, so that we have $\tau = \tau_2 = Re_0^{-1/4}$), the equations and the boundary conditions take their conventional form:

$$(Mf'')' + ff'' - \beta(g-f'^2) = 2\xi(f'\dot{f}' - \dot{f}f'')$$

$$\left(\frac{M}{\sigma}g'\right)' + fg' + \frac{1}{2}\left[M\left(1-\frac{1}{\sigma}\right)f'f''\right]' = 2\xi(f'\dot{g} - \dot{f}g') \tag{4.101}$$

$$f'(\xi,0)=0,\quad g(\xi,0)=g_w,\quad f(\xi,\infty)=g(\xi,\infty)=1,\quad f(\xi,0)=f_w(\xi)$$

$$\beta = \frac{\gamma - 1}{\gamma}\frac{d\ln p}{d\ln\xi}, \quad M = (g - f'^2)^{n-1}, \quad \mu = (g - f'^2)^n$$

where primes refer to the differentiation with respect to $\eta$ and dots to $\xi$, $\sigma$ is the Prandtl number, and $n$ is the exponent in the power-law viscosity–enthalpy dependence.

For determining $\beta(\xi)$ with the aid of Eq. (4.88), the slope of the outer boundary $d\delta_1/dx_1$ must be known

$$v_e = \frac{d\delta_1}{dx_1} = p\frac{d}{d\xi}\left[\frac{\sqrt{2\xi}}{p}\int_0^\infty (g - f'^2)\,d\eta\right] \tag{4.102}$$

Typical for the hypersonic regime is the condition $g_w \ll 1$. Because of this, for studying the regimes occurring with increase in the injection intensity, we will perform the double passage to the limit

$$g_w \to 0, \quad f_w^\circ = f_w(x = 1) \to \infty \tag{4.103}$$

In order to simplify the calculations, we will let $n = \sigma = 1$ and introduce the variables for the region located near the body surface

$$g = g_w g_3, \quad f' = f_3' g_w^{1/2}, \quad \eta = \eta_3 g_w^{-1/2}(-f_w^\circ), \quad f = (-f_w^\circ)f_3 \tag{4.104}$$

Substituting Eq. (4.104) in Eq. (4.101) and passing to limit (4.103) in the first approximation yields

$$f_3 f_3'' - \beta(\xi)(1 - f_3'^2) = 2\xi(f_3'\dot{f}_3' - \dot{f}_3 f_3''), \quad g_3 = 1$$

$$f_3(\xi, 0) = \frac{f_w(\xi)}{(-f_w^\circ)}, \quad f_3'(\xi, 0) = 0 \tag{4.105}$$

However, solution (4.105) corresponding to the inviscid injected gas flow does not make it possible to satisfy the conditions at the outer edge of the boundary layer. Therefore, it is necessary to consider region 2, for which, even at the passage to limit (4.103), the equations involve the viscous terms

$$\eta_2 = \eta - \eta_0, \quad f = f_2, \quad g = g_2 \tag{4.106}$$

where $\eta$ is the coordinate of the dividing streamline detached by injection from the body surface. At the passage to limit (4.103) for region (4.106) the equations retain the form (4.101); the conditions imposed as $\eta_2 \to +\infty$ also remain unaltered, while instead of the conditions at the body surface matching with the solution for region 3 yields

$$g_2(\xi, -\infty) = g_w \to 0, \quad f_2'(\xi, -\infty) = g_w^{1/2} f_3'(\xi, \eta_{3e}) \to 0 \tag{4.107}$$

where $\eta_{3e}$ is the coordinate of the outer boundary of region 3 determined from the condition $f_3(\eta_{3e}) = 0$, as it follows from the third equation (4.97) and (4.96). The calculation of $\delta_1$ is of chief interest, since it is necessary for determining the pressure gradient with the aid of Eqs. (4.102) and (4.95)

$$\delta_1 = \frac{\gamma - 1}{2\gamma}\frac{(2\xi)^{1/2}}{p}\left[N\int\limits_0^{\eta_{3e}(\xi)}(1 - f_3'^2)\,\mathrm{d}\eta_3 + \int\limits_{-\infty}^{\infty}(g_2 - f_2'^2)\,\mathrm{d}\eta_2\right] \tag{4.108}$$

where the similarity parameter

$$N = g_w^{1/2}(-f_w^\circ)$$

The physical meaning of the parameter $N$ is that it represents the square of the ratio of the characteristic thicknesses of regions 2 and 3. This fact is considered below in more detail. The parameter $N$ characterizes the viscosity effect on the pressure thus determining the flow pattern. If the passage to limit (4.94) is so performed that $N \to 0$, then from Eq. (4.101) it follows that, regardless of injection, the pressure distribution is completely determined by the distribution of the viscous mixing zone displacement thickness, while $\tau = \tau_2 = Re_0^{-1/4}$. In particular, for the self-similar solution (a semi-infinite body or a particularly chosen base pressure value and $f = f_w^\circ = \text{const}$) we obtain

$$p_1 = \frac{A}{\xi}, \quad A = \frac{9}{4}(\gamma + 1)^2 F^2 \tag{4.109}$$

$$F = \int\limits_{-\infty}^{+\infty}(g_2 - f_2'^2)\,\mathrm{d}\eta_2 = 0.525 \quad \sigma = n = 1$$

It is interesting to note that the same result is valid for $g_w = 0$ for finite values of $|f_w|$ greater than the value of $|f(\xi, -\infty)|$ for the viscous mixing zone obtained in solving Eqs. (4.101) under the boundary conditions (4.107). For the self-similar solution this critical value $f_w = -1.003$ for $\sigma = n = 1$. In Fig. 4.16 we have plotted the calculated dependences of the friction $f_w''$ and the quantity $F = \int_{-\infty}^{\infty}(g - f'^2)\,\mathrm{d}\eta$ on the injection intensity $f_w$; a square denotes the value of $F$ obtained for the viscous mixing zone. For $g_w = 0$ we have $f_w'' = 0$ at $f_w = -1.003$, while for $f_w < -1.003$ there is no solution within the framework of boundary layer theory. This means that zone 3 starts to develop. A similar result was obtained for a supersonic flat-plate flow when the pressure gradient has no effect in the first approximation on the original boundary layer. If for $M \sim 1$ the accelerating pressure gradient does exist, then transition to the strong injection regime accompanied by the formation of the inviscid flow region 3 takes place only when $|f_w| \to \infty$, that is, for injected gas flow rates considerably greater than that in the viscous boundary layer. But in our case transition to the strong injection regime occurs at finite $f_w$, in spite of the presence of $\beta \neq 0$.

The most general and complicated is the solution of the problem for the flow regimes in which the thicknesses of the viscous and inviscid flow regions are of the same order, that

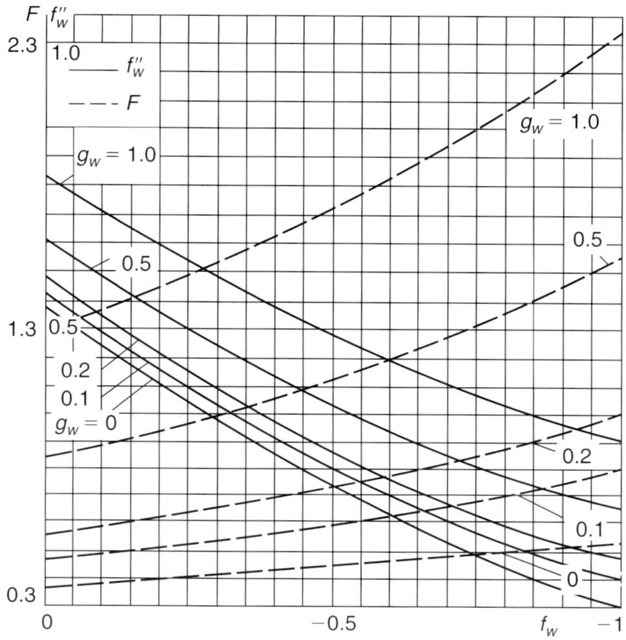


Fig. 4.16.

is, $N = O(1)$ for (4.103). In this case, the pressure distribution takes the form:

$$A = \frac{9}{4}(\gamma + 1)^2(N\Phi + F)^2, \quad \Phi = 1.957, \ F = 0.525 \ \text{for}\ \sigma = n = 1 \tag{4.110}$$

Finally, for $N \to \infty$ we obtain the flow pattern for which the pressure distribution over the body surface is related only with the inviscid flow in region 3. We note that the gas flows with $g_w = O(1)$ for $f_w^\circ \to -\infty$ pertain to the same case ("hot" wall and strong injection). It might be expected in advance that the flow in region 3 and the pressure distribution must be independent of $Re_0$, while the normalization must correspond to formulas (4.97). To demonstrate this, we introduce variables (4.104), together with the following variables:

$$\tau_3 = Re_0^{-1/4}N^{1/2}, \quad p = Np_3, \quad \rho = N\rho_3, \quad \xi = N\xi_3, \quad y = N^{1/2}y_3, \quad \delta = N^{1/2}\delta_3 \tag{4.111}$$

Substituting Eqs. (4.104) and (4.111) in Eqs. (4.110)–(4.112) and passing to the limit $N \to \infty$, Eq. (4.103) leads to the equations and boundary conditions (4.105). Instead of Eq. (4.108) we obtain

$$\delta_3 = \frac{\gamma - 1}{2\gamma}\frac{(2\xi_3)^{1/2}}{p_3}\int_0^{\eta_{3e}} (1 - f_3'^2)\,\mathrm{d}\eta_3 \tag{4.112}$$

where $f_3(\eta_{3e}) = 0$.

Finally, substituting Eqs. (4.104) and (4.111) in the original formulas (4.96), which were used for deriving Eqs. (4.110)–(4.112) (we recall that in that case the subscript 2 was omitted) leads to variables (4.104). The layer thickness $\tau_3$ and the pressure are $Re_0$-independent:

$$\tau_3 = \left\{ \frac{\sigma_w (2H_w)^{1/2}}{\rho_\infty u_\infty^2 \ell [2\xi_3(x=1)]^{1/2}} \right\}^{1/3} \tag{4.113}$$

This expression is in agreement with Eq. (4.99), taking into account that in this regime $\tau \approx \tau_3$.

It is interesting to note that for $g_w = 1$ the solution of Eq. (4.105) describing the inviscid flow in region 3 yields $f'_{3e} = g_3 = 1$ at the outer boundary. This means that there are no discontinuities in the velocity and the stagnation enthalpy between regions 1 and 3, that is, the mixing zone 2 of the conventional type does not appear. Viscosity has an effect only on the smoothing of discontinuities in the values of derivatives. This result can also be obtained from physical considerations, without solving Eq. (4.105).

In conclusion, we note that the formulations of the boundary value problems for different flow regimes obtained within the framework of the first approximation for the complete Navier–Stokes equations in Section 2 and for the Prandtl equations in Section 3 completely coincide. However, this would not be the case for the second approximation.

### *4.6.4 Analysis of the* N = O(*1*) *regime*

The equations governing in the first approximation the gas flows in regions 1, 2, and 3 are either hyperbolic or parabolic. However, the ellipticity of the original Navier–Stokes equations manifests itself in the fact that in the joint solution of the problem of the interaction between different flow regions the boundary condition preassigned at $x = 1$ should be taken into account for determining the solution near the body nose. A similar situation was studied for other flows in the works of Kozlova and Mikhailov (1970), Matveeva and Neiland (1970), and Neiland (1970*b*).

To demonstrate this fact we will consider the problem of strong injection for the most general regime $N = O(1)$ assuming $f_w$ to be constant. For region 2 the equations coincide with Eqs. (4.101) with the only difference that the conditions on the body surface are replaced by Eq. (4.107). Region 3 is described by the equations and boundary conditions (4.105). Finally, the pressure distribution is determined by formulas (4.108) and (4.95). We will now demonstrate that, apart from the above self-similar solution, already for $\xi \to 0$ the problem possesses a family of solutions dependent on a single parameter; to determine this parameter, one more boundary condition should be preassigned, for example, the value $p(x = 1)$.

It is useful to make first the change of variables

$$\zeta = \xi^a, \quad p = \zeta^{-1/a} P, \quad v_e = \zeta^{-1/2a} V \tag{4.114}$$

Then the equations and the boundary conditionstake the form:

$$f_2''' + f_2 f_2'' - \beta (g_2 f_2'^2) = 2\zeta a (f_2' \dot{f}_2' - \dot{f}_2 f_2'')$$

$$g_2'' + f_2 g_2' = 2\zeta a(f_2' \dot{g}_2 - \dot{f}_2 g_2')$$

$$f_3 f_3'' - \beta(1 - f_3'^2) = 2\zeta a(f_2' \dot{f}_3' - \dot{f}_2 f_3'')$$

$$\beta = \frac{\gamma - 1}{\gamma}\left(a\frac{\zeta}{P}\dot{P} - 1\right), \quad P = \frac{\gamma + 1}{2}V^2$$

$$V = a\sqrt{2}\zeta(\dot{F} + N\dot{\Phi}) + \frac{3}{\sqrt{2}}(F + N\Phi) - a\sqrt{2}(F + N\Phi)\frac{\mathrm{d}\ln P}{\mathrm{d}\ln\zeta} \tag{4.115}$$

$$F = \int_{-\infty}^{+\infty} (g_2 - f_2'^2)\,\mathrm{d}\eta_2, \quad \Phi = \int_0^{\eta_{3e}} (1 - f_3'^2)\,\mathrm{d}\eta_3$$

$$f_2'(\zeta, +\infty) = g_2(\zeta, +\infty) = 1, \quad f_2(\zeta, 0) = 0$$

$$f_2'(\zeta, -\infty) = g_2(\zeta, -\infty) = 0$$

$$f_3'(\zeta, 0) = 0, \quad f_3(\zeta, 0) = -1, \quad f_3(\zeta, \eta_{3e}) = 0$$

Here and in what follows dots refer to the differentiation with respect to $\eta$.

The solution in the vicinity of $\zeta = 0$ is sought in the form of a power series in $\zeta$, for example

$$f_2(\zeta, \eta_2) = f_{20}(\eta_2) + \zeta f_{21}(\eta_2) + \cdots, \quad V(\zeta) = V_0 + \zeta V_1 + \cdots \tag{4.116}$$

Collecting the terms with the same powers in $\zeta$ we obtain the following recurrent systems of equations:

$$f_{20}''' + f_{20} f_{20}'' - \beta_0(g_{20} - f_{20}'^2) = 0, \quad g_{20}'' + f_{20} g_{20}' = 0$$

$$f_{30} f_{30}'' - \beta_0(1 - f_{30}'^2) = 0, \quad V_0 = \frac{3}{\sqrt{2}}(F_0 + N\Phi_0)$$

$$P_0 = \frac{\gamma + 1}{2}V_0^2, \quad F_0 = \int_{-\infty}^{\infty} (g_{20} - f_{20}'^2)\,\mathrm{d}\eta_2, \quad \Phi_0 = \int_0^{\eta_{3e}} (1 - f_{30}'^2)\,\mathrm{d}\eta_3 \tag{4.117}$$

$$f_{20}'(+\infty) = g_{20}(+\infty) = 1, \quad f_{20}(0) = 0, \quad f_{20}'(-\infty) = g_{20}(-\infty) = 0$$

$$f_{30} = -1, \quad f_{30}'(0) = 0, \quad f_3(\eta_{3e0}) = 0, \quad \beta_0 = -\frac{\gamma - 1}{\gamma}$$

$$f_{21}''' + f_{20} f_{21}'' + f_{21} f_{20}'' - \beta_1(g_{20} - f_{20}'^2) + \frac{\gamma - 1}{\gamma}(g_{21} - 2f_{20}' f_{21}')$$
$$= 2a(f_{20}' f_{21}' - f_{21} f_{20}'')$$

$$g''_{21} + f_{20}g'_{21} + f_{21}g'_{20} = 2a(f'_{20}g_{21} - f_{21}g'_{20}), \quad \Phi_1 = -2\int_0^{\eta_{3e}} f'_{30}f'_{31}\,d\eta_3$$

$$f_{30}f''_{31} + f_{31}f''_{30} - \beta_1(1 - f'^2_{30}) - \frac{\gamma-1}{\gamma}2f'_{30}f'_{31} = 2a(f'_{30}f'_{31} - f_{31}f''_{30})$$

$$f'_{21}(\infty) = g_{21}(+\infty) = f_{21}(0) = f'_{21}(-\infty) = g_{21}(-\infty) = 0 \qquad (4.118)$$

$$f_{31}(0) = f'_{31}(0) = 0, \quad F_1 = \int_{-\infty}^{+\infty} (g_{21} - 2f'_{20}f'_{21})\,d\eta_2$$

$$\beta_1 = \frac{\gamma-1}{\gamma}a\left(\frac{P_1}{P_0}\right), \quad P_1 = (\gamma+1)V_0V_1 \qquad (4.119)$$

$$V_0P_1 + V_1P_0 = a\sqrt{2}[P_0F_1 - P_1F_0 + NF_0 + N(P_0\Phi_1 - \Phi_0P_1)] + \frac{3}{\sqrt{2}}[P_0F_1 + P_1F_0 + N(P_0\Phi_1 - P\Phi_0)]$$

The boundary value problem (4.117) determines the self-similar solution with the pressure distribution presented above.

For given $a$ the system of equations for $f_{21}$, $g_{21}$, $f_{31}$, $F_1$, $\beta_1$, $P_1$, $V_1$, and $\Phi_1$ is linear and homogeneous. Its solutions exist only for eigenvalues of $a$ and are determined correct to an arbitrary constant. One can convince oneself in this noting that for each value of $a$, if, say, $\beta$ is arbitrarily preassigned, $f_{21}$, $g_{21}$, and $f_{31}$ and then $F_1$ and $\Phi_1$ can uniquely be determined. The first two equations (4.119) give $P_1$ and $V_1$. Then it remains to verify whether the last equation (4.119) is satisfied.

These calculations were performed for the limiting case $N \to \infty$. One eigenvalue, equal approximately to 0.23, was determined in the domain $a > 0$; therefore, the effect of the boundary conditions on the flow in the upstream region must be very strong. For next terms of expansion (4.116) the system of equations is also linear but inhomogeneous; therefore, for a given value of the arbitrary constant for system (4.118) the solution is also uniquely determined in the next approximations.

Thus, apart from the self-similar solution corresponding to formulas (4.117), the boundary value problem (4.115) has a single-parameter family of solutions.

### *4.6.5 Dependence of the solution on the base pressure difference*

We note that all boundary value problems for the flows with strong injection accompanied by the formation of region 3 cannot have steady-state separationless solutions with adverse pressure gradients. This follows from the condition $u_w = 0$ and the Bernoulli equation. Then, the condition $p > k > 0$ is fulfilled for any $0 \leq x \leq 1$, since otherwise the inviscid flow region thickness would increase without bound ($\delta_3 \to \infty$) over a finite length as $p \to 0$ owing to divergence of supersonic streamtubes; however, in this case the pressure should increase starting from a certain point (cf. Eq. (4.95)).

If there is a single-parameter family of the solutions for the forward part of the body, that is, for $x \sim 1$, and the base pressure $p_4$ is given, then it is natural to try to choose the required solution from the condition $p(x=1) = p_4$. However, for this purpose it is necessary that near the base section, at distances $\Delta x \ll 1$, a region with $\Delta p \sim O(1)$ could not arise. Let us establish the conditions under which the occurrence of this region is possible. To do this, we will assume that this region does exist. Since in this region $\Delta p \sim p$, the subsonic streamtube thicknesses decrease (rarefaction flow) by the leading order, while the supersonic streamtube thicknesses increase. In the first approximation, these variations must compensate exactly each other, that is, $\mathrm{d}\delta/\mathrm{d}p = 0$. In fact, if this is not the case, the total thickness $\delta$ of the region varies by $\Delta\delta \sim \delta$ over a length $\Delta x \ll 1$. In view of Eq. (4.95), this must lead to $\Delta p \sim (\Delta\delta/\Delta x)^2 \gg p$, which is impossible. Thus, for $\mathrm{d}\delta/\mathrm{d}p \neq 0$ we have $\Delta p \ll p$ over lengths $\Delta x \ll 1$. Therefore, near the base section $x = 1$ a finite pressure difference can arise at small distances $\Delta x \ll 1$ only provided $\mathrm{d}\delta/\mathrm{d}p = 0$. The fulfillment of this condition corresponds to the appearance of a singular point on the integral curve of the boundary value problem governing the flow on the main part of the body. From Eq. (4.95) there follows

$$\mathrm{d}x = \sqrt{\frac{\gamma+1}{2p}}\left(\frac{\mathrm{d}\delta}{\mathrm{d}p}\right)\mathrm{d}p \tag{4.120}$$

For small $x$ (in the vicinity of the self-similar solution) we have $\mathrm{d}\delta/\mathrm{d}p < 0$, $\mathrm{d}p < 0$, and $\mathrm{d}x > 0$. The solution cannot be continued through the singular point $\mathrm{d}\delta/\mathrm{d}p = 0$, since to do this requires that $\mathrm{d}p > 0$ for $\mathrm{d}x > 0$, which is impossible. We note that the equations and boundary conditions (4.115) are invariant with respect to the following single-parameter group of transformations:

$$\xi = b\overline{\xi}, \quad \eta = \overline{\eta}, \quad p = \frac{1}{b}\overline{p}, \quad \rho = \frac{1}{b}\overline{\rho}, \quad v = \frac{1}{\sqrt{b}}\overline{v} \tag{4.121}$$

$$y = b^{3/2}\overline{y}, \quad x = b^2\overline{x}, \quad f_l = \overline{f}_l, \quad P = \overline{P}, \quad V = \overline{V}, \quad \xi = b^a\overline{\xi}$$

where $b$ is an arbitrary constant.

Let us assume that by preassigning $\beta_1$ in system (4.118) we obtain the entire integral curve by, for example, numerical integration up to the singular point $\mathrm{d}\delta/\mathrm{d}p = 0$. Then all other curves could be obtained with the aid of the group of transformations (4.121). The curves, at which $\mathrm{d}\delta/\mathrm{d}p = 0$ for $x < 1$, have no physical meaning. If the singular point is located at $x > 1$, then a part of the integral curve describes the flow for which the base pressure $p_4 = p(x=1)$. Finally, the integral curve having the singular point at $x = 1$ (we designate the corresponding pressure as $p(x=1) = p_5$) corresponds to all flows for which the base pressure $p_4 \leq p_5$. In this case, near the base section a flow region is formed, in which the pressure decreases by $\Delta p \sim O(1)$ over a length $\Delta x \sim \tau \ll 1$, while the transverse and longitudinal pressure gradients are of the same order. This is possible and necessary precisely due to the condition $d\delta/dp = 0$. We note that in this short region the surface pressure $p_w$ cannot be reduced lower than $[(\gamma+1)/2]^{-\gamma/(\gamma-1)}p_5$, since at the end of the body $M_w = 1$ and further flow expansion can occur only in the wave beginning at the corner point of the body, since the flow has no longer convergent subsonic streamtubes. Therefore, for all base pressures $p_4 < [(\gamma+1)/2]^{-\gamma/(\gamma-1)}p_5$ the pressure distributions over the body are the same everywhere including a small vicinity of the base section of length $\Delta x \approx \tau$.

Thus, for $p_4 \geq p_5$ the flow is completely described by a part of the integral curve of the boundary value problem for the main part of the body and $p(x=1) = p_4$, while a region with high local pressure gradients is not formed. For $p_5 > p_4 > [2/(\gamma+1)]^{\gamma/(\gamma-1)}p_5$ a local flow region with high local pressure gradients is formed, but at its end we have $p = p_4$ and $M_w < 1$. Finally, for $p_4 \leq [2/(\gamma+1)]^{\gamma/(\gamma-1)}p_5$ at the local region end $p = [2/(\gamma+1)]^{\gamma/(\gamma-1)}p_5$ and $M_w = 1$. Further decrease of the base pressure has no effect on the upstream flow. As for the main part of the body ($x \sim 1$), there for all $p_4 \leq p_5$ the flow is described by the entire integral curve up to the singular point $d\delta/dp = 0$ at $x = 1$.

### *4.6.6 Base pressure difference effect on the flow past an impermeable surface*

The qualitative analysis of the physical features of the flow near the base section drawn in Section 4.6.5 holds entirely true for flows with no injection at strong boundary layer/hypersonic flow interaction considered in the paper of Neiland (1970*b*).

Let us consider the flow presented in Fig. 4.15 but in the absence of injection. In the paper of Neiland (1970*b*) it was shown that with variation of the base pressure the flow on the main part of the body so restructures itself as the pressure $p(x=1)$ is equal to the base pressure. However, this cannot be realized at very low values of the base pressure. Firstly, as $p \to 0$, the displacement thickness $\delta \to \infty$, which, in accordance with Eq. (4.95), must lead to a pressure increase. Secondly, on the integral curve there is a singular point at which $d\delta/dp \to 0$ and $dp/dx \to -\infty$ and the solution cannot be continued through this point in view of the reasons presented in the preceding section. For this flow the limiting minimum value of the pressure at the body end in the region, where $p_y = 0$, was calculated and turned out to be equal to $p = 0.636\rho_\infty u_\infty^2 Re_0^{-1/2}$. Due to the formation of the region with high longitudinal and transverse pressure gradients at distances $x \sim \tau = Re_0^{-1/4}$ the body pressure can decrease still further by a factor of $[(\gamma+1)/2]^{\gamma/(\gamma-1)}$. For lower base pressures further flow expansion takes place in the expansion wave centered at the corner point of the body and has no effect on the upstream flow.

## 4.7 Gas injection into a hypersonic flow (moderate injection)

Distributed surface injection is used for reducing heat fluxes to the surfaces of flight vehicles moving at high supersonic velocities, cooling the blades of turbines, etc. The injection changes the effective shape of the surface and can be used for producing aerodynamic forces and moments. In the last case the normal velocity components are usually greater than the vertical velocity in the boundary layer on an impermeable surface. The regimes of the flows with intense injection were studied in many works reviewed in the paper of Levin, et al. (1980). At the same time, for the purpose of thermal protection, this is injection of a gas flow rate comparable with that in the boundary layer on an impermeable surface that turns out to be optimal; this injection intensity ensures a reduction of the heat flux in the leading order. In this case, the flow near a permeable surface is described by the system of boundary layer equations.

The greatest heat fluxes are typical of hypersonic flows, in particular, of the strong hypersonic interaction regime. The study of the flows for the strong hypersonic interaction regime in the presence of injection was chiefly restricted to the consideration of problems which admit the reduction of the system of boundary layer equations to the system of ordinary differential equations. At the same time, the injection intensity distributions realized in practice generate a need for solving non-self-similar problems. An example of the solution of the problems of this kind is given in the present section.

There is one more fact that imparts much importance to the study of flows with injection. In the classical boundary layer theory there are two known types of singularities in the solution associated with the surface friction vanishing and the flow restructuring. In the former case, a decrease in the friction to zero and the formation of a return flow region (boundary layer separation) are due to an adverse pressure gradient, while in the latter case the friction vanishing and the formation of an inviscid wall flow region (boundary layer detachment) are caused by distributed injection. The boundary layer flow structure is determined by vorticity diffusion and convection. At high Reynolds numbers the distance through which the vorticity diffuses away from the body surface is much smaller than the distance through which it is convected along the surface for the same time interval (Batchelor, 1967). Flow deceleration under the action of the adverse pressure gradient leads to the generation of the convective mechanism of vorticity transfer from the body surface and the boundary layer flow restructuring; the convective processes of this kind are also typical for flows with distributed injection. The mathematical description of the solution of the system of boundary layer equations near zero friction points is presented in the papers of Landau and Lifshitz (1944), Goldstein (1948), and Catherall (1965). An analysis of these solutions indicates that in the vicinity of zero surface friction points there appears an adverse pressure gradient induced by the displacement thickness in the outer flow. Taking account of the boundary layer/outer flow interaction made it possible to obtain smooth solutions passing through the separation point in both supersonic (Neiland, 1969*a*; Stewartson and Williams, 1969) and subsonic (Sychev, 1972) flows. Later it turned out that taking account for the induced pressure gradient in the composite system of boundary layer equations makes it possible to eliminate the singularity also for the solution describing the flows with distributed injection (Lipatov, 1977). The solution obtained in the paper of Lipatov (1977) corresponded to the weak interaction regime; in this case the effect of the induced pressure gradient began to have an effect only after the surface friction had been decreased almost to zero. Characteristic of the strong interaction regime is that the mutual influence of the boundary layer flow and the outer inviscid flow takes place along the whole body surface. Thus, the boundary layer detachment phenomenon must possess some distinctive features which differentiate it from the similar phenomenon in the weak interaction regime. These features are analyzed in this section.

### 4.7.1 Formulation of the problem and boundary conditions

We will consider the hypersonic flow of a viscous, heat-conducting gas over a plane wedge surface. The origin of a Cartesian coordinate system is at the leading edge, the $x\ell$ axis is aligned with the surface, and the $y\ell$ axis is normal to it (Fig. 4.17). In accordance with Hayes and Probstein (1966), the strong interaction regime is realized when

$$M_\infty \to \infty, \quad M_\infty \tau \to \infty$$

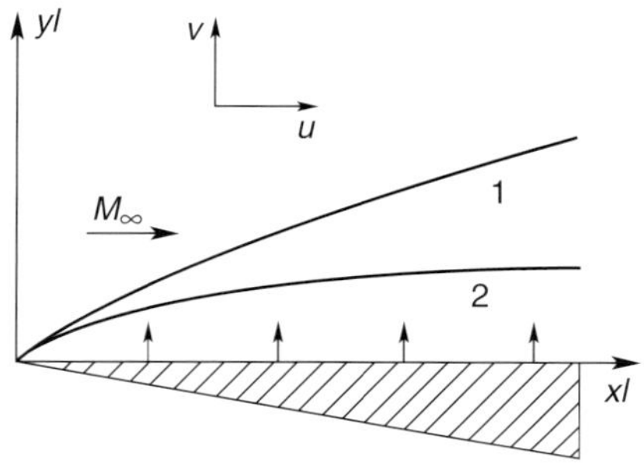


Fig. 4.17.

where $M_\infty$ is the freestream Mach number and $\tau$ is the boundary layer thickness. In accordance with the conventional estimates of theory of strong interaction in the boundary layer (region 2 in Fig. 4.17), here, as in the preceding sections, we introduce the following designations for the velocity vector components, the density, the pressure, the total enthalpy, and the dynamic viscosity: $u_\infty u_1$, $\tau u_\infty v_1$, $\tau^2 \rho_\infty \rho_1$, $\tau^2 \rho_\infty u_\infty^2 p_1$, $g_1 u_\infty^2/2$, and $\mu_0 \mu_1$. Here, the subscript $\infty$ refers to the freestream parameters and the subscript 0 to the viscosity corresponding to the stagnation temperature. The parameter $\tau$ characterizing the boundary layer thickness is expressed in terms of the Reynolds number: $\tau = Re_0^{-1/4}$, $Re_0 = \rho_\infty u_\infty \ell / \mu_0$. It is assumed that the plate surface is permeable and a gas of the same composition as in the oncoming flow is injected normal to the surface at a velocity $\tau u_\infty v_w$.

The system of boundary layer equations written in the Dorodnitsyn variables takes the form:

$$\frac{\partial u_1}{\partial x} + \frac{\partial v^*}{\partial \eta} = 0, \quad \frac{\partial p_1}{\partial \eta} = 0, \quad v^* = \rho_1 v_1 + u_1 \frac{\partial \eta}{\partial x} \tag{4.122}$$

$$u_1 \frac{\partial u_1}{\partial x} + v^* \frac{\partial u_1}{\partial \eta} + \frac{1}{\rho_1} \frac{\partial p_1}{\partial x} = \frac{\partial}{\partial \eta} \left( \rho_1 \mu_1 \frac{\partial u_1}{\partial \eta} \right)$$

$$u_1 \frac{\partial g_1}{\partial x} + v^* \frac{\partial g_1}{\partial \eta} = \frac{\partial}{\partial \eta} \left\{ \rho_1 \mu_1 \left[ \frac{1}{\sigma} \frac{\partial g_1}{\partial \eta} + \left(1 - \frac{1}{\sigma}\right) \frac{\partial}{\partial \eta} \left( \frac{u_1^2}{2} \right) \right] \right\}$$

$$p_1 = \frac{\gamma-1}{2\gamma}(g_1 - u_1^2)\rho_1, \quad \eta = \int_0^y \rho_1\,dy, \quad \delta_1 = \int_0^\infty \frac{d\eta}{\rho_1}$$

$$\rho_1\mu_1 = \frac{2\gamma}{(\gamma-1)}p_1(g_1 - u_1^2)^{n-1}, \quad u_1(x,0) = 0$$

$$g_1(x,0) = g_w, \quad v^*(x,0) = \rho_1(x,0)v_w, \quad u_1(x,\infty) = 1, \quad g_1(x,\infty) = 1$$

where $\sigma$ is the Prandtl number.

In accordance with strong interaction theory (Hayes and Probstein, 1966), the pressure distribution $p(x)$ entering in the boundary value problem is dependent on the boundary layer displacement thickness $\delta(x)$. For determining this dependence it is necessary to investigate the inviscid flow in region 1 (Fig. 4.17) located between the shock and the outer edge of the boundary layer. The flow in region 1 is described by hypersonic small perturbation theory (Chernyi, 1966). For further analysis we will use the approximate expression

$$p_1 = \frac{\gamma+1}{4}\left(\frac{d\delta}{dx}\right)^2 \tag{4.123}$$

given by the tangent wedge method.

The change of variables

$$\xi = x^{1/4}, \quad \lambda = \frac{\eta}{\xi}\left[\frac{(\gamma-1)}{8\gamma p(0)}\right]^{1/2}, \quad u = \frac{\partial\psi}{\partial\eta}, \quad v^* = -\frac{\partial\psi}{\partial x}$$

$$\psi = \xi f\left[\frac{8\gamma p(0)}{(\gamma-1)}\right]^{1/2}, \quad p_1 = \frac{p}{\xi^2}, \quad \rho_1 = \frac{\rho}{\xi^2}, \quad \delta_1 = \xi^3\delta, \quad g_1 = g$$

brings the boundary value problems (4.122) and (4.123) into the form:

$$(Nf'')' + ff'' + \frac{(\gamma-1)}{\gamma}\left(1 - \frac{\xi\dot{p}}{2p}\right)(g - f'^2) = \xi(f'\dot{f}' - \dot{f}f'') \tag{4.124}$$

$$\left(\frac{N}{\sigma}g'\right)' + fg' + \left[N\left(1 - \frac{1}{\sigma}\right)\frac{f'^2}{2}\right]' = \xi(f'\dot{g} - \dot{f}g')$$

$$f'(\xi,0) = 0, \quad g(\xi,0) = g_w, \quad f(\xi,0) = f_w, \quad f'(\xi,\infty) = 1, \quad g(\xi,\infty) = 1$$

$$\delta = 2\left[\frac{(\gamma-1)p(0)}{2\gamma p^2(\xi)}\right]^{1/2}\int_0^\infty (g - f'^2)\,d\lambda, \quad N = \frac{p(\xi)}{p(0)}$$

$$p = \frac{(\gamma+1)}{2}\left[\frac{3}{4}\delta + \frac{1}{4}\xi\frac{d\delta}{d\xi}\right]^2$$

$$f_w = -\left[\frac{2\gamma}{\gamma-1}\right]^{1/2}\frac{2v_w}{g_w\xi[p(0)]^{1/2}}\int\limits_0^{\xi} p(\xi)\xi\,d\xi, \quad f' = \frac{\partial f}{\partial\lambda}, \quad \dot{f} = \frac{\partial f}{\partial\xi}$$

The distribution $p(\xi)$ entering in Eq. (4.124) is not known beforehand and must be determined in solving the problem. The presence of the induced pressure gradient imparts new properties to the parabolic system of boundary layer equations; they are related with the upstream disturbance transfer and the associated nonuniqueness of the solution, as described in the paper of Neiland (1970*b*) and above in this chapter. An additional boundary condition imposed, for example, on the base section, $p(\xi=1)=B$, makes it possible to obtain a unique solution of the boundary value problem (4.124). For numerically solving the boundary value problems of this type the method developed in the work of Dudin and Lyzhin (1983) is used. The solution procedure consists in preassigning certain pressure and velocity fields in the region $(0 \le \xi \le 1; 0 \le \lambda < \infty)$. Then the linearized boundary value problem (4.124) is solved at the known pressure gradient, pressure distribution, and displacement thickness $\delta^i(\xi)$; as a result, a new, different from $\delta^i(\xi)$, displacement thickness distribution $\delta(\xi)$ is obtained. The next stage of calculations is related with the determination of the correction $\Delta(\xi)$ to the displacement thickness distribution. For this purpose, a linear differential second-order equation is used in which the nonhomogeneous term is proportional to the difference $\delta^i(\xi) - \delta(\xi)$. The calculation procedure is repeated for the new displacement thickness distribution $\delta^{i+1}(\xi) = \delta^i(\xi) + \Delta(\xi)$ and the corresponding pressure and pressure gradient distributions until the difference $\delta^i(\xi) - \delta(\xi)$ is not fairly small. This technique can also be applied for calculating boundary layers with return flows using oriented differences for approximating the convective derivatives.

### *4.7.2 Results of the solution*

Numerical solutions of the boundary value problem (4.124) were obtained for the following values of the parameters: $\sigma = n = 1$ and $\gamma = 1.4$. Figure 4.18 presents the calculated function $p(\xi)$ corresponding to the fixed value of the parameter $B = 1.02$ (proportional to the base pressure difference) and some values of the parameter $v_w$. Solid curves relate to the temperature factor $g_w = 1$ and dashed curves to $g_w = 0.5$. The qualitative difference between the solutions corresponding to the flows over the impermeable and permeable surfaces should be noted. The solution in the absence of injection is characterized by the constancy of the function $p(\xi)$ almost everywhere, except for a region adjoining the base section. Typical for the flows with injection are regions of rapid growth near the leading edge, almost constant values in the central part, and variations in the vicinity of the base section. Clearly, in the plateau region the function $p(\xi)$ is weakly dependent on the injection intensity and is determined by the temperature factor. Thus, with decrease in the temperature factor the maximum of $p(\xi)$ is lowered. Of course, these conclusions pertain only to the base pressure range considered, on which the flow does not contain return flow regions.

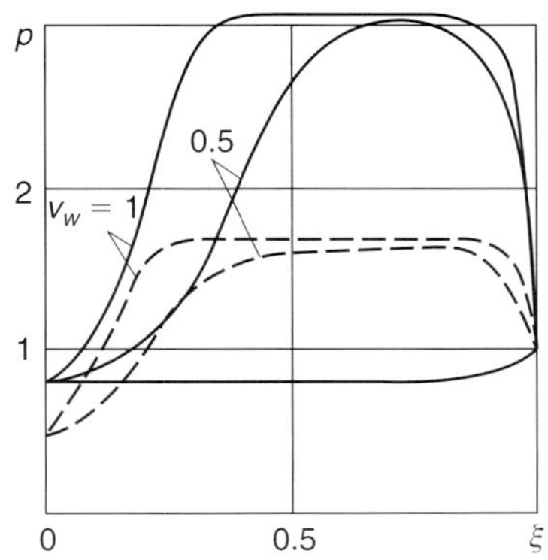


Fig. 4.18.

The results of the investigation of the parameter $B$ effect on the flow characteristics obtained for $v_w = 1$ and $g_w = 1$ are presented in Fig. 4.19. It should be noted that with variation in the base pressure difference the function $p(\xi)$ varies only near the base section. As shown in the paper of Neiland (1970), the solution of the boundary value problem (4.124), for example, for the function $p(\xi)$ can be represented, as $\xi \to 0$, in the form of a series involving an eigenfunction of the form $C\xi^a$, where $a$ is an eigenvalue and the constant $C$ is determined by the condition $p(\xi = 1) = B$. The results of the study of flows with injection preassigned by the condition $f_w(\xi) = -F$, obtained in the paper of Kovalenko (1974), indicate that an increase in the parameter $F$ leads to a decrease in the eigenvalue $a$. Correspondingly, the upstream disturbance transfer becomes more intense; in particular, at a fixed base pressure difference the pressure increases at any point near the leading edge. A weak dependence of the solution on the base pressure difference (Fig. 4.19) can be attributed to the fact that for $\xi \to 0$ and $f_w(\xi) = O(\xi)$ the injection regime under consideration ($v_w = \text{const}$) is characterized by a considerably smaller injection velocity than the regime corresponding to the condition $f_w = \text{const}$.

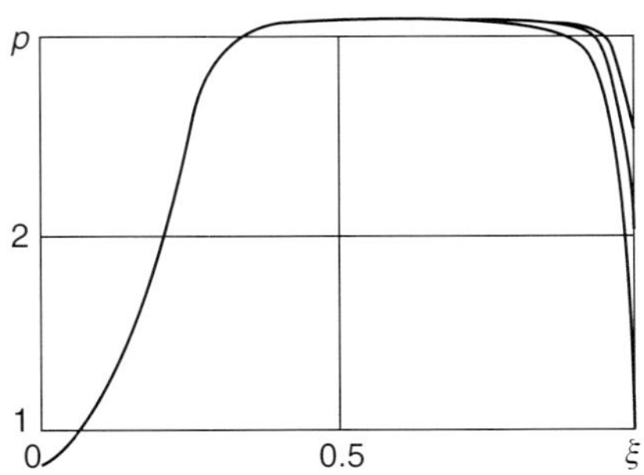


Fig. 4.19.

The results of calculations allow us to conclude that on the parameter range considered with increase in the base pressure difference the return flow region first appears near the

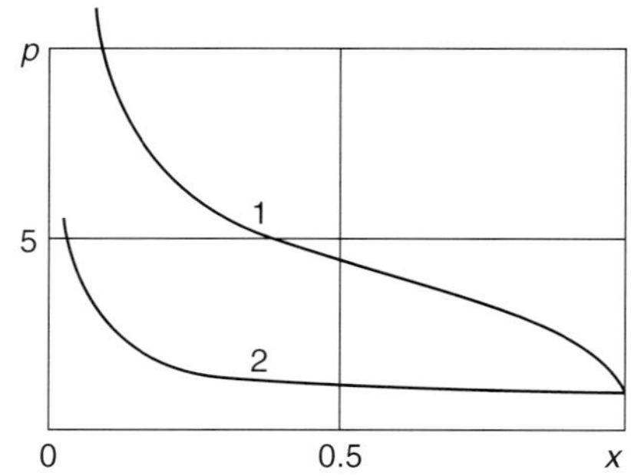


Fig. 4.20.

base section. To provide the explanation to this fact we must return to the determination of the function representing the ratio of the pressure distribution to the self-similar distribution corresponding to the flow over the impermeable surface. The pressure distributions $p_1(x)$ are plotted in the similarity variables in Fig. 4.20, in which curve 2 relates to the self-similar solution ($v_w = 0$, $g_w = 1$) and curve 1 is obtained for $v_w = 1$ and $g_w = 1$. Clearly, the pressure distributions are monotonic and associated with a negative pressure gradient on the entire surface. An increase in the base pressure difference leads to a change in the pressure distribution near the trailing edge, in particular, the occurrence of zero or negative surface friction is related with the reversion of the sign of the pressure gradient at $\xi = 1$.

Comparing the results of this study with the results of the investigation of the flows with uniformly distributed injection in the weak interaction regime (Lipatov, 1977) we note that typical of the strong interaction regime is the displacement of the pressure growth region toward the base section due to the effect of the boundary layer displacement thickness distribution ensuring a favorable pressure gradient.

In Fig. 4.21 we have presented the calculated results for the function $f_w''(\xi)$. Here, as for $p(\xi)$ (Fig. 4.19), three characteristic regions can be separated out, namely, those of a rapid decrease of the function $f_w''(\xi)$ near the leading edge, near-constant low values in the middle part, and a variation near the base section. A similar form is characteristic of the calculated

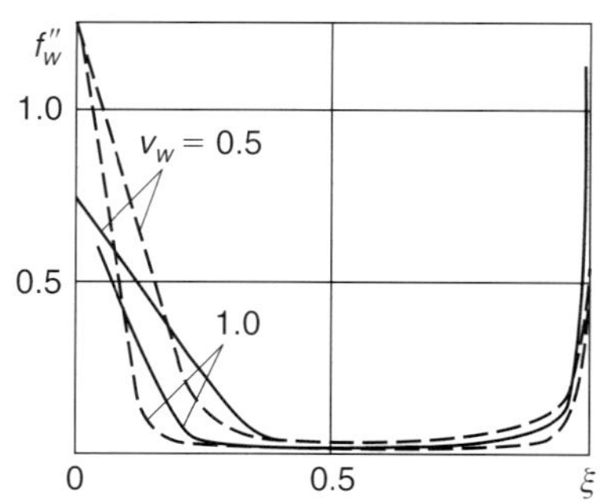


Fig. 4.21.

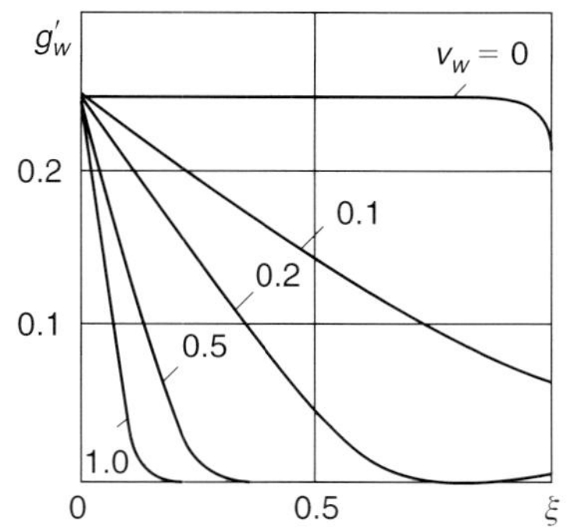


Fig. 4.22.

dimensionless heat fluxes $g'_w(\xi)$ obtained for $g_w = 0.5$ (Fig. 4.22). The existence of three characteristic regions in the flow over the permeable surface is related with different effects of diffusion, convection, and external forces (pressure gradient) on the flow in these regions. With increase in the boundary layer thickness (the distance from the zero streamline to the surface) the viscosity effect on the variation of the longitudinal momentum in the wall flow region is attenuated. In this case, the longitudinal momentum of the gas injected normal to the surface is produced to an increasingly larger degree by the action of the favorable pressure gradient. Though the surface friction produced by gas acceleration also decreases with increase in the injection intensity, the rate of the decay of the viscosity-produced friction turns out to be greater, which leads to the occurrence of the characteristic flow regions. As noted above, the existence of the third region is related with the base pressure difference effect.

# 5 Three-Dimensional Hypersonic Viscous Flows

This chapter is devoted to three-dimensional boundary layers on thin bodies in hypersonic flow in different viscous interaction regimes. Considerable attention is given to problems related with upstream disturbance transfer and the search for self-similar solutions, where it is possible.

## 5.1 Viscous flow over a low-aspect-ratio wing in the weak interaction regime (longitudinal–transverse interaction)

### *5.1.1 Special features of the formulation of the boundary value problem*

Within the framework of the classical theory of boundary layer (Prandtl, 1904) the problem of the asymptotic state of a viscous high-Reynolds-number flow past a rigid body leads to the study of the outer inviscid flow region and the boundary layer. The boundary layer is described by a parabolic system of partial differential equations, while a supersonic inviscid flow by a hyperbolic system. The solutions of boundary value problems for such systems possess the property that the distributions of the unknown functions over a certain spatial domain are determined by the boundary conditions imposed on a boundary located "upstream" of this domain. This situation takes place, for example, in the case of a flow past a thin body at moderate supersonic velocities or in the case of hypersonic flow if only the interaction between the boundary layer and the outer flow is weak. However, if the boundary conditions are not known beforehand and must be determined by joint solution of the problems for both regions, the situation changes. This pertains, in particular, to the flow with "free interaction" in the region located ahead of a separation point (Neiland, 1969*a*; see also Chapter 1) or the base section of a body (Matveeva and Neiland, 1967; see also Chapter 3), as well as the hypersonic flow past a finite-length plate (Neiland, 1970*b*) and the flow past a delta wing in the strong interaction regime (Kozlova and Mikhailov, 1970). In these problems, the outer flow and hence the pressure in the boundary layer is determined by the boundary layer displacement thickness distribution which is expressed in terms of the unknown functions of the layer in the form of an integral relation. A consequence of the integro-differential nature of the problem is that disturbances generated in the plane of symmetry of a delta wing may propagate upstream, up to its leading edges.

If in a hypersonic flow past a finite-span wing (plate) the outer inviscid flow is disturbed only slightly, then the problem reduces to a set of two-dimensional problems in each plane orthogonal to the plate and parallel to the freestream velocity. In this case, the boundary conditions for both the inviscid flow and the boundary layer are given. In fact, a variation of the pressure induced by the boundary layer in the direction transverse to the flow causes a negligibly small secondary flow in the boundary layer, so that the mechanism of the disturbance propagation toward the leading edge is absent. However, a decrease in the transverse dimension of the plate and the density in the boundary layer lead to an increase in the secondary flow velocity. As a result, a situation in which the secondary flow has an effect on the longitudinal flow, becomes possible. Then we arrive at a problem with an integral condition of the interaction over the entire surface of the plate under consideration.

### *5.1.2 Original relations and estimates*

In accordance with the paper of Ruban and Sychev (1973), we will consider a three-dimensional viscous hypersonic flow over a plate set at zero incidence (Fig. 5.1). The $x$ axis of a rectangular coordinate system is aligned with the freestream velocity $V_\infty$ in the plane of the plate, the $y$ axis is directed normal to the plate surface, and the $z$ axis is perpendicular to the above axes. We will denote the characteristic longitudinal dimension of the plate by $\ell$ and assume that the transverse dimension is of the order of $z_0\ell$, where $z_0$ is a parameter characterizing the wing aspect ratio.

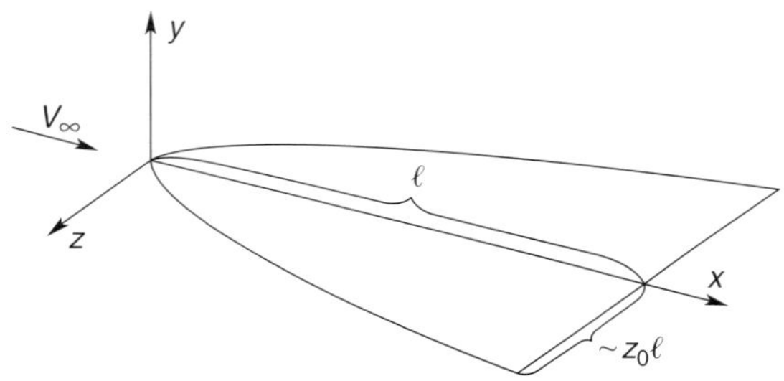


Fig. 5.1.

We will also introduce the designations: $\delta\ell$ is the characteristic thickness of the boundary layer; $u$, $v$, and $w$ are the velocity components in the $x$, $y$, and $z$ axes; $p$ is the pressure; $\rho$ is the density; and $M$ is the Mach number. The subscript $\infty$ refers to the freestream conditions.

In a hypersonic flow the outer edge of the boundary layer is, as is known, a quite definite quantity. We will write its equation in the form:

$$y_\delta = \ell\delta Y_0(X, Z) \tag{5.1}$$

where the dimensionless independent variables of the boundary layer are determined by the relations $\ell X$, $\ell z_0 Z$, and $\ell\delta Y$.

In the case under consideration, for a hypersonic flow it can be assumed

$$z_0 \gg \frac{1}{M_\infty}, \quad M_\infty \gg 1 \tag{5.2}$$

From these conditions it follows that the outer inviscid flow is two dimensional. Moreover, we will assume that the outer flow is weakly disturbed, that is

$$\varepsilon = M_\infty\delta \ll 1 \tag{5.3}$$

Condition (5.3) means that for the outer flow the linear theory holds true; from the theory there follows an expression for the pressure at the outer edge of the boundary layer for $M_\infty \gg 1$

$$p = p_\infty + p_\infty(M_\infty\delta)\left[\gamma\frac{\partial Y_0}{\partial X}(X,Z)\right] \tag{5.4}$$

This is the hypersonic counterpart of the Ackeret formula.

We will now consider the equations of the three-dimensional boundary layer (see, e.g., Struminskii, 1957; Loytsianskii, 1962). From the first momentum equation there follows:

$$\rho u\frac{\partial u}{\partial x} \sim \frac{\partial}{\partial y}\left(\mu\frac{\partial u}{\partial y}\right) \tag{5.5}$$

Using obvious estimates $H \sim V_\infty^2$ and $p \sim p_\infty$, where $H$ is the gas enthalpy in the boundary layer, we obtain $\rho \sim \rho_\infty/M_\infty^2$. Since in the boundary layer $\mu \sim \mu_0$, from relation (5.5) it follows that $\delta = M_\infty/\sqrt{Re_0}$, where $Re_0 = \rho_\infty V_\infty \ell/\mu_0$ and $\mu_0$ is the viscosity at the stagnation temperature.

From the third momentum equation and Eq. (5.4) we conclude that $\rho u(\partial w/\partial x) \sim \partial p/\partial z$ and hence $w \sim V_\infty M_\infty \delta/z_0$.

Let us consider the case in which the flow three dimensionality is important, that is, the quantity $z_0$ is so small that in the first momentum equation we have $u(\partial u/\partial x) \sim w(\partial u/\partial z)$. From this relation it follows that $z_0 = \sqrt{M_\infty\delta}$. We will present the pressure in the boundary layer as $p = p_\infty + p_\infty(M_\infty\delta)P$. We note that, in accordance with the second momentum equation, $P$ does not change across the boundary layer.

### *5.1.3 Equations and boundary conditions*

Thus, provided that relations (5.2) and (5.3) are fulfilled in the flow, introducing the dimensionless dependent variables $V_\infty U$, $V_\infty \delta V$, $V_\infty\sqrt{M_\infty\delta}W$, $V_\infty^2 g$, $(\rho_\infty/M_\infty^2)R$, $\mu_0\mu$, and $\sigma$ (the Prandtl number which is assumed here to be constant) we arrive at the system of equations

$$\begin{aligned}
RU\frac{\partial U}{\partial X} + RV\frac{\partial U}{\partial Y} + RW\frac{\partial U}{\partial Z} &= \frac{\partial}{\partial Y}\left(\mu\frac{\partial W}{\partial Y}\right)\\
RU\frac{\partial W}{\partial X} + RV\frac{\partial W}{\partial Y} + RW\frac{\partial W}{\partial Z} &= -\frac{\partial^2 Y_0}{\partial X\partial Z} + \frac{\partial}{\partial Y}\left(\mu\frac{\partial W}{\partial Y}\right)\\
RU\frac{\partial g}{\partial X} + RV\frac{\partial g}{\partial Y} + RW\frac{\partial g}{\partial Z} &= \frac{1}{\sigma}\frac{\partial}{\partial Y}\left(\mu\frac{\partial g}{\partial Y}\right) + \left(1-\frac{1}{\sigma}\right)\frac{\partial}{\partial Y}\left[\mu\frac{\partial}{\partial Y}\left(\frac{U^2}{2}\right)\right]\\
\frac{\partial(RU)}{\partial X} + \frac{\partial(RV)}{\partial Y} + \frac{\partial(RW)}{\partial Z} &= 0\\
g = \frac{1}{\gamma-1}\frac{1}{R} + \frac{U^2}{2}
\end{aligned} \tag{5.6}$$

and the corresponding system of boundary conditions

$$U = V = W = 0, \quad \frac{\partial g}{\partial Y} = 0 \quad (\text{or } g = g_w) \text{ for } Y = 0$$

$$U = 1, \quad W = 0, \quad g = \frac{1}{2} \text{ for } Y = Y_0(X, Z) \tag{5.7}$$

The problems (5.6) and (5.7) thus obtained can be written in the designations of the dimensional variables; then, using dimensionality theory, it can be shown that the flow past a wing with the planform governed by the equation $z = cx^{3/4}$ is self-similar. In this case, making in systems (5.6) and (5.7) written in the dimensional variables the following change of variables

$$s = 1 - \frac{z}{cx^{3/4}}, \quad \lambda = s^{-3/5} \frac{y}{\sqrt{x}} \sqrt{\frac{\gamma p_\infty}{\mu_0 V_\infty}}$$

$$u = V_\infty U, \quad v = V_\infty \sqrt{\frac{\mu_0 V_\infty}{\gamma p_\infty}} \frac{s^{-3/5}}{\sqrt{x}} V(s, \lambda) \tag{5.8}$$

$$w = V_\infty c \frac{s^{-1/5}}{\sqrt[4]{x}} W(s, \lambda), \quad \rho = \frac{\gamma p_\infty}{V_\infty^2} R(s, \lambda), \quad H = V_\infty^2 g(s, \lambda)$$

for the case $\sigma = 1$ and a thermally insulated wall we obtain the system of equations

$$R\left\{\left[\frac{3}{4}s^{1/5}(1-s)U - W\right]\left(s\frac{\partial U}{\partial s} - \frac{3}{5}\lambda\frac{\partial U}{\partial \lambda}\right) + \left(V - \frac{s^{6/5}}{2}\lambda U\right)\frac{\partial U}{\partial \lambda}\right\} = \frac{\partial}{\partial \lambda}\left(\mu\frac{\partial U}{\partial \lambda}\right)$$

$$R\left\{\left[\frac{3}{4}s^{1/5}(1-s)U - W\right]\left(s\frac{\partial W}{\partial s} - \frac{3}{5}\lambda\frac{\partial W}{\partial \lambda} - \frac{W}{5}\right) + \left(V - \frac{s^{6/5}}{2}\lambda U\right)\frac{\partial W}{\partial \lambda}\right.$$
$$\left. - \frac{1}{4}s^{6/5}UW\right\} = \frac{\partial}{\partial \lambda}\left(\mu\frac{\partial W}{\partial \lambda}\right) + \frac{s^{7/5}}{4\zeta^2}\left[3(1-s)(\delta s^{3/5})'' - (\delta s^{3/5})'\right] \tag{5.9}$$

$$\frac{3}{4}s^{1/5}(1-s)\left[s\frac{\partial(RU)}{\partial s} - \frac{3}{5}\lambda\frac{\partial(RU)}{\partial \lambda}\right] + \frac{\partial(RV)}{\partial \lambda} - \frac{s^{6/5}}{2}\lambda\frac{\partial(RV)}{\partial \lambda}$$
$$- \left[s\frac{\partial(RW)}{\partial s} - \frac{3}{5}\lambda\frac{\partial(RW)}{\partial \lambda} - \frac{RW}{5}\right] = 0$$

$$\frac{1}{R} = \frac{\gamma - 1}{2}(1 - U^2)$$

and the system of boundary conditions

$$U = V = W = 0 \quad \text{for } \lambda = 0 \tag{5.10}$$

$$U = 1, \quad W = 0 \quad \text{for } \lambda = \delta(s)$$

In the change of variables (5.8) the coefficients, which are some functions of $s$, are so chosen that the corresponding dimensionless variables are of the order of unity as the leading edge is approached.

In this case there are two similarity criteria

$$\zeta^2 = c^2\sqrt{\frac{\rho_\infty V_\infty}{\mu_0}\frac{1}{M_\infty^2}}, \quad \gamma = \frac{c_p}{c_v}$$

### 5.1.4 Eigenvalue problem

For the sake of simplicity, we will assume that the viscosity–enthalpy dependence is linear

$$\mu = 2h = (1 - U^2) \tag{5.11}$$

Let us determine the asymptotic expansion of the solution of systems (5.9) and (5.10) as $s \to 0$. We will present the required expansion in the form:

$$\begin{aligned}
U &= U_0(\eta) + s^{1/5}U_1(\eta) + \cdots + s^{n/5}U_n(\eta) + s^{\alpha}U_\alpha(\eta) + \cdots \\
V &= V_0(\eta) + s^{1/5}V_1(\eta) + \cdots + s^{n/5}V_n(\eta) + s^{\alpha}V_\alpha(\eta) + \cdots \\
W &= W_0(\eta) + s^{1/5}W_1(\eta) + \cdots + s^{n/5}W_n(\eta) + s^{\alpha}W_\alpha(\eta) + \cdots \\
R &= R_0(\eta) + s^{1/5}R_1(\eta) + \cdots + s^{n/5}R_n(\eta) + s^{\alpha}R_\alpha(\eta) + \cdots \\
\mu &= 2h_0(\eta) + s^{1/5}2h_1(\eta) + \cdots + s^{n/5}2h_n(\eta) + s^{\alpha}2h_\alpha(\eta) + \cdots \\
\lambda &= \lambda_0(\eta) + s^{1/5}\lambda_1(\eta) + \cdots + s^{n/5}\lambda_n(\eta) + s^{\alpha}\lambda_\alpha(\eta) + \cdots \\
\delta &= \delta_0 + s^{1/5}\delta_1 + \cdots + s^{n/5}\delta_n + s^{\alpha}\delta_\alpha + \cdots
\end{aligned}$$

where the independent variable $\eta$ is introduced by the equation $\eta = \frac{1}{A}\int_0^\lambda R(s,\lambda)\,d\lambda$ ($A$ is an arbitrary constant).

Here, it is assumed that $n/5 < \alpha < (n+1)/5$, where $n$ is a nonnegative integer.

From Eq. (5.11) it follows that $2h_0 = 1 - U_0^2$; ...; $2h_\alpha = -2U_0U_\alpha$; ... From the definition of the independent variable $\eta$ we obtain that $\lambda = A\int_0^\eta \frac{d\eta}{R(s,\eta)}$; thus, we have

$$\lambda_0(\eta) = A\int_0^\eta \frac{d\eta}{R_0(\eta)}, \cdots, \quad \lambda_\alpha(\eta) = -A\int_0^\eta \frac{R_\alpha(\eta)}{R_0^2(\eta)}\,d\eta, \cdots$$

Thus, substituting the expansions of the functions in systems (5.9) and (5.10) we obtain

$$\begin{aligned}
&\varphi_0 U_0' = \frac{5}{A^2(\gamma-1)}U_0'' \\
&(\varphi_0')^2 + 2\varphi_0\varphi_0'' + \frac{9}{20}\frac{\delta_0}{\zeta^2}(\gamma-1)(1-U_0^2) = \frac{10}{A^2(\gamma-1)}\varphi_0''' \\
&\varphi_0(0) = \varphi_0'(0) = \varphi_0'(\infty) = U_0(0) = 0, \quad U_0(\infty) = 1
\end{aligned} \tag{5.12}$$

where $\delta_0 = \frac{A}{2}(\gamma-1)\int_0^\infty (1-U_0^2)\,d\eta$

$$-\alpha\varphi_0' U_\alpha + \frac{2}{5}\varphi_0 U_0' + \left(\alpha + \frac{2}{5}\right)\varphi_\alpha U_0' = \frac{2}{A^2(\gamma-1)} U_\alpha''$$

$$\left(\frac{2}{5}-\alpha\right)\varphi_0'\varphi_\alpha' + \frac{2}{5}\varphi_0\varphi_\alpha'' + \left(\alpha+\frac{2}{5}\right)\varphi_0''\varphi_\alpha$$

$$= \frac{9}{50}(\gamma-1)\frac{\delta_0}{\zeta^2}U_0U_\alpha + \frac{3}{8}\frac{\delta_\alpha}{\zeta^2}(\gamma-1)\left(\alpha+\frac{3}{5}\right)\left(\alpha-\frac{2}{5}\right)(1-U_0^2) + \frac{2}{A^2(\gamma-1)}\varphi_\alpha''' \tag{5.13}$$

$$U_\alpha(0) = U_\alpha(\infty) = \varphi_\alpha(0) = \varphi_\alpha'(0) = \varphi_\alpha'(\infty) = 0$$

where $\delta_\alpha = -A(\gamma-1)\int_0^\infty U_0 U_\alpha\,d\eta$.

In expressions (5.12) and (5.13) it is assumed that $W_0(\eta) = \varphi_0'(\eta)$ and $W_\alpha(\eta) = \varphi_\alpha'(\eta)$. We note that Eqs. (5.12) determine uniquely the limiting velocity profile in the boundary layer as the leading edge of the wing is approached.

Equations (5.13) are homogeneous, that is, the functions $U_\alpha$ and $\varphi_\alpha$ can be multiplied by an arbitrary constant. This means that the expansion of the solution in the vicinity of the leading edge is nonunique.

Then we make the change of variables

$$\varphi_0 = \frac{4}{A^2(\gamma-1)}\Phi_0, \quad \varphi_\alpha = \frac{4}{A^2(\gamma-1)}\frac{\delta_\alpha}{\delta_0}\Phi_\alpha, \quad U_\alpha = \frac{\delta_\alpha}{\delta_0}\overline{U}_\alpha$$

and define the constant $A$ by the equation

$$\left[\frac{A^2(\gamma-1)}{4}\right]\frac{9}{20}(\gamma-1)\frac{\delta_0}{\zeta^2} = 1$$

Then from Eqs. (5.12) and (5.13) we obtain

$$\frac{4}{5}\Phi_0 U_0' = U_0'' \tag{5.14}$$

$$(\Phi_0')^2 + 2\Phi_0\Phi_0'' + (1-U_0^2) = \frac{5}{2}\Phi_0'''$$

$$\Phi_0(0) = \Phi_0'(0) = \Phi_0'(\infty) = U_0(0) = 0, \quad U_0(\infty) = 1$$

$$-\alpha\Phi_0'\overline{U}_\alpha + \frac{2}{5}\Phi_0\overline{U}_\alpha' + \left(\alpha+\frac{2}{5}\right)\Phi_\alpha U_0' = \frac{1}{2}\overline{U}_\alpha'' \tag{5.15}$$

$$\left(\frac{2}{5}-\alpha\right)\Phi_0'\Phi_\alpha' + \frac{2}{5}\Phi_0\Phi_\alpha'' + \left(\alpha+\frac{2}{5}\right)\Phi_0''\Phi_\alpha$$
$$= \frac{2}{5}U_0\overline{U}_\alpha + \frac{5}{6}\left(\alpha+\frac{3}{5}\right)\left(\alpha-\frac{2}{5}\right)(1-U_0^2) + \frac{1}{2}\Phi_\alpha'''$$

$$\overline{U}_\alpha(0) = \overline{U}_\alpha(\infty) = \Phi_\alpha(0) = \Phi_\alpha'(0) = \Phi_\alpha'(\infty) = 0$$

The eigenvalue $\alpha$ is determined from the integral condition

$$\int_0^\infty U_0\overline{U}_\alpha\,\mathrm{d}\eta + \frac{1}{2}\int_0^\infty (1-U_0^2)\,\mathrm{d}\eta = 0 \tag{5.16}$$

Thus, the functions $U_\alpha$ and $W_\alpha$ depend on the arbitrary constant $\delta_\alpha/\delta_0$ which can be determined only from an additional boundary condition imposed on the flow functions for $s = s_0 > 0$. This means that in the flow under consideration disturbances propagate upstream up to the leading edge. Of course, this reasoning is true only provided that there exists a positive number $\alpha$ satisfying system (5.15) and the integral condition (5.16).

Numerical integration of problem (5.14) was carried out using the sweeping technique with preliminary linearization of the equations. The calculated results are presented in Fig. 5.2 (solid curves).

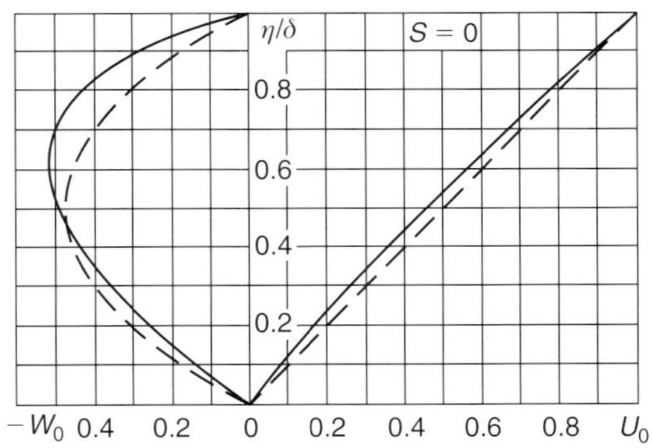


Fig. 5.2.

As can be seen from Eqs. (5.14)–(5.16), the eigenvalue $\alpha$ depends on neither the similarity criterion $\zeta$ nor the specific heat ratio $\gamma$. The eigenvalue $\alpha \approx 1.19$ was obtained from the calculations.

### 5.1.5 *Approximate calculation of the flow past a wing in the self-similar case*

The method of integral relations was used for determining approximately the flow parameters in the self-similar case. The condition of the flow symmetry about the wing axis, that is, $W|_{s=1} = 0$ was taken as an additional condition.

We will multiply the third equation of system (5.9) by $(1-U)$ and subtract the first equation from the result; we also sum up the second equation and the third equation multiplied by $W$. The equations thus obtained are integrated over the variable $\lambda$ from the wall to the outer edge of the boundary layer.

Approximating the longitudinal and transverse velocity profiles by the relations $\partial U/\partial\eta=(1-U)/\alpha(s)$, $W(s)=\alpha(s)U(1-U)$, where $\eta=\int_0^\lambda R\,d\lambda$, and performing the standard procedure of the method of integral relations under the assumption that the viscosity–enthalpy relation is linear, we obtain a system of ordinary differential equations.

The calculations were carried out for $\gamma=7/5$; their results corresponding to the case $\zeta=1$ are presented in Fig. 5.3. Here, the following designations are introduced

$$\bar{y}_0=\frac{y_\delta}{\sqrt{x}}\left(\sqrt{\frac{\mu_0 M_\infty^2}{\rho_\infty V_\infty}}\right)^{-1},\qquad \tau_u=\frac{\sqrt{x}}{V_\infty}\sqrt{\frac{\mu_0 M_\infty^2}{\rho_\infty V_\infty}}\left.\frac{\partial u}{\partial y}\right|_{y=0}$$

$$\tau_w=-\frac{x^{3/4}}{V_\infty}\sqrt[4]{\frac{\mu_0}{\rho_\infty V_\infty}}\left.\frac{\partial w}{\partial y}\right|_{y=0}$$

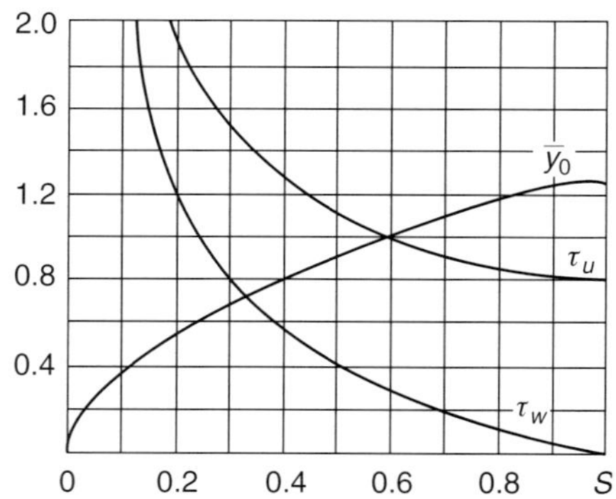


Fig. 5.3.

In Figs. 5.4 and 5.5 the calculated results for $\zeta=1$ and 1/3 (solid curves) are compared with the results of the calculation using the integral method with no account for longitudinal–transverse interaction (dashed curves). It should be noted that the distributions of the flow parameters over the entire wing surface are appreciably dependent on the chosen value of the ratio $\delta_\alpha/\delta_0$.

In Fig. 5.2 the dashed curves present the results of the calculation of the limiting velocity profile, as the wing edge is approached, by the integral method. Their comparison with the results of the solution of problem (5.14) indicates a fairly high accuracy of the integral method.

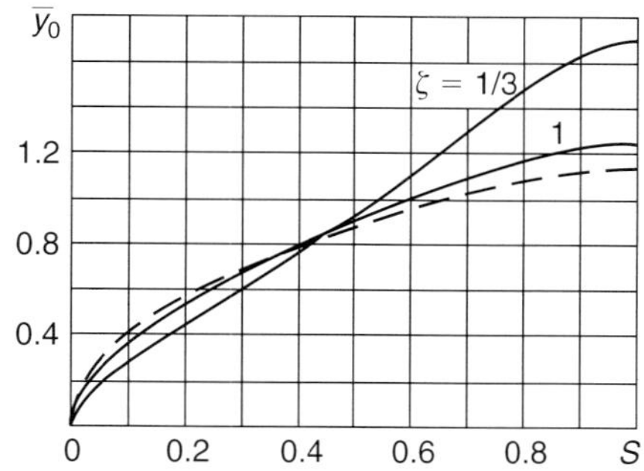


Fig. 5.4.

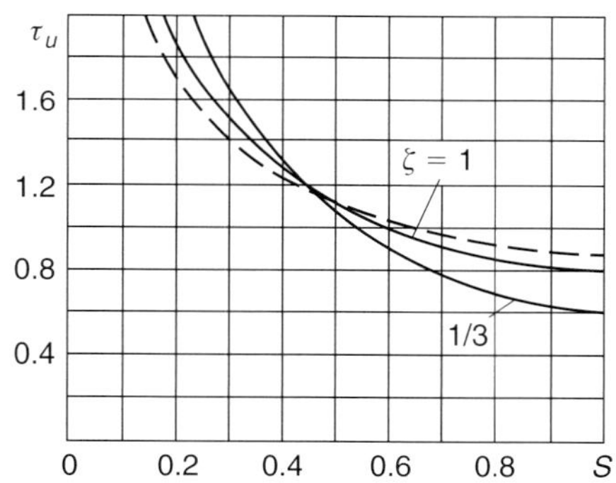


Fig. 5.5.

We note in conclusion that the conditions of applicability of relations (5.2) and (5.3) can be represented in a more illustrative form:

$$M_\infty \gg 1, \quad \frac{M_\infty^2}{\sqrt{Re_0}} \ll 1, \quad \frac{M_\infty^4}{\sqrt{Re_0}} \gg 1$$

### *5.1.6 Finite-difference method for solving the problem*

For solving the problem formulated in Section 5.1.3. Korolev (1980) proposed a stabilization method using an implicit approximation of the interaction condition in which the displacement thickness and pressure distributions satisfied this condition in each iteration stage. Using this method the boundary value problem was solved for a thin plate of planform $z = cx^{3/4}$ in the weak interaction regime over a wide range of similarity parameters. In the cited paper variables (5.8) are replaced by the following variables:

$$u = V_\infty U(s, Y_1), \quad v = V_\infty x^{-1/2} \left( \frac{\mu_0 M_\infty^2}{\rho_\infty V_\infty} \right)^{1/2} V_1(s, Y_1) \tag{5.17}$$

$$w = V_\infty c x^{-1/4} W_1, \quad \rho = \rho_\infty M_\infty^{-2} R, \quad H = V_\infty^2 g$$

$$s = 1 - \frac{z}{cx^{3/4}}, \quad Y_1 = yx^{-1/2}\left(\frac{\rho_\infty V_\infty}{M_\infty^2 \mu_0}\right)^{1/2}$$

Then instead of the system of partial differential equations (5.9) we obtain the following, more general system of equations

$$\left(\frac{3}{4}(1-s)U - W_1\right)\frac{\partial U}{\partial s} + \left(V_1 - \frac{Y_1 U}{2}\right)\frac{\partial U}{\partial Y_1} = \frac{1}{R}\frac{\partial}{\partial Y_1}\left(\mu \frac{\partial U}{\partial Y_1}\right) \tag{5.18}$$

$$\left(\frac{3}{4}(1-s)U - W_1\right)\frac{\partial W_1}{\partial s} + \left(V_1 - \frac{Y_1 U}{2}\right)\frac{\partial W_1}{\partial Y_1} - \frac{1}{4}UW_1$$
$$= \frac{1}{a_0^2 R}\left(\frac{1}{2}\delta_1(s) + \frac{3}{4}(1-s)\delta_1'(s)\right)' + \frac{1}{R}\frac{\partial}{\partial Y_1}\left(\mu\frac{\partial W_1}{\partial Y_1}\right)$$

$$\frac{3}{4}(1-s)\frac{\partial RU}{\partial s} - \frac{\partial RW_1}{\partial s} + \frac{\partial}{\partial Y_1}R\left(V_1 - \frac{YU_1}{2}\right) + \frac{R}{2}U = 0$$

$$\left(\frac{3}{4}(1-s)U - W_1\right)\frac{\partial g}{\partial s} + \left(V_1 - \frac{Y_1 U}{2}\right)\frac{\partial g}{\partial Y_1}$$
$$= \frac{1}{R}\left[\frac{1}{\sigma}\frac{\partial}{\partial Y_1}\left(\mu\frac{\partial g}{\partial Y_1}\right) + \left(1 - \frac{1}{\sigma}\right)\frac{\partial}{\partial Y_1}\left(\mu\frac{\partial}{\partial Y_1}\frac{U^2}{2}\right)\right]$$

$$g = \frac{1}{(\gamma-1)R} + \frac{U^2}{2}, \quad \gamma = \frac{c_p}{c_v}, \quad a_0^2 = c^2\left(\frac{\rho_\infty V_\infty}{\mu_0 M_\infty^4}\right)^{1/2}$$

and the corresponding system of boundary conditions

$$U = V_1 = W_1 = 0, \quad \frac{\partial g}{\partial Y_1} = 0 \,(\text{or } g = g_\omega) \text{ for } Y_1 = 0 \tag{5.19}$$

$$U = 1, \quad W = 0, \quad g = \frac{1}{2} \text{ for } Y_1 = \delta_1(s) \tag{5.20}$$

Here, $s$ and $Y_1$ are self-similar variables.

The system of equations (5.18) with the boundary conditions (5.19) and (5.20) is solved by a time-dependent method with an implicit scheme of the approximation of the interaction condition. For the sake of simplicity, it is assumed that $\sigma = 1$, $\gamma = 1.4$, and viscosity is a linear function of the enthalpy

$$\mu = 1 - U^2$$

The solution is sought in the region $0 < s \leq 1$. The condition of no flow in the transverse direction in the plane of symmetry of the wing is taken for the condition of the matching between the right and left halves of the wing, that is

$$\int_0^{Y_1} RW_1 \, \mathrm{d}Y_1 = 0$$

for $s = 1$.

In the vicinity of the leading edge the expansions for the functions $U$, $W_1$, $V_1$, and $\delta_1$ take the form:

$$\begin{aligned} U &= U_0(\xi) + s^{1/5}U_1(\xi) + O(s^{2/5}) \\ V_1 &= s^{-3/5}[V_{10}(\xi) + s^{1/5}V_{11}(\xi) + O(s^{2/5})] \\ W_1 &= s^{-1/5}[W_{10}(\xi) + s^{1/5}W_{11}(\xi) + O(s^{2/5})] \\ \delta_1 &= s^{3/5}[\delta_{10} + s^{1/5}\delta_{11} + O(s^{2/5})], \quad \xi = Y_1 s^{-3/5} \end{aligned}$$

Let us make the change of variables

$$U = U, \quad W_1 = s_1^{-1}W, \quad V_1 = s_1^{-3}\widetilde{V}$$

$$s^{1/5} = s_1, \quad \eta = s_1^{-3}\int_0^{Y_1} R \, \mathrm{d}Y_1$$

Then in the numerical solution the functions $U$, $\widetilde{V}$, and $W$, together with their derivatives, are regular as $s_1 \to 0$. In this case, the system of equations (5.18) takes the form:

$$\left[\frac{3}{4}s_1(1 - s_1^5)U - W\right]\frac{s_1}{5}\frac{\partial U}{\partial s_1} + V\frac{\partial U}{\partial \eta} = \frac{2}{\gamma - 1}\frac{\partial^2 U}{\partial \eta^2} \tag{5.21}$$

$$\left[\frac{3}{4}s_1(1 - s_1^5)U - W\right]\left(\frac{s_1}{5}\frac{\partial W}{\partial s_1} - \frac{W}{5}\right) + V\frac{\partial W}{\partial \eta} - \frac{1}{4}UWs_1^6$$
$$= \frac{2}{\gamma - 1}\frac{\partial^2 W}{\partial \eta^2} + \frac{1}{100a_0^2 R}[3(1 - s_1^5)s_1^2\delta_2'' + (6 - 11s_1^5)s_1\delta_2' - \delta_2(18 - 3s_1^5)] \tag{5.22}$$

$$\frac{\partial V}{\partial \eta} = \frac{s_1}{5}\frac{\partial W}{\partial s_1} - \frac{3}{20}s_1^2(1 - s_1^5)\frac{\partial U}{\partial s_1} + \frac{2}{5}W - \left[\frac{9}{20}s_1(1 - s_1^5) + \frac{s_1^6}{2}\right]U \tag{5.23}$$

where the following designations are introduced:

$$R = \frac{2}{\gamma - 1}\frac{1}{1 - U^2}$$

$$\delta_2(s) = \frac{\gamma - 1}{2}\int_0^{\infty}(1 - U^2)\,\mathrm{d}\eta \tag{5.24}$$

$$V = R\widetilde{V} - \frac{s_3^5 UY_1 R}{2} + s_1\frac{\partial \eta}{\partial s}\left[\frac{3}{4}(1 - s)s_1^5 U - W\right]$$

The boundary conditions are as follows:

$$U = V = W = 0 \quad \text{for } \eta = 0 \tag{5.25}$$

$$U \to 1 \quad W \to 0 \quad \text{for } \eta \to \infty \tag{5.26}$$

$$\int_0^\infty W \, \mathrm{d}\eta = 0 \quad \text{for } s_1 = 1$$

Substituting in systems (5.21)–(5.26) in the vicinity of the leading edge the expansions for the parameters $U$, $V$, and $W$ in the form:

$$\begin{aligned} U &= U_0(\eta) + s_1 U_1(\eta) + \cdots, \qquad V = V_0(\eta) + s_1 V_1(\eta) + \cdots \\ W &= W_0(\eta) + s_1 W_1(\eta) + \cdots, \qquad \delta_2 = \delta_0 + s_1 \delta_{21} + \cdots \end{aligned}$$

and passing to the limit, as $s_1 \to 0$, we obtain the system of ordinary differential equations for the functions $U_0$, $V_0$, and $W_0$

$$V_0 U_0' = \frac{2}{\gamma - 1} U_0' \tag{5.27}$$

$$\frac{W_0^2}{5} + V_0 W_0' = -\frac{18\delta_0}{100 a_0^2 R_0} + \frac{2}{\gamma - 1} W_0''$$

$$V_0' - \frac{2}{5} W_0 = 0, \qquad \frac{1}{R_0} = \frac{\gamma - 1}{2}(1 - U_0^2)$$

$$\delta_0 = \frac{\gamma - 1}{2} \int_0^\infty (1 - U_0^2) \, \mathrm{d}\eta$$

with the following boundary conditions

$$U_0 = V_0 = W_0, \qquad \eta = 0$$

$$U_0 \to 1, \qquad W_0 \to 0 \quad \text{for } \eta \to \infty$$

The solution of system (5.27) for $U_0$ and $W_0$ is taken as the initial profile of the boundary layer. The difference scheme for solving Eqs. (5.21)–(5.24) is constructed as follows. We add the terms $\partial U/\partial t$ and $\partial W/\partial t$ in Eqs. (5.21) and (5.22), respectively. We introduce a uniform grid $\{t^i, s_{1j}, \eta_k\}$ and designate the velocities at the gridpoints by $U_{j,k}^i$ and $W_{j,k}^i$. The maximum values of $j$ and $k$ are designated by $N$ and $M$, respectively. Let there be a certain velocity distribution at the time layer $t^i$. From Eq. (5.23) we determine the distribution of $V_{j,k}^i$.

Then we solve the time-dependent boundary layer equations. The difference scheme is as follows:

$$\frac{U_{j,k}^{i+1} - U_{j,k}^{i}}{\Delta t} + \left[\frac{3}{4}s_1(1 - s_1^5)U_{j,k}^{i} - W_{j,k}^{i+1}\right]\frac{M_1}{5}\frac{U_{j,k}^{i} - U_{j-M_1,k}}{\Delta s_1} + \left|V_{j,k}^{i}\right|\frac{U_{j,k}^{i+1} - U_{j,k-1}^{i+1}}{\Delta \eta}$$
$$= \frac{2}{\gamma - 1}\frac{U_{j,k+1}^{i+1} - 2U_{j,k}^{i+1} + U_{j,k-1}^{i+1}}{\Delta \eta^2} \tag{5.28}$$

$$\frac{W_{j,k}^{i+1} - W_{j,k}^{i}}{\Delta t} + \left[\frac{3}{4}s_1(1 - s_1^5)U_{j,k}^{i} - W_{j,k}^{i}\right]$$
$$\times\left(\frac{M_1}{5}\frac{W_{j,k}^{i} - W_{j-M_1,k}^{i}}{\Delta s_1} - \frac{W_{j,k}^{i}}{5}\right) + \left|V_{j,k}^{i}\right|\frac{W_{j,k}^{i+1} - W_{j,k-1}^{i+1}}{\Delta \eta} - \frac{s_1^6}{4}U_{j,k}^{i+1}W_{j,k}^{i}$$
$$= \frac{2}{\gamma - 1}\frac{W_{j,k+1}^{i+1} - 2W_{j,k}^{i+1} + W_{j,k-1}^{i+1}}{\Delta \eta^2} + \frac{PS_j^{i+1}}{100a_0^2R} \quad j = 2, N;\ k = 2, M - 1$$

Here, following the work of Klineberg and Steger (1974), the left-side or right-side approximation in $s_1$ is used depending on the sign of the expression

$$UW = \frac{3}{4}s_1(1 - s_1^5)U_{j,k}^{i+1} - W_{j,k}^{i+1}$$

The left-side approximation ($M_1 = 1$) is used when $UW \geq 0$ and the right-side one ($M_1 = -1$) when $UW < 0$. Similarly, the lower ($l = 1$) and upper ($l = -1$) approximations in $\eta$ are used depending on the sign of $V_{j,k}^{i}$. Here, $PS_j^{i+1}$ denotes the as yet unknown value of the pressure gradient

$$PS_j^{i+1} = [3s_1^2(1 - s_1^5)\delta_2'' + (6 - 11s_1^5)\delta_2's_1 - (18 - 3s_1^5)\delta_2]_j^{i+1} \tag{5.29}$$

As a result, from system (5.28) on each vertical line $j$ = const we obtain the system of equations in $U_{j,k}^{i+1}$ and $W_{j,k}^{i+1}$

$$a_kU_{j,k+1}^{i+1} + b_kU_{j,k}^{i+1} + c_kU_{j,k-1}^{i+1} + d_kW_{j,k}^{i+1} = f_k \tag{5.30}$$
$$g_kW_{j,k+1}^{i+1} + m_kW_{j,k}^{i+1} + l_kW_{j,k-1}^{i+1} + n_kU_{j,k}^{i+1} = r_kPS_j^{i+1} + rl_k$$

The solution of system (5.30) can be obtained by the matrix sweeping method

$$U_{j,k}^{i+1} = P_kU_{j,k+1}^{i+1} + Q_kW_{j,k+1}^{i+1} + R_kPS_j^{i+1} + \Phi_k$$
$$W_{j,k}^{i+1} = F_kU_{j,k+1}^{i+1} + T_kW_{j,k+1}^{i+1} + E_kPS_j^{i+1} + H_k \tag{5.31}$$

Using the conditions $U_{j,1}^{i+1}=0$ and $W_{j,1}^{i+1}=0$ we obtain the following values of the sweeping coefficients:

$$P_k = -\frac{a_k\Delta_4}{\Delta}, \quad Q_k = \frac{g_k\Delta_2}{\Delta}, \quad F_k = \frac{a_k\Delta_3}{\Delta}, \quad T_k = -\frac{g_k\Delta_1}{\Delta}$$

$$R_k = -\frac{\Delta_4 B1 + \Delta_2 B2}{\Delta}, \quad \Phi_k = \frac{\Delta_4 B4 - \Delta_2 B3}{\Delta}$$

$$E_k = \frac{\Delta_1 B2 + \Delta_3 B1}{\Delta}, \quad H_k = \frac{\Delta_1 B3 - \Delta_3 B4}{\Delta}$$

$$\Delta_1 = b_k + c_k P_{k-1}, \quad \Delta_2 = d_k + c_k Q_{k-1}$$

$$\Delta_3 = n_k + l_k F_{k-1}, \quad \Delta_4 = m_k + l_k T_{k-1}$$

$$\Delta = \Delta_1\Delta_4 - \Delta_2\Delta_3$$

$$B1 = c_k R_{k-1}, \quad B2 = r_k - l_k E_{k-1}$$

$$B3 = r_k - l_k H_{k-1}, \quad B4 = f_k - c_k \Phi_{k-1}$$

$$P_1 = Q_1 = F_1 = T_1 = R_1 = \Phi_1 = E_1 = H_1 = 0$$

Assuming that for $k=M$ the conditions $U_{j,M}^{i+1}=1$ and $W_{j,M}^{i+1}=0$ are fulfilled, we can reduce system (5.31) to a simple form:

$$\begin{aligned} U_{j,k}^{i+1} &= \alpha_{j,k} PS_j^{i+1} + \beta_{j,k} \\ W_{j,k}^{i+1} &= \gamma_{j,k} PS_j^{i+1} + \xi_{j,k} \end{aligned} \tag{5.32}$$

where $\alpha_{j,M}=\gamma_{j,M}=\xi_{j,M}=0$, $\beta_{j,M}=1$, and

$$\begin{aligned} \alpha_{j,k} &= R_k + P_k\alpha_{j,k+1} + Q_k\gamma_{j,k+1}, \quad \beta_{j,k} = \Phi_k + P_k\beta_{j,k+1} + Q_k\xi_{j,k+1} \\ \gamma_{j,k} &= E_k + F_k\alpha_{j,k+1} + T_k\gamma_{j,k+1}, \quad \xi_{j,k} = H_k + F_k\beta_{j,k+1} + T_k\xi_{j,k+1} \end{aligned}$$

For determining the unknown function $PS_j^{i+1}$ we will use the integral relation (5.24)

$$\delta_{2j}^{i+1} = \frac{\gamma-1}{2}\sum_{k=1}^{M}\left[1-\left(U_{j,k}^{i+1}\right)^2\right]\Delta\eta = \frac{\gamma-1}{2}\sum_{k=1}^{M}\left[1-(\alpha_{j,k}PS_j^{i+1}+\beta_{j,k})^2\right]\Delta\eta$$

Summing the expression on the right-hand side we obtain the relation between the unknown distributions $\delta_{2j}^{i+1}$ and $PS_j^{i+1}$:

$$\delta_{2j}^{i+1} = A_j(PS_j^{i+1})^2 + B_j PS_j^{i+1} + C_j \tag{5.33}$$

Thus, taking the interaction conditions into account, we obtain the following additional problem for determining the quantities $PS_j^{i+1}$ and $\delta_{2j}^{i+1}$:

$$PS_j^{i+1} = [3(1 - s_1^5)s_1^2\delta_2'' + (6 - 11s_1^5)\delta_2' s_1 - (18 - 3s_1^5)\delta_2]_j^{i+1} \quad (5.34)$$

At gridpoints Eq. (5.33) is used. The boundary conditions for $\delta_2$ are as follows:

$$\delta_2|_{s_1=0} = \delta_0, \quad \delta_2|_{s_1=1} = \delta_{2N}^{i+1}$$

The value of $\delta_{2N}^{i+1}$ is determined from the condition of no flow in the transverse direction. To determine it, it is sufficient to sum up the second equation (5.32) from $k = 1$ to $M$ for $j = N$ and equate the result to zero. Thence $PS_N^{i+1}$ is determined and then $\delta_{2N}^{i+1}$ from Eq. (5.33). Systems (5.33) and (5.34) are solved by the relaxation method. Given a certain distribution $PS_j^{i+1}, j = 2, \ldots, N-1$, it is substituted in the first term of the right-hand side of Eq. (5.33), while expression (5.34) is substituted in the next term for $PS_j^{i+1}$. The equation thus obtained can be solved using the sweeping technique, thus determining $\delta_{2N}^{i+1}$ and, therefore, a new distribution of $\widetilde{PS}_j^{i+1}$. The new approximation is taken with the lower relaxation

$$PS_j^{i+1} = PS_j^{i+1}(1 - os) + os \times \widetilde{PS}_j^{i+1}$$

The iteration procedure is repeated until it converges at a needed accuracy. Here, $os$ is a relaxation parameter. Thus, after the new distribution of $PS_j^{i+1}$ has been determined, new distributions of $U_{j,k}^{i+1}$ and $W_{j,k}^{i+1}$ satisfying the interaction condition can be obtained using Eq. (5.32). Then we go to a new time layer.

This numerical method can be used in solving time-dependent problems with both strong and weak interaction and separation.

### *5.1.7 Numerical results*

In numerical calculations a uniform grid with steps (0.03, 0.016, 0.4) was used; the relaxation parameter was taken to be equal $os = 0.8$. The numerical solution of the system was assumed to be completed if the difference between the values of the velocity at layer $i+1$ and the preceding time layer was less than $10^{-5}$. The same value determined the degree of convergence of the intermediate problems (5.33) and (5.34).

In Fig. 5.6 we have plotted the dimensionless displacement thickness and pressure versus the coordinate $s$ for different values of the similarity parameter $a_0$. In Figs. 5.7 and 5.8 the $s$-dependences of the longitudinal and transverse friction coefficients are presented. Here, the following designations are introduced

$$p_1 = (p - p_\infty)p_\infty^{-1}x^{1/2}M_\infty^{-2}\sqrt{\frac{\rho_\infty V_\infty}{\mu_0}}$$

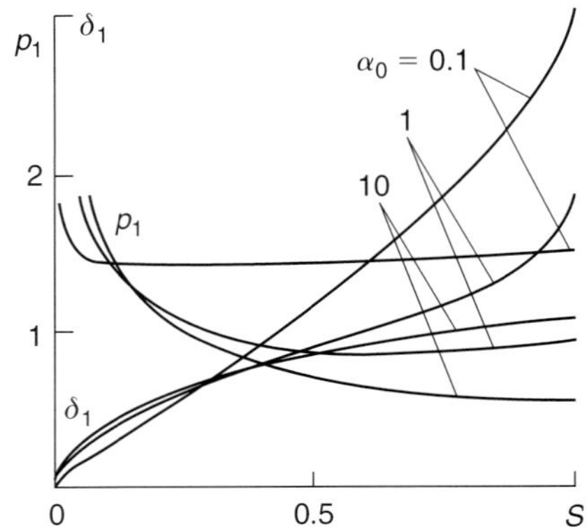


Fig. 5.6.

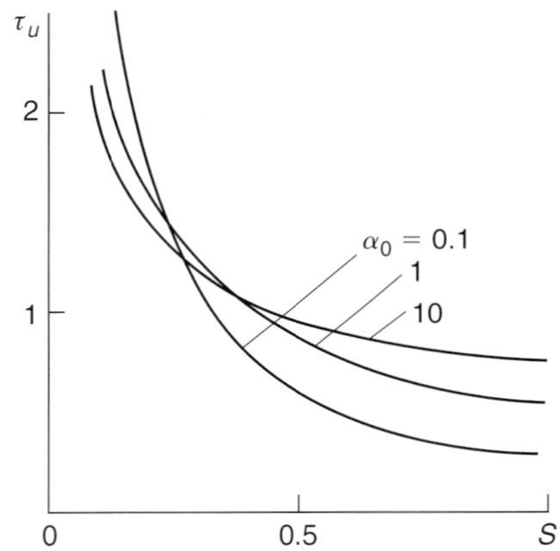


Fig. 5.7.

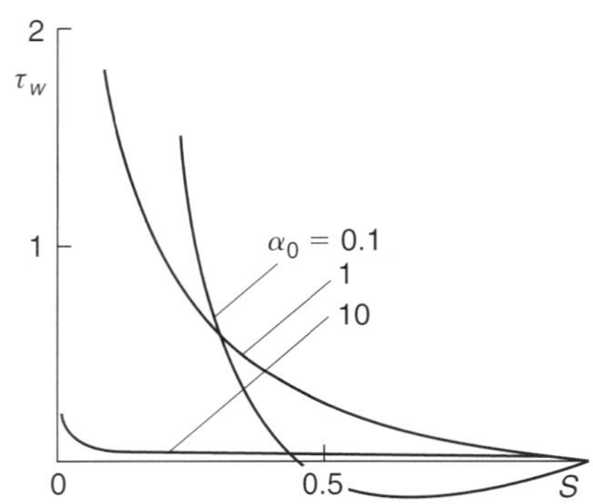


Fig. 5.8.

$$\tau_u = \frac{\partial U}{\partial y} x^{1/2} V_\infty^{-1} \sqrt{\frac{\mu_0 M_\infty^2}{\rho_\infty V_\infty}}\Bigg|_{y=0}$$

$$\tau_w = -\frac{\partial w}{\partial y} x^{3/4} c^{-1} V_\infty^{-1} M_\infty^2 \sqrt{\frac{\mu_0}{\rho_\infty V_\infty}}\Bigg|_{y=0}$$

The results of calculations show that near the axis of symmetry of the wing the derivative of the pressure with respect to $s$ vanishes, whereas the derivative of the displacement thickness increases without bound, as $(1-s)^{-1/3}$. However, the range of inapplicability of the boundary layer equations, where the slope of the displacement thickness is of the order of unity, has a relative value $\varepsilon^2 M_\infty^{-3}$, that is, it is negligibly small in the above formulation of the problem.

We call attention to the appearance of a pressure plateau at small values of $a_0$, which responds to a slow return flow in the transverse direction. In Fig. 5.9 the numerical solution thus determined is compared with the solution obtained by the integral method by Rubam and Sychev (1973) (dashed curves).

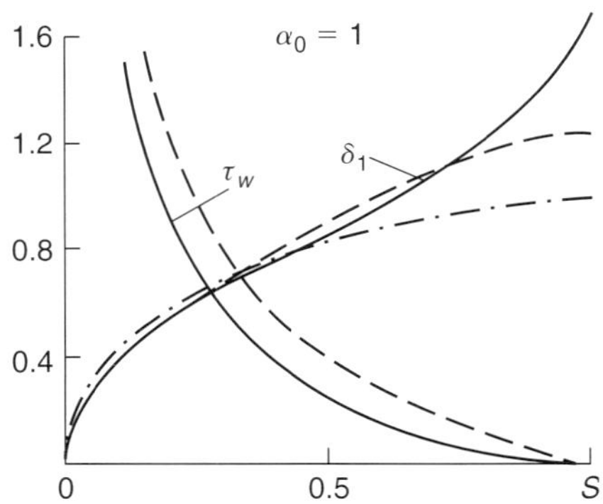


Fig. 5.9.

From the data presented it follows that for moderate values of $a_0$ the results are in fairly good agreement. The dot-and-dash curve in Fig. 5.9 presents the results of the calculation with no account for longitudinal–transverse interaction. Clearly, with decrease in $a_0$ the longitudinal–transverse interaction effect becomes more appreciable; for $a_0 = 0.1$ the flow pattern is presented in Fig. 5.10.

Thus, the numerical results show that longitudinal–transverse interaction has a noticeable effect on the gasdynamic function distributions in the boundary layer.

## 5.2 Formation of secondary flows on thin semi-infinite wings

### *5.2.1 Estimation of secondary flow parameters in boundary layers on thin wings*

Below in analyzing the hypersonic flow past thin wings we will assume that the gas is perfect, with constant specific heats ($\gamma = \text{const}$) and Prandtl number ($\sigma = \text{const}$), and the dynamic

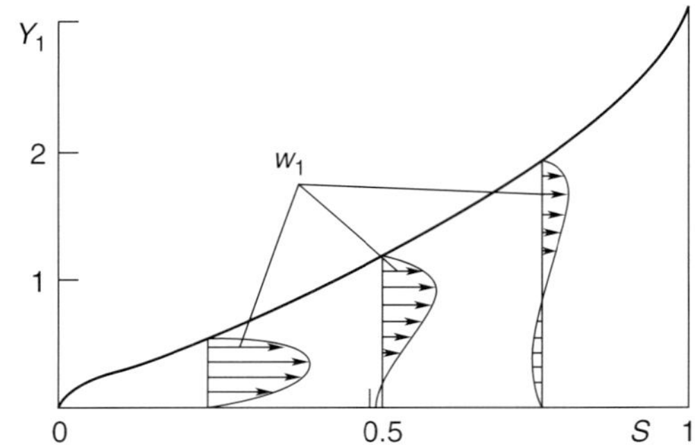


Fig. 5.10.

viscosity depends on the temperature in accordance with a power law, $\mu \sim T^n$ $(0.5 \leq n \leq 1)$. The assumption of hypersonic small perturbation theory is also fulfilled: $M_\infty \tau \geq O(1)$, where $\delta$ and $\tau$ are the characteristic dimensionless thicknesses of the boundary layer and the wing, respectively.

For solving the problem of the hypersonic flow past a thin body by the method of matched outer and inner asymptotic expansions, the Navier–Stokes equations are written in the corresponding dimensionless variables and the triple passage to the limit is performed

$$M_\infty \to \infty, \quad Re \to \infty, \quad \tau \to 0$$

Here, $M_\infty$ is the freestream Mach number and $Re = \rho_\infty u_\infty \ell / \mu_0$ is the Reynolds number based on the freestream density and velocity, a scale length of the body, and the viscosity at the freestream stagnation temperature.

As noted above, this procedure results in a double-layer flow pattern for the leading terms of the expansion: the outer inviscid flow region is governed by hypersonic small perturbation theory (Van Dyke, 1964; Hayes and Probstein, 1966) and the inner viscous region is governed by the boundary layer equations.

The systems of equations for the outer and inner flow regions should generally $(\delta \geq \tau)$ be solved jointly, since the pressure distribution on the outer edge of the boundary layer is not known beforehand and must be determined in the process of solving the problem.

An analysis of the higher-order approximations in theory of strong interaction (Matveeva and Sychev, 1965; Bush, 1966; Lee and Cheng, 1969) showed that the region of transition from the boundary layer to the inviscid shock layer has an effect on the main flow only in higher-order approximations.

The specific features of the hypersonic flow past thin wings make it possible to simplify the differential equations governing the gas flows in the outer and inner regions. For this purpose, we will derive certain estimates for each flow region.

In accordance with the conventional assumptions of hypersonic small perturbation theory, in the inviscid flow $M_\infty \tau \geq O(1)$. Thus, the relative value of the pressure disturbance $\Delta p^0 / p_\infty \geq O(1)$, since, in accordance with the work of Chernyi (1966), we have

$$\frac{p^0}{p_\infty} \sim M_\infty^2 (\tau + \delta)^2 \geq 1 \tag{5.35}$$

where $\delta$ is the dimensionless boundary layer thickness divided by $\ell$, $p_\infty$ is the static pressure in the undisturbed flow, and $p^0$ is the pressure in the disturbed flow.

In accordance with strip theory (Hayes and Probstein, 1966; Lunev, 1975) in the flow past thin wings with $\tau/z_0 \ll 1$ the velocity component directed along the wing span is of the order

$$\frac{w_0 \approx u_\infty \tau^2}{z_0} \ll 1 \tag{5.36}$$

where $z_0$ is the aspect ratio characterizing the relation between the transverse and longitudinal wing dimensions.

The perfect gas flow in a three-dimensional laminar boundary layer in an arbitrary coordinate system $\xi$, $\eta$, $\zeta$, in which the coordinate surface $\eta = 0$ coincides with the body surface, is described by the following system of equations (see, e.g., Struminskii, 1957; Loytsianskii, 1966)

$$\frac{\partial}{\partial \xi}(h_2 \rho^0 u^0) + \frac{\partial}{\partial \zeta}(h_1 \rho^0 w^0) + \frac{\partial}{\partial \eta}(h_1 h_2 \rho^0 v^0) = 0 \tag{5.37}$$

$$\frac{u^0}{h_1}\frac{\partial u^0}{\partial \xi} + \frac{w^0}{h_2}\frac{\partial u^0}{\partial \zeta} + v^0 \frac{\partial u^0}{\partial \eta} + \frac{u^0 w^0}{h_1 h_2}\frac{\partial h_1}{\partial \zeta} - \frac{w^{02}}{h_1 h_2}\frac{\partial h_2}{\partial \xi} = -\frac{1}{\rho^0 h_1}\frac{\partial p^0}{\partial \xi} + \frac{1}{\rho^0}\frac{\partial}{\partial \eta}\left(\mu^0 \frac{\partial u^0}{\partial \eta}\right)$$

$$\frac{u^0}{h_1}\frac{\partial w^0}{\partial \xi} + \frac{w^0}{h_2}\frac{\partial w^0}{\partial \zeta} + v^0 \frac{\partial w^0}{\partial \eta} + \frac{u^0 w^0}{h_1 h_2}\frac{\partial h_2}{\partial \xi} - \frac{u^{02}}{h_1 h_2}\frac{\partial h_1}{\partial \zeta}$$
$$= -\frac{1}{\rho^0 h_2}\frac{\partial p^0}{\partial \zeta} + \frac{1}{\rho^0}\frac{\partial}{\partial \eta}\left(\mu^0 \frac{\partial w^0}{\partial \eta}\right)$$

$$\frac{u^0}{h_1}\frac{\partial H^0}{\partial \xi} + \frac{w^0}{h_2}\frac{\partial H^0}{\partial \zeta} + v^0 \frac{\partial H^0}{\partial \eta} = \frac{1}{\rho^0}\frac{\partial}{\partial \eta}\left\{\mu^0\left[\frac{1}{\sigma}\frac{\partial H^0}{\partial \eta} - \frac{1-\sigma}{\sigma}\frac{\partial}{\partial \eta}\left(\frac{u^{02}+w^{02}}{2}\right)\right]\right\}$$

$$\frac{\partial p^0}{\partial \eta} = 0$$

with the boundary conditions

$$u^0 = w^0 = 0, \quad v^0 = v_w^0, \quad H^0 = H_w^0 \ \text{for}\ \eta = 0$$

$$u^0 \to u_e, \quad w^0 \to w_e, \quad H^0 \to H_e^0 \ \text{for}\ \eta \to \infty$$

Here, $u^0$, $w^0$, and $v^0$ are the velocity components directed along the $\xi$, $\eta$, and $\zeta$ coordinate axes, respectively, $\rho^0$ is the density, $p^0$ is the pressure, $H^0$ is the total enthalpy, $\mu^0$ is dynamic viscosity, $\sigma$ is the Prandtl number, and $h_1 = h_1(\xi, \zeta)$ and $h_2 = h_2(\xi, \zeta)$ are the scale factors. The subscript $e$ refers to the gasdynamic parameters at the outer edge of the boundary layer and $w$ to those at the body surface.

The system of equations (5.37) represents a system of nonlinear partial differential equations, whose solution can be obtained using methods of numerical analysis.

For thin wings with the relative thickness $\tau$, whose surfaces are described by smooth functions with continuous principal curvatures and their derivatives, the principal radii of curvature are of the order $O(\ell/\tau, z_0\ell/\tau)$. In this case, the terms proportional to $K_2 = (1/h_1h_2)\partial h_1/\partial\eta$ and $K_1 = (1/h_1h_2)\partial h_2/\partial\eta$ can be dropped out from the momentum equations of system (5.37), since $K_1$ and $K_2$ are the local curvatures of the lines $\xi = \text{const}$ and $\eta = \text{const}$ and are of the order $O(\tau\ell, \tau/z_0\ell)$.

Within the boundary layer the gas temperature is of the order of the stagnation temperature $T_0$ (Hayes and Probstein, 1966); then from the equation of state, with account for relation (5.35), we obtain the estimate for the density

$$\frac{\rho^0}{\rho_\infty} \sim \frac{p^0}{p_\infty}\frac{T_\infty}{T_0} \sim \frac{(\tau+\delta)^2 M_\infty^2}{\left[1+\left(\frac{\gamma-1}{2}\right)M_\infty^2\right](\tau+\delta)^2} \tag{5.38}$$

In estimating the transverse flow velocity $w^0$ (Dudin and Neiland, 1976; 1977) directed along the wing span, the fact that in the inviscid flow region this velocity component is small (5.36) and in the boundary layer is induced only by the pressure gradient in this direction (if the terms proportional to $K_1$ and $K_2$ are neglected) must be taken into account. Then from the transverse momentum equation there follows:

$$\frac{u^0}{h_1}\frac{\partial w^0}{\partial\xi} + \frac{w^0}{h_2}\frac{\partial w^0}{\partial\zeta} \sim \frac{1}{\rho^0 h_2}\frac{\partial p^0}{\partial\zeta} + \cdots \tag{5.39}$$

Using relations (5.35) and (5.38) and the estimate for the longitudinal velocity component in the disturbed flow $u^0 \sim u_\infty$ (Hayes and Probstein, 1966) and considering the case in which the scale factors $h_1, h_2 \approx 1$, from Eq. (5.39) we obtain

$$u_\infty w^0 + \frac{w^{0^2}}{z_0} \sim \frac{p_\infty}{\rho_\infty}\frac{M_\infty^2}{z_0} \sim \frac{u_\infty^2}{z_0} \tag{5.40}$$

Therefore, $w^0/u_\infty \sim 1$ for $z_0 \le O(1)$.

We note that for $z_0 \approx 1$ all the terms of Eq. (5.40) are of the same order; however, for $z_0 \ll 1$ (low-aspect-ratio wing) the first term turns out to be less in the order than the two other terms. The large value of the transverse flow velocity in the boundary layer is attributed to small values of the gas density. It is useful to recall that in the inviscid flow region, where the density is high, the pressure gradient $\partial p^0/\partial\zeta^0$ of the same order induces only small transverse velocities of the order $u_\infty\tau^2/z_0$ (5.36).

Going over to the case of low-aspect-ratio wings ($z_0 \ll 1$) and taking Eq. (5.40) into account, we note that in all boundary layer equations (5.37) the terms incorporating derivatives with respect to $\xi$ for $z_0 \ll 1$ turn out to be small

$$u^0\frac{\partial}{\partial\xi} \ll w^0\frac{\partial}{\partial\zeta}, \qquad \frac{\partial\rho^0 u^0}{\partial\xi} \ll \frac{\partial\rho^0 w^0}{\partial\zeta}$$

As shown below, this fact makes it possible to reduce the equations of the three-dimensional boundary layer to two-dimensional equations.

We note that if the main condition (5.35) is violated, then, in accordance with Section 5.1 (see also Ruban and Sychev, 1973), the transverse flow velocities are small $[w^0/u_\infty \sim M_\infty(\delta+\tau)/z_0]$ and for $z_0 \sim [M_\infty(\tau+\delta)]^{1/2}$ the boundary layer equations remain three dimensional. However, for $z_0 \ll [M_\infty(\tau+\delta)]^{1/2}$ the equations of the problem considered in the work of Ruban and Sychev (1973) also degenerate to two-dimensional ones.

It is important to note that the above estimates do not suggest a laminar flow and lean upon neither assumptions on the relation between the characteristic thicknesses of the wing and the boundary layer. For this reason, they are valid for all types of interaction, from weak to strong, provided condition (5.35) is fulfilled.

For the laminar boundary layer flows of the type under consideration, the interaction intensity is obviously determined by the parameter

$$N^* = \frac{1}{\tau Re^{1/4}} \tag{5.41}$$

characterizing the boundary-layer-to-wing thickness ratio. In fact, on the basis of the equality of the orders of the leading terms of the longitudinal momentum equation and with account for Eq. (5.38), the following estimate for $\delta$ can easily be derived

$$\delta \sim (\tau^2 + Re^{-1/2})^{1/2} - \tau = Re^{-1/4}\left[\left(\frac{1}{N^{*2}}+1\right)^{1/2} - \frac{1}{N^*}\right] \tag{5.42}$$

Formula (5.42) is valid for the moderate interaction regime ($N^* \sim O(1)$). For the limiting strong ($N^* \to \infty$) and weak ($N^* \to 0$) interaction regimes it can be simplified and takes the form:

$$\delta \sim Re^{-1/4} \ \text{for}\ N^* \to \infty, \quad \delta \sim N^* Re^{-1/4} \sim \tau^{-1} Re^{-1/2} \ \text{for}\ N^* \to 0$$

### 5.2.2 *Thin semi-infinite wing at zero incidence*

The problem is solved in a Cartesian coordinate system (Fig. 5.11) with the $x^0$ axis aligned with the freestream velocity $u_\infty$ and the spanwise $z^0$ axis. The velocity vector components $u^0$, $v^0$, and $w^0$ are directed along the $x^0$, $y^0$, and $z^0$ axes. The wing airfoil shape is given by the equation

$$y^0 = \delta_w^0(x^0, z^0) \tag{5.43}$$

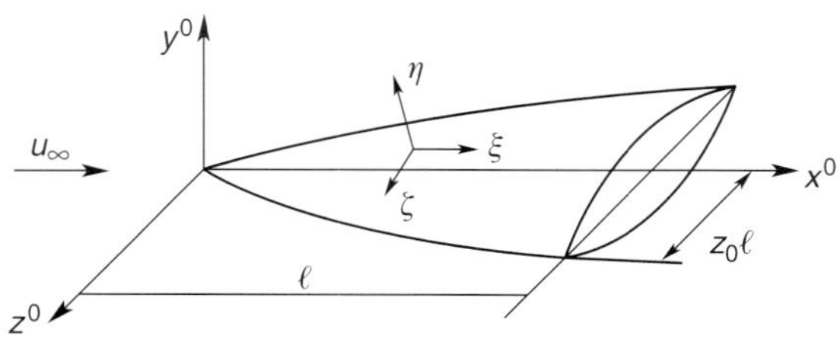


Fig. 5.11.

*5.2.2.1 Inner flow region*

In the boundary layer the perfect gas flow is described by the system of Navier–Stokes equations, which in the coordinate system presented above takes the form:

$$\frac{\partial(\rho^0 u^0)}{\partial x^0} + \frac{\partial(\rho^0 v^0)}{\partial y^0} + \frac{\partial(\rho^0 w^0)}{\partial z^0} = 0$$

$$\rho^0 u^0 \frac{\partial u^0}{\partial x^0} + \rho^0 v^0 \frac{\partial u^0}{\partial y^0} + \rho^0 w^0 \frac{\partial u^0}{\partial z^0} + \frac{\partial p^0}{\partial x^0} = \frac{\partial}{\partial y^0}\left(\mu^0 \frac{\partial u^0}{\partial y^0}\right) + \cdots$$

$$\frac{\partial p^0}{\partial y^0} = 0 + \cdots$$

$$\rho^0 u^0 \frac{\partial w^0}{\partial x^0} + \rho^0 v^0 \frac{\partial w^0}{\partial y^0} + \rho^0 w^0 \frac{\partial w^0}{\partial z^0} + \frac{\partial p^0}{\partial z^0} = \frac{\partial}{\partial y^0}\left(\mu^0 \frac{\partial w^0}{\partial y^0}\right) + \cdots$$

$$\rho^0 u^0 \frac{\partial H^0}{\partial x^0} + \rho^0 v^0 \frac{\partial H^0}{\partial y^0} + \rho^0 w^0 \frac{\partial H^0}{\partial z^0} = \frac{\partial}{\partial y^0}\left\{\mu^0\left[\frac{1}{\sigma}\frac{\partial H^0}{\partial y^0} - \frac{1-\sigma}{2\sigma}\frac{\partial(u^{02}+w^{02})}{\partial y^0}\right]\right\} + \cdots$$

where $H^0 = \frac{\gamma}{\gamma-1}\frac{p^0}{\rho^0} + \frac{u^{02}+w^{02}}{2}$

Here, the terms, which are small in the flow region under consideration, are omitted. For thin wings variables fitted to the body surface can be introduced as follows:

$$(x^0, y^0, z^0) \to (x^0, y_w^0, z^0), \quad y_w^0 = y^0 - \delta_w^0(x^0, z^0) \tag{5.44}$$

Let us also introduce the following transformation of the velocity components

$$(u^0, v^0, w^0) \to (u^0, v_w^0, w^0), \quad v_w^0 = v^0 - u^0\frac{\partial \delta_w^0}{\partial x^0} - w^0\frac{\partial \delta_w^0}{\partial z^0} \tag{5.45}$$

The change of variables (5.44) and (5.45) does not alters the form of the written-down terms of the system of Navier–Stokes equations, if $(v_w^0, y_w^0)$ are written everywhere for $(v^0, y^0)$.

In accordance with the conventional estimates for the boundary layer in a hypersonic flow (Hayes and Probstein, 1966) and with account for Eq. (5.40), we introduce the dimensionless variables of the order of unity

$$\rho^0 = \rho_\infty \Theta^2 \rho, \quad p^0 = \rho_\infty u_\infty^2 \Theta^2 p, \quad H^0 = \frac{u_\infty^2 g}{2}, \quad u^0 = u_\infty u \tag{5.46}$$

$$v_w^0 = u_\infty \delta z_0^{-1} v, \quad w^0 = u_\infty w, \quad \mu^0 = \mu_0 \mu, \quad x^0 = \ell x$$

$$y_w^0 = \ell\delta y, \quad z^0 = \ell z_0 z, \quad \delta_w^0 = \tau\ell\delta_w, \quad \delta_\varepsilon^0 = \delta\ell\delta_\varepsilon, \quad \delta = \frac{1}{\Theta}\sqrt{\frac{z_0}{Re}}$$

where $\delta_\varepsilon^0$ is the dimensionless boundary layer thickness, $\Theta=\tau$ for weak and moderate interaction, and $\Theta=\delta$ for strong interaction. Going over to the Dorodnitsyn variables

$$(x,\, y,\, z) \to (x,\, \lambda,\, z), \quad \lambda = \int_0^y \rho\, \mathrm{d}y$$

$$(u,\, v,\, w) \to (u,\, v_\delta,\, w), \quad v_\delta = \rho v + w\frac{\partial \lambda}{\partial z} + z_0 u \frac{\partial \lambda}{\partial x}$$

and dropping out the terms of the first and higher orders of smallness, from the system of the Navier–Stokes equations we obtain the following system

$$z_0 \frac{\partial u}{\partial x} + \frac{\partial v_\delta}{\partial \lambda} + \frac{\partial w}{\partial z} = 0 \tag{5.47}$$

$$z_0 u \frac{\partial u}{\partial x} + v_\delta \frac{\partial u}{\partial \lambda} + w\frac{\partial u}{\partial z} = \frac{z_0}{\rho}\frac{\partial p}{\partial x} + \frac{\partial}{\partial \lambda}\left(\rho\mu \frac{\partial u}{\partial \lambda}\right)$$

$$z_0 u \frac{\partial w}{\partial x} + v_\delta \frac{\partial w}{\partial \lambda} + w\frac{\partial w}{\partial z} = -\frac{1}{\rho}\frac{\partial p}{\partial z} + \frac{\partial}{\partial \lambda}\left(\rho\mu \frac{\partial w}{\partial \lambda}\right)$$

$$\frac{\partial p}{\partial \lambda} = 0$$

$$z_0 u \frac{\partial g}{\partial x} + v_\delta \frac{\partial g}{\partial \lambda} + w\frac{\partial g}{\partial z} = \frac{\partial}{\partial \lambda}\left\{\rho\mu\left[\frac{1}{\sigma}\frac{\partial g}{\partial \lambda} - \frac{1-\sigma}{\sigma}\frac{\partial (u^2+w^2)}{\partial \lambda}\right]\right\}$$

with the boundary conditions (with account for estimates (5.36))

$$u = v_\delta = w = 0, \quad g = g_w \ \text{ for } \lambda = 0 \tag{5.48}$$

$$u \to 1, \quad w \to 0, \quad g \to 1 \ \text{ for } \lambda \to \infty$$

where $g = \frac{2\gamma}{\gamma-1}\frac{p}{\rho} + u^2 + w^2$ and $\mu = (g - u^2 - w^2)^\omega$

For determining the boundary layer thickness it is necessary to take into account that $\lambda = \int_0^y \rho\, \mathrm{d}y$. Then we have $\int_0^\delta \mathrm{d}y = \int_0^\infty \frac{\mathrm{d}\lambda}{\rho}$ and, therefore

$$\delta_e = \frac{\gamma - 1}{2\gamma p}\int_0^\infty (g - u^2 - w^2)\, \mathrm{d}\lambda \tag{5.49}$$

The outer edge of the hypersonic boundary layer is exactly determined in the first approximation, since at the outer edge the temperature vanishes and the density increases without bound (Bush, 1966; Lee and Cheng, 1969).

### *5.2.2.2 Outer flow region*

Solving the system of equations (5.47) with the boundary conditions (5.48) requires the knowledge of the pressure distribution which is not given and must be determined in the process of the solution of the boundary value problem jointly with the equations for the outer flow. In accordance with hypersonic small perturbation theory ( Chernyi, 1966; Hayes and Probstein, 1966), for the outer flow region the following coordinate transformations and expansions of the flow functions are introduced

$$x^0 = \ell x, \quad y^0 = \Theta \ell y, \quad z^0 = z_0 \ell z$$

$$u^0 = u_\infty(1 + U\Theta^2 + \cdots), \quad v^0 = u_\infty \Theta V + \cdots, \quad w^0 = \frac{u_\infty \Theta^2 W}{z_0} + \cdots$$

$$p^0 = \rho_\infty u_\infty^2 \Theta^2 p + \cdots, \quad \rho^0 = \rho_\infty R + \cdots \tag{5.50}$$

Then in each $z = \text{const}$ plane the continuity and momentum equations reduce with a relative error $O(\Theta^2/z_0^2)$ to equations analogous to those for the time-dependent one-dimensional gas flow

$$\frac{\partial R}{\partial x} + \frac{\partial RV}{\partial y} = 0, \quad R\left(\frac{\partial V}{\partial x} + V\frac{\partial V}{\partial y}\right) + \frac{\partial P}{\partial y} = 0, \quad \left(\frac{\partial}{\partial x} + V\frac{\partial}{\partial y}\right)\left(\frac{P}{R^\gamma}\right) = 0 \tag{5.51}$$

The required solution is subject to the following boundary conditions at the shock ($y = \Psi(x, z)$):

$$V = \frac{2}{\gamma+1}\frac{\mathrm{d}\Psi}{\mathrm{d}x}\left[1 - \frac{1}{M_\infty^2 \Theta^2 \left(\frac{\mathrm{d}\Psi}{\mathrm{d}x^2}\right)}\right] \tag{5.52}$$

$$P = \frac{2}{\gamma+1}\left(\frac{\mathrm{d}\Psi}{\mathrm{d}x}\right)^2\left[1 - \frac{\gamma-1}{2\gamma M_\infty^2 \Theta^2 \left(\frac{\mathrm{d}\Psi}{\mathrm{d}x}\right)^2}\right]$$

$$R = \frac{\gamma+1}{\gamma-1}\left(1 + \frac{2}{(\gamma-1)M_\infty^2 \Theta^2 \left(\frac{\mathrm{d}\Psi}{\mathrm{d}x}\right)^2}\right)^{-1}$$

and the boundary layer edge ($y = \delta_e(x, z)$)

$$V = \frac{\partial \delta_e}{\partial x} \tag{5.53}$$

We note that within the framework of hypersonic theory the longitudinal velocity disturbance is of the second order of smallness. Because of this, the system of equations for determining the unknown disturbances can be split and the solution of the equation governing the behavior of the function $U$ can be sought separately, in accordance with the known solution of the boundary value problems (5.51)–(5.53). Since we are interested in the solution of the problem in the leading terms, here we do not consider the equation for $U$.

The joint solution of systems of equations (5.47) and (5.51) with the corresponding boundary conditions (5.48), (5.52), and (5.53) makes it possible to obtain exact solutions within the framework of hypersonic small perturbation and boundary layer theories.

Quite often in solving particular problems the system of equations (5.51) is not used but replaced by approximate relations between the pressure at the outer edge of the boundary layer and the local slope of the "effective body" surface. Most frequently it is the "tangent wedge" formula in its simplest form appropriate for $M_\infty \Theta \gg 1$ that is used for calculating the pressure distribution

$$p^0 = \frac{\gamma+1}{2}\rho_\infty u_\infty^2 \left[\frac{\partial(\delta_w^0 + \delta_e^0)}{\partial x}\right]^2$$

In the dimensionless form the formula looks as follows:

$$p = \frac{\gamma+1}{2}\left[\frac{\tau}{\Theta}\frac{\partial \delta_w}{\partial x} + \frac{\delta}{\Theta}\frac{\partial \delta_e}{\partial x}\right]^2 \tag{5.54}$$

In the case of the flow past thin wings in the weak and moderate interaction regimes ($\Theta = \tau$) formula (5.54) takes the form:

$$p = \frac{\gamma+1}{2}\left[\frac{\partial \delta_w}{\partial x} + N_1\frac{\partial \delta_e}{\partial x}\right]^2, \quad N_1 = \frac{\delta}{\tau} = \frac{1}{\tau^2}\sqrt{\frac{z_0}{Re}} \tag{5.55}$$

and in the strong interaction regime ($\Theta = \delta$)

$$p = \frac{\gamma+1}{2}\left[N_2\frac{\partial \delta_w}{\partial x} + \frac{\partial \delta_e}{\partial x}\right]^2, \quad N_2 = N_1^{-1} \tag{5.56}$$

It should be noted that formula (5.54) not only ensures good accuracy in many cases (Chernyi, 1966) but also makes it possible to obtain the same solutions of Eq. (5.47) as hypersonic small perturbation theory when preassigning power-law distributions of the wing thickness $\delta_w(x, z)$ (Hayes and Probstein, 1966). However, in this case, the factor $(\gamma+1)/2$ must be replaced by another coefficient ensuring, for example, a correct pressure distribution over a plane power-law body (Stewartson, 1955*a*, *b*).

### *5.2.3 Plane cross-section law*

As shown above, the system of equations (5.47) describes the perfect gas flow in the boundary layer on the surface of a thin wing in a hypersonic viscous flow. It involves the geometric parameter $z_0$ representing the wing aspect ratio and possesses a certain specific structure. In passing to the $z_0 \to 0$ limit for wings of general form with $\partial p/\partial z = O(1)$ the system of equations of the three-dimensional boundary layer degenerates into a two-dimensional system of equations describing the flow in a cross-section ($x = \text{const}$) and including the longitudinal coordinate as a parameter (Dudin and Neiland, 1976; 1977).

This result constitutes the essence of the plane cross-section law for the three-dimensional boundary layer: in hypersonic flow past thin wings of infinitely low aspect ratio with a finite pressure gradient in the transverse direction the system of equations of the three-dimensional boundary layer degenerates into a system dependent on two variables, describing the flow in the transverse direction, and containing the longitudinal coordinate as a parameter.

In considering the flow past bodies of relative thickness $\tau \sim Re_0^{-1/4}$, in accordance with the conditions of applicability of strip theory (Hayes and Probstein, 1966; Lunev, 1975), the passage to the limit $z_0 \to 0$ requires the fulfillment of the condition $\tau^2/z_0^2 \ll 1$ or $Re^{-1/4} \ll z_0$.

Moreover, near the leading edge of the wing, where in the presence of a pressure maximum the boundary layer can "begin," the values of $w$ are small at insufficiently large $\partial p/\partial z$, so that a local expansion, different from that obtained from Eq. (5.47) as $z_0 \to 0$, must be introduced. However, this additional expansion also obeys the plane cross-section law, that is, it admits a solution containing $x$ as a parameter in the zeroth approximation. Below all these questions are studied in detail.

The plane cross-section law is inapplicable to wings of special form with $\partial p/\partial z \leq O(z_0)$.

The application of the plane cross-section law for determining the characteristics of a three-dimensional boundary layer considerably simplifies the problem, since it is the two-dimensional rather than three-dimensional differential equations that are solved, fairly effective methods having been developed for solving such equations.

The specific structure of the system of equations (5.47) makes it possible to represent its solution in the form of a series in the small parameter $z_0$

$$\begin{aligned}
u(x,\eta,z) &= u_0(x,\eta,z) + u_1(x,\eta,z)z_0 + \cdots \\
w(x,\eta,z) &= w_0(x,\eta,z) + w_1(x,\eta,z)z_0 + \cdots \\
v(x,\eta,z) &= v_0(x,\eta,z) + v_1(x,\eta,z)z_0 + \cdots \\
g(x,\eta,z) &= g_0(x,\eta,z) + g_1(x,\eta,z)z_0 + \cdots \\
p(x,z) &= p_0(x,z) + p_1(x,z)z_0 + \cdots \\
\rho(x,\eta,z) &= \rho_0(x,\eta,z) + \rho_1(x,\eta,z)z_0 + \cdots \\
\mu(x,\eta,z) &= \mu_0(x,\eta,z) + \mu_1(x,\eta,z)z_0 + \cdots
\end{aligned} \tag{5.57}$$

After expansions (5.57) have been substituted in system (5.47) and the coefficients of the same powers of $z_0^n$ have been equated, we obtain a system of two-dimensional equations of the $n$th approximation. In analyzing applied problems the solution is usually restricted to the zeroth and first approximations, since the amount of calculations increases with the number of the approximation. The applicability of the approximate solution of the problem, together with an analysis of certain distinctive features of this approach, is considered below with reference to particular configurations of thin wings.

## 5.3 Thin power-law wings in weak viscous–inviscid interaction

Above we derived the equations governing the perfect gas flow in the laminar interacting three dimensional boundary layer on the surface of a thin wing of arbitrary shape. Generally, these equations are essentially three dimensional and their numerical analysis presents severe computational difficulties. For a particular case of low-aspect-ratio wings ($z_0 \ll 1$) the solution of the problem can be represented in the form of a series in a small parameter. As a result, the solution of the problem is reduced to the integration of two-dimensional systems of equations.

Within the framework of the classical boundary layer theory the equations of the three-dimensional boundary layer are reduced to two-dimensional equations for particular classes of bodies, for example, for the case of conical external flow. In a similar fashion, also in the case of an interacting boundary layer the three-dimensional problem can under certain conditions be reduced to the integration of a system of two-dimensional equations (self-similar solution).

In this section we derive the condition of existence of self-similar flows in the hypersonic "noninteracting" boundary layer ($N_1 = 0$) and study the salient features of the self-similar solutions depending on the relevant parameters of the problem. The self-similar solutions thus obtained make it possible to derive the applicability conditions and to determine the accuracy of the approximate solution in the form of a series in a small parameter.

The knowledge of all these distinctive features provides the foundation for an analysis of the existence of flow self-similarity and its characteristics in the case of an interacting boundary layer.

First theoretical studies of the three-dimensional flow in the hypersonic boundary layers on delta wings were apparently conducted by Ladyzhenskii (1964; 1965). In those studies viscous interaction was not taken into account; as a result the impermeability condition in the plane of symmetry of the wing was usually violated.

### 5.3.1 *Formulation of the boundary value problem*

We will consider the flow past thin power-law wings

$$z_e = x^m, \quad \delta_w = x^l \Delta_w\left(\frac{z}{z_e}\right) \tag{5.58}$$

where $z_e$ is the leading edge coordinate and $m$ and $l$ are the wing form parameters.

For studying the flow past thin power-law wings (5.58) we introduce the variables

$$x = x^*, \quad z = x^m z^*, \quad \lambda = x^k \lambda^*, \quad \mu = \mu^* \tag{5.59}$$

$$u = u^*, \quad w = w^*, \quad v_\delta = x^n v^*, \quad g = g^*$$

$$p = x^{2(l-1)} p^*, \quad \rho = x^{2(l-1)} \rho^*, \quad n_* = \frac{2l - m - 2}{2}, \quad k = \frac{2l + m - 2}{2}$$

We note that, according to Eqs. (5.58) and (5.55), for $N_1 = 0$ $p^*$ is a function of $z^*$ only.

In the absence of viscous interaction ($N_1 = 0$) the boundary value problems (5.47), (5.48), and (5.55) in variables (5.59) takes the form:

$$\frac{\partial v^*}{\partial \lambda^*} + \frac{\partial w^*}{\partial z^*} + z_0 x^{*m-1}\left(x^*\frac{\partial u^*}{\partial x^*} - mz^*\frac{\partial u^*}{\partial z^*} - k\lambda^*\frac{\partial u^*}{\partial \lambda^*}\right) = 0 \tag{5.60}$$

$$v^* \frac{\partial u^*}{\partial \lambda^*} + w^* \frac{\partial u^*}{\partial z^*} + z_0 x^{*m-1} u^* \left( x^* \frac{\partial u^*}{\partial x^*} - m z^* \frac{\partial u^*}{\partial z^*} - k\lambda^* \frac{\partial u^*}{\partial \lambda^*} \right)$$
$$= -z_0 x^{*m-1} \frac{\gamma - 1}{2\gamma p^*} (g^* - u^{*2} - w^{*2}) \left[ 2(l-1)p^* - m z^* \frac{\mathrm{d}p^*}{\mathrm{d}z^*} \right] + \frac{\partial}{\partial \lambda^*} \left( \rho^* \mu^* \frac{\partial u^*}{\partial \lambda^*} \right)$$

$$v^* \frac{\partial W^*}{\partial \lambda^*} + w^* \frac{\partial W^*}{\partial z^*} + z_0 x^{*m-1} u^* \left( x^* \frac{\partial W^*}{\partial x^*} - m z^* \frac{\partial W^*}{\partial z^*} - k\lambda^* \frac{\partial W^*}{\partial \lambda^*} \right)$$
$$= -\frac{\gamma - 1}{2\gamma p^*} (g^* - u^{*2} - w^2) \frac{\mathrm{d}p^*}{\mathrm{d}z^*} + \frac{\partial}{\partial \lambda^*} \left( \rho^* \mu^* \frac{\partial W^*}{\partial \lambda^*} \right)$$

$$\frac{\partial p^*}{\partial \lambda^*} = 0$$

$$v^* \frac{\partial g^*}{\partial \lambda^*} + w^* \frac{\partial g^*}{\partial z^*} + z_0 x^{*m-1} u^* \left( x^* \frac{\partial g^*}{\partial x^*} - m z^* \frac{\partial g^*}{\partial z^*} - k\lambda^* \frac{\partial g^*}{\partial \lambda^*} \right)$$
$$= \frac{\partial}{\partial \lambda^*} \left\{ \rho^* \mu^* \left[ \frac{1}{\sigma} \frac{\partial g^*}{\partial \lambda^*} - \frac{1-\sigma}{\sigma} \frac{\partial (u^{*2} + w^{*2})}{\partial \lambda^*} \right] \right\}$$

$$\mu^* = (g^* - u^{*2} - w^{*2})^{\omega}, \quad \rho^* = \frac{2\gamma}{\gamma - 1} \frac{p^*}{g^* - u^{*2} - w^{*2}}$$

$$p^*(z^*) = \frac{\gamma + 1}{2} \left[ l \Delta_w(z^*) - m z^* \frac{\mathrm{d}\Delta_w}{\mathrm{d}z^*} \right]^2 \tag{5.61}$$

with the boundary conditions

$$u^* = w^* = v^* = 0, \quad g^* = g^*_w \ \text{ for } \lambda^* = 0 \tag{5.62}$$

$$u^* \to 1, \quad w^* \to 0, \quad g^* \to 1 \ \text{ for } \lambda^* \to \infty$$

For $m = 1$ (wing of delta planform) the coordinate $x^*$ drops out from the above system of equations and the boundary value problems (5.60)–(5.62) admits a self-similar solution described by the two-dimensional system of equations

$$\frac{\partial v^*}{\partial \lambda^*} + \frac{\partial w^*}{\partial z^*} - z_0 \left( z^* \frac{\partial u^*}{\partial z^*} + k\lambda^* \frac{\partial u^*}{\partial \lambda^*} \right) = 0 \tag{5.63}$$

$$(v^* - k z_0 u^* \lambda^*) \frac{\partial u^*}{\partial \lambda^*} + (w^* - z_0 u^* z^*) \frac{\partial u^*}{\partial z^*}$$
$$= -z_0 \frac{\gamma - 1}{2\gamma p^*} (g^* - u^{*2} - w^{*2}) \left[ 2(l-1)p^* - z^* \frac{\mathrm{d}p^*}{\mathrm{d}z^*} \right] + \frac{\partial}{\partial \lambda^*} \left( \rho^* \mu^* \frac{\partial u^*}{\partial \lambda^*} \right)$$

$$(v^* - kz_0u^*\lambda^*)\frac{\partial W^*}{\partial \lambda^*} + (w^* - z_0u^*\lambda^*)\frac{\partial W^*}{\partial z^*}$$
$$= -\frac{\gamma - 1}{2\gamma p^*}(g^* - u^{*2} - w^2)\frac{dp^*}{dz^*} + \frac{\partial}{\partial \lambda^*}\left(\rho^*\mu^*\frac{\partial W^*}{\partial \lambda^*}\right)$$

$$\frac{\partial p^*}{\partial \lambda^*} = 0$$

$$(v^* - kz_0u^*\lambda^*)\frac{\partial g^*}{\partial \lambda^*} + (w^* - z_0u^*\lambda^*)\frac{\partial g^*}{\partial z^*}$$
$$= \frac{\partial}{\partial \lambda^*}\left\{\rho^*\mu^*\left[\frac{1}{\sigma}\frac{\partial g^*}{\partial \lambda^*} - \frac{1-\sigma}{\sigma}\frac{\partial(u^{*2} + w^{*2})}{\partial \lambda^*}\right]\right\}$$

$$\mu^* = (g^* - u^{*2} - w^{*2})^{\omega}, \quad \rho^* = \frac{2\gamma}{\gamma - 1}\frac{p^*}{g^* - u^{*2} - w^{*2}}$$

$$p^*(z^*) = \frac{\gamma + 1}{2}\left[l\Delta_w(z^*) - mz^*\frac{\mathrm{d}\Delta_w}{\mathrm{d}z^*}\right]^2 \tag{5.64}$$

with the boundary conditions (5.62), where $k = (2l - 1)/2$. The existence of the self-similar solution for a thin delta wing was noted in the paper of Kozlova and Mikhailov (1970).

The solution of the problem for a delta wing ($m = 1$) in the exact formulation, that is, including the calculation of the flowfield in the inviscid shock layer (5.51)–(5.43), imposes more rigid constraints on the existence of a self-similar solution. An analysis showed that the outer problem can be self-similar only for $l = 3/4$. In fact, introducing in the outer flow region the variables

$$z = xz^1, \quad y = x^{3/4}y^1, \quad V = x^{-1/4}V^1(z^1, y^1), \quad R = R^1(z^1, y^1)$$
$$P = x^{-1/2}P^1(z^1, y^1), \quad \Psi = x^{3/4}\Psi^1(z^1), \quad \delta_e = x^{3/4}\delta_e^1$$

from system (5.51)–(5.53) we obtain the following system of equations valid as $M_\infty \to \infty$

$$\left(V^1 - \frac{3}{4}y^1\right)\frac{\partial R^1}{\partial y^1} + R^1\frac{\partial V^1}{\partial y^1} - z^1\frac{\partial R^1}{\partial z^1} = 0$$
$$R^1\left(V^1 - \frac{3}{4}y^1\right)\frac{\partial V^1}{\partial y^1} - R^1z^1\frac{\partial V^1}{\partial z^1} - \frac{R^1V^1}{4} + \frac{\partial P^1}{\partial y^1} = 0$$
$$\left(V^1 - \frac{3}{4}y^1\right)\frac{\partial}{\partial y^1}\left(\frac{P^1}{R^{1\gamma}}\right) - z^1\frac{\partial}{\partial z^1}\left(\frac{P^1}{R^{1\gamma}}\right) - \frac{P^1}{2R^{1\gamma}} = 0$$

with the boundary conditions

$$V^1 = \frac{2}{\gamma + 1}\left(\frac{3}{4}\Psi^1 - z^1\frac{\mathrm{d}\Psi^1}{\mathrm{d}z^1}\right), \quad P^1 = \frac{2}{\gamma + 1}\left(\frac{3}{4}\Psi^1 - z^1\frac{\mathrm{d}\Psi^1}{\mathrm{d}z^1}\right)^2$$

$$R^1 = \frac{\gamma+1}{\gamma-1} \text{ for } y^1 = \Psi^1(z^1)$$

$$V^1 = \frac{3}{4}\delta_e - z^1 \frac{d\delta_e^1}{dz^1} \text{ for } y^1 = \delta_e^1(z^1)$$

### 5.3.2 On the nature of the pressure distribution

Before proceeding to further analysis, it is useful to consider the nature of the pressure distribution over thin power-law wings at zero incidence. For this purpose, we will choose a class of symmetric wings with the following cross-sectional shape

$$\Delta_w = (1 - z^{*2})^\beta \tag{5.65}$$

where $\beta$ is the form parameter. The pressure distribution over this wing is determined by the formula

$$p^* = \frac{\gamma+1}{2}(1 - z^{*2})^{2(\beta-1)}[l(1 - z^{*2}) + 2m\beta z^{*2}]^2 \tag{5.66}$$

According to the results of calculations, the pressure maximum and minimum can be located either at the edge, or in the plane of symmetry, or within the domain of variation of the variable $z^*$, depending on the values of the parameters $m$, $l$, and $\beta$ (Fig. 5.12).

### 5.3.3 Certain features of the solution of boundary value problems

We will consider certain distinctive features of the solution of the boundary value problem for power-law wings in the absence of viscous–inviscid interaction. For simplifying the analysis, we will restrict ourselves to the case of a delta wing ($m = 1$), in which the boundary layer flowfield is described by the two-dimensional system of equations (5.63) with $m = 1$. In this system the coefficient of the derivatives with respect to $z^*$ takes the form:

$$w^* - z_0 z^* u^* \tag{5.67}$$

The direction of the integration of the parabolic system of equations (5.63) must be chosen with account for the sign of expression (5.67) (Dudin and Neiland, 1977).

Thus, if the pressure decreases monotonically from the edge to the wing center (quadrant III in Fig. 5.12), then $w^* < 0$ everywhere. Since at the outer edge of the boundary layer $u^*$ is equal to unity, the sign of coefficient (5.67) remains the same for all values of $\lambda^*$. In this case, the system of boundary layer equations is integrated from the wing edges toward the wing axis.

However, for wings for which $p^*_{\max}$ is not reached at the edge (e.g., quadrants I and II in Fig. 5.12) the sign of $w^*$ may be plus. For example, if $p^*_{\max}$ is reached in the plane of symmetry, we have $w^* > 0$, at least, in the inner part of the boundary layer. In exact formulation, the integration of the problem is made difficult by the fact that at a certain value of $\lambda^*$ the sign of expression (5.67) must change for any small but finite $z_0$. In this case, the

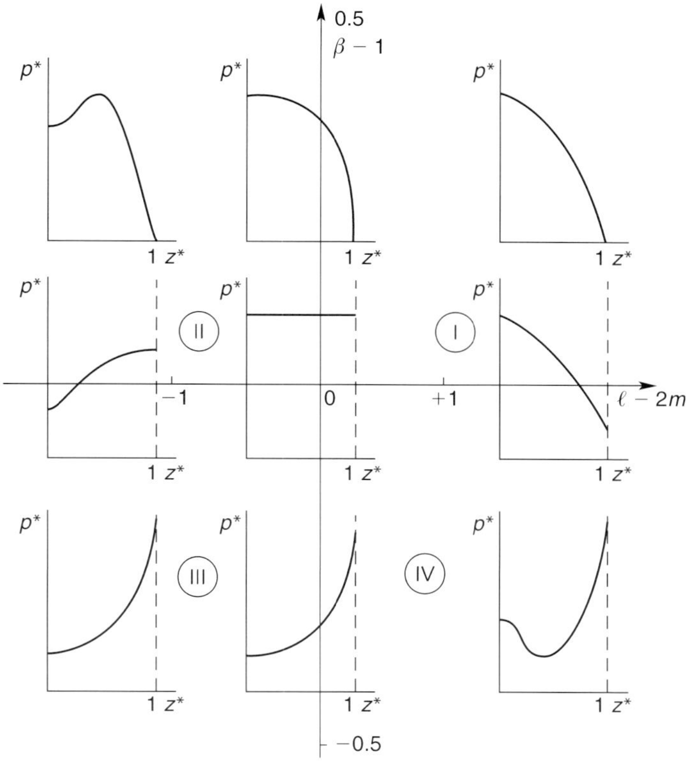


Fig. 5.12.

integration direction must be opposite in different parts of the boundary layer. This feature is similar to that occurring in return flows within a two-dimensional boundary layer. However, in the limiting case $z_0 \to 0$ the solution of the problem can be obtained in a conventional way, since the second term is dropped out from expression (5.67) and the marching method can be applied in the calculations. Thus, neglecting the disturbance transfer "upstream in the transverse flow" leads to an error $O(z_0)$. All the same, this approximation can be useful for a comparative analysis of the wing characteristics in a hypersonic viscous flow.

The second distinctive feature of the solution of the boundary value problems is related with the behavior of the solution in the vicinity of the leading edges in the case in which $p^*_{\max}$ is reached at the leading edge.

For studying this feature, new variables are introduced

$$\xi = \frac{2\gamma}{\gamma - 1}\int_{z^*}^{1} p^*(z^*)\,dz^*, \quad \eta^* = \frac{\lambda^*}{\sqrt{2\xi}}, \quad w^1 = -w^* \tag{5.68}$$

$$v^1 = \frac{\gamma - 1}{2\gamma}\sqrt{2\xi}\frac{v^*}{p^*}, \quad \widetilde{N} = \frac{\gamma - 1}{2\gamma}\frac{\rho^*\mu^*}{p^*} = (g^* - u^{*2} - w^{*2})^{\omega - 1}$$

The derivatives are transformed by the formulas

$$\frac{\partial}{\partial \lambda^*} = \frac{1}{\sqrt{2\xi}} \frac{\partial}{\partial \eta^*}, \quad \frac{\partial}{\partial z^*} = -\frac{2\gamma p^*}{\gamma - 1} \frac{\partial}{\partial \xi} + \frac{\gamma}{\gamma - 1} \frac{\eta^* p^*}{\xi} \frac{\partial}{\partial \eta^*} \tag{5.69}$$

Rewriting system (5.63) in terms of Eqs. (5.68) and (5.69) yields

$$\frac{\partial v^1}{\partial \eta^*} = \eta^* \frac{\partial w^1}{\partial \eta^*} - 2\xi \left( \frac{\partial w^1}{\partial \xi} + z_0 z^* \frac{\partial u^*}{\partial \xi} \right) + z_0 \eta^* \left( z^* + \frac{\gamma - 1}{\gamma} k \frac{\xi}{p^*} \right) \frac{\partial u^*}{\partial \eta^*} \tag{5.70}$$

$$(w^1 + z_0 z^* u^*) 2\xi \frac{\partial u^*}{\partial \xi} + \left[ v^1 - w^1 \eta^* - z_0 \eta^* u^* \left( z^* + \frac{\gamma - 1}{\gamma} k \frac{\xi}{p^*} \right) \right] \frac{\partial u^*}{\partial \eta^*}$$

$$= \frac{\partial}{\partial \eta^*} \left( \widetilde{N} \frac{\partial u^*}{\partial \eta^*} \right) + z_0 \left( \frac{\gamma - 1}{\gamma} \right)^2 (g^* - u^{*2} - (w^1)^2) \left[ \frac{z^* \xi}{2p^{*2}} \frac{\mathrm{d}p^*}{\mathrm{d}z^*} - (l - 1) \frac{\xi}{p^*} \right]$$

$$(w^1 + z_0 z^* u^*) 2\xi \frac{\partial w^1}{\partial \xi} + \left[ v^1 - w^1 \eta^* - z_0 \eta^* u^* \left( z^* + \frac{\gamma - 1}{\gamma} k \frac{\xi}{p^*} \right) \right] \frac{\partial w^1}{\partial \eta^*}$$

$$= \frac{\partial}{\partial \eta^*} \left( \widetilde{N} \frac{\partial w^1}{\partial \eta^*} \right) + \left( \frac{\gamma - 1}{\gamma} \right)^2 (g^* - u^{*2} - (w^1)^2) \frac{\xi}{2p^{*2}} \frac{\mathrm{d}p^*}{\mathrm{d}z^*}$$

$$(w^1 + z_0 z^* u^*) 2\xi \frac{\partial g^*}{\partial \xi} + \left[ v^1 - w^1 \eta^* - z_0 \eta^* u^* \left( z^* + \frac{\gamma - 1}{\gamma} k \frac{\xi}{p^*} \right) \right] \frac{\partial g^*}{\partial \eta^*}$$

$$= \frac{\partial}{\partial \eta^*} \left[ \widetilde{N} \left( \frac{1}{\sigma} \frac{\partial g^*}{\partial \eta^*} - \frac{1 - \sigma}{\sigma} \frac{\partial (u^{*2} + (w^1)^2)}{\partial \eta^*} \right) \right]$$

$$k = \frac{2l - 1}{2}$$

with the boundary conditions

$$u^* = v^1 = w^1 = 0, \quad g^* = g_w^* \ \text{ for } \eta^* = 0 \tag{5.71}$$

$$u^* \to 1, \quad w^1 \to 0, \quad g^* \to 1 \ \text{ for } \eta^* \to \infty$$

A physical singularity in the problem solution appears for wings for which in the vicinity of the leading edges the pressure gradients are fairly small and cannot lead to the formation of a transverse flow at a velocity $w^1 \sim 1$. For studying this situation we will consider a particular case in which $\beta = 1$; then the pressure distribution is determined by the formula

$$p^* = \frac{\gamma + 1}{\gamma} [l(1 - z^*)^2 + 2z^{*2}]^2 \tag{5.72}$$

After the pressure in the vicinity of $z^* = 1$ has been expanded in a series in the powers of $(1 - z^{*2})$ we obtain

$$p^* \approx 2(\gamma + 1) + 2(\gamma + 1)(l - 2)(1 - z^{*2}) + O[(1 - z^{*2})^2] \qquad (5.73)$$

$$\frac{dp^*}{dz^*} \approx -4(\gamma + 1)(l - 2)z^* + O(1 - z^{*2})$$

Let a pressure maximum $p^*_{\max}$ be reached at the edge; for this purpose, we should let $l < 2$ (see Fig. 5.12). Then at the leading edge the pressure is finite, together with its gradient. In view of this fact, from the transverse momentum equation it follows that $w^1(\xi = 0, \eta) = 0$, that is, at the leading edge there is located a spreading line. This result is understandable, since a finite pressure gradient cannot lead to an "instantaneous" appearance of $w^1 = O(1)$. Therefore, as $\xi \to 0$, for any small $z_0$ there is always a domain of fairly small $\xi$ in which $z_0 z^* u^*$ cannot be neglected as compared with $w^1$. For this reason Eqs. (5.70) with $\xi = 0$ in the limiting case $z_0 \to 0$ lead to erroneous results, so that for $\beta = 1$ the series expansion of the solution in $z_0$ is not uniformly accurate and an additional expansion is required as $\xi \to 0$.

For example, in the case described by the system of equations (5.70) the "inner" variables are determined by the formulas

$$\xi = z_0^2 \xi_*, \quad \eta^* = \eta_* z_0^{-1/2}, \quad v^1 = z_0^{1/2} v_*, \quad w^1 = z_0 w_*$$

Substituting these variables in system (5.70) and passing to the limit $z_0 \to 0$ yields the equations governing the flow near the leading edge

$$\frac{\partial v_*}{\partial \eta_*} = \eta_* \frac{\partial (w_* + u^*)}{\partial \eta_*} - 2\xi_* \frac{\partial (w_* + u^*)}{\partial \xi_*}$$

$$2\xi_*(w_* + u^*)\frac{\partial u^*}{\partial \xi_*} + (v_* - \eta_* w_* - \eta_* u_*)\frac{\partial u^*}{\partial \eta_*} = \frac{\partial}{\partial \eta_*}\left(\widetilde{N}\frac{\partial u^*}{\partial \eta_*}\right)$$

$$2\xi_*(w_* + u^*)\frac{\partial w_*}{\partial \xi_*} + (v_* - \eta_* w_* - \eta_* u_*)\frac{\partial w_*}{\partial \eta_*}$$

$$= \frac{\partial}{\partial \eta_*}\left(\widetilde{N}\frac{\partial w_*}{\partial \eta_*}\right) + \left(\frac{\gamma - 1}{\gamma}\right)^2 \frac{g^* - u^{*2}}{2}\frac{\xi_* p_{*1}}{p_{*0}^2}$$

$$2\xi_*(w_* + u^*)\frac{\partial g_*}{\partial \xi_*} + (v_* - \eta_* w_* - \eta_* u_*)\frac{\partial g_*}{\partial \eta_*} = \frac{\partial}{\partial \eta_*}\left[\widetilde{N}\left(\frac{1}{\sigma}\frac{\partial g^*}{\partial \eta_*} - \frac{1 - \sigma}{\sigma}\frac{\partial u^{*2}}{\partial \eta_*}\right)\right]$$

where $p_{*0} = 2(\gamma + 1)$ and $p_{*1} = -4(\gamma + 1)(l - 2)$, while the boundary conditions are as follows:

$$u^* = v_* = w_* = 0, \quad g^* = g^*_w \ \text{for } \eta_* = 0$$

$$u^* \to 1, \quad w_* \to 0, \quad g^* \to 1 \ \text{for } \eta_* \to \infty$$

The solution of this system of equations makes it possible to determine the initial conditions for the zeroth approximation ($z_0 = 0$) of system (5.70), as $\xi \to 0$. However, it should be noted that the above singularity at the wing edge does not appear if the boundary layer/inviscid flow interaction is taken into account, since in this case the pressure gradient is always large.

### 5.3.4 *Characteristics of the self-similar solution*

For studying the aerodynamic characteristics of thin delta wings the calculations of system (5.70) were performed over a wide range of the relevant parameters of the problem (Dudin and Neiland, 1976; 1977).

As an illustration, we will consider the flow past a wing with the aspect ratio $z_0 = 0.2$ ($l = 1.1$) at the following values of the parameters: $\gamma = 1.4$, $\sigma = 0.71$, and $g_w^* = 0.5$. In Figs. 5.13–5.15 the calculated longitudinal ($\tau_u$) and transverse ($\tau_w$) viscous stress coefficients are plotted, together with the heat flux to the body $\tau_g$ (solid curves).

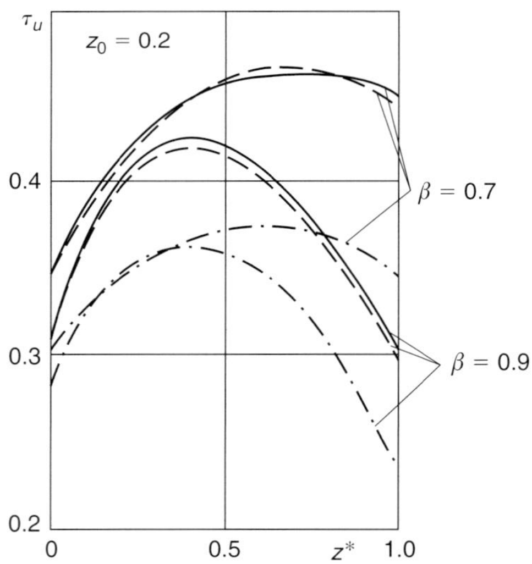


Fig. 5.13.

It should be noted that these calculations do not satisfy the condition $w^1 = 0$ for $z^* = 0$ in the plane of symmetry of the wing (Fig. 5.14). Since the calculated transverse velocities in the boundary layer turn out to be finite, an adverse pressure gradient caused by the interaction between the viscous and inviscid parts of the flow must appear in the vicinity of the plane of symmetry of the wing in order for the impermeability condition could be satisfied. The general study of flows of this type was conducted in the papers of Neiland (1974*a*, *b*).

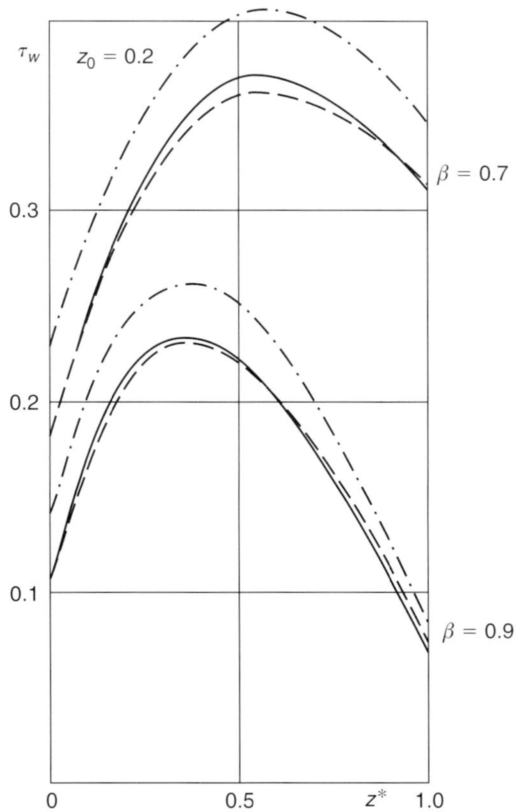


Fig. 5.14.

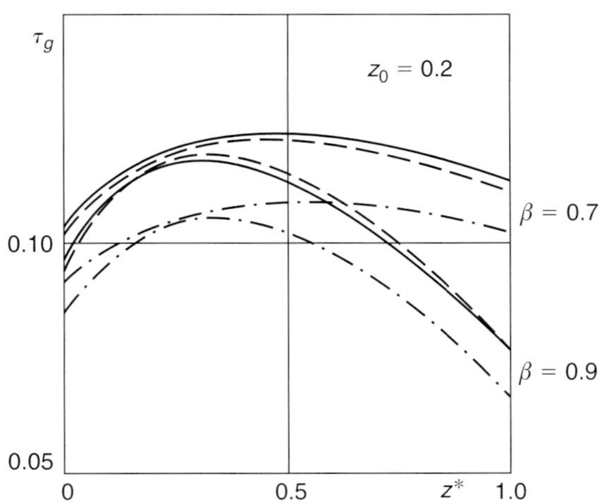


Fig. 5.15.

### 5.3.5 Approximate solution of the problem for delta wings

As noted above, the structure of the equations governing the laminar boundary layer on a thin low-aspect-ratio wing is such that it makes it possible to represent the solution in the form of a functional series in a small parameter $z_0$, thus reducing the three-dimensional problem to successive integration of systems of two-dimensional equations. Naturally, this approach remains valid for a noninteracting boundary layer if $\partial p/\partial z = O(1)$.

We will study the efficiency of the above approach with reference to the case of delta wings ($m = 1$, self-similar problem). For the sake of simplicity, we will assume a linear viscosity–temperature dependence ($\omega = 1$).

For obtaining the zeroth and first approximations we expand the functions $u^*$, $v^1$, $w^1$, and $g^*$ in powers of $z_0$

$$\begin{aligned} u^*(\xi,\eta^*,z_0) &= u_0(\xi,\eta^*) + u_1(\xi,\eta^*)z_0 + \cdots \\ w^1(\xi,\eta^*,z_0) &= w_0(\xi,\eta^*) + w_1(\xi,\eta^*)z_0 + \cdots \\ v^1(\xi,\eta^*,z_0) &= v_0(\xi,\eta^*) + v_1(\xi,\eta^*)z_0 + \cdots \\ g^*(\xi,\eta^*,z_0) &= g_0(\xi,\eta^*) + g_1(\xi,\eta^*)z_0 + \cdots \end{aligned} \tag{5.74}$$

Substituting expansions (5.74) in the system of equations (5.70) and the boundary conditions we obtain

1. *The zeroth approximation equations*

$$\frac{\partial v_0}{\partial \eta^*} = \eta^* \frac{\partial w_0}{\partial \eta^*} - 2\xi \frac{\partial w_0}{\partial \xi} \tag{5.75}$$

$$2\xi w_0 \frac{\partial u_0}{\partial \xi} + (v_0 - \eta^* w_0)\frac{\partial u_0}{\partial \eta^*} = \frac{\partial^2 u_0}{\partial \eta^{*2}}$$

$$2\xi w_0 \frac{\partial w_0}{\partial \xi} + (v_0 - \eta^* w_0)\frac{\partial w_0}{\partial \eta^*} = \frac{\partial^2 w_0}{\partial \eta^{*2}} + \left(\frac{\gamma-1}{\gamma}\right)^2 \frac{g_0 - u_0^2 - w_0^2}{2} \frac{\xi}{p^{*2}} \frac{dp^*}{dz^*}$$

$$2\xi w_0 \frac{\partial g_0}{\partial \xi} + (v_0 - \eta^* w_0)\frac{\partial g_0}{\partial \eta^*} = \frac{\partial}{\partial \eta^*}\left[\frac{1}{\sigma}\frac{\partial g_0}{\partial \eta^*} - \frac{1-\sigma}{\sigma}\frac{\partial (u_0^2 + w_0^2)}{\partial \eta^*}\right]$$

with the boundary conditions

$$\begin{aligned} u_0 &= v_0 = w_0 = 0, \quad g_0 = g_w \ \text{ for } \eta^* = 0 \\ u_0 &\to 1, \quad w_0 \to 0, \quad g_0 \to 1 \ \text{ for } \eta^* \to \infty \end{aligned} \tag{5.76}$$

2. *The first-order equations*

$$\frac{\partial v_1}{\partial \eta^*} = \eta^* \frac{\partial w_1}{\partial \eta^*} - 2\xi\left(\frac{\partial w_1}{\partial \xi} + z^* \frac{\partial u_0}{\partial \xi}\right) + \eta^*\left(z^* + \frac{\gamma-1}{\gamma} k \frac{\xi}{p^*}\right)\frac{\partial u_0}{\partial \eta^*} \tag{5.77}$$

$$2\xi w_0 \frac{\partial u_1}{\partial \xi} + (v_0 - \eta^* w_0)\frac{\partial u_1}{\partial \eta^*} = \frac{\partial^2 u_1}{\partial \eta^{*2}} + \left(\frac{\gamma-1}{\gamma}\right)^2 (g_0 - u_0^2 - w_0^2)$$
$$\times \left[\frac{z^*\xi}{2p^{*2}}\frac{\mathrm{d}p^*}{\mathrm{d}z^*} - (l-1)\frac{\xi}{p^*}\right] - 2\xi(w_1 + z^* u_0)\frac{\partial u_0}{\partial \xi}$$
$$-\left[v_1 - \eta^* w_1 - \eta^* u_0\left(z^* + \frac{\gamma-1}{\gamma}k\frac{\xi}{p^*}\right)\right]\frac{\partial u_0}{\partial \eta^*}$$

$$2\xi w_0 \frac{\partial w_1}{\partial \xi} + (v_0 - \eta^* w_0)\frac{\partial w_1}{\partial \eta^*} = \frac{\partial^2 w_1}{\partial \eta^{*2}} + \left(\frac{\gamma-1}{\gamma}\right)^2 (g_1 - 2u_0 u_1 - 2w_0 w_1)$$
$$\times \frac{\xi}{2p^{*2}}\frac{\mathrm{d}p^*}{\mathrm{d}z^*} - 2\xi(w_1 + z^* u_0)\frac{\partial w_0}{\partial \xi}$$
$$-\left[v_1 - \eta^* w_1 - \eta^* u_0\left(z^* + \frac{\gamma-1}{\gamma}k\frac{\xi}{p^*}\right)\right]\frac{\partial w_0}{\partial \eta^*}$$

$$2\xi w_0 \frac{\partial g_1}{\partial \xi} + (v_0 - \eta^* w_0)\frac{\partial g_1}{\partial \eta^*} = \frac{\partial}{\partial \eta^*}\left[\frac{1}{\sigma}\frac{\partial g_1}{\partial \eta^*} - \frac{1-\sigma}{\sigma}\frac{\partial^2(u_0 u_1 + w_0 w_1)}{\partial \eta^*}\right]$$
$$- 2\xi(w_1 + z^* u_0)\frac{\partial g_0}{\partial \xi}$$
$$-\left[v_1 - \eta^* w_1 - \eta^* u_0\left(z^* + \frac{\gamma-1}{\gamma}k\frac{\xi}{p^*}\right)\right]\frac{\partial g_0}{\partial \eta^*}$$

$$k = \frac{2l-1}{2}$$

with the boundary conditions

$$\begin{aligned} u_1 &= v_1 = w_1 = 0, \quad g_1 = 0 \ \text{ for } \eta^* = 0 \\ u_1 &\to 0, \quad w_1 \to 0, \quad g_1 \to 0 \ \text{ for } \eta^* \to \infty \end{aligned} \tag{5.78}$$

In solving numerically the system of equations (5.70) in region III (Fig. 5.12) it is necessary to take into consideration that at the wing edge $p^*(z^*) \to \infty$; therefore, integral in Eq. (5.68) is improper and analytical expressions for $\xi$, $\xi/p^*$, and $(\xi/p^{*2})\,(\mathrm{d}p^*/\mathrm{d}z^*)$ should be obtained for $z^* \to 1$.

Substituting the expression for the pressure distribution (5.66) for $m = 1$ in integral (5.68) we obtain

$$\xi|_{z^*\to 1} = 2^{2\beta}\gamma\frac{\gamma+1}{\gamma-1}(1-z^*)^{2\beta-1}\left[\frac{\beta^2}{2\beta-1} + \frac{2l-3\beta-\beta^2}{2}(1-z^*) + O(1-z^*)^2\right] \tag{5.79}$$

Then we have

$$\left.\frac{\xi}{p^*}\right|_{z^*\to 1} = \frac{\gamma}{\gamma-1}\frac{2}{2\beta-1}(1-z^*) + O(1-z^*)^2 \tag{5.80}$$

$$\left.\frac{\xi}{p^{*2}}\frac{\mathrm{d}p^*}{\mathrm{d}z^*}\right|_{z^*\to 1} = -4\gamma\frac{\beta-1}{\beta^2}\left[\frac{\beta^2}{2\beta-1} + \frac{2l-3\beta-\beta^2}{2}(1-z^*) + O(1-z^*)^2\right] \tag{5.81}$$

For $z^* < 0.99995$ integral (5.68) was calculated from the Simpson formula and for $z^* \geq 0.99995$ from formula (5.79).

The exact and approximate solutions are compared in Figs. 5.13–5.15 for wings at zero incidence at the following values of the parameters: $z_0 = 0.2$ (sweep angle $\sim 79°$), $l = 1.1$, and $\beta = 0.7$ and 0.9. In the figures solid curves present the exact solutions, dashed curves the first approximation $\left( \frac{\partial(f_0 + f_1 z_0)}{\partial \eta^*} \Big|_w \right)$, and dot-and-dash curves the zeroth approximation $\left( \frac{\partial f_0}{\partial \eta^*} \Big|_w \right)$, where $f$ stands for the functions $u$, $w$, and $g$.

Analyzing the results of these calculations, as well as calculations carried out for other values of the parameters $\beta$ and $l$ (Dudin and Neiland, 1977) it can be concluded that using the plane cross-section law for calculating the characteristics of the three-dimensional boundary layer on power-law wings in hypersonic flow actually provides an error $O(z_0)$, while using the first approximation gives $O(z_0^2)$.

## 5.4 Strong viscous interaction regime on delta and swept wings

### *5.4.1 Formulation of the problem*

Following the work of Kozlova and Mikhailov (1970) we will consider the viscous $M_\infty \to \infty$ flow past an infinite triangular flat plate. The $x$ axis of a rectangular coordinate system coincides with the axis of symmetry of the plate and is aligned with the freestream, while the $y$ axis is perpendicular to the plate surface (Fig. 5.16). We introduce the following designations: $u_\infty$ and $\rho_\infty$ are the freestream velocity and density; $\ell$ is a scale length; $\mu_0$ is the viscosity at the adiabatic stagnation temperature; $uu_\infty$, $vu_\infty$, and $wu_\infty$ are the velocity vector projections onto the $x$, $y$, and $z$ axes, respectively; $x\mu_0/(\rho_\infty u_\infty)$, $y\mu_0/(\rho_\infty u_\infty)$, and $z\mu_0/(\rho_\infty u_\infty)$ are the coordinates along the above-mentioned axes; $p\rho_\infty u_\infty^2$ is the pressure; $\rho\rho_\infty$ is the density; $hu_\infty^2$ is the enthalpy; $\mu\mu_0$ is viscosity; $\sigma$ is the Prandtl number; $\gamma$ is the adiabatic exponent; $Re_0 = \rho_\infty u_\infty \ell/\mu_0$ is the Reynolds number; and $H = h + 0.5(u^2 + v^2 + w^2)$ is the total enthalpy.

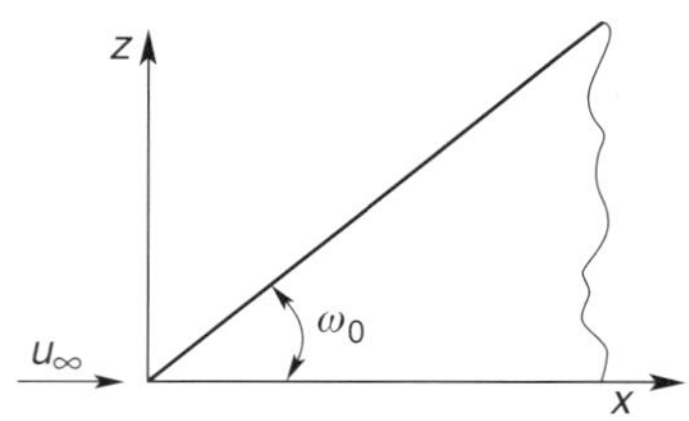


Fig. 5.16.

If for fairly high Reynolds numbers $Re_0$ the flow under consideration can be described by the equations for the boundary layer and the outer inviscid flow, then it must be self-similar, that is, a transformation reducing the number of unknown variables is possible. Such a transformation reducing the original problem to a two-dimensional one was suggested in the work of Ladyzhenskii (1964). The equations thus obtained were applied to the calculation of the flow past a delta wing (Ladyzhenskii, 1965). However, as shown in the paper of Kozlova and Mikhailov (1970), the flow pattern adopted in the work of Ladyzhenskii (1965) is contradictory, so that the flow in the boundary layer on a delta wing must be different from that on a swept plate.

### 5.4.2 *Equations and boundary conditions*

Following the work of Kozlova and Mikhailov (1970), we will derive the equations governing the viscous gas flow on a thin wing, triangular in plan, for $M_\infty = \infty$ and $\tan\omega_0 = O(1)$, where $\omega_0$ is the angle which the leading edge makes with the oncoming flow.

With a relative error of the order $\theta^2$, where $\theta$ is the greatest of the characteristic relative thicknesses of the wing and the boundary layer, the equations of the three-dimensional boundary layer on the wing can be written in the form somewhat different from system (5.47) owing to the introduction of different dimensionless parameters

$$u\frac{\partial u}{\partial x} + v\frac{\partial u}{\partial y} + w\frac{\partial u}{\partial z} + \frac{1}{\rho}\frac{\partial p}{\partial x} = \frac{1}{\rho}\frac{\partial}{\partial y}\left(\mu\frac{\partial u}{\partial y}\right) \tag{5.82}$$

$$u\frac{\partial w}{\partial x} + v\frac{\partial w}{\partial y} + w\frac{\partial w}{\partial z} + \frac{1}{\rho}\frac{\partial p}{\partial z} = \frac{1}{\rho}\frac{\partial}{\partial y}\left(\mu\frac{\partial w}{\partial y}\right)$$

$$u\frac{\partial H}{\partial x} + v\frac{\partial H}{\partial y} + w\frac{\partial H}{\partial z} = \frac{1}{\rho}\frac{\partial}{\partial y}\left[\mu\frac{\partial}{\partial y}\left(\frac{h}{\sigma} + \frac{u^2+w^2}{2}\right)\right]$$

$$\frac{\partial \rho u}{\partial x} + \frac{\partial \rho v}{\partial y} + \frac{\partial \rho w}{\partial z} = 0, \quad H = h + 0.5(u^2+w^2), \quad p = p(x,z)$$

Assuming that $\mu$ is a linear function of the temperature

$$\mu = \frac{p}{\rho\varepsilon}, \quad \varepsilon = \frac{\gamma-1}{2\gamma} \tag{5.83}$$

we introduce the stream function $\psi$ by the equation

$$u\frac{\partial\psi}{\partial x} + v\frac{\partial\psi}{\partial y} + w\frac{\partial\psi}{\partial z} = 0$$

Then, replacing $y$ by $\psi$, we introduce the following transformations for the dependent and independent variables

$$x = x, \quad \lambda = \left(\frac{\varepsilon}{px(1-t)}\right)^{1/2}\int_0^{\psi}\rho\frac{\partial y}{\partial\psi}\,d\psi, \quad t = \frac{\zeta}{\zeta_0}$$

$$L = x^{-0.5m}(1-t)^{-0.5n}p^{0.5}, \quad W = w\zeta_0, \quad \zeta = \frac{z}{x}, \quad \zeta_0 = \tan\omega_0$$

We assume that the pressure distribution over the wing is such that $L = L(t)$. In this case, system (5.82) can be brought into the following self-similar form:

$$Vu_\lambda + 2\varepsilon nth + (1-t)\left[2\varepsilon mh + \left(\frac{W}{\zeta_0^2 - ut}\right)u_t - \frac{4\varepsilon thL_t}{L}\right] = u_{\lambda\lambda} \tag{5.84}$$

$$VW_\lambda - 2\varepsilon nh + (1-t)\left[\left(\frac{W}{\zeta_0^2 - ut}\right)W_t + \frac{4\varepsilon hL_t}{L}\right] = W_{\lambda\lambda}$$

$$VH_\lambda + (1-t)\left(\frac{W}{\zeta_0^2} - ut\right)H_t = \frac{H_{\lambda\lambda}}{\sigma} + \frac{1}{2}\left(1 - \frac{1}{\sigma}\right)\left(u^2 + \frac{W^2}{\zeta_0^2}\right)_{\lambda\lambda}$$

$$V_\lambda - \frac{1}{2}(n+1)\left(\frac{W}{\zeta_0^2} - ut\right) + (1-t)\left[\frac{1}{2}(m+1)u + \left(\frac{W}{\zeta_0^2} - ut\right)\frac{L_t}{L} - tu_t + \frac{W_t}{\zeta_0^2}\right] = 0$$

Here, $V$ is the counterpart of the velocity component $v$ and vanishes for $\lambda = 0$ if $v = 0$ on the wing surface; the subscripts $t$, $\lambda$, and $\lambda\lambda$ refer to the derivatives with respect to $t$ and the first and second derivatives with respect to $\lambda$, respectively.

In solving the problem with no account for viscous interaction we will assume, following the work of Kozlova and Mikhailov (1970), that the wing shape is described by the equation

$$y_w = x^{1+m/2}(1-t)^{1+n/2}\Phi(t) \tag{5.85}$$

Assuming the inviscid flow in each $z = \text{const}$ plane to be two dimensional (stripe theory), which is true with the adopted relative error $O(\theta^2)$, we can derive equations relating $L(t)$ and $\Phi(t)$. Following the work of Ladyzhenskii (1964), we will use the approximate tangent wedge method, in accordance with which

$$p = \frac{1}{2}(1+\gamma)\cos^2(\overline{n}x)$$

Here, $\overline{n}x$ is the angle the normal to the body makes with the oncoming flow. Instead of the factor $\frac{1}{2}(1+\gamma)$, other coefficient can be chosen which would provide the correct pressure distribution over the plane power-law body with an exponent $1 + m/2$.

Then using Eq. (5.85) we obtain

$$L = \sqrt{\frac{1}{2}(1+\gamma)}\left\{(1-t)\left[\left(1 + \frac{1}{2}m\right)\Phi - t\Phi_t\right] + t\left(1 + \frac{1}{2}n\right)\Phi\right\} \tag{5.86}$$

The function $\Phi(t)$ must be fairly smooth and the wing must be thin, as required by the condition $m \le 0$. Though in this case the wing surface slope can be large near the leading edges, we will assume that fairly far from the edges the error of the solution is small.

### *5.4.3 Strong viscous interaction on a delta wing*

We are now coming to the case of strong viscous interaction on a delta wing. In this flow regime, the body thickness is negligibly small as compared with the boundary layer thickness. Thus, it is the upper edge of the boundary layer that is taken for the body surface. Calculating the boundary layer thickness $\delta$ from the formula

$$\delta = x^{(m+1)/2}(1-t)^{(n+1)/2}\varepsilon^{-1/2}L\int_0^\infty \frac{\mathrm{d}r}{\rho}, \qquad \rho = \frac{L^2x^m(1-t)^n}{2\varepsilon h}$$

and substituting $\delta$ for $y_w$ in Eq. (5.85) we obtain

$$\Phi L = 2\varepsilon^{1/2}x^{-m-1/2}(1-t)^{-n-1/2}\int_0^\infty h\,\mathrm{d}r \tag{5.87}$$

For obtaining a self-similar solution and obviating a singularity in the equations at $t=1$ we let $m=n=-1/2$.

Then Eqs. (5.84), (5.86), and (5.87) are brought into the form:

$$Vu_\lambda - \varepsilon h + (1-t)\left[\left(\frac{W}{\zeta_0^2 - ut}\right)u_t - \frac{4\varepsilon thL_t}{L}\right] = u_{\lambda\lambda} \tag{5.88}$$

$$VW_\lambda + \varepsilon h + (1-t)\left[\left(\frac{W}{\zeta_0^2 - ut}\right)W_t + \frac{4\varepsilon hL_t}{L}\right] = W_{\lambda\lambda}$$

$$VH_\lambda + (1-t)\left(\frac{W}{\zeta_0^2 - ut}\right)H_t = \frac{H_{\lambda\lambda}}{\sigma} + \frac{1}{2}\left(1-\frac{1}{\sigma}\right)\left(u^2 + \frac{W^2}{\zeta_0^2}\right)_{\lambda\lambda}$$

$$V_\lambda - \frac{1}{4}\left(\frac{W}{\zeta_0^2} - ut\right) + (1-t)\left[\left(\frac{W}{\zeta_0^2} - ut\right)\frac{L_t}{L} - tu_t + \frac{W_t}{\zeta_0^2}\right] = 0$$

$$L = \sqrt{\frac{1}{2}(1+\gamma)\left[\frac{3}{4}\Phi - t(1-t)\Phi_t\right]}$$

$$\Phi L = 2\varepsilon^{1/2}\int_0^\infty h\,\mathrm{d}r, \qquad h = H - \frac{1}{2}\left(u^2 + \frac{W^2}{\zeta_0^2}\right)$$

On the wing surface ($\lambda=0$) we have $u=W=V=0$ and $H=H_w=\text{const}$. At the outer edge of the boundary layer $u\to 1$, $W\to 0$, and $H\to 1/2$, as $\lambda\to\infty$.

It can be seen that the $t$-independent solution of system (5.88) satisfies the boundary conditions but does not ensure the matching of the solutions obtained for the left and right halves of the wing. It can be shown that this solution coincides with the solution for a swept plate which was, in essence, used in the work of Ladyzhenskii (1965).

### 5.4.4 Solution in the vicinity of the leading edge

We will study system (5.88) near $t = 1$ under the assumption that the solution is nonunique. We expand $L$ in a series

$$L(t) = L_0 + L_1(1-t)^a + \sum_2^{\infty} L_i(1-t)^{\alpha_i} \tag{5.89}$$

Other functions entering in system (5.88) can be represented in the series form, as follows:

$$f(\lambda, t) = f_0(\lambda) + f_1(\lambda)\frac{L_1}{L_0}(1-t)^a + \sum_2^{\infty} f_i(\lambda)(1-t)^{\alpha_i} \tag{5.90}$$

Here, the subscript 0 refers to the expansion terms corresponding to the self-similar solution for a swept plate of infinite span and it is assumed that $0 < a < \alpha_i < \alpha_{i+1}$. If the problem under consideration can be solved at a needed accuracy on a finite, though fairly large interval of the values of $\lambda$ and the functions $f_i$ are bounded on this interval, then as $t \to 1$ each term of the expansion can be assumed to be smaller in the order than the previous term. In this case, assuming that the remainder of the series can be neglected as $t \to 1$, for zeroth terms of the expansion we obtain

$$V_0 u_{0\lambda} - \varepsilon h_0 = u_{0\lambda\lambda}, \quad V_0 W_{0\lambda} + \varepsilon h_0 = W_{0\lambda\lambda} \tag{5.91}$$

$$V_0 H_{0\lambda} = \frac{H_{0\lambda\lambda}}{\sigma} + 0.5\left(1 - \frac{1}{\sigma}\right)\left(u_0^2 + \frac{W_0^2}{\zeta_0^2}\right)_{\lambda\lambda}$$

$$V_{0\lambda} - 0.25\left(\frac{W_0}{\zeta_0^2} - u_0\right) = 0, \quad L_0 = 0.75\sqrt{0.5(\gamma+1)}\,\Phi_0$$

$$\Phi_0 L_0 = 2\varepsilon^{1/2}\int_0^{\infty} h_0\,\mathrm{d}r, \quad h_0 = H_0 - \frac{1}{2}\left(u_0^2 + \frac{W_0^2}{\zeta_0^2}\right)$$

$$u_0 = W_0 = V_0 = 0, \quad H_0 = H_w \ \text{ for } \lambda = 0$$

$$u_0 \to 1, \quad W_0 \to 0, \quad H_0 \to \frac{1}{2} \ \text{ for } \lambda \to \infty$$

Correspondingly, for the first terms of the expansion we obtain

$$V_1 u_{0\lambda} + V_0 u_{1\lambda} - \varepsilon h_1 - a\left(\frac{W_0}{\zeta_0^2 - u_0}\right) u_1 + 4a\varepsilon h_0 = u_{1\lambda\lambda} \tag{5.92}$$

$$V_1 W_{0\lambda} + V_0 W_{1\lambda} + \varepsilon h_1 - a\left(\frac{W_0}{\zeta_0^2 - u_0}\right) W_1 - 4a\varepsilon h_0 = W_{1\lambda\lambda}$$

$$V_1 H_{0\lambda} + V_0 H_{1\lambda} - a\left(\frac{W_0}{\zeta_0^2 - u_0}\right) H_1 = \frac{H_{1\lambda\lambda}}{\sigma} + \left(1 - \frac{1}{\sigma}\right)\left(u_0 u_1 + \frac{W_0 W_1}{\zeta_0^2}\right)_{\lambda\lambda}$$

$$V_{1\lambda} - (a + 0.25)\left(\frac{W_1}{\zeta_0^2} - u_1\right) - a\left(\frac{W_0}{\zeta_0^2} - u_0\right) = 0$$

$$\frac{4a+6}{4a+3}\int_0^\infty h_0\,\mathrm{d}r = \int_0^\infty h_1\,\mathrm{d}r, \quad h_1 = H_1 - u_1 u_0 - \frac{W_1 W_0}{\zeta_0^2}$$

$$u_1 = W_1 = H_1 = V_1 = 0 \text{ for } \lambda = 0$$

$$u_1 \to 0, \quad W_1 \to 0, \quad H_1 \to 0 \text{ for } \lambda \to \infty$$

Thus, for existence of other-than-zero first terms of expansions (5.89) and (5.90) ($L_1/L_0 \neq 0$) there must exist certain eigenvalues $a$ for which the overdetermined system of equations (5.92) would have a solution. Numerical integration of systems (5.91) and (5.92) showed that on the interval $0 < a < 70$ for each value of $\zeta_0$ there is only one eigenvalue $a$. For $\omega_0 = 90°$, $74°$, $45°$, and $30°$, $\sigma = 1$, and $H_w = 0.5$ this value, to two-place accuracy, is equal to $a = 24$, 22, 12, and 7.7. Numerical integration of system (5.88) with the initial conditions corresponding to the zeroth terms of expansions (5.89) and (5.90) showed (Kozlova and Mikhailov, 1970) that this problem is ill-posed. A departure from the edge $t = 1$ using two terms of series (5.89) (accuracy is checked using two last equations of system (5.88)) gives a single-parameter family of solutions dependent on the ratio $L_1/L_0$.

Thus, it is shown that near the leading edge the solution of the problem of strong viscous interaction is nonunique. Numerical results of the calculation for a delta wing are presented below, in Section 5.4.7.

### *5.4.5 Strong viscous interaction on a swept plate*

If we consider strong viscous interaction on a swept plate and seek a solution in the forms (5.89) and (5.90), then the problem for the first two terms of the expansion reduces to the previous problem. It is assumed that the swept plate is of infinite span and, therefore, the system of equations (5.82) depends on two variables only, namely, $y$ and $x - z\cot\omega_0$ (Fig. 5.16).

As before, we introduce the streamfunction and new variables $\lambda$, $t$, $\Phi$, and $L$ by the formulas

$$t = 1 - x + \frac{z}{\zeta_0}, \quad \lambda = \left(\frac{\varepsilon}{p(1-t)}\right)^{1/2} \int_0^{\psi} \rho \frac{\partial y}{\partial \psi} \mathrm{d}\psi$$

$$L = (1-t)^{1/4} p^{1/2}, \quad \delta = (1-t)^{3/4} \Phi(t), \quad \zeta_0 = \tan \omega_0$$

After transformations analogous to those made above we arrive at the following system of equations

$$Vu_\lambda - \varepsilon h + (1-t)\left[\left(\frac{W}{\zeta_0^2 - u}\right) u_t - \frac{4\varepsilon h L_t}{L}\right] = u_{\lambda\lambda} \tag{5.93}$$

$$VW_\lambda + \varepsilon h + (1-t)\left[\left(\frac{W}{\zeta_0^2 - u}\right) W_t + \frac{4\varepsilon h L_t}{L}\right] = W_{\lambda\lambda}$$

$$VH_\lambda + (1-t)\left(\frac{W}{\zeta_0^2} - u\right) H_\lambda = \frac{H_{\lambda\lambda}}{\sigma} + \frac{1}{2}\left(1 - \frac{1}{\sigma}\right)\left(u^2 + \frac{W^2}{\zeta_0^2}\right)_{\lambda\lambda}$$

$$V_\lambda - \frac{1}{4}\left(\frac{W}{\zeta_0^2} - u\right) + (1-t)\left[\left(\frac{W}{\zeta_0^2} - u\right)\frac{L_t}{L} - u_t + \frac{W_t}{\zeta_0^2}\right] = 0$$

$$L = \sqrt{\frac{1}{2}(1+\gamma)}\left[\frac{3}{4}\Phi - (1-t)\Phi_t\right], \quad \Phi L = 2\varepsilon^{1/2} \int_0^\infty h \,\mathrm{d}r$$

The conditions at the upper and lower edges of the boundary layer coincide with the conditions for system (5.88). If a solution of system (5.93) is sought near the leading edge of the plate using expansions of type (5.89) and (5.90), then for the zeroth and first terms of expansions we obtain systems (5.91) and (5.92). Thus, near the leading edge the solution for strong interaction on a swept wing, including the case $\omega_0 = 90°$, is also nonunique. The difference of the solution from the self-similar one can be regarded as the wing's trailing edge effect. The condition which must be satisfied by the solution near the trailing edge is obviously dependent on the body geometry.

### *5.4.6 Propagation of disturbances from the trailing edge of a swept plate*

As shown above, the equations of the three-dimensional boundary layer on a swept plate of infinite length can be written in form (5.93). For further presentation, we will make, following the paper of Kozlova and Mikhailov (1971), the change of variable

$$t = 1 - \xi \tag{5.94}$$

Then expansions (5.89) and (5.90) for swept and triangular plates coincide correct to the second term of the series and can be written in the form:

$$L(\xi) = L_0 + L_1\xi^a + \cdots, \quad f(\xi,\lambda) = f_0(\lambda) + \left(\frac{L_1}{L_0}\right) f_1(\lambda)\xi^a + \cdots \tag{5.95}$$

Here, $a = a(\omega_0)$, $f$ is any function entering in relations (5.93), $L_0$ and $f_0(\lambda)$ are the self-similar solutions for the case of the flow past a swept plate, and $L_1$ is a certain arbitrary parameter, on which the expansion is dependent.

By a proper choice of the parameter $L_1$ some condition on a certain line $\xi = \text{const}$ can be satisfied. If $L_1 = 0$, the flow is that past an infinite swept plate.

The effect of disturbances propagating upstream from the trailing edge will be evaluated with reference to the case of the flow past a swept plate. It is sufficient to carry out the calculations for two values of $L_1/L_0$ equal, for example, to $+1$ and $-1$, since all other solutions can be obtained from the above-mentioned ones by linear transformation of the coordinate $\xi$. This follows from the conclusion of Section 5.4.4 and directly from relations (5.95) which admit the change of variable $\xi_* = (\pm L_1/L_0)^{1/a}\xi$ reducing the problem to one of the cases $L_1/L_0 = \pm 1$.

The calculation presented below was carried out for a thermally insulated surface, since in this case the pressure gradient effect on the boundary layer is maximum. The Prandtl number $\sigma$ was taken to be equal to unity. Near the edge ($\xi = 0$) two terms of the series for the function $L$ were used and, starting from a certain value of $\xi$, the value of $L_1/L_0$ was chosen at each step in $\xi$ from the conditions of satisfaction of two last equations of system (5.88). Other equations were solved using the technique developed in the work of Petukhov (1964). In plotting the curves the $\xi$ coordinate was so normalized that for $\xi = 1$ the dimensionless pressure $L^2 = p\xi^{1/2}/(\rho_\infty u_\infty^2)$ differed by $-20\%$ from the value of this parameter in the self-similar problem if $L_1/L_0 = -1$ and by $+50\%$ if $L_1/L_0 = +1$.

The purpose of the calculations was the qualitative estimation of the effect of disturbances occurring at the trailing edge of thin bodies on the flows with viscous interaction.

In Fig. 5.17 we have plotted the dimensionless displacement thickness $\delta^0 = \delta Re_x^{3/4}/x$ and pressure $p^0 = pRe_x^{1/2}/(\rho_\infty u_\infty^2)$. In Fig. 5.18 the distributions of the longitudinal $c_u = \mu_0(\partial u/\partial y)_w Re_x^{3/4}/(\rho_\infty u_\infty^2)$ and transverse $c_w = \mu_0(\partial w/\partial y)_w Re_x^{3/4}/(\rho_\infty u_\infty^2 \cot \omega_0)$ friction coefficients are presented. In the above relations for $\delta^0$, $p^0$, $c_u$, and $c_w$ the quantities $\delta$, $p$, $u$, $w$, and $y$ are the dimensional displacement thickness, pressure, velocity components, and coordinate, respectively, $x$ is the distance from the leading edge of the body, and $Re_x = \rho_\infty u_\infty x/\mu_0$. The data presented in Figs. 5.17 and 5.18 can be regarded as the distributions of the corresponding parameters in sections $z = \text{const}$ of the swept plate if the coordinate $x$ is measured from the leading edge. In Figs. 5.17 and 5.18 solid curves relate to the angle $\omega = 90°$ and dot-and-dash curves to $\omega = 30°$.

For $\omega = 90°$ the maximum distance at which the trailing edge effect is appreciable amounts to about 15% of the plate length for the preseparated flow and about 10% for the case of expansion at the trailing edge. The calculations show that the trailing edge effect becomes more noticeable with increase in the sweep angle. The preseparated flow at $\omega = 30°$ occupies already half the plate.

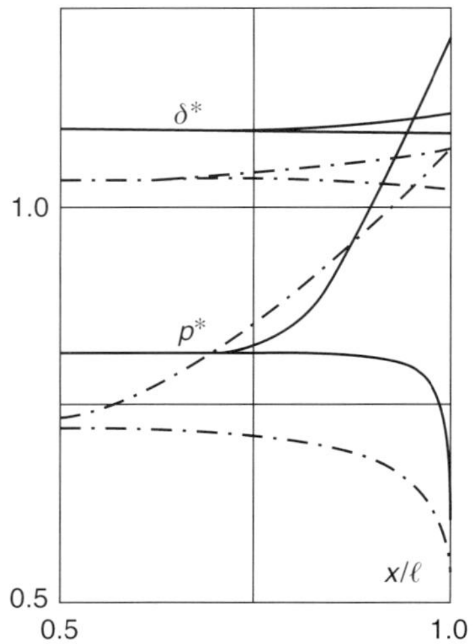


Fig. 5.17.

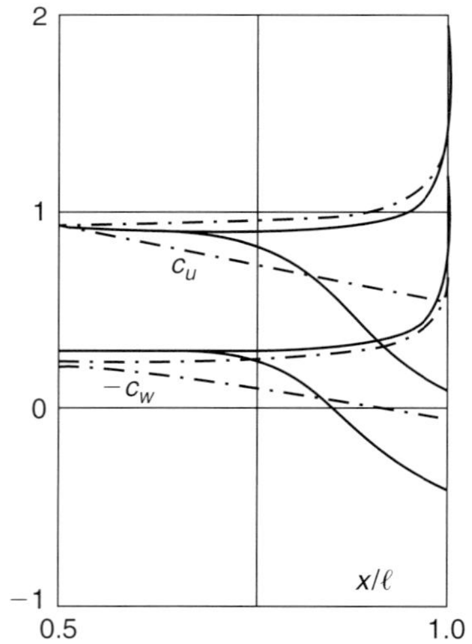


Fig. 5.18.

## *5.4.7 Delta wing*

In the case of a triangular plate of infinite length near the axis of symmetry there must exist an elevated-pressure region. The pressure rise suppresses the momentum of the gas flowing from the edges toward the axis. In this case, it is impossible to dispense with the solution of a boundary value problem, since the impermeability condition $t \equiv 0$ must necessarily be satisfied at $\xi = 1$ $(z = 0)$.

The calculation of the flow over a triangular plate was carried out using the integral method; it is described in detail in the paper of Kozlova and Mikhailov (1971). The Prandtl number was $\sigma = 1$ and the surface was thermally insulated. In Figs. 5.19 and 5.20 the results of calculations for $\delta^0$, $p^0$, $c_u$, and $c_w$ are presented. The dashed curves present the results for a swept plate with $\omega = 30°$, while the dot-and-dash and solid curves are the results for delta wings with $\omega = 30°$ and $60°$. The results can be regarded as the parameter distributions in sections $x = \text{const}$; they show that the difference between the flows on triangular and swept plates increases with decrease in the angle the leading edge makes with the oncoming flow. For $\omega = 30°$ near the axis of the triangular plate the pressure is even about 70% higher than that on the swept plate. The results obtained show the importance of taking account of the effect of boundary layer disturbances on flows with strong viscous interaction. A decrease of the local pressure in these flows leads to an increase of the friction and heat transfer coefficients and an increase results in a decrease of these coefficients. Disturbances induced by a pressure increase propagate at comparatively longer distances. The effect of boundary layer disturbances on the integral characteristics of the flow past a body becomes more appreciable with increase in the sweep angle (as shown in the paper of Kozlova and

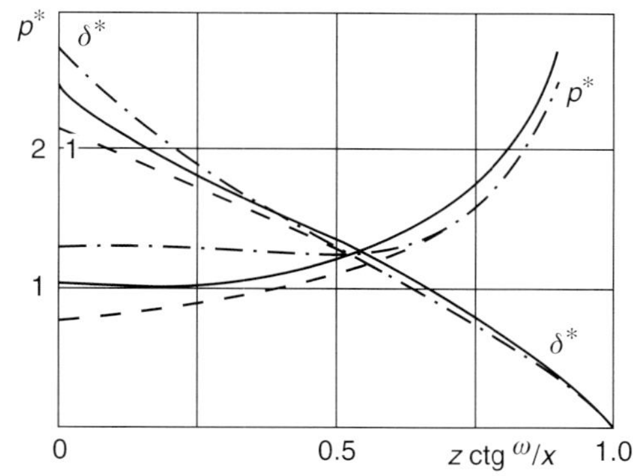


Fig. 5.19.

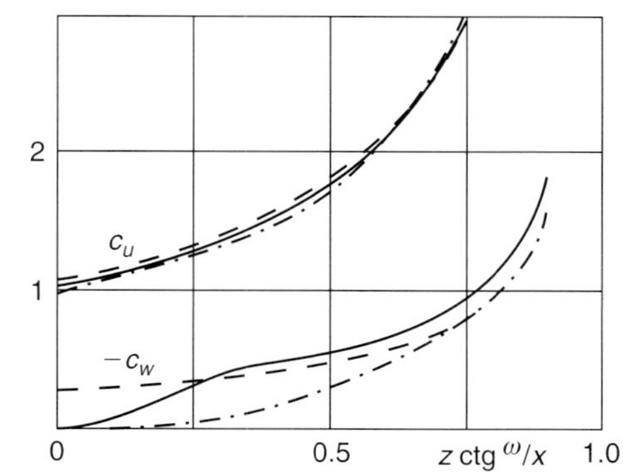


Fig. 5.20.

Mikhailov (1971), the order of the variation of these characteristics can increase from 3% to 15% when the angle $\omega_0$ varies from 90° to 30°).

## 5.5 Distinctive features of the symmetric flow over a thin triangular plate in the strong interaction regime

### 5.5.1 *Equations and boundary conditions*

Following the paper of Neiland (1974*b*), we will consider the symmetric flow over a thin triangular plate in the regime of strong interaction between the boundary layer and the outer hypersonic flow. As noted above, in this case the first important results were obtained in the work of Ladyzhenskii (1965), where it was shown that the solution depends only on two independent variables. In the same paper it was noted that, due to the existence of two counter-streaming flows inside the boundary layer, directed at a certain angle from the edges toward the center, the flow in the plane of symmetry is in no way simple and requires a special study. However, the flow pattern considered in the work of Ladyzhenskii (1965), which consisted of the boundary layer in a region adjoining the edges and a considerably thicker central region, is impossible. The boundary layer streamtubes could not inflow to the central region, where the static pressure is by an order higher than the total pressure in the boundary layer.

This fact was noted in the study of Kozlova and Mikhailov (1970). In that paper and the paper of Kozlova and Mikhailov (1971) it was assumed that the solution of the boundary layer equations is valid over the entire wing and, therefore, a solution satisfying the "impermeability" condition in the plane of symmetry was sought. In the work of Kozlova and Mikhailov (1970) the impossibility of the appearance of a locally inviscid region near the axis of symmetry was specially mentioned. In the work of Neiland (1974*b*) a possible internal contradiction in this boundary value problem formulation was noted. In fact, near the leading edge the solution involves only one arbitrary constant, while in the plane of symmetry a constraint is imposed on a function: the equality of the transverse velocity component to zero over the entire boundary layer thickness is required. The solution of Kozlova and Mikhailov (1971) was obtained in the integral approximation for which the whole profile of the transverse velocity is determined by a single constant and therefore does not make it possible to conclude whether there exist solutions of the required type for the boundary layer equations.

Following the work of Neiland (1974*b*) we will consider the solution of the problem for the general case in which the transverse velocity is nonzero in the entire profile corresponding the plane of symmetry of the wing. It can be shown that in this case, contrary to the assertion made in the work of Kozlova and Mikhailov (1970), a locally inviscid flow zone with narrow viscous flow sublayers can be constructed near the axis of symmetry. The matching of the solutions yields an additional boundary condition which makes it possible to choose the unique solution for the boundary layer. It is imposed on the gas flow rate in the transverse direction and can, therefore, be satisfied by the proper choice of the single arbitrary constant at our disposal.

The boundary layer equations can conveniently be written in the cylindrical coordinates shown in Fig. 5.21, rather than in the variables introduced in Section 5.4; here $uu_\infty$, $wu_\infty$,

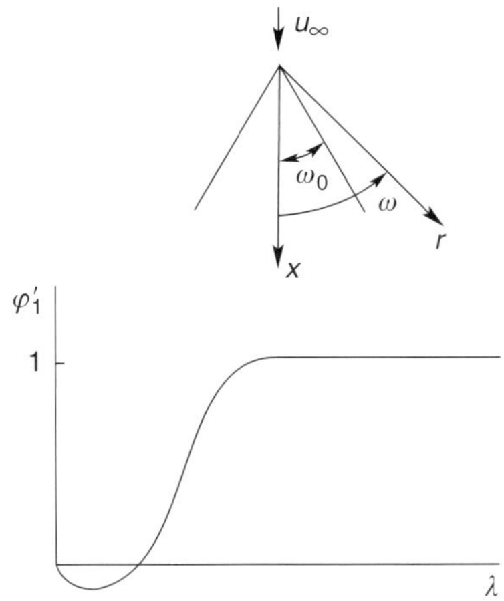


Fig. 5.21.

and $vu_\infty\tau$ are the velocity components along the radius vector $r\ell$, the transversal, and $\ell y$, respectively. The length $\ell$ has no particular geometric content; it is introduced for convenience and is automatically dropped out from final results. The angle between the leading edge and the freestream direction is designated by $\omega_0$. We introduce the variables

$$\lambda = \frac{\eta}{(r\zeta)^{1/4}}, \quad \zeta = \cos\frac{\pi\theta}{2\omega_0}, \quad \eta = \int_0^y \rho\,\mathrm{d}y \tag{5.96}$$

$$f = r(r\zeta)^{1/4} f_*(\zeta,\lambda), \quad \varphi = -r(r\zeta)^{1/4}\varphi_*(\zeta,\lambda)$$

$$p = (r\zeta)^{-1/2} p_*(\zeta), \quad \rho = (r\zeta)^{-1/2}\rho_*(\zeta,\lambda), \quad \delta = (r\zeta)^{3/4}\delta_*(\zeta)$$

$$g = g_*(\zeta,\lambda), \quad \mu = \mu_*(\zeta,\lambda)$$

where $f$ and $\varphi$ are the streamfunctions defined by formulas

$$ur = \frac{\partial f}{\partial \eta}, \quad wr = \frac{\partial \varphi}{\partial \eta} \tag{5.97}$$

We will drop out asterisks at variables in the boundary layer equations and the boundary conditions

$$(\rho\mu\varphi'')' + \frac{\pi}{8\omega_0}(1-\zeta^2)^{1/2}\left[\varphi\varphi'' + \frac{\gamma-1}{\gamma}(g - f'^2 - \varphi'^2) - \frac{4\zeta}{\rho}\dot{p}\right] + \zeta\left(\frac{4}{5}f\varphi'' - f'\varphi'\right)$$
$$= \zeta\frac{\pi}{2\omega_0}(1-\zeta^2)^{1/2}(\varphi'\dot{\varphi}' - \dot{\varphi}\varphi'')$$

$$(\rho\mu f'')' + \frac{\pi}{8\theta_0}(1-\zeta^2)^{1/2}\varphi f'' + \zeta\left[\frac{4}{5}ff'' + \varphi'^2 + \frac{\gamma-1}{4\gamma}(g - f'^2 - \varphi'^2)\right]$$
$$= \zeta\frac{\pi}{2\omega_0}(1-\zeta^2)^{1/2}(\varphi'\dot{f} - \dot{\varphi}f'')$$

$$\left\{\rho\mu\left[\frac{g'}{\sigma} - \frac{1-\sigma}{\sigma}(f'f'' + \varphi'\varphi'')\right]\right\}' + \frac{\pi}{8\omega_0}(1-\zeta^2)^{1/2}\varphi g' + \zeta\frac{5}{4}fg'$$
$$= \zeta\frac{\pi}{8\omega_0}(1-\zeta^2)^{1/2}(\varphi'\dot{g} - \dot{\varphi}g') \tag{5.98}$$

$$p = \frac{\gamma-1}{2\gamma}\rho(g - f'^2 - \varphi'^2), \quad \delta = \frac{\gamma-1}{2\gamma p}\int_0^\infty (g - f'^2 - \varphi'^2)\,\mathrm{d}\lambda$$

$$\mu = (g - f'^2 - \varphi'^2)^n$$

$$f'_w = \varphi'_w = f_w = \varphi_w = 0, \quad g(\lambda = 0) = g_w$$

$$f'_e = \cos\theta, \quad \varphi'_e = \sin\theta, \quad g_e = 1, \quad 0 \le \zeta < 1$$

Here, primes and dots refer to the differentiation with respect to $\lambda$ and $\zeta$, respectively, while the subscripts $w$ and $e$ refer to the values of the variables at the plate and the outer edge of the boundary layer. We will add a formula for calculating the pressure

$$p = \frac{\gamma+1}{2}\left\{\frac{3\pi}{8\omega_0}(1-\zeta^2)^{1/2}\delta\sin\omega_0 + \zeta\left[\frac{3}{4}\delta\cos\theta + \frac{\pi}{2\omega_0}\dot{\delta}(1-\zeta^2)^{1/2}\sin\theta\right]\right\}^2 \tag{5.99}$$

Equations (5.98) are very similar with the equations for the two-dimensional boundary layer written in the Dorodnitsyn–Lees variables. The first equation written for the transverse velocity component shows that in the region adjoining the leading edge of the wing ($\zeta = 0$) for $\varphi' > 0$ the integration should be carried out toward increasing $\zeta$, that is, toward the axis of symmetry. Contrariwise, in the flow regions, in which $\varphi' < 0$, the integration must be performed toward decreasing $\zeta$. The lines $\varphi' = 0$ or $\varphi''_w = 0$ are singular. On these lines the matching of the solutions determined separately for regions with opposite signs of $\varphi'$ can be required. Thus, there is a profound analogy with the two-dimensional separated flow, since the transverse flow streamlines with opposite signs of $\varphi'$ carry the information on different initial conditions. The analogy with the two-dimensional flow is incomplete, since the existence of the radial velocity component for the flow in a cross-sectional plane is equivalent to the flowfield produced by sources and sinks. In particular, owing to this fact, in the three-dimensional case singular lines are not always related with separation from the body surface.

We will assume that the flow is described by Eqs. (5.98) up to $\zeta = 1$. Then, using the symmetry of the flow, we can obtain local solutions of the problem in the form of the following expansions near $\theta = 0$

$$\zeta \approx 1 - \frac{1}{2}\left(\frac{\pi}{2\omega_0}\right)^2\theta^2 + \cdots, \quad \frac{\mathrm{d}\zeta}{\mathrm{d}\theta} \approx -\left(\frac{\pi}{2\omega_0}\right)^2\theta + \cdots$$

$$p \approx p_0 + \theta^2 p_2 + \cdots, \quad \rho \approx \rho_0 + \theta^2 \rho_2 + \cdots \quad \mu \approx \mu_0 + \theta^2 \mu_2 + \cdots \tag{5.100}$$

$$f \approx f_0 + \theta^2 f_2 + \cdots, \quad \varphi \approx \theta\varphi_1 + \theta^3 \varphi_3 + \cdots, \quad g \approx g_0 + \theta^2 g_2 + \cdots$$

$$\delta \approx \delta_0 + \theta^2 \delta_2 + \cdots$$

For $\theta \to 0$ we obtain the following system of ordinary differential equations and the boundary conditions for the leading terms of the expansions

$$(\rho_0\mu_0 f_0'')' + \frac{5}{4} f_0 f_0'' + \frac{\gamma - 1}{4\gamma}(g_0 - f_0'^2) - \varphi_1 f_0'' = 0$$

$$(\rho_0\mu_0 \varphi_1'')' + \frac{5}{4} f_0 \varphi_1'' + \frac{\gamma - 1}{4\gamma}(g_0 - f_0'^2)\left[\frac{4p_2}{\rho_0} + \left(\frac{\pi}{2\omega_0}\right)^2\right] - \varphi_1\varphi_1'' + \varphi_1'^2 - f_0'\varphi_1' = 0$$

$$\left[\rho_0\mu_0\left(\frac{1}{\sigma} g_0 - \frac{1-\sigma}{\sigma} f_0' f_0''\right)\right]' + \frac{5}{4} f_0 g_0' - g_0'\varphi_1 = 0 \tag{5.101}$$

$$\rho_0\mu_0 = \frac{2\gamma}{\gamma - 1} p_0 (g_0 - f_0'^2)^{n-1}, \quad p_0 = \frac{\gamma + 1}{2}\frac{9}{16}\delta_0^2, \quad \delta_0 = \frac{\gamma - 1}{2\gamma p_0}\int_0^\infty (g_0 - f_0'^2)\mathrm{d}\lambda$$

$$f_0(0) = f_0'(0) = 0, \quad f_0'(\infty) = 1, \quad \varphi_1(0) = \varphi_1'(0) = 0, \quad \varphi_1'(\infty) = 1$$

$$g_0(0) = g_w, \quad g_0(\infty) = 1$$

Problem (5.101) involves one arbitrary parameter $p_2$ determining the permissible types of solutions and dependent on the matching of the local and global solutions, as it is always the case for boundary layer flows near critical points and lines. Since $\varphi_1'(\infty) = 1$ for any $p_2$, at least, near the outer edge of the boundary layer there is always a region, in which $\varphi_1' > 0$. We will draw the further reasoning for the case

$$\frac{4\rho_2}{\rho_0} + \left(\frac{\pi}{2\theta_0}\right)^2 < 0$$

since at the axis of symmetry the pressure must be maximum. The $\varphi_1'$ profile takes the form shown in Fig. 5.21. (If $\varphi_1' > 0$ everywhere, then even without further reasoning it is obvious that the matching of solutions (5.101) and (5.98) is generally impossible and could be realized only under artificial boundary conditions, such as the presence of gas suction, the addition of a distributed body thickness in a certain region, etc.) We assume that on the main part of the wing the solution is constructed up to the point at which the $\varphi' = 0$ line goes away from the body surface. In accordance with the work of Kozlova and Mikhailov (1970), this solution depends on only one arbitrary constant. For integrating the equations in the $\varphi_1' > 0$ region the parameters

$$\lambda_0(\zeta), \quad \varphi_0(\zeta), \quad g_0(\zeta), \quad f_0(\zeta), \quad f_0'(\zeta) \tag{5.102}$$

must be preassigned at the line $\varphi' = 0$. Then, after numerical integration of (5.98), at $\zeta = 1$ we obtain certain profiles $\varphi^+(1,\lambda), f^+(1,\lambda)$, and $g^+(1,\lambda)$ (pluses refer to the variables in the $\varphi' > 0$ region). In order for these profiles to correspond to any solution (5.98), three of as yet five arbitrary functions (5.102) can be used. As a result of this procedure, the initial conditions at $\zeta = 1$ for the region $\varphi' < 0$ are determined. The boundary conditions for this region are the conditions at the body surface (5.98) and (5.102)

$$\varphi_0' = 0, \quad g_0' = g_0^+, \quad f_0'^+ = f_0'^- \text{ for } \lambda = \lambda_0(\zeta)$$

However, the following conditions are not fulfilled

$$\varphi_0^+ = \varphi_0^-, \quad \varphi_0''^+ = \varphi_0''^-, \quad g_0'^+ = g_0'^-, \quad f_0''^+ = f_0''^-, \quad f_0^+ = f_0^- \tag{5.103}$$

Moreover, conditions (5.103) cannot generally be satisfied, since only two arbitrary functions remain free. Of course, one can try to introduce some additional "degrees of freedom" assuming, for example, that in this region $\varphi_w = \varphi_w(\zeta), f_w = f_w(\zeta)$, and $g_w = g_w(\zeta)$; however, generally solution (5.98) cannot be continued to the axis.

The above reasoning shows only that the solution of problem (5.98) cannot be continued up to $\zeta = 1$. However, it is not proved that Eq. (5.98) has no solutions characterized by the fulfillment of the condition $\varphi'(\zeta \to 1, \lambda) \to 0$, which should be matched – at small distances from the axis of symmetry – with a solution for a local viscous flow with low transverse velocity components which is not described by Eqs. (5.98).

If $\varphi'$ does not tend to zero, then a region of local inapplicability of the boundary layer equations (5.98) should be considered in a small vicinity of the plane of symmetry assuming that the velocities $\theta \ll w \le 1$. We introduce the designation for the order of the velocity $w$ in the solution of Eqs. (5.98), as $\zeta \to 1$

$$w \sim O(\alpha), \quad \theta \ll \alpha \le O(1) \tag{5.104}$$

From the solution matching conditions, in the local region the velocity $v$ is of the same order. Their orders are also retained by $u \sim 1, p \sim \tau^2, \rho \sim \tau^2$, and $\delta \sim \tau$. Taking into account Eq. (5.104) and the condition $\theta \ll \alpha$ we obtain

$$\frac{1}{r}\frac{\partial w}{\partial \theta} \gg \frac{\partial u}{\partial r} \tag{5.105}$$

However, then we have $w \sim v \sim \alpha$, since it is necessary that the boundary conditions are satisfied in each section $r = \text{const}$. The above estimates show that in the region $r\theta \sim \tau$ the following variables must be introduced

$$\theta = \tau\Theta, \quad y = Y, \quad p = P, \quad \rho = R, \quad u = U$$

$$w = W, \quad v = \tau^{-1}V, \quad G = g + V^2, \quad \delta = \Delta \tag{5.106}$$

Here, for the sake of brevity, we will consider only the case $\alpha \sim 1$. However, similar results could also be obtained for the simpler case $\tau \ll \alpha \ll 1$.

Substituting Eq. (5.106) in the Navier–Stokes equations and passing to the limit $M_\infty \to \infty$ and $\tau \to 0$ leads to the inviscid flow equations involving $r$ as a parameter

$$\frac{1}{r}\frac{\partial RW}{\partial \Theta} + \frac{\partial RV}{\partial Y} = 0, \quad RW = \frac{\partial \psi}{\partial Y}, \quad RV = -\frac{1}{r}\frac{\partial \psi}{\partial \Theta}$$

$$\frac{W}{r}\frac{\partial W}{\partial \Theta} + V\frac{\partial W}{\partial Y} = \frac{1}{Rr}\frac{\partial P}{\partial \Theta}, \quad \frac{W}{r}\frac{\partial V}{\partial \Theta} + V\frac{\partial V}{\partial Y} = -\frac{1}{R}\frac{\partial P}{\partial Y} \tag{5.107}$$

$$\frac{2\gamma}{\gamma - 1}\frac{P}{R} + U^2 + V^2 + W^2 = G(\psi), \quad U = U(\psi)$$

Due to the smallness of the derivatives with respect to $r$, the longitudinal velocity component $U$ and $G$ are transferred along "streamlines" $\psi = \text{const}$. The boundary conditions of impermeability are as follows:

$$V(r, \Theta, 0) = 0, \quad V_e - W_e \left.\frac{\mathrm{d}\Delta}{r\,\mathrm{d}\Theta}\right|_{\Psi \to \infty} \to 0, \quad W(r, O, Y) = 0 \tag{5.108}$$

In new variables the condition of compatibility with the outer hypersonic flow (5.99) as $\tau \to 0$ and $\zeta \to 1$, takes the form:

$$P = \frac{\gamma + 1}{2}\left(\frac{3}{4}\Delta - \Theta\frac{\mathrm{d}\Delta}{\mathrm{d}\Theta}\right)^2$$

Integrating the continuity equation over the region thickness with account for the last condition (5.108) yields the condition which must be fulfilled for all $\Theta$

$$\int_0^\Delta RW\,\mathrm{d}Y = 0 \tag{5.109}$$

Matching solutions (5.98) and (5.107) yields the conditions necessary for integrating (5.98) in the region, in which $\varphi' < 0$. They are analogous to those obtained for the two-dimensional flow and, for the sake of brevity, are not presented here. As distinct from the two-dimensional flow, in the local region viscous layers can be introduced not only on the body surface, where the no-slip conditions are violated, but also for $z = \Delta$, if the condition of the tangential velocity matching is not satisfied there. In fact, in the outer flow $V \sim w \sim O(\tau)$, while inside the region $V \sim w \sim O(\alpha) \gg \tau$. By virtue of the matching principle and in view of condition (5.109), from all solutions (5.98) it is a solution for which the flow in the transverse direction vanishes, as $\zeta \to 1$, that should be chosen. This condition can be satisfied by a proper choice of one arbitrary constant which is contained in expansions (5.98) for $\zeta \to 0$ derived in the paper of Kozlova and Mikhailov (1970).

From the above analysis it follows also that in the locally inviscid flow regions the streamtubes, which were previously in the upper part of the boundary layer and carried large values of the stagnation enthalpy and longitudinal velocity component, turn and approach the body surface. For this reason, large heat fluxes and longitudinal components of the viscous stress should be expected in the central part of the wing.

We note in conclusion that, as shown above, boundary value problems for the boundary layers on a delta wing and ahead of a plane step turn out to be noncontradictory in the presence of secondary or return flows and regions of local inapplicability of the boundary layer equations. However, apart from the flow patterns proposed above and including locally inviscid zones with finite velocities, one more type of solutions of the boundary layer equations on the main part of the body should be studied, together with the corresponding slow flows in the regions of local inapplicability of the boundary layer equations. The solutions for the boundary layer should be chosen from the condition that on a delta wing the transverse dimension of the $w = O(1) > 0$ zone tends to zero as the plane of symmetry is approached and $u = O(1)$ vanishes on the $\Psi = 0$ streamline as the step is approached. In the solutions of this type the boundary condition at the beginning of the return flow is $u = 0$ for the two-dimensional case and $w = 0$ for the delta wing. Since these conditions are imposed for the return flow, this does not lead to contradictions in the original boundary value problem.

## 5.6 Finite-length wings in the strong viscous interaction regime

Above in this chapter we considered the flow past semi-infinite delta wings in different viscous–inviscid interaction regimes and showed that under certain conditions the three-dimensional boundary layer equations can be reduced to two-dimensional equations by introducing self-similar variables; for the latter equations corresponding methods of calculation were developed. The questions related with the flow past semi-infinite delta wings, both planar and nonplanar, were presented in detail in the book of Bashkin and Dudin (2000) and, therefore, are not considered here.

Generally, as shown below in considering the hypersonic viscous flow past planar delta wings of finite length, in the strong viscous interaction regime there remains a dependence on the longitudinal coordinate, so that the boundary value problem is not reduced to the self-similar one.

### *5.6.1 Mathematical formulation of the problem*

We will consider the hypersonic viscous flow past a planar delta wing of finite length $\ell$ on the basis of the two-layer scheme (Dudin, 1982*a*, *b*). This flow pattern is obtained from the boundary value problem for the Navier–Stokes equations at the triple passage to limit $M_\infty \to \infty$, $Re \to \infty$, and $\delta \to 0$, where $\delta$ is the characteristic dimensionless boundary layer thickness.

In solving the problem we will use a Cartesian coordinate system with origin at the delta wing vertex (Fig. 5.22).

In accordance with the conventional estimates for the boundary layer in a hypersonic flow (Hayes and Probstein, 1966) we introduce the following dimensionless variables: $uu_\infty$, $wu_\infty$, and $v\tau z_0^{-1} u_\infty$ are the velocity projections onto the axes $x^0 = x\ell$, $z^0 = zz_0\ell$, and $y^0 = y\tau\ell$, respectively; $\rho\tau^2\rho_\infty$ is the density; $p\tau^2\rho_\infty u_\infty^2$ is the pressure; $gu_\infty^2/2$ is the stagnation enthalpy; $\mu\mu_0$ is dynamic viscosity; and $\delta_e\tau\ell$ is the boundary layer thickness. Here,

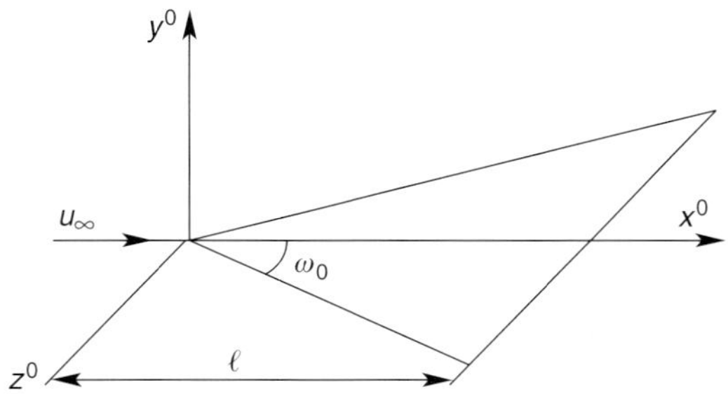


Fig. 5.22.

the parameter $\tau = (z_0/Re)^{1/4}$ and $z_0 = \tan\omega_0$. Writing the Navier–Stokes equations in these variables and passing to the limit $Re \to \infty$ we arrive at the equations of the three-dimensional laminar boundary layer (5.47) with the corresponding boundary conditions (5.48).

For solving the system of boundary layer equations (5.47) it is necessary to know the pressure distribution, which is not given and must be determined in the process of the solution of the boundary value problem (5.47) jointly with the equations for the outer inviscid flow. Since in this chapter we consider the flow past wings with the aspect ratio $z_0 = O(1)$ and stripes theory (Hayes and Probstein, 1966) is valid for the outer inviscid flow, the approximate tangent wedge formula can be applied for calculating the pressure; it can be taken, for example, in the form valid for $M_\infty\tau \gg 1$ (5.56) which in this case can be reduced to the form:

$$p = \frac{\gamma+1}{\gamma}\left(\frac{\partial\delta_e}{\partial x}\right)^2 \tag{5.110}$$

where the displacement thickness is determined by expression (5.49).

The solution of the complete boundary value problem includes taking account for the flow in the wake formed downstream of the wing (Neiland, 1974*c*); however, in this section, in order not to consider this flow, we will impose a boundary condition at the trailing edge of the wing. In this case it should be borne in mind that in the flow past a noncold plate the series expansion of the solution in the vicinity of the leading edge includes an arbitrary function, since the flow is subcritical (see Neiland, 1974; Chapter 7); therefore, a function must be preassigned at the trailing edge for choosing uniquely the solution. In this chapter, a given pressure distribution over the trailing edge is used as such a function.

The questions related with the flow past planar triangular and diamond-shaped wings in the strong viscous interaction regime were considered in the paper of Denisenko (1978*a*).

We note that at the trailing edge of a wing ($x = 1$) with a given pressure distribution $p_k(z)$ the thickness $\delta_e(x, z)$ determined as a result of the solution of the boundary value problem must satisfy the equation

$$\frac{\gamma+1}{2}\left(\frac{\partial\delta_e}{\partial x}\right)^2 = p_k(z) \tag{5.111}$$

For taking account for the singular behavior of the unknown functions in the vicinity of the delta wing vertex we introduce the similarity variables

$$z = xz^*, \quad \lambda = x^{1/4}\lambda^*, \quad \delta_e = x^{3/4}\delta_e^*(x, z^*), \quad \mu = \mu^*(x, \lambda^*, z^*) \tag{5.112}$$
$$u = u(x, \lambda^*, z^*), \quad w = w(x, \lambda^*, z^*), \quad g = g(x, \lambda^*, z^*), \quad p = x^{-1/2}p^*(x, z^*)$$
$$\rho = x^{-1/2}\rho^*(x, \lambda^*, z^*), \quad v = x^{-3/4}\left(v^* - xuz_0\frac{\partial \lambda^*}{\partial x}\right)$$

Moreover, the singularities in the behavior of the unknown functions in the vicinity of the leading edges of the delta wing ($z^* = \pm 1$) must also be taken into account. This is done by introducing the similarity variables

$$z^* = \theta, \quad \lambda^* = \sqrt{\frac{2\gamma}{\gamma - 1}\sqrt{1 - z^{*2}}}\,\eta^*, \quad p^* = \frac{p_0(x, \theta)}{\sqrt{1 - z^{*2}}}, \quad \delta_e^* = (1 - z^{*2})^{3/4}\Delta_e(x, \theta) \tag{5.113}$$
$$v^* = \left[v_0\frac{p_0}{1 - z^{*2}} - (w - z_0uz^*)\frac{\partial \eta^*}{\partial z^*}\right]\sqrt{\frac{2\gamma}{\gamma - 1}\sqrt{1 - z^{*2}}}$$

As a result of all these transformations in system (5.47), we arrive at the following boundary value problem:

$$\frac{\partial v_0}{\partial \eta^*} = (w - z_0u\theta)\frac{\theta}{2p_0} - \frac{1 - \theta^2}{p_0}\left[z_0x\frac{\partial u}{\partial x} - z_0\theta\frac{\partial u}{\partial \theta} + \frac{\partial w}{\partial \theta} + \frac{z_0u}{4}\right] \tag{5.114}$$
$$z_0ux\frac{1 - \theta^2}{p_0}\frac{\partial f}{\partial x} + (w - z_0u\theta)\frac{1 - \theta^2}{p_0}\frac{\partial f}{\partial \theta} + v_0\frac{\partial f}{\partial \eta^*} = G$$
$$f = \begin{Bmatrix} u \\ w \\ g \end{Bmatrix}$$
$$G = \begin{Bmatrix} -z_0\frac{\gamma - 1}{2\gamma p_0}(g - u^2 - w^2)\left[(1 - \theta^2)\left(\frac{x}{p_0}\frac{\partial p_0}{\partial x} - 0.5\right) - \theta\left(\theta + \frac{1 - \theta^2}{p_0}\frac{\partial p_0}{\partial \theta}\right)\right] \\ + \frac{\partial}{\partial \eta^*}\left(N\frac{\partial u}{\partial \eta^*}\right) \\ -\frac{\gamma - 1}{2\gamma p_0}(g - u^2 - w^2)\left(\theta + \frac{1 - \theta^2}{p_0}\frac{\partial p_0}{\partial \theta}\right) + \frac{\partial}{\partial \eta^*}\left(N\frac{\partial w}{\partial \eta^*}\right) \\ \frac{\partial}{\partial \eta^*}\left\{N\left[\frac{1}{\sigma}\frac{\partial g}{\partial \eta^*} - \frac{1 - \sigma}{\sigma}\frac{\partial(u^2 + w^2)}{\partial \eta^*}\right]\right\} \end{Bmatrix}$$

$$\Delta_e = \frac{1}{p_0}\sqrt{\frac{\gamma-1}{2\gamma}}\int_0^\infty (g-u^2-w^2)\,\mathrm{d}\eta^*, \quad N=(g-u^2-w^2)^{n-1}$$

$$p_0 = \frac{\gamma+1}{2}\left[0.75(1-\theta^2)\Delta_e + x(1-\theta^2)\frac{\partial \Delta_e}{\partial x} - \theta(1-\theta^2)\frac{\partial \Delta_e}{\partial \theta} - 1.5\theta\Delta_e\right]^2 \quad (5.115)$$

with the boundary conditions at $|\theta| \leq 1$ and $0 \leq x \leq 1$:

$$u = v_0 = w = 0, \quad g = g_w \quad \text{for } \eta^* = 0$$
$$u \to 1, \quad w \to 0, \quad g \to 1 \quad \text{for } \eta^* \to \infty \quad (5.116)$$

The system of partial differential equations (5.117) governs the flow in the three-dimensional boundary layer on a flat triangular plate of finite length in the strong viscous interaction regime. It should be noted that at the delta wing vertex ($x = 0$) the terms containing the variable $x$ drop out from system (5.114), so that the boundary value problem turns out to be dependent only on two independent variables $\theta$ and $\eta^*$.

As noted above, the system of partial differential equations thus obtained describes also the flow over a semi-infinite delta wing (Dudin, 1978*b*). At the leading edges of the wing for the values of the transverse coordinate $\theta = \pm 1$ the system of equations (5.114) degenerates to a system of ordinary differential equations; for a delta wing the terms containing the variable $x$ also drop out from these equations, so that their solutions are valid for all values of the coordinate $x$, $0 \leq x \leq 1$.

The domain of integration of system (5.114) represents a rectilinear parallelepiped. For solving the complete boundary value problem it is necessary to solve first the system of ordinary differential equations at the leading edges of the wing and then, taking the solution thus obtained for initial conditions, to solve the system of partial differential equations in two independent variables describing the flow at the delta wing vertex. Finally, imposing a boundary condition on the trailing edge of the wing, for example, a pressure distribution, and using the previously determined solutions at the wing vertex and its leading edges, we solve the system of the three-dimensional boundary layer (5.114). The method of the solution of the boundary value problem is presented in the book of Bashkin and Dudin (2000).

### *5.6.2 Aerodynamic characteristics of finite-length wings at zero incidence*

The effect of the finite dimensions of a wing on the wing flowfield and its aerodynamic characteristics is studied (Dudin, 1983*a*) with reference to the flow past a delta wing at zero incidence with the pressure distribution over the trailing edge

$$p_K(\theta) = p_0(x=1,\theta) = p_{0ABT}(\theta)\left[1 + C\left(\frac{1+\cos\pi\theta}{2}\right)^\nu\right] \quad (5.117)$$

where $p_{0ABT}$ is the pressure distribution in the section $x = 1$ in the flow past a semi-infinite delta wing. The exponent $\nu$ was chosen to be equal to 3 in order to ensure a smooth transition to the pressure at the leading edges, as $\theta \to \pm 1$. For $C = 0$ the pressure realized at the trailing edge corresponds to the self-similar solution. The coefficient $C$ determines maximum deviation of the pressure at the wing axis from the self-similar pressure, while its sign (plus or minus) corresponds to the compression or expansion flow pattern at the trailing edge, respectively.

The calculations were carried out for the following conditions: $z_0 = 2$ (sweep angle equal to about 27°), $\gamma = 1.4$, $\sigma = 0.71$, $g_w = 0.5$, and $C = -0.5$ to 0.2.

In Figs. 5.23 and 5.24 we have plotted the calculated profiles of the pressure $p_0$ and the longitudinal velocity $u_2$ at the second layer ($\eta = 0.15$) in the plane of symmetry $\theta = 0$ of the delta wing along the $x$ axis in the dimensionless similarity variables in which the boundary value problems (5.114)–(5.116) was solved.

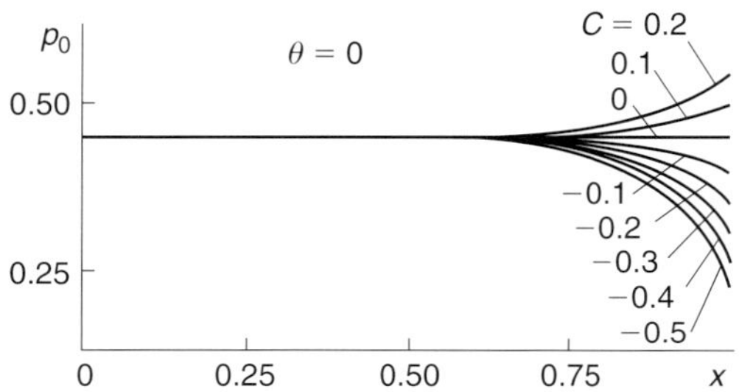


Fig. 5.23.

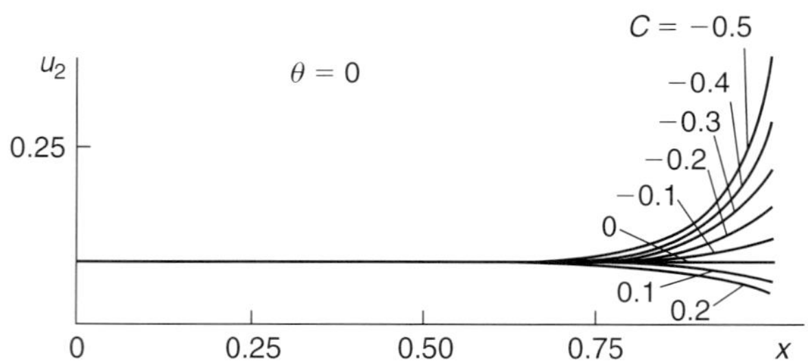


Fig. 5.24.

It should be noted that the pressure variation at the trailing edge has an effect on the flow parameters throughout an upstream distance equal to about 30–40% of the wing chord. Nearer to the wing vertex the solutions coincide with the self-similar one, that is, the solution of the problem for the flow past the semi-infinite wing. In Fig. 5.25 the longitudinal velocity profiles $u(x, \eta, \theta)$ are plotted in the same variables for the case of flow expansion at the trailing edge with the coefficient $C = -0.5$. It is visible how the profiles become more convex as $x \to 1$ and $\theta \to 0$.

In Figs. 5.26 and 5.27 we have plotted the calculated profiles of the pressure $p_0$ and the longitudinal viscous stress coefficient $\tau_u = \frac{\partial u}{\partial y}\Big|_w$ in the plane of symmetry $\theta = 0$ of the wing

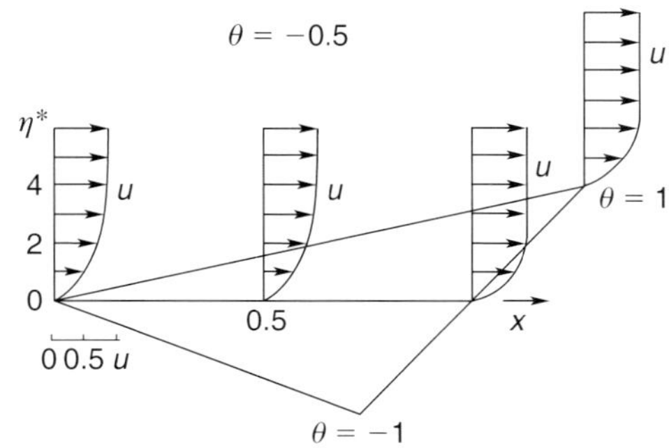


Fig. 5.25.

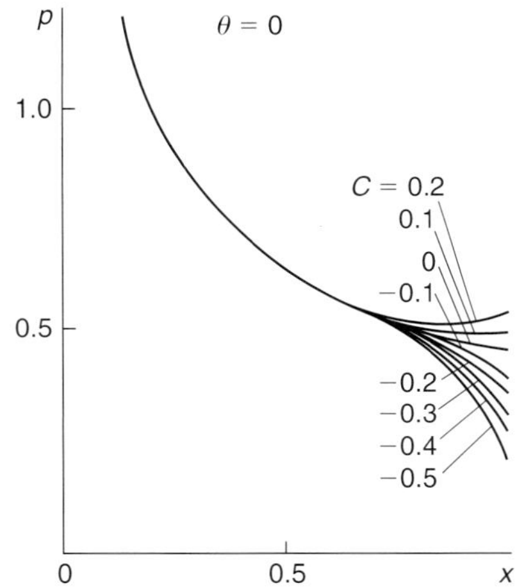


Fig. 5.26.

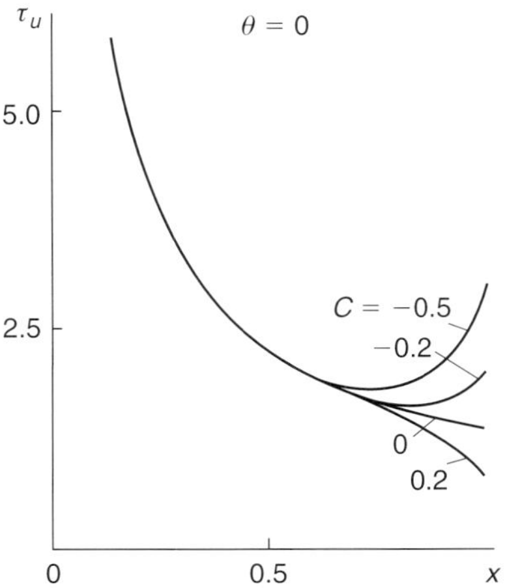


Fig. 5.27.

along the $x$ axis in the physical dimensionless variables. An appreciable effect of the pressure variation at the trailing edge on the $p$ and $\tau_u$ distributions over the wing should be noted.

We note that when the pressure distribution (5.117) is taken with the coefficient $C = -0.5$, a maximum heat flux $\tau_g$ at the axis of symmetry is observed somewhat ahead of the trailing edge rather than at point $x = 1$ itself, as it is the case for the $\tau_u$ maximum, which is apparently due to the three-dimensional nature of the boundary layer flow.

From an analysis of the boundary layer thickness $\delta_e$ distribution it can be concluded that it varies only slightly. Thus, a 2.5-fold variation of the trailing edge pressure at the axis of symmetry results in only 6% variation of the thickness $\delta_e$.

In Fig. 5.28 the distributions of the pressure $p$ and thickness $\delta_e$ over the wing are presented for the case of the calculated boundary layer with expansion at the trailing edge of the wing ($C = -0.5$). The results are presented in the dimensionless physical variables. The distribution of the dimensionless longitudinal viscous stress coefficient $\tau_u$ is presented in Fig. 5.29. It can be seen how the monotonic $\tau_u$ variation over the wing span goes over to a nonmonotonic one as the trailing edge is approached.

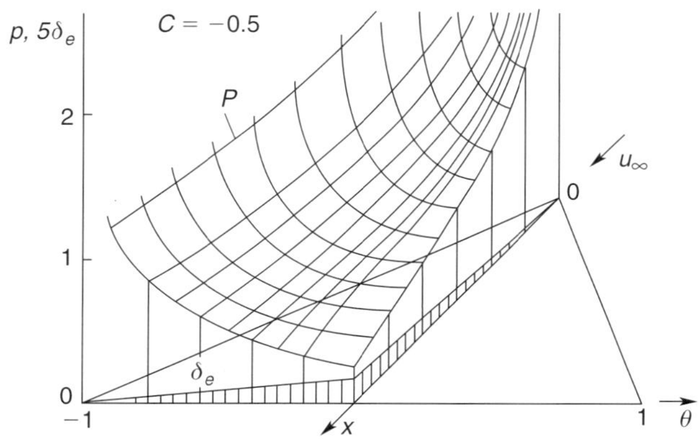


Fig. 5.28.

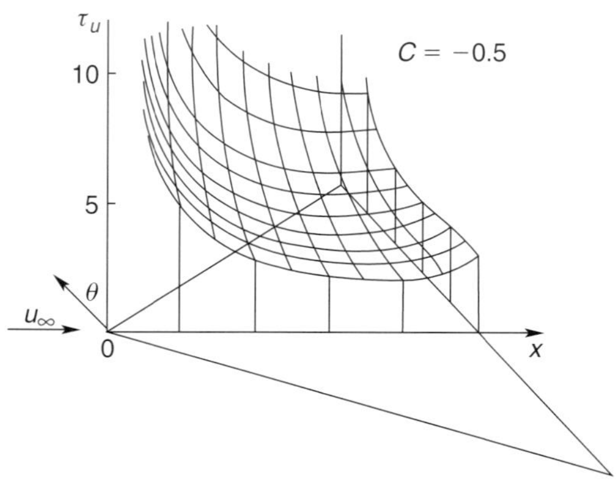


Fig. 5.29.

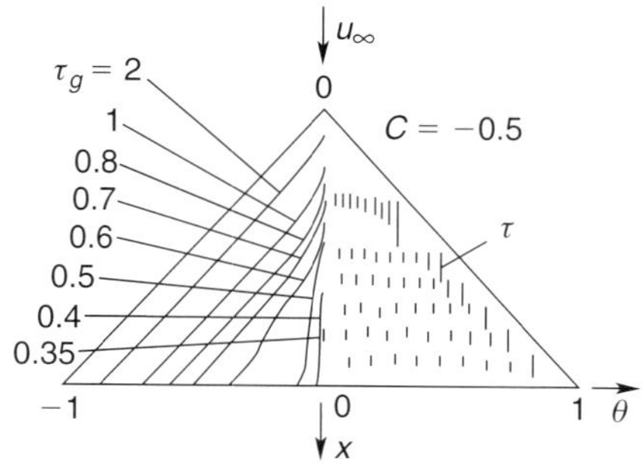


Fig. 5.30.

In Fig. 5.30 we have plotted the heat flux contours on the left half of the wing for the same coefficient $C = -0.5$. In the vicinity of the leading edge this distribution is near-linear, while in the vicinity of the plane of symmetry it becomes appreciably nonlinear. We note that a heat flux minimum is reached in the plane of symmetry at $x = 0.8$. On the right half of the wing dashes represent the direction and the value of the total viscous stress coefficient $\tau = \sqrt{\tau_u^2 + \tau_w^2}$. A $\tau$ minimum is reached in the region, in which $x \approx 0.8$ and $|z| < 0.3$.

### 5.6.3 *Wings of finite length at an angle of attack*

We will consider the hypersonic viscous flow past a finite-length delta wing at an angle of attack $\alpha^0$ (Dudin, 1983*a*). It is assumed that the angle of attack is small ($\alpha^0 < \delta$) and such that the assumption of hypersonic small perturbation theory is fulfilled

$$M_\infty(\delta \pm \alpha^0) \geq O(1) \tag{5.118}$$

where $\delta = (z_0/Re)^{1/4}$ is the characteristic dimensionless boundary layer thickness ($z_0 = \tan\omega_0$, where $\omega_0$ is the wing semi-vertex angle). In expression (5.118) plus sign corresponds to the flow over the lower (windward) wing surface and minus sign to the upper (leeward) surface. A Cartesian coordinate system with origin at the delta wing vertex is presented in Fig. 5.31. It is assumed that the boundary layer/outer hypersonic flow interaction is strong ($\chi \gg 1$) over the entire wing. The solution of the complete boundary value problem includes

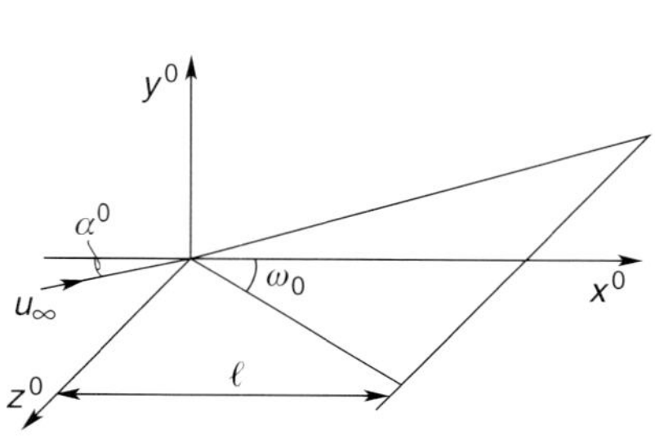


Fig. 5.31.

the flow in the wake formed downstream of the wing; however, in this, as in the preceding subsection the consideration of this flow is obviated by imposing a boundary condition at the trailing edge of the wing.

In this case, the system of boundary layer equations in dimensionless variables takes the form (5.47), while the pressure is calculated using the tangent wedge formula subject to the condition $M_\infty(\delta \pm \alpha^0) \gg 1$ and the no-slip condition

$$p = \frac{\gamma+1}{2}\left(\frac{\partial \delta_e}{\partial x} \pm \alpha\right)^2 \tag{5.119}$$

where $\delta_e$ is the boundary layer thickness and $\alpha^0 = \alpha\delta$.

After the introduction of variables (5.112) which take account for the singular behavior of the flow functions at the wing vertex, the expression for the pressure (5.119) takes the form:

$$p^*(x,\theta) = \frac{\gamma+1}{2}\left(\frac{3}{4}\delta_e^* + x\frac{\partial \delta_e^*}{\partial x} - \theta\frac{\partial \delta_e^*}{\partial \theta} \pm \alpha x^{1/4}\right)^2 \tag{5.120}$$

Due to the presence of the term $\alpha x^{1/4}$ in the above expression, on the wing surface for $\alpha \neq 0$ and $x > 0$ the pressure turns out to be a function of $x$ and $\theta$, while the unknown functions in the boundary layer are dependent on three variables: $x$, $\theta$, and $\lambda$. Thus, due to the presence of the angle of attack, the boundary value problem in variables (5.112) remains three dimensional even for a semi-infinite wing, as distinct from the case of the flow past a wing at zero incidence, in which the system of the three-dimensional boundary layer equations reduced to a system dependent on only two independent variables $\theta$ and $\lambda^*$.

Introducing then variables (5.113) which take account for the flow singularities in the vicinity of the leading edges, we obtain system of equations (5.114), while the pressure is determined by the following formula

$$\begin{aligned} p_0 = \frac{\gamma+1}{2}\Bigg\{&\frac{3}{4}(1-\theta^2)\Delta_e + x(1-\theta^2)\frac{\partial \Delta_e}{\partial x} \\ &-\theta\left[(1-\theta^2)\frac{\partial \Delta_e}{\partial \theta} - \frac{3}{2}\theta\Delta_e\right] \pm \alpha x^{1/4}(1-\theta^2)^{1/4}\Bigg\}^2 \end{aligned} \tag{5.121}$$

$$\Delta_e = \frac{1}{p_0}\sqrt{\frac{\gamma-1}{2\gamma}}\int_0^\infty (g - u^2 - w^2)\, d\eta$$

The boundary conditions for $|\theta| \leq 1$ and $0 \leq x \leq 1$ are as follows:

$$u = w = v_0 = 0, \quad g = g_w \quad \text{for } \eta^* = 0$$

$$u \to 1, \quad w \to 0, \quad g \to 1 \quad \text{for } \eta^* \to \infty$$

The system of partial differential equations (5.114), together with formula (5.121), describes the flow in the three-dimensional boundary layer on a delta wing of finite length at an angle of attack in the strong viscous interaction regime. It should be noted that at the wing vertex ($x = 0$) the terms containing the variable $x$ drop out from this system and the boundary value problem turns out to be dependent only on two independent variables $\theta$ and $\eta^*$. At the leading edges, for the values of the transverse coordinate $\theta = \pm 1$, the system degenerates to a system of ordinary differential equations.

The method for solving the system of equations thus obtained was developed in the book of Bashkin and Dudin (2000). As an illustration, we will consider the flow past a delta wing, at whose trailing edge the pressure is identically equal to that corresponding to the flow past a semi-infinite delta wing at zero angle of attack for the coordinate $x = 1$.

In the numerical calculations we took $z_0 = 2$, $\gamma = 1.4$, $\sigma = 0.71$, $g_w = 0.5$, and $\alpha = 0$ and 0.3. In Fig. 5.32 the calculated pressure is plotted along the axis of symmetry of the wing ($z = 0$). The values of $p$ with $\alpha = -0.3$ correspond to the pressure distribution over the upper surface of the wing (curve 3) and those with $\alpha = 0.3$ to the lower surface (curve 1). Curve 2 ($\alpha = 0$) corresponds to the flow past the wing at zero incidence. As it might be expected, the pressure is considerably higher on the windward than on the leeward side; thus, for $x = 0.5$ the pressure on the lower surface is almost twice as large as that on the upper surface, which is in qualitative agreement with the solution of the Navier–Stokes equations (Dudin, 1988*a*). Numerical calculations showed that the pressure variation at the trailing edge has an effect on the upstream flow only at distances of about 30–40% of the wing chord. Thus, from the vertex to the value of the longitudinal coordinate $x = 0.6$ the pressure distribution over the wing surface depends only on the angle of attack (and the parameters $z_0$, $\gamma$, $\sigma$, and $g_w$) but is independent of the pressure distribution preassigned at the trailing edge if the given pressure is not too high to produce boundary layer separation from the wing.

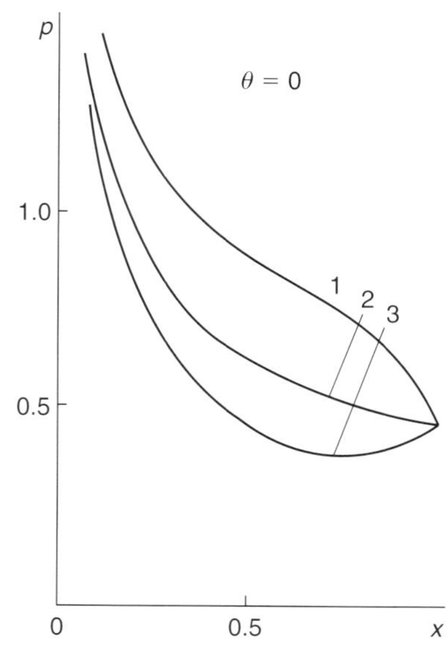


Fig. 5.32.

Figure 5.33 presents the boundary layer thickness $\delta_e(x, z)$ distribution over the wing in the flow at an angle of attack $\alpha = 0.3$. As compared with the flow at zero incidence, the thickness $\delta_e$ considerably increases on the leeward and decreases on the windward side;

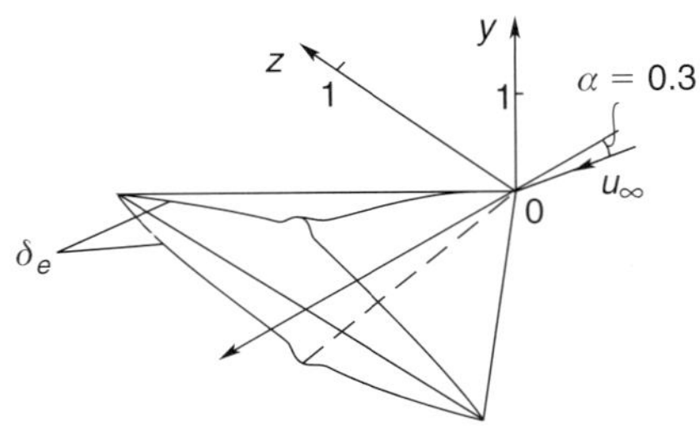


Fig. 5.33.

this is in agreement with the data of experimental studies conducted in the strong viscous interaction regime (Cross and Hankey, 1968). The variation of the boundary layer thickness is particularly strong in the vicinity of the plane of symmetry.

The calculated profiles of the longitudinal viscous stress coefficient $\tau_u$ and the heat flux $\tau_g$ to the wing surface along the axis of symmetry ($z=0$) are plotted in Fig. 5.34. The viscous stress is considerably greater on the windward than on the leeward side. On the lower surface of the wing a sharp increase in $\tau_u$ is observable in the vicinity of the trailing edge, which is due to flow acceleration. A slight effect of the trailing edge pressure on the heat flux distribution should also be noted.

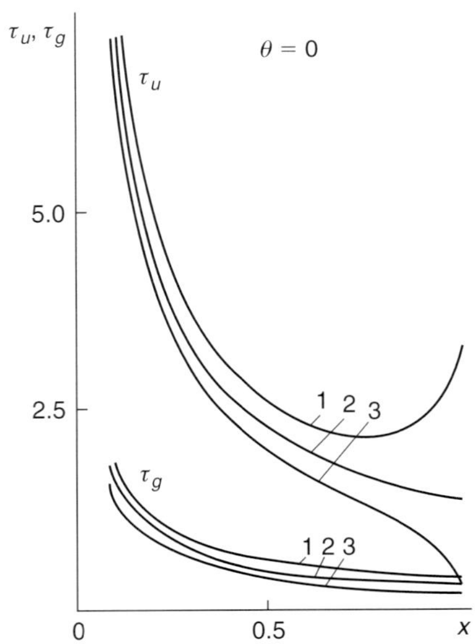


Fig. 5.34.

In Figs. 5.35 and 5.36 we have plotted the spanwise distributions of $p$, $\tau_u$, $\tau_g$, and the transverse viscous stress coefficient $\tau_w$ at the longitudinal coordinate $x=0.6$. This value of $x$ is chosen from the conditions noted above. In Fig. 5.35 the dashed curve presents the

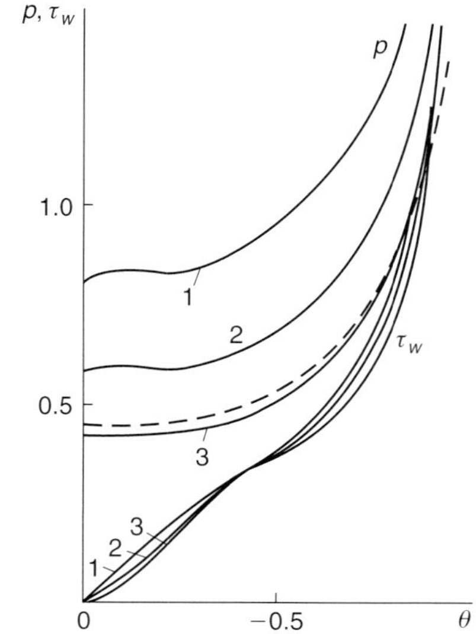


Fig. 5.35.

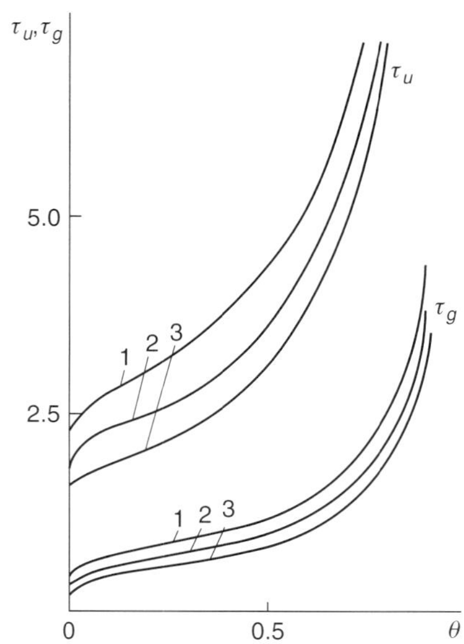


Fig. 5.36.

pressure distribution over the trailing edge $p(x = 1)$, for which all the calculations presented in this section were carried out. It should be noted that the effect of the angle of attack on the transverse viscous stress coefficient $\tau_w$ is comparatively weak, at least, for $|\theta| > 0.2$. However, near the plane of symmetry the value $|\theta| \leq 0.1$ on the windward side is considerably greater than its value on the leeward side.

As a second example, we will consider the flow past a delta wing for the following parameters: $z_0 = 1$, $\gamma = 1.4$, $\sigma = 0.71$, $g_w = 0.05$, and various angles of attack (Dudin, 1985). In Fig. 5.37 we have plotted the dependence of the aerodynamic coefficients $C_F = C_F^* Re^{3/4}$, $C_p = C_p^* Re^{1/2}$, and $m_z = m_z^* Re^{1/2}$ on the angle of attack. In order to eliminate the effect of the trailing edge pressure distribution on the aerodynamic coefficients, the integration in the longitudinal coordinate was performed to the value $x = 0.7$. In Fig. 5.38 we have plotted the profiles of the transverse velocity $w$ for $\lambda = 0.2$ on the leeward side of the wing in the vicinity of the plane of symmetry at an angle of attack $\alpha = 0.1$. The appearance of transverse return flows directed downward from the wing vertex in the region $0 < x < 0.4$ should be noted.

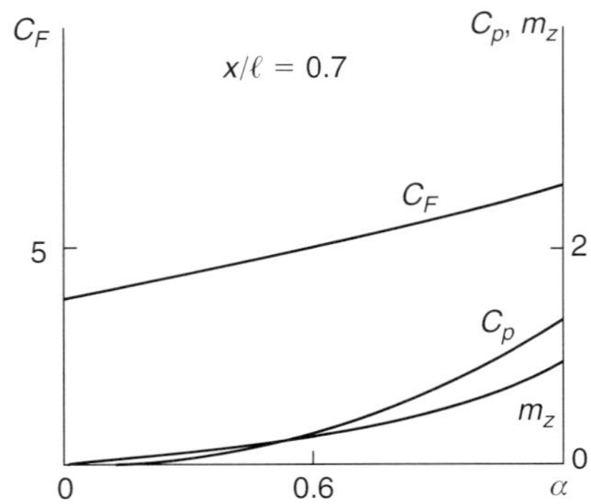


Fig. 5.37.

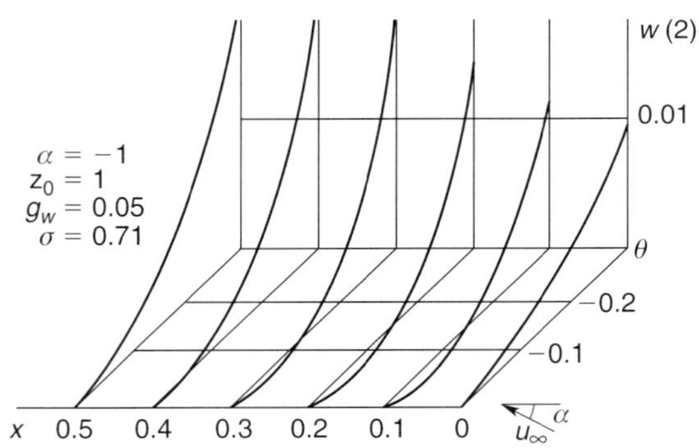


Fig. 5.38.

## 5.7 Wings of finite length in the moderate viscous interaction regime

As noted in preface, the nature of the hypersonic boundary layer flow depends considerably on the hypersonic interaction parameter $\chi = M_\infty^2 Re_0^{-1/2}$. In previous sections we considered the strong viscous interaction regime over the entire wing. However, actually, if the hypersonic interaction parameter based on the length of the region under consideration $\chi_x = M_\infty^2 Re_{0x}^{-1/2} \gg 1$, the strong viscous interaction regime is usually realized in the vicinity of the wing vertex and its leading edges and is followed by the moderate interaction region ($\chi = O(1)$). As shown below, in this case the system of governing equations remains three dimensional even for the flow past a planar triangular plate at zero incidence.

### *5.7.1 Mathematical formulation of the problem*

We will consider the hypersonic viscous perfect gas flow past a planar delta wing at zero incidence at a given surface temperature; the flow is governed by systems of equations (5.47) for the inner problem and (5.51) for the outer problem (Dudin, 1991*a*).

The joint solution of the system of equations for the viscous and inviscid flows presents severe difficulties even in the two-dimensional case. For this reason, in analyzing most problems the outer problem is usually replaced by an approximate relation for determining the pressure at the outer edge of the boundary layer.

In preceding chapters the tangent wedge formula for determining the pressure was used in its simplest form (5.55) or (5.56), appropriate for the values of the local hypersonic similarity parameter $M_\infty \theta \gg 1$, where $\theta$ is the characteristic angle of streamline deflection caused by the viscous layer or the slope of the body surface. In actual flows Mach numbers are always finite, so that in the shock layer there is always a point at which the assumption $M_\infty \theta \gg 1$ is violated (Hayes and Probstein, 1966). In this case, the tangent wedge formula can be used in a more general form (Hayes and Probstein, 1966; Stollery, 1972), valid over a wider range of the parameter $M_\infty \theta$ and yielding fairly accurate results in practical calculations (Neiland, 1974*a*).

Since in this chapter we consider the flow past delta wings with the aspect ratio $z_0 = O(1)$, the outer inviscid $M_\infty \gg 1$ flow can be described by strip theory (Hayes and Probstein, 1966; Lunev, 1975) and the pressure is determined from the tangent wedge formula, which in dimensionless variables takes the form:

$$p = \frac{1}{\gamma \chi_*^2} + \frac{\gamma+1}{4}\left(\frac{\partial \delta_e}{\partial x}\right)^2 + \frac{\partial \delta_e}{\partial x}\left[\frac{1}{\chi_*^2} + \left(\frac{\gamma+1}{4}\frac{\partial \delta_e}{\partial x}\right)^2\right]^{1/2} \tag{5.122}$$

where $\delta_e = (\gamma-1)/2\gamma p \int_0^\infty (g - u^2 - w^2)\mathrm{d}\lambda$ and $\chi_* = M_\infty \delta = M_\infty z_0^{1/4} Re_0^{-1/4}$ is the interaction parameter. The hypersonic interaction parameter $\chi$ is related with the parameter $\chi_*$ by the formula $\chi_*^2 = \chi z_0^{1/2}$.

The pressure $p_k(z)$ having been preassigned at the trailing edge of the wing, Eq. (5.122) makes it possible to close the problem.

To take account for the singular behavior of the flow functions in the vicinity of the wing vertex, we can again introduce variables (5.112), since in the vicinity of the leading edges the strong viscous interaction regime is realized. In the case under consideration rewriting the expression for the pressure (5.122) in terms of variables (5.112) yields

$$p^* = \frac{\sqrt{x}}{\gamma \chi_*^2} + \frac{\gamma+1}{4}R^2 + R\left[\frac{\sqrt{x}}{\chi_*^2} + \left(\frac{\gamma+1}{4}R\right)^2\right]^{1/2} \tag{5.123}$$

$$R_* = \frac{3}{4}\delta_e^* + x\frac{\partial \delta_e^*}{\partial x} - z^*\frac{\partial \delta_e^*}{\partial z}$$

For $\chi_* \neq \infty$ it is a function of the coordinates $z^*$ and $x$. Thus, for $\chi_* \neq \infty$ the system of equations (5.47) and (5.122) remains three dimensional even for a semi-infinite wing.

For solving numerically the boundary value problem it is also necessary to take account for the singular behavior of the pressure $p^*$ and the boundary layer thickness $\delta^*$ in the vicinity of the leading edges of the wing ($z^* = \pm 1$). In this case it is assumed that the pressure distribution $p_k(z^*)$ given at the trailing edge ($x = 1$) for $z^* = \pm 1$ coincides with the pressure corresponding to the flow past a semi-infinite wing in the strong viscous interaction regime. After the introduction of variables (5.113) the system of equations takes the form (5.114) in which the pressure is determined from the formula

$$p_0 = \frac{\sqrt{x(1-\theta^2)}}{\gamma\chi_*^2} + \frac{\gamma+1}{4}N_*^2 + N_*\left[\frac{\sqrt{x(1-\theta^2)}}{\chi_*^2} + \left(\frac{\gamma+1}{4}N_*\right)^2\right]^{1/2} \tag{5.124}$$

$$N_* = \frac{3}{4}(1-\theta^2)\Delta + x(1-\theta^2)\frac{\partial\Delta}{\partial x} - \theta\left[(1-\theta^2)\frac{\partial\Delta}{\partial\theta} - \frac{3}{2}\theta\Delta\right]$$

while the boundary conditions for $|\theta| \leq 1, 0 \leq x \leq 1$ are as follows:

$$u = v_0 = w = 0, \quad g = g_w \ \text{for } \eta^* = 0$$

$$u \to 1, \quad w \to 0, \quad g \to 1 \ \text{for } \eta^* \to \infty$$

$$p_0(x,\theta) = p_{k0}(\theta) \ \text{for } |\theta| \leq 1, \ x = 1 \tag{5.125}$$

The system of partial differential equations (5.114), together with relations (5.124) and (5.125), describes the three-dimensional boundary layer flow over a planar delta wing in the regime of viscous interaction with the outer hypersonic flow with a given pressure distribution $p_{k0}(\theta)$ at the trailing edge.

It should be noted that at the delta wing vertex the terms containing the variable $x$ drop out from the system under consideration, so that the boundary value problem turns out to be dependent only on two independent variables, $\theta$ and $\eta^*$. The system of partial differential equations obtained describes also the flow over a semi-infinite planar delta wing.

At the leading edges of the wing, at the transverse coordinate $\theta = \pm 1$, the system degenerates to a system of ordinary differential equations. In the vicinity of the wing vertex and its leading edges the strong viscous interaction regime is realized, since in expression (5.124) the terms proportional to $\chi_*^{-2}$ vanish. The domain of integration of the system of equations represents a rectilinear parallelepiped.

We note that at the trailing edge of the wing with a given pressure distribution $p_{k0}(\theta)$ the boundary layer thickness $\Delta(\theta)$ obtained as a result of the solution of the complete problem must satisfy the equation

$$\frac{\sqrt{1-\theta^2}}{\gamma\chi_*^2} + \frac{\gamma+1}{4}N_1^2 + N_1\left[\frac{\sqrt{1-\theta^2}}{\chi_*^2} + \left(\frac{\gamma+1}{4}N_1\right)^2\right]^{1/2} = p_{k0}(\theta) \tag{5.126}$$

$$N_1 = \frac{3}{4}(1-\theta^2)\Delta + (1-\theta^2)\frac{\partial\Delta}{\partial x} - \theta\left[(1-\theta^2)\frac{\partial\Delta}{\partial\theta} - \frac{3}{2}\theta\Delta\right]$$

The method of the solution of the system of equations thus obtained is described in the book of Bashkin and Dudin (2000).

### *5.7.2 Aerodynamic characteristics of a wing at zero incidence*

Below we will consider the flow past a delta wing, at whose trailing edge the pressure is identically equal to the pressure in the section $x = 1$ of a semi-infinite wing in the strong viscous interaction regime. Thus, $p_{k0}(\theta)$ is preassigned to be equal to the pressure $p_0(x, \theta)$ obtained as a result of the solution of the system of equations governing the flow at the delta wing vertex for $x = 0$.

In numerical calculations we took $z_0 = 1$ (the sweep angle is equal to 45°); $\gamma = 1.4$; $\sigma = 0.71$; $g_w = 0.05$, 0.1, and 0.2; and $\chi_* = 1$, 2, 5, $10^2$, and $10^5$.

In Fig. 5.39 we have plotted the calculated profiles of the dimensionless pressure $p$ (solid curves) and longitudinal viscous stress coefficient $\tau_u$ along the $x$ axis in the plane of symmetry ($z = 0$) for $g_w = 0.05$ and the values of the interaction parameter $\chi_* = 1$, 2, and $\infty$ which are presented by curves 1–3.

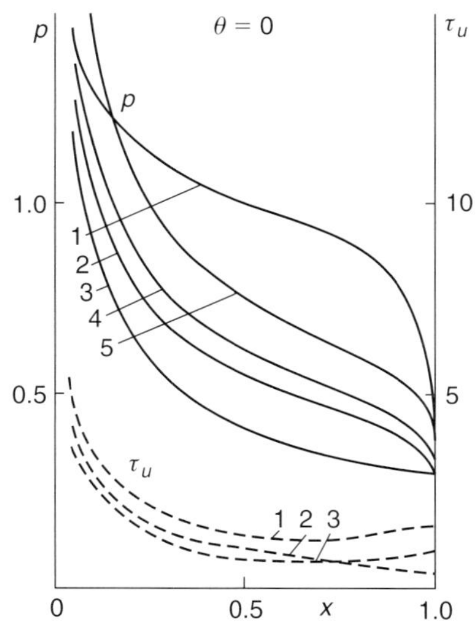


Fig. 5.39.

A considerable variation in the pressure and coefficient $\tau_u$ distributions with the parameter $\chi_*$ should be noted. As noted above, the boundary layer calculations were performed for the same dimensionless pressure distribution over the trailing edge. As in the strong interaction regime, in this case the trailing edge effect is noticeable upstream of the edge at distances of about 30–40% of the wing chord. A considerable decrease of the pressure $p$ in the vicinity of $x = 1$ for $\chi_* = 1$ leads to flow acceleration and an increase in the viscous stress coefficient $\tau_u$. The heat flux $\tau_g$ varies in a similar fashion.

In Fig. 5.40 the spanwise distributions of the pressure $p$ (solid curves) and the boundary layer displacement thickness $\delta_e$ (dashed curves) are plotted for the longitudinal coordinate

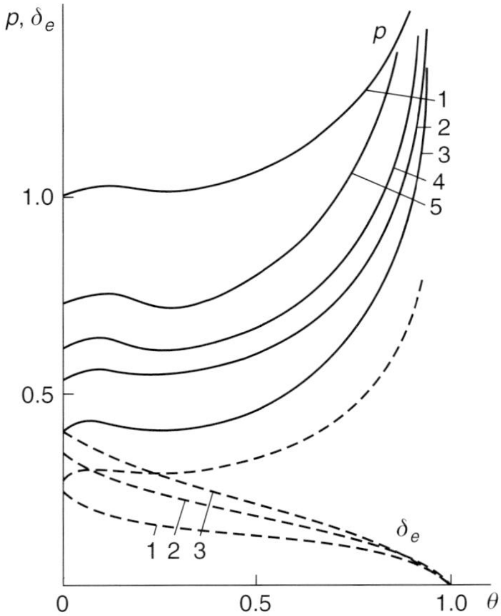


Fig. 5.40.

$x = 0.5$. The dot-and-dash curve represents the given pressure distribution $p_k(\theta)$ at the trailing edge. It should be borne in mind that the dimensionless quantities presented in the figures are related with the dimensional ones as follows:

$$\frac{p^0}{p_\infty} = \gamma\chi\sqrt{z_0}p = \gamma\chi_*^2 p, \quad \delta_e^0 = \left(\frac{z_0}{Re_0}\right)^{1/4} L\delta_e$$

where $p_\infty$ is the freestream pressure. Therefore, the pressure on the delta wing surface increases considerably with increase in the interaction parameter $\chi_*$. We note that on transition from the strong to the moderate viscous interaction regime the boundary layer displacement thickness decreases appreciably and, as a consequence, the spanwise pressure gradient also decreases everywhere, except for the plane of symmetry.

The spanwise distributions of the viscous stress coefficients $\tau_u$ (solid curves) and $\tau_w$ (dashed curves) are presented in Fig. 5.41 for $x = 0.5$. On transition to the moderate interaction regime the longitudinal friction

$$\left.\frac{\partial u}{\partial y}\right|_w = \frac{u_\infty}{\ell}\left(\frac{Re_0}{z_0}\right)^{1/4}\tau_u$$

increases, which is in a considerable degree due to a decrease in the boundary layer displacement thickness in these regimes. On the main part of the wing ($0.2 \leq |\theta| \leq 1$) the viscous stress coefficient $\tau_w$ decreases which is due to a decrease of the gradient $\partial p/\partial z$ in this region. The behavior of $\tau_w$ in the vicinity of the plane of symmetry of the wing, where the pressure gradient $\partial p/\partial z$ increases with decrease in $\chi_*$ and the velocity $w$ profile becomes more

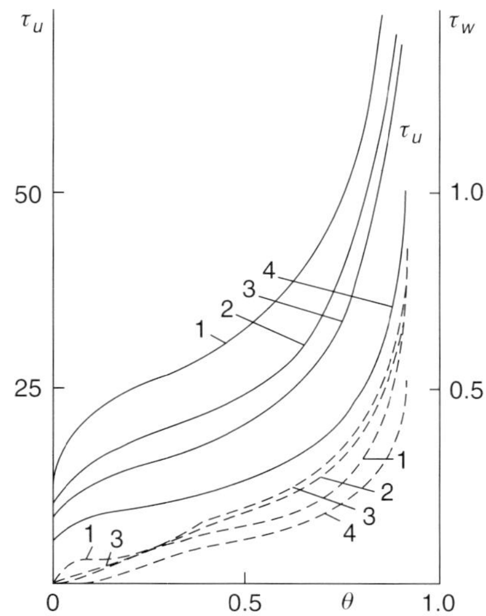


Fig. 5.41.

convex, is just the reverse of that described above. In all cases considered a flow streaming smoothly toward the plane of symmetry was realized.

In Figs. 5.39–5.41 we have also presented certain results for the flow parameters calculated for $\chi_* = 2$ and $g_w = 0.1$ and 0.2 (curves 4 and 5, respectively). An increase in the temperature factor has a considerable effect on the flow parameters.

Figure 5.42 presents the results of the calculation of the aerodynamic coefficients

$$C_F = C_F^* Re_0^{3/4} = \frac{2}{z_0^{1/4}} \sqrt{\frac{2\gamma}{\gamma - 1}} \int_0^1 \int_{-1}^1 \frac{x^{1/4} p_0(x, \theta)}{(1 - \theta^2)^{3/4}} \left. \frac{\partial u}{\partial \eta^*} \right|_w \mathrm{d}x \, \mathrm{d}\theta \tag{5.127}$$

$$C_p = C_p^* Re_0^{1/2} = \frac{2}{z_0^{1/2}} \int_0^1 \int_{-1}^1 \frac{x^{1/4} p_0(x, \theta)}{(1 - \theta^2)^{1/2}} \mathrm{d}x \, \mathrm{d}\theta$$

$$m_z = m_z^* Re_0^{1/2} = \frac{2}{z_0^{1/2}} \int_0^1 \int_{-1}^1 \frac{x^{3/2} p_0(x, \theta)}{(1 - \theta^2)^{1/2}} \mathrm{d}x \, \mathrm{d}\theta$$

calculated for one side of the wing as functions of the interaction parameter $\chi_*$ for $g_w = 0.05$. The dashed curves present the dependences of the aerodynamic characteristics calculated from the flow parameters for $0 \leq x \leq 0.7$ in order to eliminate the trailing edge effect. In Fig. 5.42 the results of the calculation of the wing with $g_w = 0.1$ are marked by symbols $a$ and those for $g_w = 0.2$ by symbols $b$.

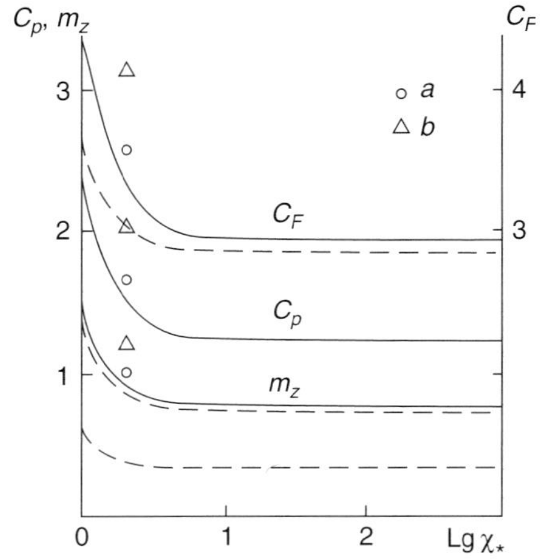


Fig. 5.42.

A considerable increase of the total aerodynamic characteristics with decrease in the parameter $\chi_*$ from 10 to 1 should be noted. For flows with $\chi_* > 10$ the aerodynamic coefficients do not change and almost coincide with those corresponding to the strong viscous interaction regime.

### 5.7.3 Angle-of-attack effect of the aerodynamic characteristics

In the experimental study of Whitehead et al. (1972) it was established that the boundary layer flow nature depends considerably on the angle of attack $\alpha^0$, at which the body is set in the flow, and the hypersonic viscous interaction parameter $\chi$. However, in the vicinity of the wing vertex and its leading edges in many cases it is the strong viscous interaction regime that is realized if the hypersonic interaction parameter based on the length of the region under consideration $\chi_x = M_\infty^2 Re_{0x}^{-1/2} \gg 1$. The hypersonic boundary layer flow over a wing in the presence of an angle of attack for the case in which the strong interaction regime ($\chi = \infty$) is realized over the entire wing, was considered in Section 5.6.3. It was shown that on the leeward side of the wing the boundary layer displacement thickness increases considerably with increase in the angle of attack which is in agreement with the experimental data of Cross and Hankey (1968). The assumption of the realization of the strong viscous interaction regime on the leeward side of a delta wing requires the fulfillment of the condition $M_\infty^2[(\partial\delta_e^0/\partial x^0) - \alpha^0]^2 \gg 1$ (Hayes and Probstein, 1966), where $\delta_e^0$ is the boundary layer displacement thickness; however, this condition is not always fulfilled. Moreover, in separate regions on the leeward side of the wing the situations, in which $\partial\delta_e^0/\partial x^0 < 0$ or $(\partial\delta_e^0/\partial x^0) - \alpha^0 < 0$, are possible, so that a local expansion regime is realized.

In what follows, we will consider the hypersonic viscous flow past a thin delta wing of finite length $\ell$ at a low angle of attack $\alpha^0 \sim Re_0^{-1/4}$ (Fig. 5.43) (Dudin, 1991*b*; 1992). It

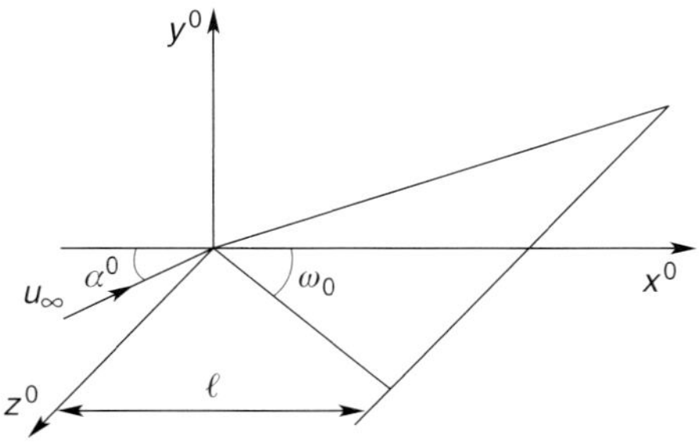


Fig. 5.43.

is assumed that in the flow past the wing the regime of viscous interaction between the three-dimensional boundary layer and the inviscid outer flow is realized.

The flow under consideration is described by the system of equations (5.47) in which the pressure is determined by the formula

$$p = \frac{1}{\gamma\chi_*^2} + \frac{\gamma+1}{4}\left(\frac{\partial\delta_e}{\partial x} \pm \alpha\right)^2 + \left(\frac{\partial\delta_e}{\partial x} \pm \alpha\right)\left[\frac{1}{\chi_*^2} + \left(\frac{\gamma+1}{4}\left(\frac{\partial\delta_e}{\partial x} \pm \alpha\right)\right)^2\right]^{1/2} \quad (5.128)$$

$$\delta_e = \frac{\gamma-1}{2\gamma p}\int\limits_0^\infty (g - u^2 - w^2)\,\mathrm{d}\lambda$$

$$\alpha = \frac{\alpha^0}{\tau}, \quad \chi_* = M_\infty\tau = M_\infty\left(\frac{z_0}{Re_0}\right)^{1/4}$$

The hypersonic interaction parameter $\chi$ is related with the parameter $\chi_*$ by the formula $\chi_*^2 = \chi z_0^{1/2}$. Plus and minus signs ahead of $\alpha$ relate to the windward and leeward wing sides, respectively. For a given pressure $p_k(z)$ at the trailing edge, Eq. (5.128) makes it possible to close the boundary value problem (5.47). After the introduction of variables (5.112) and (5.113) we arrive at the system of equations (5.114) in which the pressure is determined as follows:

$$p_0 = \frac{\sqrt{x(1-\theta^2)}}{\gamma\chi_*^2} + \frac{\gamma+1}{4}N_*^2 + N_*\left[\frac{\sqrt{x(1-\theta^2)}}{\chi_*^2} + \left(\frac{\gamma+1}{4}N_*\right)^2\right]^{1/2} \quad (5.129)$$

$$N_* = \frac{3}{4}(1-\theta^2)\Delta + x(1-\theta^2)\frac{\partial\Delta}{\partial x} - \theta\left[(1-\theta^2)\frac{\partial\Delta}{\partial\theta} - \frac{3}{2}\theta\Delta\right] \pm \alpha x^{1/4}(1-\theta^2)^{1/4}$$

The boundary conditions are as follows:

$$u = v_0 = w = 0 \quad g = g_w \ \text{ for } \eta = 0$$

$$u \to 1, \quad w \to 0, \quad g \to 1 \ \text{ for } \eta \to \infty$$

The system of partial differential equations (5.114), together with Eq. (5.129), governs the flow in the three-dimensional boundary layer on a finite-length delta wing at an angle of attack in the regime of viscous interaction with the outer hypersonic flow at a given pressure distribution $p_{k0}(\theta)$ at the trailing edge. It should be noted that for $|K| = \left|\frac{\partial \delta_e}{\partial x} \pm \alpha\right| \ll 1$ the tangent wedge formula actually goes over to the Ackeret formula (Loytsianskii, 1966). Thus, the formula for the pressure is also valid for expansion flows with small negative values of $K$.

At the leading edges of the wing, $\theta = \pm 1$, the system of equations reduces to a system of ordinary differential equations and at the wing vertex, $x = 0$, to a system dependent only on two variables, $\theta$ and $\eta$. The system of partial differential equations obtained also describes the flow over a semi-infinite planar delta wing in the strong viscous interaction regime. In the vicinity of the wing vertex and its leading edges the strong viscous interaction regime is realized, since in the expression for $p_0$ (5.129) all the terms proportional to $1/\chi_*^2$ and $\alpha$ vanish.

At the trailing edge ($x = 1$), at which a pressure distribution $p_{k0}(\theta)$ is given, the boundary layer thickness determined as a result of the solution of the complete boundary value problem must satisfy the equation

$$\frac{\sqrt{1-\theta^2}}{\gamma\chi_*^2} + \frac{\gamma+1}{4}N_1^2 + N_1\left[\frac{\sqrt{1-\theta^2}}{\chi_*^2} + \left(\frac{\gamma+1}{4}N_1\right)^2\right]^{1/2} = p_{k0}(\theta) \tag{5.130}$$

$$N_1 = \frac{3}{4}(1-\theta^2)\Delta + (1-\theta^2)\frac{\partial \Delta}{\partial x} - \theta\left[(1-\theta^2)\frac{\partial \Delta}{\partial \theta} - \frac{3}{2}\theta\Delta\right] \pm \alpha(1-\theta^2)^{1/4}$$

The method of the solution of system of equations (5.114) and (5.129) with condition (5.130) is presented in detail in the book of Bashkin and Dudin (2000). As distinct from the strong interaction regime, in the calculations with $\chi_* \approx 1$ presented below the checking of the value $N_*$ (see Eq. (5.129)) turned out to be necessary in order not to carry out calculations at such parameters when the formula for determining the pressure $p$ (5.128) becomes inapplicable.

As an illustration, we will consider the flow on the leeward side of a planar finite-length delta wing at whose trailing edge the pressure is identically equal to the dimensionless pressure corresponding to the flow past a semi-infinite delta wing at zero incidence in the strong viscous interaction regime at $x = 1$. Thus, $p_{k0}(\theta)$ was preassigned to be equal to the dimensionless pressure obtained as a result of the solution of the system of equations governing the flow at the delta wing vertex, $x = 0$. In principle, any other pressure distribution can be preassigned, if only it coincides with the pressure at the leading edges for $\theta = \pm 1$.

In the numerical calculations we took $z_0 = 1$, $\gamma = 1.4$, and $\sigma = 0.71$. The aerodynamic coefficients were calculated from formula (5.127).

The calculated results for the leeward wing side are presented as solid curves for $g_w = 0.2$ and $\chi_* = 2$ ($\chi = 4$, moderate interaction) and dashed curves for $g_w = 0.05$ and $\chi_* = 0.8$ ($\chi = 0.64$, weak interaction). Incident angles $\alpha = 0, 0.2, 0.4, 0.6, 1.0$ relate numbers 1–5. Dot-and-dash lines correspond to the dimensionless pressure in the freestream $p_\infty = 1/\gamma\chi_*^2$. Lines marked by number 6 correspond to $\chi_* = 2$ ($p_\infty = 0.178$) and number 7 to $\chi_* = 0.8$ ($p_\infty = 1.115$).

It should be noted that with increase in the angle of attack the pressure decreases sharply on the leeward side. However, for $\chi_* = 2$ the disturbed pressure is appreciably higher than the freestream pressure (line 6) and in the vicinity of the trailing edge the flow nature changes from expansion to compression at the given pressure distribution.

The distributions of the viscous stress coefficient $\tau_u$ and the thickness $\delta_e$ along $x$ are presented in Fig. 5.44. The trailing edge effect on the boundary layer flow is noticeable at upstream distances of about 30% of the wing chord.

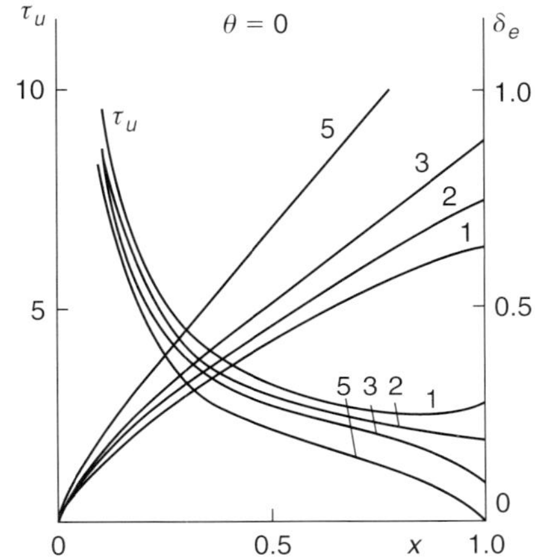


Fig. 5.44.

A considerable difference between the calculations with $\chi_* = 0.8$ (dashed curves) and $\chi_* = 2$ lies in the fact that already for $\alpha > 0.2$ the freestream pressure ($p_\infty = 1.115$) is greater than the disturbed pressure on a considerable part of the wing. Strong rarefaction at the trailing edge of the wing leads to flow acceleration in the longitudinal direction and a decrease in the boundary layer displacement thickness and a nonmonotonic behavior of the longitudinal viscous stress coefficient $\tau_u$ in this region (Fig. 5.45). The heat flux distribution over the wing surface is in qualitative agreement with the distribution of $\tau_u$.

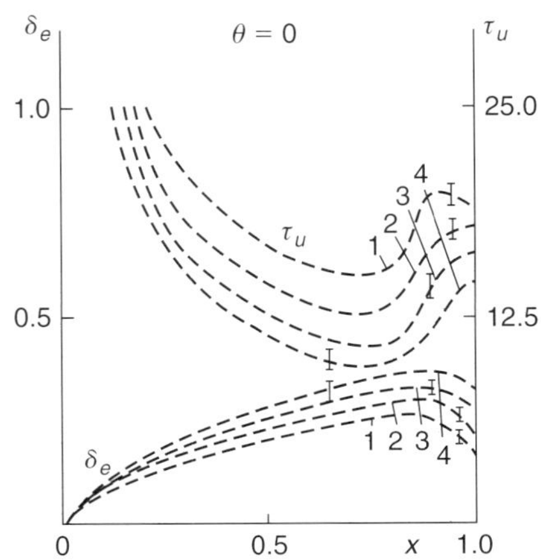


Fig. 5.45.

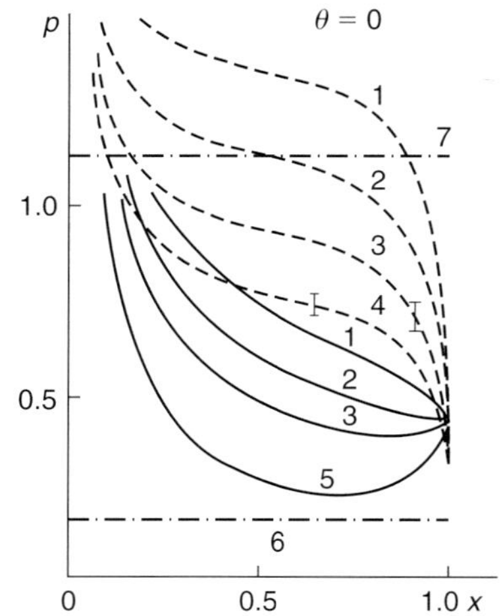


Fig. 5.46.

As noted above, in calculations in the weak interaction regime ($\chi < 1$) the value of $N_*$ must be checked at each calculated point in order to be sure in the applicability of formula (5.129). In Figs. 5.44 and 5.46 vertical dashes at dashed curves mark the boundaries to the right of which the value $n_* < -0.25$, so that using formula (5.129) can result in errors greater than about 6%.

In the weak viscous interaction regime the boundary layer displacement thickness decreases on the leeward side, though more slowly than in the strong interaction regime; therefore, the formation of aerodynamic shadow is possible.

The spanwise distribution of the transverse viscous stress coefficient $\tau_w$ is presented in Fig. 5.47 for the longitudinal coordinate $x = 0.8$. For $\chi_* = 2$ the angle-of-attack effect

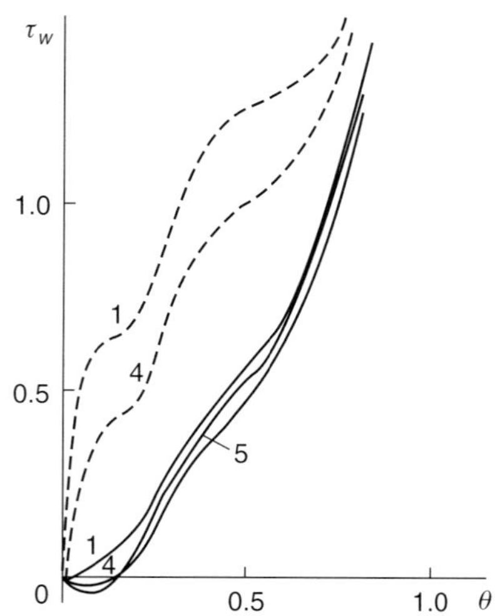


Fig. 5.47.

on $\tau_w$ is slight, at least, for $|\theta| > 0.2$. However, near the plane of symmetry ($|\theta| < 0.2$) a considerable variation of $\tau_w$ occurs, such that for $\alpha > 0.6$ in the lower part of the boundary layer there arise transverse return flows and $\tau_w$ becomes negative. For $\chi_* = 0.8$ the variation of the angle of attack has an appreciable effect on the $\tau_w$ distribution over almost the entire wing ($|\theta| < 0.8$).

For the relevant parameters considered above, the flow regime on the leeward side of the delta wing is, in accordance with the terminology of the work of Dunawant et al. (1976), of type A.

The aerodynamic coefficients $C_F$, $C_p$, and $m_z$ calculated on the leeward side of the wing are presented in Fig. 5.48 (marked 1, 2, and 3, respectively). The calculated results are shown as solid curves for $\chi_* = 2$ and dashed curves for $\chi_* = 0.8$. The angle-of-attack-dependence of the coefficients $C_F$, $C_p$, and $m_z$ is stronger for $\chi_* = 0.8$ than for $\chi_* = 2$, which is attributable to the stronger variation of the pressure distribution over the wing with variation of the angle of attack in this case.

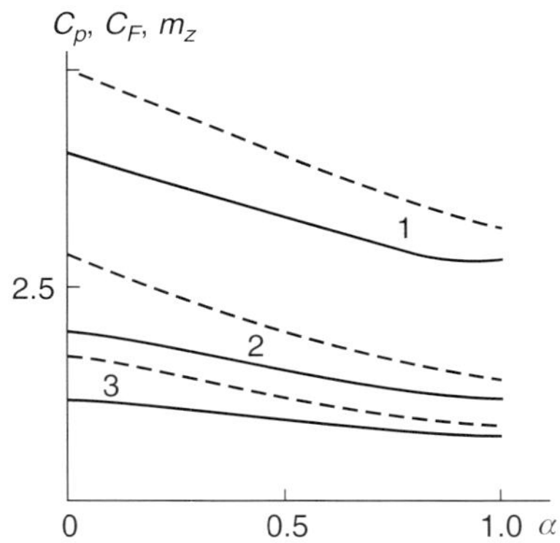


Fig. 5.48.

# 6 Supercritical and Transcritical Interaction Regimes: Two-Dimensional Flows

The asymptotic theory of the interaction between an inviscid flow and a boundary layer is an important part of viscous gas dynamics at high Reynolds numbers. It is based on the fundamental idea of Prandtl on the possibility of dividing the entire flow region into an inviscid flow and a thin boundary layer (Prandtl, 1904). This idea appeared in connection with an attempt to provide an explanation for the phenomenon of flow separation from the body surface. We note that Prandtl's idea turned out to be very fruitful not only in viscous flow dynamics but also in many other branches of applied mathematics. The original formulation of boundary layer theory includes an assumption that the problem can first be solved for the outer inviscid flow and then for the boundary layer at the previously determined pressure distribution. Later Prandtl (1939) noted the possibility of refining the solution by taking account for the displacing effect of the boundary layer on the outer flow. In the next approximation, the effect of outer flow variations on the boundary layer flow must be taken into account, etc. Actually, the concept of weak interaction theory was thus formulated.

However, further studies, both experimental and theoretical, showed that in many cases taking the weak interaction into account does not solve all resulting problems and, in particular, does not enable one to describe the flow near the boundary layer separation point. An analysis of the solutions of the boundary layer equations at a given outer pressure distribution leads often to the appearance of an "impassable" singular point (Landau and Lifshitz, 1944; Goldstein, 1948). The key role played by strong local interaction between the outer flow and the boundary layer leading to the removal of the singularity of solutions and the possibility of continuing it through the separation point was detected in the works of Neiland (1968, 1969*a*) and Stewartson and Williams (1969). In the vicinity of the laminar separation point in a supersonic flow the asymptotic representation of the solutions of the Navier–Stokes equations is of three-layer nature. Later, the use of the three-layer limiting structure of the solution turned out to be very fruitful. Thus, in the study of Sychev (1972) it was employed for solving the problem of laminar incompressible separation and in the works of Ryzhov (1977) and Schneider (1974) for studying time-dependent flows.

Moreover, the asymptotic approach to the solution of problems of supersonic viscous flow dynamics made it possible to examine a wide class of problems, which are not necessarily reduced to the three-layer flow pattern and are not governed by the classical boundary layer theory. These include, for example, flow near the shock incidence onto a boundary layer, flows past corner points on bodies and various steps on the boundary layer bottom, and other flows including strong local interaction regions, as well as the problems in which the interaction is strong over the entire body surface, which can lead to upstream disturbance transfer when the body is in a hypersonic flow. Apparently, the most complete survey of the studies in these lines of inquiry for supersonic gas flows is presented in the paper of Neiland (1981).

Thus, the range of problems described on the basis of the search for limiting solutions of the Navier–Stokes equations at the passage to the limit $Re \to \infty$ has been expanded widely. Now it includes a large number of problems, important for both theory and practice, which cannot be described by the classical boundary layer theory. It is particularly important also for the reason that determining numerical solutions for high $Re$ is difficult. Moreover, the asymptotic approach provides a rational explanation to the physical properties of the flows and helpful approximate similarity laws.

The structure of the supersonic flow/boundary layer interaction regions, the direction of disturbance transfer, and the scale of distances through which disturbances can propagate can be dependent on the Mach number profile in the undisturbed boundary layer. Crocco (1955) was the first who detected this fact. He developed the idea of subcritical and supercritical boundary layers, which are either capable of transferring pressure disturbances upstream through considerable distances, as compared with the boundary layer thickness, or do not possess this property. Crocco noted that, in order for a supercritical boundary layer to acquire this property, it is necessary that the boundary layer Mach number profile restructures itself sharply, or, in accordance with Crocco's terminology, supercritical jump occurs. These ideas were substantiated on the basis of general physical considerations and the integral method for describing the interaction.

However, earlier studies on the asymptotic theory of separation and interaction preceding the work of Neiland (1973) considered actually only subcritical regimes. It was even suggested that the supercriticality property is a consequence of inaccurate description of the phenomenon when using the integral boundary layer equations rather than a physical property of the flows (Brown and Stewartson, 1969). However, in the studies of Neiland (1973, 1974*b*, 1987) an asymptotic theory of two- and three-dimensional supercritical flows was developed and a fundamental analogy between the properties of subsonic and supersonic inviscid gas flows, on the one hand, and subcritical and supercritical boundary layers in an outer supersonic flow, on the other hand, was established. In this case, the description of transcritical flows, similar to the transonic velocity regime in conventional gas dynamics, was of fundamental interest.

Below in this chapter, the appearance of supercritical and transcritical properties in the problems of strong local interaction between the boundary layer and the supersonic flow will be considered. In this case, the body temperature must be considerably lower than the stagnation temperature of the inviscid flow, while outside a local zone the interaction must be weak.

## 6.1 Distinctive features of boundary layer separation on a cold body and its interaction with a hypersonic flow

### *6.1.1 Formulation of the problem*

We will consider the flow in the vicinity of a laminar boundary layer separation point on a smooth region of the surface of a body in a hypersonic viscous flow. It is assumed that near the outer edge of the boundary layer the flow is either hypersonic ($M_e \gg 1$) or supersonic ($M_e \sim 1$) depending on the body shape and thickness. On the main part of the body ahead of the separation zone the pressure gradient induced by the flow past an "effective body" formed by the boundary layer displacement thickness is small and should be taken into account only in the boundary layer theory of second approximation. However, a strong disturbance due, for example, to flap deflection, shock incidence from outside, etc., can lead to the formation of a separation zone ahead of it. Then the pressure distribution depends on the interaction between the viscous and inviscid flow regions. Even in the earlier experimental studies (see, e.g., Chapman et al., 1958) it was noted that if a separation region is developed and contains a zone of near-constant pressure, then the flow near the separation point is almost independent of the type of the separation-producing disturbance. The flow is determined by the local characteristics of the undisturbed boundary layer and the inviscid flow ahead of the beginning of the interaction region (of course, the rear part of the flow has an effect on the separation point position). A slight dependence of the flow nature in the vicinity of the separation point on the other flow regions is attributable to the fact that large local pressure gradients induced in this region are chiefly dependent on the form of the flow parameter profiles in the undisturbed boundary layer ahead of the strong local interaction and on a small region of the outer inviscid flow. Precisely this makes it possible to construct local limiting solutions of the Navier–Stokes equations in the vicinity of separation points (Neiland, 1968; Stewartson and Williams, 1969; Neiland, 1971*a*).

### *6.1.2 Starting estimates*

Following the studies of Neiland (1969*a*, 1973), we will estimate the flow functions near the separation point. We will first assume the flow at the outer edge of the boundary layer to be hypersonic, $M_e \gg 1$ and $Re \gg 1$. The Reynolds number $Re = \rho_e u_e \ell / \mu_0$ is based on the local velocity and density at the outer edge of the boundary layer, a characteristic distance from the separation point $\ell$, and viscosity $\mu_0$ calculated at the stagnation temperature of the inviscid flow. Below all the variables are assumed to be dimensionless, scaled on the values at the outer edge of the boundary layer ahead of the beginning the strong interaction region, except for viscosity $\mu$ and the stagnation enthalpy normalized on their values at the stagnation temperature of the outer inviscid flow. The coordinates are nondimensionalized by $\ell$.

In a hypersonic flow, in the weak interaction regime the boundary layer thickness $\delta \sim M_e / Re^{1/2}$ (see, e.g., Hayes and Probstein, 1966).

Even for very small values of the temperature factor ($g_w \ll 1$, where $g_w$ is the dimensionless stagnation enthalpy calculated at the body surface temperature) the viscous stress and the heat flux directly at the body are of the same order as in the boundary layer. Taking this fact into account one can easily determine the behavior of the flow functions in the undisturbed boundary layer near the body surface correct to unessential constant factors

$$y \sim \delta[(u + g_w)^{1+\omega} - g_w^{1+\omega}], \quad \rho \sim \frac{\rho_0}{g_w + u} \tag{6.1}$$

where $u$ is the dimensionless velocity. Here, $\omega$ is the exponent in the enthalpy dependence of $\mu$. Since we consider a short-size region $\Delta x \ll 1$, a small pressure difference $\Delta p \ll 1$ is sufficient for producing separation (though the pressure gradient must be large).

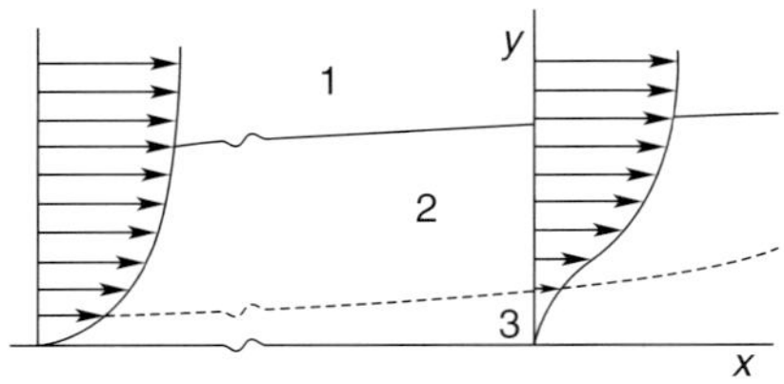


Fig. 6.1.

Using the momentum and continuity equations, together with the equation of state, we can derive the estimates for disturbances in the streamtubes of the main part of the boundary layer (region 2 in Fig. 6.1)

$$\Delta u^{(2)} \sim \Delta p \sim \frac{\Delta \rho^{(2)}}{\rho^{(2)}} \sim \frac{\Delta \delta^{(2)}}{\delta} \tag{6.2}$$

Here and in what follows, superscripts in parentheses refer to the numbers of regions in which the corresponding functions are calculated.

However, according to Eq. (6.1), in the wall part of the undisturbed boundary layer, where $u \ll g_w$, there is a region in which $\rho u^2$ is smaller in the order than in the main part of the boundary layer. Because of this, on the basis of the same equations, as were used in deriving estimates (6.2), together with formulas following from relations (6.1), the following estimates can be derived for region 3 in Fig. 6.1:

$$\delta^{(3)} \sim \delta g_w^{\omega} u^{(3)}, \quad \rho^{(3)} \sim \frac{1}{g_w M_e^2}, \quad \mu^{(3)} \sim g_w^{\omega} \text{ for } u \ll g_w \tag{6.3}$$

$$\delta^{(3)} \sim \delta g_w^{\omega+1/2} \Delta p^{1/2} \tag{6.4}$$

Since the vicinity of the separation point must include both separated and separationless flow regions, in region 3 the leading inertial and viscous terms must be of the same order:

$$\frac{\rho^{(3)}(u^{(3)})^2}{\Delta x} \sim \frac{\mu^{(3)} u^{(3)}}{(\delta^{(3)})^2 Re} \tag{6.5}$$

Comparing Eqs. (6.2) and (6.4) it can be seen that for the flows studied in the works of Neiland (1969*a*), Stewartson and Williams (1969), Neiland (1971*a*), and Sychev (1972), in which $g_w \sim 1$, we have always $\delta^{(3)} \gg \Delta\delta^{(2)}$. In the flow past a body with a cold surface, according to estimate (6.4), there can exist regimes, for which both layers, 2 and 3, take part in the interaction with the outer inviscid flow. We will consider the general case, in which $\delta^{(3)} \sim \Delta\delta^{(2)}$. In accordance with hypersonic small perturbation theory, we obtain

$$\Delta p \sim \frac{M_e \delta^{(3)}}{\Delta x} \tag{6.6}$$

Relations (6.3)–(6.6) give seven conditions for determining the scales of the parameters $\rho^{(3)}$, $u^{(3)}$, $\Delta x$, $\mu^{(3)}$, $\delta^{(3)}$, $\Delta\delta^{(2)}$, and $\Delta p$:

$$\Delta p \sim M_e Re^{-1/4}, \quad \Delta x \sim M_e^{3/2} Re^{-3/8} g_w^{\omega+1/2} \tag{6.7}$$

$$\delta^{(3)} \sim M_e^{3/2} Re^{-5/8} g_w^{\omega+1/2}, \quad \Delta\delta^{(2)} \sim M_e^2 Re^{-3/4}$$

In estimating the range of applicability of the theory an important role is played by the following equations

$$\frac{\Delta\delta^{(2)}}{\delta^{(3)}} = \frac{M_e^{1/2}}{g_w^{\omega+1/2} Re^{1/8}} = N, \quad \frac{\delta}{\Delta x} = \frac{N}{M_e} \tag{6.8}$$

For $N = O(1)$ the temperature factor $g_w \sim (M_e\delta)^{1/(2(2\omega+1))} \sim (M_e^2/Re^{1/2})^{1/(2(2\omega+1))}$.

The theory developed in the works of Neiland (1969), Stewartson and Williams (1969*a*), and Neiland (1971*a*) is applicable only when $N \ll 1$, that is, for $g_w^{\omega+1/2} \gg (M_e^2/Re^{1/2})^{1/4}$.

The distinctive features of the flow regimes corresponding to $g_w \ll (M_e^2/Re^{1/2})^{1/(2(2\omega+1))}$ will be considered in Section 6.2.

The general regime, in which the whole boundary layer takes part in the free interaction process, corresponds to $N = O(1)$. The case $N \to \infty$ must be considered separately. Moreover, for $M_e \sim 1$ and $N \sim 1$ or $N \sim M_e$ considerable transverse pressure differences might be expected, since $\Delta x_e \sim \delta$.

### 6.1.3 *Solution for the hypersonic regime of weak viscous interaction*

Let $N \sim O(1)$, $M_e \gg 1$, and $Re \gg 1$. On the basis of the above estimates, we will introduce the following coordinates and flow functions for region 3

$$x = \frac{M_e^{3/2} g_w^{\omega+1/2} (\gamma-1)^{1/2}}{Re^{3/8} 2^{5/4} a^{5/4}} x_3, \quad y = \frac{M_e^{3/2} g_w^{\omega+1/2} (\gamma-1)^{1/2}}{Re^{5/8} 2^{7/4} a^{3/4}} y_3 \tag{6.9}$$

$$p - 1 = \frac{\gamma M_e a^{1/2}}{Re^{1/4} 2^{1/2}} \xi, \quad \psi = \frac{g_w^{\omega}}{Re^{3/4} 2^{1/2} a^{1/2}} f, \quad y_3 = \xi^{1/2} \eta$$

$$\rho = \frac{2}{(\gamma-1) M_e^2 g_w}, \quad \mu = g_w^{\omega} + \cdots, \quad a = \frac{M_e g_w^{\omega}}{Re^{1/2}} \left(\frac{\partial u}{\partial y}\right)_w$$

Here, the subscripts refer to the dimensionless variables retaining the order $O(1)$ in the corresponding region. The dimensionless streamfunction $\psi$ is based on the density, velocity, and length scales adopted above, while the origin is at the separation point.

Substituting Eqs. (6.9) in the Navier–Stokes equations and passing to the limit

$$Re \to \infty, \quad M_e \to \infty, \quad g_w \to 0, \quad N = O(1) \tag{6.10}$$

leads to the conventional system of incompressible boundary layer equations. In the variables introduced above we obtain

$$f''' = \xi^{1/2}\frac{d\xi}{dx_3}\left[1+\frac{1}{2}f'^2 - ff'' + \xi(f'\dot{f}' - \dot{f}f'')\right] \tag{6.11}$$

where primes and dots refer to the differentiation with respect to $\eta$ and $\xi$, respectively. Two boundary conditions correspond to the equality of the velocity vector to zero. The third boundary condition is obtained by matching the solutions for regions 3 and 2 at the outer boundary of region 3. The initial condition is obtained by matching the solution for region 3 with the solution for the undisturbed boundary layer; they are derived in the same way as in Chapter 1

$$f(\xi, 0) = f'(\xi, o) = 0, \quad f''(\xi, \eta \to \infty) = 1, \quad f(0, \eta) = \frac{1}{2}\eta^2 \tag{6.12}$$

For the subsequent presentation we will need the expression for the displacement thickness of region 3

$$\delta_3^* = g_w^{\omega+1/2} M_e^{3/2}(\gamma - 1)^{1/2} 2^{-7/4} a^{-3/4} Re^{-5/8} \lim_{\eta\to\infty}\{2\xi^{1/2}[\eta - (2f)^{1/2}]\} \tag{6.13}$$

Let us introduce the variables for region 2

$$x = M_e^{3/2} Re^{-3/8} g_w^{\omega+1/2} x_2, \quad \psi = \frac{\psi_2}{M_e Re^{1/2}} \tag{6.14}$$

$$p - 1 = \gamma M_e Re^{-1/4} p_2(x_2, \psi_2) + \cdots$$

$$\rho = M_e^{-2}[\rho_0(\psi_2) + M_e Re^{-1/4}\rho_2(x_2, \psi_2) + \cdots], \quad H = H_0(\psi_2) + \cdots$$

$$y = M_e Re^{-1/2}[y_0(\psi_2) + M_e Re^{-1/4} y_2(x_2, \psi_2) + \cdots]$$

$$u = u_0(\psi_2) + M_e Re^{-1/4} u_2(x_2, \psi_2) + \cdots, \quad v = M_e Re^{-1/4} v_2(x_2, \psi_2) + \cdots$$

Here, the subscript 0 refers to the flow parameter profiles corresponding to the undisturbed boundary layer directly ahead of the beginning of the interaction region. Substituting Eqs. (6.14) in the Navier–Stokes equations and passing to limit (6.10) leads to the following system of equations:

$$\rho_0 u_0 \frac{\partial u_2}{\partial x_2} = -\frac{\partial p_2}{\partial x_2}, \quad \frac{\partial p_2}{\partial \psi_2} = 0, \quad \frac{\partial y_0}{\partial \psi_2} = \frac{1}{\rho_0 u_0} \tag{6.15}$$

$$\frac{\partial y_2}{\partial \psi_2} = -\frac{1}{\rho_0 u_0}\left(\frac{\rho_2}{\rho_0} + \frac{u_2}{u_0}\right), \quad v_2 = N u_0 \frac{\partial y_2}{\partial x_2}, \quad \frac{\partial H_0}{\partial x_2} = 0$$

In order to determine the displacement thickness induced by region 2, we will integrate the fourth equation (6.15) using an expression for $u_2$ which can easily be obtained by integrating the first equation (6.15). Moreover, it is convenient to introduce the Mach number profile in the undisturbed boundary layer $M_0(y_0)$ noting that $M_0 = u_0\rho_0^{1/2}$. Finally, system (6.15) admits the isentropicity condition for $\rho_2$, so that $\rho_2 = \rho_0 p_2$. Thus, we have

$$y_2(x_2, \psi_2) = p_2(x_2) \int_0^{y_0} (M_0^{-2} - 1)\,\mathrm{d}y_0 = p_2(x_2)L(y_0) \qquad (6.16)$$

At the outer edge of the hypersonic boundary layer, that is, as $\psi_2 \to \infty$, the quantity $y_0$ is finite, since there the density $\rho_0 \to \infty$. Thus, integral in formula (6.16) is convergent (in what follows it will be designated by $L$).

The integral $L$ has the same importance for interaction theory, as the Mach number for gas dynamics. If $L < 0$, then region 2 behaves as a supersonic streamtube, that is, it is convergent for $\Delta p > 0$, whereas for $L > 0$ it behaves as a subsonic streamtube. Since region 3 is always "subsonic", the appearance of $L < 0$ can in certain cases lead to a "supercritical" (supersonic) behavior of the problem. This will be shown below.

In region 1 the flow is described by hypersonic small perturbation theory which in this case leads to the Ackeret formula relating the pressure disturbance and the slope of the effective body induced by the variation of displacement thicknesses of regions 2 and 3. Omitting cumbersome though simple algebra we obtain

$$\beta = \xi^{1/2}\frac{\mathrm{d}\xi}{\mathrm{d}x_3} = \xi^{3/2}\left(\frac{\mathrm{d}}{\mathrm{d}\xi}\left\{2\xi^{1/2}\lim_{\eta\to\infty}[\eta - (2f)^{1/2}]\right\} + K\right)^{-1} \qquad (6.17)$$

$$[K = (2a)^{5/4}(\gamma - 1)^{-1/2}NL]$$

Thus, now the solution of the boundary layer equation (6.11) subject to the boundary and initial conditions (6.12) must be determined. The pressure distribution $\xi(x_3)$ is not given but must be determined in the process of the solution from formula (6.17) relating pressure disturbances with the variation of the boundary layer 3 displacement thickness and that of the region 2 of inviscid slightly disturbed flow.

### *6.1.4 Discussion of the results*

The boundary value problem obtained differs from that solved in the works of Neiland (1969*a*), Stewartson and Williams (1969), and Neiland (1971*a*) only in that in formula (6.17) for the dimensionless pressure gradient $K \neq 0$. The quantity $K$ allows for the effect of the variation of the region 2 thickness. Clearly, for $g_w = O(1)$ in the limit $K = 0$. This follows from the first formula (6.8), that is, the fact that $N \to 0$ in passing to the limit $M_e Re^{-1/2} \to 0$, which is characteristic in developing boundary layer theory. The physical reason for the appearance of region 2 when cooling a body is a decrease in the characteristic thickness of the wall layer of the locally viscous flow 3.

Numerical solution of the problem was obtained using the method described in Chapter 1. The calculated pressure distributions are plotted in Fig. 6.2 for different values of

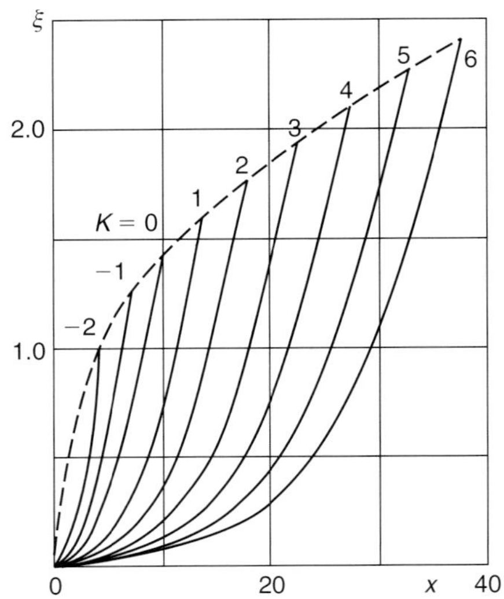


Fig. 6.2.

the parameter $K$. The right ends of the integral curves correspond to separation points. The $K = 0$ curve relates to the case of separation from a body with $g_w = O(1)$ considered above.

The results presented indicate that the value of $\xi$ at the separation point and the interaction region scale change when cooling a body (with increase in $|K|$) in quite different ways depending on sign of integral $L$ (6.16). For giving a clearer idea of the behavior of the solution with variation of $K$ and $N$ we will study the asymptotics of upstream disturbance decay.

For this purpose, we will construct a solution in the form of a series in $\xi$ as $x_3 \to -\infty$ or $\xi \to 0$

$$(\xi, \eta) \to (\xi, n_3), \quad n_3 = \eta\xi^{1/2} \tag{6.18}$$

$$f(\xi, n_3) \sim \frac{n_3^2}{2\xi} + \Phi_0(n_3) + \xi\Phi_1(n_3) + \cdots$$

Substituting Eq. (6.18) in Eq. (6.17) yields

$$\beta = \xi^{3/2}[K - 2\Phi'(\infty)]^{-1} + O(\xi^{5/2}) \tag{6.19}$$

Using Eqs. (6.18) and (6.19) from Eqs. (6.11) and (6.12) we obtain

$$\Phi_0''' = \frac{(1 + n_3\Phi_0' - \Phi_0)}{K - 2\Phi_0'(\infty)} \tag{6.20}$$

$$\Phi_0(0) = \Phi_0'(0) = \Phi_0''(\infty) = 0$$

From the solution of problem (6.20) we obtain

$$-\Phi_0'(\infty) = \frac{1.3667}{[K - 2\Phi_0'(\infty)]^{1/3}} \tag{6.21}$$

As $\xi \to 0$ and $x_3 \to -\infty$, in accordance with Eq. (6.19), the pressure disturbance distribution takes the form:

$$\xi \sim \exp(x_3 b^{-1}), \quad b = K - 2\Phi_0'(\infty) \tag{6.22}$$

Thus, in accordance with the first formula (6.9) and Eq. (6.22), the characteristic scale for the upstream disturbance transfer is as follows:

$$\Delta x \sim \frac{(\gamma - 1)^{1/2} M_e^{3/2} g_w^{\omega+1/2} b}{(2a)^{5/4} Re^{3/8}} \tag{6.23}$$

If the body temperature decreases and $N \to \infty$, then for $L > 0$, in accordance with Eq. (6.21), as $K \to \infty$, we obtain

$$b \approx K, \quad \Phi_0'(\infty) \to -1.3667K^{-1/3}, \quad \Delta x \sim M_e^2 Re^{-1/2} \tag{6.24}$$

Thus, in this case disturbances are chiefly transferred by the subsonic part of the locally inviscid flow 2. The effect of the viscous sublayer 3 becomes minimal. In the case of a hypersonic outer flow, the subsonic sublayer with a thickness of the order $\delta^{(2)} \sim M_e Re^{-1/2}$ transmits disturbances through distances of the order $M_e \delta^{(2)} \sim M_e^2 Re^{-1/2}$. Let us study the behavior of the solution of the complete nonlinear problem as $K \to +\infty$. For this purpose, we will make the change of variables

$$f(\xi, \eta) = \bar{f}(\bar{\xi}, \bar{\eta}), \quad \bar{\eta} = \eta, \quad \xi = K^{2/3}\bar{\xi}$$

Then in the limit instead of Eqs. (6.11), (6.12), and (6.17) we obtain

$$\bar{f}''' = \bar{\xi}^{1/2} \frac{d\bar{\xi}}{dx_3} \left[ 1 + \frac{1}{2}\bar{f}''^2 - \bar{f}\bar{f}'' + \bar{\xi}(\bar{f}'\dot{\bar{f}}' - \dot{\bar{f}}\bar{f}'') \right]$$

$$\bar{f}(\bar{\xi}, 0) = \bar{f}'(\bar{\xi}, 0) = 0, \quad \bar{\bar{f}}''(\bar{\xi}, \bar{\eta} \to \infty) = 1, \quad \bar{f}(0, \bar{\eta}) = \frac{1}{2}\bar{\eta}^2$$

Solving numerically the problem gives the asymptotic behavior of $\xi_0$ at the separation point

$$\xi_0 \to 0.51K^{2/3}, \quad K \to +\infty$$

Hence follows that at the separation point the pressure difference increases as

$$\Delta p \sim g_w^{-2/3(\omega+1/2)} \tag{6.25}$$

Previously it was shown that, in essence, the flow in region 2 containing the main part of streamtubes of the undisturbed boundary layer behaves locally as an inviscid flow. The case

$K > 0$ and $L > 0$ considered above corresponds to such original profiles $M_0(y_0)$, for which the displacement thickness of the locally inviscid flow region increases with the pressure, while the transverse pressure difference is small.

This boundary layer behaves as a subsonic streamtube; following Crocco's terminology, we will call it subcritical. We note that for $L > 0$ and $|K| \gg 1$ the estimate for the length of the region of upstream disturbance transfer can easily be obtained from estimates (6.2) for $\Delta\delta^{(2)}$ and a counterpart of the Ackeret formula (6.6); taking into account that in this case $\Delta\delta^{(2)} \gg \delta^{(3)}$, we obtain $\Delta x \sim M_e\delta$.

Let us now consider the $L < 0$ regime. In this case, the total thickness of streamtubes of region 2 decreases in the absence of a transverse pressure gradient and with increase in the pressure, since the major part of the original boundary layer is supersonic.

Disturbance transfer through distances larger than $M_e Re^{-1/2}$ due to the free interaction between the boundary layer and the outer flow for $L < 0$ and $N = O(1)$ is possible only thanks to the existence of viscous flow region 3. With decrease in $g_w$ the parameter $K \to -\infty$. According to Eqs. (6.21) and (6.22), we have

$$b \sim (-K)^{-3}, \quad \Phi_0'(\infty) \to +\frac{1}{2}K \tag{6.26}$$

In this case, region 2 behaves as a supersonic streamtube and the scale of the upstream disturbance propagation region decreases. By analogy with supersonic flows, the flows of this type are called supercritical. In what follows it will be shown that for $\Delta\delta^{(2)} \gg \delta^{(3)}$ the downstream scale can be $\Delta x \sim M_e\delta$. However, in this case relations (6.26) cannot be considered as limiting, since for $\Delta x \sim M_e Re^{-1/2}$ the longitudinal and transverse dimensions of region 2 are of the same order. But then the longitudinal and transverse pressure gradients are also of the same order and the above estimates and the formulation of the boundary value problem no longer hold.

### *6.1.5 Supercritical regime of incipient separation*

Above we considered the general interaction regime in which the parameter $N$ (6.8) proportional to $\Delta\delta^{(2)}/\delta^{(3)}$ was of the order $O(1)$. It was shown that the involvement of supersonic streamtubes has a considerable effect on the flow pattern. In this section, we will consider a limiting regime in which $\Delta\delta^{(2)} \gg \delta^{(3)}$ and the nature of the interaction is independent of the subsonic flow in region 3.

For obtaining the estimates we can, as before, use formula (6.3) to (6.5) and replace formula (6.6) by the result obtained above

$$\Delta x \sim M_e^2 Re^{-1/2} \tag{6.27}$$

Then we arrive at the following scales for the flow functions

$$\Delta p \sim M_e^{4/3} Re^{-1/3} g_w^{-2/3(\omega+1/2)}, \quad \Delta x \sim M_e^2 Re^{-1/2}$$

$$\Delta\delta^{(2)} \sim M_e^{7/3} Re^{-5/6} g_w^{-2/3(\omega+1/2)} \tag{6.28}$$

We note that for $g_w \sim (M_e\delta)^{1/(2(2\omega+1))}$ scales (6.28) go over to (6.7).

We will now consider the flow past a concave corner with a given inclination angle $\theta$

$$-1 \leq x < 0, \quad \theta = 0, \quad x > 0, \quad \theta = \theta_0 \tag{6.29}$$

We introduce the variables for region 2

$$x = M_e^2 Re^{-1/2} x_2, \quad \psi = \frac{\psi_2}{M_e Re^{1/2}} \tag{6.30}$$

$$p - 1 = \gamma M_e^{4/3} Re^{-1/3}\, g_w^{-2(\omega+1/2)/3}\, p_2(x_2, \psi_2) + \cdots$$

$$\rho = M_e^{-2}[\rho_0(\psi_2) + M_e^{4/3} Re^{-1/3} g_w^{-2(\omega+1/2)/3} \rho_2(x_2, \psi_2) + \cdots], \quad H = H_0(\psi_2) + \cdots$$

$$y = M_e Re^{-1/2}[y_0(\psi_2) + M_e^{4/3} Re^{-1/3} g_w^{-2(\omega+1/2)/3} y_2(x_2, \psi_2) + \cdots]$$

$$u = u_0(\psi_2) + M_e^{4/3} Re^{-1/3} g_w^{-2(\omega+1/2)/3} u_2(x_2, \psi_2) + \cdots$$

Substituting Eq. (6.30) in the Navier–Stokes equations and passing to the limit

$$M_e \to \infty, \quad Re \to \infty, \quad g_w \to 0 \tag{6.31}$$

leads to the system of equations

$$\rho_0 u_0 \frac{\partial u_2}{\partial x_2} + \frac{\partial p_2}{\partial x_2} = 0, \quad \frac{\partial p_2}{\partial \psi_2} = 0, \quad \frac{\partial y_0}{\partial \psi_2} = \frac{1}{\rho_0 u_0} \tag{6.32}$$

$$\frac{\partial y_2}{\partial \psi_2} = -\frac{1}{\rho_0 u_0}\left(\frac{\rho_2}{\rho_0} + \frac{u_2}{u_0}\right)$$

But in this case the required result for $y_2(x_2, \infty)$ is given by formula (6.16).

Using the Ackeret formula, together with Eq. (6.16), we obtain

$$p_2 = \theta^* + L\frac{\mathrm{d}p_2}{\mathrm{d}x_2} m \tag{6.33}$$

where $\theta^*$ is the "flap" deflection angle

$$\theta^* = \frac{Re^{1/3} g_w^{2(\omega+1/2)/3}}{M_e^{1/3}} \theta \tag{6.34}$$

If the deflection leads to flow compression, then $\theta^* > 0$.

For a while, we will assume that $\theta^* > 0$ but the value of $\theta^*$ is insufficient in order to lead to separation of the viscous nonlinear layer 3.

In a separationless flow it is natural to seek solutions for which disturbances decay far away from the corner point, such that $p_2 \to 0$ as $x_2 \to -\infty$ and $p_2 \to \theta^*$ as $x_2 \to \infty$.

However, for $L > 0$ solutions of the required form $(p_2(-\infty) = 0)$ exist only in the region $x < 0$ $(\theta^* = 0)$, whereas behind the corner point $(\theta^* > 0)$ there are no solutions leading to $p_2 = \theta^*$.

For $L > 0$ the only type of solution (subcritical boundary layer) is as follows:

$$\begin{aligned} x_2 < 0: &\quad p_2 = \theta^* e^{x_2/L}, \\ x_2 > 0: &\quad p_2 = \theta^* \end{aligned} \tag{6.35}$$

Contrariwise, for $L < 0$ (supercritical boundary layer) the solution having a physical meaning takes the form:

$$\begin{aligned} x_2 < 0: &\quad p_2 = 0, \\ x_2 > 0: &\quad p_2 = \theta^*(1 - e^{x_2/L}) \end{aligned} \tag{6.36}$$

This behavior of limiting solutions is natural, since the subsonic $(L > 0)$ boundary layer transmits disturbances upstream, while in the supersonic $(L < 0)$ boundary layer they are carried away downstream.

We note that with increase in $\theta^*$ the boundary layer starts to separate ahead of the corner point for $L > 0$ and behind it for $L < 0$.

It remains to study the problem of possible generation of separation in an immediate vicinity of the corner point, at distances $\Delta x \sim \delta_0$. This region can appear for $L < 0$, since in the presence of fairly large $\partial p/\partial y$, which is possible due to the scale $\Delta x$, the thicknesses of supersonic streamtubes can compensate each other and the body thickness can vary due to the presence of the corner.

In the system of the estimates presented above, in this case the scale

$$\Delta x \sim \delta \tag{6.37}$$

should be used instead of the Ackeret formula.

Solving the system of equations we can obtain the pressure disturbance amplitude leading to generation of separation in region 3

$$\Delta p \sim \gamma M_e^{2/3} g_w^{-2(\omega+1/2)/3} Re^{-1/3} \tag{6.38}$$

Comparing Eqs. (6.38) and (6.28) and using the formula for the flow past the corner $\theta$ we can see that in the region with $\Delta x \sim M_e \delta$ separation is generated for

$$M_e \theta \sim M_e^{1/3} Re^{-1/3} g_w^{-2(\omega+1/2)/3}$$

However, on the $\Delta x \sim \delta$ scale, due to $M_2 \sim 1$, the pressure difference can be estimated as follows:

$$\Delta p_w \sim \theta \sim \gamma M_e^{1/3} g_w^{-2(\omega+1/2)/3} Re^{-1/3} \tag{6.39}$$

Let us compare Eqs. (6.39) and (6.38)

$$\frac{\Delta p_w}{\Delta p} \sim M_e^{-1/3} \ll 1 \tag{6.40}$$

This means that for $L < 0$ flow separation is generated at a distance $\Delta x \sim M_e \delta$ downstream of the corner point, whereas in the subcritical flow with $L > 0$ separation starts to form at a distance $\Delta x \sim M_e \delta$ upstream of the corner point.

## 6.2 Distinctive features of interaction and separation of a transcritical boundary layer

We will consider the asymptotic theory of the interaction between a hypersonic flow and a laminar boundary layer in flow regimes with low values of the temperature factor. We present a complete classification of the flows of this type unifying the previously considered subcritical and supercritical flows. Emphasis is placed on the transitional, or transcritical, flows, whose properties are similar from the physical standpoint to those of inviscid transonic flows. We develop a theory of discontinuous flows including a relation at a discontinuity which, to a certain degree, is analogous to the Rankine–Hugoniot relation at a shock. Certain solutions describing flow past some forms typical of aerodynamic applications are derived. It is shown that at fairly large angles of flow turning overexpansion and a transcritical shock necessarily occur, that is to say, the solution is discontinuous on the main scale of the disturbed region, though, of course, it is continuous on a lesser scale.

### *6.2.1 Equations and boundary conditions*

We will first study the hypersonic viscous flow around a vicinity of a corner point on the contour of a very simple body shown in Fig. 6.3.

We will consider the weak interaction regime on the main part of the body, up to a certain vicinity of the corner point $O$. In this case, the undisturbed boundary layer parameter profiles are, as is known (see, e.g., Hayes and Probstein (1966)), determined by the equations

$$u = u_e f'(\xi, \eta), \quad H = \left(\frac{u_e^2}{2}\right) g(\xi, \eta), \quad N = \frac{\rho\mu}{\rho_w \mu_w}$$

$$\xi = \int_{-1}^{0} \rho_w \mu_w u_e \, \mathrm{d}x, \quad \eta = \left(\frac{u_e}{\sqrt{2\xi}}\right) \int_0^y \rho \, \mathrm{d}y, \quad f'(\xi, \infty) = 1, \quad g(\xi, \infty) = 1 \tag{6.41}$$

$$(Nf'')' + ff'' = 0, \quad \left(\frac{N}{\sigma} g'\right) + fg' = 0, \quad f(\xi, 0) = f'(\xi, 0) = 0, \quad g(\xi, 0) = g_w$$

The Cartesian coordinate system is shown in Fig. 6.3; its origin is at the corner point, $x = -\ell$ relates to the beginning of the boundary layer, that is, $\ell$ is the length of the forward

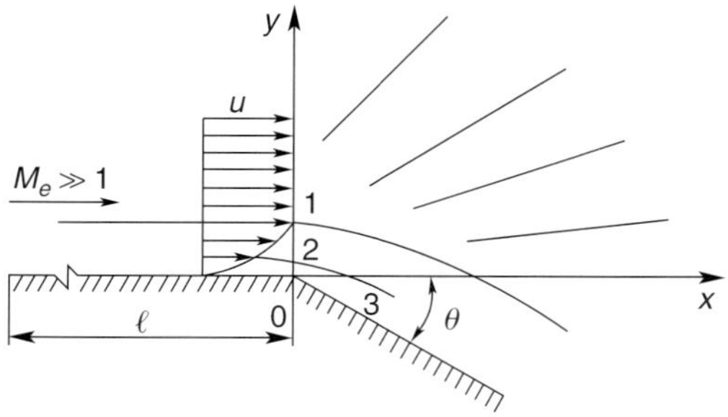


Fig. 6.3.

part of the plate, and $u$, $H$, $\rho$, and $\mu$ are the longitudinal velocity, the stagnation enthalpy, the density, and dynamic viscosity. The subscripts $e$ and $w$ refer to the parameters at the outer edge of the boundary layer and at the wall.

The power-law viscosity-enthalpy dependence with an exponent $\omega$ is adopted.

Prior to deriving limiting asymptotic solutions of the Navier–Stokes equations by the matched asymptotic expansion method (see, e.g., Van Dyke, 1964), we will derive the estimates of the flow parameter scales in the region of the strong local interaction between the outer hypersonic flow and the boundary layer near point $O$.

In accordance with Eq. (6.41), the corresponding estimates for the boundary layer not disturbed by the interaction ahead of point $O$ are as follows:

$$u \sim u_e, \quad \rho \sim \rho_0 \sim \frac{\rho_e}{M_e^2}, \quad H \sim \frac{u_e^2}{2}, \quad \delta_0 \sim l\,Re^{-1/2}, \quad Re_0 \sim \frac{\rho_0 u_e l}{\mu_0}, \quad \rho_0 u_e^2 \sim p \tag{6.42}$$

where $\delta_0$ is the undisturbed boundary layer thickness at $x=0$, the subscript $e$ refers to the parameters at the outer edge of the undisturbed boundary layer and the subscript 0 to the values of the density and viscosity at the stagnation temperature of the outer inviscid flow.

In accordance with hypersonic small perturbation theory, for $M_e \gg 1, \theta \ll 1$, and $M_e\theta \ll 1$ we have the Ackeret formula

$$M_e(p - p_e) = \rho_e u_e^2 \left( \frac{d\delta^*}{dx} + \theta \right) \tag{6.43}$$

where $\delta^*$ is the boundary layer displacement thickness and $\theta$ is the body contour deflection angle (Fig. 6.3).

If $\delta^* = 0$ (there is no boundary layer), the pressure on the body changes jumpwise in a centered expansion wave. Due to the presence of the boundary layer, the discontinuity in the pressure distribution over the body surface is smeared; however, pressure gradients remain large, though the pressure disturbance itself is small for $M\theta \ll 1$, in accordance with Eq. (6.43)

$$\frac{\Delta p}{p} \sim M_e\theta \tag{6.44}$$

In the main part of the boundary layer, where $u \sim u_e$, using the momentum equation, together with the equation of state, we obtain the conventional estimates

$$\Delta u \sim u_e \left(\frac{\Delta p}{p}\right), \quad \Delta\rho \sim \rho_0 \left(\frac{\Delta p}{p}\right), \quad \frac{\Delta\delta}{\delta_0} \sim \frac{\Delta p}{p} \tag{6.45}$$

This part of the boundary layer is associated with region 2 in Fig. 6.3. However, there exists always region 3 in which

$$\rho u^2 \sim \Delta p \tag{6.46}$$

Therefore, if in what follows it will turn out that in region 3 the viscous terms of the Navier–Stokes equations are small, then velocity disturbances in this region will be of the leading order (Neiland (1968, 1969a)). Generally, in region 3 with low dynamic heads it is required to construct separate asymptotic expansions.

Thus, we will seek the solution of the Navier–Stokes equations in passing to the following limits:

$$Re_0 \to \infty, \quad M_e \to \infty, \quad \theta \to 0 \quad \text{for } M_e\theta \to 0, \; g_w \ll O(1) \tag{6.47}$$

Obviously, the solution in the outer region 1 is described by small perturbation theory for the outer hypersonic flow. For further presentation we shall need only the formula for the pressure distribution over the boundary formed by the boundary layer displacement thickness, or region 2 (6.43).

In region 2 of a weakly disturbed vortex flow including the major part of the streamtubes of the original undisturbed boundary layer, the Navier–Stokes equations can conveniently be written in the von Mises variables

$$\begin{aligned}
\rho u\frac{\partial u}{\partial x} + \frac{\partial p}{\partial x} - \rho v\frac{\partial p}{\partial \psi} &= \rho u\frac{\partial}{\partial \psi}\left[\rho u\mu\frac{\partial u}{\partial \psi} + \mu\left(\frac{\partial v}{\partial x} - \rho v\frac{\partial v}{\partial \psi}\right)\right] \\
&\quad + \left(\frac{\partial}{\partial x} - \rho v\frac{\partial}{\partial \psi}\right)\left[\frac{4}{3}\mu\left(\frac{\partial u}{\partial x} - \rho v\frac{\partial u}{\partial \psi}\right) - \frac{2}{3}\rho u\mu\frac{\partial u}{\partial \psi}\right] \\
\rho u\left(\frac{\partial v}{\partial x} + \frac{\partial p}{\partial \psi}\right) &= \rho u\frac{\partial}{\partial \psi}\left[\frac{4}{3}\rho u\mu\frac{\partial v}{\partial \psi} - \frac{2}{3}\mu\left(\frac{\partial u}{\partial x} - \rho v\frac{\partial u}{\partial \psi}\right)\right] \\
&\quad + \left(\frac{\partial}{\partial x} - \rho v\frac{\partial}{\partial \psi}\right)\left[\rho u\mu\frac{\partial u}{\partial \psi} + \mu\left(\frac{\partial v}{\partial x} - \rho u\frac{\partial v}{\partial \psi}\right)\right]
\end{aligned} \tag{6.48}$$

$$\frac{\partial n}{\partial x} = \frac{v}{u'} \quad \frac{\partial n}{\partial \psi} = \frac{1}{\rho u}$$

$$\begin{aligned}
\rho u\frac{\partial h}{\partial x} - u\frac{\partial p}{\partial x} &= \rho u\frac{\partial}{\partial \psi}\left[\rho\mu u\frac{\partial h}{\partial \psi}\right] + \left(\frac{\partial}{\partial x} - \rho v\frac{\partial}{\partial \psi}\right)\left(\mu\frac{\partial h}{\partial x} - \rho v\mu\frac{\partial h}{\partial \psi}\right) \\
&\quad + \mu\left[\left(\rho u\frac{\partial v}{\partial \psi} + \frac{\partial v}{\partial x} - \rho v\frac{\partial v}{\partial \psi}\right)^2 + 2\left(\frac{\partial u}{\partial x} - \rho v\frac{\partial u}{\partial \psi}\right)^2\right. \\
&\quad \left. + 2\left(\rho u\frac{\partial v}{\partial \psi}\right)^2 - \frac{2}{3}\left(\frac{\partial u}{\partial x} - \rho v\frac{\partial u}{\partial \psi} + \rho u\frac{\partial v}{\partial \psi}\right)^2\right]
\end{aligned}$$

Let us introduce the following independent variables and the functions in region 2:

$$x = \ell \Delta x_2 x_2, \quad \psi = \rho_0 u_e \ell \Delta \psi_2 \psi_2$$

$$p = \rho_0 u_e^2 (p_{20} + \Delta p_2 p_{21} + \cdots), \quad \rho = \rho_0 (p_{20} + \Delta \rho_2 \rho_{21} + \cdots)$$

$$u = u_e (u_{20} + \Delta u_2 u_{21} + \cdots), \quad v = u_e \Delta v_2 (v_{21} + \cdots) \tag{6.49}$$

$$n = \ell Re_0^{-1/2} [(n_{20} + \Delta n_2 n_{21} + \cdots) + \Delta n_3 (n_3 - n_{30})_\infty]$$

where $p_{20}$, $p_{21}$, $\rho_{20}$, $\rho_{21}, \ldots$ are dimensionless functions of the order $O(1)$, while $\Delta x_2$, $\Delta n_3, \ldots$ are dimensionless scales dependent on the form of the passage to limit (6.47)

$$p_{20} = \frac{(\gamma - 1)}{2\gamma}, \quad \rho_0 = \rho_e \left[ \frac{2}{(\gamma - 1) M_e^2} \right]$$

$$\Delta p_2 = \Delta u_2 = \Delta \rho_2 = \Delta n_2 = M_e \theta, \quad \Delta \psi_2 = Re_0^{-1/2} \tag{6.50}$$

Here, $\Delta x_2$ is determined separately in each individual case; however, it will be shown that $Re_0^{-1/2} \ll \Delta x_e \ll 1$. Substituting Eq. (6.49) in the Navier–Stokes equations (6.48) and passing to limit (6.47) we arrive at the following system of equations:

$$\rho_{20} u_{20} \frac{\partial u_{21}}{\partial x_2} + \frac{\partial p_{21}}{\partial x_2} = 0, \quad \frac{\partial p_{21}}{\partial \psi_2} = 0, \quad \frac{\partial}{\partial x} \left( \frac{p_2}{\rho_2^\gamma} \right) = 0$$

$$\frac{dn_{20}}{d\psi_2} = \frac{1}{\rho_{20} u_{20}}, \quad \frac{\partial n_{21}}{\partial \psi_2} = \frac{1}{\rho_{20} u_{20}} \left( -\frac{\rho_{21}}{\rho_{20}} - \frac{u_{21}}{u_{20}} \right) \tag{6.51}$$

We will now match the solution for region 2, as $x_2 \to -\infty$, with the solution for the undisturbed boundary layer, thus determining the profiles of the functions $\rho_{20}(\psi_2)$ and $u_{20}(\psi_2)$; the value $\rho_{21}(x_{21} \to -\infty) \to 0$. The system of equations (6.51) can be integrated thus obtaining the contribution of the flow in region 2 to the variable part of the displacement thickness $\delta^*$ in the flow region under consideration: for $g_w = 0$ we have

$$n_{21}(x_2) = \frac{1}{\gamma} \frac{p_{21}(x_2)}{p_{20}} \int_0^{n_{20e}} \left( \frac{1}{M_0^2} - 1 \right) dn_{20} \tag{6.52}$$

Clearly, $n_{21}$ is linearly dependent on the pressure disturbance $p_{21}$, while the nature of the variation of $u_{21}$ is determined by the $M_0$ profile in the undisturbed boundary layer ahead of the beginning of the interaction region. We note the $M_0$ profile in the flat-plate boundary layer at the temperature factor $g_w = 0$ is such, that integral in Eq. (6.52) is negative. This means that the flow in region 2 behaves as in a supersonic streamtube, that is, with decrease in the pressure ($p_{21} < 0$) its thickness grows ($n_{21} > 0$).

### *6.2.2 Flow in region 3*

We will now consider the flow in region 3, in which relation (6.46) is fulfilled and, therefore, disturbances may be nonlinear.

We introduce the following scales of the arguments and functions:

$$x = \ell \Delta x_3 x_3, \quad \psi = \rho_0 u_e \ell \, Re_0^{-1/2} \Delta\psi_3 \psi_3 \tag{6.53}$$

$$p = \rho_0 u_e^2 (p_{30} + \Delta p_3 \cdot p_{31} + \cdots), \quad \rho = \rho_0 \Delta\rho_3 (\rho_3 + \cdots)$$

$$u = u_e \Delta u_3 (u_3 + \cdots), \quad n = \ell \, Re_0^{-1/2} \Delta n_3 (n_3 + \cdots)$$

$$v = u_e \Delta v_3 (v_3 + \cdots), \quad \mu = \mu_0 \Delta\mu_3 (\mu_3 + \cdots)$$

$$h = \left(\frac{u_e^2}{2}\right) \Delta h_3 (h_3 + \cdots)$$

where

$$\Delta p_3 = \Delta p_2 = \Delta\rho_3 \cdot \Delta u_3^2, \quad p_{30} = p_{20}, \quad p_{31} = p_{21}, \quad \Delta\psi_3 = \Delta\rho_3 \cdot \Delta u_3 \cdot \Delta n_3$$

Substituting Eq. (6.53) in the Navier–Stokes equations (6.48) and passing to limit (6.47) we obtain the following system of equations:

$$\rho_3 u_3 \frac{\partial u_3}{\partial x_3} = -\frac{\partial p_3}{\partial x_3} + K \rho_3 u_3 \frac{\partial}{\partial \psi_3}\left(\rho_3 u_3 \mu_3 \frac{\partial u_3}{\partial \psi_3}\right)$$

$$\frac{\mathrm{d}n_3}{\mathrm{d}\psi_3} = \frac{1}{\rho_3 u_3}, \quad \rho_3 h_3 = \frac{2\gamma}{\gamma - 1} p_{30}, \quad K = \frac{\Delta\mu_3 \Delta x_3}{\Delta n_3^2 \Delta\rho \Delta u_3} \tag{6.54}$$

$$\frac{\partial h_3}{\partial x_3} = K \frac{\partial}{\partial x_3}\left(\rho_3 u_3 \mu_3 \frac{\partial h_3}{\partial \psi_3}\right), \quad \frac{\partial p_{31}}{\partial \psi_3} = 0$$

Depending on the relation between small quantities $g_w$ and $M_e\theta$, the relative dimensionless scales of the variables $\Delta x$, $\Delta p$, and $\Delta\rho$ can take different values. Then passing to limit (6.47), including the case $g_w \to 0$, can lead to equations of different type resulting from system (6.54).

### *6.2.3 Classification of flow regimes*

We will carry out the classification of flow regimes in accordance with the conventional procedure of estimating the orders of the flow functions. For this purpose, it is convenient to return partially (only in this section) to dimensional variables referring by subscripts 0, 1, 2, and 3 their characteristic values in the undisturbed boundary layer and in the regions 1, 2, and 3 of the disturbed part of the flow, respectively.

Using Eq. (6.1) we obtain the following estimates for the behavior of the dimensionless velocity $u/u_e$ and the dimensionless enthalpy $g = 2h/u_e^2$ at the bottom of the undisturbed boundary layer, with the characteristic temperature $g_w \leq O(1)$

$$g \sim (y + g_w^{\omega+1})^{1/(\omega+1)}, \quad u \sim (y + g_w^{\omega+1})^{1/(\omega+1)} - g_w \tag{6.55}$$

In relations (6.55) we have dropped factors $O(1)$ unessential for deriving the estimates of the orders of the flow functions; the coordinate $y$ is divided by the boundary layer thickness $\delta_0$.

We will first consider the case in which the flow in region 3 characterized by the thickness $y \sim \delta_3$ ($\delta_3$ is also dimensionless and scaled on $\delta_0$) is near-isothermal, that is

$$g_w^{\omega+1} \gg \delta_3 \tag{6.56}$$

Then, in accordance with Eq. (6.55) we obtain the estimates

$$g_3 \sim g_w, \quad \rho_3 \sim \rho_w \sim \frac{\rho_0}{g_w}, \quad \mu_3 \sim \mu_0 g_w^{\omega}, \quad u_3 \sim \frac{u_e \delta_3}{g_w^{\omega}} \tag{6.57}$$

where $\rho_0$ and $\mu_0$ are the density and viscosity at the stagnation temperature. The last estimate (6.57) is obtained from relations (6.55) with account for inequality (6.56).

By definition, in region 3 the nonlinear disturbances are

$$\Delta p_3 \sim \rho_3 u_3^2 \tag{6.58}$$

Then the layer thickness is of the order

$$\delta_3 \sim g_w^{\omega+1/2} \left(\frac{\Delta p}{p}\right)^{1/2} \tag{6.59}$$

In accordance with relation (6.58), if viscous stresses are not predominant in region 3, the velocity varies in the leading order, so that we have

$$\Delta\delta_3 \sim \delta_3 \tag{6.60}$$

We consider flows with nonlinear disturbances in region 3. Depending on the relation between the values of $g_w$ and $\Delta p/p$, the following flow regimes are possible.

If $g_w$ is not too small or, generally, $g_w = 0(1)$, then the following relation is fulfilled:

$$\delta_3 \gg \Delta\delta_2 \sim \frac{\Delta p}{p} \tag{6.61}$$

This means that the main contribution of the derivative of the boundary layer displacement thickness $\delta^*$ (6.43) is given by the subsonic flow in region 3. Then, as in the classical three-layer scheme, the flow pattern is necessarily subcritical, since in region 3 the flow is subsonic.

The sufficient condition of subcritical flow follows also immediately from the comparison of Eqs. (6.59)–(6.61) and (6.45)

$$\frac{\Delta p}{p} \ll g_w^{2\omega+1} \tag{6.62}$$

If we assume that region 2 is supercritical, that is, the integral in formula (6.16) (we designate it by $L$) is, due to the passage to the limit $g_w \to 0$, convergent and negative, the flow as a whole can become supercritical if

$$\Delta\delta_2 \gg \Delta\delta_3 \sim \delta_3 \tag{6.63}$$

Thus, for $L < 0$ the flow is supercritical provided that

$$\frac{\Delta p}{p} \gg g_w^{2\omega+1} \tag{6.64}$$

Thus, transcritical flows, in which the boundary layer can be both subcritical and supercritical in different flow regions, can exist provided that

$$\frac{\Delta p}{p} \sim g_w^{2\omega+1}, \quad \Delta\delta_2 \sim \delta_3 \tag{6.65}$$

Fundamental differences between the subcritical and supercritical flows are similar to those between subsonic and supersonic flows and will be illustrated below with reference to particular problems.

Considering necessarily subcritical regimes ($\Delta\delta_3 \gg \Delta\delta_2$) and possible transcritical regimes ($\Delta\delta_3 \sim \delta_3 \sim \Delta\delta_2$) and using Eq. (6.43) for deriving the estimates, one can determine the scale $\Delta x$ of the region in which disturbances can be transferred upstream

$$\frac{\Delta x}{\ell} \sim \left(M_e \frac{\delta_0}{\ell}\right) g_w^{\omega+1/2} \left(\frac{\Delta p}{p}\right)^{-1/2} \tag{6.66}$$

Comparing estimates (6.65) and (6.66) it can be seen that for all subcritical and transcritical interaction regimes we have $\Delta x \gg \delta_0$, which means the absence of transverse pressure differences from regions 2 and 3 in the main approximation, except for singular points of the solutions considered below.

We are now coming to the comparison of the characteristic values of the viscous-to-inertial term ratio in the momentum equation for region 3. In order for the flow in region 3 to remain nonlinear, it is necessary and sufficient that the following condition is fulfilled:

$$\frac{\rho_3 u_3^2 \delta_3^2}{\Delta x \mu_3 u_3} \geq O(1) \tag{6.67}$$

Using the estimates derived above, from relation (6.27) we obtain

$$\left(\frac{\Delta p}{p}\right)^2 \geq O(M_e Re_0^{-1/2}) \sim O\left(\frac{M_e \delta_0}{\ell}\right) \tag{6.68}$$

Thus, the flow in region 3 is described by the complete Prandtl equations, that is, it is "viscous and nonlinear" when the order of pressure disturbances is as follows:

$$\frac{\Delta p}{p} \sim M_e^{-1/2} Re_0^{-1/4} \tag{6.69}$$

which is in agreement with the results of the studies of Neiland (1968, 1969b) and Stewartson and Williams (1969) extended by Neiland (1970*a*) to hypersonic flows.

If the order of the pressure disturbances is smaller, then in region 3 the disturbances are linear. If

$$1 \gg \frac{\Delta p}{p} \gg M_e Re_0^{-1/4} \tag{6.70}$$

then the flow in region 3 must be nonlinear and inviscid in the main approximation (then a thinner viscous sublayer should be introduced for satisfying the no-slip conditions at the body).

In what follows, emphasis is placed on the transcritical regimes with nonlinear inviscid flow in region 3 for which conditions (6.65) and (6.70) are fulfilled. In Fig. 6.4 these regimes correspond to curve $AB$ in the plane $O(g_w), O(\Delta p/p)$. Subcritical flow regimes with nonlinear disturbances in region 3 lie in domain $ABC$. Point $C$ is associated with the conventional three-layer flow model (Neiland, 1968; 1969*b*; Stewartson and Williams, 1969), while the disturbance regimes lying to the right of $AB$ but above $AC$, that is, those with inviscid nonlinear layer 3, were studied in the works of Matveeva and Neiland (1967) and Ollsson and Messiter (1969).

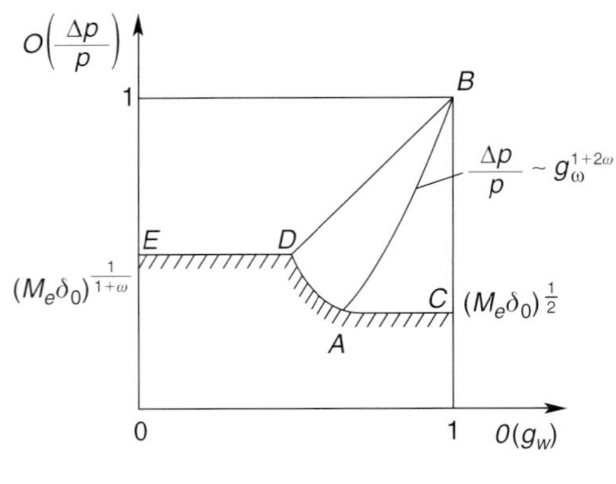


Fig. 6.4.

In all subcritical and transcritical nonlinear problems corresponding to triangle $ABC$ in Fig. 6.4, the flow in region 3 remains isothermal in the main approximation, that is $g_3 \sim g_w$. Therefore, it might be expected that transcritical interaction regimes can be found to the left of curve $AB$ (for $L < 0$).

In order for supercritical properties of the interaction could manifest themselves, it is necessary that the main part of $\delta^*$ variation is produced by supersonic streamtubes from region 2. For this purpose it is necessary that condition (6.64) is fulfilled. From Eq. (6.43) and estimate (6.61) for $\Delta\delta_2$ there follows the estimate:

$$\Delta x \sim M_e \delta_0 \tag{6.71}$$

Thus, with decrease in $g_w$ at a given value of the pressure disturbance the scale of $\Delta x$ in region *ABC* (Fig. 6.4) decreases, in accordance with relation (6.66), to the transcritical curve *AB* and then, upon intersection of this curve, ceases to change in the order, since it is no longer influenced in the main approximation by region 3. Using condition (6.67) we obtain

$$\frac{\Delta p}{p} \sim \left(M_e Re_0^{-1/2}\right)^{2/3} g_w^{-2(\omega+1/2)/3} \tag{6.72}$$

and determine how the critical pressure difference producing boundary layer separation varies in the order to the left of *AB* . Clearly, at point *A* (Fig. 6.4), at which the transcriticality condition (6.65) is fulfilled, the value of the critical pressure difference satisfies relation (6.69).

Moving along the diagram of Fig. 6.4 in accordance with relation (6.72) toward decreasing $g_w$ we can see that the thickness of the nonlinear viscous disturbed flow layer 3 continuously decreases, whereas the pressure disturbance increases until condition (6.69), in accordance with which estimates for layer 3 were carried out, is violated. Using relation (6.59) it can be shown that inequality (6.56) is violated at the following values of the temperature factor:

$$g_w \sim \frac{\Delta p}{p} \ll \left(\frac{\Delta p}{p}\right)^{1/(1+2w)} \tag{6.73}$$

In Fig. 6.4 transition to the nonisothermal region 3 is associated with the leftward crossing of line *BD*, at which condition (6.73) is satisfied. For smaller $g_w$, in accordance with Eq. (6.53), the orders of the flow parameters in region 3 with nonlinear disturbances are as follows:

$$g_3 \sim \delta_3^{1/(1+w)}, \quad u_3 \sim \delta_3^{1/(1+w)}, \quad \mu_3 \sim \delta_3^{w/(1+w)}, \quad \rho_3 \sim \delta_3^{1/(1+w)} \tag{6.74}$$

As above, the comparison of the orders of the leading viscous and inertial terms gives the order of the critical pressure difference which can produce boundary layer separation in compression flows

$$\frac{\Delta p}{p} \sim (M_e^2 Re_0^{-1/2})^{1/3} \tag{6.75}$$

In Fig. 6.4 it is line *ED* that corresponds to formula (6.75).

Thus, Fig. 6.4 gives an idea of possible flow patterns with strong local interaction up to boundary layer separation. We note that small-size regions with $\Delta x \sim \delta_0$, in which the Ackeret formula (6.43) is not fulfilled, can also appear; these will be considered in Section 6.2.5. A major part of the existing studies is devoted to subcritical regimes associated with region *ABC* and, particularly, those at point *C*. There are a few studies on supercritical regimes (Neiland, 1973; 1974*b*); see also Section 6.1). For this reason, below we study the salient features of the transcritical regime which is in a sense similar to the transonic flow regime in conventional gas dynamics.

### 6.2.4 *Properties of transcritical flows corresponding to curve* AB

We will now consider in more detail the properties of transcritical flows corresponding to curve *AB* in Fig. 6.4. In accordance with Eqs. (6.54), the flow in region 3 is described by the inviscid boundary layer equations with

$$\Delta\delta_2 \sim \delta_3 \tag{6.76}$$

In this case, the variable part of the boundary layer displacement thickness $\Delta\delta^*$ can particularly simply be calculated using the unified formula for regions 2 and 3 following from the equation of the gas flow rate:

$$\Delta\delta^* = \int_0^{\delta_0} \left( \frac{\rho_0}{\rho} \frac{u_0}{u} - 1 \right) dn_0 \tag{6.77}$$

where subscript 0 refers to the parameters ahead of the interaction region. Since in both regions under consideration the flows are locally inviscid, using the isentropicity equation and the Bernoulli equation we can obtain the following expression:

$$\Delta\delta^* = \int_0^{\delta_0} \left[ \frac{\left(\frac{p_0}{p}\right)^{1/\gamma}}{\left\{1 + \frac{2}{((\gamma-1)M_0^2)} \left[1 - \left(\frac{p}{p_0}\right)^{(\gamma-1)/\gamma}\right]\right\}^{1/2}} - 1 \right] dy_0 \tag{6.78}$$

Thus, the profile of the original undisturbed boundary layer enters in Eq. (6.78) in the form of the $M_0(y_0)$ dependence, where $y_0$ is the coordinate of the corresponding streamline in the undisturbed boundary layer.

Using notation (6.41) we obtain

$$M_0 = \left[ \frac{2}{(\gamma - 1)} \right]^{1/2} \frac{u_0}{(g - u_0^2)^{1/2}} \tag{6.79}$$

In the regime under consideration $\Delta p/p = O(g_w^{2\omega+1})$ the velocity $u_0(y_0)$ and stagnation enthalpy $g(y_0)$ profiles near the body can be represented in the form:

$$u_0 = \frac{\tau_w \mu_0 u_\infty^2}{2\sigma q_w} \left\{ \left[ g_w^{\omega+1} + 2(\omega+1) \frac{g_w^\sigma y_0}{\mu_0 u_\infty^2} \right]^{1/(\omega+1)} - g_w \right\} + \cdots$$

$$g = \left[ g_w^{\omega+1} + 2(\omega+1) \frac{g_w^\sigma y_0}{\mu_0 u_\infty^2} \right]^{1/(\omega+1)} + \cdots \tag{6.80}$$

where $\tau_w$ and $q_w$ are the viscous stress and the heat flux in the undisturbed boundary layer ahead of the local interaction region and $\sigma$ is the Prandtl number.

Since $p/p_0 = 1 + \Delta p/p_0$, where $\Delta p/p_0 \ll 1$, the second term under the radicand sign in the denominator of formula (6.78) can be expanded in series in $\Delta p/p_0 \ll 1$ everywhere, except for a region, in which $M_0$ is also small. The $M_0 = O(1)$ region is precisely region 2, while the nonlinearity region 3 corresponds to small values of $y_0$ for which the second term is not small.

In sublayer 3 we have

$$u_0 \approx \left(\frac{AB}{2}\right)\left(\frac{y_0}{q_w}\right) + \cdots, \quad g \approx g_w + \cdots$$

$$B = \left(\frac{4q_w}{\mu_0 u_\infty^2}\right) \quad A = \frac{\tau_\infty u_e}{2q_w} \tag{6.81}$$

We introduce an intermediate limit of integration

$$0 < \delta_1(\varepsilon) \ll \delta_0, \quad \varepsilon = -\frac{\Delta p}{p} \tag{6.82}$$

Taking integral (6.78) from 0 to $\delta_1$ and using representations (6.73) to (6.81) we obtain

$$\Delta\delta_3^* = \left(\frac{p_0}{p}\right)^{1/\gamma}\left[\sqrt{\delta_1^2 + \frac{\gamma-1}{\gamma}\frac{4g_w^{1+2w}}{A^2B^2}\varepsilon} - \sqrt{\frac{\gamma-1}{\gamma}\frac{4g_w^{1+2w}}{A^2B^2}\varepsilon}\right] \tag{6.83}$$

In region 2 with "linear disturbances", that is, for $\delta_1(\varepsilon) \le y_0 \le \delta_0$ we obtain

$$\Delta\delta_2^* = \left[\left(\frac{p_0}{p}\right)^{1/\gamma} - 1\right](\delta_0 - \delta_1) - \left(\frac{p_0}{p}\right)^{1/\gamma}\left(\frac{\gamma-1}{2\gamma}\right)\varepsilon\int\limits_{\delta_1(\varepsilon)}^{\delta_0}\frac{g\,\mathrm{d}y_0}{u_0^2} \tag{6.84}$$

It should be remembered that as $\varepsilon \to 0$, we have $\delta_1(\varepsilon) \to 0$ and integral (6.84) is divergent. Therefore, the results should be matched accurately, taking relation (6.69) into account

$$\varepsilon \sim g_w^{2\omega+1}, \quad \varepsilon < \delta_1(\varepsilon) \ll 1 \tag{6.85}$$

Using Eqs. (6.83)–(6.85) in the vicinity of $\varepsilon = 0$ we arrive at the expression

$$\Delta\delta^* = \Delta\delta_2^* + \Delta\delta_3^* = \varepsilon\left[\frac{1}{\gamma}\delta_0^* - \left(\frac{\gamma-1}{\gamma}\right)\delta_0\int\limits_0^1\left(\frac{g}{\overline{u}_0^2} - \frac{4g_w^{1+2w}}{A^2B^2\overline{y}_0^2}\right)\mathrm{d}y_0\right]$$
$$-\sqrt{\frac{\gamma-1}{\gamma}\frac{2\delta_0}{A\overline{B}}}(\varepsilon g_w^{1+2w})^{1/2}$$

$$y_0 = \delta_0\overline{y}_0, \quad \overline{B} = \delta_0 \; B = O(1) \tag{6.86}$$

The structure of the integrand in Eq. (6.86) is such that the integral converges as $y_0 \to 0$, since, by virtue of Eq. (6.41) $\overline{u}_0'(0) = \overline{u}_0''(0) = 0$.

In order to understand the result thus obtained, we note, firstly, that in view of Eq. (6.81), the quantity $A$ is dimensionless but $B$ is dimensional, its dimension being the inverse of length. Secondly, from Eq. (6.86) it follows that $\Delta\delta^*/\delta_0 \sim \varepsilon \sim g_w^3$. The second term on the right-hand side of Eq. (6.86) has appeared in integrating over the nonlinear, always subsonic sublayer. For this reason, for expansion flows, that is, for $\varepsilon > 0$, this term is always negative. Of course, the sign of the first term depends on the form of the Mach number profile in the main part of the boundary layer. However, it is the supersonic part of the profile that is almost always predominant, so that for $g_w \ll 1$ the first term in the brackets is positive. We note that for flow regimes with an inviscid flow in region 3, by virtue of the Bernoulli equation, the case $\varepsilon < 0$ has no physical meaning.

We will use Eqs. (6.43) and (6.86) for obtaining the equation governing the pressure distribution

$$\frac{\mathrm{d}P}{\mathrm{d}\zeta} = \frac{P^{1/2}(P + \widetilde{\theta})}{P_*^{1/2} - P^{1/2}} \tag{6.87}$$

where the following notation is introduced

$$\widetilde{\theta} = \frac{\gamma M_e \theta}{g_w^{1+2w}}, \quad \varepsilon = g_w^{1+2w} P, \quad \zeta = \frac{x}{\gamma M_e \delta_0 T}, \quad P_*^{1/2} = \frac{R}{2T}$$

$$T = \left(\frac{1}{\gamma}\right)\left[1 - \frac{\gamma - 1}{2}\int_0^1 \left(\frac{g}{\overline{u}_0^2} - \frac{4g_w^{1+2w}}{A^2B^2\overline{y}_0^2}\right)\mathrm{d}\overline{y}_0\right], \quad R = \sqrt{\frac{\gamma - 1}{\gamma}\frac{2}{A\overline{B}}} \tag{6.88}$$

For a sufficiently convex profile in the undisturbed boundary layer, when transition to supercritical flow is possible, we have $T > 0$.

### 6.2.5 *Properties of integral curves*

Analyzing Eq. (6.87) we can construct a family of integral curves for the surfaces $\widetilde{\theta}(\zeta)$ for different values of the critical pressure difference $P_*$. We recall that we consider transcritical flows, locally inviscid in region 3, associated with curve $AB$ in Fig. 6.4. Because of this, only the case $P > 0$ has a meaning. In variables (6.88) the dependence of the variable part of the displacement thickness (6.86) takes the form:

$$\Delta\delta^* = \delta_0 g_w^3\left(TP - RP^{1/2}\right), \quad T > 0, \ R > 0 \tag{6.89}$$

Clearly, at very small disturbances it is the second term corresponding to sublayer 3 that is predominant in Eq. (6.89), so that $\Delta\delta^*$ decreases with growing expansion. This is the subsonic, or subcritical type of behavior. For $P > P_*$ $\Delta\delta^*$ starts to increase as in an one-dimensional supersonic flow. The critical point, analogous to that appearing in the solution for a convergent-divergent nozzle for $M = 1$, corresponds to

$$\left(\frac{d\delta^*}{dP}\right) = 0, \quad P = P_* = \left(\frac{R}{2T}\right)^2 \tag{6.90}$$

The explicit dependence of $P_*$ on $g_w$ is eliminated by normalization (6.88), while the dependence on the form of the original undisturbed boundary layer profile is clear from formulas (6.86) and (6.88).

For a qualitative analysis of possible solutions of the problem, we will present, using Eq. (6.87), a diagram of integral curves for body regions $\widetilde{\theta} = \text{const}$ corresponding to different relations between $\widetilde{\theta}$ and $P_*$ (Fig. 6.5).

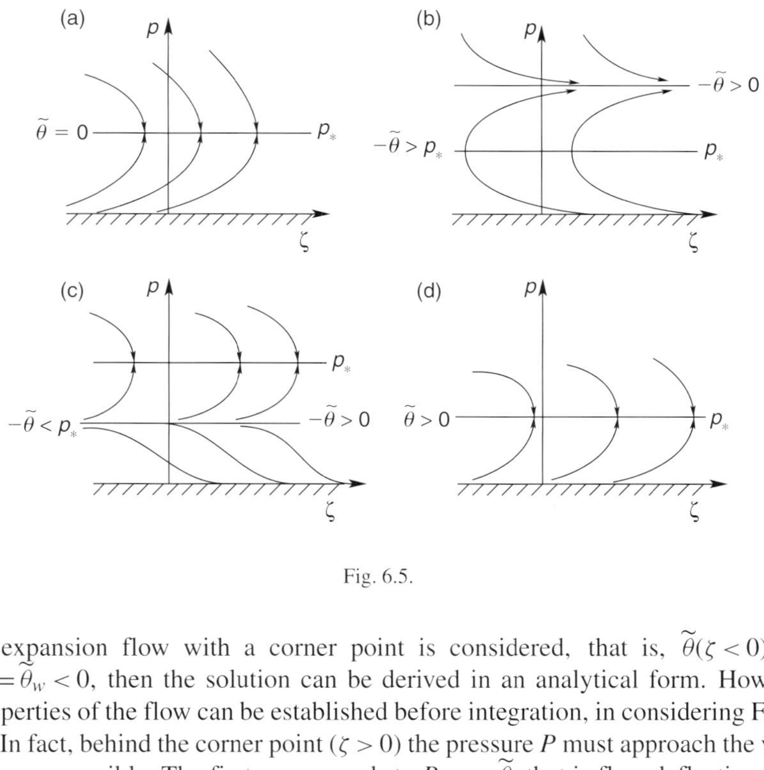


Fig. 6.5.

If an expansion flow with a corner point is considered, that is, $\widetilde{\theta}(\zeta < 0) = 0$ and $\widetilde{\theta}(\zeta > 0) = \widetilde{\theta}_w < 0$, then the solution can be derived in an analytical form. However, the basic properties of the flow can be established before integration, in considering Fig. 6.5, a, b, and c. In fact, behind the corner point ($\zeta > 0$) the pressure $P$ must approach the value $-\widetilde{\theta}$. Two cases are possible. The first corresponds to $P_* > -\widetilde{\theta}$, that is flow deflection by a subcritical angle. As can be seen from Fig. 6.5b, for $\zeta > 0$ there are no integral curves arriving at the line $P = -\widetilde{\theta}$; therefore, the flow turn up to $P = -\widetilde{\theta}$ is realized ahead of the corner point $\zeta = 0$, in the flow region $\zeta < 0$, where $\widetilde{\theta} = 0$, and is described by one of the integral curves in Fig. 6.5a.

The second case is that of supercritical deflection, $P_* > \widetilde{\theta}$. From Fig. 6.5c it is visible that integral curves proceeding from the critical point $P = P_*$ arrive at line $P = -\widetilde{\theta}$, as $\zeta \to +\infty$. Since for $\zeta < 0$ the flow is described by integral curves of Fig. 6.5a, it is the flow in which the "sonic", or critical, point corresponds to $P(0) = P_*$, that is, the critical point is the corner point, that turns out to be uniquely possible.

Thus, in the flow past a body with a deflection angle larger than the critical value $|\widetilde{\theta}| > P_*$, the whole subsonic part of the flow is realized ahead of the corner point, while the whole supersonic part is realized behind it. In particular, at a supercritical original state, in which $R \ll T$, the flow turn is realized almost completely behind the corner point.

The solution of many problems requires the existence of discontinuous solutions of the type of supercritical jump proposed by Crocco (1955). We note that within the framework of the asymptotic theory such solutions are possible. We recall that for the supercritical flows and the transcritical flows under consideration $\Delta x \sim M_e \delta_0$. For this reason, variations occurring on short-scale distances $\Delta x \sim \delta_0$ represent discontinuities on the main integral curves. Of course, the consideration of flow details on distances $\Delta x \sim \delta_0$ with rapid pressure variations is a special problem, which for a while is not considered. We can only claim that on distances $\Delta x \sim \delta_0$ the quantity $\Delta\delta^*$ must not vary in the leading term

$$[\Delta\delta^*] = 0 \tag{6.91}$$

The solution for these regions can be written down; however, the structure of this flow plays no part for the purposes of this study, in the same fashion as the Becker solution (Becker, 1921/22) is unnecessary for conventional gas dynamics. Using Eq. (6.83) we will write down a counterpart of the Hugoniot relation following from Eq. (6.91)

$$P_2^{1/2} = 2P_*^{1/2} - P_1^{1/2} \tag{6.92}$$

where subscripts 1 and 2 refer to the values of the function ahead of and behind the discontinuity provided condition (6.91) is fulfilled. In Fig. 6.6 we have plotted the dependence $\Delta\delta^*(P)$. The flow can accelerate from point $A$ to point $B$ through the critical point $O$ of the continuous expansion flow. The way back from $B$ to $A$ is possible across a shock determined by relation (6.92).

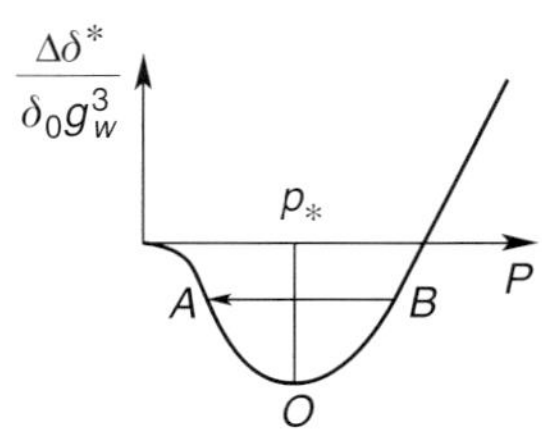


Fig. 6.6.

Using the existence of continuous integral curves and the relation at discontinuity (6.92) (Fig. 6.6), we can construct certain fairly complicated flows possessing characteristic properties of transcriticality, similar to a certain degree to the properties of inviscid transonic flows.

We will consider the flow over an oblique step shown in Fig. 6.7. Let first $|\widetilde{\theta}| < P_*$. The flow region $\zeta > \zeta_B$ corresponds to Fig. 6.5a. Clearly, there are no integral curves at

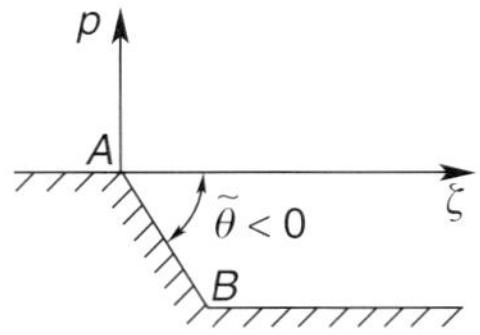


Fig. 6.7.

which $P \to 0$. Then it is necessary that $P(\zeta_B) = 0$. In region $AB$ the flow is described by the integral curves in Fig. 6.5b. All integral curves for $P < -\tilde{\theta}$ are defined correct to a constant displacement along the $\zeta$ axis. Since $P(\zeta_B) = 0$, we can take any of these curves and going along it a distance equal to $\zeta_A - \zeta_B$ in the similarity variables, corresponding to the length of interval $AB$ in Fig. 6.7, obtain the value $P(\zeta_A) < P_*$. It remains to describe in an obvious manner the disturbance decay in the region $\zeta < \zeta_A$ using the integral curves of Fig. 6.5a. Thus, the solution has been completely determined and does not require any discontinuities.

The case $|\tilde{\theta}| > P_*$ is more complicated. As in the former case, $P(\zeta_B) = 0$. The solution in this region is described by Fig. 6.5c. We designate the distance between the points $P = P_*$ and $P = 0$ by $\Delta\zeta_*$. Let in the similarity variables the distance $AB$ be $\Delta\zeta_{AB} \leq \Delta\zeta_*$. In this case, from Fig. 6.5c we determine $P(\zeta_A)$ measuring the distance $\Delta\zeta_{AB}$ from point $P = 0$ along the integral curve. Obviously, this value $P < P_*$. Then using Fig. 6.5a we construct a part of the solution on the interval $\zeta < \zeta_A$. Thus, for a small length $\Delta\zeta_{AB} < \Delta\zeta_*$ the solution can be smooth and subcritical even for a supercritical deflection angle $|\tilde{\theta}| > P_*$.

However, for $\Delta\zeta_{AB} > \Delta\zeta_*$ there is no such possibility. In this case, it is necessary to consider a flow in which $P(\zeta_A) = P_*$, that is to say, the supercritical behavior exhibits itself already in the first region. Having passed along the whole integral curve in Fig. 6.5a, after point $A$ we go upward along the integral curve in Fig. 6.5c; in order to arrive at point $B$ at which $P = 0$ discontinuity (6.92) is required. Since in this case $\Delta\zeta > \Delta\zeta_*$, the position of discontinuity (6.92) can be so chosen that we arrive at such a point of the subcritical integral curve that the pressure on the remaining part of the curve up to point $B$ is zero. Obviously, it is always possible, since in going along the integral curve $P_* < P < -\tilde{\theta}$ in Fig. 6.5c any distance can be passed. The curve reaches the $P = -\tilde{\theta}$ line only asymptotically, as $\zeta \to \infty$. Thus, an interesting result is obtained. If the flow deflection angle is supercritical, $-\tilde{\theta} > P_*$, then the flow becomes supercritical only at a fairly long distance $AB$ and is decelerated across discontinuity (6.92). However, if the length $AB$ in the similarity variables is less than the critical value $\Delta\zeta_{AB} < \Delta\zeta_*$, then the flow remains smooth and subcritical everywhere.

In conclusion, we will consider the flow past the "airfoil" presented in Fig. 6.8. In this case, we have again $P(\zeta_D) = 0$. If $-\tilde{\theta} < P_*$, then the construction of the solution presents no difficulty. We measure the pressure at a distance $\Delta\zeta = \Delta\zeta_{CD}$ from the point $P = 0$ at the integral curve in Fig. 6.5b, thus determining $P(\zeta_C)$. Then, $\Delta\zeta_{BC}$ having been known, we determine $P(\zeta_B)$ in Fig. 6.5a. Then, using Fig. 6.5d and $\Delta\zeta_{AB}$ having been known, we determine $P(\zeta_A)$ and, finally, using Fig. 6.5a determine the solution on the interval $\zeta < \zeta_A$.

Let now there be $-\tilde{\theta} < P_*$. As before, $P(\zeta_D) = 0$. Region $CD$ is described by the integral curves in Fig. 6.5c. Two cases are possible. The simpler case corresponds to $\Delta\zeta_{CD} < \Delta\zeta_*$. Then from Fig. 6.5c we determine $P(\zeta_C) < P_*$. Region $BC$ corresponds to Fig. 6.5a; $\Delta\zeta_{BC}$

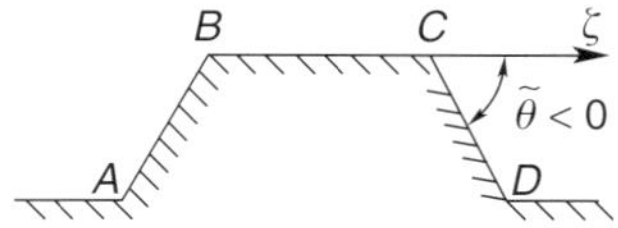


Fig. 6.8.

having been known, we determine $P(\zeta_B) < P_*$. Then using Fig. 6.5c we determine $P(\zeta_A)$ and from Fig. 6.5a the pressure distribution upstream of point $A$. Thus, if the lengths of the inclined intervals $AB$ and $CD$ in the similarity variables are shorter than the critical value $\Delta\zeta_*$, then the flow remains subcritical even for $-\widetilde{\theta} < P_*$.

The case in which $-\widetilde{\theta} > P_*$ and $\Delta\zeta_{CD} > \Delta\zeta_*$ (Fig. 6.5a) is more complicated. Here, we have again $P(\zeta_D) = 0$; according to Fig. 6.5c, in view of the inequality $\Delta\zeta_{CD} > \Delta\zeta_*$, discontinuity (6.92) must exist on interval $CD$. We designate the discontinuity coordinate by $\zeta_1$ and the pressure ahead of it by $P_1$; then the pressure behind the discontinuity $P_2(P_1)$ can be determined from Eq. (6.92). Moreover, according to Fig. 6.5c, there exists a relation between $P_2(\zeta_D - \zeta_1)$ and $P_1(\zeta_1 - \zeta_C)$, since at point $C$ we have necessarily $P = P_*$. Thus, we obtain three relations for determining three unknown quantities $P_1$, $P_2$, and $\zeta_2$; for $\Delta\zeta_{CD} > \Delta\zeta_*$ these equations are always resolvable. Then for a given length $\Delta\zeta_{BC}$ the flow in region $BC$ can be calculated from the integral curves in Fig. 6.5a, thus determining $P(B) < P_*$. Then at a known length $\Delta\zeta_{AB}$ from Fig. 6.5d we determine $P(\zeta_A)$ and from Fig. 6.5a an integral curve describing the disturbance decay upstream of point $A$.

Now the flow has been completely described and it turns out again that for deflection angles $-\widetilde{\theta} > P_*$ a supercritical flow occurs only at fairly long acceleration regions and is terminated by discontinuities of type (6.92) in which the flow decelerates. A detailed study of the flow structure inside the discontinuity region (6.92) at lengths $\Delta x \sim \delta_0$ is of great interest, since in the region $\Delta x \sim \delta_0$, in which discontinuity (6.92) is localized, thin viscous sublayer formed at the bottom of locally inviscid flow in region 3 can be locally separated; local separation of this sublayer can also occur near points $B$ (Fig. 6.7) and $D$ (Fig. 6.8).

## 6.3 Study of time-dependent processes of transcritical interaction between the laminar boundary layer and a hypersonic flow

In this section the theory of the viscous–inviscid interaction in the transcritical regime presented in the previous section is generalized to the case of time-dependent flows.

An asymptotic analysis on the basis of the matching asymptotic expansion method is applied to derive a fundamental equation of hyperbolic type which reduces in the limiting case to the Burgers equation.

### *6.3.1 Estimates of the scales*

We will consider the flow over a flat plate with a region of time-dependent disturbances located at a distance $\ell$ from the leading edge; the disturbances can be due to the incidence of a shock on the laminar boundary layer, base pressure fluctuations, or other reasons. The following notation is used for Cartesian coordinates measured along the body surface and normal to it, time, the velocity components, the density, the pressure, the total enthalpy, and dynamic viscosity:

$$\ell x,\ \ell y,\ \frac{\ell t}{u_e},\ u_e u,\ u_e v,\ \rho_0\rho,\ \rho_0 u_e^2 p,\ \frac{u_e^2 H}{2},\ \mu_0\mu$$

Here, the subscript $e$ refers to the parameters of the undisturbed inviscid flow above the boundary layer, while the subscript 0 refers to the dimensional values of the density and dynamic viscosity calculated at the stagnation temperature and the static pressure in the inviscid flow. It is assumed that the perfect gas is characterized by a constant specific heat ratio $\gamma$, while the Prandtl number $\sigma = 1$. The flow pattern studied below is characterized by small spatial scales, as compared with the body length. It is also assumed that the Reynolds number is high but not higher than the critical value at which laminar–turbulent transition occurs. As is known (Chapman et al., 1958), for supersonic flows the transition Reynolds numbers are high which justifies the above assumption.

Let us estimate the effect of a certain external action characterized by a pressure disturbance amplitude $\Delta p$ on the flow. We assume that $\Delta p$ is small, whereas the pressure gradient induced in the vicinity of the disturbance source is large. This infers that the interaction region is local, or the extent of the disturbed region is much smaller than the body scale-length. It is assumed that on the main body length the weak hypersonic interaction regime (Hayes and Probstein, 1966) is realized. It is characterized by the following values of the relevant parameters:

$$\chi = M_e\tau \ll 1,\quad M_e \to \infty,\ \tau \to 0 \tag{6.93}$$

$$\tau = O(Re_0^{-1/2}),\quad Re = \frac{\rho_0 u_e \ell}{\mu_0}$$

where $\tau$ is the dimensionless boundary layer thickness near the interaction region; it is also assumed that viscosity is linearly dependent on the temperature. The following limiting relations are assumed to hold:

$$M_\infty\tau \to 0,\quad M_\infty \to \infty,\ \tau \to 0$$

$$g_w \to 0,\quad \Delta p \sim O(g_w^3 M_\infty^{-2}) \tag{6.94}$$

As shown in the previous section, the estimates following from Eq. (6.94) lead to the following structure of the disturbed flow regions: the outer inviscid slightly disturbed flow (region 1), the boundary layer flow (region 2), and the nonlinearly disturbed wall flow (region 3). Thus, the disturbed flows in regions 1 and 2 are in the first approximation inviscid and weakly disturbed, respectively, while typical of region 3 are nonlinear variations of the flow functions. The viscosity effect is absent from region 3; therefore, in this region the solution does not satisfy the no-slip condition. For this reason, a boundary layer should be introduced at the bottom of region 3. The flow in this region can have an effect on the flow in regions 1 to 3 only if local separation occurs. In what follows, we consider only such situations or such time intervals in which the flow in the wall boundary layer remains separationless.

The characteristic longitudinal extent of all the regions is equal in the order to $O(M_e\tau)$ (Fig. 6.4). The transverse dimensions of the regions are different. Region 1 has the largest transverse dimension: for a given length its thickness is determined by the inclination of the characteristics in the outer inviscid flow $O(M_e^{-1})$ and is equal to $O(\tau)$. The transverse dimension of region 1 is in the order the same as that of region 2 (boundary layer thickness). At the same time, different gas flow rates are typical of these regions, since, though the thicknesses and velocities in these regions are of the same order, their densities are asymptotically different.

For each region the time scale is determined either from the condition that the corresponding convective derivatives are of the same order or the ratio of the scalelength to the scale velocity. The lengths of all the regions being the same, the least longitudinal velocity is characteristic of region 3, so that the characteristic time is largest for this region. Using the estimates obtained in the previous section we can derive an estimate for the characteristic time in the wall region: $\Delta t \sim M_e g_w^{-2}$. Below we consider time-dependent flows characterized by time $\Delta t \sim M_e g_w^{-2}$, that is, a time-dependent flow in region 3 and quasi-stationary flows in regions 1 and 2.

### 6.3.2 Formulation and solution of the boundary value problem

The results obtained in Section 6.2 show that in analyzing the flow in wall regions 2 and 3 we can proceed, at least, for first approximations, from the system of boundary layer equations rather than from the Navier–Stokes system. The former system governs the flow over the entire plate surface up to point $O$ and is as follows:

$$\frac{\partial \rho}{\partial t} + \frac{\partial(\rho u)}{\partial x} + \frac{\partial(\rho v)}{\partial y_2} = 0, \qquad \frac{\partial p}{\partial y_2} = 0$$

$$\rho\frac{\partial u}{\partial t} + \rho u\frac{\partial u}{\partial x} + \rho v_2\frac{\partial u}{\partial y_2} + \frac{\partial p}{\partial x} = \frac{\partial}{\partial y_2}\left(\mu\frac{\partial u}{\partial y_2}\right) \tag{6.95}$$

$$\rho\frac{\partial g}{\partial t} + \rho u\frac{\partial g}{\partial x} + \rho v_2\frac{\partial g}{\partial y_2} = \frac{\partial}{\partial y_2}\left(\mu\frac{\partial g}{\partial y_2}\right) + 2\frac{\partial p}{\partial t}$$

The boundary conditions are as follows:

$$u(x,0) = v_2(x,0) = 0, \quad g(x,0) = g_w$$

$$u(x,\infty) = g(x,\infty) = 1 \tag{6.96}$$

$$y = y_2 Re^{1/2}, \quad v = v_2 Re^{1/2}, \quad Re = \frac{\rho_0 u_e \ell}{\mu_0}$$

For further analysis it is convenient to use the generalization of the Dorodnitsyn–Lees variables to the time-dependent case

$$x, y_2, t \to x, \quad \lambda = \int\limits_0^{y_2} \rho \, \mathrm{d}y_2, \quad u = \frac{\partial f}{\partial \lambda}, \quad g = H$$

$$\rho = \frac{2\gamma p}{(\gamma - 1)(g - f'^2)}, \quad f' = \frac{\partial f}{\partial \lambda}, \quad \dot{f} = \frac{\partial f}{\partial x} \tag{6.97}$$

Then the continuity equation takes the form:

$$\left.\frac{\partial}{\partial t}\right|_{x,y_2} \int\limits_0^{y_2} \rho \, \mathrm{d}y_2 + \left.\frac{\partial f}{\partial x}\right|_{x,y_2} + \rho v = 0$$

or

$$\left.\frac{\partial \lambda}{\partial t}\right|_{x,y_2} + \left.\frac{\partial f}{\partial x}\right|_{x,y_2} + \rho v = 0$$

$$\left.\frac{\partial \lambda}{\partial t}\right|_{x,y_2} + \left.\frac{\partial f}{\partial x}\right|_{x,\lambda} + u\left.\frac{\partial \lambda}{\partial x}\right|_{x,y_2} + \rho v = 0$$

Hence follows

$$-\left.\frac{\partial f}{\partial x}\right|_{x,\lambda} = -\rho v - u\left.\frac{\partial \lambda}{\partial x}\right|_{x,y_2} - \left.\frac{\partial \lambda}{\partial t}\right|_{x,y_2} = v^*$$

and

$$\frac{\partial u}{\partial x} + \frac{\partial v^*}{\partial x} = 0$$

After the change of variables the boundary value problems (6.95) and (6.96) takes the form:

$$\frac{\partial^2 f}{\partial \lambda \partial t} + \frac{\partial f}{\partial \lambda}\frac{\partial^2 f}{\partial x \partial \lambda} - \frac{\partial f}{\partial x}\frac{\partial^2 f}{\partial \lambda^2} + \frac{1}{\rho}\frac{\partial p}{\partial x} = \frac{\partial^3 f}{\partial \lambda^3} \tag{6.98}$$

$$\frac{\partial g}{\partial t} + \frac{\partial f}{\partial \lambda}\frac{\partial g}{\partial x} - \frac{\partial f}{\partial x}\frac{\partial g}{\partial \lambda} = \frac{\partial^2 g}{\partial \lambda^2} + \frac{2}{\rho}\frac{\partial p}{\partial t}$$

$$\frac{\partial f}{\partial \lambda} = f = 0, \quad g = g_w, \ \lambda = 0$$

$$\frac{\partial f}{\partial \lambda} = g = 1, \ \lambda = \infty \tag{6.99}$$

The system is valid provided the viscosity–temperature dependence is linear and the Prandtl number is unity.

The flat-plate boundary layer flow upstream of the local region of influence of the disturbances is governed by the self-similar solution of problem (6.98)

$$f = f_0(\eta)x^{1/2}, \quad g = g_0(\eta), \quad \eta = \lambda x^{1/2} \tag{6.100}$$

$$2f_0''' + f_0 f_0'' = 0, \quad g_0'' + f_0 g_0' = 0$$

$$f_0(0) = f_0'(0) = g_0(0) = 0, \quad f_0'(\infty) = g_0(\infty)$$

The solution of problem (6.100) is taken below as the initial conditions for solutions in regions 2 and 3.

In region 3 the following variables are introduced

$$(x, \lambda, t) = (1 + M_e \tau x_3, \ g_w^2 \lambda_3, \ M_e \tau g_w^{-2} t_3) \tag{6.101}$$

$$(f, g, p, \rho) = (g_w^4 f_3, \ g_w, \ \gamma_1 \gamma^{-1} + g_w^3 \gamma_1 p_3, \ g_w^{-1}) + \cdots$$

Substituting expressions (6.101) in the boundary value problems (6.95) and (6.96) and passing to limit (6.94) yields the following result for the first-approximation functions

$$\frac{\partial^2 f_3}{\partial \lambda_3 \partial t_3} + \frac{\partial f_3}{\partial \lambda_3}\frac{\partial^2 f_3}{\partial x_3 \partial \lambda_3} - \frac{\partial f_3}{\partial x_3}\frac{\partial^2 f_3}{\partial \lambda_3^2} + \frac{\partial p_3}{\partial x_3} = 0 \tag{6.102}$$

The solution of problem (6.102) must satisfy the impermeability condition

$$f_3(x_3, 0, t_3) = 0 \tag{6.103}$$

and the condition of matching with the solution for the wall region of the undisturbed boundary layer (6.100) as $\lambda \to 0$

$$f_3'(x_3 \to -\infty, \lambda) = a\lambda_3 \tag{6.104}$$

The solution of problems (6.102)–(6.104) can be sought in the form:

$$f_3 = 2^{-1} a \lambda_3^2 + A_3(x_3, t_3)\lambda_3 \tag{6.105}$$

which follows from the condition of constant vorticity conservation (6.104) in the two-dimensional inviscid flow under consideration and the possibility of representing the longitudinal velocity in the form of a sum of two terms. One term is due to vorticity of the undisturbed oncoming wall flow (6.104) while the second represents the velocity at the surface.

For the function $A_3(x_3, t_3)$ we obtain the following equation:

$$\frac{\partial A_3}{\partial t_3} + A_3 \frac{\partial A_3}{\partial x_3} + \gamma_1 \frac{\partial p_3}{\partial x_3} = 0, \quad \gamma_1 = \frac{\gamma - 1}{2} \tag{6.106}$$

From formula (6.105) it follows that the asymptotic expansion of the function $f_3(x_3, \lambda_3, t_3)$ as $\lambda_3 \to \infty$ takes the form:

$$f_3 = \frac{a\lambda_3^2}{2} + A_3\lambda_3 + O(1) \tag{6.107}$$

This expansion will be used for deriving the solution in region 2.

For region 2 the asymptotic representations are as follows:

$$(x, \lambda, t) = (1 + M_\infty \tau_1 x_3, \lambda_2, g_w^{-2} M_\infty \tau_1 t_3) \tag{6.108}$$

$$\begin{aligned} f(x, \lambda, t) &= f_0(\lambda_2) + g_w^3 f_2(x_2, \lambda_2, t_2) + \cdots \\ g(x, \lambda, t) &= g_0(\lambda_2) + g_w^3 g_2(x_2, \lambda_2, t_2) + \cdots \\ p(x, \lambda, t) &= \frac{(\gamma - 1)}{2} + \frac{(\gamma - 1)}{2} g_w^3 p_2(x_2, t_2) + \cdots \end{aligned}$$

Substituting expressions (6.108) in the original boundary value problem (6.98) leads to the following problem for region 2:

$$\frac{\partial f_0}{\partial \lambda_2}\frac{\partial^2 f_2}{\partial x_2 \partial \lambda_2} - \frac{\partial f_2}{\partial x_2}\frac{\partial^2 f_0}{\partial \lambda_2^2} + \frac{(\gamma - 1)}{2}\left[g_0 - \left(\frac{\partial f_f}{\partial \lambda_2}\right)^2\right]\frac{\partial p_2}{\partial x_2} = 0 \tag{6.109}$$

$$\frac{\partial f_0}{\partial \lambda_2}\frac{\partial g_2}{\partial x_2} - \frac{\partial f_2}{\partial x_2}\frac{\partial g_0}{\partial \lambda_2} = 0$$

The solution of problem (6.109) can be sought in the form:

$$f_2 = \overline{f}_2(\lambda_2, t_2) p_2(x_2, t_2), \quad g_2 = \overline{g}_2(\lambda_2, t_2) p_2(x_2, t_2) \tag{6.110}$$

For the functions $\overline{f}_2$ and $\overline{g}_2$ we obtain

$$\begin{aligned} f_0' \overline{f}_2' - \overline{f}_2 f_0'' + \gamma_1 (g_0 - f_0'^2) &= 0 \\ f_0' \overline{g}_2 - \overline{f}_2 g_0' &= 0 \end{aligned} \tag{6.111}$$

The solution of system (6.111) satisfying the impermeability condition at the surface is as follows:

$$f_2 = p_2 f_0' \left[\gamma_1 \int_0^{\lambda_2} (g_0 - f_0'^2) f_0'^{-2} \, d\lambda_2 + A_3 g_w^2 \alpha^{-2}\right] \tag{6.112}$$

$$g_2 = p_2 f_{20} g_0' f_0'^{-1}$$

The determination of the function $C_1(x_2, t_2)$ requires matching solutions (6.112) and (6.105) hence follows:

$$C_1(x_2, t_2) = -\frac{A_3(x_3, t_3)g_w^2}{a} \tag{6.113}$$

In region 1 the representations of the coordinates and flow functions are as follows:

$$(x, y, t) = (1 + M_e\tau x_1,\ \tau y_1,\ M_e\tau g_w^{-2}t_1) \tag{6.114}$$

$$\begin{aligned}
u(x, y, t) &= 1 + 2g_w^3 M_e^{-2}\gamma_1^{-1}u_1(x_1, y_1, t_1) + \cdots \\
v(x, y, t) &= 2g_w^3 M_e^{-2}\gamma_1^{-1}v_1(x_1, y_1, t_1) + \cdots \\
p(x, y, t) &= \frac{\gamma - 1}{2\gamma} + g_w^3 p_1(x_1, y_1, t_1) + \cdots \\
\rho(x, y, t) &= \frac{\gamma - 1}{2}M_e^2 + g_w^3\rho_1(x_1, y_1, t_1) + \cdots
\end{aligned}$$

Substituting these expressions in the system of Navier–Stokes equations yields

$$\begin{aligned}
\frac{\partial\rho_1}{\partial x_1} + \frac{\partial v_1}{\partial y_1} + \frac{\partial u_1}{\partial x_1} &= 0 \\
\frac{\partial v_1}{\partial x_1} + M_e^2\frac{\partial p_1}{\partial y_1} &= 0 \\
M_e^2\frac{\partial p_1}{\partial x_1} - \frac{\partial\rho_1}{\partial x_1} &= 0 \\
\frac{\partial u_1}{\partial x_1} + \frac{\partial p_1}{\partial x_1} &= 0
\end{aligned} \tag{6.115}$$

From system (6.115), which can be reduced to the wave equation, the Ackeret formula relating the boundary layer displacement thickness and the induced pressure disturbance can be derived

$$p_1(x_1, 0, t_1) = M_e^{-2}v_{1w}(x_1, 0, t_1) \tag{6.116}$$

$$v_1(x_1, 0, t_1) = \frac{\mathrm{d}}{\mathrm{d}x_1}(\delta_2 + \Delta_w g_w^{-3}\tau^{-1})$$

where $\Delta_w$ is the dimensionless variation of the thickness of a local roughness which can be a source of disturbances.

The expression for the variation of the boundary layer thickness $\delta_2(x_2,\ t_2)$ can be obtained using formulas (6.97) and (6.107)

$$\delta_2(x_1, t_1) = \int_0^\infty (g_2 - 2f_0' f_2')\,\mathrm{d}\lambda_2 \tag{6.117}$$

Using expressions (6.112) for the functions $f_2'$ and $g_2$ formula (6.117) can be brought into the form:

$$\delta_2 = p_2\left[\frac{(\gamma-1)}{2}J_1 - J_2\right] - \frac{A_3}{a} = \overline{L}p_2 - \frac{A_3}{a} \tag{6.118}$$

where

$$J_1 = \int_0^\infty \frac{(g_0 - f_0'^2)^2}{f_0'^2}\,\mathrm{d}\lambda_2, \quad J_2 = \int_0^\infty (g_0 - f_0'^2)\,\mathrm{d}\lambda_2, \quad \overline{L} = \frac{\gamma-1}{2}y_1 - y_2 \tag{6.119}$$

Then the system of equations governing the disturbed flow in region 3 takes the form following from formulas (6.106), (6.116), and (6.118):

$$\frac{\partial A_3}{\partial t_3} + A_3\frac{\partial A_3}{\partial x_3} + \frac{(\gamma-1)}{2}\frac{\partial p_3}{\partial x_3} = 0 \tag{6.120}$$

$$\frac{2}{(\gamma-1)}p_3 = \left[\frac{(\gamma-1)}{2}J_1 - J_2\right]\frac{\partial p_3}{\partial x_3} - \frac{1}{a}\frac{\partial A_3}{\partial x_3} + \frac{\partial \Delta_w}{\partial x_3}$$

The change of variables

$$A = A_3\frac{4a^2}{(\gamma-1)^2}, \quad P = p_3\frac{8a^2}{(\gamma-1)^3}, \quad T = t_3\frac{(\gamma-1)^2}{4a}$$

$$\overline{\Delta}_w = D\frac{(\gamma-1)^2}{4a^2}, \quad \overline{L} = L\frac{(\gamma-1)}{2}$$

brings the system of equations (6.120) into the following form:

$$\frac{\partial A}{\partial T} + A\frac{\partial A}{\partial x} + \frac{\partial P}{\partial x} = 0 \tag{6.121}$$

$$P = L\frac{\partial P}{\partial x} - \frac{\partial A}{\partial x} + \frac{\partial D}{\partial x}$$

For $L = 0$ the system of equations (6.121) reduces to the Burgers equation and describes the nonlinear regime of the interaction between the boundary layer and the outer hypersonic flow at the temperature factor values larger than the critical value $g_w \sim O(\Delta p^{1/3}M_e^{2/3})$. In the general case, for $L \neq 0$ the system of equations (6.121) reduces to an equation of the hyperbolic type, as distinct from the parabolic Burgers equation.

The system of equations (6.121) can be brought into the form:

$$\frac{\partial A}{\partial T} + \left(A + \frac{1}{L}\right)\frac{\partial A}{\partial x} = -P + \frac{\Theta}{L} \tag{6.122}$$

$$\frac{\partial}{\partial x}(A - LP) = -P + \Theta, \quad \Theta = \frac{\partial D}{\partial x}$$

The characteristics of system (6.122) are described by the following equations:

$$\frac{\mathrm{d}\tau_1}{\mathrm{d}x} = 0, \quad \frac{\mathrm{d}\tau_2}{\mathrm{d}x} = -\left(A + \frac{1}{L}\right)^{-1} \tag{6.123}$$

The system of equations (6.123) has discontinuous solutions which appear as a result of the intersection of characteristics of the same family. For determining the conditions satisfied by the solution at a discontinuity line, the Green formula should be applied and a contour adjoining this line and enclosing a certain interval of it should be considered. From the Green formula there follows the expression for the velocity of the discontinuity displacement:

$$\frac{\mathrm{d}x}{\mathrm{d}T} = \frac{(A_1 + A_2)}{2} + \frac{1}{L} \tag{6.124}$$

where $A_1$ and $A_2$ are the values of a function to the left and to the right of the discontinuity, respectively. The condition of stationarity of discontinuity $\mathrm{d}x/\mathrm{d}T = 0$ reduces to that derived by Neiland (1987).

The system of equations (6.121) was numerically integrated. We studied the flow over a body with a bend in the contour generator, whose configuration at the initial moment $T = 0$ was specified by the formulas

$$\Theta(x, 0) = -2^{-1}H(x), \quad \Theta(x < 1, T > 0) = H(x)$$

$$\Theta(x > 1,\ T > 0) = -\exp(-T), \quad L = -1$$

where $H(x)$ is the Heaviside function. In the process of solution an artificial viscosity of the form $\varepsilon(\partial^2 A/\partial x^2)$ was introduced into the system of equations. The calculated results are presented in Fig. 6.9, in which the pressure distribution is plotted at different moments. The dashed curve corresponds to the stationary solution of the problem.

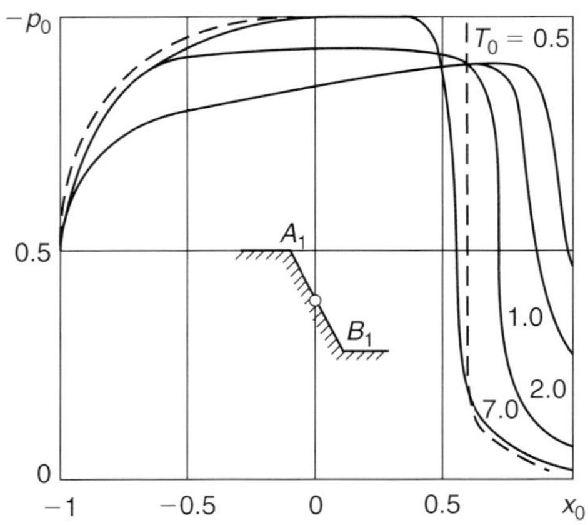


Fig. 6.9.

### *6.3.3 Conclusion*

Above it is shown that transition from the supercritical to the subcritical interaction regime occurs with the formation of a discontinuity in the distribution of the pressure and other flow functions. Therefore, an analogy between one-dimensional inviscid flows (subsonic and supersonic) and boundary layer flows (subcritical and supercritical) is maintained. At the same time, there is a considerable difference between these flows. The processes in a gas dynamic shock are governed by kinetic equations, while the shock thickness is about a few free path lengths. In the problem under consideration, the characteristic longitudinal dimension is equal in the order to the boundary layer thickness ($\Delta x = O(\tau)$) which is considerably larger than the shock thickness.

Results important for understanding these phenomena can be obtained on the basis of the study of the transcritical interaction regime, for which the role played by viscosity forces in the nonlinear disturbance region is important. In particular, transcritical transition in the boundary layer can have an effect on the local and total aerodynamic characteristics of hypersonic flight vehicles.

## 6.4 Analysis of the boundary layer flow near the trailing edge of a flat plate and in its wake in the strong hypersonic interaction regime

In this section we analyze the flow patterns in the region adjoining the traling edge of a flat plate in a hypersonic outer flow. Possible generalizations of the results obtained to other problems are also considered.

### *6.4.1 Formulation of the problem*

We will consider the hypersonic viscous flow past a flat plate. The plate of length $\ell$ is set at zero incidence (Fig. 6.10). It is assumed that the flow is laminar both near the plate and in the wake region. The notation and the choice of dimensionless variables are the same as in Section 4.2, in which the strong hypersonic interaction regime was studied. This flow regime is governed by the following system of equations describing the laminar boundary layer flow

$$\rho u \frac{\partial u}{\partial x} + \rho v \frac{\partial u}{\partial y} + \frac{\mathrm{d}p}{\mathrm{d}x} = \frac{\partial}{\partial y}\left(\mu \frac{\partial u}{\partial y}\right), \quad \frac{\partial p}{\partial y} = 0 \tag{6.125}$$

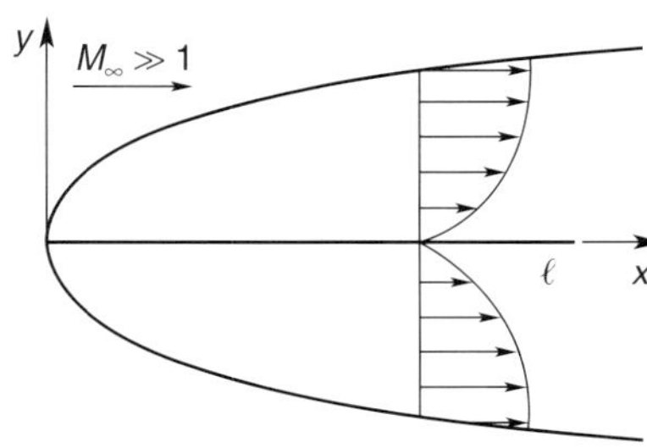


Fig. 6.10.

$$\rho u \frac{\partial g}{\partial x} + \rho v \frac{\partial g}{\partial y} = \frac{\partial}{\partial y}\left(\frac{\mu}{\sigma}\frac{\partial g}{\partial y}\right) + \frac{\partial}{\partial y}\left[\mu\left(1 - \frac{1}{\sigma}\right)\frac{\partial u^2}{\partial y}\right]$$

$$\frac{\partial \rho u}{\partial x} + \frac{\partial \rho v}{\partial y} = 0, \;\; p = \frac{(\gamma - 1)}{2\gamma}\rho(g - u^2)$$

where $\sigma$ is the Prandtl number. The boundary conditions are as follows:

$$u = v = 0, \quad g = g_w, \quad 0 < x \leq 1, \quad y = 0 \tag{6.126}$$

$$\frac{\partial u}{\partial y} = v = \frac{\partial g}{\partial y} = 0, \quad x > 1, \quad y = 0$$

$$u = g = 1, \quad y = \delta$$

The pressure distribution entering in the system of equations is unknown beforehand and must be determined in the process of solving the problem. The inviscid flow in the region between the shock and the outer edge of the boundary layer is described by hypersonic small perturbation theory. For determining the relation between the boundary layer thickness $\delta$ and the pressure disturbance $p$ the approximate equation given by the tangent wedge method can be used

$$p = \frac{(\gamma + 1)}{2}\left(\frac{\mathrm{d}\delta}{\mathrm{d}x}\right)^2 + O(\chi^{-1}) \tag{6.127}$$

For further analysis it is convenient to go over to the Dorodnitsyn–Lees variables

$$\eta = x^{-1/4}\int_0^y \rho\, \mathrm{d}y, \quad \delta = x^{1/4}\int_0^\infty \frac{\mathrm{d}\eta}{\rho}$$

$$u = \frac{\partial f}{\partial \eta}, \quad R = x^{1/2}\rho, \quad P = x^{1/2}p, \quad \delta = x^{3/4}\Delta$$

in which the boundary value problems (6.125)–(6.127) takes the form:

$$(Nf'')' + \frac{1}{4}ff'' + \frac{(\gamma - 1)}{4\gamma}\left(1 - \frac{2x\dot{P}}{P}\right)(g - f'^2) = x(f'\dot{f}' - \dot{f}f'')$$

$$\left(\frac{N}{\sigma}g'\right)' + \frac{1}{4}fg' + \left[N\left(1 - \frac{1}{\sigma}\right)(f'^2)'\right]' = x(f'\dot{g} - \dot{f}g')$$

$$N = \frac{2\gamma}{\gamma - 1}P(g - f'^2)^{n-1}, \quad \Delta = \frac{(\gamma - 1)}{2\gamma P}\int_0^\infty (g - f'^2)\,\mathrm{d}\eta \tag{6.128}$$

$$P = \frac{(\gamma + 1)}{2}\left[\frac{3}{4}\Delta + x\dot{\Delta}\right]^2$$

$$f = f' = 0, \quad g = g_w, \quad 0 < x \leq 1, \quad \eta = 0$$

$$f'' = f = g' = 0, \quad x > 1, \quad \eta = 0$$

$$f' = g = 1, \quad \eta \to \infty$$

In deriving Eq. (6.128) the power-law viscosity–temperature dependence was assumed

$$\mu = C_\mu T^n \tag{6.129}$$

### 6.4.2 *Investigation of the plate wake flow in the vicinity of the point of subcritical-to-supercritical transition*

Due to the replacement of the no-slip condition at the plate surface by the symmetry condition in the wake and owing to momentum transfer, near the wake axis the longitudinal velocity increases downstream and at a certain distance from the trailing edge becomes comparable with the velocity of the outer inviscid flow. Thus, the problem under consideration always involves a point in the wake $x_0 = 1 + \xi_0$ corresponding to transition from the subcritical to the supercritical interaction regime, if only the flow ahead of $x_0 = 1$ is subcritical (the coordinate $x$ is measured from the leading edge of the plate and the coordinate $\xi$ from its trailing edge, Fig. 6.10). We will also introduce a coordinate $s$ measured from the point of transition from the subcritical to the supercritical interaction regime.

Near point $x_0$ the solution for $\dot{t}(s) \gg 1$ can be represented in the series-form:

$$\begin{aligned} f &= F_0(\eta) + t(s)F_\alpha(\eta) + sF_1 + \cdots \\ g &= G_0(\eta) + t(s)G_\alpha(\eta) + sG_1 + \cdots \\ P &= P_0 + t(s)P_\alpha + sP_1 + \cdots \\ \Delta &= D_0 + t(s)D_\alpha + sD_1 + \cdots \end{aligned} \tag{6.130}$$

where $t(s) = O(1)$, $s = O(1)$, $s = x_0 - x$, $F_0 = f(x_0, \eta)$, $G_0 = g(x_0, \eta)$, $D_0 = \Delta(x_0)$, and $P_0 = P(x_0)$.

Substituting expansions (6.130) in the boundary value problem (6.128) leads to the following problem for the functions $F_\alpha$ and $G_\alpha$:

$$-\frac{(\gamma - 1)}{2\gamma}\frac{P_\alpha}{P_0}(G_0 - F_0'^2) = (F_0'F_\alpha' - F_\alpha F_0'') \tag{6.131}$$

$$F_0'G_\alpha - F_\alpha G_0' = 0, \quad F_\alpha(0) = 0$$

Solution (6.131) can be written in explicit form:

$$F_\alpha = -\frac{(\gamma - 1)}{2\gamma}\frac{P_\alpha}{P_0}F_0' \int_0^\eta \frac{(G_0 - F_0'^2)}{F_0'^2}\,\mathrm{d}\eta \tag{6.132}$$

$$G_\alpha = -\frac{(\gamma - 1)}{2\gamma}\frac{P_\alpha}{P_0}G_0' \int_0^\eta \frac{(G_0 - F_0'^2)}{F_0'^2}\,\mathrm{d}\eta$$

The variation of the displacement thickness corresponding to this approximation can be written in the form:

$$D_\alpha = \frac{(\gamma - 1)}{2\gamma P_0}\left[\int_0^\infty (G_\alpha - 2F_0'F_\alpha')\,\mathrm{d}\eta - \frac{P_\alpha}{P_0}\int_0^\infty (G_0 - F_0'^2)\,\mathrm{d}\eta\right] \tag{6.133}$$

In view of solutions (6.132), formula (6.133) can be brought into the form:

$$D_\alpha = \frac{(\gamma - 1)}{2\gamma^2 P_0}\frac{P_\alpha}{P_0}L, \quad L = \frac{(\gamma - 1)}{2}J_\alpha - J_0 \tag{6.134}$$

where

$$J_\alpha = \int_0^\eta \frac{(G_0 - F_0'^2)^2}{F_0'^2}\mathrm{d}\eta \quad \text{and} \quad J_0 = \int_0^\eta (G_0 - F_0'^2)\mathrm{d}\eta$$

An analysis of the relation between the displacement thickness and the pressure disturbance

$$P_\alpha = -\frac{3(\gamma + 1)}{4}D_0 x_0 \frac{\dot{t}}{t} D_\alpha$$

shows that solution (6.130) exists if

$$D_\alpha = 0$$

or

$$\frac{(\gamma - 1)}{2}J_\alpha - J_0 = 0$$

This relation is reduced to that derived in the paper of Neiland (1974*b*) for determining the choking position in the wake. Formula (6.134) shows that the sign of the displacement thickness variation is the same as the sign of the pressure disturbance if integral $L$ is positive. In this case the flow is subcritical (the wake displacement thickness variation is of the same sign as the variation of the subsonic flow streamtube).

If integral $L(s)$ is regular and monotonic near point $s=0$, that is, it can be expanded as follows:

$$L = H_0 s + o(s)$$

then formula (6.134) can be written in the form:

$$D_\alpha = \frac{(\gamma-1)}{4\gamma^2 P_0}\frac{P_\alpha}{P_0} H_0 s$$

Hence follows that $t = s^\alpha$. The exponent $\alpha$ satisfies the equation

$$\alpha = -\frac{16\gamma^2}{3(\gamma+1)(\gamma-1)D_0 x_0 H_0} \tag{6.135}$$

Thus, if in the global solution of problem (6.128) the parameter $P_\alpha$ is nonzero, then the pressure distribution has a singularity as $x \to x_0$, since $\alpha < 0$. A possibility of smooth transition to the supercritical flow region is related with the choice of a solution in which $P_\alpha = 0$. This choice can be ensured by the choice of the constant $P_a$ entering in expansion (6.130) in the vicinity of the leading edge. In other words, the condition for the choice of a unique solution consists in the elimination of a singular eigen solution at the boundary of the domain of influence (near the point of transition from the subcritical to the supercritical flow regime).

### 6.4.3 *Investigation of the flow in the vicinity of the transition point for a near-supercritical regime*

It is assumed that the distance from the transition point to the trailing edge of the plate is small ($\xi = o(1)$). The factors favoring the transition point displacement toward the trailing edge are presented below. If the distance from the trailing edge is small but greater in the order than the wake thickness, the functions $F_0(\eta)$ and $G_0(\eta)$ can be expressed in terms of the velocity and total enthalpy profiles at the trailing edge of the plate and in the wake.

The incompressible flow near the trailing edge of a flat plate was studied in the paper of Goldstein (1930), where it was shown that the thickness of the viscous flow region in the wake depends on the distance from the trailing edge as $\xi^{1/3}$. The minimum value of the velocity at the wake axis varies in the same fashion (Fig. 6.11). In our case we deal with a compressible flow; however for nonzero values of the temperature factor of the plate surface the flow near the wake axis is characterized in the first approximation by a constant density. It can be shown that at nonzero surface temperatures the density variation near the wake axis induced by flow acceleration is small at small distances from the trailing edge. For this reason, in the case under consideration the functional form of the expansions in coordinates in the viscous flow region near the wake axis is the same as in the paper of Goldstein (1930).

In the main part of the flow, for $\eta = O(1)$ variations for $\xi = o(1)$ are small; in particular, for the integral entering in formula (6.134) the following estimate is valid:

$$J_\alpha = I_0 + O(\xi^{1/3}), \quad f = \varphi_0(\eta) + \xi^{1/3}\varphi_1(\eta), \quad g = z_0(\eta) + \xi^{1/3} z_1(\eta)$$

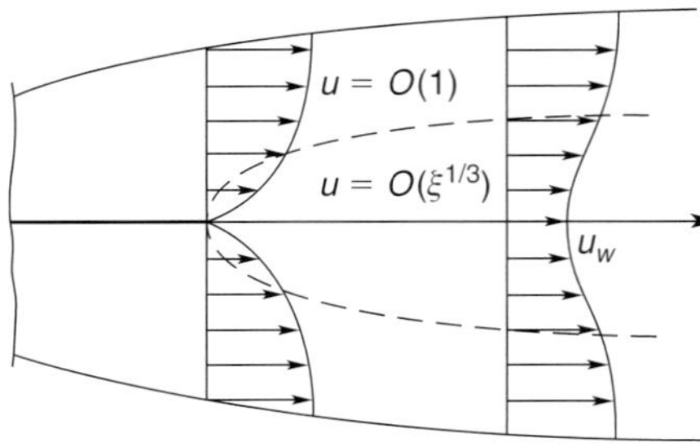


Fig. 6.11.

Integral $J_\alpha$ can be represented as the sum

$$J_\alpha = \int_0^{\eta_0} \frac{(G_0 - F_0'^2)}{F_0'^2}\, d\eta + \int_{\eta_0}^{\infty} \frac{(G_0 - F_0'^2)}{F_0'^2}\, d\eta \tag{6.136}$$

For small values of $\eta_0$ the second integral on the right-hand side of Eq. (6.136) is brought into the form:

$$\int_{\eta_0}^{\infty} \frac{\left(G_0 - F_0'^2\right)^2}{F_0'^2}\, d\eta = \int_0^1 \left[\frac{(z_0 - \varphi_0'^2)^2}{\varphi_0'^2} - \frac{g_w^2}{a^2\eta}\right] d\eta + \int_1^{\infty} \frac{(z_0 - \varphi_0'^2)^2}{\varphi_0'^2}\, d\eta - \frac{g_w^2}{a^2\eta_0} + \frac{g_w^2}{a^2} + O(\eta_0^2) \tag{6.137}$$

where $a = \varphi_{0w}'$.

In the first integral on the right-hand side of Eq. (6.136) the functions can be represented in the form:

$$G_0 = g_w + \xi^{1/3} q_1(\lambda) + \cdots, \quad F_0 = \xi^{2/3}\psi_1(\lambda) + \cdots, \quad \lambda = \frac{\eta}{\xi^{1/3}} \tag{6.138}$$

In view of Eq. (6.138) the above-mentioned integral can be written in the form:

$$\int_0^{\eta_0} \frac{(G_0 - F_0'^2)^2}{F_0'^2}\, d\eta = \xi^{-1/3} \int_0^{\infty} \frac{(g_w + \xi^{1/3} q_1)^2}{\psi_1'^2}\, d\lambda - \xi^{-1/3} \int_{\eta_0/\xi^{1/3}}^{\infty} \frac{(g_w + \xi^{1/3} q_1)^2}{\psi_1'^2}\, d\lambda \tag{6.139}$$

In the $\lambda_1 = \eta_0/\xi^{1/3} \to \infty$ limit we obtain

$$\psi_1' = a\lambda_1 + O(e^{-\lambda_1}) + \cdots$$

Formula (6.139) is transformed as follows:

$$\int_0^{\eta_0} \frac{(G_0 - F_0'^2)^2}{F_0'^2}\,\mathrm{d}\eta = \xi^{-1/3} g_w^2 \int_0^{\infty} \frac{\mathrm{d}\lambda}{\psi_1'^2} - \frac{g_w^2}{a^2 \eta_0} \tag{6.140}$$

In view of Eqs. (6.137) and (6.140), formula (6.136) in the limit $\eta_0 \to 0$, $\xi \to 0$, and $\lambda_1 = \eta_0/\xi^{1/3} \to \infty$ takes the form:

$$J_\alpha = \int_0^1 \left[ \frac{(z_0 - \varphi_0'^2)^2}{\varphi_0'^2} - \frac{g_w^2}{a^2 \eta} \right] \mathrm{d}\eta + \int_1^{\infty} \frac{(z_0 - \varphi_0'^2)^2}{\varphi_0'^2}\,\mathrm{d}\eta + \frac{g_w^2}{a^2} + \xi^{-1/3} g_w^2 \int_0^{\infty} \frac{\mathrm{d}\lambda}{\psi_1'^2}$$

or

$$J_\alpha = J_{12} + \xi^{-1/3} g_w^2 J_{11}$$

The position of the point of transition from the subcritical to the supercritical regime is determined in explicit form:

$$\xi_0 = x_0 - 1 = \left[ \frac{\frac{(\gamma-1)}{2} g_w^2 J_{11}}{I_0 - \frac{(\gamma-1)}{2} J_{12}} \right]^3$$

provided that near the trailing edge the flow over the plate is subcritical

$$I_0 - \frac{(\gamma - 1)}{2} J_{12} > 0$$

The parameter $H_0$ is as follows:

$$H_0 = \frac{(\gamma - 1)}{2} \frac{g_w^2 J_{11}}{3\xi_0^{4/3}}$$

Hence, in view of Eq. (6.135), there follows the formula for the exponent

$$\alpha = -\frac{16\gamma^2 P_0^2 \xi_0^{1/3}}{3(\gamma^2 - 1) D_0 g_w J_{11}}$$

Thus, the distance to the choking point is small, if the temperature factor $g_w$ is small. For zero value of the temperature factor formulas (6.140) and (6.137) take the form:

$$\int_0^{\eta_0} \frac{(G_0 - F_0'^2)^2}{F_0'^2}\,\mathrm{d}\eta = \xi^{1/3} g_w^2 \int_0^{\infty} \left( \frac{q_1}{\psi_1'^2} - \frac{b^2}{a^2} \right) \mathrm{d}\lambda + \frac{b^2}{a^2} \eta_0$$

where $b = q'_{1w}$ and

$$\int_{\eta_0}^{\infty} \frac{(G_0 - F_0'^2)^2}{F_0'^2} \mathrm{d}\eta = \int_0^{\infty} \frac{(z_0 - \varphi_0'^2)^2}{\varphi_0'^2} \mathrm{d}\eta - \frac{b^2}{a^2}\eta_0^2$$

In this case, the condition determining the transition point position is brought into the form:

$$\frac{(\gamma-1)}{2}\xi^{1/3}J_{20} + J_{21}\frac{(\gamma-1)}{2} - I_0 = 0 \tag{6.141}$$

where

$$J_{20} = \int_0^{\infty} \left(\frac{q_1}{\psi_1'^2} - \frac{b^2}{a^2}\right) \mathrm{d}\lambda_0, \quad J_{21} = \int_0^{\infty} \frac{(z_0 - \varphi_0'^2)^2}{\varphi_0'^2} \mathrm{d}\eta$$

### *6.4.4 Analysis of the flow in the vicinity of the trailing edge of a flat plate in the subcritical and supercritical regimes*

The change of the no-slip boundary condition on the plate $f'_w = 0$ for the symmetry condition in the wake $f''_w = 0$ violates the regularity of the solution of the system of the boundary layer equations. An analysis of the solution with the violation of the continuity in the boundary conditions was drawn in the paper of Goldstein (1930), in accordance with which the discontinuity leads to the necessity of introducing an additional subdomain near the wake axis. As noted above, for further analysis it is important that the geometry of this subdomain (dependence of the thickness on the distance from the trailing edge) and the dependence of the longitudinal velocity on the same distance are determined from the condition of the balance between the viscosity and inertia forces and are independent in the first approximation of both the nature of the outer flow and the temperature factor. The variables characteristic of this subdomain can be written in the form:

$$\lambda = \eta\xi^{1/3}, \quad f = \xi^{2/3}\psi_1(\lambda) + \cdots, \quad g = g_w + \xi^{1/3}q_1(\lambda) + \cdots \tag{6.142}$$

Substituting Eq. (6.142) in the boundary value problem (6.128) yields for the first-approximation functions

$$\psi_1''' + \frac{2}{3}\psi_1\psi_1'' - \frac{1}{3}\psi_1'^2 = \beta_1 g_w \tag{6.143}$$

$$q_1'' + \frac{2}{3}\psi_1 q_1' - \frac{1}{3}\psi_1' q_1 = 0$$

$$\psi_1 = \psi_1'' = q_1' = 0, \quad \lambda = 0$$

$$\psi_1'' = a, \quad q_1' = b, \quad \lambda \to \infty$$

$$\beta_1 = \frac{(\gamma - 1)}{2\gamma} \frac{\dot{P}}{P} \xi^{1/3}$$

The corresponding expansions and the system of equations governing the disturbed flow in the main region, where $\eta = O(1)$, depend on the temperature factor $g_w$.

We will first consider the case of a finite temperature factor (the body temperature is nonzero). Then for the main flow region, that is, the boundary layer giving way to the wake, the following representations hold

$$f = \varphi_0(\eta) + \xi^{1/3}\varphi_1(\eta) + \cdots \tag{6.144}$$

$$g = z_0(\eta) + \xi^{1/3} z_1(\eta) + \cdots$$

$$P = P_0 + \xi^{2/3} P_1 + \cdots, \quad P_0 = P(1)$$

The form of the expansions for the functions $f$ and $g$ is determined by the conditions of matching with solutions (6.142). The form of the expansion for the function $P$ is discussed below. The system of equations for the functions $\varphi_1$ and $z_1$ is obtained after the substitution of expansions (6.144) in system (6.128)

$$\varphi_0'\varphi_1' - \varphi_0''\varphi_1 = 0 \tag{6.145}$$

$$\varphi_0' z_1 - z_0' \varphi_1 = 0$$

The solution of system (6.145) is as follows:

$$\varphi_1 = A\varphi_0', \quad z_1 = Az_0'$$

where $A$ is a constant determined as a result of the matching with the solution for the sublayer. The expansion for the displacement thickness takes the form:

$$\Delta = \Delta_0 + \Delta_1 \xi^{1/3}, \quad \Delta_0 = \Delta(1) \tag{6.146}$$

$$\Delta_1 = \frac{(\gamma - 1)}{2\gamma P_0} A \int_0^\infty (z_0' - 2\varphi_0'\varphi_0')\, d\eta = -\frac{(\gamma - 1)}{2\gamma P_0} A g_w$$

The irregular variation of the displacement thickness results in the appearance of an infinitely large induced pressure gradient. In order for the self-consistent structure of the solution to be conserved, it is necessary that the solution of problem (6.143) satisfies the condition

$$\psi_1'(\infty) = a\lambda + o(1) \tag{6.147}$$

Then $A = 0$ and $\Delta_1 = 0$. Condition (6.147) can be satisfied by a proper choice of the parameter $\beta_1$. If $\beta_1 = O(1)$, then the expansion for the function $P$ must include a term of the form $\xi^{2/3} P_1$. The boundary value problems (6.143) and (6.147) was solved in the work of Hakkinen and Rott (1965), in which an incompressible flow was studied. Thus, for a nonzero temperature factor the known interaction pattern is realized.

For zero temperature factor ($g_w = 0$) the coefficient $\Delta_1 = 0$; this follows from Eq. (6.146). The contribution of the flow in the $\Delta\delta_1$ subdomain to the displacement thickness variation is equal in the order to $O(\xi^{2/3})$. This displacement thickness variation would also lead to a singularity in the pressure distribution and should, therefore, be compensated by an opposite-in-sign variation of the displacement thickness $\Delta_2$ of the main region. Using formula (6.134) for determining the displacement thickness variation we obtain that also in the case under consideration the pressure distribution is described by expansion (6.144). The compensation condition must be applied for determining the coefficient $P_1$

$$\Delta\delta_1 + \Delta_2 = 0 \tag{6.148}$$

$$\frac{P_1}{P_0} L + \int_0^\infty (q_1 - b\lambda)\, \mathrm{d}\lambda = 0$$

In the first approximation the pressure gradient has no effect on the flow in the subdomain $\beta_1 = O(\xi^{1/3})$. The integral entering in formula (6.148) is determined by numerical integration of problem (6.142)

$$\int_0^\infty (q_1 - b\lambda)\, \mathrm{d}\lambda \approx 0.4 \frac{b}{a^{2/3}}$$

As follows from Eq. (6.148), the sign of the pressure gradient in the local flow near the trailing edge is determined by the sign of integral $L$. For $L > 0$, that is, in the subcritical flow regime, the pressure gradient is negative for $\xi > 0$, whereas for $L < 0$, that is, in the supercritical flow regime, the pressure gradient is positive. It should also be noted that for $\beta_1 = 0$ the solution of problem (6.143) satisfies the condition $q_1 = (b/a)\psi_1'$. In this case, in Eq. (6.141) integral $J_{20} = 0$, so that the higher approximations must be investigated for determining the position of point $\xi_0$.

### 6.4.5 *Analysis of the flow in the vicinity of the trailing edge of a flat plate in the transcritical interaction regime*

As follows from Eq. (6.148), the pressure disturbance $\Delta P = O(\xi^{2/3})$ increases without bound as integral $L$ vanishes; this indicates a change in the functional form of the solution. For $L = 0$ the contribution of the main region $\eta = O(1)$ in the displacement thickness variation is zero, while the contribution of the subdomain $\eta = O(\xi^{1/3})$ is equal in the order to $O(\xi^{2/3})$.

In order to construct a self-consistent solution, the next term of the expansion for the main flow region in the boundary layer (wake), proportional to $\Delta P^2$, should be taken into account.

The fulfillment of the compensation condition gives an estimate for the pressure disturbance $\Delta P = O(\xi^{1/3})$ which determines the functional form of the coordinate expansions of the functions $f$ and $g$ in the main region

$$\begin{aligned} f &= \varphi_0(\eta) + \xi^{1/3}\varphi_1(\eta) + \xi^{2/3}\varphi_2(\eta) + \cdots \\ g &= z_0(\eta) + \xi^{1/3}z_1(\eta) + \xi^{2/3}z_2(\eta) + \cdots \\ P &= P_0 + \xi^{1/3}P_1 + \xi^{2/3}P_2 + \cdots \\ \Delta &= \Delta_0 + \xi^{1/3}0 + \xi^{2/3}\Delta_2 + \cdots \end{aligned} \tag{6.149}$$

The first and second terms of expansions (6.149) satisfy the following system of equations:

$$\begin{aligned} &-\frac{(\gamma-1)}{2\gamma}\frac{P_1}{P_0}(z_0 - \varphi_0'^2) = \varphi_0'\varphi_1' - \varphi_0''\varphi_1 \\ &\varphi_0' z_1 - z_0'\varphi_1 = 0 \\ &-\frac{(\gamma-1)}{2\gamma}\left(\frac{2P_2}{P_0} - \frac{P_1^2}{P_0^2}\right)(z_0 - \varphi_0'^2) - \frac{(\gamma-1)}{2\gamma}\frac{P_1}{P_0}(z_1 - 2\varphi_0'\varphi_1') \\ &\quad = 2(\varphi_0'\varphi_2' - \varphi_2\varphi_0'') + (\varphi_1'^2 - \varphi_1\varphi_1'') \\ &2(\varphi_0' z_2 - z_0'\varphi_2) + (\varphi_1' z_1 - \varphi_1 z_1') = 0 \end{aligned} \tag{6.150}$$

The solution of system (6.150) can be represented in the form:

$$\varphi_1 = C_1\varphi_0', \quad \varphi_2 = C_2\varphi_0', \quad z_1 = C_1 z_0', \quad z_2 = C_2 z_0' - \frac{(\varphi_1' z_1 - z_1'\varphi)_1}{2\varphi_0'}$$

The functions $C_1$ and $C_2$ satisfy the equations

$$C_1' = -\frac{(\gamma-1)}{2\gamma}\frac{P_1}{P_0}\frac{A_0}{\varphi_0'^2}, \quad A_0 = z_0 - \varphi_0'^2 \tag{6.151}$$

$$C_2' = -\frac{(\gamma-1)}{4\gamma}\left(2\frac{P_2}{P_0} - \frac{P_1^2}{P_0^2}\right)\frac{A_0}{\varphi_0'^2} - \frac{1}{2}\frac{P_1}{P_0}\frac{(z_1 - 2\varphi_0'\varphi_1')}{\varphi_0'^2} - \frac{1}{2}\frac{(\varphi_1'^2 - \varphi_1\varphi_1'')}{\varphi_0'^2}$$

Using the formula for the displacement thickness in Eq. (6.128) we can derive the following expressions for the first and second coefficients in the expansion for $\Delta$

$$\Delta_1 = 0 = \frac{(\gamma-1)}{2\gamma P_0}\left[\int_0^\infty (z_1' - 2\varphi_0'\varphi_1')\,\mathrm{d}\eta - \frac{P_1}{P_0}J_0\right], \quad J_0 = \int_0^\infty A_0\,\mathrm{d}\eta \tag{6.152}$$

$$\Delta_2 = \frac{(\gamma-1)}{2\gamma P_0}\left[\int_0^\infty (z_2' - 2\varphi_0'\varphi_2')\,d\eta - \frac{P_1}{P_0}\int_0^\infty (z_1 - 2\varphi_0'\varphi_1')\,d\eta - \int_0^\infty \varphi_1'^2\,d\eta + \left(\frac{P_1^2}{P_0^2} - \frac{P_2}{P_0}\right)J_0\right]$$

The equation for $\Delta_2$ includes integrals divergent as $\eta \to 0$; their principal values will be determined below. In Eq. (6.152) the value $\eta = \eta_0 = o(1)$ is taken for the lower limit for these integrals.

In order to single out the contribution of nonhomogeneous terms it is convenient to represent the second-approximation functions in the form of a sum

$$C_1 = C_{10}\frac{P_1}{P_0}, \quad z_1 = z_{10}\frac{P_1}{P_0}, \quad \varphi_1 = \varphi_{10}\frac{P_1}{P_0}$$

$$C_2 = C_{20} + C_{21}\frac{P_1^2}{P_0^2}, \quad C_{20} = C_1\frac{P_0}{P_1}\left(\frac{P_2}{P_0} - \frac{1}{2}\frac{P_1^2}{P_0^2}\right)$$

$$\varphi_2 = \varphi_{20} + \varphi_{21}\frac{P_1^2}{P_0^2}, \quad z_2 = z_{20} + z_{21}\frac{P_1^2}{P_0^2}$$

The formula for $\Delta_2$ takes the form:

$$\Delta_2 = \frac{(\gamma-1)}{2\gamma P_0}\left[\int_0^\infty (z_2' - 2\varphi_0'\varphi_2')\,d\eta - \frac{P_1^2}{P_0^2}\int_0^\infty \varphi_1'^2\,d\eta - \frac{1}{2}\frac{P_1^2}{P_0^2}\right] \tag{6.153}$$

The integrals entering in Eq. (6.153) can be brought into the form:

$$\int_{\eta_0}^\infty \varphi_0'\varphi_{21}'\,d\eta = \frac{C_{21}(\eta_0)\varphi_0'^2(\eta_0)}{2} + \frac{B}{4}\int_{\eta_0}^\infty (z_{10} - 2\varphi_0'\varphi_{10}')\,d\eta + \frac{1}{2}\varphi_{10}(\eta_0)\varphi_{10}'(\eta_0) - \frac{1}{2}\int_{\eta_0}^\infty \varphi_{10}'^2\,d\eta$$

$$\int_{\eta_0}^\infty z_{21}'\,d\eta = C_{21}(\eta_0)z_0(\eta_0) - \frac{B}{2}\Pi_1 + \frac{1}{2}\Pi_2 - \frac{1}{2}\Pi_3, \quad B = -\frac{(\gamma-1)}{2\gamma} \tag{6.154}$$

where the following notation is used

$$\Pi_1 = \int_{\eta_0}^\infty \frac{(z_{10} - 2\varphi_0'\varphi_{10}')\,d\eta}{\varphi_0'^2} = \int_{\eta_0}^\infty \frac{C_{10}z_0(z_0' - 2\varphi_0'\varphi_0'')\,d\eta}{\varphi_0'^2} - 2B\int_{\eta_0}^\infty \frac{(z_0 - \varphi_0'^2)\,d\eta}{\varphi_0'^2}$$

$$\Pi_2 = \int_{\eta_0}^{\infty} \frac{(\varphi_{10}^2 - \varphi_{10}\varphi_{10}'')\mathrm{d}\eta}{\varphi_0'^2} = -\frac{C_{10}(\eta_0)[C_{10}(\eta_0)\varphi_0']'z_0(\eta_0)}{\varphi_0'^2} + 2B^2 \int_{\eta_0}^{\infty} \frac{(z_0 - \varphi_0'^2)z_0\,\mathrm{d}\eta}{\varphi_0'^4}$$

$$+ 4B \int_{\eta_0}^{\infty} \frac{(z_0 - \varphi_0'^2)C_{10}(\eta)\varphi_0''z_0\mathrm{d}\eta}{\varphi_0'^3} + 2 \int_{\eta_0}^{\infty} \frac{C_{10}^2\varphi_0''z_0\,\mathrm{d}\eta}{\varphi_0'^2} + \int_{\eta_0}^{\infty} \frac{C_{10}C_3\,\mathrm{d}\eta}{\varphi_0'^2}$$

$$+ C_3 \left[ \frac{B(z_0 - {\varphi_0'}^2)}{\varphi_0'} + C_{10}\varphi_0'' \right] (z_0'\varphi_0' - 2z_0\varphi_0'')$$

$$\Pi_3 = \int_{\eta_0}^{\infty} \frac{(z_{10}\varphi_{10}' - z_{10}'\varphi_{10})\,\mathrm{d}\eta}{\varphi_0'} = -C_1^2(\eta_0)z_0'(\eta_0) + \int_{\eta_0}^{\infty} \frac{C_{10}^2\varphi_0''z_0'\,\mathrm{d}\eta}{\varphi_0'} + 2B \int_{\eta_0}^{\infty} \frac{C_{10}A_0z_0'\,\mathrm{d}\eta}{\varphi_0'^2}$$

The domain of integration can be subdivided into two subdomains

$$\Pi_i = \int_1^{\infty} T_i\,\mathrm{d}\eta + \int_{\eta_0}^{1} T_i\,\mathrm{d}\eta$$

where $T_i$ is the corresponding integrand.

Formulas (6.151), together with the formulas for the asymptotic representation of the functions $\varphi_0(\eta)$ and $z_0(\eta)$ for $\eta \to 0$, make it possible to single out the components of the integrands $T_{i2}$, the integrals of which are divergent

$$T_i = T_{i1} + T_{i2}, \quad S_{i2}' = T_{i2}$$

$$\Pi_{i1} = \int_1^{\infty} T_i\,\mathrm{d}\eta + \int_0^{1} T_{i1}\,\mathrm{d}\eta + S_{i2}(1) \quad \text{is the principal value of } \Pi_i$$

In view of the definition of the principal part of integrals, formula (6.153) takes the form:

$$\Delta_2 = \frac{(\gamma - 1)}{4\gamma P_0}\frac{P_1^2}{P_0^2}[-\Pi_{11} + \Pi_{21} - \Pi_{31} - (B+1)J_0 + S_{12}(\eta_0) + S_{22}(\eta_0) + S_{32}(\eta_0)] + O(\eta_0) \tag{6.155}$$

The boundary value problem for the subdomain near the wake axis takes the form:

$$\psi_1''' + \frac{2}{3}\psi_1\psi_1'' - \frac{1}{3}\psi_1'^2 = \frac{(\gamma - 1)}{6\gamma}\frac{P_1}{P_0}q_1 \tag{6.156}$$

$$q_1'' + \frac{2}{3}\psi_1 q_1' - \frac{1}{3}\psi_1' q_1 = 0$$

$$\psi_1 = \psi_1'' = q_1' = 0, \ \ \lambda = 0$$

$$\psi_1'' = a, \quad q_1' = b, \ \ \lambda \to \infty$$

where the coordinate $\lambda$ and the functions $\psi_1$ and $q_1$ were defined above (Eq. (6.142)). The subdomain displacement thickness can be expressed as follows:

$$\Delta\delta_2 = \frac{(\gamma-1)}{2\gamma P_0}\int^{\lambda_1} q_1 \, \mathrm{d}\lambda \tag{6.157}$$

Since integral (6.157) diverges as $\lambda \to \infty$, the value $\lambda = \lambda_1 \gg 1$ should be taken for the upper limit. In order to single out the principal value of integral (6.157) it is necessary to use the asymptotic representations for the functions $\psi_1$ and $q_1$ as $\lambda \to \infty$

$$\psi_1 = \frac{a\lambda^2}{2} + a_1\lambda\ln\lambda + a_2\lambda + a_3\ln^2\lambda + a_4\ln\lambda + a_5 + a_6\frac{\ln^2\lambda}{\lambda} + a_7\frac{\ln\lambda}{\lambda} + \frac{a_8}{\lambda} + O\left(\frac{\ln^m\lambda}{\lambda^2}\right)$$

$$q_1 = b\left[\lambda + \frac{a_1\lambda}{a} + \frac{a_2}{a} + \left(\frac{a_4}{a} - \frac{a_1a_2}{a^2}\right)\frac{\ln\lambda}{\lambda} + \left(\frac{a_5}{a} - \frac{a_2^2}{2a^2}\right)\frac{1}{\lambda} + \left(\frac{a_6}{a} - \frac{a_1a_4}{a^2} - \frac{a_2a_3}{a_2} + \frac{3a_1^2a_2}{a^3}\right)\frac{\ln^2\lambda}{\lambda} + \left(\frac{a_7}{a} - \frac{a_1a_5}{a^2} - \frac{a_2a_4}{a^2} + \frac{3a_1a_2^2}{2a^3}\right)\frac{\ln\lambda}{\lambda} + \left(\frac{a_1a_2^3}{2a^3} - \frac{a_2a_5}{a^2} + \frac{a_8}{a}\right) + O\left(\frac{\ln^m\lambda}{\lambda^3}\right)\right]$$

where the expansion coefficients are expressed in terms of the parameters $a$, $b$, and $\beta = [(\gamma-1)/6\gamma](P_1/P_0)$ as follows:

$$a_1 = -\frac{3\beta b}{a}, \quad a_3 = \frac{9\beta^2b^2}{2a^3}, \quad a_4 = \frac{3\beta b}{a^2}\left(\frac{3\beta b}{a} - a_2\right)$$

$$a_5 = \frac{1}{2a}\left(\frac{27\beta^2b^2}{a^2} - \frac{3\beta b a_2}{a} + a_2^2\right), \quad a_6 = -\frac{27\beta^3b^3}{2a^5}$$

$$a_7 = \frac{9\beta b^2}{a^4}\left(-\frac{3\beta b}{a} + a_2\right), \quad a_8 = \frac{3\beta b}{a^3}\left(-\frac{9\beta^2b^2}{a^2} + \frac{3\beta b a_2}{a} - \frac{a_2^2}{2}\right)$$

The coefficient $a_2$ must be determined by numerical integration of the problem. In the main region the first-approximation solution for $\varphi_1$ and $z_1$ is determined correct to a constant which can be expressed in terms of the coefficient $a_2$. In integral (6.157), apart from isolating the singularity, we must also take account for the contribution to the displacement thickness, which is already present in the expressions for $\Delta_2$ and $\Delta_1$

$$\Delta\delta_2 \frac{2\gamma P_0}{(\gamma-1)} = \int_0^\infty (q_1 - b\lambda - z_{1w})\,\mathrm{d}\lambda$$
$$= \int_0^1 (q_1 - b\lambda - z_{1w})\,\mathrm{d}\lambda + \int_1^\infty (q_1 - b\lambda - z_{1w})\,\mathrm{d}\lambda - Q_1(1) + Q_1(\lambda_1) \tag{6.158}$$

where

$$z_{1w} = b\lambda + \frac{ba_1 \ln\lambda}{a} + \frac{ba_2}{a}$$
$$Q_1' = \left(\frac{a_4}{a} - \frac{a_1 a_2}{a^2}\right)\frac{\ln\lambda}{\lambda} + \left(\frac{a_5}{a} - \frac{a_2^2}{2a^2}\right)\frac{1}{\lambda}$$

Rewriting the compensation condition in terms of Eqs. (6.158) and (6.155) yields

$$\Delta\delta_2 + \frac{P_1^2}{2P_0^2}[-\Pi_{11} + \Pi_{21} - \Pi_{31} - (B+1)J_0] = 0 \tag{6.159}$$

In deriving Eq. (6.159) it was taken into account that

$$\lim_{\eta_0\to 0} \frac{P_1^2}{2P_0^2}(S_{12} + S_{22} + S_{32}) + \lim_{\lambda\to\infty_1} Q_1 = 0, \quad \lambda_1 = \frac{\eta_0}{\xi^{1/3}}$$

As an illustration we will present the results of the numerical solution of the problem for the case in which the zeroth approximation for the flat-plate boundary layer flow is described by the following problem:

$$\varphi_0''' + \frac{1}{4}\varphi_0\varphi_0'' + \beta_0(z_0 - \varphi_0'^2) = 0$$
$$z_0'' + \frac{1}{4}\varphi_0 z_0' = 0$$
$$\varphi_0' = z_0 = 1, \quad \eta \to \infty$$

As a result of numerical integration, the value of the parameter $\beta_0$, for which the transcriticality condition is fulfilled, was determined

$$L = \frac{(\gamma-1)}{2}\int_0^\infty \frac{(z_0 - \varphi_0'^2)^2}{\varphi_0'^2}\,\mathrm{d}\eta - \int_0^\infty (z_0 - \varphi_0'^2)\,\mathrm{d}\eta = 0$$
$$\beta_0 \approx 0.13$$

The values of the parameters $P_1/P_0 = -0.36$, $\Pi_{11} = -0.46$, $\Pi_{21} = -265$, $\Pi_{31} = 0.97$, and $\Delta\delta_2 = -0.037$ were determined by numerical integration of problem (6.156) and calculation of integrals $\Pi_{i1}$ followed by an iteration procedure for satisfying the compensation condition (6.159).

This regime of the interaction between the inviscid flow in the main region and the viscous flow in the subdomain can also be realized in other situations. In the paper of Neiland (1974*b*) it was shown that on a cold delta wing in a hypersonic flow on a certain sweep angle range there exists a plane in which an integral, analogous to integral $L$, reverses sign. Disturbances can propagate in the subcritical flow region located between this plane and the plane of symmetry. The eigenfunctions governing the disturbance propagation near the region of transition to the subcritical flow were determined in the study of Dudin and Lipatov (1985). The procedure of determining these solutions was based on the assumption of the linearity of disturbances in the viscous flow subdomain. An increase in the intensity of the disturbances transferred (those of the pressure in the plane of symmetry) must ultimately lead to appearance of nonlinear disturbances in both the viscous flow subdomain and the main region. As a result, the surface friction and the heat flux will vary in the leading order near the plane of transition to the subcritical flow. In particular, transition to the region of return flow occurring due to an elevated pressure in the plane of symmetry can be ensured. In order to maintain the existence of this regime, the existence of a three-dimensional problem analogous to problem (6.156) and the fulfillment of the three-dimensional compensation condition, analogous to condition (6.159) must be proven.

Another situation, in which the nonlinear interaction regime considered above can be realized, pertains to a disturbed flow in the boundary layer on a cold surface whose parameters satisfy the condition $L = 0$. The estimation performed shows that the nonlinear interaction zone length is less in the order than the boundary layer thickness. Therefore, in this case local disturbances can give rise to the compensation regime of interaction for a nonisothermal wall flow. The study of this flow presents an individual problem.

## 6.5 Global solution for the hypersonic flow over a finite-length plate with account for the wake flow

### *6.5.1 Formulation of the problem*

Two facts noted in the preceding sections of this chapter are important for formulating the problem of the symmetric flow over a finite-length plate in the regime of subcritical strong viscous–inviscid interaction and choosing a method for solving it.

The first is related with the effect of upstream disturbance propagation over distances comparable with the streamwise body dimension, typical of this flow regime. This results in the flow pattern in which the parts of the flow on the upper and lower surfaces of the plate are subjected to mutual ejecting influence leading to flow acceleration in the vicinity of the trailing edge. In this connection, using the self-similar solution of the hypersonic boundary layer equations (Hayes and Probstein, 1966), valid for the flow over a semi-infinite plate, is unjustified in calculating the aerodynamic characteristics of a finite-length plate. A correct solution can be obtained only with account for the wake flow.

The second is related with the presence of a "saddle"-type singularity in the solution. We mean the wake cross-section in which the streamtube acceleration leads to the change of the subcritical for the supercritical interaction regime. Hence follows the necessity of developing a numerical method which would make it possible to continue the solution in the region located downstream of this cross-section called the "choking" section.

We will consider the flow over a plate and in its wake (Fig. 6.12). It is assumed that the plate is set at zero incidence in a hypersonic perfect-gas flow. It is also assumed that the Reynolds number is high but not higher than the critical value associated with laminar–turbulent transition and the strong viscous–inviscid interaction regime is realized.

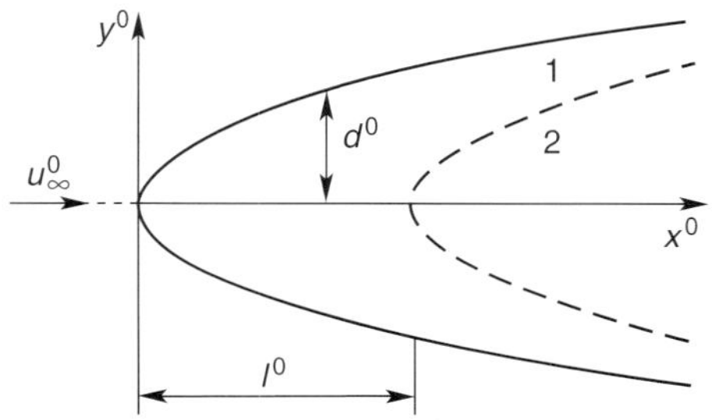


Fig. 6.12.

The system of boundary layer equations written in the variables introduced in Section 4.2 takes the form:

$$\frac{\partial}{\partial x}(\rho u) + \frac{\partial}{\partial x}(\rho v) = 0, \quad p = \frac{(\gamma - 1)}{2\gamma}\rho(g - u^2), \;\; \mu = g - u^2 \tag{6.160}$$

$$\rho u \frac{\partial u}{\partial x} + \rho v \frac{\partial u}{\partial y} + \frac{\partial p}{\partial x} = \frac{\partial}{\partial y}\left(\mu \frac{\partial u}{\partial y}\right), \quad \frac{\partial p}{\partial y} = 0$$

$$\rho u \frac{\partial g}{\partial x} + \rho v \frac{\partial g}{\partial y} = \frac{\partial}{\partial y}\left(\frac{\mu}{\sigma}\frac{\partial g}{\partial y}\right) + \frac{\partial}{\partial y}\left[\mu\left(1 - \frac{1}{\sigma}\right)\frac{\partial}{\partial y}(u^2)\right]$$

$$y = 0, \;\; v = u = 0, \;\; g = g_w \quad (0 \le x \le 1)$$

$$y = 0, \;\; v = \frac{\partial u}{\partial y} = \frac{\partial g}{\partial y} = 0 \quad (x > 1)$$

$$y = \delta, \;\; u = g = 1 \quad (x \ge 0)$$

Here $\gamma$ is the specific heat ratio of the perfect gas and $\sigma$ is the Prandtl number.

The equation determining the induced pressure is obtained using the tangent wedge method

$$p = \frac{(\gamma + 1)}{2}\left(\frac{d\delta}{dx}\right)^2 \tag{6.161}$$

### 6.5.2 Transformation of variables

For numerically solving the problem we will introduce independent variables which allow us not only to exclude the density from Eqs. (6.160) but also to make the computation domain rectangular. From the computational point of view it is also convenient to use variables which take account for the form of the well-known self-similar Lees–Stewartson solution (Hayes and Probstein, 1966) in the vicinity of the leading edge. These properties are typical of the variables

$$X = x, \quad Y = \frac{(\gamma - 1)}{2\gamma} x^{-1/4} \int^{y} \rho \, dy, \quad V = xu\frac{\partial Y}{\partial x} + \frac{(\gamma - 1)}{2\gamma} x^{3/4} \rho v \tag{6.162}$$

$$U = u, \quad H = g, \quad \Delta = x^{-3/4}\delta$$

In these variables the equations and boundary conditions take the form:

$$V_Y + \frac{1}{4}U + XU_X = 0 \tag{6.163}$$

$$XUU_X + VU_Y + \frac{(\gamma - 1)}{2\gamma}\left(X\frac{P_X}{P} - \frac{1}{2}(H - U^2)\right) = \frac{(\gamma - 1)}{2\gamma} PU_{YY} \tag{6.164}$$

$$XUH_X + VH_Y = \frac{(\gamma - 1)}{2\gamma\sigma} PU_{YY} + \frac{(\gamma - 1)}{2\gamma}\left(1 - \frac{1}{\sigma}\right) P(U^2)_{YY} \tag{6.165}$$

$$Y = 0, \quad V = U = 0, \quad H = H_w \quad (0 \le X \le 1)$$

$$Y = 0, \quad V = U_Y = H_Y = 0 \quad (x > 1)$$

$$Y = \infty, \quad U = H = 1 \quad (x \ge 0) \tag{6.166}$$

$$\Delta = \frac{1}{P}\int_0^\infty (H - U^2)\, dY, \quad P = \frac{(\gamma + 1)}{2}\left(X\Delta_X + \frac{3}{4}\Delta\right)^2 \tag{6.167}$$

In the problem formulated the main role is played by the procedure of searching an effective body thickness distribution which would be consistent with the pressure distribution. The procedure of searching the self-consistent thickness is presented in the paper of Kovalenko (1989).

### 6.5.3 Results of calculations

The graphical representation of the calculated results is so structured that it demonstrates clearly the difference between the flow over semi-infinite and finite-length plates. The

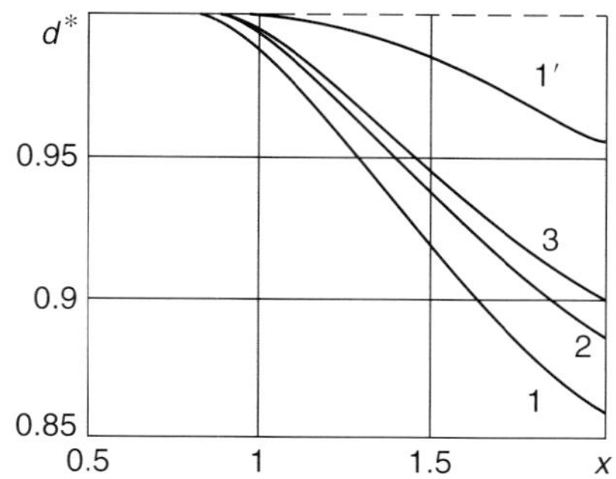


Fig. 6.13.

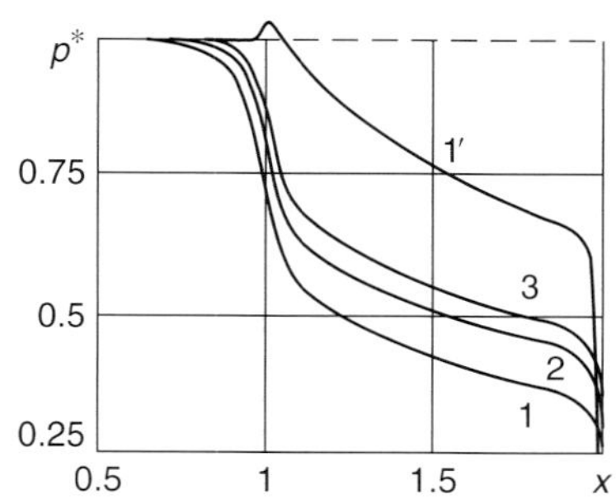


Fig. 6.14.

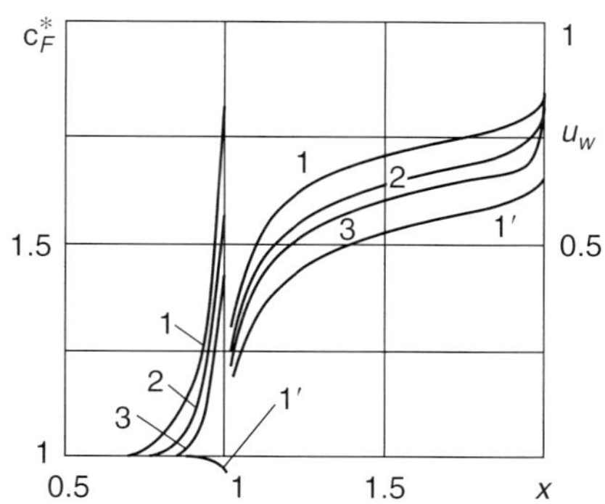


Fig. 6.15.

calculations were carried out for monatomic, diatomic, and triatomic perfect gases with $\gamma = 5/3$, 7/5, and 4/3, respectively. In Figs. 6.13–6.15 the corresponding curves are marked by numbers 1, 2, and 3. Moreover, for a monatomic gas ($\gamma = 5/3$) calculations were performed for the purpose of studying the surface cooling effect on the solution. In this set of calculations the temperature factor was zero. The corresponding curves are marked by 1′.

Dashed lines present the self-similar solutions, that is, the solutions for the flow over a semi-infinite plate. The functions divided by their self-similar values are marked by asterisks. All the calculations were carried out for $\sigma = 1$. The coordinate $X = 1$ corresponds to the trailing edge of the plate. On the right boundary of the computation domain $X = 2$ the pressure was kept constant: $P^*(2) = 0.25$. This choice of a fairly low pressure ensures against undesirable occurrence of return flow streamtubes near this boundary. Of course, this leads to a local flow acceleration but has no effect on the flow near the plate and along a considerable part of the wake length.

From an analysis of the solutions presented in Fig. 6.13 it follows that the change of the no-slip condition for the symmetry condition in the section $X = 1$ has a slight effect on the displacement thickness. Thus, in this section its deviation from the corresponding self-similar value is not greater than 1% for the triatomic gas and 2% for the monatomic gas. The static pressure is much more sensitive to the change of the boundary condition near the trailing edge (Fig. 6.14). The pressure starts to decrease relative to its value for the semi-infinite plate already at a distance from the trailing edge equal to one fourth of the plate length. This is due to the local flow acceleration caused by the mutual ejecting influence of the flows on the upper and lower surfaces. Naturally, the streamtube acceleration is accompanied by an increase in the local surface friction coefficient $c_F^*$ (Fig. 6.15). In the same figure the distributions of the velocity $U_w$ along the wake axis are presented. Its growth is most intense when the plate is in the monatomic gas flow.

Evaluating the integral determining the flow type (subcritical or supercritical) made it possible to establish that the flow downstream of a noncold plate becomes supercritical already at a small distance from the trailing edge. The flow over a cold plate is everywhere supercritical.

## 6.6 Strong interaction of the boundary layer with a hypersonic flow under local disturbances of boundary conditions

In this section we present the study of the effect of sharp variations in boundary conditions on the local and global flow characteristics at strong global interaction between the original boundary layer and the outer hypersonic flow. It is shown that at fairly large disturbance amplitudes most of the boundary layer (outside a narrow viscous wall layer) behaves as a locally inviscid flow. The flow regimes are classified versus the disturbance amplitude, the similarity parameters are determined, and the corresponding boundary value problems are formulated. Of particular interest is the flow with large pressure disturbances, for which the range of separationless flow over an upstream-facing step is established, together with the rule of the choice of the solution for the main part of the body. As distinct from the flows with discontinuous boundary conditions considered in the previous chapters, in this formulation the sharp variations in the boundary conditions have not only local but also global effect on the boundary layer flow from the disturbed region up to the leading edge.

### *6.6.1 Formulation of the problem*

The schematics of the flow under consideration are shown in Fig. 6.16. A hypersonic viscous perfect-gas flow streams over a flat plate. At a distance $L$ from the plate nose there is a small

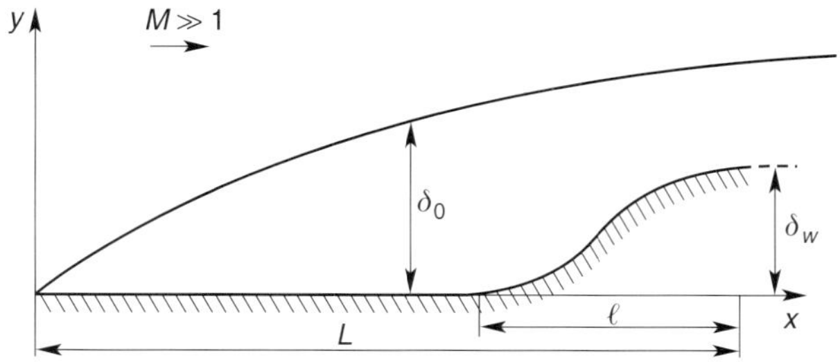


Fig. 6.16.

bump, depression, or some other variation of the boundary condition with a scale length $\ell \ll L$.

Let the Reynolds number $Re_0 = \rho_\infty u_\infty L/\mu_0$, where $\rho_\infty$ and $u_\infty$ are the freestream density and velocity and $\mu_0$ is the viscosity at the freestream stagnation temperature, be such that the boundary layer/outer hypersonic flow interaction is strong. Then for the boundary layer flow parameters the following estimates are valid

$$u \sim u_\infty, \quad \rho \sim \rho_\infty \tau^2, \quad p \sim \rho_\infty u_\infty \tau^2$$
$$\tau = \frac{\delta_0}{L}, \quad \tau = Re_0^{-1/4}, \quad M_\infty^2 \tau^2 \gg 1 \tag{6.168}$$

where $\delta_0$ is the boundary layer thickness.

We will restrict our consideration to disturbances for which

$$\tau L \ll \ell \ll L \tag{6.169}$$

while the body thickness $\delta_w$ can be both positive (bump) and negative (depression).

We will estimate the flow parameter disturbances preassigning the pressure disturbance $\Delta p$ and determining the scale lengths $\delta_w$ and the flow patterns associated with this values of $\Delta p$.

### *6.6.2 Estimates of the orders of the flow parameters*

In an undisturbed boundary layer the viscous and inertial terms are of the same order and the pressure gradient (in the strong interaction regime) is as follows:

$$\frac{\partial p}{\partial x} \sim \frac{p}{L} \tag{6.170}$$

We will consider a disturbance $\Delta p \le p$ assuming that the pressure gradient corresponding to this $\Delta p$ is greater in the order than that in the undisturbed boundary layer and, therefore, greater than viscous stresses. Then almost the entire boundary layer flow over a distance $\Delta x/\ell \sim O(1)$ is in the first approximation inviscid, except for a thin wall layer.

Let there be first $\Delta p/p \sim O(1)$ and the flow be separationless. Then, in accordance with the Bernoulli equation, together with the equations of continuity and state, in the boundary layer disturbances are of the same order as the undisturbed parameters

$$\frac{\Delta\delta}{\delta_0} \sim \frac{\Delta\rho}{\rho_0} \sim \frac{\Delta u}{u_e} \sim \frac{\Delta p}{p} \sim O(1) \tag{6.171}$$

Here, the subscript 0 refers to the boundary layer parameters ahead of the region, where the boundary conditions are disturbed, while $e$ and $w$ refer to the values of the variables at the outer edge of the boundary layer and the body surface, respectively.

However, along the length $\ell$ the outer edge of the boundary layer $y_e$ cannot be displaced by an amount $\delta_0$, since, in accordance with hypersonic small perturbation theory, we have

$$p \sim \rho_\infty u_\infty^2 \left(\frac{\mathrm{d}y_e}{\mathrm{d}x}\right)^2 \tag{6.172}$$

so that if $\Delta y_e \sim \delta_0$ along the length $\Delta x \sim \ell$, then $\Delta p/p \sim (L/\ell)^2 \gg 1$.

Therefore, it is necessary to assume that $\Delta\delta + \delta_w \approx 0$. In other words, on lengths (6.169) the pressure disturbances (6.171) can occur only due to variations in the boundary layer and body thicknesses

$$\delta_w \sim \Delta\delta \sim \delta_0 \sim L\tau \tag{6.173}$$

Then in this flow regime the values $\Delta p > 0$ cannot be obtained in a separationless flow, since near the wall there are streamtubes with low velocities which, in accordance with the Bernoulli equation, cannot arrive to the region with finite $\Delta p > 0$. If $\Delta p < 0$, then in an effectively inviscid flow near the body the gas velocity reaches a value $u \sim u_e$, so that the no-slip boundary conditions are no longer fulfilled. Therefore, near the body surface we must consider a viscous sublayer in which velocities are finite, while the pressure gradient and the inertial terms of the Navier–Stokes equations are of the same order as the leading viscous terms. This condition determines the viscous sublayer thickness $\delta_v$

$$\frac{\rho_0 u_e^2}{\ell} \sim \frac{\mu_0 u_e}{\delta_v^2}, \quad \frac{\delta_v}{\delta_0} \sim \left(\frac{\ell}{L}\right)^{1/2} \tag{6.174}$$

With decrease in the order of the pressure disturbances $\Delta p/p \ll 1$, in accordance with the equation for the longitudinal momentum component (with account for the equations of continuity and state), in the main part of the boundary layer, where the undisturbed velocities are of the order of $u_e$, the relative values of the disturbances decrease

$$\frac{\Delta\delta}{\delta_0} \sim \frac{\Delta\rho}{\rho_0} \sim \frac{\Delta u}{u_e} \sim \frac{\Delta p}{p} \ll 1$$

However, since in the original undisturbed boundary layer the velocity vanishes as the body surface is approached, there can exist a region in which dynamic heads are low and the relative values of the disturbances are large. This region is determined by the conditions

$$\rho_0 u_n^2 \sim \Delta p, \quad \left(\frac{u_n}{u_e}\right) \sim \left(\frac{\delta_n}{\delta_e}\right) \tag{6.175}$$

where the subscript $n$ refers to the variables in the region with nonlinear disturbances. The former condition indicates that in this region the dynamic head must be of the same order as the pressure disturbance, while the latter follows from the form of the velocity profile in the undisturbed boundary layer near the body. From relations (6.175) we obtain the estimate for the nonlinear disturbance zone thickness

$$\left(\frac{\delta_n}{\delta_0}\right) \sim \left(\frac{\Delta p}{p}\right)^{1/2}, \quad \frac{u_n}{u_e} \sim \left(\frac{\Delta p}{p}\right)^{1/2} \tag{6.176}$$

At the bottom of this zone a viscous layer must be introduced using the condition of the equality of the orders of the viscous and inertial terms in the momentum equation

$$\left(\frac{\delta_v}{\delta_0}\right) \sim \left(\frac{\ell}{L}\right)^{1/2} \left(\frac{\Delta p}{p}\right)^{1/4}, \quad u_v \sim u_n \tag{6.177}$$

The comparison of Eqs. (6.176) and (6.177) shows that the thicknesses of the regions, in which nonlinear and viscous effects are important, are the same for pressure disturbances

$$\frac{\Delta p}{p} \sim \left(\frac{\ell}{L}\right)^{2/3}$$

Thus, the flow pattern, for which three local flow regions should be considered, namely, two regions of locally inviscid flow with linear and nonlinear disturbances and a viscous sublayer, corresponds to the following disturbance values

$$\left(\frac{\ell}{L}\right)^{2/3} \ll \frac{\Delta p}{p} \ll 1, \quad \left(\frac{\ell}{L}\right)^{1/3} \ll \left(\frac{\delta_w}{\tau L}\right) \ll 1 \tag{6.178}$$

If condition (6.177) is fulfilled, the main part of the boundary layer behaves as a weakly disturbed, locally inviscid flow, while in the region of thickness

$$\left(\frac{\delta_v}{\delta_0}\right) \sim \left(\frac{\Delta p}{p}\right)^{1/2} \sim \left(\frac{\ell}{L}\right)^{1/3} \tag{6.179}$$

the viscous and inertial terms are of the same order. Therefore, in this region the flow is governed by the Prandtl equations. The thickness of the wall region with nonlinear disturbances (6.179) varies in the leading order, while the variation of the thickness of the outer part of the boundary layer remains small. However, then from Eq. (6.172) it follows that the compatibility of the boundary layer flow and the outer hypersonic flow is ensured if in the leading term

$$\delta_w + \Delta\delta_v \approx 0 \tag{6.180}$$

We note that precisely in this flow regime incipient separation can appear for $\Delta p > 0$ due to nonlinearity of the disturbances of the viscous part of the flow.

Finally, for even smaller pressure disturbances, $\Delta p/p \ll (\ell/L)^{1/3}$, disturbances become small even in the viscous sublayer. In this case, the flow can be separationless and consist of two regions with linear disturbances: a relatively thick ($\sim\delta_0$) locally inviscid flow region and a thin viscous sublayer.

### *6.6.3 Flow regime with finite pressure disturbances*

On the basis of condition (6.169), we will proceed from the boundary layer equations, since the longitudinal to transverse dimension ratio of the disturbed flow region is large and the equation for the momentum component across the boundary layer degenerates. We will assume that the Prandtl number is unity, the viscosity–temperature relation is linear, and the body is thermally insulated. Then in the Dorodnitsyn–Lees variables the equations and boundary conditions take the form:

$$f''' + ff'' - \beta(1 - f'^2) = 2\xi(f'\dot{f}' - \dot{f}f'') \tag{6.181}$$

$$f'(0,\eta) = 1, \quad f'(\xi,\infty) = 1, \quad f(\xi,0) = f'(\xi,0) = 0, \quad F = \int_0^\infty (1 - f'^2)\,\mathrm{d}\eta \tag{6.182}$$

where the approximate tangent wedge formula

$$\beta = \frac{(\gamma-1)}{\gamma}\frac{\xi}{P}\frac{\mathrm{d}P}{\mathrm{d}\xi}, \quad P = \frac{(\gamma+1)}{2}\left[P\frac{\mathrm{d}}{\mathrm{d}\xi}\left(\frac{\sqrt{2\xi}}{P}F + \frac{2\gamma}{\gamma-1}\Delta_w\right)\right]^2 \tag{6.183}$$

is used for determining the pressure and $\gamma$ is the specific heat ratio.

The Cartesian coordinate system $(x, y)$ is shown in Fig. 6.16. In Eqs. (6.181)–(6.183) the dimensionless variables are related by the formulas

$$u = u_\infty f'(\xi,\eta), \quad \xi = \frac{2\gamma}{\gamma-1}\int_0^X P\,\mathrm{d}X, \quad Y = \Delta_w + \sqrt{2\xi}\,\frac{\gamma-1}{2\gamma P}\int_0^\eta (1 - f'^2)\,\mathrm{d}\eta \tag{6.184}$$

$$p = \rho_\infty u_\infty^2 \tau^2 P, \quad x = LX, \quad y = \tau LY$$

Here, primes refer to the differentiation with respect to $\xi$, while $\Delta_w \tau L = \delta_w$ is the obstacle thickness, positive values corresponding to bumps and negative to depressions.

We will seek the solution for the following distribution $\Delta_w(\xi)$:

$$\Delta_w = 0, \quad 0 \le \xi < \xi_0, \quad \Delta = \delta \Delta_w(\xi_1), \quad \xi_1 = \frac{(\xi - \xi_0)}{\varepsilon} \tag{6.185}$$

where $\tau \ll \varepsilon \ll 1$ is the smallness parameter characterizing the relative length of the disturbed region. The function $\Delta_w(\xi_1)$ is assumed to be continuous and differentiable.

In this section we assume that $\delta = 1$; below we will consider small-perturbation regimes with $\delta(\varepsilon) \ll 1$.

We will first consider the local solution of the problem for $\eta \sim O(1)$, assuming that $\varepsilon \to 0$. In this case, the asymptotic expansions are as follows:

$$f \sim f_1(\xi_0, \xi_1, \eta) + \cdots, \quad P = P_1(\xi_0, \xi_1) + \cdots \tag{6.186}$$

$$F \sim F_1(\xi_0, \xi_1, \eta) + \cdots$$

Substituting Eq. (6.186) in Eqs. (6.181)–(6.183) and passing to the limit

$$\varepsilon \to 0, \quad \eta = O(1)$$

yields in the first approximation

$$-\frac{\gamma - 1}{\gamma} \frac{\xi_0}{P_1} \frac{dP_1}{d\xi_1}(1 - f_1'^2) = 2\xi_0 \left( f_1'' \frac{\partial f_1'}{\partial \xi_1} - f_1'' \frac{\partial f_1}{\partial \xi_1} \right) \tag{6.187}$$

$$\beta = \frac{1}{\varepsilon} \frac{\gamma - 1}{\gamma} \frac{\xi_0}{P_1} \frac{dP_1}{d\xi} + \cdots$$

$$\frac{d}{d\xi} \left( \frac{\sqrt{2\xi_0}}{P_1} F_1 + \frac{2\gamma}{\gamma - 1} \Delta_w \right) = 0 \tag{6.188}$$

Equation (6.187) admits the Bernoulli integral: along each streamline we have

$$f_1 = \text{const}, \; f_1' = \sqrt{1 - (1 - f_0'^2) \left( \frac{P_1}{P_{10}} \right)^{(\gamma - 1)/\gamma}} \tag{6.189}$$

where $f_0'(\eta)$ and $P_{10}$ are the solutions of Eqs. (6.181)–(6.183) for $\xi = \xi_0$ for the region in which $\Delta_w \equiv 0$. From Eq. (6.188) we obtain the equation determining $\Delta_w(P_1)$ in explicit form

$$\frac{\sqrt{2\xi_0}}{P_1} F_1 + \frac{2\gamma}{\gamma - 1} \Delta_w = \frac{\sqrt{2\xi_0}}{P_{10}} F_{10} \tag{6.190}$$

since, by virtue of Eq. (6.189), $F_1$ is uniquely determined by the last formula (6.182) as a function of $P_1$.

Solutions (6.189) and (6.190) has a simple physical meaning. In the region $\xi \sim O(1)$ a locally inviscid vortical flow, across which the pressure is constant ($\varepsilon \ll 1$), has the same initial parameter profiles as the original boundary layer for $\xi = \xi_0$. The layer thickness varies only due to the wall displacement ($\Delta_w \neq 0$). However, this flow must have a minimum thickness $Y - \Delta_w$ for a certain critical value $P_1$. We will demonstrate this property proceeding from the condition

$$\frac{\partial}{\partial P_1}\left(\frac{F_1}{P_1}\right) = 0 \tag{6.191}$$

since for a given velocity profile the quantity $F_1$ depends only on $P_1$ (cf. Eq. (6.189)).

Using Eqs. (6.189) and (6.190), together with the equation for the Mach number $M$

$$\frac{\gamma - 1}{2}(1 - f_0'^2)P_1^{(\gamma-1)/\gamma}M^2 = 1 - (1 - f_0'^2)P_1^{(\gamma-1)/\gamma} \tag{6.192}$$

we obtain the choking condition

$$\int_{\Delta_w}^{Y_e} (M^{-2} - 1)\,\mathrm{d}Y = 0 \tag{6.193}$$

In order to verify that the extremum condition (6.191) actually corresponds to a minimum of $F_1(P_1)$, we note that

$$\frac{F_1}{P_1} \sim P_1^{-1/\gamma} \int_0^\infty \frac{(1 - f_0'^2)f_0'\,\mathrm{d}\eta}{\sqrt{1 - (1 - f_0'^2)P_1^{(\gamma-1)/\gamma}}} \tag{6.194}$$

From Eq. (6.194) it follows that the thickness of the region increases without bound as $P_1$ vanishes (for $P_1 > 1$ there are no separationless solutions). The critical value of $P_1$, for which the cross-sectional area is minimum, corresponds, in accordance with Eq. (6.193), to the mean-integral value of the Mach number $M$ equal to unity. An analogy with the inviscid gas flow in a convergent–divergent nozzle is obvious.

Differentiating Eq. (6.194) with respect to $P_1$ it can be shown that for positive sign of integral (6.193) a streamtube narrowing (increase of $\Delta_w$ in Eq. (6.190)) corresponds to a decrease in $P_1$, whereas for negative sign the same increase of $\Delta_w$ corresponds to an increase in $P_1$. Thus, the problem of the subcritical ("subsonic") boundary layer flow over the smoothed step shown in Fig. 6.15 (when integral (6.193) is negative at $\xi = \xi_0$) admits solutions for $\Delta_{w\,\max}$ smaller than a certain fixed value dependent on the choice of the solution of the boundary layer equations in the region $0 < \xi < \xi_0$.

Let us consider in more detail the problem of the flow over a flat plate at the rear of which there is a gently sloping step followed by the base section downstream of which the pressure $P_1$ is preassigned.

In Fig. 6.17 we have plotted the curves $P(X)$ for the main part of the body, where $\Delta_w = 0$. The pressure is divided by its value corresponding to the well-known self-similar solution

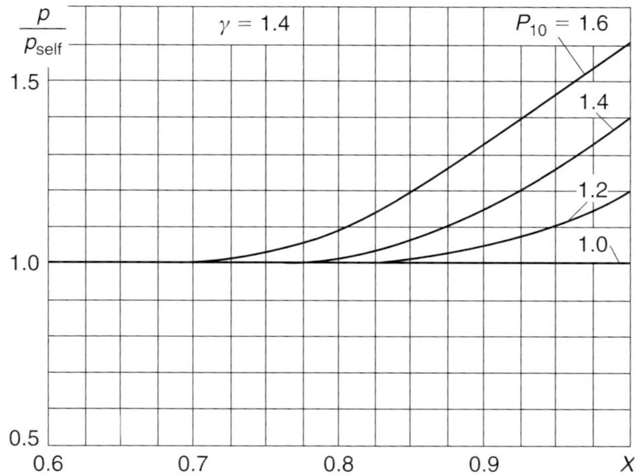


Fig. 6.17.

(Hayes and Probstein, 1966). Each curve of the family can be characterized by a value of the parameter $P_{10} = P(X = 1)$. In accordance with the studies of Bogolepov and Neiland (1971) and Neiland (1971*a*), for all values of $P_{10}$, except for a minimum value $P^0_{10}$, integral (6.193) is negative and the boundary layer is subcritical.

Thus, each fixed solution for the forward part of the body is associated with a set of solutions with different values of $0 \le \Delta_w < \Delta_{\max}$, $P_{10}$, and the base pressure $P_{11\,\min} \le P_{11} \le P_{10}$, where $P_{11\,\min}$ and $\Delta_{w\,\max}$ correspond to the choking regime. In Fig. 6.18 we have plotted the dependences of the step height on the base pressure $P_1$ for fixed values of $P_{10}$ determining the flow patterns at the forward part of the body. These dependences are presented as solid curves, where the numbers adjacent to the curves indicate the values of $P_{10}$. The dashed curve *BC* corresponds to choking regimes, while the rightmost solid curve *BD* relates to the base pressure $P'_{11}$ (point *D*) at which a separation point appears on a plate with no step.

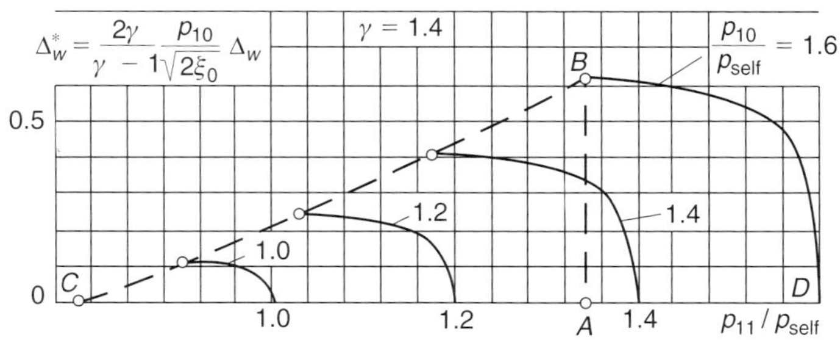


Fig. 6.18.

Using the data in Fig. 6.18 we can construct any solution of the separationless flow problem for fixed values of the step height $\Delta_w$ and the base pressure $P_{11} \le P'_{11}$. Let, for example, the base pressure $P_{11} \le P'_{11}$ and $\Delta_w$ increase starting from zero. Clearly, at a fixed

$P_{11}$ an increase in $\Delta_w$ is associated with an increase in $P_{10}$, that is, the pressure rise at the main part of the body. For the values of $P_{11}$ to the left of point $A$ the choking condition (6.193) is fulfilled for a certain $\Delta_w$ determined by curve $BC$ and the flow is no longer dependent on the base pressure.

With further increase in $\Delta_w$ the change in the flow pattern (value of $P_{10}$) takes place provided condition (6.193) is fulfilled and the $P_{10}(\Delta_w)$ dependence is determined by the dashed curve $BC$. For $\Delta_w > \Delta_B$ the flow is necessarily separated. If the base pressure $P_{11} \leq P_{10}^0$, where $P_{10}^0$ corresponds to point $C$, then, in accordance with the paper of Neiland (1970*a*), it has no effect on the flow, since the chocking condition has been already fulfilled for $\Delta_w = 0$. In this case, the $P_{10}(\Delta_w)$ dependence is also determined by the dashed curve $BC$. For the values of $P_{11}$ to the right of point $A$ the choking condition is not fulfilled for separationless flows, since an increase in $\Delta_w$ leads to separation at a certain value of $\Delta_w$ determined by curve $BD$.

Thus, if the step length $\ell/L \ll 1$, then for any base pressure the flow is separated if the following condition is fulfilled:

$$\frac{\delta_w}{L} > 0.17\frac{2\gamma}{\gamma-1}Re_0^{-1/4} \tag{6.195}$$

We recall that, in accordance with Eq. (6.189), in a locally inviscid expansion flow region the velocity at the body surface is nonzero. For this reason, in accordance with the general theory (Neiland and Sychev, 1966), we must introduce a viscous sublayer in which the viscous and inertial terms (6.181) are of the same order and the velocity $f' = o(1)$. For this purpose it is sufficient to use the internal variables

$$\xi_1 = \frac{(\xi - \xi_0)}{\varepsilon}, \quad \eta_1 = \frac{\eta^{1/2}}{\varepsilon}, \quad f = \varepsilon^{1/2} f_1 \tag{6.196}$$

Then substituting them in Eq. (6.181) and performing the internal passage to the limit

$$\varepsilon \to 0, \quad \eta_1 = O(1), \quad f_1 = O(1) \tag{6.197}$$

we obtain the equation

$$\frac{\partial^3 f_1}{\partial \eta_1^3} - \frac{\gamma-1}{\gamma}\frac{\xi_0}{P_1}\frac{dP_1}{d\xi_1}\left[1 - \left(\frac{\partial f_1}{\partial \eta_1}\right)^2\right] = 2\xi_0\left(\frac{\partial f_1}{\partial \eta_1}\frac{\partial^2 f_1}{\partial \xi_1 \partial \eta_1} - \frac{\partial f_1}{\partial \xi_1}\frac{\partial^2 f_1}{\partial \eta_1^2}\right) \tag{6.198}$$

On the basis of Eqs. (6.182) and (6.189) the boundary conditions at $f_0' = 0$ can be written in the form:

$$f_1(\xi_1, 0) = \frac{\partial f_1}{\partial \eta_1}(\xi_1, 0) = 0, \quad \frac{\partial f_1}{\partial \eta_1}(\xi_1, \infty) = \left[1 - \left(\frac{P_1}{P_{10}}\right)^{(\gamma-1)/\gamma}\right]^{1/2} \tag{6.199}$$

A general approach to the solution of these problems was developed in the paper of Neiland and Sychev (1966).

### *6.6.4 Flow patterns with small pressure differences*

The solutions obtained for the case of a finite pressure difference are determined by conditions (6.187) (viscous effects have no effect on the pressure distribution). However, the results can be considerably simplified using the condition $\delta \ll 1$ in Eq. (6.185). In this case

$$P \sim P_{10} + \delta^2 \Delta P_1 + \cdots, \quad \beta \approx \frac{\delta^2}{\varepsilon} \frac{\gamma - 1}{\gamma} \frac{\xi_0}{P_{10}} \frac{\mathrm{d}\Delta P_1}{\mathrm{d}\xi_1} + \cdots \tag{6.200}$$

Then, in accordance with Eq. (6.189), in the $\eta \sim \delta$ region the velocity disturbances are of the order of the undisturbed velocity, $\sim\delta$, while $\Delta f' = O(\delta^2)$ for $\eta \sim 1$. We will first introduce the internal variables

$$\eta = \delta\eta_2, \quad f = \delta^2 f_2(\xi_1, \eta_2), \quad \varepsilon^{1/3} \ll \delta \ll 1 \tag{6.201}$$

In these variables, after passing to the internal limit and going over to the variables $\xi_1$ and $f_1$ Eq. (6.187) takes the form:

$$-\frac{(\gamma - 1)}{2\gamma} \frac{1}{P_{10}} \frac{\mathrm{d}\Delta P_1}{\mathrm{d}\xi_1} = u_2 \frac{\partial u_2}{\partial \xi_1}, \quad u_2 = \left(\frac{\partial f_2}{\partial \eta_2}\right) \tag{6.202}$$

Integrating Eq. (6.202) with account for $u_2 = \sqrt{2f''_{0w} f_2} + \ldots$ yields

$$\eta_2(\xi_1, f_2) = \frac{1}{f''_{0w}} \left( \sqrt{2f''_{0w} f_2 - \frac{\gamma - 1}{\gamma} \frac{\Delta P_1}{P_{10}}} - \sqrt{-\frac{\gamma - 1}{\gamma} \frac{\Delta P_1}{P_{10}}} \right) \tag{6.203}$$

$$f_2(\xi_1, \eta_2) \approx \frac{1}{2} f''_{0w} \eta_2^2 + \left( -\frac{\gamma - 1}{\gamma} \frac{\Delta P_1}{P_{10}} \right)^{1/2} \eta_2 + \cdots$$

In the outer variables $(\eta, f)$ the latter formula takes the form:

$$f = \delta^2 f_2 \approx \frac{1}{2} f''_{0w} \eta^2 + \delta \left( -\frac{\gamma - 1}{\gamma} \frac{\Delta P_1}{P_{10}} \right)^{1/2} \eta + \cdots$$

Therefore, any outer solution $(\eta \sim 1, f \sim 1)$ must be sought in the form:

$$f \approx f_0(\eta) + \delta f_1(\xi_1, \eta) + O(\delta^2) \tag{6.204}$$

Substituting Eqs. (6.204) and (6.200) in Eq. (6.187) and passing to the limit $\delta \to 0$, $\varepsilon \to 0$ for $\eta = O(1)$ leads to the equation

$$f'_0 \dot{f}'_1 - f''_0 \dot{f}_1 = 0$$

whose solution is as follows:

$$f_1(\xi_1, \eta) = \Phi_1(\xi_1) f_0'(\eta) \tag{6.205}$$

Matching solutions (6.205) and (6.203) makes it possible to obtain two terms of the outer solution

$$f \approx f_0(\eta) + \frac{\delta}{f_{0w}''} \left( -\frac{\gamma - 1}{\gamma} \frac{\Delta P_1}{P_{10}} \right)^{1/2} f_0'(\eta) + \cdots \tag{6.206}$$

The second term on the right side of Eq. (6.206) allows for the outer region displacement due to the narrowing of the near-wall streamtubes, while the natural variation of the outer streamtubes is of the order of $\delta^2$.

Using the second formula (6.203) and Eq. (6.206) for calculating the function $F(\bar{\xi})$ in Eq. (6.190) we obtain the dependence of the pressure on the body shape in explicit form:

$$\frac{\sqrt{2\xi_0}}{P_{10} f_{0w}''} \sqrt{-\frac{\gamma - 1}{\gamma} \frac{\Delta P_1}{P_{10}}} + \frac{2\gamma}{\gamma + 1} \Delta_w = 0 \tag{6.207}$$

The simple form of Eq. (6.207) is due to the fact that the main part of streamtubes in the boundary layer has no effect on the pressure distribution which depends only on the thin wall layer, in which the velocity profile is determined by the value of $f_{0w}''$.

The solution of the problem for the third flow regime, in which viscosity starts to have an effect on the wall layer with nonlinear disturbances, is more complicated. In accordance with Eq. (6.178), in this case $\delta = \varepsilon^{1/3}$ in Eq. (6.185). Studying this flow pattern is important, since it corresponds to transition from separationless to separated flows for $\Delta P > 0$. However, due to the intricacy of the problem, universal results cannot be obtained, so that we are led to determine a numerical solution for any given shape $\Delta_w(\xi)$. For this reason, we will restrict ourselves to the formulation of the corresponding boundary value problem.

For determining the internal expansion we will let $\delta = \varepsilon^{1/3}$ in Eqs. (6.200) and (6.201); then instead of Eq. (6.202) we obtain the boundary layer equation

$$f_2''' + f_2 f_2'' - \beta(1 - f_2'^2) = 2\xi_0 (f_2' \dot{f}_2' - \dot{f}_2 f_2'') \tag{6.208}$$

with the no-slip and impermeability conditions imposed on the body surface

$$f_2(\xi, 0) = f_2'(\xi, 0) = 0 \tag{6.209}$$

Formulas (6.204) and (6.205), though with $\delta = \varepsilon^{1/3}$, remain valid for the outer solution. For this reason, matching with the outer solution gives the third boundary condition for Eq. (6.208) and the function $\Phi_1(\bar{\xi})$ in Eq. (6.205)

$$f_2''(\xi_1, \infty) = f_{0w}'', \quad \Phi(\xi_1) = \lim_{\eta_2 \to \infty} \left( \frac{f_2' - f_{0w}'' \eta_2}{f_{0w}''} \right) \tag{6.210}$$

Then Eq. (6.190) takes the form:

$$-\frac{\sqrt{2\xi_0}}{P_{10}}\Phi_1(\xi_1)+\frac{2\gamma}{\gamma+1}\Delta_w=0 \tag{6.211}$$

In solving the boundary value problems (6.208)–(6.210) Eq. (6.211) is used for determining the pressure.

Though weaker flow disturbances occurring for $\delta \ll \varepsilon^{1/3}$ can give rise to large local pressure gradients in the flow, they cannot lead to boundary layer separation. The formal solution of the outer problem ($\eta \sim 1$) for not-too-small $\delta$ retains the forms (6.204) and (6.205). However, everywhere up to the body surface the velocity disturbances become small as compared with the undisturbed velocity $f_0'$ at the same streamline. Therefore, the equations for the viscous sublayer admit linearization. The solution of this problem is obtained from solutions (6.208)–(6.211) for $\Delta_w \ll 1$ and is of little interest.

### *6.6.5 Concluding remarks*

The question about the temperature factor effect on the flow over a step and boundary layer separation is important. In the previous sections it is shown that the boundary layer on a cold plate is supercritical in the strong interaction regime, so that there is no upstream disturbance transfer. Moreover, in this case the integral in Eq. (6.193) is positive. Therefore, the typical flow behavior is the reverse of that for the subcritical flow regime considered above. For the subcritical and supercritical cases, small bumps and depressions can give rise to disturbances of opposite sign. For this reason, in experimental investigation of the flows under consideration the modeling of the temperature factor, as well as thermodynamic and transfer gas properties, takes on great significance.

# 7 Three-Dimensional Hypersonic Viscous Flows with Supercritical and Subcritical Regions

A fundamental analogy between a three-dimensional boundary layer (or an entropy layer) flow in an interaction regime and a two-dimensional inviscid supersonic flow is found. As the equations governing supersonic flow, the equations for cold-body boundary layer flows and wake flows possess, apart from streamsurfaces, two families of characteristic surfaces confining disturbance transfer domains. For a subcritical regime, analogous to a subsonic flow, the solution in the vicinity of the leading edge involves an arbitrary function which can be determined from the conditions on a singular line analogous to the sonic line in an inviscid flow. The equations of characteristics and "sonic" lines, as well as disturbance blocking and unblocking conditions, are derived. In particular, supercritical flows over delta wings with subcritical and supercritical leading edges are studied (an analogy with subsonic and supersonic leading edges for a wing in an inviscid supersonic flow).

## 7.1 Strong interaction between a hypersonic flow and the boundary layer on a cold delta wing

### *7.1.1 Equations and boundary conditions*

We will consider the distinctive features of disturbance propagation and the formulation of the boundary value problem for the boundary layer on a delta wing in the case in which the interaction with the outer hypersonic flow is nonweak, while the body temperature is low as compared with the freestream stagnation temperature ($g_w \ll 1$). The results obtained make it possible to draw some interesting conclusions concerning the general three-dimensional flow.

Let us introduce a Cartesian coordinate system with origin at the wing vertex, the $xl$ and $zl$ axes directed normal to one of the wing edges and along this edge in the wing plane, respectively, and the $yl\tau$ axis normal to the wing plane (Fig. 7.1).

In accordance with the conventional estimates for the boundary layer in a hypersonic flow (Hayes and Probstein, 1966), we will use the notation

$$u_\infty u, \quad u_\infty w, \quad \tau u_\infty v, \quad \tau^2 \rho_\infty \rho$$

$$\tau^2 \rho_\infty u_\infty^2 p, \quad \left(\frac{u_\infty^2}{2}\right) g, \quad \mu_0 \mu$$

for the velocity components, the density, the pressure, the stagnation enthalpy, and viscosity. The parameter $\tau = Re_0^{-1/4}$, where the Reynolds number $Re_0 = \rho_\infty u_\infty \ell / \mu_0$ is based on the

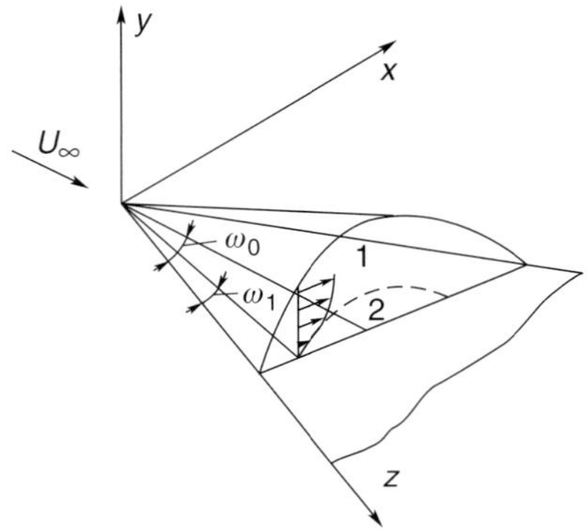


Fig. 7.1.

freestream density and velocity, the viscosity at the freestream stagnation temperature, and a scale length $\ell$ which in the self-similar case drops out of final results.

In the Dorodnitsyn variables the boundary layer equations and the boundary conditions take the form:

$$\frac{\partial u}{\partial x} + \frac{\partial v^*}{\partial \eta} + \frac{\partial w}{\partial z} = 0, \quad \frac{\partial p}{\partial \eta} = 0, \quad v^* = \rho v + u\frac{\partial \eta}{\partial x} + w\frac{\partial \eta}{\partial z} \tag{7.1}$$

$$u\frac{\partial u}{\partial x} + v^*\frac{\partial u}{\partial \eta} + w\frac{\partial u}{\partial z} = -\frac{1}{\rho}\frac{\partial p}{\partial x} + \frac{\partial}{\partial \eta}\left(\mu\rho\frac{\partial u}{\partial \eta}\right)$$

$$u\frac{\partial w}{\partial x} + v^*\frac{\partial w}{\partial \eta} + w\frac{\partial w}{\partial z} = -\frac{1}{\rho}\frac{\partial p}{\partial z} + \frac{\partial}{\partial \eta}\left(\mu\rho\frac{\partial w}{\partial \eta}\right)$$

$$u\frac{\partial g}{\partial x} + v^*\frac{\partial g}{\partial \eta} + w\frac{\partial g}{\partial z} = \frac{\partial}{\partial \eta}\left\{\rho\mu\left[\frac{1}{\sigma}\frac{\partial g}{\partial \eta} + \left(1 - \frac{1}{\sigma}\right)\frac{\partial}{\partial \eta}(u^2 + w^2)\right]\right\}$$

$$p = \frac{\gamma - 1}{2\gamma}(g - u^2 - w^2)\rho, \quad \eta = \int_0^y \rho\,\mathrm{d}y, \quad \delta = \int_0^\infty \frac{\mathrm{d}\eta}{\rho}$$

$$\rho\mu = \frac{2\gamma}{\gamma - 1}p(g - u^2 - w^2)^{n-1}, \quad p = \frac{\gamma + 1}{2}\left(\frac{\partial \delta}{\partial x}\sin\omega_0 + \frac{\partial \delta}{\partial z}\cos\omega_0\right)^2$$

$$u_w = v_w^* = w_w = g_w = 0, \quad u_e = \sin\omega_0, \quad w_e = \cos\omega_0, \quad g_e = 1$$

Here, $\omega_0$ is the angle between the $z$ axis (wing edge) and the freestream direction.

For further analysis we will make the change of variables

$$(x,\ \eta,\ z) \to \left(x,\ \zeta = \frac{1}{\omega_0}\arctan\frac{x}{z},\ \lambda = \frac{\eta}{x^{1/4}}\right) \tag{7.2}$$

$$u = \frac{\partial \psi}{\partial \eta}, \quad w = \frac{\partial \varphi}{\partial \eta}, \quad v^* = -\left(\frac{\partial \psi}{\partial x} + \frac{\partial \varphi}{\partial z}\right)$$

$$\psi = x^{1/4} f_*, \quad \varphi = x^{1/4} \varphi_*, \quad p = x^{-1/2} p_*, \quad \rho = x^{-1/2} \rho_*, \quad \delta = x^{3/4} \delta_*$$

We will write down the equations and the boundary conditions in the new variables omitting asterisks

$$(Nf'')' + \frac{1}{4} ff'' + \frac{\gamma - 1}{4\gamma}\left(1 - \frac{\sin 2\omega \times \dot{p}}{\omega_0 p}\right)(g - f'^2 - \varphi'^2)$$
$$= \frac{\sin 2\omega}{2\omega_0}(f'\dot{f}' - f''\dot{f}) - \frac{\sin^2 \omega}{\omega_0}(\varphi'\dot{f}' - f''\dot{\varphi}) \tag{7.3}$$

$$(N\varphi'')' + \frac{1}{4} f\varphi'' + \frac{\gamma - 1}{2\gamma}\frac{\sin^2 \omega}{\omega_0}\frac{\dot{p}}{p}(g - f'^2 - \varphi'^2)$$
$$= \frac{\sin 2\omega}{2\omega_0}(f'\dot{\varphi}' - \varphi''\dot{f}) - \frac{\sin^2 \omega}{\omega_0}(\varphi'\dot{\varphi}' - \varphi''\dot{\varphi})$$

$$\left(\frac{N}{\sigma} g'\right)' + \frac{1}{4} gf' + \left[N\left(1 - \frac{1}{\sigma}\right)(f'^2 + \varphi'^2)'\right]'$$
$$= \frac{\sin 2\omega}{2\omega_0}(f'\dot{g} - g'\dot{f}) - \frac{\sin^2 \omega}{\omega_0}(\varphi'\dot{g} - g'\dot{\varphi})$$

$$N = \frac{2\gamma}{\gamma - 1} p(g - f'^2 - \varphi'^2)^{n-1}, \quad \omega = \omega_0 \zeta$$

$$\delta = \frac{\gamma - 1}{2\gamma p}\int_0^\infty (g - f'^2 - \varphi'^2)\,\mathrm{d}\lambda, \quad F = \int_0^\infty (g - f'^2 - \varphi'^2)\,\mathrm{d}\lambda$$

$$f_w = \varphi_w = f'_w = \varphi'_w = g_w = 0, \quad f'_e = \sin \omega_0, \quad \varphi'_e = \cos \omega_0, \quad g_e = 1$$

$$p = \frac{\gamma + 1}{2}\left[\frac{3}{4}\delta \sin \omega_0 + \frac{1}{\omega_0}\frac{\mathrm{d}\delta}{\mathrm{d}\zeta}\left(\frac{1}{2}\sin 2\omega \sin \omega_0 - \sin^2 \omega \cos \omega_0\right)\right]^2$$

### 7.1.2 *Solution near the leading edge*

As usually (Kozlova and Mikhailov, 1970; Neiland, 1970*b*), we can construct an expansion for the solution near the leading edge

$$p(\zeta) = p_0 + p_1 \zeta^a + \cdots, \quad f(\zeta, \lambda) = f_0(\lambda) + \frac{p_1}{p_0}\zeta^a f_1(\lambda) + \cdots \tag{7.4}$$

$$\varphi(\zeta,\lambda) = \varphi_0(\lambda) + \frac{p_1}{p_0}\zeta^a\varphi_1(\lambda) + \cdots, \quad g(\zeta,\lambda) = g_0(\lambda) + \frac{p_1}{p_0}\zeta^a g_1(\lambda) + \cdots$$

$$N(\zeta,\lambda) = N_0(\lambda) + \frac{p_1}{p_0}\zeta^a N_1(\lambda) + \cdots, \quad \delta(\zeta) = \delta_0 + \frac{p_1}{p_0}\zeta^a\delta_1 + \cdots$$

$$F(\zeta) = F_0 + \frac{p_1}{p_0}\zeta^a F_1 + \cdots$$

An equation for the zeroth approximation can be derived from Eq. (7.3) by letting $\omega = 0$ in this equation. The next terms of the expansion are determined from the system of linear equations

$$(N_1 f_0'' + N_0 f_1'')' + \frac{1}{4}(f_0 f_1'' + f_1 f_0'') + \frac{\gamma-1}{4\gamma}[g_1 - 2f_0' f_1' - 2\varphi_0'\varphi_1' - 2a(g_0 - f_0'^2 - \varphi_0'^2)] = a(f_0' f_1' - f_0'' f_1) \tag{7.5}$$

$$(N_1\varphi_0'' + N_0\varphi_1'')' + \frac{1}{4}(f_0\varphi_1'' + f_1\varphi_0'') = a(f_0'\varphi_1' - \varphi_0'' f_1)$$

$$\left(\frac{N_0}{\sigma}g_1' + \frac{N_1}{\sigma}g_0'\right)' + \frac{1}{4}(f_0 g_1' + f_1 g_0') + \left\{\left(1 - \frac{1}{\sigma}\right)[N_1(f_0'^2 + \varphi_0'^2)' + 2N_0(f_0' f_1' + \varphi_0'\varphi_2')']\right\}' = a(f_0' g_1 - g_0' f_1)$$

$$N_1 = N_0\left[1 - \frac{(n-1)(g_1 - 2f_1' f_0' - 2\varphi_1'\varphi_0')}{g_0 - f_0'^2 - \varphi_0' 2}\right]$$

The boundary conditions for system (7.5) are trivial. After substitution of Eq. (7.4) in formulas for $\delta$ and $p$ in Eq. (7.3), they give two independent relations determining $\delta_1$. The constant $p_1$ drops out from the consideration owing to the special form of expansion (7.4) involving $p_1$ in the form of a factor of all functions. If $p_1$ is introduced only in the expansion for the function $p(\zeta)$, then we obtain a system of linear homogeneous equations resolvable only for an eigenvalue $a$ determined from the same condition as for Eq. (7.5), namely:

$$a = \frac{3}{4}\left[\frac{F_0}{2(F_1 - F_0)} - 1\right] \tag{7.6}$$

Of course, solution (7.4)–(7.6) is determined correct to the arbitrary constant $p_1$, if a nontrivial solution of the problem exists. However, numerical investigations showed that for $g_w = 0$ eigensolutions exist only for wings with a large sweep of the leading edge (small values of $\omega_0$). The dependence of the eigenvalue $a$ on the sweep angle $(\pi/2) - \omega_0$ is plotted in Fig. 7.2 for $\sigma = n = 1$. Thus, for wings with fairly large values of $\omega_0$ there exists a region adjoining the leading edge within which a self-similar flow, which coincides with the flow past a semi-infinite swept plate, is realized, as it was suggested by Ladyzhenskii (1965), though for almost the whole wing. An analogous phenomenon of disturbance blocking on

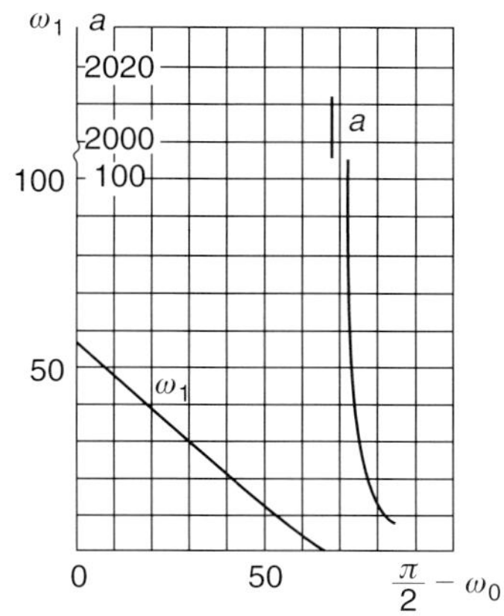


Fig. 7.2.

cold bodies in supersonic flows with free interaction was found by Neiland (1973). The extent of the supercritical flow region $\omega_1$ is presented in Fig. 7.2.

Below, in Section 7.1.6, it is shown that for $g_w = 0$ in the wing boundary layer there appear characteristic surfaces of a new type, similar to the Mach lines (characteristics) of two-dimensional inviscid flows. Regions adjoining the leading edge, in which self-similar solutions are realized for large values of $\omega_0$ (small sweep angles), lie outside a Mach cone whose vertex coincides with that of the delta wing. Then the leading edges may naturally be called supercritical, by analogy with the supersonic edges in conventional gas dynamics. Below it is shown that within the characteristic cone proceeding from the wing vertex the effect of the boundary conditions imposed on the plane of symmetry of the wing begins again to manifest itself. In what follows, these domains of disturbance transfer within Mach cones may naturally be called m-subcritical regions. As in conventional inviscid supersonic flows, they confine the domain of influence of the outer boundary conditions. Of course, this is not the upstream disturbance transfer, which exists for $g_w = O(1)$ (subcritical regions) and is similar to the subsonic flow behavior.

### *7.1.3 Flow regimes*

The nature of the flow over a cold delta wing depends on the regime of the boundary layer/outer hypersonic flow interaction (Neiland, 1974*b*). In m-subcritical regimes disturbances can propagate from the plane of symmetry toward the edges. For choosing an unique solution an additional boundary condition should be taken, for example, the value of the pressure at a certain surface $\zeta = \text{const}$. In the supercritical flow region disturbances do not propagate upstream and the flow is governed by the self-similar solution of system (7.3). On a cold wing displacement thickness variation is produced by the main part of the boundary layer and this variation depends linearly on the pressure disturbance (Neiland, 1974). The flow regime is determined by the sign of the derivative $d\delta/dp$. The supercritical regime is realized for $d\delta/dp < 0$ and the subcritical regime for $d\delta/dp > 0$.

With decrease in the wing sweep angle $\chi = (\pi/2) - \omega_0$ the derivative $\frac{d\delta}{dp}|_{\zeta=0}$ decreases monotonically, vanishes at a certain critical value $\chi^* = (\pi/2) - \omega^*$, and remains negative with further decrease in the sweep angle to zero. Thus, for $\omega_0 > \omega^*$ in the delta-wing boundary layer there exist the regions of both supercritical and m-subcritical flow and transition from the flow of one type to the other takes place at a certain value $\omega^* < \omega_1 < \omega_0$ (Neiland, 1974*b*). The flow in the region between the leading edge and the $\omega = \omega_1$ surface is governed by the self-similar solution of system (7.3). In the region between the $\omega = \omega_1$ surface and the plane of symmetry of the wing the effect of the upstream disturbance transfer should be taken into account.

We note that in proving the principle of impossibility of disturbance localization in the flows with nonweak interaction between the boundary layer and the outer hypersonic flow (Neiland, 1970*b*) the flow pattern in a thin wall layer, in which velocity disturbances induced by a small pressure difference are of the same order as the velocity itself, is important. Precisely a large value of $\Delta\delta \sim g_w \Delta p^{1/2}$ produced by this sublayer at small pressure disturbances $\Delta p$ leads to the conclusion on inevitable upstream disturbance transfer through finite distances. However, for $g_w \to 0$ (one more case of disturbance blocking under the action of an intense expansion wave was studied by Neiland, 1972) the quantity $\Delta\delta$ depends completely on the deformation of the flow parameter profiles in the main part of the boundary layer. There $\Delta\delta \sim \Delta p$, while the sign of $d\delta/dp$ depends on the Mach number profile across the boundary layer. If $d\delta/dp < 0$, then, in accordance with the terminology introduced for plane flows by Crocco (1955), the layer is supercritical and the disturbance transfer is absent. The subcritical case $d\delta/dp > 0$ corresponds to upstream disturbance propagation. The rough integral approach adopted in the work of Crocco (1955) did not take account for the effect of sublayers and led to a qualitative discrepancy with rigorous solutions when studying the flows with $g_w \sim O(1)$. However, for the flows with $g_w \to 0$ there appears a qualitative agreement: the flow, "subsonic" in the mean ($d\delta/dp > 0$) transmits disturbances, whereas the "supersonic" flow does not.

On transition to three-dimensional flows the analogy with the subsonic and supersonic types of the pressure disturbance transfer is also conserved. The solution of Kozlova and Mikhailov (1970) with $g_w = O(1)$ corresponds to the subsonic case, while the above solution with $g_w = 0$ to the supersonic case. However, in an inviscid supersonic flow (we note that even for $g_w = 0$ the characteristic value of the Mach number within the boundary layer is not hypersonic, $M \sim 1$) within Mach cones there are directions, along which disturbances are transferred. For this reason, we will consider a delta wing with a large $\omega_0$ and $g_w = 0$ and seek a value of $\omega_1$ at which the eigensolutions of the problem, which make it possible to take account for disturbances proceeding from the plane of symmetry or neighboring regions with large $\zeta$, appear at a certain ray $0 < \zeta_1 < 1$, unknown beforehand.

In the region $0 < \zeta < \zeta_1$ there is a self-similar solution corresponding to the flow over a flat plate with a leading edge which is not necessarily perpendicular to the freestream velocity. This solution corresponds to the zeroth term of expansion (7.4).

### *7.1.4 Analysis of the solution in the vicinity of the critical section*

For $\lambda = O(1)$ the viscosity forces have no effect on the disturbed boundary layer flow near the $\omega = \omega_1$ plane. An analysis of the system of equations governing this flow shows (Dudin and Lipatov, 1985) that it can be once integrated with respect to $\zeta$; hence follows that the disturbances of the functions $f$ and $g$ are proportional to the pressure disturbance $p_1(\zeta_1)$. To the right of the $\omega = \omega_1$ plane, we will represent the solution in the form of the expansions for the flow functions

$$p = p_0 + p_1(\zeta_1) + \cdots, \quad \delta = \delta_0 + \delta_1 + \cdots, \quad \zeta_1 = \zeta - \frac{\omega_1}{\omega_0} \tag{7.7}$$

$$f = f_0(\lambda) + \frac{p_1(\zeta_1)}{p_0} f_1(\lambda) + \cdots, \quad \varphi = \varphi_0(\lambda) + \frac{p_1(\zeta_1)}{p_0} \varphi_1(\lambda) + \cdots$$

$$g = g_0(\lambda) + \frac{p_1(\zeta_1)}{p_0} g_1(\lambda) + \cdots$$

where functions with the subscript 0 correspond to the self-similar solution. Substituting Eq. (7.7) in the system of equations (7.3) yields a linear system of equations for the first approximation, the solution of which was derived by Neiland (1974*b*). We note that for $\lambda = O(1)$ this solution is independent of the form of the function $p_1(\zeta_1)$.

The equations for zeroth terms correspond to Eq. (7.3) for $\omega = 0$ (self-similarity in the undisturbed flow region)

$$-\frac{\gamma - 1}{4\gamma} \sin 2\omega_1 (g_0 - f_0'^2 - \varphi_0'^2) = \frac{\sin 2\omega_1}{2} (f_0' f_1' - f_0'' f_1) - \sin^2 \omega_1 (\varphi_0' f_1' - f_0'' \varphi_1) \tag{7.8}$$

$$\frac{\gamma - 1}{2\gamma} \sin^2 \omega_1 (g_0 - f_0'^2 - \varphi_0'^2) = \frac{\sin 2\omega_1}{2} (f_0' \varphi_1' - \varphi_0'' f_1) - \sin^2 \omega_1 (\varphi_0' \varphi_1' - \varphi_0'' \varphi_1)$$

$$\frac{1}{2} \sin 2\omega_1 (f_0' g_1 - g_0' f_1) - \sin^2 \omega_1 (\varphi_0' g_1 - g_0' \varphi_1) = 0$$

$$\omega_1 = \omega_0 \zeta_1, \quad (\delta_1 + \delta_0) \frac{2\gamma}{\gamma - 1} p_0 = F_1$$

$$F_1 = \int_0^\infty (g_1 - 2f_1' f_0' - 2\varphi_1' \varphi_0')\, \mathrm{d}\lambda, \quad \delta_1 = 0$$

These expansions are invalid in the vicinity $\lambda \to 0$, where it is necessary to consider a region $\bar{\lambda} = \lambda/\sigma(\Delta\zeta) = O(1)$ with the scale $\sigma(\Delta\zeta) \to 0$ as $\Delta\zeta \to 0$, while $\sigma(\Delta\zeta)$ must be so introduced that the viscous terms of Eqs. (7.3) are retained and the solution admits matching with solution (7.8).

Equation (7.8) can be integrated as follows:

$$f_1' = \frac{\gamma - 1}{2\gamma} \left[ -\frac{\cos \omega_1 (g_0 - f_0'^2 - \varphi_0'^2)}{\cos \omega_1 f_0' - \sin \omega_1 \varphi_0'} - f_0'' \int_0^\lambda \frac{g_0 - f_0'^2 - \varphi_0'^2}{(\cos \omega_1 f_0' - \sin \omega_1 \varphi_0')^2} \mathrm{d}\lambda \right] \tag{7.9}$$

$$\varphi_1' = \frac{\gamma-1}{2\gamma}\left[\frac{\sin\omega_1(g_0 - f_0'^2 - \varphi_0'^2)}{\cos\omega_1 f_0' - \sin\omega_1\varphi_0'} - \varphi_0''\int\limits_0^\lambda \frac{g_0 - f_0'^2 - \varphi_0'^2}{(\cos\omega_1 f_0' - \sin\omega_1\varphi_0')^2}\,\mathrm{d}\lambda\right]$$

$$g_1 = -\frac{\gamma-1}{2\gamma}g_0'\int\limits_0^\eta \frac{g_0 - f_0'^2 - \varphi_0'^2}{(\cos\omega_1 f_0' - \sin\omega_1\varphi_0')^2}\,\mathrm{d}\lambda$$

According to the last formulas (7.8), in order for an eigensolution to exist the quantity $\zeta_1$ must satisfy the equation

$$F_0 = \frac{\gamma-1}{2}\int\limits_0^\infty \left(\frac{g_0 - f_0'^2 - \varphi_0'^2}{\cos\omega_1 f_0' - \sin\omega_1\varphi_0'}\right)^2 \mathrm{d}\lambda \tag{7.10}$$

where

$$F_0 = \int\limits_0^\infty (g_0 - f_0'^2 - \varphi_0'^2)\,\mathrm{d}\lambda$$

Then in region $0 \le \zeta \le \zeta_1$ the solution is self-similar, while the disturbed part of the flow region and the eigensolutions determined correct to an arbitrary constant $p_1$ appear for $\zeta = \zeta_1$. The dependence of $\omega_1$ on the leading edge sweep angle $(\pi/2) - \omega_0$ is presented in Fig. 7.2.

The expression for the displacement thickness formed in the $\lambda = O(1)$ region is as follows:

$$\delta_1 = \frac{\gamma-1}{2\gamma p_0}\Delta p\left[\frac{\gamma-1}{2}\int\limits_0^\infty \left(\frac{g_0 - f_0'^2 - \varphi_0'^2}{f_0'\cos\omega_1 - \varphi_0'\sin\omega_1}\right)^2 \mathrm{d}\lambda - \int\limits_0^\infty (g_0 - f_0'^2 - \varphi_0'^2)\,\mathrm{d}\lambda\right] \tag{7.11}$$

where $\Delta p = p_1/p_0$.

The derivative $\mathrm{d}\delta_1/\mathrm{d}\Delta p$ is linearly dependent on the coordinate $\zeta$. Differentiating Eq. (7.11) with respect to the coordinate $\zeta$ gives the following expression:

$$\frac{\mathrm{d}\delta_1}{\mathrm{d}\Delta p} = \frac{(\gamma-1)^2}{4\gamma p_0}\omega_0 J_1\zeta_1, \quad J_1 = 2\int\limits_0^\infty (g_0 - f_0'^2 - \varphi_0'^2)\frac{f_0'\sin\omega_1 + \varphi_0'\cos\omega_1}{(f_0'\cos\omega_1 - \varphi_0'\sin\omega_1)^3}\,\mathrm{d}\lambda \tag{7.12}$$

The solution for the first approximation is not uniformly accurate, since it does not take the viscosity effect into account. In order to satisfy the boundary conditions imposed on the wing surface, it is necessary to bring into consideration region 2 in which the influence of the viscosity and inertia forces is the same in the first approximation. The wall region 2 (Fig. 7.1) induces a variation of the displacement thickness $\Delta_1$. The estimates for the region 2 thickness and the scales of the functions in this region are obtained by equating the orders of the terms of the system of equations responsible for the effects of the viscosity and inertia forces. Finally, the requirement of the matching of solutions in regions 1 and 2 leads to the equality of the orders of the pressure gradient and the inertial terms thus making it possible to determine – at the known region 2 thickness – the scales of the disturbed functions $f$, $\varphi$, and $g$. The expression for the variation of the displacement thickness formed in region 2 is as follows:

$$\Delta_1 = c_1 p_1 \left(\frac{p_1}{\dot{p}_1}\right)^{1/3}, \quad c_1 = \text{const} \tag{7.13}$$

For determining the function $p_1(\zeta_1)$ the interaction condition must be used. The assumption that the main contribution to the displacement thickness variation is formed in region 1 leads, in view of the interaction condition, to a function of the form $p_1 = c\zeta_1^{\alpha}$. This expression can be fairly simply obtained by substituting $\delta_1$ (7.11) in the expression for $p_1 = c\zeta_1^{\alpha}$ obtained by substitution of the expansions for the flow functions (7.7) in the expression for the pressure (see Eq. (7.3)) with account for Eq. (7.12). Then the flow in region 2 induces the thickness variation $\Delta_1 \approx \zeta_1^{1/3} p_1$ which is greater in the order than $\delta_1 \approx \zeta_1 p_1$ for all permissible values $\alpha > 0$. The assumption that the main contribution to the displacement thickness is formed in region 2 also does not lead to a self-consistent scheme. In this case it turns out that the induced pressure disturbance is greater than the original disturbance.

An analysis of the interaction condition shows that a solution of Eq. (7.7) exists if in the first approximation the total variation of the displacement thickness is equal to zero

$$\Delta\delta = \delta_1 + \Delta_1 = 0 \tag{7.14}$$

The interaction of analogous nature was described in the papers of Stewartson (1969), in which the flow in the vicinity of the leading edge of a plate was studied, and Bogolepov and Neiland (1976), in which the flow over small roughnesses at the bottom of a laminar boundary layer was investigated. The expression for the function $p_1(\zeta_1)$ follows from Eqs. (7.12) and (7.13) and takes the form:

$$p_1(\zeta_1) = c\exp\left(-\frac{\alpha}{\zeta_1^2}\right) \tag{7.15}$$

The total displacement thickness variation can be derived from Eq. (7.3)

$$\Delta\delta = \frac{\gamma - 1}{2\gamma p_0}\Delta p\left[\int_0^\infty (g_1 - 2f_0'f_1' - 2\varphi_0'\varphi_1' - g_0 + f_0'^2 + \varphi_0'^2)d\lambda + \zeta_1\int_0^\infty (g_2 - g_1(0))\,d\eta\right] + O\left(\zeta_1^2\frac{p_1}{p_0}\right), \quad \eta = \frac{\lambda}{\sqrt{\zeta_1}} \tag{7.16}$$

The first integral on the right side of Eq. (7.16) represents the contribution of region 1 and, in accordance with Eq. (7.12), can be written in the form $\delta_1 = [(\gamma - 1)^2/4\gamma p_0]\Delta p\,\omega_0 J_1\zeta_1$.

The second integral in Eq. (7.16) is the contribution of the wall region to the displacement thickness variation; for determining this integral a solution for region 2 should preliminarily be found.

In region 2 we introduce the following expansions of the functions:

$$
\begin{aligned}
f &= \frac{1}{2}a\eta^2\zeta_1 + \frac{c}{p_0}\sqrt{\zeta_1}\exp\left(-\frac{\alpha}{\zeta_1^2}\right)f_2(\eta) + \cdots \\
\varphi &= \frac{1}{2}b\eta^2\zeta_1 + \frac{c}{p_0}\sqrt{\zeta_1}\exp\left(-\frac{\alpha}{\zeta_1^2}\right)\varphi_2(\eta) + \cdots \\
g &= \mathrm{d}\eta^2\zeta_1^{1/2} + \frac{c}{p_0}\exp\left(-\frac{\alpha}{\zeta_1^2}\right)g_2(\eta) + \cdots \\
p &= p_0 + c\exp\left(-\frac{\alpha}{\zeta_1^2}\right) + \cdots, \qquad \eta = O(1)
\end{aligned}
\tag{7.17}
$$

where $c$ is an arbitrary constant. The first terms of the expansions for the functions $f$, $\varphi$, and $g$ are the asymptotic representations of the functions $f_0$, $\varphi_0$, and $g_0$ as $\lambda \to 0$. The parameters $a$, $b$, and $d$ are determined from the self-similar solution. Substituting Eq. (7.17) in the system of equations (7.13) leads after certain transformations to the following first-approximation system of equations

$$
\begin{aligned}
& z_1''' - \eta_1 = \eta_1 z_1', \quad z_2''' - \eta_1 = \eta_1 z_2' - z_2, \quad z_3''' - \eta_1 = \eta_1 z_3 - z_3 \\
& z_1(0) = z_1'(0) = z_2(0) = z_2'(0) = z_3(0) = 0 \\
& z_1'(\infty) = -1 + O(1), \quad z_2'(\infty) = -\ln\eta_1 + O(1), \quad z_3'(\infty) = -\ln\eta_1 + O(1)
\end{aligned}
$$

where

$$
\begin{aligned}
z_1 &= 2^{4/3}\alpha^{1/3}\gamma\omega_0 T^{4/3}\frac{bf_2 - a\varphi_2}{\mathrm{d}(\gamma-1)\sin\omega_1(b\cos\omega_1 + a\sin\omega_1)} \\
z_2 &= 2^{4/3}\alpha^{1/3}\gamma\omega_0 T^{4/3}\frac{f_2\cos\omega_1 - \varphi_2\sin\omega_1}{\mathrm{d}(\gamma-1)\sin\omega_1} \\
z_3 &= 2\gamma\omega_0^2 T^2\frac{g_2}{\mathrm{d}(\gamma-1)\sin^2\omega_1} \\
\eta &= \eta_1 T^{-1/3}2^{-1/3}\alpha^{-1/3}, \quad T = \frac{\sin\omega_1}{\omega_0(a\cos\omega_1 - b\sin\omega_1)}
\end{aligned}
$$

The solution of the system of equations governing the flow in region 2 leads, in view of Eqs. (7.14) and (7.16), to an expression for the eigenvalue $\alpha$

$$\alpha = \frac{1}{2}\left(\frac{d^2 J_2 \sin^2 \omega_1}{\gamma \omega_0^3}\right)^3, \quad J_2 = \int_0^\infty [z_3(\eta_1) - z_3(\infty)]\,\mathrm{d}\eta$$

The dependence $\alpha(\chi)$ is presented in Fig. 7.3. Clearly, with increase in the leading edge sweep angle $\chi = (\pi/2) - \omega_0$ the eigenvalue $\alpha(\chi)$ increases monotonically and tends to infinity as the sweep angle tends to the value $\chi \to \chi^* = (\pi/2) - \omega^*$. The increase of the eigenvalue $\alpha(\chi)$ is due to a decrease in the intensity of the upstream disturbance transfer. This variation of the nature of the disturbance transfer can be attributed to the fact that in region 2 the thickness displacement variation is inverse proportional to the exponents of both the eigenvalue $\alpha(\chi)$ and the parameter $\omega_1$.

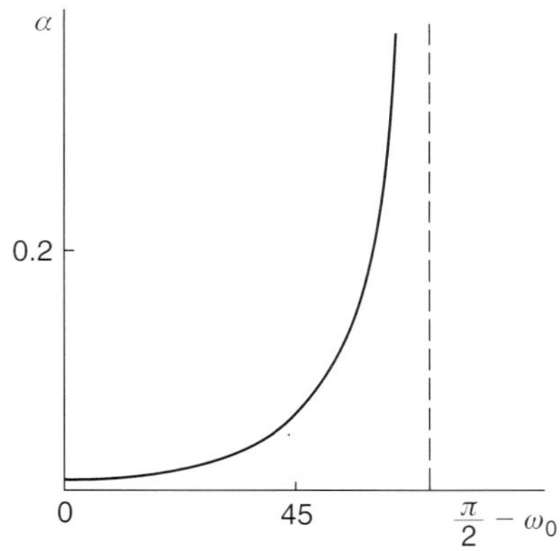


Fig. 7.3.

Condition (7.14) requires the conservation of the quantity $\Delta_1$; therefore, with decrease of the distance from the leading edge to the plane, in which the disturbed flow begins, the eigenvalue $\alpha(\chi)$ must increase.

The physical meaning of the growth of the disturbance transfer intensity (decrease of $\alpha(\chi)$) with decrease in the wing sweep is related with the fact that the supercritical flow region length increases (in $\zeta$) with increasing $\omega_0$; therefore, the flow function gradients decrease, the velocity profiles become less convex, and the thickness of the layer formed by subsonic streamtubes increases.

As $\omega_0 \to \omega^*$, the functions $\sin^2 \omega$ and $\sin 2\omega$ entering in the system of equations (7.3) can be approximated by the first terms of the series expansion, since $\omega = \omega_1 + \omega_0 = O(1)$. Then for estimating the region 2 thickness we obtain:

$$\lambda_2 \sim \left[\frac{p_1}{(\omega_1 + \omega_0 \zeta_1)} \dot{p}_1\right]^{1/3}$$

Accordingly, the displacement thickness variation formed in region 2 is determined as follows:

$$\Delta_1 = Ap_1 \left[ \frac{p_1}{(\omega_1 + \omega_0 \zeta_1)\dot{p}_1} \right]^{1/3}, \quad A = \left( \frac{\gamma - 1}{2\gamma} \right)^2 \frac{d^2 \omega_0^{1/3} J_2}{p_0 a^{7/3}} < 0 \tag{7.18}$$

In view of the expression for $\Delta_1$, condition (7.14) takes the form:

$$B\zeta_1 p_1 + Ap_1 \left[ \frac{p_1}{(\omega_1 + \omega_0 \zeta_1)} \dot{p}_1 \right]^{1/3}, \quad B = \frac{(\gamma - 1)^2}{4\gamma p_0} \omega_0 J_1 \tag{7.19}$$

The solution of the differential equation (7.19) is as follows:

$$p_1 = c \exp \left\{ \frac{A^3}{\omega_0 B^3} \left( \frac{\omega_0}{\omega_1} \right)^3 \left[ \ln \left( \frac{\omega_0 \zeta_1}{\omega_1 + \omega_0 \zeta_1} \right) + \frac{\omega_1}{\omega_0 \zeta_1} - \frac{\omega_1^2}{2\omega_0^2 \zeta_1^2} \right] \right\} \tag{7.20}$$

For $\omega_1 = O(1)$ expression (7.19) coincides in the leading term with Eq. (7.15). As $\omega_1 \to 0$ and for a small but fixed value $\zeta_1$, the right side of Eq. (7.20) having been expanded in a series in the small parameter $\omega_1/\omega_0\zeta_1$ leads to the function

$$p_1 = c \exp \left( \frac{A^3}{3B^3 \omega_0 \zeta_1^3} \right) \tag{7.21}$$

As $\omega_0 \to \omega^*$ $(\omega_0 - \omega^* < 0)$ the variation of the displacement thickness of region 1 can be presented in the form $\delta_1 = B[\zeta_1 - 1 + (\omega^*/\omega_0)]p_1$. The estimate for the region 2 thickness is obtained from Eq. (7.18) letting there $\omega_1 = 0$. Accordingly, condition (7.14) takes the form:

$$B \left( \zeta_1 - 1 + \frac{\omega^*}{\omega_0} \right) p_1 + Ap_1 \left( \frac{p_1}{\omega_0 \zeta_1 \dot{p}_1} \right)^{1/3} = 0 \tag{7.22}$$

Then the solution of Eq. (7.22) is as follows:

$$p_1 = c \exp \left\{ \left( \frac{A}{B(\omega^* - \omega_0)} \right)^3 \left[ \ln \left( \frac{\omega_0 \zeta_1}{\omega^* - \omega_0 + \omega_0 \zeta_1} \right) + \frac{\omega^* - \omega_0}{\omega^* - \omega_0 + \omega_0 \zeta_1} + \frac{1}{2} \left( \frac{\omega^* - \omega_0}{\omega^* - \omega_0 + \omega_0 \zeta_1} \right)^2 \right] \right\} \tag{7.23}$$

As $\zeta_1 \to 0$, for finite values of the parameter $\omega^* - \omega_0$ we obtain a power-law eigenfunction of the form $p = c\zeta_1^\alpha$, which is in agreement with the results of Neiland (1974*b*) and formula (7.4). The passage to the limit $\zeta \to 0$ and $\omega_0 \to \omega^*$ leads to function (7.21). Thus, in the subcritical flow regime continuous transition of the coordinate expansions near the leading edge to the expansions describing the beginning of the interaction in the flow containing the regions of both subcritical and supercritical flows is ensured.

### *7.1.5 Aerodynamic characteristics of delta wings*

The numerical solution of the system of equations for the three-dimensional boundary layer on a delta wing in the strong viscous interaction regime was obtained using the method described in the book of Bashkin and Dudin (2000). It should be noted that in the calculations the system of boundary layer equations (5.114)–(5.116) written in the coordinate system fitted to the axis of symmetry of the wing was used (in considering semi-infinite wings, $x = 0$ should be let in these equations).

In numerically integrating the system of equations the vicinity of $\omega = \omega_1$ was not specially singled out, so that the boundary value problem was solved from one leading edge up to the other.

Figures 7.4–7.6 present the calculated results which allow one to investigate the temperature factor effect; in these figures the spanwise pressure distributions are plotted in variables (5.113) at $z_0 = 2$, 1, and 0.5 for $\sigma = 1$, $n = 1$, $\gamma = 1.4$, and $g_w = 0.5$, 0.4, 0.3, 0.2, 0.1, and 0. It should be noted that with decrease in the body surface temperature the pressure also decreases, which is due to a decrease in the boundary layer displacement thickness, whereas the region, in which a near-self-similar flow is realized, enlarges. Crosses on the curves corresponding to the flow over a cold wing ($g_w = 0$) indicate points of transition from the supercritical to the m-subcritical flow regime obtained in accordance with the paper of Neiland (1974*b*).

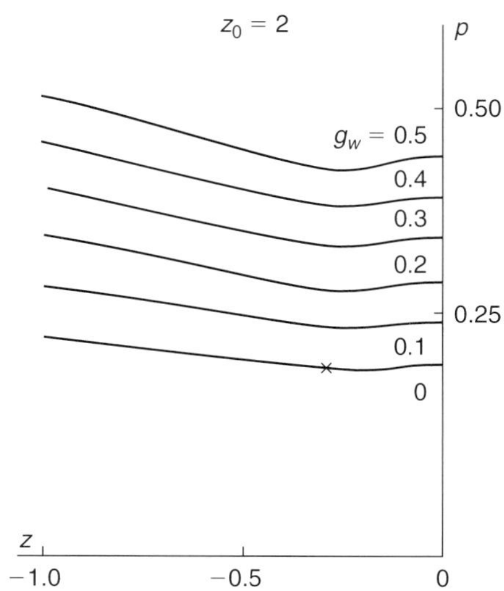


Fig. 7.4.

In Fig. 7.7 curves 1 to 4 relate to the pressure distributions in the coordinate $\omega$ for different wing aspect ratios $z_0 = \tan\omega_0 = 0.5$, 0.8, 1, and 2 at $g_w = 0$. Here, crosses indicate the values of the coordinate $\omega_1$ corresponding to transition from the supercritical to the m-subcritical flow regime in accordance with the vanishing of expression (7.11) for the displacement thickness variation. The results of the numerical solution of the boundary value problem

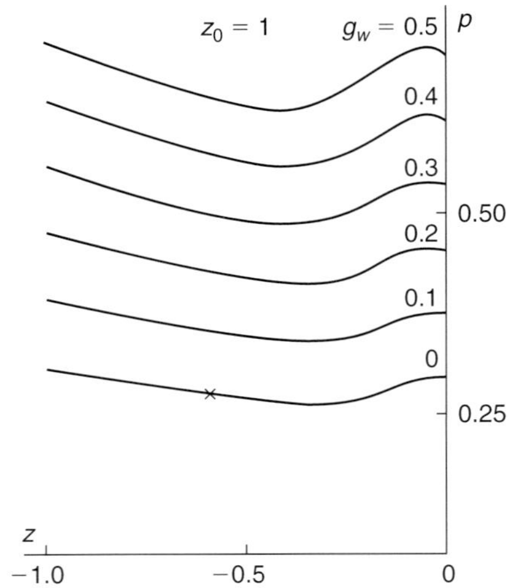


Fig. 7.5.

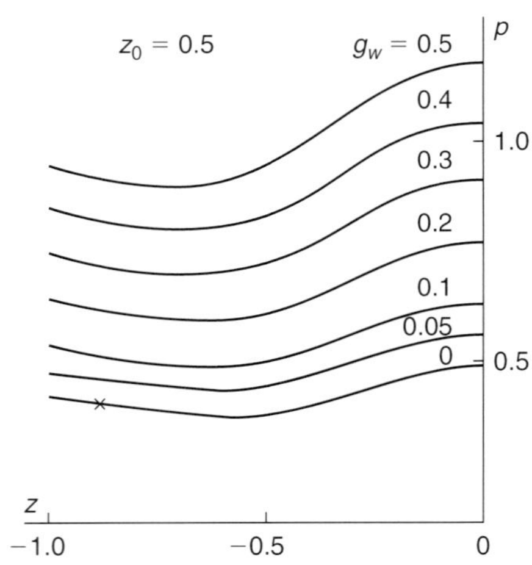


Fig. 7.6.

show that the departure from self-similar solutions, that is, transition from the supercritical to the m-subcritical flow regime, takes place in accordance with the values of $\omega_1$ determined by Eq. (7.11). Thus, disregarding the fine structure of the flow in the vicinity of $\omega = \omega_1$ in solving the global problem does not result in appreciable errors in determining the pressure distribution.

The flows with regions of subcritical and supercritical interaction studied above are realized under fairly smooth boundary conditions in the plane of symmetry of the wing. As noted in the paper of Neiland (1974*b*), other types of flows with m-subcritical and supercritical

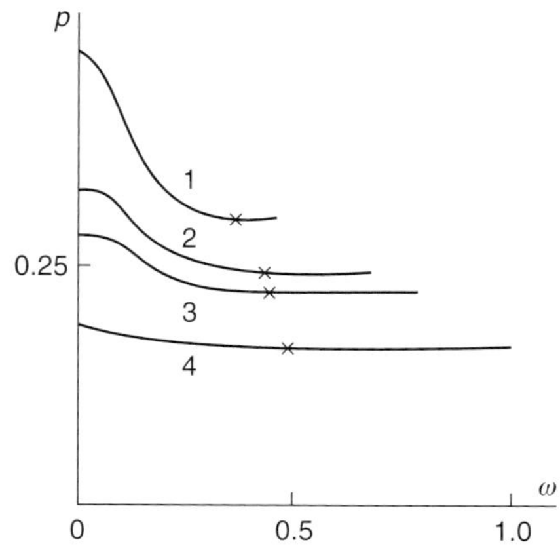


Fig. 7.7.

regions are also possible; in this case, transition from one region to the other occurs over small distances due to large disturbances preassigned downstream.

### *7.1.6 Characteristics for supercritical boundary layers and wakes for an arbitrary wing planform*

In the previous section an analogy between disturbance propagation in a supercritical three-dimensional boundary layer and a supersonic inviscid flow was found with reference to the particular example of a delta wing. It was shown that by varying the wing sweep we can derive an analogy with the supersonic inviscid flow over wings with supersonic or subsonic leading edges. In the case of the strong viscous hypersonic interaction it is the presence of supercritical regions near the leading edge for small sweep angles or their absence for large sweep angles. It is natural to try to construct characteristic surfaces and the corresponding relations for the general case (apart from the characteristics related with streamsurfaces, e.g., see Wang, 1971).

Following the conventional technique, we will seek the surfaces at which the normal derivatives of the flow functions are not uniquely determined by preassigning initial data on these surfaces.

We introduce a curvilinear coordinate system $(\xi, \eta, \zeta)$. The $\eta = 0$ surface coincides with the body surface or a certain surface within the wake downstream of the body; the coordinate $\eta$ is measured along the normal to this surface. Some initial data are given at a cylindrical surface $\xi = \text{const}$ normal to the body surface, that is, the flow functions are preassigned as functions of $\zeta$ and $\eta$ ($\zeta = \text{const}$ are cylindrical surfaces orthogonal to the $\xi$ and $\eta$ families). It is required to determine the derivatives with respect to the coordinate $\xi$.

We will present the boundary layer equations in this coordinate system

$$\frac{u}{h_1}\frac{\partial u}{\partial \xi} + v^* \frac{\partial u}{\partial \eta^*} + \frac{w}{h_3}\frac{\partial u}{\partial \zeta} = -\frac{1}{\rho h_1}\frac{\partial p}{\partial \xi} + \frac{\partial \tau_1}{\partial \eta^*} - l_1 \tag{7.24}$$

$$\frac{u}{h_1}\frac{\partial w}{\partial \xi} + v^* \frac{\partial w}{\partial \eta^*} + \frac{w}{h_3}\frac{\partial w}{\partial \zeta} = -\frac{1}{\rho h_3}\frac{\partial p}{\partial \zeta} + \frac{\partial \tau_3}{\partial \eta^*} - l_3$$

$$\frac{u}{h_1}\frac{\partial g}{\partial \xi} + v^* \frac{\partial g}{\partial \eta^*} + \frac{w}{h_3}\frac{\partial g}{\partial \zeta} = \frac{\partial G}{\partial \eta^*}, \quad \frac{\partial p}{\partial \eta^*} = 0$$

$$l_1 = \frac{w}{h_1 h_3}\left(u\frac{\partial h_1}{\partial \zeta} - \frac{w}{h_1 h_3}\frac{\partial h_3}{\partial \xi}\right), \quad l_3 = \frac{u}{h_1 h_3}\left(w\frac{\partial h_3}{\partial \xi} - \frac{\partial h_1}{\partial \zeta}\right)$$

where $h_1$ and $h_3$ are the scale factors. The quantities $\tau_1$ and $\tau_3$ are the viscous stresses in $\xi$ and $\eta$ and $G$ is the heat flux. We will not specify the form of these functions, since further calculations are valid for not only laminar flows but also for inviscid flows occurring at strong injection or entropy layers from which these terms are absent. The consideration presented is also applicable for certain types of approximate equations governing turbulent flows. The quantities $v^*$ and $\eta^*$ are the Dorodnitsyn variables; in what follows, the asterisks at these variables will be dropped out, as in Eqs. (7.1) and (7.2).

We also have

$$\delta = \int_{\eta_1}^{+\infty} \frac{d\eta}{\rho}, \quad g - u^2 - w^2 = \frac{2\gamma}{\gamma - 1}\frac{p}{\rho} \tag{7.25}$$

$$p = \frac{\gamma + 1}{2}\left(\frac{1}{h_1}\frac{\partial \delta}{\partial \xi}\cos\omega - \frac{1}{h_3}\frac{\partial \delta}{\partial \zeta}\sin\omega\right)^2$$

where $\omega$ is the angle which the freestream velocity vector makes with the plane tangent to the $\xi$ axis, normal to $\eta = 0$. The quantity $\eta_1$ is equal to 0 or $-\infty$ for the flows over the body and in the wake, respectively.

The boundary conditions are as follows:

$$\eta \to \pm\infty, \quad u \to \cos\omega, \quad w \to \sin\omega, \quad g \to 1 \tag{7.26}$$

$$\eta = 0, \quad u = v = w = g = 0$$

The condition for $\eta \to -\infty$ is written down for the wake and that for $\eta = 0$ for the flow past the body surface.

From Eqs. (7.25) we obtain

$$\frac{1}{\delta\cos\omega}\left(\sqrt{\frac{2p}{\gamma + 1}} + \frac{1}{h_3}\frac{\partial \delta}{\partial \zeta}\sin\omega\right) = \frac{1}{F h_1}\frac{\partial F}{\partial \xi} - \frac{1}{p h_1}\frac{\partial p}{\partial \xi} \tag{7.27}$$

$$F = \int_{\eta_1}^{+\infty} (g - q^2)\, d\eta, \quad q^2 = u^2 + w^2$$

Differentiating under the integral sign in $F$ and substituting the expression for the derivatives of $u$, $w$, and $g$ with respect to $\xi$ from Eq. (7.24) we obtain

$$A = \int_{\eta_1}^{\infty} \left[ \frac{1}{u}\frac{\partial G}{\partial \eta} - \frac{w}{uh_3}\frac{\partial g}{\partial \zeta} + 2\left(\frac{w}{h_3}\frac{\partial u}{\partial \zeta} - \frac{\partial \tau_1}{\partial \eta} + l_1\right) + \frac{2w}{u}\left(\frac{w}{h_3}\frac{\partial w}{\partial \zeta} + \frac{1}{\rho h_3}\frac{\partial p}{\partial \zeta} - \frac{\partial \tau_3}{\partial \eta} + l_3\right)\right] \mathrm{d}\eta - \frac{F}{\delta \cos\omega} \times \left(\sqrt{\frac{2p}{\gamma+1}} + \frac{\sin\omega}{h_3}\frac{\partial \delta}{\partial \zeta}\right) \tag{7.28}$$

$$A = \int_{\eta}^{+\infty} \frac{v}{u}\frac{\partial}{\partial \eta}(g - q^2)\,\mathrm{d}\eta + \frac{F}{\gamma p h_1}\frac{\partial p}{\partial \xi}$$

Introducing streamfunctions in such a way that $\psi(\xi, 0, \zeta) = 0$, from the first two equations (7.24) we obtain

$$\frac{1}{h_3}\frac{\partial \psi}{\partial \xi} = u\int_0^{\eta} h_1\left(\frac{\partial \varphi}{\partial \zeta}\frac{\partial u}{\partial \eta}\frac{1}{h_1 h_3} - \frac{w}{h_3}\frac{\partial w}{\partial \zeta} + \frac{\partial \tau_1}{\partial \eta} + \frac{u^2}{h_1 h_3}\frac{\partial h_3}{\partial \xi} - l_1\right)\frac{\mathrm{d}\eta}{u^2} - \frac{\gamma - 1}{2\gamma}\frac{u}{p}\frac{\partial p}{\partial \xi}\int_0^{\eta}\frac{g - q^2}{u^2}\,\mathrm{d}\eta \tag{7.29}$$

$$h_3 u = \frac{\partial \psi}{\partial \eta}, \quad h_1 w = \frac{\partial \varphi}{\partial \eta}, \quad h_1 h_3 v = -\left(\frac{\partial \psi}{\partial \xi} + \frac{\partial \varphi}{\partial \zeta}\right)$$

Using Eq. (7.29) for eliminating $v$ from Eq. (7.28) we obtain an explicit expression for the pressure gradient normal to the $\xi = \mathrm{const}$ surface

$$\frac{1}{\gamma p h_1}\frac{\partial p}{\partial \xi} = \frac{A - B - D}{F - \frac{\gamma-1}{2}F_1}, \quad F_1 = \int_{\eta_1}^{+\infty}\left(\frac{g - q^2}{u}\right)^2 \mathrm{d}\eta$$

$$B = -\int_{\eta_1}^{+\infty}\frac{\partial (g - q^2)}{\partial \eta}\frac{1}{h_3 u h_1}\frac{\partial \varphi}{\partial \zeta}\,\mathrm{d}\eta \tag{7.30}$$

$$D = \int_{\eta_1}^{+\infty}\frac{g - q^2}{u^2}\left(\frac{1}{h_3 h_1}\frac{\partial \varphi}{\partial \zeta}\frac{\partial u}{\partial \eta} - \frac{w}{h_3}\frac{\partial u}{\partial \zeta} - l_1 + \frac{\partial \tau_1}{\partial \eta} + \frac{u^2}{h_3 h_1}\frac{\partial h_3}{\partial \xi}\right)\mathrm{d}\eta$$

In particular, for a two-dimensional wake flow we have

$$\frac{\mathrm{d}p}{\mathrm{d}x} = \frac{\gamma - 1}{2}\left[\frac{F}{\delta}\sqrt{\frac{2p}{\gamma + 1}} - \int\limits_{-\infty}^{+\infty}\left(\frac{1}{u}\frac{\partial G}{\partial \eta} - 2\frac{\partial \tau}{\partial \eta}\right)\mathrm{d}\eta \right.$$
$$\left. + \int\limits_{-\infty}^{+\infty}\frac{\partial \tau}{\partial \eta}\frac{g - u^2}{u^2}\frac{\mathrm{d}p}{\mathrm{d}x}\,\mathrm{d}\eta\right] \Bigg/ \int\limits_{-\infty}^{+\infty}\left(\frac{1}{M^2} - 1\right)\frac{\mathrm{d}\eta}{\rho} \tag{7.31}$$

Using Eq. (7.29) and the original equations (7.24) we can obtain the normal derivatives for the other functions. A surface is characteristic if the denominator in Eq. (7.30) vanishes. This condition can be brought into the form which admits an obvious physical interpretation

$$F - \frac{\gamma - 1}{2}F_1 = \frac{2\gamma p}{\gamma - 1}\int\limits_{\delta_1}^{\delta_2}\left(1 - \frac{a^2}{u^2}\right)\mathrm{d}y = 0 \tag{7.32}$$

where $\delta_1$ and $\delta_2$ are the upper and lower edges of the boundary layer or the wake which are exactly determined for the hypersonic flow regime under consideration.

The characteristic surfaces are cylindrical surfaces, the projection of the mean-integral Mach number onto which is equal to unity. This property is analogous to the property of inviscid supersonic flows in which the Mach number projection on the normal to a characteristic is equal to unity. If at a given point on the surface the maximum value of integral (7.32) remains negative when projecting the velocity onto any arbitrary direction, then the boundary layer is subcritical. In this case, a pressure disturbance is transferred in all directions over finite distances. If there is a direction, the projection of integral (7.32) onto which is positive, then, as in the case of a supersonic inviscid flow, pressure disturbances propagate within “Mach cones” whose boundaries are determined by Eq. (7.32) (of course, there is also a mechanism of information transfer along streamsurfaces).

In accordance with Eq. (7.30), along the characteristics we have

$$A - B - D = 0 \tag{7.33}$$

Other relations can be derived by a conventional way from Eqs. (7.24) and (7.29). Formula (7.32) also makes it possible to establish the existence of a counterpart of “sonic surfaces” in inviscid flows, on which the nature of the disturbance transfer can change. This role is played by surfaces that consist of points at which the maximum values of integral (7.32) of the Mach number profile vanish.

This moment corresponds to subcritical-to-supercritical transition and vice versa. For example, in the flow over a thin noncold plate with a smooth leading edge the series expansions of the solution in the vicinity of the leading edge involve an arbitrary function, since the flow is subcritical (this is direct generalization of the results of the work of Neiland (1970*b*) to the case of three-dimensional flows). However, as the boundary layer is shed from the trailing edge of the wing, the velocities in the wake increase and on a certain "sonic" surface (a line in the $(\xi, \zeta)$ plane) there occurs transition to the supercritical flow. The arbitrary function must be chosen from the condition of the simultaneous fulfillment of equalities (7.32) (for a maximum value of the integral) and (7.33). Thus, the domain of influence is closed.

An inverse, supercritical-to-m-subcritical, transition can also be considered. Condition (7.32) indicates that the "sonic lines" are associated with arbitrary functions. A particular case of the transition was considered at the beginning of this chapter. However, in the presence of strong counter-traveling disturbances transition can take place across supercritical jumps, that is, regions of local inapplicability of the boundary layer equations (Crocco, 1955). For a supersonic flow and the weak interaction regime these regions were studied in the paper of Neiland (1973).

## 7.2 Propagation of disturbances in three-dimensional time-dependent boundary layers

### *7.2.1 Formulation of the problem*

We will consider the hypersonic viscous flow over a plane surface with a rectilinear leading edge at zero angle of attack in the strong hypersonic viscous–inviscid interaction regime under the following conditions:

$$M_\infty \to \infty, \quad M_\infty \tau \to \infty \tag{7.34}$$

where $M_\infty$ is the Mach number of the undisturbed inviscid flow and $\tau$ is the dimensionless boundary layer thickness.

We will use the following notation for Cartesian coordinates measured in the direction perpendicular to the leading edge in the wing plane, normal to the surface, and along the leading edge: time; the corresponding velocity components; the density; the pressure; the total enthalpy; and the dynamic viscosity

$$\ell x, \ \ell y, \ \ell z, \ \ \frac{\ell t}{u_\infty}, \ \ u_\infty u, \ \ u_\infty v, \ \ u_\infty w, \ \ \rho_\infty \rho, \ \ \rho_\infty u_\infty^2 p, \ \ u_\infty^2 \frac{H}{2}, \ \ \mu_0 \mu$$

The parameter $\ell$ is a scale length (e.g., the length of the surface generator in the longitudinal direction); $\tau = (\rho_\infty u_\infty \ell / \mu_0)^{-1/4}$, where the subscript $\infty$ refers to the dimensional parameters of the oncoming flow; and $\mu_0$ is the dynamic viscosity determined at the stagnation temperature. It is assumed that the gas is thermodynamically perfect and characterized by a constant specific heat ratio $\gamma$. The Reynolds number is high but not higher than the critical value corresponding to laminar-turbulent transition. It is known that for supersonic and hypersonic flows the transition Reynolds number is fairly large (e.g., see Chapman et al., 1958).

In accordance with theory of strong interaction, the disturbed flow region can be divided into two subdomains, namely, the shock layer I and the boundary layer II (Fig. 7.8). In this section, singular domains located near the leading edge and in the transitional temperature layer near the outer edge of the boundary layer are not considered, since in the first approximation the flow in these regions has no effect on the boundary layer flow.

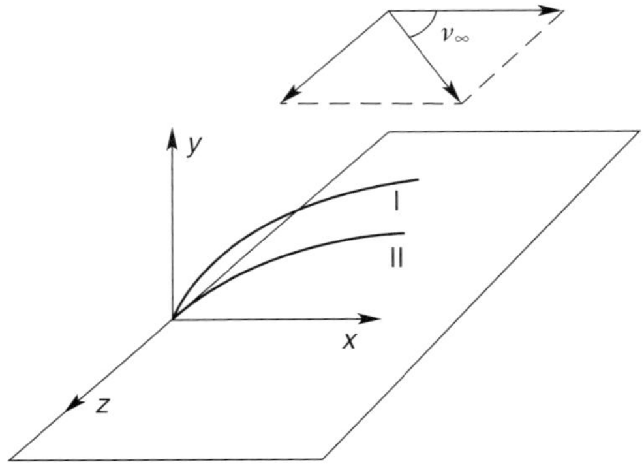


Fig. 7.8.

In subdomain I the flow functions and the coordinates can be represented as follows (Hayes and Probstein, 1966):

$$(x, y, z, t) = (x_1, \tau y_1, z_1, t_1)$$

$$(u, v, w, p, \rho, h) = (1 + \cdots, \tau v + \cdots, w_1 + \cdots, \tau^2 p_1 + \cdots, \rho_1 + \cdots, H_1 + \cdots)$$

Substituting these expansions in the system of Navier–Stokes equations and passing to limit (7.34) yields a system of nonlinear equations describing the inviscid shock-layer flow; in this section, this system is omitted.

For further analysis it is necessary to obtain a relation between the boundary layer thickness (or the vertical velocity at the outer edge of the boundary layer) and the induced pressure disturbance. We will use the approximate equation

$$p_1 = \frac{\gamma + 1}{2} v_1^2 \tag{7.35}$$

which represents the generalization of the tangent wedge formula to the time-dependent case presented in the book of Hayes and Probstein (1966).

The following asymptotic representations are characteristic of subdomain II:

$$(x, y, z, t) = (x_1, \tau y_2, z_1, t_1) \tag{7.36}$$

$$\begin{aligned}(u, v, w) &= (u_2 + \cdots, \tau v_2 + \cdots, w_2 + \cdots)\\ (p, \rho, H) &= (\tau^2 p_2 + \cdots, \tau^2 \rho_2 + \cdots, H_2 + \cdots)\end{aligned} \tag{7.37}$$

Substituting Eqs. (7.36) and (7.37) in the system of Navier–Stokes equations and passing to limit (7.34) leads to the system of equations for a time-dependent three-dimensional boundary layer

$$X\frac{\partial U}{\partial T} + X\left(U\frac{\partial U}{\partial X} - \frac{\partial F}{\partial X}\frac{\partial U}{\partial Y} - \frac{\partial \Phi}{\partial Z}\frac{\partial U}{\partial Y}\right) - \frac{F}{4}\frac{\partial U}{\partial Y} + \left(-\frac{1}{2} + \frac{X}{P}\frac{\partial P}{\partial X}\right)\frac{(\gamma - 1)}{2\gamma}Q = \left(\frac{P}{C_0}\right)\frac{\partial^2 U}{\partial Y^2}$$

$$X\frac{\partial W}{\partial T} + X\left(U\frac{\partial W}{\partial X} - \frac{\partial F}{\partial X}\frac{\partial W}{\partial Y} - \frac{\partial \Phi}{\partial Z}\frac{\partial W}{\partial Y}\right) - \frac{F}{4}\frac{\partial W}{\partial Y} + \frac{X}{P}\frac{\partial P}{\partial Z}\frac{(\gamma-1)}{2\gamma}Q = \left(\frac{P}{C_0}\right)\frac{\partial^2 W}{\partial Y^2}$$

$$X\frac{\partial G}{\partial T} + X\left(U\frac{\partial G}{\partial X} - \frac{\partial F}{\partial X}\frac{\partial G}{\partial Y} - \frac{\partial \Phi}{\partial Z}\frac{\partial G}{\partial Y}\right) - \frac{F}{4}\frac{\partial G}{\partial Y} = XQ\frac{\gamma-1}{\gamma}\frac{\partial P}{\partial T} + \left(\frac{P}{C_0}\right)\frac{\partial^2 G}{\partial Y^2} \tag{7.38}$$

$$\frac{\partial P}{\partial Y} = 0, \quad \beta = -1 + \frac{2X}{P}\frac{\partial P}{\partial X}, \quad \Delta = \left[\frac{(\gamma-1)C_0}{2\gamma P^2}\right]^{1/2}\int_0^\infty Q^2\, \mathrm{d}Y$$

$$P = \frac{\gamma+1}{2}\left[\frac{3\Delta}{4} + X\left(\frac{\partial \Delta}{\partial X} + \frac{\partial \Delta}{\partial T}\right)\right]^2, \quad Q = G - U^2 - W^2$$

where

$$Y = \left(\frac{\gamma-1}{2\gamma C_0}\right)^{1/2} x_1^{-1/4}\int_0^{y_2} R\,\mathrm{d}y_2, \quad u_2 = U = \frac{\partial F}{\partial Y}, \quad w_2 = W = \frac{\partial \Phi}{\partial Y}, \quad \delta = x_1^{3/4}\Delta$$

$$u_2 = U = \frac{\partial F}{\partial Y}, \quad w_2 = W = \frac{\partial \Phi}{\partial Y}, \quad X = x_1, \quad Z = z_1, \quad T = t_1$$

$$p_2 = x_1^{-1/2}P, \quad \rho_2 = x_1^{-1/2}R, \quad C_0 = P(0,T), \quad G = H_2$$

We note that in the boundary layer approximation the equation for the transverse momentum component is degenerated; hence follows that the functions $P$ and $\beta$ are independent of the transverse coordinate $Y$. It is assumed that the viscosity–temperature dependence is linear and the Prandtl number is unity. The solution of this system must satisfy the boundary conditions at the surface and the outer edge of the boundary layer

$$U = F = \Phi = 0, \quad G = g_w, \quad Y = 0 \tag{7.39}$$

$$U = G = 1, \quad W = 0, \quad Y = \infty$$

where $\alpha$ is the angle between the oncoming flow direction and the $OX$ axis. For determining the unique solution it is necessary to impose an additional condition on a certain line downstream of the leading edge, for example a base pressure at the trailing edge (Neiland, 1970*b*). The appearance of the additional condition is due to possible upstream disturbance transfer and their effect on the boundary layer flow

$$P[X, Z = \beta(X), T] = \varphi(X, T) \tag{7.40}$$

### 7.2.2 *Determining subcharacteristic surfaces in time-dependent three-dimensional flows*

A characteristic (subcharacteristic) surface $\Omega(X, Z, T)$ associated with a function $P(X, Z, T)$ is a surface on which some initial data are preassigned and the derivative $\partial P/\partial\Omega$ is not determined. The procedure used below is applicable only under conditions in which the pressure distribution in the boundary layer equations is not known beforehand and is determined in the process of solution. For this purpose, the system must involve the additional interaction condition (7.35) and the additional boundary conditions (7.40).

After introduction of new variables $X, Y, Z, T \to \Omega(X, Z, T), Y, Z, T$ the boundary value problem (7.38)–(7.40) takes the form:

$$\frac{\partial U}{\partial \Omega}\left(\frac{\partial \Omega}{\partial T} + U\frac{\partial \Omega}{\partial X} + W\frac{\partial \Omega}{\partial Z}\right) - \frac{\partial U}{\partial Y}\left(\frac{\partial F}{\partial \Omega}\frac{\partial \Omega}{\partial X} + \frac{\partial \Phi}{\partial \Omega}\frac{\partial \Omega}{\partial Z}\right) + A_1\frac{\partial P}{\partial \Omega}\frac{\partial \Omega}{\partial X} = B_u \quad (7.41)$$

$$\frac{\partial W}{\partial \Omega}\left(\frac{\partial \Omega}{\partial T} + U\frac{\partial \Omega}{\partial X} + W\frac{\partial \Omega}{\partial Z}\right) - \frac{\partial W}{\partial Y}\left(\frac{\partial F}{\partial \Omega}\frac{\partial \Omega}{\partial X} + \frac{\partial \Phi}{\partial \Omega}\frac{\partial \Omega}{\partial Z}\right) + A_2\frac{\partial P}{\partial \Omega}\frac{\partial \Omega}{\partial Z} = B_w$$

$$\frac{\partial G}{\partial \Omega}\left(\frac{\partial \Omega}{\partial T} + U\frac{\partial \Omega}{\partial X} + W\frac{\partial \Omega}{\partial Z}\right) - \frac{\partial G}{\partial Y}\left(\frac{\partial F}{\partial \Omega}\frac{\partial \Omega}{\partial X} + \frac{\partial \Phi}{\partial \Omega}\frac{\partial \Omega}{\partial Z}\right) + A_3\frac{\partial P}{\partial \Omega}\frac{\partial \Omega}{\partial T} = B_g$$

subject to additional conditions

$$P = \frac{\gamma+1}{2}\left[X\frac{\partial \Delta}{\partial \Omega}\left(\frac{\partial \Omega}{\partial X} + \frac{\partial \Omega}{\partial T}\right) + \frac{3\Delta}{4} + X\frac{\partial \Delta}{\partial T}\right]^2 \quad (7.42)$$

$$A_1 = A_2 = \frac{(\gamma-1)Q}{2\gamma P}, \quad A_3 = \frac{(\gamma-1)Q}{\gamma P}$$

Below we do not use the expressions for the right-hand sides of the equations. Using the definition of the boundary layer displacement thickness the derivative on the right-hand side of Eq. (7.42) can be expressed as follows:

$$\frac{\partial \Delta}{\partial \Omega} = \left[\frac{(\gamma-1)C_0}{2\gamma P^2}\right]^{1/2}\left[\int_0^\infty \frac{\partial Q}{\partial \Omega}\,\mathrm{d}Y - \frac{1}{P}\frac{\partial P}{\partial \Omega}\int_0^\infty Q\,\mathrm{d}Y\right]$$

For determining the derivative $\partial Q/\partial \Omega$ we will transform the system of equations (7.41) into a single equation for the function $D = (\partial F/\partial \Omega)(\partial \Omega/\partial X) + (\partial \Phi/\partial \Omega)(\partial \Omega/\partial Z)$. Summing the first equation multiplied by $\partial \Omega/\partial X$ and the second equation multiplied by $\partial \Omega/\partial Z$ we obtain the equation

$$A_0\frac{\partial D}{\partial Y} + D\frac{\partial A_0}{\partial Y} + B_1 Q\frac{\gamma-1}{2\gamma P}\frac{\partial P}{\partial \Omega} = B_0 \quad (7.43)$$

where

$$A_0 = \frac{\partial \Omega}{\partial T} + U\frac{\partial \Omega}{\partial X} + W\frac{\partial \Omega}{\partial Z}, \quad B_1 = \left[\left(\frac{\partial \Omega}{\partial X}\right)^2 + \left(\frac{\partial \Omega}{\partial Z}\right)^2\right]$$

Equation (7.43) has the following solution

$$D = -B_1 A_0 \frac{\gamma - 1}{2\gamma P} \frac{\partial P}{\partial \Omega} \int_0^Y \frac{Q}{A_0^2} \mathrm{d}Y + A_0 \int_0^Y \frac{B_0}{A_0^2} \mathrm{d}Y$$

which yields the following expressions for the derivatives

$$\frac{\partial U}{\partial \Omega} = \frac{1}{A_0} \left( D \frac{\partial U}{\partial Y} - A_1 \frac{\partial P}{\partial \Omega} \frac{\partial \Omega}{\partial X} + B_u \right)$$

$$\frac{\partial W}{\partial \Omega} = \frac{1}{A_0} \left( D \frac{\partial W}{\partial Y} - A_2 \frac{\partial P}{\partial \Omega} \frac{\partial \Omega}{\partial Z} + B_w \right)$$

$$\frac{\partial G}{\partial \Omega} = \frac{1}{A_0} \left( D \frac{\partial G}{\partial Y} + A_3 \frac{\partial P}{\partial \Omega} \frac{\partial \Omega}{\partial T} + B_g \right)$$

Finally, we obtain the expression for the derivative of the induced pressure in the direction perpendicular to the subcharacteristic surface

$$\frac{\partial P}{\partial \Omega} = \frac{P B_p}{N \left( \frac{\partial \Omega}{\partial X} + \frac{\partial \Omega}{\partial Z} \right)}, \quad N = \frac{\gamma - 1}{2} \left[ \left( \frac{\partial \Omega}{\partial X} \right)^2 + \left( \frac{\partial \Omega}{\partial Z} \right)^2 \right] I_1 - I_0 \tag{7.44}$$

$$I_0 = \int_0^\infty (G - U^2 - W^2)\, \mathrm{d}Y, \quad I_1 = \int_0^\infty \frac{(G - U^2 - W^2)^2}{\left( \frac{\partial \Omega}{\partial T} + U \frac{\partial \Omega}{\partial X} + W \frac{\partial \Omega}{\partial Z} \right)^2} \mathrm{d}Y$$

After the introduction of the subcharacteristic surface displacement velocity

$$a_x = a \times \cos \omega = -\frac{1}{B_1} \frac{\partial \Omega}{\partial X} \frac{\partial \Omega}{\partial T}, \quad a_z = a \times \sin \omega = -\frac{1}{B_1} \frac{\partial \Omega}{\partial Z} \frac{\partial \Omega}{\partial T}$$

where $\omega$ is the angle between the *OX* axis and the disturbance propagation direction in the *XZ* plane, the expression for $N$ in Eq. (7.44) takes the form:

$$N = \frac{\gamma - 1}{2} \int_0^\infty \frac{(G - U^2 - W^2)^2}{(a - U \cos \omega - W \sin \omega)^2}\, \mathrm{d}Y - \int_0^\infty (G - U^2 - W^2)\, \mathrm{d}Y$$

The subcharacteristic surface $\Omega(X, Z, T)$ is determined by the condition

$$N = 0 \tag{7.45}$$

Equation (7.45) determines the mean velocity of disturbance propagation if the velocity and enthalpy profiles are known. Though formula (7.45) was derived under certain assumptions of the interaction conditions, it can be shown that it is also valid for other interaction regimes. Moreover, this expression determines the disturbance propagation velocity in turbulent boundary layers.

Thus, if the surface $\Omega(X, Z, T)$ exists, then it divides the flow into regions in which disturbances can (subcritical flow region) and cannot (supercritical flow region) travel upstream.

We note that the common subcharacteristic surface for all unknown functions $\Omega_1(X, Y, Z, T)$ can be determined on the basis of an analysis of the system of equations (7.41), (7.42), and (7.45) written in the form:

$$\|E\| \frac{\partial \mathbf{S}}{\partial \Omega} = \mathbf{B}$$

where the matrix $\|E\|$ and the vector $\mathbf{S}$ are as follows:

$$\|E\| = \begin{pmatrix} A_0 & 0 & 0 & 0 & A_1 \\ 0 & A_0 & 0 & 0 & A_2 \\ C_1 & C_2 & C_3 & 0 & 0 \\ 0 & 0 & 0 & A_0 & A_3 \\ 0 & 0 & 0 & 0 & A_4 \end{pmatrix}, \quad \mathbf{S} = \begin{vmatrix} U \\ W \\ V \\ G \\ P \end{vmatrix}$$

$$C_1 = \frac{\partial \Omega}{\partial X}, \quad C_2 = \frac{\partial \Omega}{\partial Z}, \quad C_3 = \frac{\partial \Omega}{\partial Y}, \quad A_4 = N\left(\frac{\partial \Omega}{\partial X} + \frac{\partial \Omega}{\partial Z}\right)$$

Then the subcharacteristic surface is determined by the equation

$$\det \|E\| = 0$$

### 7.2.3 Results of the numerical analysis

For determining the disturbance propagation velocity it is necessary to know the velocity and enthalpy profiles in the boundary layer. These profiles can be found by solving stationary two-dimensional or three-dimensional boundary layer equations.

Below, for determining the disturbance propagation velocity we use the results of the solution of a model problem describing the boundary layer flow at a given pressure distribution

$$\xi(x) = \int_0^x \rho_w \mu_w u_\delta \, d\tilde{x}, \quad \eta(x, y) = \frac{u_\delta}{\sqrt{2\xi \cos\alpha}} \int_0^{y_2} \rho \, d\tilde{y} \tag{7.46}$$

$$f = \int_0^\eta \frac{u}{u_\delta} d\eta, \quad g = \frac{H}{H_\delta}, \quad w = \int_0^\eta \frac{w}{w_\delta} d\eta$$

$$f_{\eta\eta\eta} + f f_{\eta\eta} + \beta(g - \cos^2\alpha \times f_\eta^2 - \sin^2\alpha \times \varphi_\eta^2) = 0 \tag{7.47}$$

$$\beta = -\frac{\gamma - 1}{\gamma} \frac{m}{m+1} \frac{1}{\cos^2\alpha}, \quad \tilde{\varphi}_{\eta\eta} + f\tilde{\varphi}_\eta = 0$$

where $\tilde{\varphi} = \varphi_\eta$ and $g_{\eta\eta} + f g_\eta = 0$.

The boundary conditions are as follows:

$$\eta = 0: \quad f = f_w, \quad f_\eta = \widetilde{\varphi} = 0, \quad g = g_w \tag{7.48}$$

$$\eta = \infty: \quad f_\eta = \widetilde{\varphi} = 1, \quad g = 1$$

Here, $\gamma$ is the adiabatic exponent, $m$ is the exponent in the pressure distribution law ($p = c_1 X^n$), and $\alpha$ is the angle between the freestream velocity vector and the $OX$ axis.

The system of equations (7.47) is written in the linearized form:

$$f_k''' + f_{k-1} f_k'' + \beta(g_{k-1} - \cos^2\alpha \times f_{k-1} f_k - \sin^2\alpha \times \varphi_{k-1}^2) = 0$$

$$\widetilde{\varphi}_{k-1}'' + f_{k-1}\widetilde{\varphi}_{k-1}' = 0 \quad g_{k-1}'' + f_{k-1} g_{k-1}' = 0$$

where $k$ is the number of the approximation, and approximated by a second-order difference scheme. In the case $m = -0.5$ the first equation, subject to given boundary conditions, is solved using four-diagonal sweeping technique. The last two equations have analytical solutions

$$\widetilde{\varphi}_{k-1}(\eta) = \frac{I(\eta)}{I(\infty)}$$

where

$$I(\eta) = \int_0^\eta \exp\left(\int_0^{\widetilde{\eta}} f_{k-1}\,d\overline{\eta}\right) d\widetilde{\eta}, \quad g_{k-1}(\eta) = g_w + (1 - g_w)\widetilde{\varphi}_{k-1}(\eta)$$

In the case $n > 0$ for the solution branch corresponding to negative surface friction, $f_w'' < 0$ (Stewartson, 1954), the domain of definition of the unknown functions is subdivided into two segments $[0, \eta^*] \cup [\eta^*, \eta_\infty]$ (Fig. 7.9) and the general problem is broken into two problems: the lower boundary value problem I with the upper boundary condition for the function $f$ at

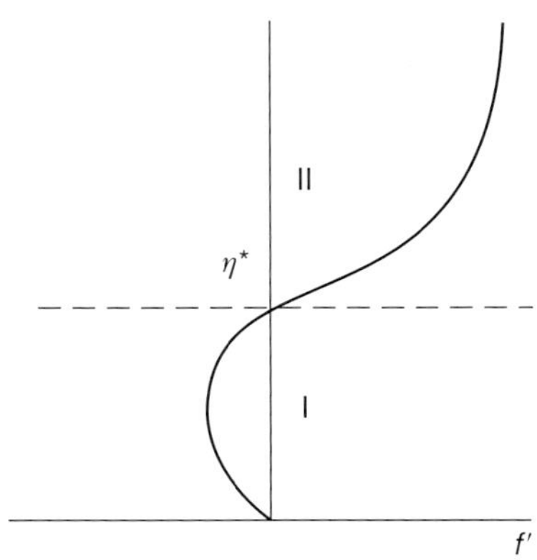


Fig. 7.9.

point $\eta^*: f' = 0$, and the upper boundary value problem II with the lower boundary condition for the function $f: f' = 0\ f = f_{\mathrm{I}}(\eta^*)$, where $f_{\mathrm{I}}(\eta^*)$ is the value of the solution in domain I at point $\eta^*$.

Then the difference of one-sided second derivatives is determined at point $\eta^*$: $\Delta_1 = f_{\mathrm{I}}''(\eta^*) - f_{\mathrm{II}}''(\eta^*)$. Using one or another iteration procedure (e.g., segment bisection) the point $\eta^*$ is so chosen that it provides a minimum of the quantity $|\Delta_1|$ with a given accuracy. After the velocity and enthalpy profiles have been obtained, Eq. (7.45) written in variables (7.46) is used for determining the disturbance propagation velocity

$$\frac{\gamma - 1}{2} \int_0^\infty \frac{(G - U^2 - W^2)^2}{(a - U\cos\omega - W\sin\omega)^2}\, \mathrm{d}\eta - \int_0^\infty (G - U^2 - W^2)\, \mathrm{d}\eta = 0 \tag{7.49}$$

where $G = g$, $U = \cos\alpha \times f_\eta$, and $W = \sin\alpha \times \widetilde{\varphi}$.

Below we present the results of the theoretical and numerical analysis of disturbance propagation under changes of the relevant parameters of the problem for two-dimensional and three-dimensional flows.

### *7.2.4 Two-dimensional flows*

Figure 7.10 presents the plots of the velocities of the upstream and downstream disturbance propagation on the parameter ranges $f_w \in [-1, 1]$ and $g_w \in [0, 1]$. A nonzero value of the streamfunction at the surface ($f_w \neq 0$) corresponds to a power-law distribution of the suction or injection velocity. An increase in the suction velocity leads to a decrease in the velocity $a_-$ due to the relative decrease of the thickness of the subsonic flow region within the boundary layer. An increase in the injection velocity gives rise to an opposite tendency. The temperature factor $g_w$ is equal to the ratio of the surface temperature to the stagnation temperature. Heating the surface leads to an increase in the boundary layer thickness and a corresponding increase in the relative thickness of the subsonic flow region. As a result, the velocity of the upstream disturbance propagation increases. Clearly that, as $g_w$ tends to zero, the velocity of the upstream disturbance propagation also vanishes.

Negative values of the parameter $\beta$ in system (7.47) are associated with the self-similar solution for the laminar boundary layer with increase in the outer-flow pressure. It is known (Stewartson, 1954) that under these conditions the solution of the problem is nonunique, one of its branches describing a flow with return currents.

The appearance of the return flows in the boundary layer ensures an additional convective mechanism of upstream disturbance propagation. It is important that in this case the characteristics of the pressure disturbance propagation also change. This fact is maintained by the calculated results. In Fig. 7.11 we have plotted the dependence of $a_-$ on the parameter $f_w$ for $g_w = 1$ and a power-law pressure distribution in the outer flow ($p = c_1 X^{0.1}$) for the branch of the solution of Eq. (7.47) corresponding to negative surface friction. It should be noted that the appearance of the return flow region increases appreciably the velocity of upstream disturbance propagation.

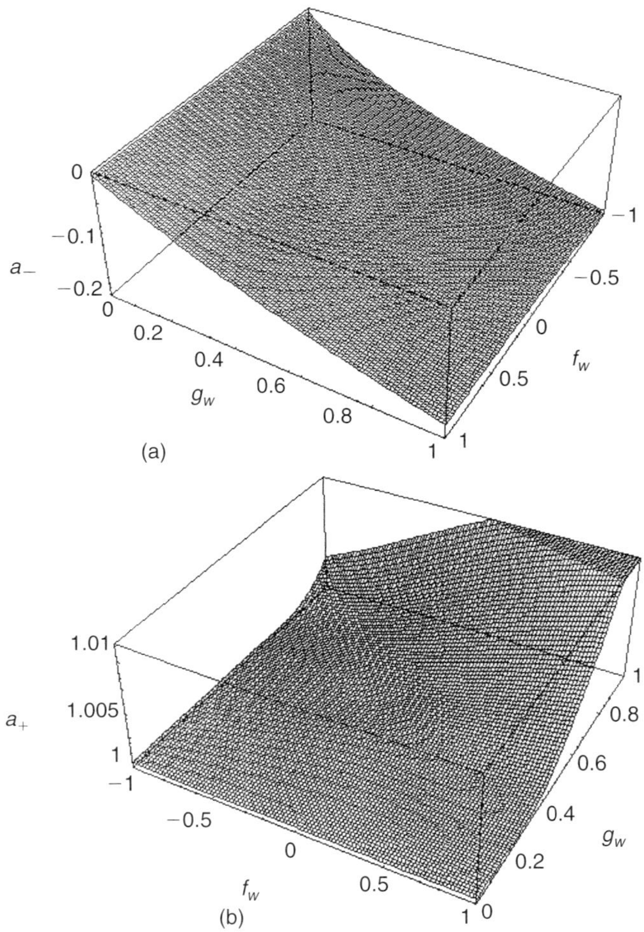


Fig. 7.10.

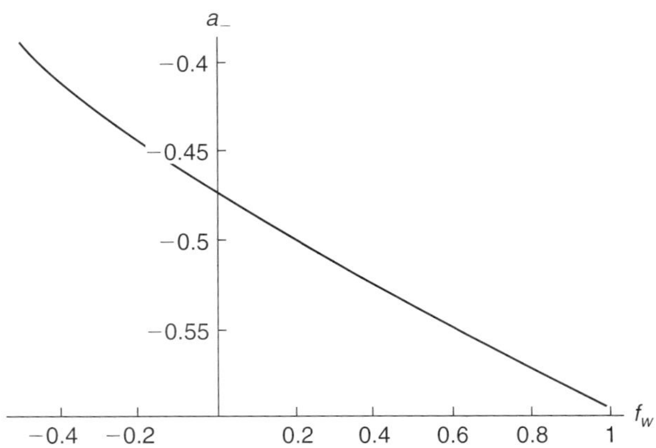


Fig. 7.11.

### 7.2.5 Three-dimensional boundary layer

The self-similar solution of Eqs (7.47) for $m = -0.5$ and $\alpha = 30°$ was used for determining the vector $\overline{a} = (a_x, a_z)$ as a function of the temperature factor $g_w$. In Fig. 7.12 the radiation pattern of the disturbance propagation velocities is presented for $\alpha = 30°$. Qualitatively, it is similar to the radiation pattern obtained for zero yaw angle (Lipatov, 1995).

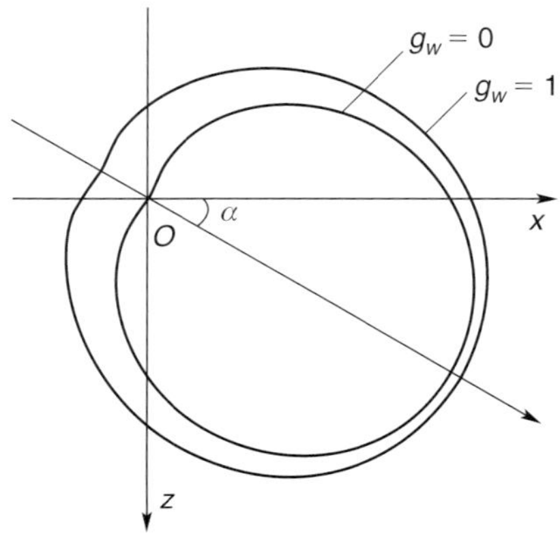


Fig. 7.12.

## 7.3 Supercritical regimes of hypersonic flow over a yawed planar delta wing

Here, we will study the yaw angle effect on the characteristics of the laminar boundary layer on a cold wing and the formation of subcritical and supercritical flow regions.

As shown in Section 7.1, generally, in symmetric flow over cold wings with the sweep angle of the wing's leading edge less than the critical value the delta-wing boundary layer includes the regions of both supercritical and subcritical flows, transition from one to another type of flow occurring at a certain value $\omega_1$. The supercritical flow is realized in the region between the leading edge and a surface determined by the angle $\omega_1$, while in the region located between the $\omega_1$ surface and the plane of symmetry of the wing the effect of the upstream disturbance transfer should be taken into account in constructing solutions. In the presence of a yaw angle the nature of the boundary layer flow considerably changes (Dudin, 1978*b*). In nonsymmetric flow over the wing the streamlines in the boundary layer proceeding from different edges of the wing converge at a certain surface. This surface divides the flow region into zones with different directions of disturbance transfer. The intersection of this surface with a plane perpendicular to the axis of symmetry of the wing determines a curve at which the direction of parabolicity of the system of boundary layer equations changes. In this case, the conditions occurring downstream the cross flow are considerably different from those for symmetric flow over the wing (Dudin, 1995).

### *7.3.1 Equations and boundary conditions*

We will consider the hypersonic viscous perfect-gas flow over a planar semi-infinite delta wing at zero incidence (Fig. 7.13). It is assumed that the wing surface temperature is constant and low as compared with the freestream stagnation temperature and that the regime of strong interaction between the boundary layer and the outer hypersonic flow is realized. As distinct from Section 7.1, we introduce a Cartesian coordinate system with origin at the wing nose, the $x\ell$ axis aligned with the axis of symmetry of the wing, the $z\ell$ axis normal to the former and lying in the wing plane, and the $y\ell$ axis normal to the $xz$ plane. The velocity components $uu_\infty$, $vu_\infty$, and $wu_\infty$ are directed along the $x$, $y$, and $z$ axes, respectively. The parameter $z_0 = \tan\omega_0$ characterizing the transverse-to-longitudinal wing size ratio is of the order $O(1)$. The yaw angle $\beta$ is the angle between the freestream velocity vector and the $x$ axis.

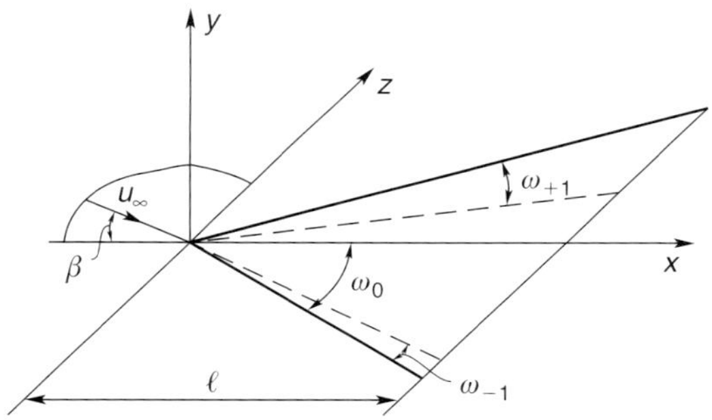


Fig. 7.13.

In accordance with the estimates for the laminar boundary layer in a hypersonic flow we introduce the following variables:

$$z_0 z^*, \quad \delta y^*, \quad \delta z_o^{-1} v^*, \quad \rho_\infty \delta^2 \rho^* \tag{7.50}$$

$$\rho_\infty u_\infty^2 \delta^2 p^*, \quad \frac{u_\infty^2 g}{2}, \quad \mu_0 \mu, \quad \ell \delta \delta_e^*$$

These are the transverse and normal coordinates, the normal component of the velocity vector, the density, the pressure, the stagnation enthalpy, the viscosity, and the boundary layer thickness. Here $\delta = z_0^{1/4} Re_0^{-1/4}$, where the Reynolds number $Re_0 = \rho_\infty u_\infty \ell / \mu_0$ is based on the undisturbed flow density and velocity, the viscosity at the stagnation temperature, and a scale length $\ell$ which drops out of the final results when considering the flow past the semi-infinite wing. Then we introduce the Dorodnitsyn variables

$$\eta = \int_0^{y^*} \rho^* \, \mathrm{d}y^*, \quad v_\delta^* = \rho^* v^* + z_0 u \frac{\partial \eta}{\partial x} + w \frac{\partial \eta}{\partial z^*}$$

In considering the flow over semi-infinite delta wings in the strong viscous–inviscid interaction regime, it is convenient to introduce the variables

$$z^* = x z_*, \quad \eta = x^{1/4} \lambda_* \tag{7.51}$$

$$v_\delta^* = x^{-3/4}\left(v_* - z_0 x u \frac{\partial \lambda_*}{\partial x}\right), \quad p^* = x^{-1/2} p_*(z_*)$$

$$\rho^* = x^{-1/2}\rho_*(\lambda_*, z_*), \quad \delta_e^* = x^{3/4}\Delta_e^*(z_*)$$

When variables (7.51) are substituted in the equations of the three-dimensional boundary layer and the boundary conditions, the coordinate $x$ drops out of the boundary value problem and the system obtained for gas dynamic variables depends only on two independent variables $z_*$ and $\lambda_*$.

Moreover, in order to take account of the special features of the behavior of the flow functions in the vicinity of the leading edges of the triangular plate ($z = \pm 1$) we will introduce the variables

$$\lambda_* = \left(\frac{2\gamma}{\gamma - 1}\right)^{1/2}(1 - z_*^2)^{1/4}\lambda, \quad p_* = (1 - z_*^2)^{-1/2} p(z_*) \tag{7.52}$$

$$\Delta_e^* = (1 - z_*^2)^{3/4}\Delta_e(z_*)$$

$$v_* = \left(\frac{2\gamma}{\gamma - 1}\right)^{1/2}(1 - z_*^2)^{1/4}\left[v\frac{p}{1 - z_*^2} - (w - z_0 z_* u)\frac{\partial \lambda}{\partial z_*}\right]$$

Rewriting the system of dimensionless boundary layer equations and boundary conditions in terms of Eqs. (7.50)–(7.52) yields (asterisks at the transverse coordinate are omitted)

$$(w - z_0 z\, u)F_1 \dot{f} + v f' = G \tag{7.53}$$

$$f = \begin{Bmatrix} u \\ w \\ g \end{Bmatrix}, \quad G = \begin{cases} z_0 \frac{\gamma - 1}{2\gamma p}(g - u^2 - w^2)F_2 + u'' \\ -\frac{\gamma - 1}{2\gamma p}(g - u^2 - w^2)F_3 + w'' \\ \frac{1}{\sigma} g'' - \frac{1-\sigma}{\sigma}(u^2 + w^2)'' \end{cases}$$

$$v' = (w - z_0 z u)F_4 - F_1(0.25 z_0 u - z_0 z \dot{u} + \dot{w})$$

$$F_1 = \frac{1 - z^2}{p}, \quad F_2 = \frac{1 + z^2}{2} + zF_1\frac{\mathrm{d}p}{\mathrm{d}z}, \quad F_3 = z + F_1\frac{\mathrm{d}p}{\mathrm{d}z}, \quad F_4 = \frac{z}{2p}$$

$$\lambda = 0: \quad u = w = v = g = 0$$

$$\lambda \to \infty: \quad u \to \cos\beta, \quad w \to -\sin\beta, \quad g \to 1$$

where $|z| \le 1$, $\sigma$ is the Prandtl number, and $\gamma$ is the specific heat ratio. Dots and primes denote the derivatives with respect to the coordinates $z$ and $\lambda$, respectively.

By virtue of the fact that at the outer edge of the boundary layer the gas density increases without bound, since the temperature vanishes (Lee and Cheng, 1969), for the boundary layer thickness in variables (7.52) we obtain

$$\Delta_e = \frac{1}{p}\left(\frac{\gamma-1}{2\gamma}\right)^{1/2}\int_0^\infty (g-u^2-w^2)\,\mathrm{d}\lambda \tag{7.54}$$

For solving the system of equations (7.53) it is necessary to know the pressure distribution produced by the displacing action of the boundary layer and the body thickness. This pressure is not given and must be determined in the process of solution of the boundary value problem (7.53), together with the equations for the outer inviscid flow obtained by applying hypersonic small perturbation theory. However, in considering the flow over thin wings with the aspect ratio $z_0 = O(1)$, strip theory (Hayes and Probstein, 1966; Lunev, 1975) is applicable for the outer inviscid $M_\infty \gg 1$ flow, while for determining the pressure under condition $M_\infty\delta \gg 1$ the approximate tangent wedge formula can be used, which, after variables (7.50)–(7.52) have been introduced, takes the form:

$$p = \frac{\gamma+1}{2}\left\{\left[\frac{3}{4}(1-z^2)\Delta_e - z\left((1-z^2)\frac{\mathrm{d}\Delta_e}{\mathrm{d}z} - \frac{3}{2}z\Delta_e\right)\right]\cos\beta + \left((1-z^2)\frac{\mathrm{d}\Delta_e}{\mathrm{d}z} - \frac{3}{2}z\Delta_e\right)\frac{\sin\beta}{z_0}\right\}^2 \tag{7.55}$$

Here, it is taken into account that, correct to $O(\delta^2/z_0)$, the boundary layer displacement thickness is equal to $\Delta_e$.

Equations (7.54) and (7.55) make it possible to close the system of equations in partial derivatives (7.53) which describes the flow in the laminar boundary layer on a cold delta wing in the strong viscous interaction regime. We note that in substituting the expression for the pressure (7.55) in the system of equations (7.53), in the latter system, due to the presence of the term $\mathrm{d}p/\mathrm{d}z$, there appears the second derivative $\mathrm{d}^2\Delta_e/\mathrm{d}z^2$ which makes it possible to take account for a downstream boundary condition, for example, the no-flow condition in the plane of symmetry of the wing for $\beta = 0$. At the leading edges of a delta wing ($z = \pm 1$) the system of equations (7.53)–(7.55) degenerates to systems of ordinary differential equations and their solutions make it possible to determine all flow functions in the boundary layer at the edges.

### *7.3.2 Results of the calculations*

The numerical solution of the system of equations was obtained using the method presented in the paper of Dudin (1985).

In numerically integrating the system of equations ($\omega_{1+}, \omega_{1-}$) vicinities of supercritical-to-subcritical transition were not specially singled out, so that the boundary value problem was solved in a united domain, from one edge to the other. The coordinates $z_k$ corresponding to supercritical-to-m-subcritical transition were determined from the condition of vanishing of the expression for the displacement thickness variation (7.11), which takes the following form in coordinates (7.51) and (7.52):

$$\frac{\gamma-1}{2}\int_0^\infty \left[\frac{g-u^2-w^2}{u\sin(\omega_0\pm\beta-\omega_1)+w\cos(\omega_0\pm\beta-\omega_1)}\right]^2 \mathrm{d}\lambda = \int_0^\infty (g-u^2-w^2)\,\mathrm{d}\lambda \tag{7.56}$$

where $z_k = \tan(\omega_0-\omega_1)/\tan\omega_0$.

Here, plus and minus signs relate to the right and left wing halves, respectively (Fig. 7.13). As an illustration, we present the calculated results for the flow over a wing for the following parameters: $\sigma=1$, $\gamma=1.4$, $z_0=1$ (the semi-vertex wing angle $\omega_0=45°$), and the yaw angles $\beta=0$, $10°$, $25°$, and $35°$ (curves 1 to 4 in Figs. 7.14–7.18). In Fig. 7.14 the spanwise

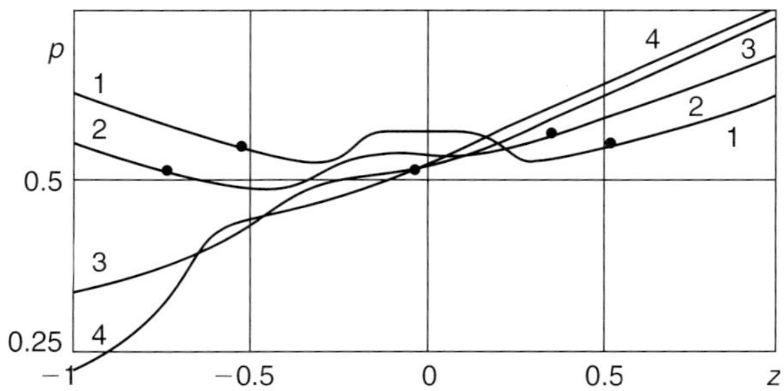


Fig. 7.14.

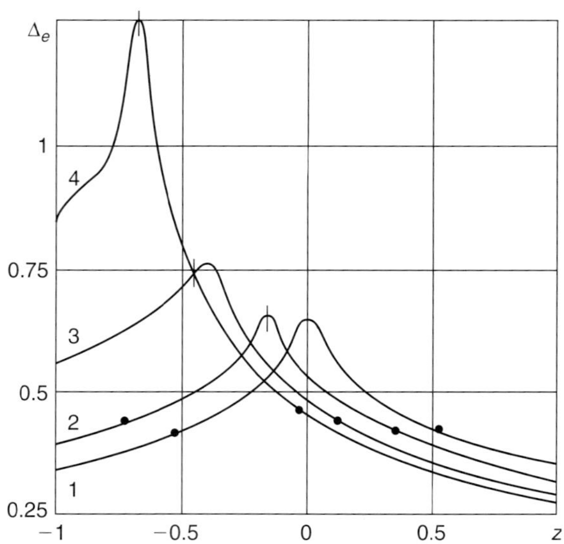


Fig. 7.15.

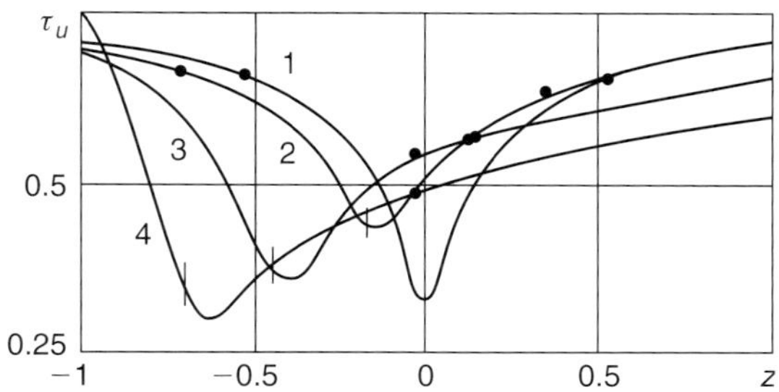


Fig. 7.16.

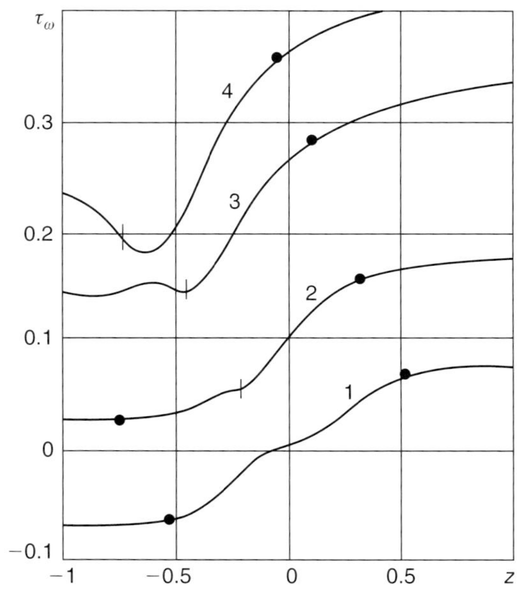


Fig. 7.17.

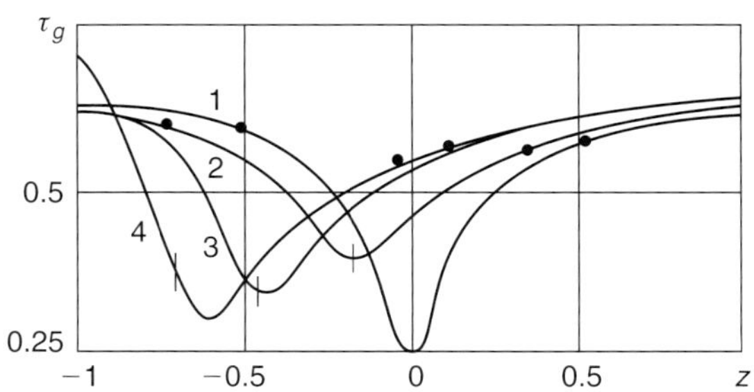


Fig. 7.18.

distribution of the induced pressure $p$ is presented. Points on the curves indicate the critical values of the coordinate $z_k$ obtained from formula (7.56).

As can be seen from the presented results, the departure from the self-similar solutions corresponding to the flow past a semi-infinite plate, that is, supercritical-to-m-subcritical flow transition, takes place in accordance with the values determined by Eq. (7.56).

We note that with increase in the yaw angle the supercritical flow region enlarges on the right wing half and diminishes on the left half; for $\beta \geq 25°$ on the left half the supercritical flow region no longer exists. With further increase in the yaw angle, $\beta \geq 35°$, the supercritical-to-subcritical transition line passes through the plane of symmetry of the wing and the supercritical flow region occupies more than a half of the wing.

In all the cases considered a pressure disturbance occurring in the region of convergence of two flows proceeding from different wing edges propagates over the wing span only within the subcritical flow region whose boundaries are determined by the corresponding values of the coordinate $z_k$. In Fig. 7.15 the spanwise distribution of the boundary layer thickness is presented. With increase in the yaw angle the location of the $\Delta_e$ maximum is displaced along the wing by almost the value of the yaw angle (in Figs. 7.15–7.18 these values are indicated by vertical dashes). In the vicinity of the boundary layer thickness maximum there is a region of convergence of the flows proceeding from the leading edges.

The results of calculation of the spanwise distributions of the longitudinal, $\tau_u = \left|\frac{\partial u}{\partial \lambda}\right|_w$, and transverse, $\tau_w = \left|-\frac{\partial w}{\partial \lambda}\right|_w$, viscous stress and heat flux, $\tau_g = \left|\frac{\partial w}{\partial \lambda}\right|_w$, coefficients are presented in Figs. 7.16–7.18. The coefficients $\tau_u$ and $\tau_g$ are minimum on the leeward wing side, $z < 0$. However, the coordinates of the $\tau_u$ and $\tau_g$ minima do not coincide with the coordinate of the ray parallel to the freestream direction. It should be noted that the variation of the coefficient $\tau_w$ is appreciably nonmonotonic; already for $\beta \geq 10°$ the coefficient $\tau_w$ becomes positive on the left wing half, so that the velocity component $w < 0$ over the entire wing.

The calculations show that, though the variation of the yaw angle has no effect on the transition coordinate (fulfillment of Eq. (7.56)), this variation has a considerable effect on the spanwise distributions of the viscous stress and heat flux coefficients, as well as the boundary layer thickness and pressure distributions. For a yawed delta wing the flow can be supercritical on one wing half and subcritical on the other. This means that disturbances which can arise in the boundary layer, for example, in the plane of symmetry of a cold wing, propagate in the subcritical region up to the leading edge, whereas on the other half a flow corresponding to that past a semi-infinite swept plate is realized.

## 7.4 Existence of self-similar solutions in the supercritical region on a nonplanar delta wing in hypersonic flow

In previous sections we considered flows over planar delta wings. In this section we will consider nonplanar wings with a power-law variation of the thickness with the exponent equal to 3/4 (Dudin, 1997).

### *7.4.1 Equations and boundary conditions*

We will consider the symmetric flow over a delta wing (the yaw angle $\beta = 0$). The coordinate system and the notation for the flow functions are the same as in Section 7.3.1. We assume that the shape of the thin wing surface is determined by the equation $y_w = \delta_w(x, z)$ and introduce the variables fitted to the wing surface

$$y_w = y - \delta_w(x, z), \quad v_w = v - u\frac{\partial \delta_w}{\partial x} - w\frac{\partial \delta_w}{\partial z} \tag{7.57}$$

Then in the dimensionless variables (7.50) only the following relations change their form:

$$y_w = \delta y^*, \quad v_w = \delta z_0^{-1} v^*, \quad \delta_w = \tau \delta_w^* \tag{7.58}$$

Here, we introduced an additional parameter, namely, the dimensionless wing thickness $\tau$. In considering the flow past semi-infinite triangular bodies with a thickness

$$\delta_w^* = x^{3/4} \Delta_w^* \left(\frac{z^*}{x}\right) \tag{7.59}$$

in the strong viscous interaction regime the boundary value problem can be brought into a more convenient form by means of transformations (7.51). In substituting these variables in the three-dimensional boundary layer equations the longitudinal coordinate $x$ drops out from the boundary value problem and the system of equations thus obtained depends only on two independent variables $z_*$ and $\lambda_*$. At the leading edges of the wing the interaction between the boundary layer and the outer flow appreciably depends on not only the edge shape $\Delta_w^*(z_* \to \pm 1)$ but also the parameter $D = \tau/\delta = \tau z_0^{-1/4} Re_0^{1/4}$ (Dudin, 1978*a*) characterizing the wing-to-boundary layer thickness ratio.

In what follows we will consider the flow over wings with the following cross-sectional shape (the asterisk at $z$ is omitted)

$$\Delta_w^*(z) = (1 - z^2)^{3/4} \Delta_w(z) \tag{7.60}$$

where the function $\Delta_w(z)$ has no singularities at the edges for $z = \pm 1$. In order to take account for the flow function behavior in the vicinities of the leading edges, we introduce variables (7.52). As a result, we arrive at the system of equations (7.53) but, in view of the flow symmetry, with different boundary conditions at the outer edge of the boundary layer, namely:

$$\lambda \to \infty: \quad u \to 1, \quad w \to 0, \quad g \to 1 \tag{7.61}$$

The expression for determining the boundary layer displacement thickness retains its form (7.54). The pressure, which in the case in question is produced by both the body and the displacement thickness, is determined by the expression

$$p = \frac{\gamma + 1}{2} \left[\frac{3}{4}(1 + z^2)(D\Delta_w + \Delta_e) - z(1 - z^2)\frac{d(D\Delta_w + \Delta_e)}{dz}\right]^2 \tag{7.62}$$

rather than by Eq. (7.55).

Equations (7.54) and (7.62) make it possible to close the system of equations in partial derivatives (7.53) with modified boundary conditions (7.61). The solution of this boundary value problem for the case of the flow over delta wings with a power-law cross-sectional shape $\Delta_w(z) = (1-z)^\alpha$, where $\alpha \geq 0$ and $g(\lambda = 0) \neq 0$, was obtained in the paper of Dudin (1988*b*). Introducing the above-mentioned variables makes it possible to solve the system of equations from one edge to the other (Dudin and Lyzhin, 1983) without specially separating the supercritical and subcritical flow regions.

At the leading edges of the wing the system obtained degenerates to systems of ordinary differential equations

$$
\begin{aligned}
v\frac{\mathrm{d}u}{\mathrm{d}\lambda} &= z_0\frac{\gamma-1}{2\gamma p}(g-u^2-w^2)J_1 + \frac{\mathrm{d}^2u}{\mathrm{d}\lambda^2} \\
v\frac{\mathrm{d}w}{\mathrm{d}\lambda} &= -\frac{\gamma-1}{2\gamma p}(g-u^2-w^2)J_2 + \frac{\mathrm{d}^2w}{\mathrm{d}\lambda^2} \\
v\frac{\mathrm{d}g}{\mathrm{d}\lambda} &= \frac{1}{\sigma}\frac{\mathrm{d}^2g}{\mathrm{d}\lambda^2} - \frac{1-\sigma}{\sigma}\frac{\mathrm{d}^2(u^2+w^2)}{\mathrm{d}\lambda^2} \\
\frac{\mathrm{d}v}{\mathrm{d}\lambda} &= (w - z_0 u J_3)\frac{J_4}{p}
\end{aligned}
\tag{7.63}
$$

$$\Delta_e = \frac{1}{p}\left(\frac{\gamma-1}{2\gamma}\right)^{1/2}\int\limits_0^\infty (g-u^2-w^2)\,\mathrm{d}\lambda$$

$$p = \frac{\gamma+1}{2}[J_5(D\Delta_w + \Delta_e)]^2$$

$$J_1 = 1, \quad J_2 = J_3 = \pm 1, \quad J_4 = \pm\frac{1}{2}, \quad J_5 = \frac{3}{2}$$

$$\lambda = 0: \quad u = w = v = g = 0$$

$$\lambda \to \infty: \quad u \to 1, \quad w \to 0, \quad g \to 1$$

where plus and minus signs relate to the wing edges with $z = \pm 1$.

It should be noted that in solving the boundary value problem throughout the entire wing in the variables $z$ and $\lambda$ even in the case $D = 0$ the flow functions $f$, $p$, and $\Delta_e$ turn out to be dependent on the transverse coordinate $z$ not only in the subcritical but also in the supercritical flow region. This means that in variables (7.52) the system of boundary layer equations in the supercritical flow region cannot be brought into the self-similar form corresponding to the flow past a semi-infinite yawed plate. Below we will introduce variables in which the equations have a self-similar form also for supercritical flows.

### *7.4.2 Self-similar solutions*

For studying whether self-similar solutions are possible for $D \neq 0$ we will consider a vicinity of one of the leading edges, for example, $z = 1$. Then, in view of the behavior of the flow functions in the vicinity of the leading edge of a planar delta wing (Neiland, 1974*b*), it can be shown that if, instead of variables (7.52) and (7.60), the variables of the form:

$$\lambda_* = \left(\frac{2\gamma}{\gamma-1}\right)^{1/2}(1-z)^{1/4}\lambda, \quad p_* = (1-z)^{-1/2}p(z) \tag{7.64}$$

$$\Delta_e^* = (1-z)^{3/4}\Delta_e(z), \quad \Delta_w^*(z) = (1-z)^{3/4}\Delta_w(z)$$

$$v_* = \left(\frac{2\gamma}{\gamma-1}\right)^{1/2}(1-z)^{1/4}\left[v\frac{p}{1-z} - (w - z_0 z u)\frac{\partial\lambda}{\partial z}\right]$$

are introduced in the region $0 \le z \le 1$ (the asterisk at the variable $z$ is omitted), then the system of equations (7.53) with the boundary conditions (7.61) and expression (7.54) for $\Delta_e$ retains its form for $0 \le z \le 1$ if we let

$$F_1 = \frac{1-z}{p}, \quad F_2 = \frac{1}{2} + zF_1\frac{\mathrm{d}p}{\mathrm{d}z}, \quad F_3 = z + F_1\frac{\mathrm{d}p}{\mathrm{d}z}, \quad F_4 = \frac{1}{4p} \tag{7.65}$$

$$p = \frac{\gamma+1}{2}\left[\frac{3}{4}(D\Delta_w + \Delta_e) - z(1-z)\frac{\mathrm{d}(D\Delta_w + \Delta_e)}{\mathrm{d}z}\right]^2$$

Let us consider the case in which $\Delta_w(z) = \text{const}$. In view of the fact that the equation involves only the product $D\Delta_w$, we may, without loss of generality, let the constant be equal to unity. Then, in accordance with Eq. (7.64), in the $0 \le z \le 1$ region we have a power-law wing thickness variation $\Delta_w^*(z)$ with the exponent equal to 3/4. Assuming that there is a supercritical flow region in the boundary layer on a part of the wing adjoining the leading edge $z = 1$ and in it the functions $f(\lambda)$, $v(\lambda)$, $p$, and $\Delta_e$ are independent of the transverse coordinate $z$, we can easily show that this system of equations in partial derivatives, with account for Eq. (7.65), reduces in this region to a system of ordinary differential equations whose solution is self-similar. The form of this system is the same as that of system (7.63) with the following coefficients $J_i$

$$J_1 = J_2 = \frac{1}{2}, \quad J_3 = 1, \quad J_4 = \frac{1}{4}, \quad J_5 = \frac{3}{4} \tag{7.66}$$

For obtaining self-similar solutions in the supercritical region adjoining the $z = -1$ edge it is necessary to replace the expressions $(1-z)$ by $(1+z)$ in variables (7.64). Then, after the substitution in the general system of equations we again arrive at a system of ordinary differential equations for $\Delta_w = 1$, though with coefficients somewhat different from (7.66). These self-similar solutions govern the flows only in the regions in which the supercritical flow regime is realized. As shown below, the existence and particular dimensions of these regions depend on the wing aspect ratio and the parameters $\gamma$ and $\sigma$. The value of the coordinate $z_k$ corresponding to supercritical-to-m-subcritical flow transition is determined by Eq. (7.56) which in the case of symmetric flow over a wing takes the form:

$$\frac{\gamma-1}{2}\int_0^\infty\left[\frac{g - u^2 - w^2}{u\sin(\omega_0 - \omega_1) + w\cos(\omega_0 - \omega_1)}\right]^2 \mathrm{d}\lambda = \int_0^\infty (g - u^2 - w^2)\,\mathrm{d}\lambda \tag{7.67}$$

where $z_k = \tan(\omega_0 - \omega_1)/\tan\omega_0$. Here, the angle $\omega_1$ is measured from the wing edge $z = 1$.

For constant $\gamma$, $\omega_0$, and $\sigma$ the quantity $z_k$ depends on the solution of the system of equations (7.63) in which $J_i$ are determined by formulas (7.66). An analysis of system (7.63) shows that this solution depends on the parameter $D$, so that it might be expected that the values of $z_k$ calculated from Eq. (7.67) would be different for different $D$. However, this is not the case. Taking into account that $p$ is constant in the supercritical region we will perform the deformation of the variables

$$\lambda = \eta_* \sqrt{p}, \quad v = \frac{v_*}{\sqrt{p}}, \quad \Delta_e = \frac{\Delta_{e*}}{\sqrt{p}} \tag{7.68}$$

As a result, rewriting system (7.63) in terms of equalities (7.66) we arrive at pressure-independent system of equations and boundary conditions, while the pressure itself is determined by the expression

$$p = \left( \frac{3}{8}\sqrt{\frac{\gamma+1}{2}}D + \frac{1}{2}\sqrt{\frac{9}{32}(\gamma+1)D^2 + 3\sqrt{\frac{\gamma+1}{2}}\Delta_{e*}} \right)^2 \tag{7.69}$$

In deriving Eq. (7.69) it was assumed that in Eq. (7.64) $\Delta_w \equiv 1$. This does not influence the result since the parameters $D$ and $\Delta_w$ enter in the equation only in the form of a product.

The flow functions $u$, $w$, and $g$ obtained by solving this system are independent of the parameter $D$. Since in Eq. (7.67) the integrand written in variables (7.68) does not change its form, the transition coordinate $z_k$ is independent of the parameter $D$ in the flow over the delta wing with the thickness $\Delta_w^* = (1-z)^{3/4}$ $(0 \le z \le 1)$. In particular, if the distributions of the flow functions $u$, $w$, $v$, $g$, and $\Delta_e$ are known in the supercritical region of a planar delta wing $(D=0)$, then, using Eqs. (7.68) and (7.69), the values of these functions can easily be obtained for an arbitrary $D$.

Thus, if a supercritical flow region exists in hypersonic flow over a planar delta wing in the strong viscous interaction regime, then this region shall exist in flow over delta wings with power-law cross-sectional shape with the exponent 3/4, the dimensions of this region being independent of the ratio of the characteristic thicknesses of the wing and the boundary layer.

### *7.4.3 Results of calculations*

Below we present the results of numerical calculations for certain values of the relevant parameters. The numerical solution of the system of three-dimensional boundary layer equations was obtained by the method described in the paper of Dudin and Lyzhin (1983). The system of ordinary differential equations was solved using the sweeping method.

In Fig. 7.19 we have plotted the profiles of the flow functions $u$, $w$, and $g$ (curves 1 to 3) along the coordinate $\eta_*$ in the supercritical flow region for the flow over a delta wing with $\omega_0 = 45°$ $(z_0 = 1)$ at $\gamma = 1.4$, $\sigma = 1$, and $D = 0.7038$. We note that for this value of $D$ the system of equations (7.63) written in variables (7.64) with account for Eqs. (7.66) does not change its form as a result of the variable deformation (7.68), since in the case under consideration we obtain the displacement thickness $\Delta_{e*} = 0.1901$ and the pressure $p = 1$. The coordinate of supercritical-to-subcritical flow regime transition is $z_k = 0.4958$.

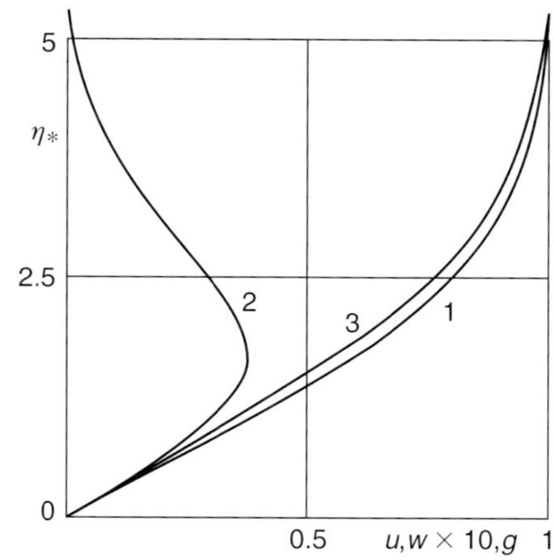


Fig. 7.19.

Figure 7.20 presents the dependence of the angle $\omega_1$ corresponding to supercritical-to-m-subcritical flow regime transition on the leading-edge sweep angle $(\pi/2) - \omega_0$. The systems of equations (7.53) and (7.63) include the wing aspect ratio $z_0 = \tan\omega_0$ as a parameter. This makes difficult the numerical solution as $\omega_0 \to 90°$, since in this case it is necessary to consider $z_0 \to \infty$. The calculations were carried out up to $z_0 = 50$ corresponding to $\omega_0 \sim 88°50'$. For $\gamma = 1.4$ the calculated results are somewhat different from the data presented in Fig. 7.2, particularly for $\omega_0 \leq 50°$. In particular, according to the present results, the supercritical flow region no longer exists at an angle $\omega_0 \sim 20°$.

The effect of the parameter $\gamma$ on the supercritical flow region dimensions is considerable. A decrease of $\gamma$ to 1.2 leads to an appreciable enlargement of this region; in this case the

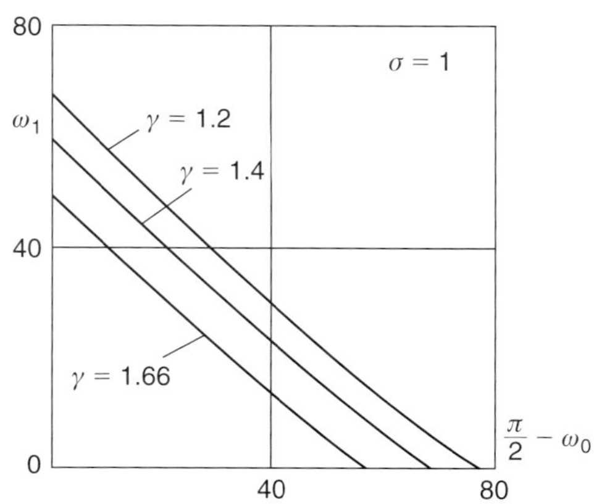


Fig. 7.20.

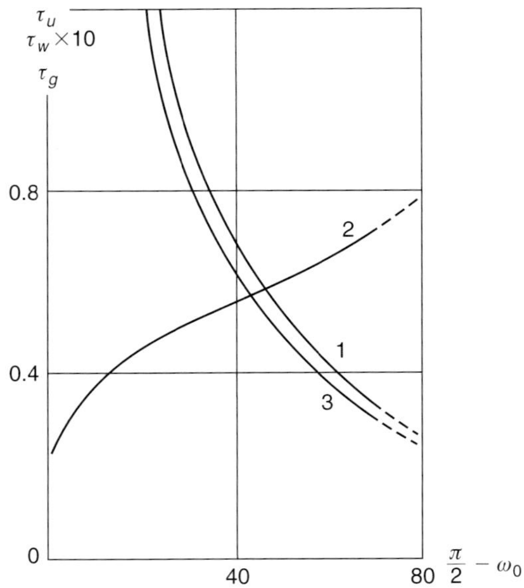


Fig. 7.21.

subcritical flow past the leading edges is realized only for wings with sweep angles $\geq 76°$. In Fig. 7.21 the calculated longitudinal and transverse viscous stress and heat flux coefficients

$$\tau_u = \left.\frac{\partial u}{\partial \eta_*}\right|_{\eta_*=0}, \qquad \tau_w = \left.\frac{\partial w}{\partial \eta_*}\right|_{\eta_*=0}, \qquad \tau_g = \left.\frac{\partial g}{\partial \eta_*}\right|_{\eta_*=0}$$

are plotted against the angle $(\pi/2) - \omega_0$ (curves 1 to 3) in the supercritical flow region for $\gamma = 1.4$ and $\sigma = 1$. The distributions of $p$ and $\Delta_e$ in variables (7.64) (curves 1 and 2) and the $\omega_0$ dependence of the transition coordinate $z_k$ (curve 3) are presented in Fig. 7.22. Dashed curves in Figs. 7.21 and 7.22 present the values of the flow functions for the angle $(\pi/2) - \omega_0$ for which the supercritical flow region does not exist, the values themselves being realized only at the wing edges. It should be noted that $z_k \to 0$ as $\omega_0 \to 90°$.

For obtaining the global solutions, in which self-similar solutions are realized in supercritical flow regions, we will solve the system of equations for the case in which the wing thickness in Eq. (7.60) is given by the expression $\Delta_w(z) = (1 + |z|)^{-3/4}$. Though in this case for $z = 0$ the derivative $d\Delta_w/dz$ does not exist, this has no effect on the problem solution, since this derivative enters only in expression (7.62) for determining $p$, where it is multiplied by $z$. However, in the presence of a yaw angle this can no longer be neglected. In Fig. 7.23 we have plotted the spanwise distributions of the induced pressure $p^1 = p(1 + z)^{-1/2}$ (solid curves) and the boundary layer thickness $\Delta_e^1 = \Delta_e(1 + z)^{3/2}$ (dashed curves) for $0 \leq z \leq 1$ in the flow over a cold wing at $z_0 = 1$, $\gamma = 1.4$, $\sigma = 1$, and $D = 0$, 0.1, 0.2, and 0.3 (curves 1 to 4). We note that in the supercritical flow region $p^1$ and $\Delta_e^1$ correspond to $p$ and $\Delta_e$ introduced by Eq. (7.64). In Fig. 7.23 points on the curves indicate the values of the coordinate $z_k$

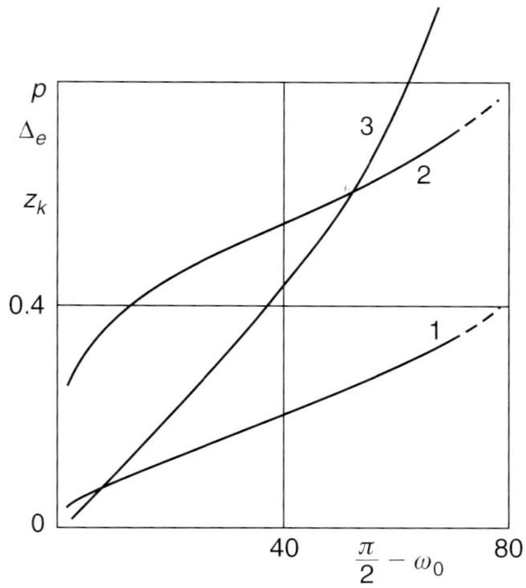


Fig. 7.22.

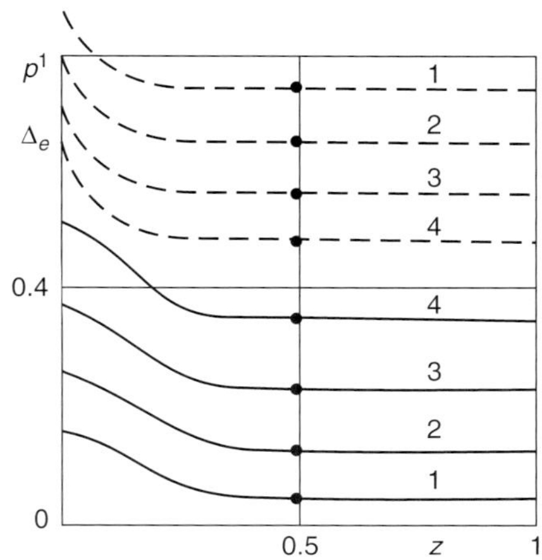


Fig. 7.23.

obtained from Eq. (7.67). The results of the numerical solution of the boundary value problem show that a self-similar solution is actually realized in the supercritical flow region and the dimensions of this region are independent of the parameter $D$. Contrariwise, in the m-subcritical flow region the flow functions depend on this parameter. Figure 7.24 presents the distributions of the quantities

$$\tau_u^1 = (1+z)^{-1/4} \left.\frac{\partial u}{\partial \eta^*}\right|_w, \quad \tau_w^1 = (1+z)^{-1/4} \left.\frac{\partial w}{\partial \eta^*}\right|_w, \quad \tau_g^1 = (1+z)^{-1/4} \left.\frac{\partial g}{\partial \eta^*}\right|_w$$

(curves 1 to 3, respectively) over the wing surface for $D = 0$ (solid curve) and $D = 0.3$ (dashed curve). In the supercritical flow region ($z_k > 0.4958$) the above functions are constant.

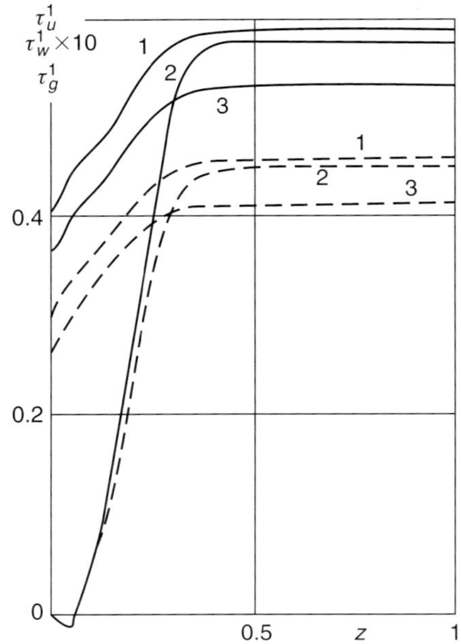


Fig. 7.24.

## 7.5 Effect of strong cooling of the surface on the hypersonic viscous flow over a nonplanar delta wing

Here, we consider the effect of strong cooling of the body surface on the parameters of the boundary layer flow on a thin delta wing in hypersonic viscous perfect-gas flow in the strong viscous–inviscid interaction regime. The influence of the cross-sectional shape and the ratio of the wing thickness to the boundary layer displacement thickness on the local and total aerodynamic characteristics is studied.

In hypersonic flow over thin delta wings the nature of the flow in the three-dimensional boundary layer interacting with the outer inviscid flow considerably depends on the cross-sectional shape of the wing and a parameter characterizing the wing to boundary layer displacement thickness ratio (Dudin, 1988*b*). A fairly intense secondary flow that occurs makes the pattern of the flow around such a wing fairly complicated as compared with the flow around a planar delta wing. In flows around wings the boundary layer/outer-flow interaction can lead to the formation of regions with developed zones of secondary return flow and, as a consequence, the appearance of local zones of elevated heat fluxes.

The flow over thin delta wings is particular in the case in which the surface temperature is asymptotically low as compared with the stagnation temperature. In Section 7.4 we considered the flow past delta wings with three-fourth leading edges (Dudin, 1997). It was shown that for sweep angles less than a critical value in the laminar boundary layer there appear supercritical and m-subcritical flow regions. In the former, disturbances do not propagate upstream and a self-similar flow is realized. If the sweep angle is greater than the critical angle, then the m-subcritical regime is realized over the entire wing and disturbances propagate from the plane of symmetry of the wing up to the leading edges.

In view of the above-said, it might be expected that in the flow over cold wings the wing thickness distribution has an effect on the boundary layer flow parameters and in supercritical flow regions there can occur flows which are not governed by self-similar solutions, so that such solutions can appear only for a special law of the wing thickness distribution (Dudin, 1998).

### *7.5.1 Equations and boundary conditions*

We will consider the symmetric hypersonic viscous perfect-gas flow over a semi-infinite thin delta wing with a characteristic thickness $\tau$ of the order of the boundary layer thickness at zero incidence. It is assumed that the wing surface temperature is constant and low as compared with the freestream stagnation temperature and the regime of strong interaction between the boundary layer and the outer inviscid flow is realized. The formulation of the problem was described in Section 7.4.1 of the preceding section; therefore, we will present only the final system of the laminar boundary layer equations and the boundary conditions which takes the form:

$$(w - z_0 z u)F_1 \dot{f} + v f' = G \tag{7.70}$$

$$f = \begin{Bmatrix} u \\ w \\ g \end{Bmatrix}, \quad G = \begin{cases} z_0 \frac{\gamma - 1}{2\gamma p}(g - u^2 - w^2)F_2 + u'' \\ -\frac{\gamma - 1}{2\gamma p}(g - u^2 - w^2)F_3 + w'' \\ \frac{1}{\sigma} g'' - \frac{1 - \sigma}{\sigma}(u^2 + w^2)'' \end{cases}$$

$$v' = (w - z_0 z u)F_4 - F_1(0.25 z_0 u - z_0 z \dot{u} + \dot{w})$$

$$F_1 = \frac{1 - z^2}{p}, \quad F_2 = \frac{1 + z^2}{2} + z F_1 \frac{\mathrm{d}p}{\mathrm{d}z}, \quad F_3 = z + F_1 \frac{\mathrm{d}p}{\mathrm{d}z}, \quad F_4 = \frac{z}{2p}$$

$$\lambda = 0: \quad u = w = v = g = 0$$

$$\lambda \to \infty: \quad u \to 1, \quad w \to 0, \quad g \to 1$$

where $|z| \leq 1$, $\sigma$ is the Prandtl number, and $\gamma$ is the specific heat ratio.

For the boundary layer thickness we obtain

$$\Delta_e = \frac{1}{p}\left(\frac{\gamma - 1}{2\gamma}\right)^{1/2} \int_0^\infty (g - u^2 - w^2)\,\mathrm{d}\lambda \tag{7.71}$$

Using the approximate tangent wedge formula, for the pressure we obtain

$$p = \frac{\gamma + 1}{2}\left[\frac{3}{4}(1 + z^2)(D\Delta_w + \Delta_e) - z(1 - z^2)\frac{\mathrm{d}(D\Delta_w + \Delta_e)}{\mathrm{d}z}\right]^2 \tag{7.72}$$

where $D = \tau/\delta$ and it is taken into account that, correct to $O(\tau^2/z_0)$, the boundary layer displacement thickness is equal to $\Delta_e$ (Bashkin and Dudin, 2000).

Equations (7.71) and (7.72) make it possible to close the system of equations in partial derivatives (7.70) which describes the flow in the laminar boundary layer on a cold delta wing in the strong viscous interaction regime. We note that in substituting expression (7.72) for the pressure in the system of equations (7.70), the latter, due to the presence of the term $dp/dz$, includes the second derivative $d^2\Delta_e/dz^2$ which makes it possible to take account for a boundary condition imposed somewhere downstream, for example, the no-flow condition in the plane of symmetry of the wing. At the leading edges of the delta wing ($z = \pm 1$) the system of equations (7.70)–(7.72) degenerates to systems of ordinary differential equations whose solutions make it possible to determine all flow functions in the boundary layer at the edges.

In accordance with the results of Section 7.4, in studying the hypersonic flow over a triangular plate and a wing with the leading edge shape given by the function $\Delta_w^* = (1 - |z|)^{3/4}\Delta_w^1$, where $\Delta_w^1 = \text{const}$, for obtaining self-similar solutions in supercritical regions it is necessary to introduce variables of type (7.64) with the terms $1 - z^2$ replaced by $1 - |z|$ rather than variables (7.52).

As noted above (in Section 7.4.3), in considering the flow over a delta wing with $\Delta_w(z) = (1 + |z|)^{-3/4}$ in the expression for the thickness (7.60), the self-similar form of the solutions of system (7.70)–(7.72) in supercritical regions is obtained if the calculated results are presented in the form:

$$p^1(z) = (1 + |z|)^{-1/2}p(z), \quad \Delta_e^1(z) = (1 + |z|)^{3/4}\Delta_e \tag{7.73}$$

$$\tau_u^1 = (1 + |z|)^{-1/4}\frac{\partial u}{\partial \lambda}(z, 0), \quad \tau_w^1 = (1 + |z|)^{-1/4}\frac{\partial w}{\partial \lambda}(z, 0)$$

$$\tau_g^1 = (1 + |z|)^{-1/4}\frac{\partial g}{\partial \lambda}(z, 0)$$

When the function $\Delta_w(z)$ in Eq. (7.60) is arbitrarily preassigned, self-similar solutions are generally absent from supercritical flow regions; nevertheless, for the sake of presentation, the calculated results can conveniently be represented in form (7.73) which will be done in discussing the results of numerical calculations.

The coordinate $z_k$ corresponding to supercritical-to-m-subcritical flow transition in the boundary layer on a cold body is determined from Eq. (7.67) which can be written down as follows:

$$\frac{\gamma - 1}{2}\int_0^\infty \left[\frac{g - u^2 - w^2}{u\sin(\omega_0 - \omega_1) + w\cos(\omega_0 - \omega_1)}\right]^2 d\lambda = \int_0^\infty (g - u^2 - w^2)\, d\lambda \tag{7.74}$$

where $z_k = \tan(\omega_0 - \omega_1)/\tan\omega_0$ and the angle $\omega_1$ is measured from the $z = 1$ wing edge.

We note that the above equation includes the values of the functions $g$, $u$, and $w$ at point $z = z_k$, as distinct from the previous section in which these functions were taken at the leading edges of the wing; this is due to the fact that in the case under consideration self-similar solutions no longer exist in the supercritical flow region. In presenting the results of numerical calculations the critical values of the coordinate $z_k$ obtained when this condition is fulfilled will be indicated by points on the curves for the pressure.

### *7.5.2 Results of calculations*

In solving the system of equations (7.70)–(7.72) it is necessary to take account for the fact that the equations of motion include the term with the second derivative of $\Delta_e$ with respect to $z$. For solving the boundary value problem obtained above we use the method developed by Dudin (1985) which allows for upstream disturbance transfer in the boundary layer and makes it possible to carry out calculations when secondary return flows occur, that is, when in the equations of motion (7.70) the sign of the coefficient $w - szu$ varies across the boundary layer.

Let us consider an example of the flow over a delta wing with a power-law cross-sectional shape, for which the function $\Delta_w(z) = (1 - z)^\alpha$, where the exponent $\alpha \geq 0$. In all numerical calculations presented below it was assumed that $\sigma = 1$, $\gamma = 1.4$, and $z_0 = 1$ (the semi-vertex wing angle $\omega_0 = 45°$). In Fig. 7.25 the profiles of the pressure $p^1$ and the boundary layer displacement thickness $\Delta_e^1$ along the coordinate $z$ are presented for $\alpha = 9.25$ and $D = 0$, 0.1, 0.5, and 0.85 (curves 1 to 4), together with the corresponding distribution of the wing thickness $\Delta_w^1$.

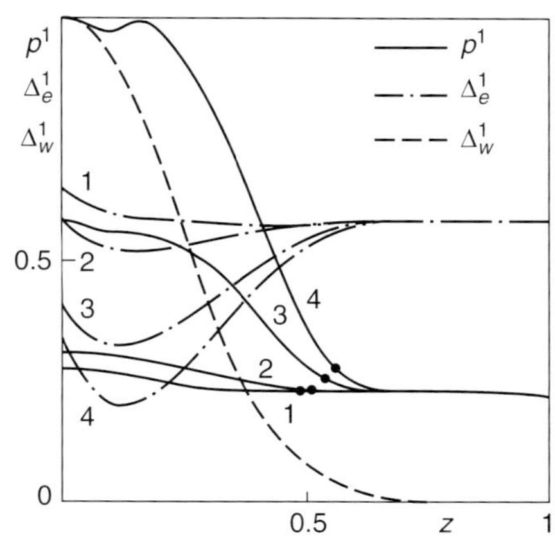


Fig. 7.25.

A considerable rise of the pressure with increase in $D$ and an earlier departure of the corresponding curve from the line corresponding to the self-similar solution should be noted. Curves 1 relate to the case of the flow past a triangular plate ($D = 0$) for which supercritical-to-m-subcritical flow transition takes place at $z_k = 0.4958$ for the flow parameters noted above. With increase in $D$ the transition coordinates $z_k$ (points on the pressure curves) are displaced toward the leading edges of the wing, that is, the supercritical flow region diminishes. In the cases in question the transition coordinate depends on the values of the function $g$, $u$, and $w$ at the transition point, these values being different from self-similar (except for the case $D = 0$). An increase in $D$ also leads to a nonmonotonic behavior of the function $\Delta_e^1$. A decrease of the quantity $\Delta_e^1$ is attributable to the fact that an increase in the pressure leads to flow deceleration in the transverse direction.

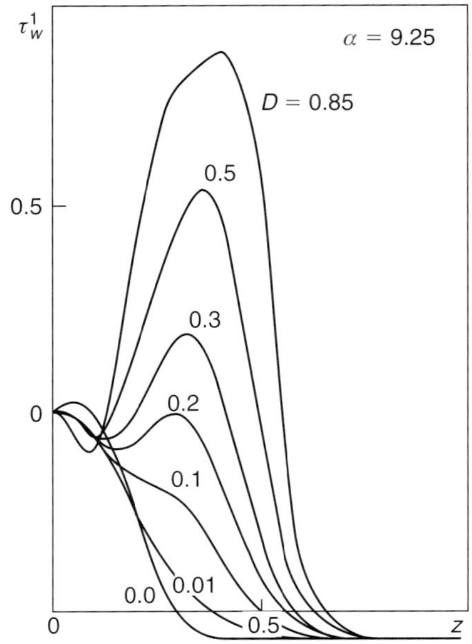


Fig. 7.26.

Figure 7.26 presents the spanwise distributions of the viscous stress in the transverse direction $\tau_w^1$, in accordance with which for $D > 0.2$ in the vicinity of the wing surface there appears a region in which the velocity component $w$ reverses sign, while an increase in $D$ leads not only to the enlargement of this region but also to an increase of the return flow velocity $w$ in absolute magnitude. We note a significant difference between the boundary layer flows in the vicinity of the plane of symmetry of the wing for $D = 0$ and $D > 0$. For $D = 0$ in this near-surface region there appears a small-sized ($\Delta z \approx 0.08$) return-flow zone. For $D > 0$ here a flow with smooth convergence is realized, which provides an explanation for the growth of the boundary layer displacement thickness $\Delta_e^1$ in this region (Fig. 7.25). For $D > 0.2$ intense flow deceleration in the transverse direction also leads to an appreciably nonmonotonic distribution of the longitudinal viscous stress coefficient $\tau_u^1$ (Fig. 7.27). In the case $D = 0.85$ in the vicinity of $z \approx 0.4$ near the wing surface there appears a small-sized region of secondary return flow, since there the coefficient $w - z_0 z u$ determining the parabolicity direction of system (7.70) reverses sign. In the order, $|\tau_w^1| \sim 0.1 |\tau_u^1|$.

For $D = 0.4$ the results of the study of the exponent $\alpha$ effect on the flow parameters are presented in Figs. 7.28–7.30. In Fig. 7.28 we have plotted the results of numerical calculations for $p^1$ and $\Delta_e^1$ at $\alpha = 2.25$, 9.25, and 19.25 (curves 1 to 3). The spanwise distributions of the wing thickness $\Delta_w$ for $\alpha = 2.25$ and 19.25 are presented as dashed curves 1 and 3. With increase in the exponent $\alpha$ the pressure $p^1$ decreases and the elevated-pressure region diminishes. The numerical calculations carried out for $D = 0.4$ showed that the departure of the curves for both $p^1$ and $\Delta_e^1$ from the straight lines corresponding to the self-similar solutions takes place at the values of the coordinate $z$ at which the value of $\Delta_w$ is equal to approximately 0.0005–0.001. For $\alpha = 2.25$ the departure from the self-similar solution

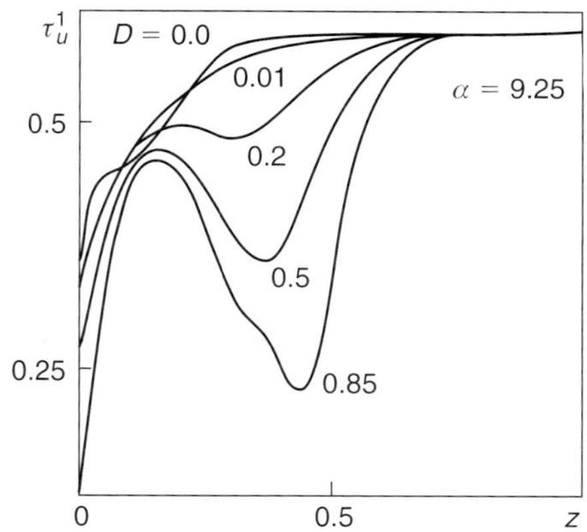


Fig. 7.27.

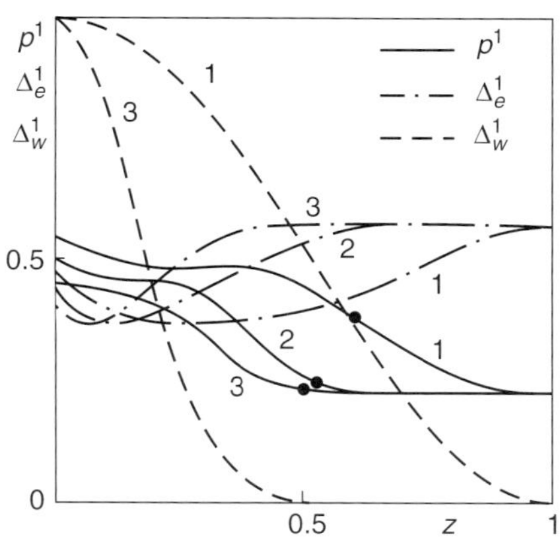


Fig. 7.28.

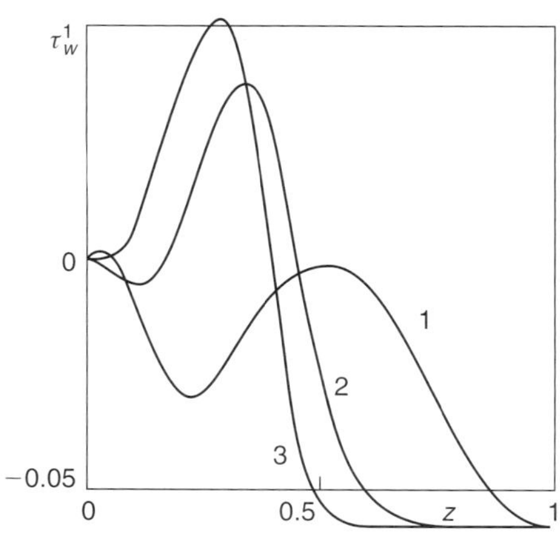


Fig. 7.29.

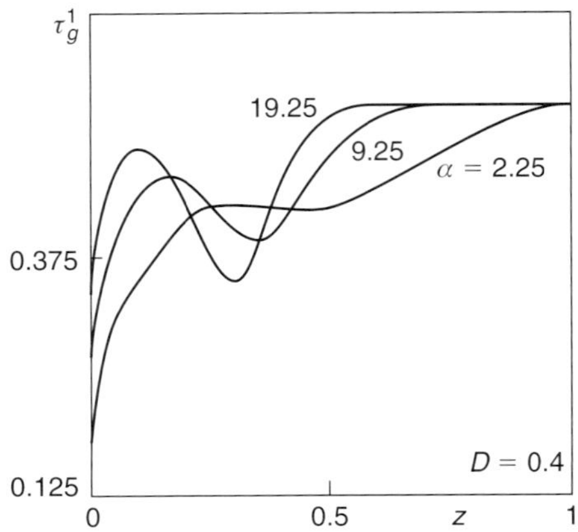


Fig. 7.30.

takes place in the immediate vicinity of the leading edge. With decrease in the exponent $\alpha$ the supercritical flow region considerably diminishes. According to the data presented, the $\Delta_e^1$ minimum remains constant and $\alpha$-independent but the region of nonmonotonic variation of $\Delta_e^1$ diminishes with increase in $\alpha$. The distribution of the parameter $\tau_w^1$ in Fig. 7.29 shows that for $\alpha = 2.25$ (curve 1) the velocity component $w$ is directed from the edges to the plane of symmetry throughout the entire boundary layer flow region. For $\alpha = 9.25$ and 19.25 (curves 2 and 3) in the boundary layer there appear developed flow zones ($\Delta z \approx 0.3$) in which the velocity component $w$ reverses sign. The distributions of both the heat flux $\tau_g^1$ in Fig. 7.30 and the longitudinal viscous stress coefficient $\tau_u^1$ are considerably nonmonotonic for $\alpha = 9.25$ and 19.25. For greater values of $\alpha$ there appear zones of elevated heat flux which are displaced toward the plane of symmetry and become narrower with increase in $\alpha$.

In Fig. 7.31 we have plotted the calculated aerodynamic coefficients

$$C_f^* = C_f Re_0^{3/4} = \frac{8}{5\sqrt[4]{z_0}}\sqrt{\frac{2\gamma}{\gamma - 1}}\int_{-1}^{1} \frac{p}{(1 - z^2)^{3/4}}\frac{\partial u}{\partial \lambda}(z, 0)\,\mathrm{d}z \tag{7.75}$$

$$C_p^* = C_p Re_0^{1/2} = \frac{4}{3}\sqrt{z_0}\int_{-1}^{1} \frac{p}{\sqrt{1 - z^2}}\,\mathrm{d}z, \quad m_z^* = m_z Re_0^{1/2} = \frac{3}{5}C_p^*$$

calculated for one wing side at $\alpha = 9.25$ and 19.25 (curves 1 and 2) versus the parameter $D$. The growth of the aerodynamic coefficients $C_p^*$ and $m_z^*$ with increase in $D$ should be noted; this is due to an increase in the pressure $p$ on the wing surface (Fig. 7.25). The relative increase of $C_f^*$ is weaker which is attributable to a decrease in $\tau_u^1$ over a considerable part of the wing with increase in $D$ (Fig. 7.27).

We note in conclusion that in the numerical calculations the dependence of the gas dynamic functions on the parameters $\alpha$ and $D$ characterizing the cross-sectional wing shape and the wing to boundary layer thickness ratio is found. It is established that an increase in

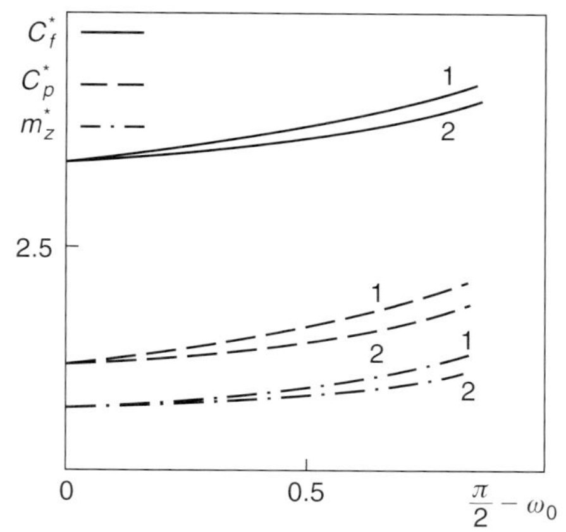


Fig. 7.31.

both $\alpha$ and $D$ leads to significantly nonmonotonic spanwise variation of the flow functions, in particular, to the appearance of local elevated-heat-flux zones on the wing surface in a region adjoining the plane of symmetry. An increase in the parameter $D$ for a given cross-sectional shape of the wing leads to a decrease in the extent of the supercritical flow region in the laminar boundary layer. An analogous phenomenon is observable with decrease in the parameter $\alpha$ at a fixed value of $D$. The effect of the parameters $\alpha$ and $D$ on the total aerodynamic coefficients is fairly weak.

## 7.6 Self-similar flows with gas injection from the triangular plate surface into a hypersonic flow

In this section we study the flow occurring at gas injection through the permeable surface of a triangular plate in the regime of strong viscous–inviscid interaction between a hypersonic flow and a laminar boundary layer. The distinctive features of flows past strongly cooled surfaces with the formation of supercritical and subcritical flow regions are considered. An injection velocity distribution, for which self-similar solutions exist in supercritical flow regions, is derived. The results of numerical calculations of the flow parameters are presented.

Gas injection through a permeable body surface is an important factor influencing the flow parameters. In particular, with considerable increase in the injection intensity the boundary layer can be detached from the body surface; in this case an inviscid (in the first approximation) flow region appears between the boundary layer and the body surface (Neiland, 1972). In this case, the pressure distribution is additionally influenced by the distribution of the injected-gas layer displacement thickness which, in turn, depends on the pressure distribution.

In hypersonic flow past a body the body surface temperature can frequently be asymptotically small as compared with the stagnation temperature and both supercritical and subcritical flow regions can occur in the laminar boundary layer. With decrease in the surface temperature the injected-gas density increases which favors an increase in the gas flow rate through the permeable surface for a given injection velocity (Neiland, 1972). As a result, even for fairly small injection velocities an inviscid flow zone can be formed near the surface of a strongly cooled body. This should be borne in mind in considering hypersonic viscous flow past bodies with mass transfer at the body surface. In this section emphasis is placed on the study of the effect of distributed surface gas injection on the parameters of the flow past a strongly cooled triangular plate in hypersonic flow (Dudin, 2000).

### *7.6.1 Equations and boundary conditions*

We will consider a symmetric hypersonic viscous perfect-gas flow past a semi-infinite triangular plate at zero incidence. We introduce a Cartesian coordinate system with origin at the plate vertex, the $x\ell$ axis aligned with the axis of symmetry, the $z\ell$ axis normal to the $x\ell$ axis in the plane of the plate, and the $y\ell$ axis normal to the $xz$ plane. The velocity components $uu_\infty$, $vu_\infty$, and $wu_\infty$ are directed along the $x$, $y$, and $z$ axes. The semi-vertex angle of the triangular plate is equal to $\omega_0$. The parameter $z_0 = \tan\omega_0$ characterizing the transverse-to-longitudinal dimension ratio is assumed to be of the order of unity. The scale length $\ell$ drops out from the final results.

Continuous distributed gas injection is performed through the permeable plate surface; the given injection velocity $v(y=0) = F(x,z)$ is of the order $O(\tau)$, where $\tau \sim \delta \sim Re_0^{-1/4}$ corresponds to the normal velocity component in the boundary layer in the absence of mass transfer. It is assumed that the injection intensity is such that an inviscid flow region is not formed near the plate surface. As a result, the hypersonic flow streams past an "effective body" formed by processes due to viscosity and heat conduction. We will consider flow regimes in which the characteristic slope of the effective body is of the order of the characteristic dimensionless boundary layer displacement thickness, $\delta \ll 1$. Then in the first approximation the inviscid shock layer on the plate is described by hypersonic small perturbation theory. In considering the strong viscous interaction regime, $M_\infty\delta \gg 1$ ($M_\infty$ is the undisturbed flow Mach number), the induced pressure difference is greater than the undisturbed flow pressure by a factor of $M_\infty^2\delta^2$ and the approximate tangent wedge formula can be used for determining the pressure.

The plate surface temperature $T_w$ is assumed to be given and constant along the surface. We will also consider the case in which $T_w$ is asymptotically small, as compared with the stagnation temperature.

In what follows, a particular form of the function $F_w(x,z) \times 2c_pT_w/u_\infty^2$, where $c_p$ is the specific heat at constant pressure and $u_\infty$ is the freestream velocity, is chosen from the condition that the three-dimensional boundary value problem can be reduced to a self-similar problem. In order for the gas injection velocity to be independent of the plate surface temperature in the boundary value problem formulated in the dimensionless variables, it is assumed that the injection velocity is proportional to the temperature factor.

In accordance with the conventional estimates for the laminar boundary layer in hypersonic flow, in the region, in which the leading viscous and inertial terms of the Navier–Stokes equations are of the same order, we introduce the dimensionless coordinates and functions (7.50) and, additionally

$$F_w = u_\infty \delta z_0^{-1} F^*, \quad T_w = \frac{u_\infty^2 g_w}{2c_p} \tag{7.76}$$

For the sake of simplicity, we will assume the linear viscosity–temperature dependence $\mu = g - u^2 - w^2$ and introduce the Dorodnitsyn variables

$$\eta = \int_0^{y^*} \rho^* \, \mathrm{d}y^*, \quad v_\delta^* = \rho^* v^* + z_0 u \frac{\partial \eta}{\partial x} + w \frac{\partial \eta}{\partial z^*}$$

At the plate surface the dimensionless injection velocity is as follows:

$$v_\delta^*(\eta = 0) = \rho_w^* F^* g_w = \frac{2\gamma}{\gamma - 1} p^* F^*, \quad \text{since } \rho^* = \frac{2\gamma}{\gamma - 1} \frac{p^*}{g - u^2 - w^2}$$

In considering the flow past semi-infinite triangular bodies with mass transfer at the surface in the strong viscous–inviscid interaction regime the boundary value problem can be reduced to a self-similar problem (Dudin, 1979) if we require that the function

$$F^* = x^{-1/4} F_* \left( \frac{z^*}{x} \right) \tag{7.77}$$

and introduce the self-similar variables (7.51).

In substituting variables (7.51) in the equations of the three-dimensional boundary layer and the boundary conditions the coordinate $x$ drops out from the boundary value problem and the system of equations in partial derivatives thus obtained describes the flow in a three-dimensional boundary layer in the $z, \lambda_*$ plane (self-similar problem). In this case, the injection velocity at the plate surface is determined by the quantity $v_*(\lambda_* = 0) = (2\gamma/\gamma - 1) p_*(z_*) F_*(z_*)$.

In what follows it is assumed that the function characterizing mass transfer is determined by the expression

$$F_* = \left( \frac{\gamma - 1}{2\gamma} \right)^{1/2} (1 - z_*^2)^{-1/4} F(z_*) \tag{7.78}$$

where the function $F(z_*)$ has no singularities at the leading edges of the triangular plate at the values of the coordinate $z_* = \pm 1$. In this case, for solving the system of equations in partial derivatives thus obtained it is convenient to introduce variables (7.52) which allow for the behavior of the flow functions in the vicinity of the leading edges.

Then the system of equations of the laminar boundary layer and the boundary conditions take the form (the asterisk at the coordinate $z$ is omitted):

$$(w - z_0 z u) F_1 \dot{f} + v f' = G \tag{7.79}$$

$$f = \begin{Bmatrix} u \\ w \\ g \end{Bmatrix}, \quad G = \begin{cases} z_0(g - u^2 - w^2)F_2 + u'' \\ -(g - u^2 - w^2)F_3 + w'' \\ \frac{1}{\sigma}g'' - \frac{1-\sigma}{\sigma}(u^2 + w^2)'' \end{cases}$$

$$v' = (w - z_0 z u)\frac{z}{2p} - F_1(0.25 z_0 u - z_0 z \dot{u} + \dot{w})$$

$$F_1 = \frac{1 - z^2}{p}, \quad F_2 = \frac{\gamma - 1}{2\gamma}[0.5(1 + z^2) + zF_1\dot{p}], \quad F_3 = \frac{\gamma - 1}{2\gamma}(z + F_1\dot{p})$$

$$\lambda = 0: \quad u = w = 0, \quad g = g_w, \quad v = F(z)$$

$$\lambda \to \infty: \quad u \to 1, \quad w \to 0, \quad g \to 1$$

where primes and dots denote the differentiation with respect to $\lambda$ and $z$, respectively, $|z| \leq 1$, $\sigma$ is the Prandtl number, and $\gamma$ is the specific heat ratio. We note that at the plate surface in the dimensional variables the injection velocity is determined by the relation

$$F^0 = \left(\frac{\gamma - 1}{2\gamma}\right)^{1/2} u_\infty z_0^{-3/4}[Re_0 \times x \times (1 - z^2)]^{-1/4} F g_w \tag{7.80}$$

Therefore, in considering the flow past bodies with $g_w \to 0$ for obtaining final injection velocities in dimensional variables it is necessary that $F \to \infty$.

Using the Dorodnitsyn variables and taking into account that the gas density at the outer edge of the boundary layer increases without bound, since the temperature vanishes, for the laminar boundary layer thickness we obtain

$$\Delta_e = \frac{1}{p}\left(\frac{\gamma - 1}{2\gamma}\right)^{1/2} \int_0^\infty (g - u^2 - w^2)\,\mathrm{d}\lambda \tag{7.81}$$

As noted above, in the flow regimes under consideration the pressure distribution over the body surface is not known beforehand and must be determined in solving jointly the problems of the boundary layer flow with mass transfer at the surface and the outer inviscid flow described in the first approximation by hypersonic small perturbation theory. In considering the flow past thin bodies with the aspect ratio $z_0 = O(1)$ strip theory is applicable for the outer inviscid flow with $M_\infty \gg 1$ and, provided that $M_\infty \delta \gg 1$, the approximate tangent wedge formula can be used for determining the induced pressure; after the Dorodnitsyn variables have been introduced, this formula takes the form:

$$p = \frac{\gamma + 1}{2}\left[\frac{3}{4}(1 + z^2)\Delta_e - z(1 - z^2)\frac{\mathrm{d}\Delta_e}{\mathrm{d}z}\right]^2 \tag{7.82}$$

Equations (7.81) and (7.82) make it possible to close the system of equations in partial derivatives (7.79) which describes the three-dimensional flow in the laminar boundary layer on a triangular plate with a given surface injection velocity in the strong viscous–inviscid interaction regime. At the leading edges of the plate $z=\pm 1$ the system of equations degenerates to systems of ordinary differential equations whose solutions make it possible to determine all the flow functions in the boundary layer at the wing edges. It should be noted that expression (7.82) can be used for determining the pressure only until an inviscid (in the first approximation) flow region begins to be formed near the plate surface with increase in the injection velocity (Neiland, 1972). Within the framework of the problem under consideration this occurs at such injection velocities $v(\eta=0)=F(z)$ at which the viscous stress coefficient $\tau=\sqrt{\tau_u^2+\tau_w^2}$ vanishes at the plate surface (Neiland, 1972), where $\tau_u=\frac{\partial u}{\partial\lambda}\big|_{\lambda=0}$ and $\tau_w=\frac{\partial w}{\partial\lambda}\big|_{\lambda=0}$ are the longitudinal and transverse viscous stress coefficients.

### *7.6.2 Reduction to self-similar form*

We will consider certain distinctive features of the flow past cold triangular plates in the case in which $g_w=0$. It should be noted that in solving the boundary value problem (7.79), (7.81), and (7.82) in the $z$, $\lambda$ variables the flow functions $f$, $p$, and $\Delta_e$ turn out to be dependent on the transverse coordinate $z$ in the supercritical flow region even in the absence of injection (Dudin, 1997). In order to bring the boundary value problem into the self-similar form, in supercritical flow regions, it is necessary to consider – instead of flow (7.78) – the flow in which the function determining mass transfer is as follows:

$$F_*=\left(\frac{\gamma-1}{2\gamma}\right)^{1/2}(1-|z|)^{-1/4}F^1 \tag{7.83}$$

where $F^1$ is constant, and to introduce – instead of variables (7.52) – variables of type (7.64) in which the terms $(1-z^2)$ are replaced by $(1-|z|)$.

It can be shown that in considering the flow past a triangular plate with distributed injection given by expression $F(z)=(1+|z|)^{1/4}F^1$ in Eq. (7.78) the self-similar form of the solution of system (7.79), (7.81), (7.82) in the supercritical flow regions is obtained if the results of calculations are represented in form (7.73). Obviously, in the case in which the function $F(z)$ in Eq. (7.78) is arbitrarily specified, there are no self-similar solutions in the supercritical flow regions.

For determining the transition coordinate $z_k$ corresponding to supercritical-to-subcritical flow transition in the boundary layer on a cold body ($g_w=0$) Eq. (7.74) is used.

We note that if in the supercritical flow region there are no self-similar solutions, then Eq. (7.74) involves the current values of the functions $g$, $u$, and $w$ at point $z=z_k$. If the function $F(z)$ is so chosen that in the supercritical region of the boundary layer self-similar solutions exist, then for $0<z\leq 1$, as in the papers of Neiland (1974*b*) and Dudin (1997), the functions are taken from the solution of the system of ordinary differential equations at the leading edge of the plate. In presenting the calculated results, the critical values of the coordinate $z_k$ obtained when the above condition is fulfilled are marked by points on the pressure curves.

### 7.6.3 Results of calculations

We will present some calculated results for the flow past triangular plates. In all calculations presented below it was assumed that $\sigma = 1$ and $\gamma = 1.4$.

Figures 7.32–7.34 present the results obtained in solving the system of ordinary differential equations at the leading edge of the plate for $z = -1$. In Fig. 7.32 we have plotted the dependence of the viscous stress coefficient $\tau$ on the injection velocity $F$ for $g_w = 0, 0.1, 0.2, 0.5$, and 1 (curves 1 to 5) for a plate with the sweep angle of the leading edge 45° ($z_0 = 1$). Clearly that in the case of the flow past a cold plate ($g_w = 0$, curve 1) the viscous stress coefficient $\tau \to 0$ as $F \to 1.1$; for $F > 1.1$ there is no longer a solution within the framework of boundary layer theory. This means that for the injection velocities greater than the limiting value an inviscid flow region starts to develop in the vicinity of the plate surface. A qualitatively similar result was obtained by Neiland (1972) in studying the two-dimensional flow past a flat plate with gas injection through the plate surface. When comparing the present results with those described in the paper of Neiland (1972) two important differences should be noted. First, in the flow past triangular plates for any $g_w > 0$ even in the vicinity of leading edges there are no self-similar solutions corresponding to the flow past a yawed plate, since the subcritical flow regime is realized over the entire triangular plate and the boundary conditions in the plane of symmetry of the plate have an effect on the flow up to the leading edge. Secondly, for $g_w = 0$ in the flow past triangular plates self-similar solutions corresponding to the flow past a yawed plate exist only at the sweep angles less than 70° (for

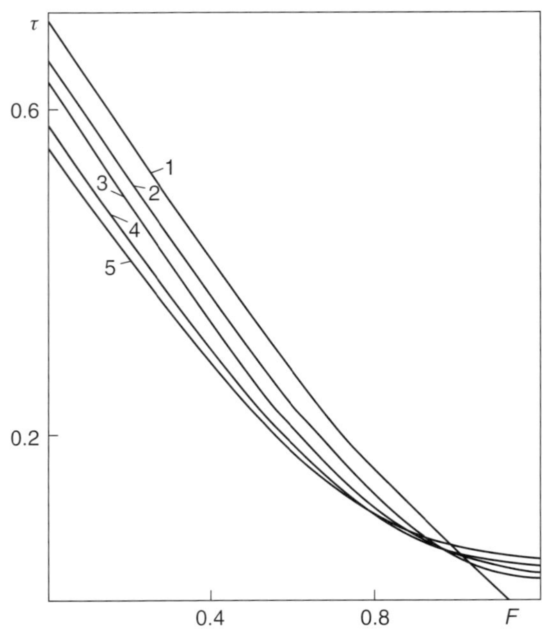


Fig. 7.32.

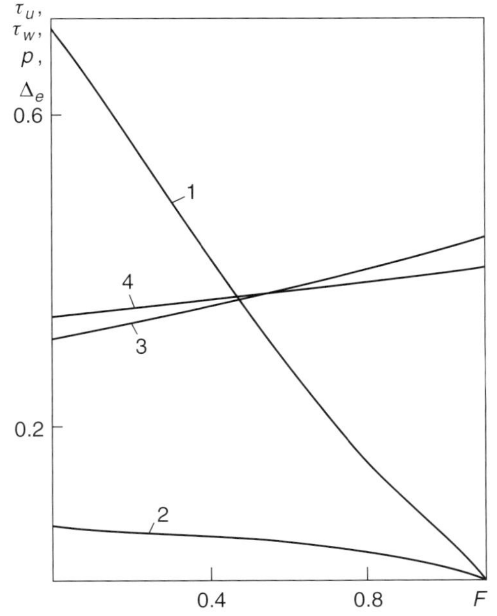


Fig. 7.33.

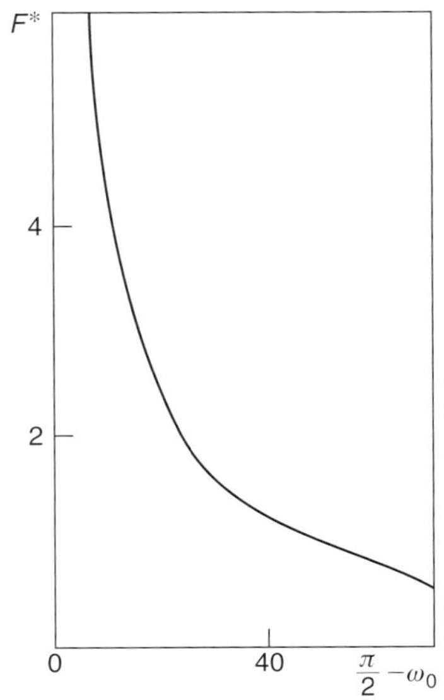


Fig. 7.34.

$\sigma = 1$ and $\gamma = 1.4$), the dimensions of the supercritical flow region depending on the sweep angle (Neiland, 1974*b*; Dudin and Lipatov, 1985). In the two-dimensional case, self-similar solutions for the flow past semi-infinite plates or for finite plates at a properly chosen base pressure exist for any $g_w \geq 0$ along the whole plate length.

In Fig. 7.33 we have plotted the flow functions $\tau_u$, $\tau_w$, $p$, and $\Delta_e$ (curves 1 to 4) versus the injection velocity $F$ for $g_w = 0$ and $z_0 = 1$ at the leading edge of the plate. As the limiting injection velocity $F^* = 1.1$ is approached, both the longitudinal and transverse viscous stress coefficients $\tau_u$ and $\tau_w$ vanish. The calculations also show that the behavior of the heat flux $\tau_g = \left.\frac{\partial g}{\partial \lambda}\right|_{\lambda=0}$ is similar to that of $\tau_u$. Calculations of the supercritical-to-subcritical flow transition coordinate $z_k$ in the flow past a triangular plate with $z_0 = 1$ show that for any injection velocities smaller than the critical value $F = 1.1$ the coordinate $z_k \approx 0.49326$, this value being independent of the injection velocity. From the consideration of the system of ordinary differential equations at the leading edge of the plate it follows that in the supercritical flow region the quantity $\Delta_e/\sqrt{p} = (2/3)\sqrt{2/(\gamma+1)}$ and hence is independent of the injection velocity $F$. Making the change of variable $\lambda = \lambda_* p^{3/2}$ in this system of equations we find that the value of the integral on the right-hand side of Eq. (7.74) $\int_0^\infty (g - u^2 - w^2)\, d\lambda_* = (4/3)\sqrt{\gamma/(\gamma-1)}$ is also independent of the mass transfer rate. Hence follows that the integral on the left-hand side of this expression is also constant if the change of variable $\lambda = \lambda_* p^{3/2}$ is made in it, though the functions themselves $g$, $u$, and $w$ depend on the quantity $F$ characterizing the injection. In this case, the absence of the injection velocity effect on transition can be attributed to the fact that for $g_w = 0$ and finite values of $F$, as follows from Eq. (7.80), the dimensional injection velocity is almost zero.

In Fig. 7.34 we have plotted the dependence of the limiting injection velocity $F^*$ on the parameter $(\pi/2) - \omega_0$, that is, the sweep angle of the leading edge of the plate, for $g_w = 0$. A decrease in the sweep angle leads to a considerable increase in the limiting injection velocity. Thus, for $z_0 = 100$ ($\omega_0 \approx 89.5^\circ$) the limiting injection velocity $F \approx 33$. The curve is calculated only up to the sweep angle of $70^\circ$, since for greater sweep angles there is no supercritical flow region in the vicinity of the leading edge (for given $\sigma$ and $\gamma$). Thus, for the cases considered there exists a minimum value of the limiting injection velocity $F \approx 62$ at which an inviscid flow zone does not appear near the plate surface, at least, in the supercritical flow region. For the injection velocities above the curve plotted in Fig. 7.34 the boundary layer is detached from the plate surface even in the supercritical flow region with the formation of an inviscid (in the first approximation) region in which the tangential momentum is produced only due to a negative induced pressure gradient.

We are now coming to the calculated results corresponding to the solution of the system of equations over the entire triangular plate for different injection velocities and the following parameter values: $z_0 = 1$, $\sigma = 1$, and $\gamma = 1.4$. In Fig. 7.35 we have plotted the profiles of the induced pressure $p^1$ and the boundary layer displacement thickness $\Delta_e^1$ (7.33) along the $z$ coordinate for the case in which $g_w = 0$ and the injection velocity is determined by the function $F(z) = (1 + |z|^{1/4})F^1$, where $F^1 = 0$ and 0.2 (curves 1 and 2) and self-similar solutions are realized in the supercritical region. We note that in this region, for example, the numerical value of the pressure $p^1$ varies only in the fifth significant digit. Points on the solid curves indicate the values of the coordinate $z_k$ obtained from formula (7.74) with the current values of the flow functions substituted in it. The results of the numerical solution

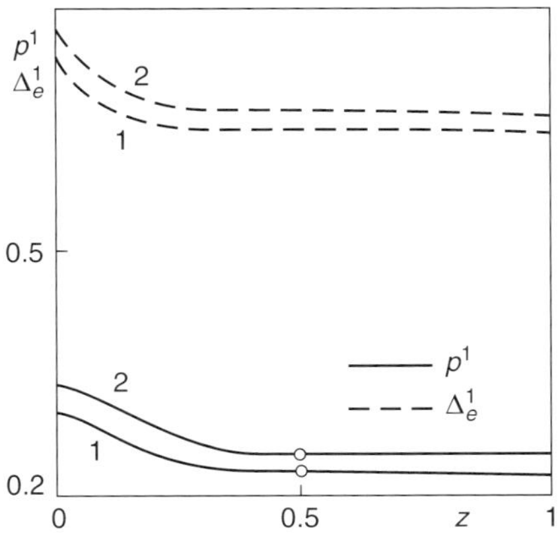


Fig. 7.35.

of the boundary value problem presented above show that a self-similar solution is actually realized in the supercritical flow region and the dimensions of the region do not vary with the parameter $F^1$ determining the injection velocity. An increase in the injection velocity leads to an increase in both the induced pressure and the boundary layer thickness.

The distributions of the quantities $\tau'_u$, $\tau'_w$, and $\tau'_g$ (7.73) (curves 1 to 3, respectively) over the plate surface (along the $z$ coordinate) are presented in Fig. 7.36 for $g_w = 0$ and $F^1 = 0$ (solid curve) and 0.2 (dashed curve). In the supercritical flow region ($z_k > 0.49326$) all the functions presented above are constant. An increase in the injection velocity leads to a

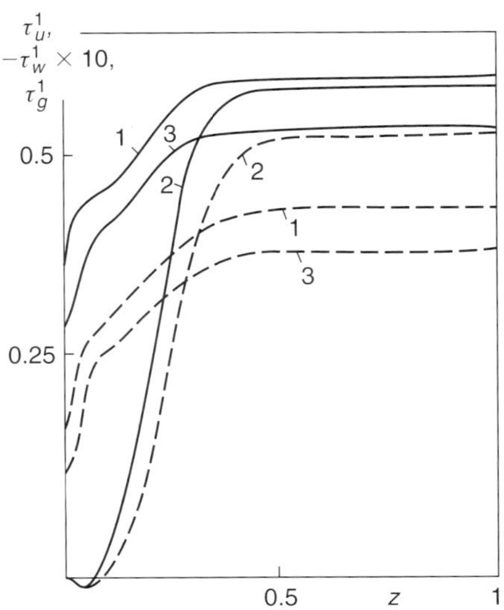


Fig. 7.36.

considerable decrease in the coefficient $\tau'_u$ in the vicinity of the convergence plane $z=0$; this means that in the subcritical flow region the solutions obtained within the framework of the boundary layer equations can turn out to be inadequate for considerably lower velocity injections than the critical injection velocities in supercritical flow regions. An increase in the injection velocity leads to a decrease of all the quantities listed above, which is attributable to a growth of the displacement thickness; however, on the parameter $F^1$ range mentioned above the coefficient $\tau'_w$ remains almost constant in the return flow region in the vicinity of the plane of symmetry $0 \leq z < 0.07$.

The results of the calculations of the boundary layer flow parameters are presented in Figs. 7.37 and 7.38 for the case in which $g_w \geq 0$. In Fig. 7.37 we have plotted the profiles of the induced pressure $p^1$ (solid curve) and the boundary layer displacement thickness $\Delta_e^1$ (dashed curve) along the $z$ coordinate for the cases in which $g_w = 0$, 0.1, and 0.2 (curves 1 to 3) and the mass transfer is determined by the quantity $F^1 = 0.1$. A point on the solid curve 1 indicates the transition coordinate. When $g_w > 0$ in the vicinities of the leading edges the flow functions are no longer constant, since there are no supercritical flow regions. Thus, the pressure $p^1$ (solid curve 3) increases from 0.356 to 0.358 as the $z$ coordinate varies from 1 to 0.5. An increase in the plate surface temperature at a constant injection velocity $F^1$ leads to a considerable increase of all the functions presented.

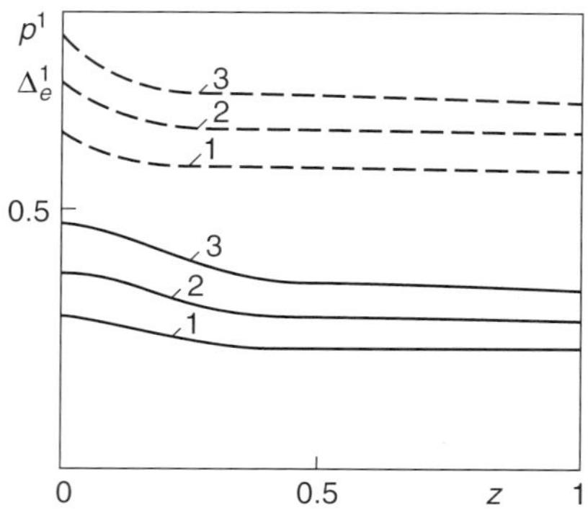


Fig. 7.37.

In Fig. 7.38 we have plotted the distributions of the parameters $\tau'_u$, $\tau'_w$, and $\tau'_g$ (curves 1 to 3) over the triangular plate surface (along the $z$ coordinate) for $F^1 = 0.1$ and $g_w = 0$ (solid curve) and 0.2 (dashed curve). The results of the calculations show that an increase in the surface temperature leads to a decrease of the parameters $\tau'_u$ and $\tau'_g$ over the entire range of the transverse coordinate, whereas the parameter $\tau'_w$ in the region $0.3 \leq z \leq 1$ is considerably greater for $g_w = 0.2$ than for $g_w = 0$. For smaller values of the $z$ coordinate this relation reverses and both the return flow region dimensions and the transverse viscous stress coefficient increase in the vicinity of the convergence plane. It should also be noted that the gradients of the functions $\tau'_u$ and $\tau'_g$ with respect to the coordinate $z$ are considerably smaller for $g_w = 0.2$ than for $g_w = 0$. This is apparently due to the fact that for $g_w > 0$ the triangular plate does not contain supercritical flow regions.

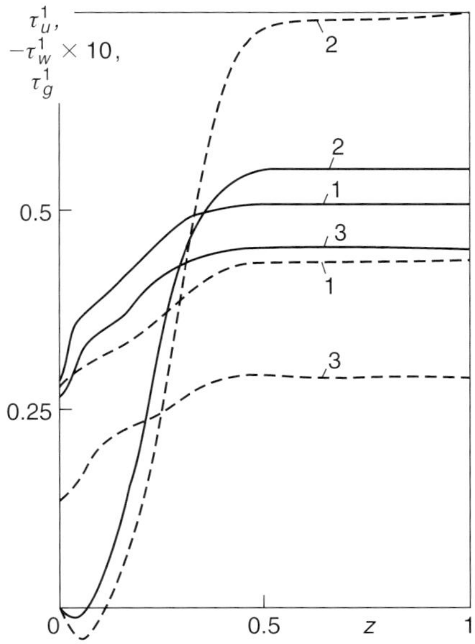


Fig. 7.38.

Numerical calculations made it possible to determine the dependence of the gas dynamic variables on the parameters $F(F')$ and $g_w$ characterizing the injected-gas velocity and the triangular plate surface temperature. It is established that in the symmetric hypersonic viscous flow past a cold triangular plate for any given value of the leading edge sweep angle less than the critical value there is a limiting gas injection velocity such that an inviscid flow region is formed near the plate surface when this value is exceeded. If the injection velocity distribution is so specified that it ensures self-similarity in the supercritical flow region, then the supercritical-to-subcritical transition coordinate is independent of the injection velocity. An increase in the plate surface at a given dimensionless injection velocity leads to a change, both quantitative and qualitative, in the boundary layer flow characteristics, particularly, in the vicinity of the line of symmetry of the triangular plate.

## 7.7 Mass transfer on a planar delta wing in the presence of a supercritical flow region in the boundary layer

In this section we study the effect of gas injection and suction and the position of a permeable region on the wing surface on the flow parameters and supercritical-to-m-subcritical flow transition.

Gas injection, or suction, through the permeable surface of a body in a hypersonic viscous flow can have a considerable effect not only on friction and heat flux but also on the boundary layer displacement thickness and, therefore, the induced pressure distribution. In this section emphasis is placed on the effect of the mass transfer region position and intensity on the parameters of the flow over a strongly cooled delta wing and the position of supercritical-to-subcritical flow transition (Dudin, 2001).

### 7.7.1 Equations and boundary conditions

We will consider the symmetric hypersonic viscous flow of a perfect heat-conducting gas over a semi-infinite planar delta wing. The formulation of the boundary value problem is actually analogous to that of Section 7.6 but it is assumed that the mass transfer region can either occupy the entire wing surface or begin at a certain distance from the leading edges and extend up to the plane of symmetry.

Introducing variables (7.76), (7.78), and (7.52) we arrive at the system of equations (7.79), (7.81), and (7.82) describing the three-dimensional flow in the laminar boundary layer on a delta wing with given mass transfer intensity $F(z)$ and the surface temperature $g = g_w$ in the strong viscous–inviscid interaction regime.

Below we will consider the flow over cold delta wings for $g_w = 0$. To bring the boundary value problem into the self-similar form in supercritical flow regions it is necessary to consider the flow past a wing with the mass transfer determining function of the form:

$$F^* = \left(\frac{\gamma - 1}{2\gamma}\right)^{1/2} x^{-1/4}(1 - |z|)^{-1/4} F^1 \tag{7.84}$$

where $F^1$ must be constant, and to introduce variables of type (7.52) in which the expression $1 - z^2$ is replaced by $1 - |z|$.

We note that in considering the flow past a delta wing with distributed mass transfer specified by the function $F(z) = (1 + |z|)^{1/4} F^1$ in Eq. (7.78) the self-similar form of the solution of the system of equations in supercritical flow regions is obtained when the calculated results are represented in form (7.73).

If the function $F(z)$ in Eq. (7.78) is arbitrarily specified, then self-similar solutions do not exist in supercritical flow regions.

The supercritical-to-m-subcritical flow transition coordinate $z_k$ in the cold-wing boundary layer for $0 < z \leq 1$ is determined from Eq. (7.74).

We note that if there are no self-similar solutions in the supercritical flow region, then Eq. (7.74) involves the current values of the functions $g$, $u$, and $w$ for $z = z_k$. In presenting the calculated results the coordinates $z_k$ are marked by crosses on the curves corresponding to pressure distributions.

### 7.7.2 Results of calculations

Below we present the results obtained in solving the system of equations for a delta wing at the following values of the parameters: $z_0 = 1$, $\sigma = 1$, $\gamma = 1.4$, and $g_w = 0$. In Figs. 7.39–7.43 the results are presented with account for transformation (7.73), the solid curves relate to the flows with $F^1 = 0$, dashed to those with injection ($F^1 = 0.2$), and dot-and-dashed to those with suction ($F^1 = -0.5$). The positions of the beginning of the mass transfer region

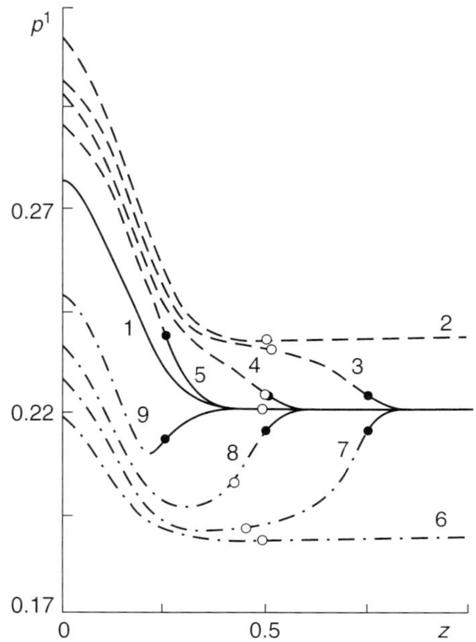


Fig. 7.39.

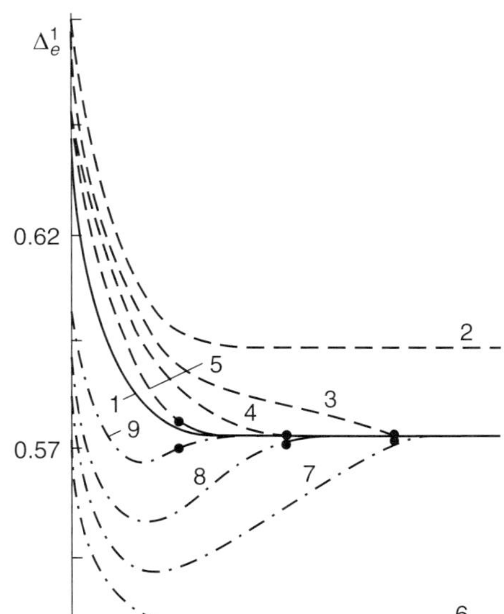


Fig. 7.40.

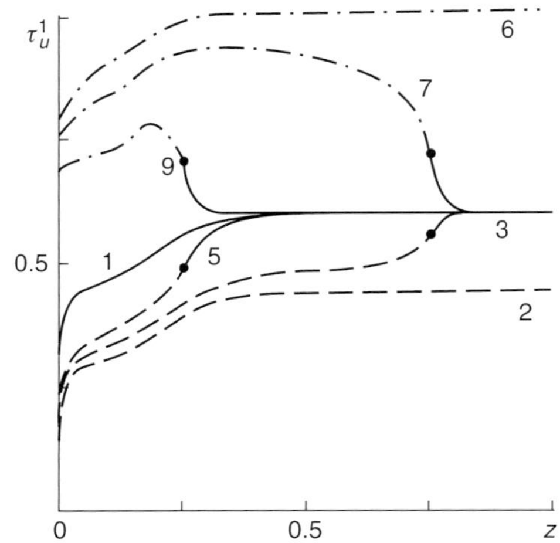


Fig. 7.41.

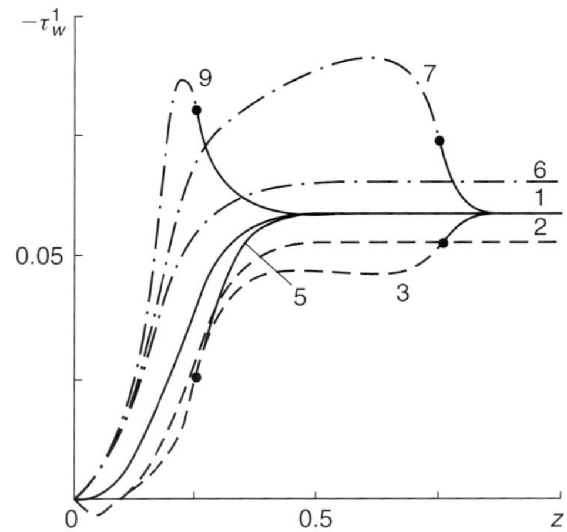


Fig. 7.42.

are marked by black points on the corresponding curves. It is assumed that in the case of variable mass transfer, when $F^1$ is a function of $z$ rather than constant, it varies linearly from zero to $F^1 = 0.2$ or $F^1 = -0.5$ in the last grid step ahead of the beginning of the mass transfer region.

In Fig. 7.39 the induced pressure $p^1$ profiles along the $z$ coordinate are presented. In the cases in which mass transfer is absent (curve 1) or $F^1$ is constant for all $z$ [0.2 (curve 2) or $-0.5$ (curve 6)], self-similar solutions are realized in the supercritical region, where the pressure $p^1$ is constant. In all three cases considered the supercritical flow region length

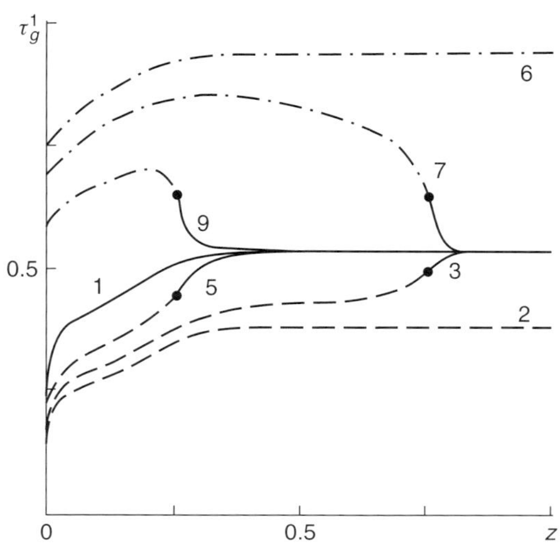


Fig. 7.43.

turned out to be independent of $F^1$. The transition coordinate $z_k = 0.49236$ is marked by light points on curves 1, 2, and 6. An increase in injection intensity (curve 2) leads to an increase in the pressure $p^1$ over the entire wing surface due to the boundary layer thickness $\Delta_e^1$ growth (Fig. 7.40), whereas an increase in the suction intensity (curve 6) leads to its decrease, the behavior of both $p^1(z)$ and $\Delta_e^1(z)$ being monotonic. For constant values of $F^1$ in the supercritical flow region on the wing the viscous stress and heat flux coefficients $\tau_u^1$, $\tau_w^1$, and $\tau_g^1 = \left.\frac{\partial g}{\partial \lambda}\right|_{\lambda=0}$ remain constant (Figs. 7.41–7.43, curves 2 and 6). An increase in the suction intensity leads to an increase in $\tau_u^1$, $\tau_w^1$, and $\tau_g^1$. In this case, in the vicinity of the plane of symmetry of the wing a flow with smooth convergence is realized (curve 6 in Fig. 7.42). With increase in the injection intensity (curve 2) the coefficients naturally decrease and the flow nature so changes that in the $0 \le z \le 0.1$ vicinity of the plane of symmetry the coefficient $\tau_w^1$ reverses sign (Fig. 7.42), so that within this region of the boundary layer a return transverse flow zone appears near the wing surface.

The distributions of the flow functions in the case in which mass transfer occurs on the plate surface for $|z| \le 0.75$, that is, begins in the supercritical flow region, are presented as curves 3 and 7 in Figs. 7.39–7.43. The "upstream" disturbance propagation from the beginning of the injection (curve 3) or suction (curve 7) region is limited by two–three steps of the difference grid ($\Delta z = 0.025$) which is natural bearing in mind the actual presence of the second derivative of the boundary layer thickness with respect to the transverse coordinate in Eqs. (7.79), (7.81), and (7.82). The distributions of the pressure and the other flow functions in the supercritical flow region toward the plane of symmetry are no longer self-similar. In these cases, transition occurs outside the self-similarity domain. In the case of injection the transition coordinate determined from Eq. (7.74) for the current flow functions is displaced toward the leading edge (a light point on curve 3). In the suction flow transition is delayed and the supercritical flow region enlarges (curve 7). The appreciably nonmonotonic nature of the variation of the functions $p^1(z)$ and $\Delta_e^1(z)$ in the case of suction (curve 7 in Figs. 7.39 and 7.40) leads to a nonmonotonic behavior of the viscous stress and heat flux coefficients in the transverse coordinate. A fairly strong variation of the parameters $\tau_u^1$, $\tau_w^1$, and $\tau_g^1$ in the vicinity of the beginning of the mass transfer region should be noted (Figs. 7.41–7.43). However, while the variations of $\tau_u^1$ and $\tau_g^1$ are not greater than the corresponding values in the flows with constant $F^1$ (Figs. 7.41 and 7.43), this is not the case for $\tau_w^1$. For example, in the case of suction (curve 7 in Fig. 7.42) in the supercritical flow region the transverse viscous stress coefficient increases in absolute magnitude to 0.091 instead of 0.0645 (curve 6). However, for injection flows this effect is not so considerable (cf. curves 3 and 2).

For flows with mass transfer at $|z| \leq 0.5$ (curves 4 and 8 in Figs. 7.39 and 7.40) the nature of the "upstream" disturbance propagation is analogous to the previous case. For injection flows (curve 4) the transition coordinate almost coincides with the beginning of the injection region. Since for suction flows this coordinate is displaced toward the plane of symmetry to $z = 0.42$ (curve 8), the beginning of the suction region is located in the supercritical flow region.

The profiles of the flow functions for the mass transfer region located at $|z| \leq 0.25$ are presented in the figures as curves 5 and 9. Numerical calculations show that in these cases disturbances propagate "upstream" up to the transition coordinate $z_k = 0.49326$ realized in the absence of mass transfer or for a constant $F^1$. As a result, the behavior of the flow functions $\tau_u^1$ and $\tau_g^1$ in the vicinity of the beginning of the injection (suction) region is smoother as compared with the previous cases and the mass transfer effect on the variation of $\tau_u^1$ and $\tau_g^1$ is smaller than that in the $|z| \leq 0.75$ region (Figs. 7.41 and 7.43). We also note a considerable change of the behavior of $\tau_w^1$ for the suction flow (curve 9 in Fig. 7.42) in which it has a peak in the vicinity of the beginning of the suction region. A sharp flow acceleration in the transverse direction in this region of the boundary layer can be attributed to a pressure decrease in it (curve 9 in Fig. 7.39). We note that in the vicinity of the plane of symmetry the flow nature is not changed. The effect of gas injection on $\tau_w^1$ is weaker which is, apparently, attributable to the monotonic nature of the induced pressure distribution $p^1(z)$ (curve 5 in Fig. 7.39).

An increase in the injection intensity (curves 2, 3, and 5 in Fig. 7.41) leads to a considerable decrease of $\tau_u^1$ in the vicinity of the plane of symmetry of the wing which means that in the subcritical flow region the solutions obtained within the framework of the boundary layer equations can turn out to be inadequate for considerably smaller injection velocities than the limiting injection velocities in supercritical flow regions.

Thus, numerical calculations made it possible to determine the dependence of the gas dynamic variables on the function $F^1(z)$ characterizing the mass transfer intensity on the surface of a cold delta wing. It is established that mass transfer beginning in the supercritical region has an effect on the supercritical-to-subcritical flow transition coordinate and has almost no effect on the "upstream" flow parameters; gas injection makes the supercritical flow region to shrink, whereas suction enlarges it. Mass transfer beginning in the m-subcritical flow region has an effect on the flow characteristics throughout the entire region. In this case, gas suction can lead to the formation of fairly narrow zones of elevated transverse viscous stress with conservation of smooth flow convergence in the vicinity of the plane of symmetry of the wing and without formation of transverse return flow zones.

## 7.8 Mass transfer on a nonplanar delta wing

In this section we study the mass transfer effect on the parameters of the laminar boundary layer on a cold delta wing in a hypersonic viscous perfect-gas flow in the strong viscous–inviscid interaction regime. The mass transfer intensity effect on the supercritical-to-m-subcritical flow transition coordinate is numerically investigated and the total aerodynamic characteristics of a delta wing with a power-law cross-sectional shape are determined.

The study of the flow over nonplanar delta wings (Dudin, 1997) showed that if the cross-sectional shape is a power-law function with the exponent equal to 3/4, then the dimensions of the supercritical flow region are the same as in the flow over a planar wing. In this case, given the parameter $D = \tau/\delta$ characterizing the wing-to-boundary-layer thickness ratio, we can, using the similarity transformation (Dudin, 1997), determine the flow parameters on the basis of those in the supercritical region on a planar delta wing.

However, the study of the flow over power-law delta wings with other values of the exponent showed (Dudin, 1998) that both the supercritical flow region dimensions and the flow parameters in it depend on the exponent and the parameter $D$. In this case the transition coordinate can be determined only if the solution of the system of equations in partial derivatives is known in the supercritical flow region.

The study of the effect of gas injection through the surface of a cold delta wing on the formation of supercritical and subcritical flow regions (Dudin, 2000) showed that if distributed injection is so specified that it ensures self-similarity in the supercritical flow region on the delta wing, then the transition coordinate is independent of the injection intensity.

In this section we study the effect of mass transfer (injection or suction) through the surface of a nonplanar delta wing on the formation of supercritical and m-subcritical flow regions in the laminar boundary layer and the flow parameters in it. It is assumed that mass transfer can occur only on a part of the wing surface (Dudin, 2002).

### 7.8.1 *Equations and boundary conditions*

We will consider the symmetric hypersonic viscous perfect-gas flow over a semi-infinite thin delta wing with the characteristic thickness $\tau$ of the order of the boundary layer thickness at zero angle of attack. Gas is injected or sucked off normal to the wing surface; the gas composition is the same as in the undisturbed flow and the temperature is equal to the surface temperature. The gas flow rate due to mass transfer is assumed to be comparable with that in the boundary layer on the impermeable surface. It is assumed that the wing surface temperature is constant and low as compared with the freestream stagnation temperature and the regime of strong interaction between the boundary layer and the outer inviscid hypersonic flow is realized. Otherwise, the formulation of the problem is analogous to that presented in Sections 7.6 and 7.7.

As shown by Dudin (1997), in the absence of mass transfer on the surface of the delta wing whose thickness varies along the $x^0$ axis following the $\delta_w^0(x^0, z^0) \sim (x^0)^{3/4}$ law, we can introduce dimensionless variables and flow functions in which the system of boundary layer equations is self-similar, that is, dependent only on two variables:

$$(w - z_0 z u)F_1 \frac{\partial f}{\partial z} + v \frac{\partial f}{\partial \lambda} = G \tag{7.85}$$

$$f = \begin{Bmatrix} u \\ w \\ g \end{Bmatrix}, \quad G = \begin{cases} z_0 \frac{\gamma-1}{2\gamma p}(g - u^2 - w^2)F_2 + \frac{\partial^2 u}{\partial \lambda^2} \\ -\frac{\gamma-1}{2\gamma p}(g - u^2 - w^2)F_3 + \frac{\partial^2 w}{\partial \lambda^2} \\ \frac{1}{\sigma}\frac{\partial^2 g}{\partial \lambda^2} - \frac{1-\sigma}{\sigma}\frac{\partial^2 (u^2+w^2)}{\partial \lambda^2} \end{cases}$$

$$\frac{\partial v}{\partial \lambda} = (w - z_0 z u)\frac{z}{2p} - F_1\left(\frac{z_0 u}{4} - z_0 z \frac{\partial u}{\partial z} + \frac{\partial w}{\partial z}\right)$$

$$F_1 = \frac{1 - z^2}{p}, \quad F_2 = \frac{1 + z^2}{2} + z F_1 \frac{\mathrm{d}p}{\mathrm{d}z}, \quad F_3 = z + F_1 \frac{\mathrm{d}p}{\mathrm{d}z}$$

Here, $\sigma$ is the Prandtl number, $\gamma$ is the specific heat ratio,

$$z_0 = \tan \omega_0, \quad h = (y^0 - \delta_w^0)/\ell\delta, \quad \lambda = \sqrt{(\gamma - 1)/2\gamma} \int_0^h \rho^0 \, \mathrm{d}y^0 / \ell \delta^3 \rho_\infty \sqrt[4]{x(1 - z^2)}$$

$$p(z) = \sqrt{x(1 - z^2)}\frac{p^0}{\delta^2 \rho_\infty u_\infty^2}, \quad \delta = z_0^{1/4} Re_0^{-1/4}$$

$p^0$ is the pressure, and $\rho^0$ is the density. In deriving system (7.85) it was assumed that the viscosity–enthalpy dependence is linear.

In accordance with the study of Dudin (2000), in the above variables self-similarity is retained if the injection, or suction, velocity $v_w^0$ varies proportional to $(x^0)^{-1/4}$. Then in system (7.85) the quantity $v$ on the wing surface, which is actually a dimensionless mass transfer intensity, is determined by the equation

$$v_w(z) = v(\lambda = 0) = z_0 \sqrt{\frac{(\gamma - 1)}{2\gamma}} \sqrt[4]{x(1 - z^2)}\, u_\infty \rho_w^0 \frac{v_w^0}{\delta p^0}$$

Here it is assumed that the function $v_w(z)$ has no singularities at the leading edges of the wing for $z = \pm 1$.

In this case, the boundary conditions for system (7.85) take the form:

$$\lambda = 0: \quad u = w = g = 0, \quad v = v_w(z) \tag{7.86}$$

$$\lambda \to \infty: \quad u \to 0, \quad w \to 0, \quad g \to 1$$

As noted in the paper of Dudin (2000), preassigning the parameter $v_w$ in the boundary conditions (7.86) makes it possible to determine the solutions for different mass transfer intensities $\Psi = \rho_w^0 v_w^0$. We note that its value turns out to be dependent on the induced pressure which in turn depends on $v_w$. In considering the flow past bodies with $g_w^0 \to 0$ at the wing surface the velocity $v_w^0 \to 0$; however, the injected or sucked off gas flow rate $\Psi = \rho_w^0 v_w^0$ is finite if the parameter $v_w \neq 0$.

Then, as in the paper of Dudin (1997), we consider the flow over delta wings with the following shape of the surface

$$\delta_w^0(z) = \ell\tau[x(1 - z^2)]^{3/4}\Delta_w(z), \quad \text{where } \Delta_w(z) = (1 - z^2)^\alpha \tag{7.87}$$

where the exponent $\alpha \geq 0$. Then in the flow over thin wings with $z_0 = O(1)$ the distribution of the induced pressure $p$ produced under the displacing action of the boundary layer

$$\Delta_e = \frac{1}{p}\left(\frac{\gamma - 1}{2\gamma}\right)^{1/2} \int_0^\infty (g - u^2 - w^2)\,d\lambda \tag{7.88}$$

and the wing thickness $\Delta_w(z)$, under the condition that $M_\infty(\tau + \delta) \gg 1$, can approximately be determined using the tangent wedge method

$$p = \frac{\gamma + 1}{2}\left[\frac{3}{4}(1 + z^2)(D\Delta_w + \Delta_e) - z(1 - z^2)\frac{d(D\Delta_w + \Delta_e)}{dz}\right]^2 \tag{7.89}$$

Equations (7.88) and (7.89) make it possible to close the system of equations in partial derivatives (7.85) with the boundary conditions (7.86) which describe the flow in the laminar boundary layer on a cold delta wing of power-law shape with a given mass transfer intensity $v_w$ in the strong viscous–inviscid interaction regime.

It can be shown that in considering the flow over a delta wing with $\Delta_w(z) = (1 + |z|)^{-3/4}$ in Eq. (7.87) and a distributed mass transfer specified by the expression $v_w(z) = \sqrt[4]{1 + |z|}\,v_w^1$, where $v_w^1 = \text{const}$, the self-similar form of the solution of the system of equations in supercritical flow regions is obtained if the calculated results are represented in form (7.73).

When specifying the functions $v_w$ and $\Delta_w(z)$ otherwise, self-similar solutions in supercritical flow regions can no longer exist.

The coordinate $z_k$ corresponding to supercritical-to-subcritical flow transition in the cold-body boundary layer is determined from condition (7.74).

### *7.8.2 Results of calculations*

As an illustration, we will consider the mass transfer effect on the flow over a delta wing with the leading edge sweep angle 45° and the cross-sectional shape $\Delta_w(z)$ (7.87) in which the exponent $\alpha = 9.25$ and the parameter $D = 0.5$. In the calculations it was also assumed that $\sigma = 1$ and $\gamma = 1.4$. In the absence of mass transfer the flow with the above parameters was considered in the paper of Dudin (1998). Below the calculated flow parameter profiles are plotted as dashed curves for the regions on the wing surface in which injection is performed as dot-and-dash curves for the suction regions. In the cases in which mass transfer is produced only on a part of the wing span, the beginning of the injection (suction) region is indicated by points on the corresponding solid curves. Since these calculations are also aimed at studying the effect of mass transfer beginning off the leading edges of the wing, the injection intensity can no longer be preassigned in the form $v_w(z) = \sqrt[4]{1+|z|}v_w^1$, where $v_w^1 = \text{const}$ and $z \in [-1, 1]$; however, the flow function distributions can conveniently be presented in the figures in form (7.73). Below, curves 2 and 5 relate to the flows with injection $v_w = 0.15$ and suction $v_w = -1$ over the entire wing span. For the flows associated with curves 3 and 6 and 4 and 7 mass transfer on the surface with the above values of $v_w$ was produced in the $|z| \leq 0.75$ and $|z| \leq 0.25$ regions, respectively, while outside these regions it was zero.

In Fig. 7.44 the induced pressure $p^1$ profiles along the $z$ coordinate are presented. Curve 1 relates to the case of the flow over the wing in the absence of mass transfer on its surface, in which supercritical-to-subcritical flow transition occurs for $z_k \sim 0.5407$; this value is marked by a cross. In the cases in question the transition coordinate depends on the functions $g$, $u$, and $w$ at the transition point. We note that, as compared with the flow over the planar delta wing ($D = 0$) with the same flow parameters, the supercritical flow region dimensions decreased by about 10% (Dudin, 1998). Continuous gas injection (curve 2) leads to a slight shift of the transition coordinate toward the leading edges, up to $z_k \sim 0.5483$, and an increase in the induced pressure $p^1$ over the entire wing span. On gas injection with an intensity $v_w = 0.15$ (curve 3) through the permeable wing surface zone $|z| \leq 0.75$, which begins in the supercritical flow region, the transition coordinate increases somewhat ($z_k \sim 0.5565$); however, the pressure distribution thus obtained turns out to be close to that for the previous case (curve 2) even for $|z| \leq 0.6$, where the difference is less than 1%. Injection in the $|z| \leq 0.25$ region (curve 4) located entirely in the subcritical flow region has no effect on the transition coordinate ($z_k \sim 0.5407$).

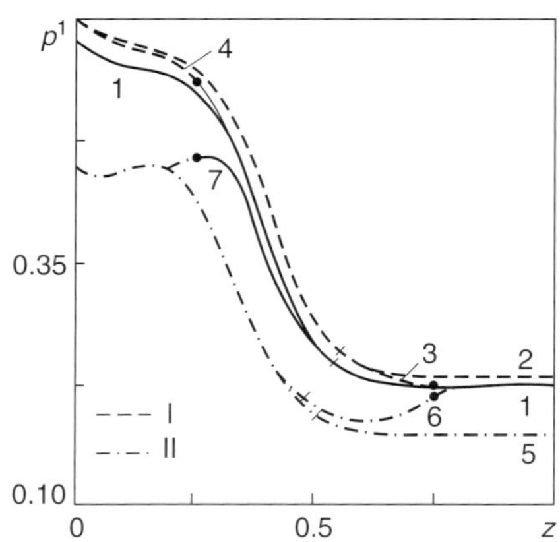


Fig. 7.44.

Contrariwise, continuous suction with an intensity $v_w = -1$ (curve 5) delays transition and enlarges the supercritical flow region ($z_k \sim 0.5125$) and also reduces considerably the pressure $p^1(z)$, particularly, at the central part of the wing; as for the function $p^1(z)$ behavior in the $|z| \leq 0.175$ region, it becomes nonmonotonic. When gas is sucked off in the $|z| \leq 0.75$

region (curve 6), the transition coordinate is further displaced toward the wing center, up to the value $z_k \sim 0.474$, but in the $|z| \leq 0.35$ region curve 6 comes closer to curve 5. An analysis of the calculated results show that for $|z| > 0.75$ the departure of curves 3 and 6 from curve 1 begins actually at a distance not greater than two or three grid steps along the $z$ axis from the beginning of the mass transfer region and, therefore, in this flow region the supercritical regime is realized and disturbances do not propagate upstream.

Gas suction in the subcritical region $|z| \leq 0.25$ (curve 7), as injection in the same region, has no effect on the transition coordinate. It is important to note that the mass transfer produced only in the subcritical region leads to the variation of the pressure distribution throughout the entire subcritical region which is clearly visible by comparing the pressure distributions associated with curves 1 and 7 for $|z| \leq 0.54$. In this region the difference in the behavior of curves 1 and 4 is not so great, but the comparison of the corresponding calculated results shows that this effect takes place also in the presence of gas injection.

In Fig. 7.45 the profiles of the quantity $\Delta_e^1$ associated with the displacement thickness are presented. A decrease in $\Delta_e^1$ (curve 1) in the region $0.13 < z < 0.75$ can be attributed to flow deceleration in the transverse direction due to a pressure increase with further formation of a flow with $w > 0$ in the lower part of the boundary layer. Continuous injection (curve 2) and suction (curve 5) lead naturally to increase and decrease of the displacement thickness. The departure of curves 3 and 6 from curve 1 is particularly descriptive of the absence of disturbance propagation toward smaller $z$ from $z = 0.75$ in the supercritical flow region. At the same time, the mass transfer preassigned in the subcritical region (curves 4 and 7) has an effect on the flow parameters throughout the entire subcritical flow region.

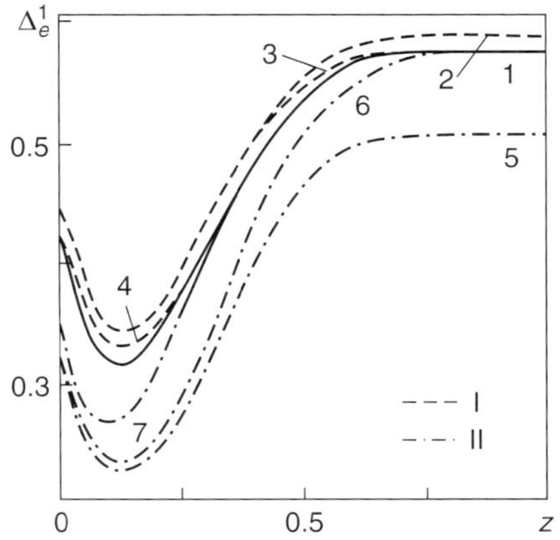


Fig. 7.45.

In Fig. 7.46 the spanwise distributions of the transverse viscous stress coefficient $\tau_w^1$ are presented. As noted in the paper of Dudin (2000), a region near the wing surface, in which the velocity component $w$ reverses sign, appears in the boundary layer already for the values of the parameter $\chi > 0.2$. In this case, an increase in the parameter $D$ leads to not only the enlargement of this region but also to an increase of the return flow velocity $w$ in absolute magnitude. In the absence of mass transfer and for $D = 0.5$ (curve 1) in the region

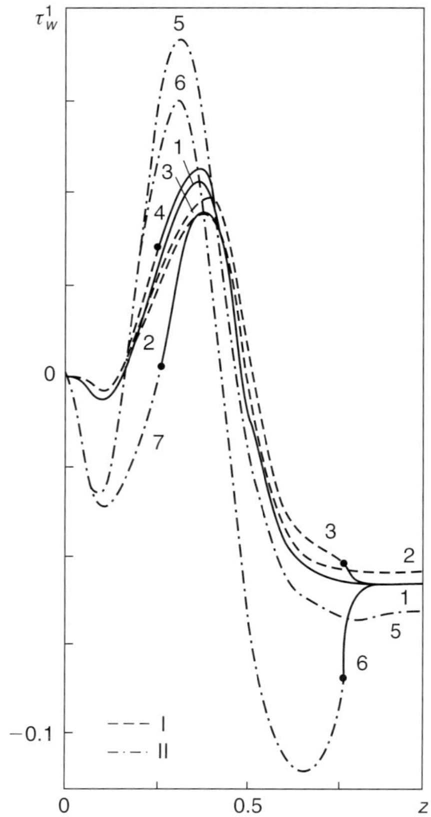


Fig. 7.46.

$0.15 \le |z| \le 0.47$ the quantity $\tau_w^1 > 0$. A considerable difference between the flows in the boundary layer in the vicinity of the plane of symmetry of the wing for $D = 0$ and $D > 0$ should be noted. For $D = 0$ a small return flow zone ($\Delta z \approx 0.1$) is formed in this region near the body surface. For $D > 0$ a flow with smooth convergence is realized which is responsible for the growth of the boundary layer displacement thickness $\Delta_e^1$ in this region (Fig. 7.45). Continuous injection (curve 2) and suction (curve 5) have a slight effect on the dimensions of the flow region with $\tau_w^1 > 0$ but influence considerably the quantity $\tau_w^1$ itself in this region. Thus, suction with $v_w = -1$ over the entire wing (curve 5) increases the transverse viscous stress coefficient in the vicinity of the $\tau_w^1$ peak ($z \approx 0.3$) by more than 80%. Gas injection with $v_w = 0.15$ beginning in both the supercritical (curve 3) and subcritical (curve 4) flow regions has a comparatively slight effect on both the nature of the variation of $\tau_w^1$ and the quantity itself, whereas suction (curves 6 and 7) reduces fairly considerably the dimensions of the region in which $\tau_w^1 > 0$. In this case, suction beginning in the supercritical region (curve 6) leads to a sharp increase in absolute magnitude of the viscous stress coefficient in the $0.65 \le |z| \le 0.8$ region.

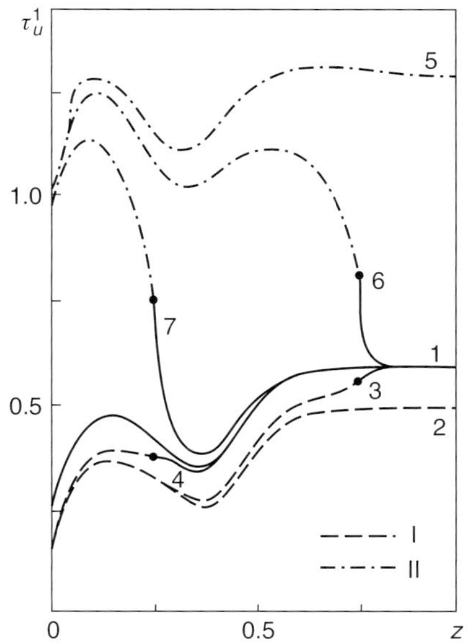


Fig. 7.47.

For $D = 0.5$ intense flow deceleration in the transverse direction also leads to considerably nonmonotonic distributions of the longitudinal viscous stress coefficient $\tau_u^1$ (curve 1 in Fig. 7.47). Continuous injection (curve 2) and suction (curve 5) result in decrease and increase of $\tau_u^1$, respectively. Mass transfer beginning in the supercritical flow region (curves 3 and 6) has a very strong effect on the longitudinal viscous stress coefficient downstream from the injection (suction) beginning and only a slight effect upstream of it. As it might be expected, mass transfer preassigned in the subcritical flow region (curves 4 and 7) has an effect on the $\tau_u^1$ distribution only within this region. In all the cases considered the behavior of $\tau_u^1$ is appreciably nonmonotonic.

In Fig. 7.48 we have plotted the heat flux $\tau_g^1$ distributions for all the cases considered; as the $\tau_u^1$ distributions, these are appreciably nonmonotonic. Suction realized in the subcritical flow region (curve 7) can lead to the formation of fairly narrow zones of elevated heat flux.

The results presented above show that the distributed mass transfer with the beginning in the supercritical flow region of the boundary layer on a delta wing with a power-law cross-sectional shape has an effect on both the downstream flow parameters and the transition coordinate. For a wider range of the gas flow rate through the wing surface the dependence of the transition coordinate on the intensity of mass transfer preassigned in the $|z| \leq 0.75$ region is presented in Fig. 7.49. An increase in the suction intensity to $v_w = -5$ leads to supercritical region enlargement ($z_k \sim 0.406$), whereas an increase in the injection intensity to $v_w = 0.4$ results in its shrinking ($z_k \sim 0.587$). On the mass transfer range considered the variation of the dimensions of this region can amount to almost 20% of the wing span. It should be noted that in increasing the injection intensity the coefficient $\tau_u^1$ decreases in the plane of symmetry of the wing and vanishes for $v_w > 0.45$. In accordance with the results

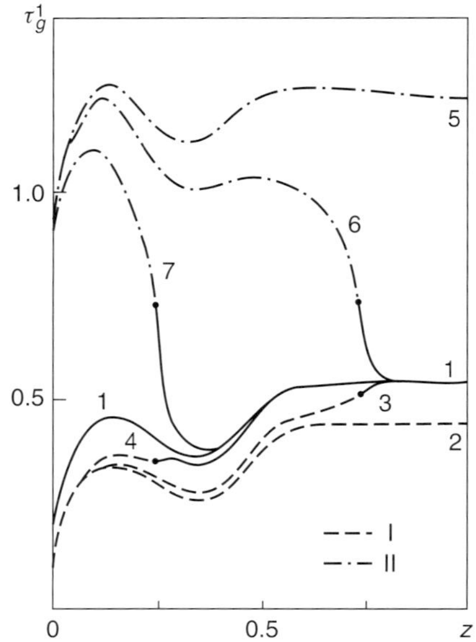


Fig. 7.48.

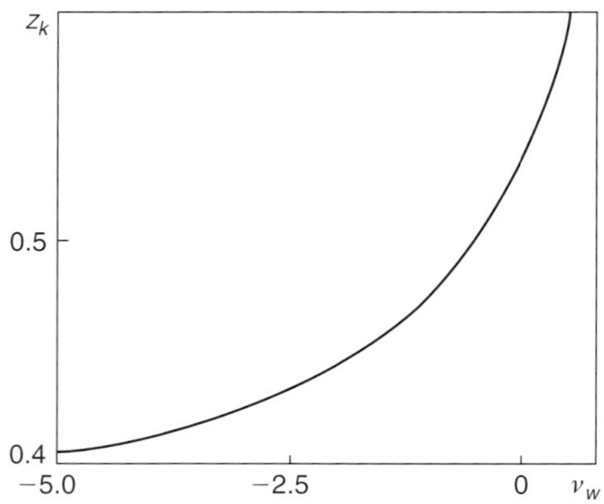


Fig. 7.49.

of the study of Dudin (2000), this means that for $v_w > 0.45$ there is no solution within the framework of boundary layer theory, at least, in the vicinity of the plane of symmetry of the wing, since for these injection velocities an inviscid flow region starts to develop near the wing surface.

In Fig. 7.50 the dependence of the aerodynamic coefficients on the mass transfer intensity in the $|z| \leq 0.75$ region is presented; they are calculated by formulas (7.75) for one side of the wing for $\alpha = 9.25$ and $D = 0.5$. With increase in the suction intensity the coefficient

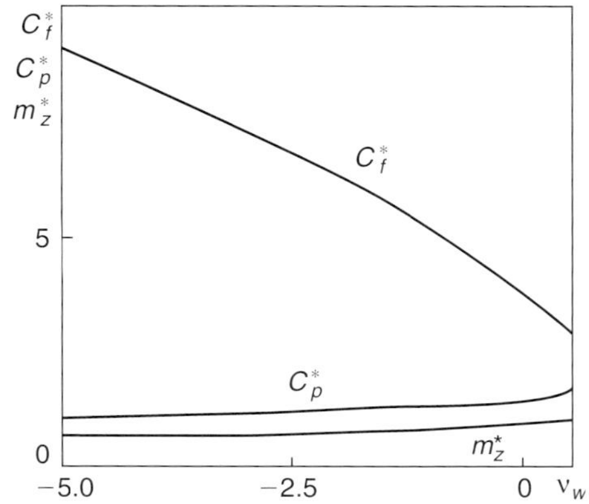


Fig. 7.50.

$C_f^*$ considerably increases which is due to an increase of $\tau_u^1$ (Fig. 7.47). In this case, the coefficients $C_p^*$ and $m_z^*$ decrease, since the induced pressure $p^1$ decreases (Fig. 7.44).

Thus, the numerical calculations made it possible to establish that mass transfer realized in the subcritical flow region on the surface of a cold delta wing with a power-law cross-sectional shape has an effect on the flow parameters throughout the entire subcritical flow region but has no effect on the transition coordinate. In the case in which the beginning of the mass transfer zone is located in the supercritical flow region, the mass transfer has an effect on the flow parameters only downstream of the mass transfer zone beginning; in this case the dependence of the transition coordinate on the mass transfer intensity is appreciably nonlinear. The most mass transfer effect is that on the aerodynamic coefficient $C_f^*$.

## 7.9 Using the Newtonian passage to limit for studying the flow over a delta wing

In this section, we study the flow over a planar delta wing in the regime of strong interaction between the boundary layer and the outer supersonic flow. An analytical investigation is performed using the "Newtonian" passage to limit when the adiabatic exponent tends to unity while the Mach and Reynolds numbers increase without bound (Dudin and Neiland, 2002). In the case of the flow over a cold wing with a fairly large aspect ratio, in which transverse currents in the boundary layer are small, we derive an analytical expression determining, correct to the second approximation, the line of supercritical-to-m-subcritical flow transition. The results obtained are compared with the experimental data.

In the hypersonic flow over a cold planar delta wing in the strong viscous interaction regime, when the sweep angle of the leading edge is smaller than the critical value, supercritical and m-subcritical flow regions arise in the boundary layer (Neiland, 1974*b*; Dudin and Lipatov, 1985). In the former region disturbances cannot propagate upstream and a self-similar flow corresponding to that past a semi-infinite swept plate is realized. With increase in the sweep angle, the extent of the supercritical flow regions located near the leading edges decreases and, as the critical value has been attained, an m-subcritical flow regime is realized over the entire wing; in this regime disturbances propagate from the plane of symmetry of the wing up to the leading edges. Generally this flow is governed by a system of equations in partial differential equations. Numerical solutions of the corresponding boundary value problem showed (Dudin, 1997) that the transition coordinate depends on not only the sweep angle but also the adiabatic exponent $\gamma = C_p/C_v$, where $C_p$ and $C_v$ are the specific heats at constant pressure and volume, respectively. A decrease of the parameter $\varepsilon = \gamma - 1$ leads to considerable enlargement of the supercritical flow regions (Dudin, 1997). In this section, we consider the flow over delta wings with the aspect ratio of the order of unity in the case in which the parameter $\varepsilon$ is asymptotically small.

### *7.9.1 Estimates of the flow parameters*

We will consider the symmetric flow over a semi-infinite delta wing in the strong viscous–inviscid interaction regime (Fig. 7.51). It is assumed that the surface temperature $T_w$ is low as compared with the freestream stagnation temperature $T_0$ and the parameter $\varepsilon$ asymptotically vanishes. The gas is assumed to be perfect with constant values of $C_p$ and $C_v$ and a linear viscosity–temperature dependence $\mu^0/\mu_\infty = C_\infty T^0/T_\infty$, where $C_\infty = \text{const}$ and the subscript $\infty$ refers to the freestream parameters. The velocity components $u^0$, $v^0$, and $w^0$ are directed along the $x^0$, $y^0$, and $z^0$ axes of a Cartesian coordinate system with origin at the vertex of a wing with half-angle $\omega_0$. The wing aspect ratio $z_0 = \tan\,\omega_0$. In the undisturbed flow the velocity $u_\infty$, the density $\rho_\infty$, and the stagnation enthalpy $g_\infty$ tend to certain constant values as the Mach number $M_\infty \to \infty$. In this case, the pressure $p_\infty$, the speed of sound $a_\infty$, and the temperature $T_\infty$ vanish.

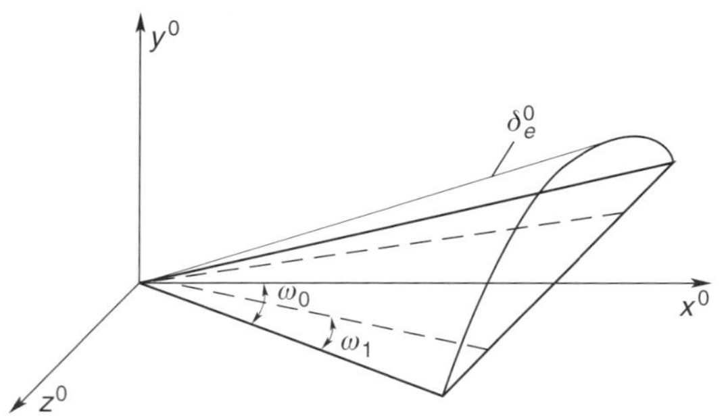


Fig. 7.51.

In accordance with the small perturbation theory, for $M_\infty \gg 1$ and the dimensionless thickness of the laminar boundary layer $\delta \ll 1$, when the assumption of the strong interaction $M_\infty\delta \gg 1$ is fulfilled, the induced pressure is of the order $p^0 \sim \rho_\infty u_\infty^2 \delta^2$. In the boundary layer the static enthalpy $h^0 \sim u_\infty^2/2$. Then from the equation of state for the gas density in the boundary layer we obtain the following estimate:

$$\frac{\rho^0}{\rho_\infty} \sim \frac{p^0 h_\infty}{p_\infty h^0} \sim \frac{\rho_\infty u_\infty^2 \delta^2}{p_\infty} \frac{a_\infty^2}{(\gamma - 1)u_\infty^2} \sim \delta^2 \varepsilon^{-1} \tag{7.90}$$

For estimating the thickness $\delta$, we equate the orders of the leading viscous and inertial terms and obtain

$$\rho^0 u^0 \frac{\partial u^0}{\partial x^0} \sim \frac{\partial}{\partial y^0}\left(\mu^0 \frac{\partial u^0}{\partial y^0}\right), \quad \frac{\rho_\infty \delta^2 u_\infty^2}{\varepsilon \ell} \sim \frac{\mu_0 u_\infty}{\delta^2 \ell^2}, \quad \delta \sim \varepsilon^{1/4} Re_0^{-1/4} \tag{7.91}$$

Here, $Re_0 = \rho_\infty u_\infty \ell/\mu_0$ is the Reynolds number. Below we consider the case in which the density in the boundary layer is small as compared with the undisturbed flow density $\rho^0/\rho_\infty \ll 1$ and, therefore, $\varepsilon \gg \delta^2$. We estimate the value of the transverse velocity component $w^0$ produced in the boundary layer by the pressure difference along the $z^0$ axis from the equation of momentum transport along this axis

$$\rho^0 u^0 \frac{\partial w^0}{\partial x^0} + \rho^0 v^0 \frac{\partial w^0}{\partial y^0} + \rho^0 w^0 \frac{\partial w^0}{\partial z^0} = -\frac{\partial p^0}{\partial z^0} + \frac{\partial}{\partial y^0}\left(\mu^0 \frac{\partial w^0}{\partial y^0}\right) \tag{7.92}$$

If the wing aspect ratio $z_0$ is fairly high, then the estimate follows from the equality of the orders of the following terms of Eq. (7.92)

$$\rho^0 u^0 \frac{\partial w^0}{\partial x^0} \sim \frac{\partial p^0}{\partial z^0}, \quad \frac{\rho_\infty \delta^2 u_\infty w^0}{\varepsilon \ell} \sim \frac{\rho_\infty u_\infty^2 \delta^2}{z_0 \ell}, \quad \frac{w^0}{u_\infty} \sim \varepsilon z_0^{-1} \tag{7.93}$$

In this case, $\rho^0 w^0 \partial w^0/\partial z^0 \times (\partial p^0/\partial z^0)^{-1} \sim \varepsilon/z_0^2$; this means that for the wing aspect ratios $z_0 \gg \sqrt{\varepsilon}$ in Eq. (7.92) the convective term $\rho^0 w^0 \partial w^0/\partial z^0$ can be neglected. We note that in the case in question ($\delta/z_0 \ll 1$) outside the boundary layer this velocity component is of the order $w_e^0 \sim z_0 \ll 1$ (Chernyi, 1966).

Another limiting case is realized for very small $z_0$. Then the estimate for $w^0$ follows from the condition of the equality of the terms

$$\rho^0 w^0 \frac{\partial w^0}{\partial z^0} \sim \frac{\partial p^0}{\partial z^0}, \quad \frac{\rho_\infty \delta^2 (w^0)^2}{\varepsilon z_0 \ell} \sim \frac{\rho_\infty u_\infty^2 \delta^2}{z_0 \ell}, \quad \frac{w^0}{u_\infty} \sim \varepsilon^{1/2} \tag{7.94}$$

This case will be considered in the next section. In this case, the longitudinal convective terms in Eq. (7.92) turn out to be small under condition $\rho^0 u^0 (\partial w^0/\partial x^0) \times (\rho^0 w^0 (\partial w^0/\partial z^0))^{-1} \ll 1$, that is, for $z_0 \ll \sqrt{\varepsilon}$. In this case, the boundary layer are degenerated: they lose their convective terms including derivatives with respect to the coordinate $x^0$.

The general case in which all the terms are of the same order is realized for wing aspect ratios $z_0 \sim \sqrt{\varepsilon}$; in this case the estimate for the transverse flow velocity component $w^0/u_\infty \sim \varepsilon^{1/2}$ (Eq. (7.94)) is retained.

Further, we consider the wings with the aspect ratio $z_0 = O(1) \gg \varepsilon^{1/2}$; under the condition $M_\infty \delta \gg 1$ the approximate tangent wedge formula (Hayes and Probstein, 1966) can be applied to determine the induced pressure produced by the displacement thickness.

In accordance with the conventional estimates for the laminar boundary layer in a hypersonic flow and in view of Eqs. (7.90), (7.91), and (7.93), we introduce the following self-similar (in the $x$ axis) variables

$$\ell x, \quad \ell\delta x^{3/4}\int_0^{\lambda^*}\frac{d\lambda^*}{\rho}, \quad \ell z_0 x z \tag{7.95}$$

$$\rho_\infty u_\infty^2 \delta^2 x^{-1/2} p^*(z), \quad \rho_\infty \delta^2 \varepsilon^{-1} x^{-1/2}\rho(\lambda^*, z)$$

$$\mu_0\mu(\lambda^*, z), \quad u_\infty^2\frac{g(\lambda^*, z)}{2}, \quad u_\infty u(\lambda^*, z)$$

$$u_\infty \delta z_0^{-1} x^{-1/4}\rho^{-1}\left(v^*(\lambda^*, z) - \frac{z_0 u\lambda^*}{4} - z_0 x u\frac{\partial\lambda^*}{\partial x} - \varepsilon w\frac{\partial\lambda^*}{\partial z}\right)$$

$$u_\infty \varepsilon w(\lambda^*, z), \quad \ell\delta x^{3/4}\delta_e^*(z), \quad \delta = \varepsilon^{1/4} z_0^{1/4} Re_0^{-1/4}$$

In variables (7.95) for a semi-infinite delta wing the system of equations of the three-dimensional boundary layer reduces to a two-dimensional system dependent on $\lambda^*$ and $z$, since the longitudinal coordinate $x$ drops out from the boundary value problem.

In order to take account for the flow function behavior in the case $\varepsilon \ll 1$, in the vicinity of the leading edges we will introduce variables analogous to those used in the paper of Dudin, (1997) which are also non-self-similar (in the $z$ axis) in the vicinity of these edges even in the presence of supercritical flow regions

$$\lambda = \frac{\lambda^*}{\sqrt{2(1-z^2)^{1/2}}}, \quad p(z) = \sqrt{1-z^2}\,p^*, \quad \Delta(z) = (1-z^2)^{-3/4}\Delta_e^* \tag{7.96}$$

$$v(\lambda, z) = \frac{1-z^2}{p}\left[(\varepsilon w - z_0 u z)\frac{\partial\lambda}{\partial z} + \frac{v^*}{\sqrt{2(1-z^2)^{1/2}}}\right]$$

In view of Eqs. (7.95) and (7.96) the system of equations of the three-dimensional boundary layer and the boundary conditions on a cold planar delta wing take the form:

$$f = (\varepsilon w - z_0 u z)(1-z^2)p^{-1} \tag{7.97}$$

$$f\frac{\partial u}{\partial z} + v\frac{\partial u}{\partial\lambda} = -\frac{z_0\varepsilon}{2p(1+\varepsilon)}(g - u^2 - \varepsilon^2 w^2)\left(-\frac{1+z^2}{2} + z\frac{1-z^2}{p}\frac{dp}{dz}\right) + \frac{\partial^2 u}{\partial\lambda^2}$$

$$f\frac{\partial w}{\partial z} + v\frac{\partial w}{\partial\lambda} = -\frac{1}{2p(1+\varepsilon)}(g - u^2 - \varepsilon^2 w^2)\left(z + \frac{1-z^2}{p}\frac{dp}{dz}\right) + \frac{\partial^2 w}{\partial\lambda^2}$$

$$f\frac{\partial g}{\partial z} + v\frac{\partial g}{\partial\lambda} = \frac{\partial}{\partial\lambda}\left[\frac{1}{\sigma}\frac{\partial g}{\partial\lambda} - \frac{1-\sigma}{\sigma}\frac{\partial(u^2+\varepsilon^2 w^2)}{\partial\lambda}\right]$$

$$\frac{\partial v}{\partial \lambda} = (\varepsilon w - z_0 u z)\frac{z}{2p} - \left(\frac{z_0 u}{4} - z_0 z\frac{\partial u}{\partial z} + \varepsilon\frac{\partial w}{\partial z}\right)\frac{1 - z^2}{p}$$

$$\Delta = \frac{1}{\sqrt{2(1+\varepsilon)p}}\int_0^\infty (g - u^2 - \varepsilon^2 w^2)\,\mathrm{d}\lambda$$

$$p = \frac{2+\varepsilon}{2}\left[\frac{3}{4}(1+z^2)\Delta - z(1-z^2)\frac{\mathrm{d}\Delta}{\mathrm{d}z}\right]^2$$

$$\lambda = 0: \quad u = w = v = g = 0$$

$$\lambda \to \infty: \quad u \to 1, \quad w \to 0, \quad g \to 1$$

Here, $\sigma$ is the Prandtl number. The system of equations (7.97) is considerably different from the corresponding system considered in the paper of Dudin (1997) in that the displacement thickness does not vanish in this system when passing to the $\varepsilon \to 0$ limit. The solution of the boundary value problem (7.97) determining the boundary layer flow over the entire wing depends generally on the parameters $z_0$, $\sigma$, and $\varepsilon = \gamma - 1$. Passing to the $\varepsilon = 0$ limit in Eq. (7.97) we obtain

$$-z_0 u z\frac{1-z^2}{p}\frac{\partial u}{\partial z} + v\frac{\partial u}{\partial \lambda} = \frac{\partial^2 u}{\partial \lambda^2} \tag{7.98}$$

$$-z_0 u z\frac{1-z^2}{p}\frac{\partial g}{\partial z} + v\frac{\partial g}{\partial \lambda} = \frac{\partial}{\partial \lambda}\left(\frac{1}{\sigma}\frac{\partial g}{\partial \lambda} - \frac{1-\sigma}{\sigma}\frac{\partial u^2}{\partial \lambda}\right)$$

$$\frac{\partial v}{\partial \lambda} = -\frac{z_0 u z^2}{2p} - \left(\frac{z_0 u}{4} - z_0 z\frac{\partial u}{\partial z}\right)\frac{1-z^2}{p}$$

$$\Delta = \frac{1}{\sqrt{2p}}\int_0^\infty (g - u^2)\,\mathrm{d}\lambda$$

$$p = \left[\frac{3}{4}(1+z^2)\Delta - z(1-z^2)\frac{\mathrm{d}\Delta}{\mathrm{d}z}\right]^2$$

$$-z_0 u z\frac{1-z^2}{p}\frac{\partial w}{\partial z} + v\frac{\partial w}{\partial \lambda} = -\frac{1}{2p}(g-u^2)\left(z + \frac{1-z^2}{p}\frac{\mathrm{d}p}{\mathrm{d}z}\right) + \frac{\partial^2 w}{\partial \lambda^2}$$

$$\lambda = 0: \quad u = w = v = g = 0$$

$$\lambda \to \infty: \quad u \to 1, \quad w \to 0, \quad g \to 1$$

System (7.98) can be considered as a system of equations for determining the leading terms of series expansions of the flow functions in the parameter $\varepsilon$. This system can be separated, since the unknown functions $u$, $g$, $p$, and $\Delta$ depend only on the parameters $z_0$ and $\sigma$ and are independent of the transverse velocity $w$; in this it differs considerably from the general case described by system (7.97). The equation for determining $w$ is linear, with zero boundary conditions; its solution is determined after the functions listed above have been found. From the fact that the coefficient of the derivative with respect to the coordinate $z$ in the transport equations of system (7.98) is proportional to $z$ and reverses sign only in the plane of symmetry of the wing ($z=0$) it follows that a flow with smooth convergence to this plane is realized and the parabolicity direction of this system is retained on each wing side. Since, in view of the symmetry, in the $z=0$ plane the pressure gradient is zero, the equation for $w$ in the plane of symmetry becomes an ordinary differential equation with the solution $w(\lambda, z=0) \equiv 0$ (Smirnov, 1958). Similar results were obtained in the paper of Neiland and Sokolov (1977) in considering the laminar boundary layer on a cone at small angles of attacks in a supersonic flow and in the book of Bashkin and Dudin (2000) in studying the flow over wings of special shape ensuring low pressure gradients.

At the leading edges of the wing $z=\pm 1$ system (7.98) degenerates to two systems of ordinary differential equations. As shown below, in the case $\sigma=1$ the solutions of these equations can be expressed in terms of the solution of the Blasius problem (Schlichting, 1968).

In studying the behavior of the flow functions in the boundary layer in the vicinity of the plane of symmetry of the wing it is assumed that the following expansions (Dudin, 1980) are valid

$$u = u_0(\lambda) + u_1(\lambda)z^2 + o(z^2), \quad g = g_0(\lambda) + g_1(\lambda)z^2 + o(z^2) \tag{7.99}$$

$$v = v_0(\lambda) + v_1(\lambda)z^2 + o(z^2), \quad w = w_1(\lambda)z + o(z^2)$$

$$\Delta = \Delta_0 + \Delta_1 z^2 + o(z^2), \quad p = p_0 + p_1 z^2 + o(z^2)$$

Substituting expansions (7.99) in Eq. (7.98) and equating the terms of the same order we obtain the system of zeroth-approximation equations

$$v_0 \frac{du_0}{d\lambda} = \frac{d^2 u_0}{d\lambda^2}, \quad v_0 \frac{dg_0}{d\lambda} = \frac{d}{d\lambda}\left(\frac{1}{\sigma}\frac{dg_0}{d\lambda} - \frac{1-\sigma}{\sigma}\frac{du_0^2}{d\lambda}\right) \tag{7.100}$$

$$\frac{dv_0}{d\lambda} = -\frac{z_0 u_0}{4p_0}, \quad \Delta_0 = \frac{1}{\sqrt{2p_0}} \int_0^\infty (g_0 - u_0^2)\, d\lambda, \quad p_0 = (0.75\Delta_0)^2$$

$$v_0 \frac{dw_1}{d\lambda} = -\frac{1}{2p_0}(g_0 - u_0^2)\left(1 + \frac{2p_1}{p_0}\right) + \frac{d^2 w_1}{d\lambda^2} + \frac{z_0 u_0 w_1}{p_0}$$

$$\lambda = 0: \quad u_0 = w_1 = v_0 = g_0 = 0$$

$$\lambda \to \infty: \quad u_0 \to 1, \quad w_1 \to 0, \quad g_0 \to 1$$

We note that introducing the transformation $w_1 = (1 + 2p_1/p_0)w_1^*$ in the equation for $w_1$ of system (7.100) we obtain an equation for determining $w_1^*$ which no longer depends on the parameter $p_1$, though, of course, its solution depends on the functions $g_0$, $u_0$, $v_0$, and $p_0$. System (7.100) can also be separated; in this system the profile $w_1(\lambda) = (dw/dz)_{z=0}$ depends on the parameter $p_1 = 0.5(\mathrm{d}^2p/\mathrm{d}z^2)_{z=0}$ which generally must be determined from the condition of the matching with the solution proceeding from the leading edge of the wing. However, in the case in question, by virtue of the special features of this system of equations, the parameter $p_1$ can be found from the solution of a part of the equations of the system for the next-order terms of the expansion

$$-\frac{2z_0u_0u_1}{p_0} + v_0\frac{\mathrm{d}u_1}{\mathrm{d}\lambda} + v_1\frac{\mathrm{d}u_0}{\mathrm{d}\lambda} = \frac{\mathrm{d}^2u_1}{\mathrm{d}\lambda^2} \tag{7.101}$$

$$-\frac{2z_0u_0g_1}{p_0} + v_0\frac{\mathrm{d}g_1}{\mathrm{d}\lambda} + v_1\frac{\mathrm{d}g_0}{\mathrm{d}\lambda} = \frac{\mathrm{d}}{\mathrm{d}\lambda}\left[\frac{1}{\sigma}\frac{\mathrm{d}g_1}{\mathrm{d}\lambda} - 2\frac{1-\sigma}{\sigma}\frac{\mathrm{d}(u_0u_1)}{\mathrm{d}\lambda}\right]$$

$$\frac{\mathrm{d}v_1}{\mathrm{d}\lambda} = \frac{7z_0u_1}{p_0} - \frac{z_0u_0}{4p_0}\left(1 - \frac{p_1}{p_0}\right)$$

$$p_1 = \frac{3}{7}\left[\frac{5\sqrt{2}}{4\sqrt{p_0}}\int_0^\infty (g_1 - 2u_0u_1)\,\mathrm{d}\lambda - 2p_0\right]$$

$$\lambda = 0: \quad u_1 = v_1 = g_1 = 0$$

$$\lambda \to \infty: \quad u_1 = 0, \quad g_1 = 0$$

Given $z_0$ and $\sigma$, the functions $u_0$, $g_0$, $v_0$, and $p_0$ are determined from system (7.100). Then from the solution of the linear system (7.101) with zero boundary conditions the functions $u_1$, $g_1$, and $w_1$ are determined, together with the parameter $p_1$; knowing this parameter makes it possible to solve the linear inhomogeneous equation for $w_1$ of system (7.100). Substituting the functions thus obtained in expansions (7.99) we can find, correct to $o(z^2)$, the solution in the vicinity of the plane of symmetry of the wing. This solution describes the flow, at least, near the plane of symmetry in the subcritical flow region, when this region is formed at fairly low values of the parameter $\varepsilon$. For constructing the solution in the entire subcritical flow region it is generally necessary to consider the next terms of expansion in system (7.97). As noted above, in the formulation considered a flow with smooth convergence is realized in the vicinity of the plane of symmetry of the wing. We note that the time it takes for streamtubes to reach this plane is of the order $t \sim z^0/w^0 \sim z_0x\ell z/(\varepsilon u_\infty w_1 z) \sim 1/\varepsilon$ and increases without bound as $\varepsilon \to 0$. It should be noted that system (7.100) was derived for the case of the flow past wings with aspect ratios $z_0 = O(1) \gg \sqrt{s}$, so that the passage to limit $z_0 \to 0$ is not permissible even formally, since in this case $v_0(\lambda) = 0$, while the solution for the velocity component $u_0(\lambda)$ is a linear function of the coordinate $\lambda$ which does not satisfy the condition at the outer edge of the boundary layer. As noted above, considering the flow past narrow delta wings with $z_0 \ll \sqrt{\varepsilon}$ or $z_0 = \sqrt{\varepsilon}$ requires other estimates for the transverse velocity (7.94).

Let us now consider system (7.100) in the case in which $\sigma = 1$. Then the Crocco integral $g_0 = u_0$ is valid. Introducing the function $u_0(\lambda) = \varphi'$, where prime denotes the derivative with respect to $\lambda$, and assuming that $\varphi(0) = 0$ from the integrated continuity equation we obtain $v_0 = -0.25 s \varphi p_0$. Making then the change of variables $\varphi(\lambda) = \sqrt{2p_0/s}\,\Phi(\lambda)$ and $\lambda = \sqrt{2p_0/s}\,\zeta$ from the equation for $u_0$ in system (7.100) we obtain the Blasius problem whose solution is tabulated (Schlichting, 1968)

$$2\Phi''' + \Phi\Phi'' = 0 \tag{7.102}$$

$$\zeta = 0: \quad \Phi = \Phi' = 0, \quad \zeta \to \infty: \quad \Phi' \to 1$$

Here, primes denote the derivatives with respect to the coordinate $\zeta$. In view of the fact that for the boundary value problem (7.102) the value of the integral $\int_0^\infty \Phi'(1-\Phi')\,\mathrm{d}\zeta \approx 0.664$ (Schlichting, 1968), the following expressions for the zeroth-approximation functions and the coordinate normal to the surface can be obtained from system (7.100)

$$p_0 = 0.498 z_0^{-1/2}, \quad u_0 = g = \Phi', \quad v_0 = -0.5(0.996)^{-1/2} z_0^{3/4} \Phi \tag{7.103}$$

$$\lambda = \sqrt{0.996}\, z_0^{-3/4} \zeta$$

In this case, the boundary value problem for determining $w_1$ can be brought into the form:

$$w_1'' + 0.5\Phi w_1' + 2z_0 \Phi' w_1 = z_0^{-1}\left(1 + 2\frac{p_1\sqrt{z_0}}{0.498}\right)\Phi'(1-\Phi')$$

$$\zeta = 0: \quad w_1 = 0, \quad \zeta \to \infty: \quad w_1 \to 0$$

where the constant $p_1$ can be determined from solution (7.101) with $\sigma = 1$.

### 7.9.2 *Self-similar variables*

The flow in supercritical regions can be studied with reference to the example of the flow near one of the leading edges, for example, $z = 1$, introducing instead of variables (7.96) the following self-similar variables

$$\lambda_a = \frac{\lambda^*}{\sqrt{2(1-z)^{1/2}}}, \quad p_a = \sqrt{1-z}\,p^*, \quad \Delta_a = (1-z)^{-3/4}\Delta_e^* \tag{7.104}$$

$$v_a = \frac{1-z}{p_a}\left[(\varepsilon w - z_0 u z)\frac{\partial \lambda_a}{\partial z} + \frac{v^*}{\sqrt{2(1-z)^{1/2}}}\right]$$

Then in the supercritical flow region in which the flow functions depend on the coordinate $\lambda_a$ ($u = u_a(\lambda_a)$, $w = w_a(\lambda_a)$, $v = v_a(\lambda_a)$, and $g = g_a(\lambda_a)$), in the case in which it exists for $z_k \leq z \leq 1$, we obtain the following system of ordinary differential equations

$$v_a \frac{du_a}{d\lambda_a} = \frac{z_0 \varepsilon}{4p_a(1+\varepsilon)}(g_a - u_a^2 - \varepsilon^2 w_a^2) + \frac{d^2 u_a}{d\lambda_a^2} \tag{7.105}$$

$$v_a \frac{dw_a}{d\lambda_a} = -\frac{1}{4p_a(1+\varepsilon)}(g_a - u_a^2 - \varepsilon^2 w_a^2) + \frac{d^2 w_a}{d\lambda_a^2}$$

$$v_a \frac{dg_a}{d\lambda_a} = \frac{d}{d\lambda_a}\left[\frac{1}{\sigma}\frac{dg_a}{d\lambda_a} - \frac{1-\sigma}{\sigma}\frac{d(u_a^2 + \varepsilon^2 w_a^2)}{d\lambda_a}\right]$$

$$\frac{dv_a}{d\lambda_a} = (\varepsilon w_a - z_0 u_a)\frac{1}{4p}$$

$$\Delta_a = \frac{1}{\sqrt{2(1+\varepsilon)p_a}} \int_0^\infty (g_a - u_a^2 - \varepsilon^2 w_a^2)\, d\lambda_a, \quad p_a = \frac{2+\varepsilon}{2}(0.75\Delta_a)^2$$

$$\lambda_a = 0: \quad u_a = w_a = v_a = g_a = 0$$

$$\lambda_a \to \infty: \quad u_a \to 1, \quad w_a \to 0, \quad g_a \to 1$$

We note that system (7.105) describes the flow on a yawed cold plate in the strong interaction regime (Neiland, 1974*b*]. On the basis of the form of Eqs. (7.105) in what follows we assume that in the supercritical region for small values of the parameter $\varepsilon$ the flow functions in the self-similar variables (7.104) can be expanded as follows:

$$F_a = F_{a0} + O(\varepsilon) \tag{7.106}$$

Substituting Eq. (7.106) in system (7.105) for determining the zeroth-approximation terms we obtain

$$v_{a0} \frac{du_{a0}}{d\lambda_a} = \frac{d^2 u_{a0}}{d\lambda_a^2} \tag{7.107}$$

$$v_{a0} \frac{dw_{a0}}{d\lambda_a} = -\frac{1}{4p_{a0}}(g_{a0} - u_{a0}^2) + \frac{d^2 w_{a0}}{d\lambda_a^2}$$

$$v_{a0} \frac{dg_{a0}}{d\lambda_a} = \frac{d}{d\lambda_a}\left(\frac{1}{\sigma}\frac{dg_{a0}}{d\lambda_a} - \frac{1-\sigma}{\sigma}\frac{du_{a0}^2}{d\lambda_a}\right)$$

$$\frac{dv_{a0}}{d\lambda_a} = -\frac{z_0 u_{a0}}{4p_a} \tag{7.108}$$

$$\Delta_{a0} = \frac{1}{\sqrt{2}p_{a0}} \int_0^\infty (g_{a0} - u_{a0}^2)\,\mathrm{d}\lambda_a, \quad p_{a0} = (0.75\Delta_{a0})^2$$

$$\lambda_a = 0: \quad u_{a0} = w_{a0} = v_{a0} = g_{a0} = 0$$

$$\lambda_a \to \infty: \quad u_{a0} \to 1, \quad w_{a0} \to 0, \quad g_{a0} \to 1$$

In system (7.107) the functions $u_{a0}, g_{a0}, v_{a0}, p_{a0}$, and $\Delta_{a0}$, as the corresponding functions in system (7.98), are independent of the transverse velocity component. The solution for $w_{a0}$ is obtained after the above-listed functions have been determined. It is important to note that the equations of systems (7.100) and (7.107), except for the equations determining the $w_1(\lambda)$ and $w_{a0}(\lambda_a)$ profiles, coincide correct to notation; therefore, for $\sigma = 1$, in view of Eqs. (7.99) and (7.100) we can immediately write

$$p_{a0} = 0.498 z_0^{-1/2}, \quad u_{a0} = g_{a0} = \Phi' \tag{7.109}$$

$$v_{a0} = -0.5(0.996)^{-1/2} z_0^{3/4} \Phi, \quad \lambda_a = \sqrt{0.996} z_0^{-3/4} \zeta$$

where the function $\Phi$ is determined by Eq. (7.102). Moreover, in view of Eqs. (7.96) and (7.104) for $z = 0$, we obtain the equality $\lambda = \lambda_a$. Thus, if in the limiting case $\varepsilon \equiv 0$ the supercritical flow region exists over the entire wing surface, then in this region the solution of system (7.107) coincides with the solution of system (7.100) determining the flow in the plane of symmetry of the wing; moreover, in view of Eq. (7.95), in this case on the entire wing we obtain $w^0 = u_\infty \varepsilon w_{a0} \equiv 0$. For determining $w_{a0}$ in the case $\sigma = 1$ we note that after the change of variable $w_{a0} = z_0^{-1} W$ in Eq. (7.106) we arrive at the following boundary value problem

$$W'' + 0.5\Phi W' - 0.5\Phi'(1 - \Phi') = 0$$

$$\zeta = 0: \quad W = 0, \quad \zeta \to \infty: \quad W \to 0$$

This equation is linear and its solution subject to the boundary conditions is as follows (Smirnov, 1958):

$$W(\zeta) = -0.5\Phi'(\zeta) \int_0^\infty \Phi''(\theta) \left[ \int_0^\theta \frac{\Phi'(\vartheta)(1 - \Phi'(\vartheta))}{\Phi''(\vartheta)} \mathrm{d}\vartheta \right] \mathrm{d}\theta$$

$$- 0.5 \int_0^\zeta \Phi''(\theta) \left[ \int_0^\theta \frac{\Phi'(\vartheta)(1 - \Phi'(\vartheta))}{\Phi''(\vartheta)} \mathrm{d}\vartheta \right] \mathrm{d}\theta \tag{7.110}$$

We note that solution (7.109) is independent of the wing aspect ratio and can be calculated only once. In Fig. 7.52 the velocity profile $W(\zeta)$ at the leading edge ($z = 1$) is presented.

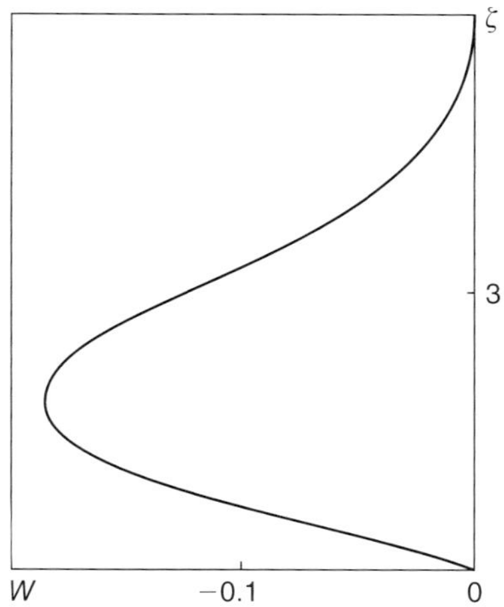


Fig. 7.52.

### *7.9.3 Conditions of supercritical-to-m-subcritical flow regime transition*

In the case of the flow past a semi-infinite delta wing the coordinate $z_k$ corresponding to supercritical-to-subcritical flow transition is determined from the condition of the vanishing of an integral (Neiland, 1974*b*) which in the self-similar variables takes the form:

$$I = \int_0^\infty \left[ \frac{\varepsilon}{2} \left( \frac{g_a - u_a^2 - \varepsilon^2 w_a^2}{u_a \sin(\omega_0 - \omega_1) - \varepsilon w_a \cos(\omega_0 - \omega_1)} \right)^2 - g_a + u_a^2 + \varepsilon^2 w_a^2 \right] \mathrm{d}\eta_a = 0$$

$$z_k = \frac{\tan(\omega_0 - \omega_1)}{\tan \omega_0} \tag{7.111}$$

Here, the angle $\omega_1$ is measured from the leading edge with $z = 1$, while the flow functions are determined from the solution of system (7.105).

In view of the fact that with decrease in the parameter $\varepsilon$ the difference of the angles $\omega_0 - \omega_1$ becomes small (Dudin, 1997), in what follows we will assume that for small $\varepsilon$ the following expansion is valid

$$\omega_0 - \omega_1 = c_1 \varepsilon^{1/2} + c_2 \varepsilon + o(\varepsilon) \tag{7.112}$$

Here, $c_1$ and $c_2$ are generally functions of $z_0$ and $\sigma$. Substituting expansions (7.106) and (7.111) in Eq. (7.110) and assuming that

$$\sin(\omega_0 - \omega_1) = \omega_0 - \omega_1 + O((\omega_0 - \omega_1)^3), \quad \cos(\omega_0 - \omega_1) = 1 + O((\omega_0 - \omega_1)^2)$$

we obtain

$$I = \int_0^\infty \left[ \frac{(g_{a0} - u_{a0}^2)^2}{2u_{a0}^2 c_1^2} - (g_{a0} - u_{a0}^2) - \frac{(g_{a0} - u_{a0}^2)^2 (u_{a0} c_2 - w_{a0})\sqrt{\varepsilon}}{u_{a0}^3 c_1^3} + O(\varepsilon) \right] \mathrm{d}\eta_a = 0$$

$$z_k = \frac{c_1}{z_0}\sqrt{\varepsilon} + \frac{c_2}{z_0}\varepsilon + O(\varepsilon) \tag{7.113}$$

We will first note that in the $\varepsilon = 0$ limit the transition coordinate $z_k = 0$ and, therefore, a supercritical flow region is formed in the boundary layer on the entire wing. From the condition of vanishing of integral (7.109) we obtain two conditions for determining the constants $c_1$ and $c_2$ which take the form:

$$c_1 = \sqrt{\int_0^\infty \frac{(g_{a0} - u_{a0}^2)^2}{u_{a0}^2} \mathrm{d}\eta_a \times \left[ 2\int_0^\infty (g_{a0} - u_{a0}^2)\, \mathrm{d}\eta_a \right]^{-1}} \tag{7.114}$$

$$c_2 = \int_0^\infty \frac{(g_{a0} - u_{a0}^2)^2 w_{a0}}{u_{a0}^3} \mathrm{d}\eta_a \times \left[ \int_0^\infty \frac{(g_{a0} - u_{a0}^2)^2}{u_{a0}^2} \mathrm{d}\eta_a \right]^{-1} \tag{7.115}$$

Thus, the solution of system (7.107) for the zeroth approximation makes it possible to determine two terms in the expansion for the transition line (7.112). In the case of an arbitrary Prandtl number the system of equations (7.107) must be numerically solved and then, as a result of numerical integration, the constants $c_1$ and $c_2$ are determined.

For further simplification we will consider the case in which $\sigma = 1$. Then from system (7.107) we obtain $g_{a0} = u_{a0}$ and from Eqs. (7.113) and (7.114) the constants are determined from the formulas

$$c_1 = \sqrt{\frac{3}{8\sqrt{2}} p_{a0}^{-3/2} \int_0^\infty (1 - u_{a0})^2\, \mathrm{d}\eta_a} \tag{7.116}$$

$$c_2 = \frac{3}{8\sqrt{2}} c_1^{-2} p_{a0}^{-3/2} \int_0^\infty (1 - u_{a0})^2 \frac{w_{a0}}{u_{a0}} \mathrm{d}\eta_a$$

In view of Eq. (7.108), Eq. (7.115) for determining $c_1$ reduces to the following form:

$$c_1 = \sqrt{\frac{1}{1.328} \int_0^\infty (1 - \Phi')^2\, \mathrm{d}\zeta}$$

Since, according to Schlichting (1968)

$$\int_0^\infty (1-\Phi')^2\,d\zeta = \int_0^\infty (1-\Phi')\,d\zeta - \int_0^\infty \Phi'(1-\Phi')\,d\zeta \approx 1.721 - 0.664 \approx 1.057$$

we have $c_1 \approx 0.8921$ and, therefore, for $\sigma = 1$ the quantity $c_1$ is independent of the wing aspect ratio. Then in the first approximation, correct to $o(\sqrt{\varepsilon})$, for the transition coordinate $z_k$ (Eq. (7.112)) and the slope of the transition line $\omega_1$ (Eq. (7.111)) we obtain

$$z_k = 0.8921\sqrt{\varepsilon}z_0^{-1}, \quad \omega_1 = \omega_0 - 0.8921\sqrt{\varepsilon} \tag{7.117}$$

For calculating the quantity $c_2$ entering in the second term of expansion (7.112) it is necessary that the transverse velocity component $w_{a0} = z_0^{-1}W$ is also known; here, $W(\zeta)$ is determined by formula (7.109). Then we have

$$c_2 = 0.946z_0^{-1}\int_0^\infty (1-\Phi'^2)\frac{W}{\Phi'}\,d\zeta \approx -0.8266z_0^{-1}$$

Therefore, the quantity $c_2$ is inverse proportional to the wing aspect ratio. We note that in the case $\sigma = 1$ using the two-term expansion (7.112) is permissible if

$$\sqrt{\varepsilon} \ll \left|\frac{c_1}{c_2}\right| \approx 1.0792z_0 \tag{7.118}$$

Obviously, this condition is violated when considering wings with aspect ratios $z_0 \sim \sqrt{\varepsilon}$ or less; contrariwise, for fairly large values of $z_0$ a satisfactory applicability of expansion (7.112) might be expected even for $\gamma$ not too close to unity. Thus, for $\sigma = 1$ for the second approximation we have

$$z_k = 0.8921\sqrt{\varepsilon}z_0^{-1} - 0.8266\varepsilon z_0^{-2} \tag{7.119}$$

$$\omega_1 = \omega_0 - 0.8921\sqrt{\varepsilon} + 0.8266\varepsilon z_0^{-1}$$

In Fig. 7.53 we have plotted the dependence of the angle $\omega_1$ corresponding to supercritical-to-subcritical flow transition in the laminar three-dimensional boundary layer on the sweep angle of the leading edge for $\gamma = 1.2$ (curves 1) and 1.05 (curves 2) for the Prandtl number $\sigma = 1$. Curves I present the exact solutions of Eq. (7.110), curves II the first approximation (7.118), and curves III the second approximation (7.118). A decrease in $\gamma$ leads to enlargement of the supercritical flow region. The first approximation (curves II) underestimates the transition angle by about 15% on the wings with the sweep angle 45° in the case of $\gamma = 1.2$ flow (curves 1). With increase in the angle $\omega_0$ this underestimation, both absolute and relative, becomes appreciably smaller. It should be particularly noted that for $\gamma = 1.2$ and sweep angles $\geq$40° the second approximation (curve III) turns out to be worse than

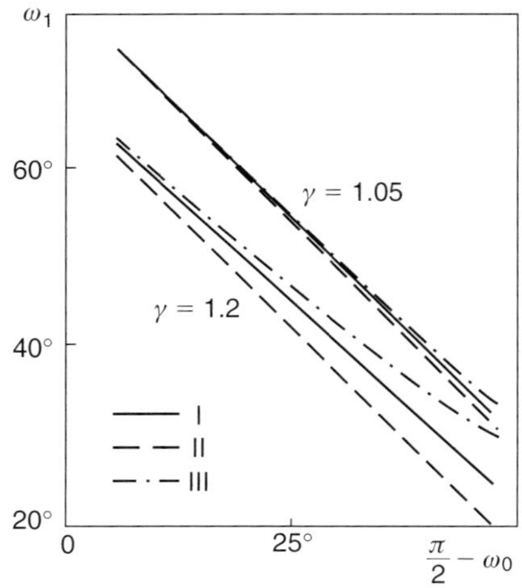


Fig. 7.53.

the first approximation. This is due to the fact that inequality (7.117) is badly fulfilled. In the $\gamma = 1.05$ wing flow the greatest deviation of the first approximation (curve II) from the exact solution (curve I) is not greater than 4%; in this case, the second approximation (curve III) is more accurate than the first for all sweep angle considered.

The dependence of the transition angle $\omega_1$ on the parameter $\varepsilon = \gamma - 1$ for $z_0 = 1$, 2, and 5 (curves 1 to 3 in Fig. 7.54) and $\sigma = 1$ is essentially nonlinear, particularly at $\varepsilon < 0.1$. An analysis of the results shows that for fairly small values of the parameter $\varepsilon$ the second

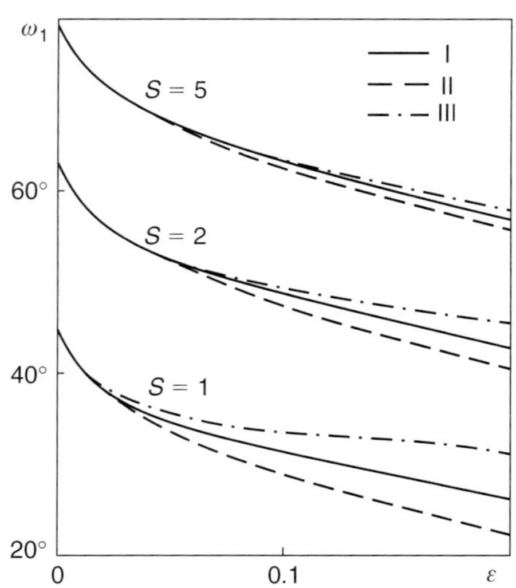


Fig. 7.54.

approximation (curves I and II) is more accurate than the first approximation (curve II) for all wing aspect ratios considered.

Thus, it is established that, as the adiabatic exponent $\gamma \to 1$, three flow patterns can be realized depending on the wing aspect ratio. It is shown that in the flow over a cold planar delta wing with the aspect ratio $z_0 = O(1) \gg \sqrt{\gamma - 1}$ and the adiabatic exponent $\gamma \to 1$ in the laminar three-dimensional boundary layer there occur secondary flows with the transverse velocity component of the order $O(\gamma - 1)$. In the limiting case $\gamma = 1$ it is established that the system of equations in partial derivatives governing the flow on the entire wing can be separated and the flow in the vicinity of the plane of symmetry of the wing is described – in the zeroth approximation – by a system of ordinary differential equations which can be closed using the next terms of the expansion. An analytical dependence of the supercritical-to-subcritical flow transition coordinate is derived in the form of a two-term expansion, whose coefficients are determined by solving a system of ordinary differential equations written in self-similar variables for the zeroth approximation in the supercritical flow region. For the perfect-gas flow with the Prandtl number equal to unity numerical values of the coefficients in the expansion of the transition coordinate are calculated and explicit analytical expressions of the flow functions on the wing aspect ratio are derived.

# 8 Boundary Layer Flow Over Roughnesses at Body Surfaces

In modern aerodynamics an important role is played by viscous high-Reynolds-number flows near body surfaces which are usually studied within the framework of Prandtl's classical boundary layer theory (Prandtl, 1904). The main assumption of this theory is the smallness of the longitudinal (streamwise) gradients of the flow functions, as compared with the vertical gradients. However, in some cases this assumption does not hold, so that an important role is acquired by boundary conditions and flow details with the scales different from those of Prandtl's boundary layer theory. Thus, in the paper of Neiland and Sychev (1966) the asymptotic expansion matching method was applied to develop an approach to the solution of problems with high longitudinal pressure gradients and locally inviscid flow regions were found (see Chapter 3), while in the works of Neiland (1968; 1969*a*) and Stewartson and Williams (1969) another type of local solutions was studied, for which the variation of the boundary layer displacement thickness has an effect on the boundary layer flow even in the first approximation (see Chapter 1). In the paper of Hunt (1971) a wake flow behind a small bump on a flat plate was considered and it was shown that in the shear layer (wall region of the undisturbed flat-plate boundary layer) disturbances decay more rapidly than in an uniform flow. In the paper of Zubtsov (1971) the solution for the flow over a small single roughness was derived, while in the work of Smith (1973) the flow over a small hump was studied in the free interaction regime (Neiland, 1968; 1969; Stewartson and Williams, 1969).

In this chapter we consider flows over small roughnesses at the surface of a body in a supersonic viscous flow. Using the asymptotic expansion matching method the main parameters determining the physical features of the flows and the types of the equations and boundary conditions are established and the flow regimes are classified. It is shown that small roughnesses can lead to the occurrence of local separation zones and a considerable increase in the viscous stress and the heat flux not only in the vicinity of the roughness itself but also in regions of larger extent (Bogolepov and Neiland, 1971).

## 8.1 Flow over two-dimensional roughnesses

### *8.1.1 General formulation of the problem and classification of flow regimes*

Let a small roughness of the type of a bank or a trench be located on the surface of a flat plate at a distance $\ell$ from its leading edge. Let also the plate be in a uniform supersonic viscous flow with a finite Mach number, $(M_\infty^2 - 1) \sim 1$, and a high but subcritical Reynolds number $Re_\infty = \rho_\infty u_\infty \ell / \mu_\infty = \varepsilon^{-2} \gg 1$ (here, $\rho$, $u$, and $\mu$ are the density, the velocity, and the dynamic viscosity, respectively, and the subscript $\infty$ refers to the oncoming uniform flow).

We will construct a steady solution of the Navier–Stokes equations for the disturbed laminar flow region as $Re_\infty \to \infty$ or $\varepsilon \to 0$. In what follows, the linear dimensions are scaled on $\ell$, the velocity components $u$ and $v$ (along the $x$ and $y$ axes of a Cartesian coordinate system) on $u_\infty$, the pressure $p$ on $\rho_\infty u_\infty^2$, the enthalpy $h$ on $u_\infty^2$, and $\rho$ and $\mu$ on $\rho_\infty$ and $\mu_\infty$, respectively; only dimensionless variables are used. The gas is assumed to be perfect, with a constant specific heat ratio $\gamma$, and a certain relation of the form $\mu = \mu(\rho, h)$ is used for determining viscosity.

As for the roughness dimensions, we will assume that is characteristic thickness $a$ is not greater in the order than the characteristic flat-plate boundary layer thickness $\delta$ at the location of the roughness, $a \leq \delta$, while its characteristic length $b$ is in the order equal to or greater than the roughness thickness $a$ and, obviously, is not greater than the distance to the plate's leading edge or a characteristic body length, $a \leq b \leq 1$.[1] If the roughness thickness $a$ is greater than its length $b$ ($a > b$), then the flows over such roughnesses can possess the same features as in the case $a \sim b$, with the difference that in this case maximum dimensions of the disturbed regions are determined by the parameter $a$. It is also known that for moderate oncoming flow Mach numbers $M_\infty$ and surface temperatures (or for $h_w \sim 1$, where the subscript $w$ refers to the values at the roughness location) the characteristic thickness of the undisturbed laminar boundary layer $\delta$ is in the order $\delta \sim \varepsilon$.

Under the above constraints on the parameters $a$, $b$, and $\delta$ the following classification of small roughnesses at the body surface can be imagined:

1. "short" roughnesses with the length less than the boundary layer thickness:
   (*a*) "short and thin" roughnesses for $1 \gg \delta \gg b \gg a$;
   (*b*) "short and fairly thick" roughnesses for $1 \gg \delta \gg b \sim a$;
2. "fairly long" roughnesses with the length comparable with the boundary layer thickness:
   (*a*) "fairly long and thin" roughnesses for $1 \gg \delta \sim b \gg a$;
   (*b*) "fairly long and fairly thick" roughnesses for $1 \gg \delta \sim b \sim a$;
3. "long" roughnesseses with the length greater than the boundary layer thickness:
   (*a*) "long" roughnesses with the thickness smaller than the boundary layer thickness for $1 \gg b \gg \delta \gg a$;
   (*b*) "long" roughness with the thickness comparable with the boundary layer thickness for $1 \gg b \gg \delta \sim a$.

As shown below, each case of the classification presented above possesses its own physical features and requires the solution of its own boundary value problem. The flow under consideration is sketched in Fig. 8.1; here, $(s, n)$ is a curvilinear orthogonal coordinate system fitted to the roughness surface.

[1] In this chapter the vertical dimension $a$ of the roughness (see Fig. 8.1) is alternately called the thickness, when it is compared with the thicknesses of the boundary layer and other layers, and the height, when the stress is laid on the body geometry.

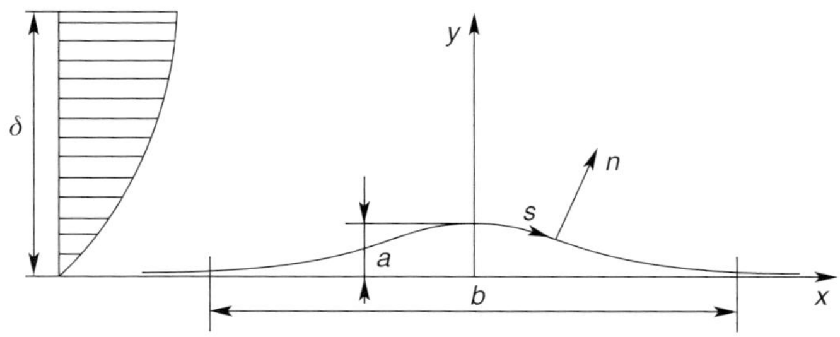


Fig. 8.1.

### *8.1.2 Flow over "short" roughnesses embedded in the wall region of the undisturbed boundary layer*

Let first the roughness length be in the order smaller than the boundary layer thickness: $b \ll \delta \sim \varepsilon$ (case 1 of the classification of roughnesses adopted in Section 8.1.1). Then the following relation is valid for $b$: $\varepsilon^2 \ll b \ll \varepsilon$ (in other words, for $b \sim \varepsilon^2$ the disturbed flow region dimensions are comparable with the molecular free path and the Navier–Stokes equations no longer hold for this region).

It is known that, as $\varepsilon \to 0$, in a region with characteristic dimensions $\Delta x \sim \Delta y \sim 1$ near a flat plate the solution of the Navier–Stokes equations is represented by the uniform oncoming flow, $u = 1$. This solution is valid everywhere, except for a narrow region near the plate, since the uniform flow cannot satisfy the no-slip conditions at the surface. Therefore, in accordance with Prandtl's theory, a region with the characteristic dimensions $\Delta x \sim 1$ and $\Delta y \sim \varepsilon$, that is, the boundary layer, should be considered. The solution for this region is also well known: this is the Blasius solution. However, it does not describe the flow in a region with characteristic dimensions $\Delta x \sim \Delta y \sim b \ll \varepsilon$, which on the scale of the boundary layer represents a point on the passage to the limit $Re_\infty \to \infty$.

In order to investigate the flow in this disturbed region it is necessary to determine the orders of the flow functions in this region and to go over to new variables which would be of the order of unity. The necessary estimates can be obtained from the consideration that the disturbed flow region is located at the "bottom" of the undisturbed boundary layer. Since the dimensionless viscous stress $\tau$ and heat flux $q$

$$\tau = \varepsilon^2 \mu \frac{\partial u}{\partial y} \sim \varepsilon, \quad q = \varepsilon^2 \left(\frac{\mu}{\sigma}\right) \frac{\partial h}{\partial y} \sim \varepsilon \tag{8.1}$$

where $\sigma$ is the Prandtl number, retain their orders throughout almost the entire boundary layer, in its wall region for $\Delta y/\varepsilon \ll 1$ and $h_w \sim 1$ we can obtain

$$u \sim \frac{\Delta y}{\varepsilon}, \quad h \sim h_w + \frac{\Delta y}{\varepsilon} \tag{8.2}$$

From these estimates it follows that the disturbed flow region with the characteristic dimensions $\varepsilon^2 \ll \Delta x \sim \Delta y \sim b \ll \varepsilon$ for $\varepsilon \to 0$ is in the first approximation isothermal, with constant density and viscosity $h_w \sim \rho_w \sim \mu_w \sim 1$, while the longitudinal velocity $u$ is in the order as follows: $u \sim b/\varepsilon \ll 1$.

If the roughness is not thin, that is, its thickness is comparable with its length, $a \sim b$ (case 1$b$, according to the classification of roughnesses), it obviously can induce nonlinear disturbances in this region (e.g., $\Delta u \sim u \sim b/\varepsilon \ll 1$). Then from the comparison of the orders of the terms of the Navier–Stokes equations and from the definition of the stream function ($\Delta p \sim \rho_w u \Delta u$, $u/\Delta x \sim v/\Delta y$, $\Delta\psi \sim \rho_w u \Delta y$) we obtain the required estimates for the orders of the flow functions in this region

$$u \sim v \sim \frac{b}{\varepsilon}, \quad \Delta p \sim \left(\frac{b}{\varepsilon}\right)^2, \quad \Delta\psi \sim \frac{b^2}{\varepsilon} \tag{8.3}$$

In the Russian literature the nonlinear disturbance region near the body surface with $\Delta u \sim u \ll 1$ is conventionally designated by the number 3. Thus, estimates (8.3) make it possible to introduce the following new independent variables and asymptotic expansions for the flow functions in this disturbed region

$$x = bx_3, \quad y = by_3 \tag{8.4}$$

$$u(x, y; \varepsilon) = \frac{b}{\varepsilon}u_3 + \cdots, \quad v(x, y; \varepsilon) = \frac{b}{\varepsilon}v_3 + \cdots$$

$$p(x, y; \varepsilon) = \frac{1}{\gamma M_\infty^2} + \left(\frac{b}{\varepsilon}\right)^2 p_3 + \cdots$$

$$\rho(x, y; \varepsilon) = \rho_w + \cdots, \quad h(x, y; \varepsilon) = h_w + \frac{b}{\varepsilon}h_3 + \cdots$$

$$\mu(x, y; \varepsilon) = \mu_w + \cdots, \quad \psi(x, y; \varepsilon) = \frac{b^2}{\varepsilon}\psi_3 + \cdots$$

Substituting expansions (8.4) in the Navier–Stokes equations and passing to the limit $\varepsilon \to 0$ for $\varepsilon^2 \ll b \ll \varepsilon$ shows that in the first approximation region 3 is described by

1. the Stokes equation for $\varepsilon^2 \ll b \ll \varepsilon^{3/2}$;
2. the Navier–Stokes equations for $b \sim \varepsilon^{3/2}$; and
3. the Euler equations for $\varepsilon^{3/2} \ll b \ll \varepsilon$.

Let first $b \sim \varepsilon^{3/2}$; this is the most general flow regime in the case $\varepsilon^2 \ll b \ll \varepsilon$. Then in the first approximation the flow in the disturbed region 3 is described by the incompressible Navier–Stokes equations

$$\frac{\partial u_3}{\partial x_3} + \frac{\partial v_3}{\partial y_3} = 0 \tag{8.5}$$

$$\rho_w\left(u_3\frac{\partial u_3}{\partial x_3} + v_3\frac{\partial u_3}{\partial y_3}\right) + \frac{\partial p_3}{\partial x_3} = \mu_w\left(\frac{\partial^2 u_3}{\partial x_3^2} + \frac{\partial^2 u_3}{\partial y_3^2}\right)$$

$$\rho_w\left(u_3\frac{\partial v_3}{\partial x_3}+v_3\frac{\partial v_3}{\partial y_3}\right)+\frac{\partial p_3}{\partial y_3}=\mu_w\left(\frac{\partial^2 v_3}{\partial x_3^2}+\frac{\partial^2 v_3}{\partial y_3^2}\right)$$

$$\rho_w\left(u_3\frac{\partial h_3}{\partial x_3}+v_3\frac{\partial h_3}{\partial y_3}\right)=\frac{\mu_w}{\sigma}\left(\frac{\partial^2 h_3}{\partial x_3^2}+\frac{\partial^2 h_3}{\partial y_3^2}\right)$$

The conventional boundary conditions of no-slip and impermeability are imposed on the small roughness surface $y_3 = f(x_3)$

$$u_3 = v_3 = h_3 = 0, \quad y_3 = f(x_3) \tag{8.6}$$

The outer boundary conditions are obtained by matching with the solution for the wall region of the undisturbed flat-plate boundary layer

$$u_3 \to Ay_3, \quad h_3 \to By_3, \quad v_3, p_3 \to 0, \quad x_3^2 + y_3^2 \to \infty \tag{8.7}$$

where $A$ and $B$ are the dimensionless vertical gradients of the longitudinal velocity and the enthalpy near the surface in the undisturbed boundary layer at the location of the small roughness. For a short, fairly thick roughness (case 1$b$ of the classification of roughnesses) in the boundary value problem (8.5)–(8.7) the following affinely similar transformation of variables can be introduced

$$x_3 = a_1\bar{x}, \ \ y_3 = a_1\bar{y}, \ \ u_3 = Aa_1\bar{u}, \ \ v_3 = Aa_1\bar{v}, \ \ p_3 = \rho_w A^2 a_1^2 \bar{p}, \ \ h_3 = Ba_1\bar{h} \tag{8.8}$$

where $a_1 \sim 1$ is the maximum vertical dimension of the small roughness ($a = \varepsilon^{3/2}a_1$), while $Aa_1$ and $Ba_1$ are the longitudinal velocity and the enthalpy increment at a distance $a_1$ from the plate surface in the undisturbed boundary layer. Then in the new variables the problem is brought into the form:

$$\frac{\partial\bar{u}}{\partial\bar{x}}+\frac{\partial\bar{v}}{\partial\bar{y}}=0, \quad \bar{u}\frac{\partial\bar{u}}{\partial\bar{x}}+\bar{v}\frac{\partial\bar{u}}{\partial\bar{y}}+\frac{\partial\bar{p}}{\partial\bar{x}}=\frac{1}{Re}\left(\frac{\partial^2\bar{u}}{\partial\bar{x}^2}+\frac{\partial^2\bar{u}}{\partial\bar{y}^2}\right) \tag{8.9}$$

$$\bar{u}\frac{\partial\bar{v}}{\partial\bar{x}}+\bar{v}\frac{\partial\bar{v}}{\partial\bar{y}}+\frac{\partial\bar{p}}{\partial\bar{y}}=\frac{1}{Re}\left(\frac{\partial^2\bar{v}}{\partial\bar{x}^2}+\frac{\partial^2\bar{v}}{\partial\bar{y}^2}\right), \quad \bar{u}\frac{\partial\bar{h}}{\partial\bar{x}}+\bar{v}\frac{\partial\bar{h}}{\partial\bar{y}}=\frac{1}{\sigma Re}\left(\frac{\partial^2\bar{h}}{\partial\bar{x}^2}+\frac{\partial^2\bar{h}}{\partial\bar{y}^2}\right)$$

$$\bar{u}=\bar{v}=\bar{h}=0, \ \ \bar{y}=f(\bar{x}); \ \ \bar{u}, \ \bar{h}\to\bar{y}, \ \ \bar{v}, \ \bar{p}\to 0, \ \ \bar{x}^2+\bar{y}^2\to\infty; \ \ Re=\frac{\rho_w A a_1^2}{\mu_w}\sim 1$$

where $Re$ is the local Reynolds number, while the shape of the normalized small roughness is given by the equation $\bar{y} = f(\bar{x})$, where $|f(\bar{x})|_{\max} = 1$.

The elliptic boundary value problem (8.9) governs the flow over a small roughness embedded in the wall "shear" region of the undisturbed boundary layer when the orders of the characteristic longitudinal and vertical dimensions of the roughness are the same: $a \sim b \sim \varepsilon^{3/2} \ll \varepsilon \sim \delta$. From expansions (8.4) it follows that in this case the disturbances of the viscous stress $\Delta\tau$ and heat flux $\Delta q$ are in the order equal to their values near the plate surface in the undisturbed boundary layer: $\Delta\tau \sim \Delta q \sim \tau \sim q \sim \varepsilon$. This means that small roughnesses can induce finite relative local variations of the viscous stress and heat flux and initiate flow separation. A similar boundary value problem was considered in the paper of Stewartson (1968) in studying the flow over the trailing edge of a thin plate.

If the roughness is "thin", that is, its thickness is less than the length, $a \ll b$, and the small roughness shape is given by the equation $y_3 = (a/b)f(x_3)$ (case 1$a$, according to the classification of roughnesses), then the solution of the boundary value problem (8.5)–(8.7) can be sought in the form of small disturbances with respect to the flow in the wall region of the undisturbed flat-plate boundary layer

$$u_3 = Ay_3 + \frac{a}{b}U + \cdots, \quad v_3 = \frac{a}{b}V + \cdots, \quad p_3 = \frac{a}{b}P + \cdots, \quad h_3 = By_3 + \frac{a}{b}H + \cdots \tag{8.10}$$

Substituting expansions (8.10) in the boundary value problem (8.5)–(8.7) yields a system of linear equations for the flow function disturbances of the type of the Oseen equations (Oseen, 1910). The internal boundary conditions are carried to the plane surface, while the outer conditions determine the disturbance decay

$$\frac{\partial U}{\partial x_3} + \frac{\partial V}{\partial y_3} = 0 \tag{8.11}$$

$$\rho_w\left(Ay_3\frac{\partial U}{\partial x_3} + AV\right) + \frac{\partial P}{\partial x_3} = \mu_w\left(\frac{\partial^2 U}{\partial x_3^2} + \frac{\partial^2 U}{\partial y_3^2}\right)$$

$$\rho_w Ay_3\frac{\partial V}{\partial x_3} + \frac{\partial P}{\partial y_3} = \mu_w\left(\frac{\partial^2 V}{\partial x_3^2} + \frac{\partial^2 V}{\partial y_3^2}\right)$$

$$\rho_w\left(Ay_3\frac{\partial H}{\partial x_3} + BV\right) = \frac{\mu_w}{\sigma}\left(\frac{\partial^2 H}{\partial x_3^2} + \frac{\partial^2 H}{\partial y_3^2}\right)$$

$$U = -Af(x_3), \quad V = 0, \quad H = -Bf(x_3), \quad y_3 = 0$$

$$U, V, P, H \to 0, \quad x_3^2 + y_3^2 \to \infty$$

This boundary value problem is less interesting than problems (8.5)–(8.7) or (8.9), since in this case the roughness generates only small disturbances of the viscous stress and heat flux $\Delta\tau \sim \Delta q \sim (a/b)\varepsilon \ll \varepsilon$ and cannot initiate flow separation.

For smaller roughnesses with $\varepsilon^2 \ll a \sim b \ll \varepsilon^{3/2}$ (this is also case 1$b$ according to the classification of roughnesses) in the disturbed flow region 3 with the characteristic dimensions $\Delta x \sim \Delta y \sim b$ the convective terms of the Navier–Stokes equations are negligibly small as compared with the viscous terms. The flow is governed by the Stokes equations (Stokes, 1851) in which the viscous stresses are balanced by the pressure force. The corresponding boundary value problem is immediately obtained if the convective terms are dropped out from Eqs. (8.5) or the formal passage to the limit $Re \to 0$ is performed in Eqs. (8.9); then the viscous stresses and the heat flux vary in the leading order. This means that possible flow separation is described by equations with no convective terms (the solution of this problem will be considered below). For thin roughnesses with $\varepsilon^2 \ll a \ll b \ll \varepsilon^{3/2}$ we obviously arrive at the boundary value problem of type (8.11), only with no convective terms; in this case, the viscous stress and heat flux disturbances are again small.

### *8.1.3 Flow over "short" roughnesses with the formation of locally inviscid disturbed flow regions*

We will now consider the flow over larger, as compared with those of the previous section, roughnesses; however, their length is again smaller in the order than the boundary layer thickness, $\varepsilon^{3/2} \ll b \ll \varepsilon \sim \delta$ (these also pertain to case 1 according to the roughness classification adopted). Substituting expansions (8.4) in the Navier–Stokes equations and passing to the limit $\varepsilon \to 0$ for $\varepsilon^{3/2} \ll b \ll \varepsilon$ shows that in the disturbed region with the characteristic dimensions $\varepsilon^{3/2} \ll \Delta x \sim \Delta y \sim b \ll \varepsilon$ in the first approximation the flow is described by the incompressible Euler equations (Eqs. (8.5) with no viscous terms)

$$\frac{\partial u_3}{\partial x_3} + \frac{\partial v_3}{\partial y_3} = 0, \quad \rho_w \left( u_3 \frac{\partial u_3}{\partial x_3} + v_3 \frac{\partial u_3}{\partial y_3} \right) + \frac{\partial p_3}{\partial x_3} = 0 \tag{8.12}$$

$$\rho_w \left( u_3 \frac{\partial v_3}{\partial x_3} + v_3 \frac{\partial v_3}{\partial y_3} \right) + \frac{\partial p_3}{\partial y_3} = 0, \quad u_3 \frac{\partial h_3}{\partial x_3} + v_3 \frac{\partial h_3}{\partial y_3} = 0$$

Obviously, the same result can be obtained from Eq. (8.9) passing to the limit $Re \to \infty$. In view of the fact that in the undisturbed flat-plate boundary layer the stream function is described by the equation $\psi_3 = A\rho_w y_3^2/2$, from Eq. (8.12) for the enthalpy disturbance $h_3$ we immediately obtain

$$h_3 = h_3(\psi_3) = B\sqrt{\frac{2\psi_3}{A\rho_w}}, \quad \rho_w u_3 = \frac{\partial \psi_3}{\partial y_3}, \quad \rho_w v_3 = -\frac{\partial \psi_3}{\partial x_3} \tag{8.13}$$

For the stream function $\psi_3 = \psi_3(x_3, y_3)$ the following boundary value problem can be formulated

$$\frac{\partial^2 \psi_3}{\partial x_3^2} + \frac{\partial^2 \psi_3}{\partial y_3^2} = A\rho_w, \quad \psi_3 \to A\rho_w \frac{y_3^2}{2}, \quad x_3^2 + y_3^2 \to \infty \tag{8.14}$$

If the characteristic thickness $a$ and length $b$ of the small roughness are of the same order, $\varepsilon^{3/2} \ll a \sim b \ll \varepsilon$ (a version of case 1*b*), then only impermeability conditions must be fulfilled on its surface

$$\psi_3 = 0, \quad y_3 = f(x_3) \tag{8.15}$$

The solution of the boundary value problem (8.14), (8.15) can be found using well-known methods for solving the incompressible Euler equations (e.g., the singularity method or the conformal mapping method). After this outer inviscid problem has been solved, a viscous sublayer 4 near the roughness surface should be considered in order to satisfy the no-slip conditions and that for the enthalpy. For this purpose, in the viscous sublayer 4 the following new independent variables and asymptotic expansions for the flow functions must be introduced as follows:

$$s = bs_4, \quad n = \varepsilon^{3/2} n_4 \tag{8.16}$$

$$u(s,n;\varepsilon) = \frac{b}{\varepsilon} u_4 + \cdots, \quad v(s,n;\varepsilon) = \varepsilon^{1/2} v_4 + \cdots$$

$$p(s,n;\varepsilon) = \frac{1}{\gamma M_\infty^2} + \left(\frac{b}{\varepsilon}\right)^2 p_4 + \cdots$$

$$\rho(s,n;\varepsilon) = \rho_w + \cdots, \quad h(s,n;\varepsilon) = h_w + \frac{b^{1/2}}{\varepsilon^{1/4}} h_4 + \cdots$$

$$\mu(s,n;\varepsilon) = \mu_w + \cdots, \quad \psi(s,n;\varepsilon) = b\varepsilon^{1/2} \psi_4 + \cdots$$

Here, $s$ and $n$ are curvilinear coordinates ($s$ is measured along the small roughness surface and $n$ normal to it) and $u$ and $v$ are the velocity components along the $s$ and $n$ axes. Substituting expansions (8.16) in the Navier–Stokes equations written in the $(s, n)$ coordinates and passing to the limit $\varepsilon \to 0$ for $\varepsilon^{3/2} \ll b \ll \varepsilon$ yields that in the first approximation the flow in the viscous sublayer 4 is governed by the incompressible Prandtl boundary layer equations

$$\frac{\partial u_4}{\partial s_4} + \frac{\partial v_4}{\partial n_4} = 0, \quad \rho_w \left( u_4 \frac{\partial u_4}{\partial s_4} + v_4 \frac{\partial u_4}{\partial n_4} \right) + \frac{\mathrm{d} p_4}{\mathrm{d} s_4} = \mu_w \frac{\partial^2 u_4}{\partial n_4^2} \tag{8.17}$$

$$\frac{\partial p_4}{\partial n_4} = 0, \quad \rho_w \left( u_4 \frac{\partial h_4}{\partial s_4} + v_4 \frac{\partial h_4}{\partial n_4} \right) = \frac{\mu_w}{\sigma} \frac{\partial^2 h_4}{\partial n_4^2}$$

At the small roughness surface the conventional conditions for the velocity components and the enthalpy must be fulfilled

$$u_4 = v_4 = h_4 = 0, \quad n_4 = 0 \tag{8.18}$$

while the outer boundary conditions are obtained by matching with the outer inviscid solution (8.13)–(8.15)

$$u_4 \to u_e(s_4), \quad h_4 \to B\sqrt{\frac{2\psi_4}{A\rho_w}}, \quad n_4 \to \infty \tag{8.19}$$

where $u_e = u_e(s_4)$ is the velocity distribution over the outer edge of the viscous sublayer 4 which is derived from the outer inviscid solution. The pressure gradient $\mathrm{d}p_4/\mathrm{d}s_4$ is conventionally determined from the Bernoulli equation

$$\frac{\mathrm{d} p_4}{\mathrm{d} s_4} = -\rho_w u_e \frac{\mathrm{d} u_e}{\mathrm{d} s_4} \tag{8.20}$$

The distinctive feature of the parabolic problem under consideration is that the integration of Eqs. (8.17) must be started from a stagnation point located at $s_4 = -\infty$.

From the form of expansions (8.16) it follows that in the case in question near the roughness the viscous stress and the heat flux are greater in the order than those in the undisturbed boundary layer near the plate surface

$$\tau \sim \varepsilon \frac{b}{\varepsilon^{3/2}} \gg \varepsilon, \quad q \sim \varepsilon \left( \frac{b}{\varepsilon^{3/2}} \right)^{1/2} \gg \varepsilon, \quad \varepsilon^{3/2} \ll b \ll \varepsilon \tag{8.21}$$

Therefore, ahead of the roughness there must be a transitional flow region in which the viscous stress and the heat flux are of the same order as in the undisturbed flat-plate boundary layer on the plate, increasing to the values corresponding to the viscous sublayer 4 (Neiland, 1969*b*). The flow in the transitional region will be considered below, when solving the boundary value problem (8.38)–(8.42).

Let now the roughness be thin for $a/b \ll 1$, $\varepsilon^{3/2} \ll b \ll \varepsilon$. Then, obviously, the flow in the region with the characteristic dimensions $\varepsilon^{3/2} \ll \Delta x \sim \Delta y \sim b \ll \varepsilon$ is inviscid and the thin roughness induces in it only small disturbances with respect to the flow in the wall region of the undisturbed boundary layer (e.g., $\Delta u \ll u \sim b/\varepsilon \ll 1$). In the Russian literature this region is conventionally designated by the number 2.

To construct the solution in this region, we can again use the variables and asymptotic expansions (8.4) with the subscript 3 replaced by the subscript 2. Substituting these expansions in the Navier–Stokes equations leads, obviously, in the first approximation, as $\varepsilon \to 0$, to the system of Euler equations (8.12), the solution for the enthalpy disturbance (8.13), and the boundary value problem (8.14) for the stream function (with the subscript 2 substituted everywhere for the subscript 3). The solution of the latter problem can be represented in the form of small disturbances with respect to the shear layer, or the wall region of the undisturbed flat-plate boundary layer

$$\psi_2(x_2, y_2) = A\rho_w \left[ \frac{y_2^2}{2} + \left( \frac{a}{b} \right)^2 \psi_{21}(x_2, y_2) + \cdots \right] \tag{8.22}$$

Then for the function $\psi_{21}(x_2, y_2)$ we obtain the following boundary value problem

$$\frac{\partial^2 \psi_{21}}{\partial x_2^2} + \frac{\partial^2 \psi_{21}}{\partial y_2^2} = 0, \quad \psi_{21} \to 0, \quad x_2^2 + y_2^2 \to \infty \tag{8.23}$$

and its analytical solution

$$\psi_{21}(x_2, y_2) = \frac{1}{\pi} \int_{-\infty}^{\infty} \frac{y_2 \psi_{21}(\xi, 0)}{(x_2 - \xi)^2 + y_2^2} \, d\xi \tag{8.24}$$

Substituting expansion (8.22) in the second equation (8.12) rewritten for the subscript 2 and integrating it once with respect to the longitudinal coordinate $x_2$, on condition that all the disturbances decay far ahead of the roughness, as $x_2 \to -\infty$, we obtain

$$\left(\frac{a}{b}\right)^2 A^2 \rho_w \left( y_2 \frac{\partial \psi_{21}}{\partial y_2} - \psi_{21} \right) + p_2 = 0 \tag{8.25}$$

Hence, as $y_2 \to 0$, there immediately follows:

$$p_2(x_2, 0) = \left(\frac{a}{b}\right)^2 A^2 \rho_w \psi_{21}(x_2, 0) \tag{8.26}$$

since $\psi_{21}(x_2, y_2)$ is a harmonic function and for $y_2 = 0$ its derivative $\partial \psi_{21}/\partial y_2$ is a limited function.

Then it is useful to consider separately the flow in a layer with a characteristic thickness of the same order as the roughness thickness $\Delta y \sim a$, in which nonlinear disturbances are induced (e.g., $\Delta u \sim u \sim a/\varepsilon \ll 1$) and which represents now region 3. In this layer the following independent variables and the asymptotic expansions for the flow functions are valid

$$x_3 = x_2 = \frac{x}{b}, \quad y_3 = \frac{y}{a} \tag{8.27}$$

$$u(x, y; \varepsilon) = \frac{a}{\varepsilon} u_3 + \cdots, \quad v(x, y; \varepsilon) = \frac{a^2}{\varepsilon b} v_3 + \cdots$$

$$p(x, y; \varepsilon) = \frac{1}{\gamma M_\infty^2} + \left(\frac{a}{\varepsilon}\right)^2 p_3 + \cdots, \quad \rho(x, y; \varepsilon) = \rho_w + \cdots$$

$$h(x, y; \varepsilon) = h_w + \frac{a}{\varepsilon} h_3 + \cdots, \quad \mu(x, y; \varepsilon) = \mu_w + \cdots, \quad \psi(x, y; \varepsilon) = \frac{a^2}{\varepsilon} \psi_3 + \cdots$$

Requiring that the characteristic dimensions of the small roughness satisfy the inequality $\varepsilon b^{1/3} \ll a \ll b \ll \varepsilon$ and substituting expansions (8.27) in the Navier–Stokes equations for the leading terms of the expansions, as $\varepsilon \to 0$, we obtain

$$\frac{\partial u_3}{\partial x_3} + \frac{\partial v_3}{\partial y_3} = 0, \quad \rho_w \left( u_3 \frac{\partial u_3}{\partial x_3} + v_3 \frac{\partial u_3}{\partial y_3} \right) + \frac{dp_3}{dx_3} = 0 \tag{8.28}$$

$$\frac{\partial p_3}{\partial y_3} = 0, \quad u_3 \frac{\partial h_3}{\partial x_3} + v_3 \frac{\partial h_3}{\partial y_3} = 0$$

These are the so-called inviscid boundary layer equations. The solution for the enthalpy disturbance can again be derived in finite form

$$h_3 = h_3(\psi_3) = B \sqrt{\frac{2\psi_3}{A \rho_w}} \tag{8.29}$$

The dynamic part of problem (8.28) can conventionally be solved in the von Mises variables $(x, \psi)$

$$\rho_w u_3 \frac{\partial u_3}{\partial x_3} + \frac{\mathrm{d}p_3}{\mathrm{d}x_3} = 0, \quad \frac{\partial y_3}{\partial x_3} = \frac{v_3}{u_3}, \quad \frac{\partial y_3}{\partial \psi_3} = \frac{1}{\rho_w u_3}, \quad \frac{\partial p_3}{\partial \psi_3} = 0 \tag{8.30}$$

$$y_3 = f(x_3), \ \ \psi_3 = 0; \quad y_3 \to \sqrt{\frac{2\psi_3}{A\rho_w}}, \ \ p_3(x_3) \to 0, \ \ f(x_3) \to 0, \quad x_3 \to -\infty$$

The solution of problem (8.30) can also be obtained in finite form:

$$y_3 = f(x_3) + \sqrt{\frac{2}{A^2\rho_w}} \left( \sqrt{A\psi_3 - p_3} - \sqrt{-p_3} \right) \tag{8.31}$$

or

$$\psi_3(x_3, y_3) = A\rho_w \frac{y_3^2}{2} + y_3 \left( \sqrt{-2\rho_w p_3} - A\rho_w f \right) + A\rho_w \frac{f^2}{2} - f\sqrt{-2\rho_w p_3} \tag{8.32}$$

Then, from the condition of the matching of the asymptotic expansions for the pressures in region 2 with the characteristic dimensions $\varepsilon^{3/2} \ll \Delta x \sim \Delta y \sim b \ll \varepsilon$ and in layer 3 with the characteristic thickness $\Delta y \sim a$ and in view of Eq. (8.26) we obtain

$$p_3(x_3) = A^2 \rho_w \psi_{21}(x_2, 0) \tag{8.33}$$

Writing solution (8.22) for $y_2 \to 0$, taking Eq. (8.33) into account, and going over to the variables of layer 3, we can derive the outer condition for the boundary value problem (8.30)

$$\psi_3 \to A\rho_w \left( \frac{y_3^2}{2} + \frac{p_3(x_3)}{A^2 \rho_w} \right), \quad y_3 \to \infty \tag{8.34}$$

It can easily be verified that solution (8.32) satisfies this condition only if

$$p_3(x_3) = -\frac{A^2 \rho_w f^2(x_3)}{2} \tag{8.35}$$

and, therefore,

$$\psi_3(x_3, y_3) = A\rho_w \frac{y_3^2 - f^2(x_3)}{2}, \quad \psi_{21}(x_2, 0) = -\frac{f^2(x_3)}{2} = \frac{p_3(x_3)}{A^2 \rho_w} \tag{8.36}$$

After the inviscid problem has been solved, a viscous heat-conducting sublayer 4 with the characteristic thickness $\Delta y \sim \varepsilon^{3/2}(b/a)^{1/2}$ must be considered in order to satisfy the no-slip conditions and the condition for the enthalpy at the small roughness surface; in this layer, the leading viscous terms of the Navier–Stokes equations and the inertial terms must be of the same order. The solution near the small roughness surface is described by the conventional Prandtl boundary layer equations with a given outer pressure distribution. It can easily be verified that in this case, as in the solution of the boundary value problem (8.17)–(8.20), the viscous stress and the heat flux are greater in the order than those in the undisturbed flat-plate boundary layer. Hence follows that ahead of this small roughness there must also be a transitional flow region in which the viscous stress and the heat flux are of the same order as those in the undisturbed boundary layer on the plate surface and increase (Neiland, 1969*b*). Mathematically, this problem coincides with the problem in which viscosity is important throughout the entire layer 3 with the characteristic thickness of the order of the roughness thickness, $\Delta y \sim a$, whereas in region 2 with the characteristic dimensions $\Delta x \sim \Delta y \sim b$ the flow is inviscid. In this case, $a \sim \varepsilon b^{1/3}$ and the outer solution (8.13) (with the subscript 2), (8.22), and (8.24) and the matching conditions (8.33) and (8.34) remain unaltered, though Eqs. (8.28) now involve the terms dependent on viscosity and thermal conductivity.

For this problem we can introduce the following affine transformation of variables

$$x_3 = a_1\bar{x}, \quad y_3 = b_1\bar{y}, \quad u_3 = Aa_1\bar{u}, \quad v_3 = \frac{Aa_1^2}{b_1}\bar{v} \tag{8.37}$$

$$p_3 = \rho_w A^2 a_1^2 \bar{p}, \quad h_3 = Ba_1\bar{h}, \quad \psi_3 = \rho_w A a_1^2 \bar{\psi}$$

where $a_1 \sim 1$ and $b_1 \sim 1$ are the thickness and length of the small roughness ($a = \varepsilon b^{1/3} a_1$, $\Delta x = bb_1$). Then in the new variables the flow over the roughness is described in the following incompressible boundary layer equations:

$$\frac{\partial \bar{u}}{\partial \bar{x}} + \frac{\partial \bar{v}}{\partial \bar{y}} = 0, \quad \bar{u}\frac{\partial \bar{u}}{\partial \bar{x}} + \bar{v}\frac{\partial \bar{u}}{\partial \bar{y}} + \frac{d\bar{p}}{d\bar{x}} = \Pi \frac{\partial^2 \bar{u}}{\partial \bar{y}^2} \tag{8.38}$$

$$\frac{\partial \bar{p}}{\partial \bar{y}} = 0, \quad \bar{u}\frac{\partial \bar{h}}{\partial \bar{x}} + \bar{v}\frac{\partial \bar{h}}{\partial \bar{y}} = \frac{\Pi}{\sigma}\frac{\partial^2 \bar{h}}{\partial \bar{y}^2}, \quad \Pi = \frac{\mu_w b_1}{\rho_w A a_1^3} \sim 1$$

The following conditions must be fulfilled at the small roughness surface

$$\bar{u} = \bar{v} = \bar{h} = 0, \quad \bar{y} = f(\bar{x}) \tag{8.39}$$

where the length and thickness of the normalized surface $\bar{y} = f(\bar{x})$ are equal to unity, the outer boundary conditions are derived by matching with the outer inviscid solution (8.13) (with the subscript 2), (8.22), and (8.24)

$$\frac{\partial \bar{u}}{\partial \bar{y}} \to 1, \quad \frac{\partial \bar{h}}{\partial \bar{y}} \to 1, \quad \bar{y} \to \infty \tag{8.40}$$

the pressure disturbance is determined from condition (8.34)

$$\bar{\psi} \to \frac{\bar{y}^2}{2} + \bar{p}(\bar{x}), \quad \bar{y} \to \infty \tag{8.41}$$

the initial conditions are as follows:

$$\overline{u}, \overline{h} \to \overline{y}, \quad \overline{p}(\overline{x}), f(\overline{x}) \to 0 \;\; \overline{x} \to -\infty \tag{8.42}$$

and the similarity parameter $\Pi$ corresponds to the inverse local Reynolds number.

The boundary value problem (8.38)–(8.42) is more informative than the inviscid problem (8.28) since it incorporates some additional terms. For $a \gg \varepsilon b^{1/3}$ the results for problem (8.28) can be obtained by passage to the limit $\Pi \to 0$ in the solution (8.38)–(8.42). Another passage to limit, $\Pi \to \infty$, corresponds to the case in which the characteristic thickness of the viscous heat-conducting layer is in the order greater than the roughness thickness, $a \ll \varepsilon b^{1/3}$. Obviously, in this case the solution of the boundary value problem (8.38)–(8.42) can be linearized with respect to the shear flow in the wall region of the undisturbed boundary layer. Clearly, this roughness can induce only small disturbances of the viscous stress and the heat flux. The lower limit for the roughness thickness $a$ can be derived from the condition that a large pressure gradient $\partial p/\partial x \gg 1$, greater in the order than that on the main part of the body, must be induced in the flow around it. For $\partial p/\partial x \leq 1$ classical Prandtl's boundary layer theory can be applied and there is no need of considering the fine structure of flow around small roughnesses. Therefore, the condition $\partial p/\partial x \gg 1$ leads to the following relation:

$$\frac{\partial p}{\partial x} \sim \left(\frac{\varepsilon b^{1/3}}{\varepsilon}\right)^2 \frac{1}{b}\frac{a}{\varepsilon b^{1/3}} \sim \frac{a}{\varepsilon b^{2/3}} \gg 1, \quad \varepsilon b^{1/3} \gg a \gg \varepsilon b^{2/3} \tag{8.43}$$

In solving the boundary value problem (8.38)–(8.42) the pressure distribution is determined in the process of solution, so that we deal with a typical problem of the interaction between an inviscid flow and a viscous sublayer. However, as distinct from the flows with free interaction (Neiland, 1969*a*), here disturbances are not transferred upstream. Mathematically, this is due to the fact that in free interaction flows the pressure disturbance is proportional to the variation of the boundary layer displacement thickness, $\Delta p \sim \mathrm{d}\delta^*/\mathrm{d}x$, and the equation for the longitudinal momentum contains, though under sign of the integral over the vertical coordinate, the second derivative with respect to the longitudinal coordinate $x$. As for system (8.38)–(8.42), the condition (8.41) for determining the pressure does not increase the order of the derivatives with respect to the longitudinal coordinate $x$ entering in the system.

Obviously, in the flow over a bump the pressure disturbance is negative, $\Delta p < 0$, since the small roughness interacts actually only with the subsonic wall region of the boundary layer and the narrowing of streamtubes results in pressure decreasing (it can be directly seen for the inviscid solution (8.35)). Flow separation can occur only at the bump rear, where the pressure increases. For a family of affinely similar roughnesses it is the parameter $\Pi$ that determines whether flow separation is involved or not in solution (8.38)–(8.42). Undoubtedly, when $\Pi \to 0$ flow separation takes place always, since a thin viscous sublayer at the bottom of the inviscid layer (8.28) cannot arrive at the stagnation point of the flow corresponding to $x \to \infty$ without separation (in this sublayer the pressure disturbance varies in the leading order). When $\Pi \to \infty$, the small roughness is in a separationless flow (in this flow the pressure disturbance varies in the next order of smallness). Obviously, in the flow over a depression the pressure disturbance is positive, $\Delta p > 0$, and flow separation occurs immediately.

In Sections 8.1.2 and 8.1.3 we considered the flows over small roughnesses when there is no interaction between the flow in the disturbed regions and the outer flow. Therefore, these regimes of the flow over small roughness are realized in both subsonic and hypersonic outer flows (they differ only by the values of the temperature factor).

### *8.1.4 Flow over roughnesses with a characteristic length equal in the order to the boundary layer thickness*

Flows over roughnesses whose vertical and longitudinal dimensions are equal in the order to the boundary layer thickness, $a \sim b \sim \delta \sim \varepsilon$, (case $2b$ according to the classification of roughnesses) pertain to the type considered generally in the paper of Neiland and Sychev (1966) and in Chapter 3 of this book. Therefore, below we will study only the flow over thin roughnesses with $\varepsilon^2 \ll a \ll b \sim \delta \sim \varepsilon$ (case $2a$). An analysis of these flows can conveniently be drawn on the basis of the Navier–Stokes equations written in the von Mises variables $x$, $\psi$.

As in the flow over "short" roughnesses (Section 8.1.2), as $\varepsilon \to 0$, in the region with the characteristic dimensions $\Delta x \sim \Delta y \sim 1$ the solution of the Navier–Stokes equations is represented in the first approximation by the undisturbed oncoming flow. Then, in accordance with the Prandtl theory, a region with the characteristic dimensions $\Delta x \sim 1$ and $\Delta y \sim \varepsilon$, that is, the boundary layer, should be considered. The well-known Blasius solution for the flat-plate boundary layer is invalid, at least, in the vicinity of a small roughness, that is, in a region with the characteristic dimensions $\Delta x \sim \Delta y \sim b \sim \delta \sim \varepsilon$.

Obviously, a slight distortion of the plate surface (the characteristic slope of the small roughness is in the order equal to $a/b \ll 1$) results in a small pressure disturbance $\Delta p \ll 1$ and a variation of the boundary layer displacement thickness $\Delta\delta^*$. Following the studies of Neiland (1969*a*, *b*), it is convenient to consider first the main part of the boundary layer with the characteristic thickness $\Delta y \sim \delta \sim \varepsilon$ (in accordance with the notation adopted in the Russian literature, this is region 2), in which the longitudinal velocity is equal in the order to $u \sim 1$, while its disturbance is naturally small, $\Delta u \ll 1$. From the equations of longitudinal momentum conservation, continuity, and state it can be obtained that along a fixed streamline we have

$$\Delta p \sim \Delta u^{(2)} \sim \Delta\rho^{(2)} \sim \frac{\Delta\delta^{*(2)}}{\varepsilon} \tag{8.44}$$

Here, the superscript 2 (in parentheses) denotes disturbances of the flow functions in the main part of the boundary layer (region 2), while $\Delta\delta^{*(2)}$ is the variation of its thickness. Then a flow region near the small roughness surface (region 3) must be considered; here, generally, viscous stresses are important, while the longitudinal velocity, owing to the no-slip conditions, tends to zero as the surface is approached, and is equal in the order to its own disturbance, $\Delta u^{(3)} \sim u^{(3)}$. Then from the equation of longitudinal momentum conservation we can obtain that in the wall region 3

$$u^{(3)} \sim \Delta u^{(3)} \sim \Delta p^{1/2} \sim \frac{\Delta\delta^{*(3)}}{\varepsilon} \tag{8.45}$$

Here, it was also used that in the wall region of the undisturbed boundary layer the flow functions vary proportional to the distance from the plate surface (cf. estimates (8.2)). Comparing relations (8.44) and (8.45) we can note that the variation $\Delta\delta^{*(3)}$ of the wall region 3 thickness is greater in the order than the variation of the thickness of the main part of the boundary layer, that is, region 2

$$\varepsilon\Delta p \sim \Delta\delta^{*(2)} \ll \Delta\delta^{*(3)} \sim \varepsilon\Delta p^{1/2} \tag{8.46}$$

Therefore, it can be assumed that here, as in the works of Neiland (1969*a*, *b*) the thickness of the boundary layer as a whole $\Delta\delta^*$ varies mainly due to the variation of the thickness of the wall region 3, in which $u^{(3)} \sim \Delta u^{(3)} \sim \Delta p^{1/2} \ll 1$, while in the first approximation the main part of the boundary layer, that is, region 2 in which the longitudinal velocity is large, $u^{(2)} \sim 1$ and the velocity disturbance is small, $\Delta u^{(2)} \sim \Delta p \ll 1$, is equidistantly displaced by the value $\Delta\delta^{*(3)}$, as $\varepsilon \to 0$ and $\varepsilon^2 \ll a \ll \varepsilon$. However, in accordance with the linear theory of supersonic flows (see, e.g., Hayes and Probstein, 1959) the following relation must hold:

$$\Delta p \sim \frac{\Delta\delta^*}{b} \sim \frac{\varepsilon\Delta p^{1/2}}{\varepsilon} \sim \Delta p^{1/2} \tag{8.47}$$

which is possible only if $\Delta p \sim 1$. Therefore, for $\Delta p \ll 1$ the variation of the wall region 3 thickness must be offset in the first approximation by the small roughness thickness

$$\Delta\delta^{*(3)} \sim \varepsilon\Delta p^{1/2} \sim a, \quad \Delta p \sim \left(\frac{a}{\varepsilon}\right)^2 \tag{8.48}$$

while the outer edge of the boundary layer is displaced only by the value $\Delta\delta^* \sim \Delta\delta^{*(2)} \sim a^2/\varepsilon$. On the basis of these estimates, in the main part of the boundary layer (region 2) the following independent variables and asymptotic expansions for the flow functions should be introduced

$$x = \varepsilon x_2, \quad \psi = \varepsilon\psi_2 \tag{8.49}$$

$$u(x,\psi;\varepsilon) = u_{20}(\psi_2) + \left(\frac{a}{\varepsilon}\right)^2 u_2 + \cdots, \quad v(x,\psi;\varepsilon) = \left(\frac{a}{\varepsilon}\right)^2 v_2 + \cdots$$

$$p(x,\psi;\varepsilon) = \frac{1}{\gamma M_\infty^2} + \left(\frac{a}{\varepsilon}\right)^2 p_2 + \cdots, \quad \rho(x,\psi;\varepsilon) = \rho_{20}(\psi_2) + \left(\frac{a}{\varepsilon}\right)^2 \rho_2 + \cdots$$

$$h(x,\psi;\varepsilon) = h_{20}(\psi_2) + \left(\frac{a}{\varepsilon}\right)^2 h_2 + \cdots, \quad y(x,\psi;\varepsilon) = \varepsilon y_{20}(\psi_2) + \frac{a^2}{\varepsilon} y_2 + \cdots$$

Here, $u_{20}(\psi_2)$, $\rho_{20}(\psi_2)$, $h_{20}(\psi_2)$, and $y_{20}(\psi_2)$ are known functions taken from the solution for the undisturbed flat-plate boundary layer. Substituting expansions (8.49) in the

Navier–Stokes equations and passing to the limit $\varepsilon \to 0$, $\varepsilon^2 \ll a \ll \varepsilon$, for the leading terms of the expansions we obtain

$$\frac{\partial y_2}{\partial x_2} = \frac{v_2}{u_{20}}, \quad \frac{\partial y_{20}}{\partial \psi_2} = \frac{1}{\rho_{20}u_{20}}, \quad \frac{\partial y_2}{\partial \psi_2} = -\frac{1}{\rho_{20}u_{20}}\left(\frac{u_2}{u_{20}} + \frac{\rho_2}{\rho_{20}}\right) \tag{8.50}$$

$$\rho_2 = M_\infty^2 \rho_{20} p_2, \quad \frac{\partial v_2}{\partial x_2} + \frac{\partial p_2}{\partial \psi_2} = 0, \quad \rho_{20}u_{20}\frac{\partial u_2}{\partial x_2} + \frac{\partial p_2}{\partial x_2} = 0, \quad \rho_{20}\frac{\partial h_2}{\partial x_2} - \frac{\partial p_2}{\partial x_2} = 0$$

In accordance with the above estimates, in the wall layer 3, in which $u^{(3)} \sim \Delta u^{(3)} \ll 1$, the new independent variables and the asymptotic expansions of the flow functions take the form:

$$x_3 = x_2 = \frac{x}{\varepsilon}, \quad \psi_3 = \frac{\varepsilon}{a^2}\psi \tag{8.51}$$

$$u(x, \psi; \varepsilon) = \frac{a}{\varepsilon}u_3 + \cdots, \quad v(x, \psi; \varepsilon) = \left(\frac{a}{\varepsilon}\right)^2 v_3 + \cdots$$

$$p(x, \psi; \varepsilon) = \frac{1}{\gamma M_\infty^2} + \left(\frac{a}{\varepsilon}\right)^2 p_3 + \cdots, \quad \rho(x, \psi; \varepsilon) = \rho_w + \cdots$$

$$h(x, \psi; \varepsilon) = h_w + \frac{a}{\varepsilon}h_3 + \cdots, \quad y(x, \psi; \varepsilon) = a y_3 + \cdots, \quad \mu(x, \psi; \varepsilon) = \mu_w + \cdots$$

After expansions (8.51) have been substituted in the Navier–Stokes equations and the passage to the limit $\varepsilon \to 0$, $\varepsilon^2 \ll a \ll \varepsilon$ has been performed, the following incompressible boundary layer equations are obtained for the leading terms of the expansions

$$\rho_w u_3 \frac{\partial u_3}{\partial x_3} + \frac{\mathrm{d}p_3}{\mathrm{d}x_3} = \frac{\varepsilon^4}{a^3}\rho_w^2 \mu_w u_3 \frac{\partial}{\partial \psi_3}\left(u_3 \frac{\partial u_3}{\partial \psi_3}\right), \quad \frac{\partial p_3}{\partial \psi_3} = 0 \tag{8.52}$$

$$\frac{\partial y_3}{\partial x_3} = \frac{v_3}{u_3}, \quad \frac{\partial y_3}{\partial \psi_3} = \frac{1}{\rho_w u_3}, \quad \frac{\partial h_3}{\partial x_3} = \frac{\varepsilon^4}{a^3}\frac{\rho_w \mu_w}{\sigma}\frac{\partial}{\partial \psi_3}\left(u_3 \frac{\partial h_3}{\partial \psi_3}\right)$$

In the most general case $a \sim \varepsilon^{3/4}$; then the wall layer 3 is viscous. Particular cases will be obtained below from the general case by passing to the corresponding limits. At the small roughness surface $y_3 = f(x_3)$ for $\psi_3 = 0$ the solution of system (8.52) must satisfy the following boundary conditions:

$$u_3(x_3, 0) = v_3(x_3, 0) = h_3(x_3, 0) = 0 \tag{8.53}$$

The outer boundary conditions are obtained by matching the solutions in the main part of the boundary layer (region 2) and the wall layer 3. For this purpose, we pass to the limit $\varepsilon \to 0$, $\varepsilon^2 \ll a \ll \varepsilon$ at a fixed value $x_2 = x_3$ and for a fixed $\psi_3$, which corresponds to expansion of the flow functions in region 2 as $\psi_2 \to 0$. For the pressure disturbance the matching yields immediately

$$p_2(x_2, 0) = p_3(x_3) \tag{8.54}$$

In view of the fact that all disturbances vanish as $x_2 \to -\infty$, from the penultimate equation of system (8.50) it can easily be obtained that

$$\rho_{20}u_{20}u_2 + p_2 = 0 \tag{8.55}$$

Since in the undisturbed boundary layer near the plate surface relations (8.2) are valid, we have

$$u_{20}(\psi_2) \approx \sqrt{\frac{2A\psi_2}{\rho_w}} + \cdots, \quad \psi_2 \to 0 \tag{8.56}$$

Then we can write the expansion for the longitudinal velocity $u$ in the main part of the boundary layer (region 2) in the variables of the wall layer 3

$$u \approx \frac{a}{\varepsilon}\left(\sqrt{\frac{2A\psi_3}{\rho_w}} - \frac{p_3(x_3)}{\sqrt{2A\rho_w\psi_3}}\right) + \cdots \tag{8.57}$$

The same formula describes the solution for $u_3(x_3, \psi_3)$ when passing to the limit $\varepsilon \to 0$, $\varepsilon^2 \ll a \ll \varepsilon$ at a fixed value of $\psi_2$, that is, as $\psi_3 \to \infty$

$$u_3 \to \sqrt{\frac{2A\psi_3}{\rho_w}}, \quad \psi_3 \to \infty \tag{8.58}$$

The outer boundary condition for the enthalpy disturbance $h_3$ is obtained in the same fashion

$$h_3 \to B\sqrt{\frac{2\psi_3}{A\rho_w}}, \quad \psi_3 \to \infty \tag{8.59}$$

From Eqs. (8.50) and (8.55) it can be obtained that

$$\frac{\partial y_2}{\partial \psi_2} = \frac{p_2}{\rho_{20}}\left(\frac{1}{\rho_{20}u_{20}^3} - \frac{M_\infty^2}{u_{20}}\right) \tag{8.60}$$

Now in the main part of the boundary layer (region 2), as $\psi_2 \to 0$, for the function $y(x, \psi)$ we can write

$$y \approx \varepsilon\sqrt{\frac{2\psi_2}{A\rho_w}} + \frac{a^2}{\varepsilon}\left[\int_{\psi_2}^{\psi_2^*} \frac{p_3(x_3)}{\sqrt{2A\rho_w}}\left(\frac{1}{2A\psi^{3/2}} - \frac{M_\infty^2}{\psi^{1/2}}\right) d\psi\right] + \cdots \tag{8.61}$$

where $\psi_2^*$ is a certain small but fixed value of $\psi_2$. For the sake of matching, only the term having a singularity, as $\psi_2 \to 0$, should be retained in the expression in the brackets. The integral in the brackets can be taken in the quadrature form, so that solution (8.61) can be expressed in terms of the variables of the wall layer 3 as follows:

$$y \sim a\left(\sqrt{\frac{2\psi_3}{A\rho_w}} - \frac{p_3(x_3)}{A\sqrt{2A\rho_w\psi_3}}\right) + \cdots \tag{8.62}$$

Hence follows that

$$y_3 \to \sqrt{\frac{2\psi_3}{A\rho_w}} - \frac{p_3}{A\sqrt{2A\rho_w\psi_3}} + \cdots, \quad \psi_3 \to \infty \tag{8.63}$$

or

$$\psi_3 \to A\rho_w\left(\frac{y_3^2}{2} + \frac{p_3(x_3)}{A^2\rho_w}\right) \quad y_3 \to \infty \tag{8.64}$$

Supplementing the system of equations (8.52), boundary conditions (8.53), (8.58), and (8.59), and condition (8.64) for determining the pressure by the initial conditions

$$u_3 \to \sqrt{\frac{2A\psi_3}{\rho_w}}, \quad h_3 \to B\sqrt{\frac{2\psi_3}{A\rho_w}}, \quad x_3 \to -\infty \tag{8.65}$$

and introducing the affine transformation of variables (8.37) we can note that the boundary value problem thus obtained is almost the same as the previously considered problem (8.38)–(8.42) with the similarity parameter $(\varepsilon^4/a^3)\Pi$ substituted for the parameter $\Pi$. In the most general case of the problem under consideration $a \sim \varepsilon^{4/3}$, so that we arrive again at the boundary value problem (8.38)–(8.42); then $a \sim \varepsilon b^{1/3}$.

It can easily be seen that the case $\varepsilon^{4/3} \ll a \ll \varepsilon$, in which layer 3, where $u^{(3)} \sim \Delta u^{(3)} \ll 1$, is inviscid, is analogous to the previously considered case (8.28) and can again be obtained as a limit of the solution as the similarity parameter $\Pi \to 0$. If $a \ll \varepsilon^{4/3}$ and viscosity is important even above the nonlinear disturbance layer, then the corresponding solution can be obtained as a limit when the similarity parameter $\Pi \to \infty$. The lower limit for the vertical roughness dimension $a$ is again determined from the condition $\partial p/\partial x \gg 1$; then $a \gg \varepsilon^{5/3}$ or $a \gg \varepsilon b^{2/3}$.

The fact that in the first approximation the roughness thickness $a$ is offset by the variation of the wall layer 3 thickness, $\Delta\delta^{*(3)} \sim \varepsilon \Delta p^{1/2} \approx -a$, while the outer edge of the boundary layer is displaced only by the value of the variation of the main part of the boundary layer (region 2), $\Delta\delta^* \sim \Delta\delta^{*(2)} \sim \varepsilon\Delta p \ll \Delta\delta^{*(3)} \sim a$ (i.e., it remains fixed in the leading order and the outer inviscid flow simply does not "feel" the presence of the roughness) makes it possible to designate this new type of the interaction flow (8.38–8.42) as compensation flow (Bogolepov and Neiland, 1971).

### *8.1.5 Flow over "long" roughnesses whose length is greater than the boundary layer thickness*

We will consider case 3, according to the classification of roughnesses, in which the characteristic length is greater in the order than the boundary layer thickness, $1 \gg b \gg \delta \sim \varepsilon \geq a$. A slight distortion of the plate surface, $a/b \ll 1$, results in a small pressure variation, $\Delta p \ll 1$, which, in turn, leads to the variation of the boundary layer displacement thickness $\Delta\delta^*$. In accordance with the linear theory of supersonic flows, the variations of the boundary layer thickness and the small roughness thickness are related with the pressure disturbance by the following equation

$$\Delta p \sim \frac{\Delta\delta^*}{b} + \frac{a}{b} \tag{8.66}$$

Following the works of Neiland (1968; 1969*a*) and repeating estimates (8.44) and (8.45) we can again obtain that the thickness of the boundary layer as a whole varies chiefly due to the variation of the thickness of the wall layer 3 in which the longitudinal velocity disturbance is of the same order as the velocity itself, $\Delta u^{(3)} \sim u^{(3)} \ll 1$, and $\Delta\delta^{*(3)} \sim \varepsilon\Delta p^{1/2}$. Since in Eq. (8.66) the last term which allows for the roughness effect must be always retained, it can be rewritten in the form:

$$\Delta p\frac{b}{a} \sim \Delta p^{1/2}\frac{\varepsilon}{a} + 1 \tag{8.67}$$

In the most general case, all terms in Eq. (8.67) must be of the same order

$$\Delta p\frac{b}{a} \sim \Delta p^{1/2}\frac{\varepsilon}{a} \sim 1 \quad \text{or} \quad \Delta p \sim \frac{a}{b}, \quad ab \sim \varepsilon^2 \tag{8.68}$$

Moreover, the thicknesses of the small roughness and the layer 3, in which $u^{(3)} \sim \Delta u^{(3)} \ll 1$, are of the same order, $\Delta\delta^{*(3)} \sim a$, and the pressure disturbance is produced by the joint action of the roughness and the wall layer on the outer supersonic flow. In the most general case layer 3 is viscous; then $a \sim \varepsilon^{5/4}$ and $b \sim \varepsilon^{3/4}$. Similar flow regimes, namely, free interaction flows, were previously studied in the works of Neiland (1968; 1969*a*) and Stewartson and Williams (1969). If viscosity is important above this layer, then it is necessary to consider a thin roughness at the viscous layer bottom; in this case the general problem admits linearization of the solution with respect to the main shear flow, that is, the wall region of the undisturbed flat-plate boundary layer. In this case $\varepsilon^{3/2} \ll a \ll \varepsilon^{5/4}$ and $b \sim \varepsilon^{3/4}$ (the lower limit for the roughness thickness $a$ is again determined from the condition $\partial p/\partial x \gg 1$).

If the pressure disturbance is produced only due to the interaction between the roughness and the outer supersonic flow, we have

$$\Delta p\frac{b}{a} \sim 1, \quad \Delta p^{1/2}\frac{\varepsilon}{a} \ll 1 \quad \text{or} \quad \Delta p \sim \frac{a}{b}, \quad ab \gg \varepsilon^2 \tag{8.69}$$

In the case in which

$$\Delta p \frac{b}{a} \ll 1, \quad \Delta p^{1/2} \frac{\varepsilon}{a} \sim 1 \quad \text{or} \quad \Delta p \sim \left(\frac{a}{\varepsilon}\right)^2, \quad ab \ll \varepsilon^2 \tag{8.70}$$

the thickness of layer 3, in which $u^{(3)} \sim \Delta u^{(3)} \ll 1$, is equal in the order to the small roughness thickness, $\Delta\delta^{*(3)} \sim a$, but they must have opposite signs and compensate one another in the first approximation. In this case the outer edge of the boundary layer is displaced by the value equal to the thickness of the main part of the boundary layer, or region 2 with $u^{(2)} \sim 1$ and $\Delta u^{(2)} \sim \Delta p \ll 1$; then we have $\Delta\delta^* \sim \Delta\delta^{(2)} \sim \varepsilon\Delta p \sim a^2/\varepsilon$, the pressure disturbance is produced due to the interaction between the subsonic wall region of the undisturbed flat-plate boundary layer and the small roughness, and there is no interaction with the outer supersonic flow. Thus, in this case we also arrive at the "compensation" regime of flow over roughnesses (8.38)–(8.42).

From estimates (8.68)–(8.70) it follows that, if near the small roughness surface there is a layer in which $u \sim \Delta u \ll 1$, then the characteristic cross-sectional area of the "long" roughness $ab$ determines the pattern of the interaction between the outer supersonic flow and the flow in the vicinity of the small roughness. Obviously, in the flow over a bump, we have a pressure rise $\Delta p > 0$ on its forward side in the case $ab \gg \varepsilon^2$, whereas for $ab \ll \varepsilon^2$ the bump produces flow expansion; the case $ab \sim \varepsilon^2$ is transitional.

Estimates (8.68)–(8.70) for $a \ll \varepsilon$ describe different versions of case 3$a$ according to the classification of roughnesses. For case 3$b$ estimates (8.69) for $a \sim \varepsilon$ are valid.

For the flow regime in which the pressure disturbance is produced only by a small roughness, from estimates (8.69) and the linear theory of supersonic flows it follows that in the outer supersonic flow we must consider a disturbed region (in the Russian literature it is conventionally designated by the number 1) with the characteristic dimensions $\Delta x \sim \Delta\psi \sim b$ in which the following new independent variables and asymptotic representations of the flow functions are valid:

$$x = bx_1, \quad \psi = b\psi_1 \tag{8.71}$$

$$u(x, \psi; \varepsilon) = 1 + \frac{a}{b}u_1 + \cdots, \qquad v(x, \psi; \varepsilon) = \frac{a}{b}v_1 + \cdots$$

$$p(x, \psi; \varepsilon) = \frac{1}{\gamma M_\infty^2} + \frac{a}{b}p_1 + \cdots, \qquad \rho(x, \psi; \varepsilon) = 1 + \frac{a}{b}\rho_1 + \cdots$$

$$h(x, \psi; \varepsilon) = \frac{1}{(1-\gamma)M_\infty^2} + \frac{a}{b}h_1 + \cdots, \qquad y(x, \psi; \varepsilon) = b\psi_1 + ay_1 + \cdots$$

Substituting expansions (8.71) in the Navier–Stokes equations and in the equation of state and passing to the limit $\varepsilon \to 0$ for $ab \gg \varepsilon^2$ and $a \gg b^2$ (the latter inequality follows from the condition $\partial p/\partial x \gg 1$), it can be obtained that the flow in the disturbed region of the supersonic flow (region 1) is described in the first approximation by the following equations:

$$\frac{\partial u_1}{\partial x_1}+\frac{\partial p_1}{\partial x_1}=0, \quad \frac{\partial y_1}{\partial x_1}=v_1, \quad \frac{\partial y_1}{\partial \psi_1}=-\rho_1-u_1 \tag{8.72}$$

$$h_1=\frac{\gamma}{\gamma-1}p_1-\frac{1}{(\gamma-1)M_\infty^2\rho_1}, \quad \frac{\partial v_1}{\partial x_1}+\frac{\partial p_1}{\partial \psi_1}=0, \quad \frac{\partial h_1}{\partial x_1}-\frac{\partial p_1}{\partial x_1}=0$$

This system of equations can be reduced to wave equations of the form:

$$\frac{\partial^2 p_1}{\partial \psi_1^2}-(M_\infty^2-1)\frac{\partial^2 p_1}{\partial x_1^2}=0, \quad \frac{\partial^2 v_1}{\partial \psi_1^2}-(M_\infty^2-1)\frac{\partial^2 v_1}{\partial x_1^2}=0 \tag{8.73}$$

The solution of Eqs. (8.73) is well known; on condition of the decay of all disturbances far ahead of the roughness, $x_1 \to -\infty$, for $\psi_1=0$ it gives

$$\sqrt{M_\infty^2-1}\,p_1(x_1,0)=v_1(x_1,0) \tag{8.74}$$

In the main region of the boundary layer (region 2) it is necessary to introduce new independent variables and asymptotic expansions for the flow functions

$$x_2=x_1=\frac{x}{b}, \quad \psi_2=\frac{\psi}{\varepsilon} \tag{8.75}$$

$$u(x,\psi;\varepsilon)=u_{20}(\psi_2)+\frac{a}{b}u_2+\cdots, \quad v(x,\psi;\varepsilon)=\frac{a}{b}v_2+\cdots$$

$$p(x,\psi;\varepsilon)=\frac{1}{\gamma M_\infty^2}+\frac{a}{b}p_2+\cdots, \quad \rho(x,\psi;\varepsilon)=\rho_{20}(\psi_2)+\frac{a}{b}\rho_2+\cdots$$

$$h(x,\psi;\varepsilon)=h_{20}(\psi_2)+\frac{a}{b}h_2+\cdots, \quad y(x,\psi;\varepsilon)=\varepsilon y_{20}(\psi_2)+af(x_2)+\cdots$$

where $y_2=f(x_2)$ is the small roughness shape, while $u_{20}(\psi_2)$, $\rho_{20}(\psi_2)$, $h_{20}(\psi_2)$, and $y_{20}(\psi_2)$ are again some known functions taken from the solution for the undisturbed flat-plate boundary layer. Substituting expansions (8.75) in the Navier–Stokes equations and passing to the limit $\varepsilon \to 0$, $\varepsilon^{4/3} \ll a \le \varepsilon$, and $\varepsilon^2/a \ll b \ll a^{1/2}$, for the leading terms of the expansions we obtain

$$\frac{\mathrm{d}f(x_2)}{\mathrm{d}x_2}=\frac{v_2}{u_{20}}, \quad \frac{\mathrm{d}y_{20}}{\mathrm{d}\psi_2}=\frac{1}{\rho_{20}u_{20}}, \quad \frac{\partial p_2}{\partial \psi_2}=0 \tag{8.76}$$

$$\rho_{20}u_{20}\frac{\partial u_2}{\partial x_2}+\frac{\mathrm{d}p_2}{\mathrm{d}x_2}=0, \quad \rho_{20}\frac{\partial h_2}{\partial x_2}-\frac{\partial p_2}{\partial x_2}=0$$

Hence immediately follows that, as $\psi_2 \to \infty$

$$v_2(x_2, \infty) = \frac{\mathrm{d}f(x_2)}{\mathrm{d}x_2} \tag{8.77}$$

In the wall layer 3, where $u^{(3)} \sim \Delta u^{(3)} \ll 1$, the following independent variables and asymptotic representations for the flow functions can be introduced

$$x_3 = x_2 = x_1 = \frac{x}{b}, \quad \psi_3 = \frac{b}{a\varepsilon}\psi \tag{8.78}$$

$$u(x, \psi; \varepsilon) = \left(\frac{a}{b}\right)^{1/2} u_3 + \cdots, \quad v(x, \psi; \varepsilon) = \left(\frac{a}{b}\right)^{3/2} v_3 + \cdots$$

$$p(x, \psi; \varepsilon) = \frac{1}{\gamma M_\infty^2} + \frac{a}{b} p_3 + \cdots, \quad \rho(x, \psi; \varepsilon) = \rho_w + \cdots$$

$$h(x, \psi; \varepsilon) = h_w + \left(\frac{a}{b}\right)^{1/2} h_3 + \cdots, \quad y(x, \psi; \varepsilon) = af(x_3) + \varepsilon \left(\frac{a}{b}\right)^{1/2} y_3 + \cdots$$

After expansions (8.78) have been substituted in the Navier–Stokes equations for $a, b, \varepsilon \to 0$ we arrive at the following system of equations for the leading terms of the expansions

$$\rho_w u_3 \frac{\partial u_3}{\partial x_3} + \frac{\mathrm{d}p_3}{\mathrm{d}x_3} = \frac{b^{5/2}}{a^{3/2}} \rho_w^2 \mu_w u_3 \frac{\partial}{\partial \psi_3}\left(u_3 \frac{\partial u_3}{\partial \psi_3}\right), \quad \frac{\partial p_3}{\partial \psi_3} = 0 \tag{8.79}$$

$$\frac{\mathrm{d}f}{\mathrm{d}x_3} = \frac{v_3}{u_3}, \quad \frac{\partial y_3}{\partial \psi_3} = \frac{1}{\rho_w u_3}, \quad \frac{\partial h_3}{\partial x_3} = \frac{b^{5/2}}{a^{3/2}} \frac{\rho_w \mu_w}{\sigma} \frac{\partial}{\partial \psi_3}\left(u_3 \frac{\partial h_3}{\partial \psi_3}\right)$$

In the most general case, in which $b \sim a^{3/5}$ and $\varepsilon^{5/4} \ll a \leq \varepsilon$, layer 3 is viscous and Eqs. (8.79) are the conventional Prandtl equations for an incompressible boundary layer; obviously that the boundary and initial conditions coincide with Eqs. (8.53), (8.58), (8.59), and (8.65). Performing the usual solution matching procedure for different flow regions we can obtain the equation for determining the pressure disturbance

$$p_3 = p_2 = p_1 = \frac{\mathrm{d}f(x_3)}{\mathrm{d}x_3} \frac{1}{\sqrt{M_\infty^2 - 1}} \tag{8.80}$$

Therefore, the problem of the flow over a small roughness with $b \sim a^{3/5}$ reduces to the well-known problem of the integration of the Prandtl equations for an incompressible boundary layer with a given pressure distribution. After the affine transformation of the variables, the problem involves a similarity parameter of the order $b^{5/2}/a^{3/2}$, which in the general case, in which the nonlinear disturbance layer 3 is viscous and $a \sim b^{5/3}$, is of the order of unity. Solutions for the flows in which the viscous layer is thinner in the order than the nonlinear disturbance layer ($b^{5/3} \ll a \le \varepsilon$) or viscosity is important in this layer and above it ($b^2 \ll a \ll b^{5/3}$) can again be obtained from the general case for small or large values of this parameter, respectively.

In considering case (8.70) we can completely repeat the reasoning of the previous Section 8.1.4 with only slight modifications. In the main part of the boundary layer (region 2) the following independent variables and the asymptotic representations of the flow functions are valid:

$$x = bx_2, \quad \psi = \varepsilon\psi_2 \tag{8.81}$$

$$u(x, \psi; \varepsilon) = u_{20}(\psi_2) + \left(\frac{a}{\varepsilon}\right)^2 u_2 + \cdots, \quad v(x, \psi; \varepsilon) = \frac{a^2}{\varepsilon b} v_2 + \cdots$$

$$p(x, \psi; \varepsilon) = \frac{1}{\gamma M_\infty^2} + \left(\frac{a}{\varepsilon}\right)^2 p_2 + \cdots, \quad \rho(x, \psi; \varepsilon) = \rho_{20}(\psi_2) + \left(\frac{a}{\varepsilon}\right)^2 \rho_2 + \cdots$$

$$h(x, \psi; \varepsilon) = h_{20}(\psi_2) + \left(\frac{a}{\varepsilon}\right)^2 h_2 + \cdots, \quad y(x, \psi; \varepsilon) = \varepsilon y_{20}(\psi_2) + \frac{a^2}{\varepsilon} y_2 + \cdots$$

As before, we will consider only the cases with $\partial p/\partial x \gg 1$. In the main part of the boundary layer the flow is again described by the inviscid equations (8.50) with the only difference that the pressure disturbance is constant across the boundary layer, $\partial p_2/\partial\psi_2 = 0$.

In the nonlinear disturbance wall layer (region 3) the following new variables and the asymptotic representations for the flow functions must be introduced

$$x_3 = x_2 = \frac{x}{b}, \quad \psi_3 = \frac{\varepsilon}{a^2}\psi \tag{8.82}$$

$$u(x, \psi; \varepsilon) = \frac{a}{\varepsilon} u_3 + \cdots, \quad v(x, \psi; \varepsilon) = \frac{a^2}{\varepsilon b} v_3 + \cdots$$

$$p(x, \psi; \varepsilon) = \frac{1}{\gamma M_\infty^2} + \left(\frac{a}{\varepsilon}\right)^2 p_3 + \cdots, \quad \rho(x, \psi; \varepsilon) = \rho_w + \cdots$$

$$h(x, \psi; \varepsilon) = h_w + \frac{a}{\varepsilon} h_3 + \cdots, \quad y(x, \psi; \varepsilon) = a y_3 + \cdots, \quad \mu(x, \psi; \varepsilon) = \mu_w + \cdots$$

After expansions (8.82) have been substituted in the Navier–Stokes equations as $a, b, \varepsilon \to 0$, we arrive again at equations of the form (8.52); after affine transformations (8.37) they involve the similarity parameter $(\varepsilon^3 b/a^3)\Pi$. Obviously, in the general case $a \sim \varepsilon b^{1/3}$.

Matching the solutions for the main part of the boundary layer and the wall sublayer (in this case all calculations of Section 8.1.4 are repeated) we obtain that, as in Section 8.1.4, the problem reduces to the previously studied boundary value problem (8.38)–(8.42).

### 8.1.6 Classification diagram of the regimes of the flow over small two-dimensional roughnesses

The studies performed in Sections 8.1.2–8.1.5 make it possible to construct a classification diagram of the regimes of the disturbed flow over small two-dimensional roughnesses on a plane surface (Bogolepov and Neiland, 1976). In Fig. 8.2 the orders of the characteristic length $b$ and thickness $a$ of the roughness are measured along the horizontal and vertical axes, respectively. Lines $CB$ ($a \sim \varepsilon$), $BF$ ($b \sim 1$), $FG$ ($a \sim \varepsilon^2$), and $GC$ ($a \sim b$) confine the domain of possible variation of the parameters $a$ and $b$.

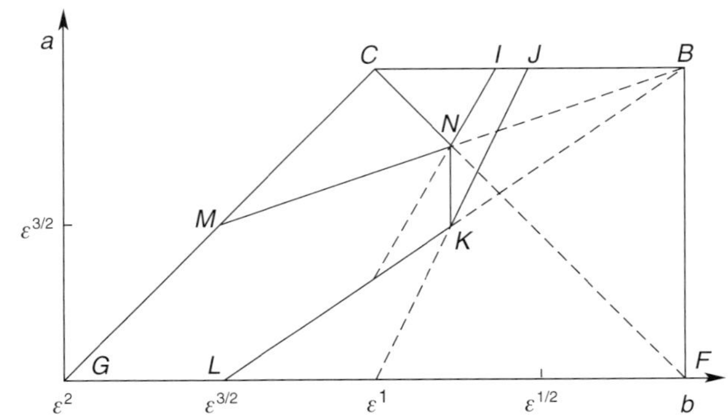


Fig. 8.2.

As shown above, the main regimes are those for which the disturbed flow near the roughness surface is viscous and nonlinear. These are lines $MN$ ($a \sim b\varepsilon^{1/3}$) and $NI$ ($a \sim b^{5/3}$); in the vicinities of such roughnesses the viscous nonlinear layer thickness is of the order $\Delta y \sim \varepsilon b^{1/3}$ and in the flow past them finite relative disturbances of the viscous stress $\Delta\tau \sim \tau \sim \varepsilon$ and the heat flux $\Delta q \sim q \sim \varepsilon$ are induced (that is to say that they can induce flow separation). Above, in domain $MCIN$, the roughnesses induce inviscid nonlinear disturbances; therefore, a viscous, heat-conducting layer must be additionally considered near the roughness surface. Below, in domain $GMNIJKL$, the roughnesses lead to viscous linear disturbances; lines $LK$ ($a \sim \varepsilon b^{2/3}$) and $KJ$ ($a \sim b^2$) cut off flow regimes in which small or finite-in-order pressure gradients, $\partial p/\partial x \leq 1$, are induced. Among all flow regimes, those at lines $MN$ and $NI$ are least degenerate; the corresponding boundary value problems involve governing similarity parameters. In what follows, emphasis is placed on the study of precisely these flows. The solutions for domains $MCIN$ and $GMNIJKL$ can be obtained from the solutions for these basic regimes using the corresponding passages to limit.

Estimate (8.68) shows that the characteristic area of the longitudinal section $ab$ of the roughness determines the pattern of the interaction between the roughness and the outer oncoming flow. For this reason, lines $CN$ ($a \sim \varepsilon^2/b$) and $NK$ ($b \sim \varepsilon^{3/4}$) separate the flow regimes in which the flow function disturbances are produced due to the interaction either only with the subsonic wall region of the undisturbed flat-plate boundary layer (to the left, in domain $GCNKL$) or only with the uniform oncoming flow (to the right, in domain $CJKN$).

At the boundary of these domains disturbances are produced due to the interaction of the roughnesses with the flat-plate boundary layer as a whole.

For $a \sim b \sim \varepsilon^{3/2}$ point $M$ corresponds to the case in which the flow past a roughness embedded in the subsonic wall region of the undisturbed flat-plate boundary layer is described by the incompressible Navier–Stokes equations; then the boundary value problem (8.9) involves a similarity parameter, namely, the local Reynolds number $Re \sim 1$. Solutions for lines $MG$ or $MC$ can be obtained from the main solution by passing to the limits $Re \to 0$ or $Re \to \infty$, respectively.

On line $MN$, for $a \sim \varepsilon b^{1/3}$ and $\varepsilon^{3/2} \ll b \ll \varepsilon^{3/4}$, the flow past a roughness embedded in the subsonic wall region of the undisturbed flat-plate boundary layer is described by the incompressible boundary layer equations with interaction (8.38)–(8.42). The pressure disturbance is produced only due to the interaction between the roughness and the wall region of the boundary layer and the roughness thickness is offset by the variation of the wall region thickness in such a way that in the first approximation the outer edge of the boundary layer is not displaced as $\varepsilon \to 0$. The interaction condition is local, that is, it is determined by the flow parameters at a particular point and does not lead to upstream disturbance propagation.

For $a \sim \varepsilon^{5/4}$ and $b \sim \varepsilon^{3/4}$ point $N$ corresponds to the classical flow regime with free interaction (Neiland, 1968; 1969*a*; Stewartson and Williams, (1969)), while the flow over roughnesses associated with line $NI$ for $a \sim b^{5/3}$ and $\varepsilon^{3/4} \ll b \ll \varepsilon^{3/5}$ is described by Prandtl's equations for an incompressible boundary layer in which the pressure disturbance is induced due to the interaction between the roughness and the uniform oncoming flow (Eqs. (8.53), (8.58), (8.59), (8.65), (8.79), and (8.80)).

### *8.1.7 Examples of solutions for the flow over two-dimensional roughnesses*

*Flow over "short" roughnesses embedded in the wall region of the undisturbed boundary layer*. The numerical solution of the boundary value problem (8.9) was obtained for roughnesses of different shape over a wide local Reynolds number $Re$ range. For the variables we take the disturbances of the stream function $\psi$, the vorticity $\omega$, and the enthalpy $g$

$$\bar{u} = \frac{\partial \psi}{\partial \bar{y}}, \quad \bar{v} = -\frac{\partial \psi}{\partial \bar{x}}, \quad \psi = \frac{\bar{y}^2}{2} + \varphi, \quad \frac{\partial^2 \varphi}{\partial \bar{x}^2} + \frac{\partial^2 \varphi}{\partial \bar{y}^2} = -\omega, \quad \bar{h} = \bar{y} + g \tag{8.83}$$

At the roughness surface $\bar{y} = f(\bar{x})$ the boundary conditions for the functions $\varphi = \varphi(\bar{x}, \bar{y})$ and $g = g(\bar{x}, \bar{y})$ are immediately obtained, while for the function $\omega = \omega(\bar{x}, \bar{y})$ an approximate condition (Tom, 1928) is taken. In deriving the outer boundary conditions it is used that far away from the roughness the flow becomes inviscid and disturbances decay. For this reason, as asymptotics of solution (8.9) we take the harmonic solution of the Laplace equation

$$\frac{\partial^2 \varphi}{\partial \bar{x}^2} + \frac{\partial^2 \varphi}{\partial \bar{y}^2} = 0 \tag{8.84}$$

which describes the inviscid flow past a roughness of chosen shape. The pressure disturbance $\bar{p}$ distribution over the roughness surface was determined by integrating Eq. (8.9).

The difference grid was refined as the roughness surface was approached; a first-order difference scheme with upwind-oriented differences was used for approximating the equations (Gosman et al., 1969). The well-known alternate-direction method was used for solving the problem.

In Figs. 8.3 and 8.4 we have plotted the streamline patterns corresponding to the flow over a semi-cylindrical bump $f(\overline{x}) = \sqrt{1-\overline{x}^2}$ on a plane surface (Bogolepov, 1975) at the Reynolds numbers $Re = 0$ and 30; the well-known dipole solution of Eq. (8.84) was used as the outer asymptotics. Clearly, for $Re = 0$ the flow is symmetric (this follows from the form of the equations and boundary conditions for $Re = 0$) and in the vicinities of the stagnation points there are small separation zones (analogous separation zones were detected in the work of Burggraf, 1966 in studying the flows in rectangular cavities). With increase in $Re$ the forward separation zone shrinks somewhat, while the rear zone considerably enlarges.

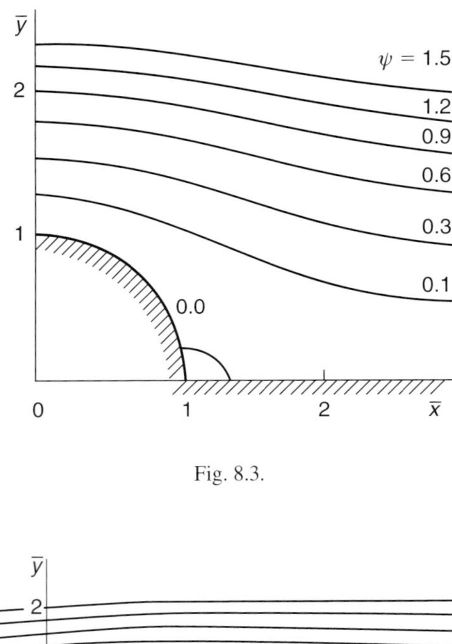


Fig. 8.3.

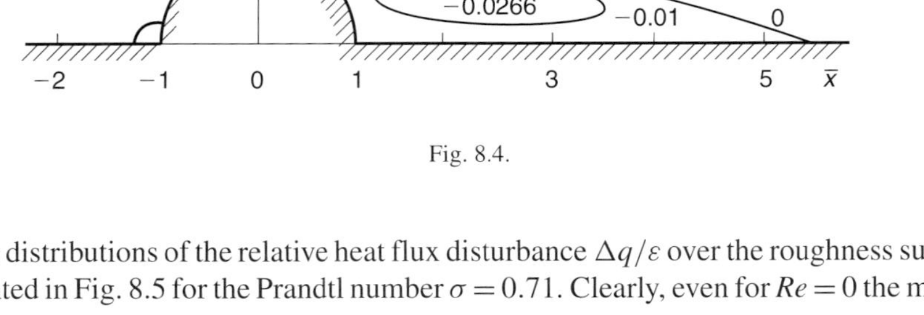


Fig. 8.4.

The distributions of the relative heat flux disturbance $\Delta q/\varepsilon$ over the roughness surface are presented in Fig. 8.5 for the Prandtl number $\sigma = 0.71$. Clearly, even for $Re = 0$ the maximum

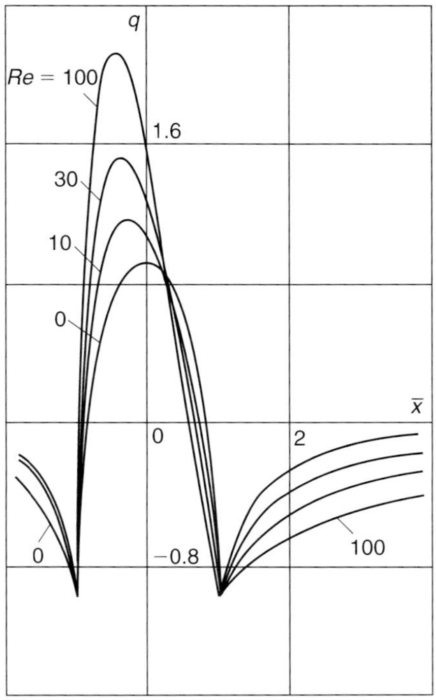


Fig. 8.5.

heat flux to the surface is almost twice as large as its value in the undisturbed boundary layer. With increase in the local Reynolds number *Re* the heat fluxes to the surface increase, their peaks are displaced forward, and the disturbance decay downstream of the roughness, as $\bar{x} \to \infty$, becomes slower. The relative disturbance of the viscous stress $\Delta\tau/\varepsilon$ varies in a similar fashion (Fig. 8.6), only the *Re* dependence is stronger. In Fig. 8.7 we have plotted the pressure disturbance $\bar{p}$ distribution over the roughness surface (for $Re = 0$ on the chosen scale it is the quantity $\bar{p}Re$ that is finite).

It is known that in the case of the uniform flow past a circular cylinder the Stokes equations have no solutions, since the errors due to the fact that the convective terms are disregarded grow rapidly with distance from the body surface. Similar errors take place in the flow under consideration, though they remain small as compared with the rapidly growing streamfunction in the outer shear flow, $\psi \to (\bar{x}^2 + \bar{y}^2)/2$ as $\bar{x}^2 + \bar{y}^2 \to \infty$. For a uniform oncoming flow the streamfunction grows much more slowly, $\psi \to \sqrt{\bar{x}^2 + \bar{y}^2}$ as $\bar{x}^2 + \bar{y}^2 \to \infty$, and the errors due to the fact that the convective terms are disregarded exhibit themselves earlier than the outer flow conditions are attained by the solution.

The results presented in Figs. 8.4–8.7 almost coincide with the data of the paper of Kiya and Arie (1973); however in that work the solution for $Re = 0$ was not obtained.

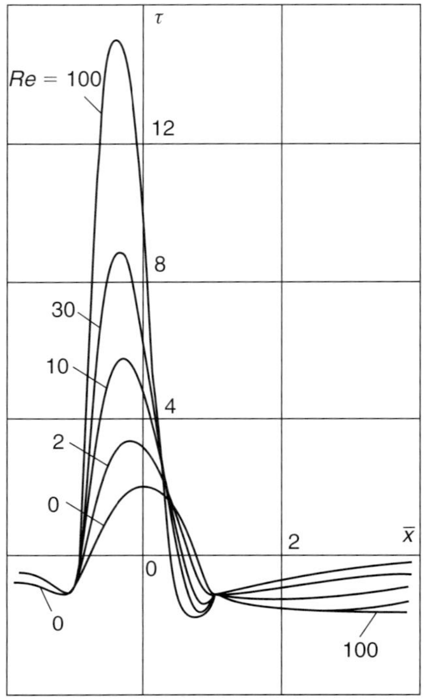


Fig. 8.6.

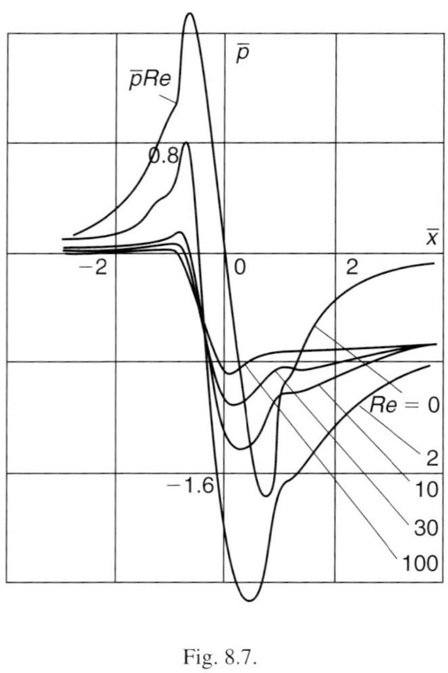


Fig. 8.7.

If the local Reynolds number is expressed in terms of the characteristic values of the small roughness height $a$ and the undisturbed boundary layer thickness $\delta$, the following relation (Bogolepov, 1977) can be derived

$$Re \sim Re_\infty^{1/2}\left(\frac{a}{\delta}\right)^2 \tag{8.85}$$

From this formula it follows that for the laminar $Re_\infty \approx 10^6$ boundary layer and the relative roughness height $a/\delta \approx 0.1$ the local Reynolds number $Re \approx 10$. This means that the solution of the boundary value problem (8.9) can be obtained using well-developed methods for solving the two-dimensional Navier–Stokes equations at quite moderate Reynolds numbers; it describes fairly reliably the flow over various local distortions of flight vehicle surfaces.

The numerical solution of the boundary value problem (8.9) over a wide range of the local Reynolds number $Re$ was also obtained for the flow over different barriers on plane surfaces (Bogolepov, 1980; 1983; 1985*b*; Bogolepov and Lipatov, 1988*a*, *b*), the flow past the trailing edge of a thin plate (Bogolepov, 1985*b*), and the flow of chemically nonequilibrium viscous gas past a small step (Bogolepov, 1984). The asymptotic laws of disturbance decay downstream of roughnesses were studied in the work of Bogolepov and Lipatov (1982).

*Flow over "short" roughnesses with the formation of locally inviscid disturbed flow regions*. As illustration of the solution of the boundary value problem (8.14), (8.15), we will consider the superposition of the flow in the wall region of the undisturbed flat-plate boundary layer and a dipole with an intensity $J$ (Bogolepov and Neiland, 1971; 1976)

$$\psi_3(x_3, y_3) = \rho_w\left(\frac{Ay_3^2}{2} - \frac{Jy_3}{x_3^2 + y_3^2}\right) \tag{8.86}$$

From the impermeability condition at the roughness surface (8.15) we can derive its shape and the total velocity $u_e^2 = u_3^2 + v_3^2$ distribution over its surface

$$x_3^2 + y_3^2 = \frac{2J}{Ay_3}, \quad u_e^2 = \frac{A^2y_3^2}{4}\left(1 + \frac{4Ay_3^3}{J}\right) \tag{8.87}$$

For further analysis it is convenient to introduce in the solution (8.87) and the boundary value problem (8.17)–(8.20) the following affine transformation of variables

$$x_3 = \left(\frac{2J}{A}\right)^{1/3}\bar{x}, \quad y_3 = \left(\frac{2J}{A}\right)^{1/3}\bar{y}, \quad s_4 = \left(\frac{2J}{A}\right)^{1/3}\bar{s}, \quad n_4 = \left(\frac{2J}{A}\right)^{1/3}\bar{n} \tag{8.88}$$

$$u_e = 3\left(\frac{A^2J}{4}\right)^{1/3}\bar{u}_e, \quad u_4 = 3\left(\frac{A^2J}{4}\right)^{1/3}\bar{u}, \quad v_4 = 3\left(\frac{A^2J}{4}\right)^{1/3}\bar{v}$$

$$p_4 = 9\rho_w\left(\frac{A^2J}{4}\right)^{2/3}\bar{p}, \quad h_4 = \sqrt{3}B\left(\frac{2J}{A}\right)^{1/3}\bar{h}, \quad \psi_4 = 3\rho_w\left(\frac{AJ^2}{2}\right)^{1/3}\bar{\psi}$$

Then in the inviscid region the roughness shape and the velocity distribution over its surface take the form:

$$\bar{y}\left(\bar{x}^2+\bar{y}^2\right)=1, \quad \bar{u}_e=\frac{\bar{y}}{3}\left(1+8\bar{y}^3\right)^{1/2} \tag{8.89}$$

In variables (8.88) the boundary value problem (8.17)–(8.20) is brought into the form:

$$\frac{\partial \bar{u}}{\partial \bar{s}}+\frac{\partial \bar{v}}{\partial \bar{n}}=0, \quad \bar{u}\frac{\partial \bar{u}}{\partial \bar{s}}+\bar{v}\frac{\partial \bar{u}}{\partial \bar{n}}-\bar{u}_e\frac{d\bar{u}_e}{d\bar{s}}=K^2\frac{\partial^2 \bar{u}}{\partial \bar{n}^2} \tag{8.90}$$

$$\bar{u}\frac{\partial \bar{h}}{\partial \bar{s}}+\bar{v}\frac{\partial \bar{h}}{\partial \bar{n}}=\frac{K^2}{\sigma}\frac{\partial^2 \bar{h}}{\partial \bar{n}^2}, \quad K^2=\frac{\mu_w}{3\rho_w}\left(\frac{2}{AJ^2}\right)^{1/3}$$

$$\bar{u}(\bar{s},0)=\bar{v}(\bar{s},0)=\bar{h}(\bar{s},0)=0; \quad \bar{u}\to\bar{u}_e, \ \bar{h}\to\sqrt{\bar{\psi}}, \ \bar{n}\to\infty$$

Following the paper of Neiland (1966) for numerical integration we introduce the new independent and dependent variables

$$\xi=\int_{-\infty}^{\bar{s}}\bar{u}_e(\bar{s})\,d\bar{s}, \quad \eta=\frac{\bar{u}_e^{3/4}(\bar{s})\bar{n}}{K}, \quad \frac{\partial \phi}{\partial \eta}=\frac{\bar{u}}{\bar{u}_e}, \quad g=\frac{h}{K^{1/2}\bar{u}_e^{1/8}} \tag{8.91}$$

in which the equations and boundary conditions (8.90) are rewritten as follows:

$$\frac{\partial^3\phi}{\partial\eta^3}+\frac{1}{4\bar{u}_e^{1/2}}\frac{d\bar{u}_e}{d\xi}\phi\frac{\partial^2\phi}{\partial\eta^2}+\frac{1}{\bar{u}_e^{1/2}}\frac{d\bar{u}_e}{d\xi}\left[1-\left(\frac{\partial\phi}{\partial\eta}\right)^2\right]=\bar{u}_e^{1/2}\left(\frac{\partial\phi}{\partial\eta}\frac{\partial^2\phi}{\partial\xi\partial\eta}-\frac{\partial^2\phi}{\partial\eta^2}\frac{\partial\phi}{\partial\xi}\right) \tag{8.92}$$

$$\frac{1}{\sigma}\frac{\partial^2 g}{\partial\eta^2}+\frac{1}{4\bar{u}_e^{1/2}}\frac{d\bar{u}_e}{d\xi}\left(\phi\frac{\partial g}{\partial\eta}-\frac{\partial\phi}{\partial\eta}\frac{g}{2}\right)=\bar{u}_e^{1/2}\left(\frac{\partial\phi}{\partial\eta}\frac{\partial g}{\partial\xi}-\frac{\partial\phi}{\partial\xi}\frac{\partial g}{\partial\eta}\right)$$

$$\phi(\xi,0)=\frac{\partial\phi}{\partial\eta}(\xi,0)=g(\xi,0)=0; \quad \frac{\partial\phi}{\partial\eta}\to 1, \ g\to\sqrt{\phi}, \ \eta\to\infty$$

In the new variables $(\xi,\eta)$ the infinite integration interval along the longitudinal coordinate $\bar{s}$ reduces to a finite interval; for $\xi=0$ ($\bar{s}\to-\infty$) Eqs. (8.92) take the self-similar form, and for small $\xi$ we have $\bar{u}_e\approx 3\xi^2$. At the roughness surface (8.21) the viscous stress and the heat flux are determined by the formulas:

$$\frac{\tau}{\varepsilon}=\frac{b}{\varepsilon^{3/2}}\frac{3}{2K}\bar{u}_e^{7/4}\frac{\partial^2\phi}{\partial\eta^2}(\xi,0), \quad \frac{q}{\varepsilon}=\frac{b^{1/2}}{\varepsilon^{3/4}}\sqrt{\frac{3}{K}}\bar{u}_e^{7/8}\frac{\partial g}{\partial\eta}(\xi,0) \tag{8.93}$$

Equations (8.92) were solved using a semi-standard program developed by Seliverstov on the basis of the method proposed by Petukhov (1964). In Fig. 8.8 we have plotted the

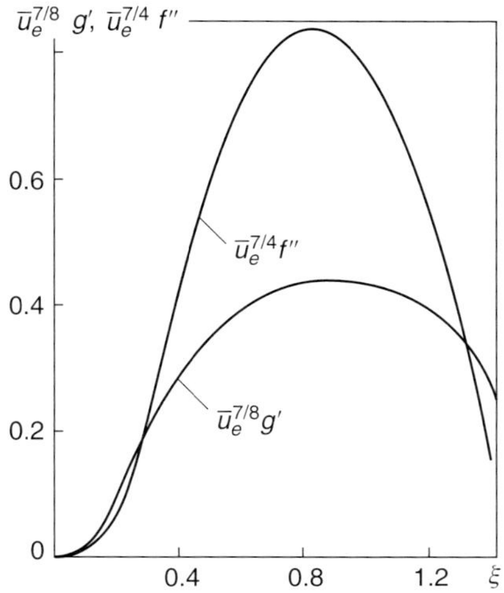


Fig. 8.8.

distributions of the quantities $\overline{u}_e^{7/4}(\xi)(\partial^2\phi/\partial\eta^2)(\xi,0)$ and $\overline{u}_e^{7/8}(\xi)(\partial g/\partial\eta)(\xi,0)$ entering in the expressions for the viscous stress and the heat flux (8.93) for the Prandtl number $\sigma=1$.

For $\xi=0$ or $\overline{s}\to-\infty$ the viscous stress and the heat flux tend to their values in the undisturbed boundary layer, which on the scales of the quantities $b/\varepsilon^{3/2}\gg 1$ and $b^{1/2}/\varepsilon^{3/4}\gg 1$ are equal to zero when $\varepsilon\to 0$, $\varepsilon^{3/2}\ll b\ll 1$. At distances of the order of the small roughness length the products $\overline{u}_e^{7/4}(\partial^2\phi/\partial\eta^2)$ and $\overline{u}_e^{7/8}(\partial g/\partial\eta)$ become of the order of unity and decrease sharply as the separation point $\xi\approx 1.4$ is approached (the calculations were performed up to the point of separation which occurs almost immediately behind the bump vertex, as the pressure begins to rise).

*Solution for the compensation regime of the flow over thin roughnesses.* As noted above, if the viscous stress and the heat flux induced in the flow over a roughness are greater in the order than those in the undisturbed flat-plate boundary layer, then ahead of this roughness there must be a transition flow region in which the viscous stress and the heat flux are of the same order as those in the undisturbed boundary layer and increase. In this transitional region the flow must be governed by the solution of the boundary value problem (8.38)–(8.42). It was obtained using the semi-standard program developed by Seliverstov on the basis of the method proposed by Petukhov (1964). In Figs. 8.9–8.11 it is presented by the pressure disturbance distributions and the relative distributions of the viscous stress and heat flux disturbances, respectively, in the flow over a sinusoidal bump $f(\overline{x})=\sin\pi\overline{x}$ for different values of the parameter $\Pi$ and the Prandtl number $\sigma=0.71$; integration was performed up to the separation point (Bogolepov, 1974). Ahead of the roughness, for $\overline{x}<0$, the flow remains undisturbed, while in the vicinity of point $\overline{x}=0$ the limiting solution of the problem yields $(-\overline{p})\sim\overline{x}^{4/3}$, $\Delta\tau\sim\overline{x}^{2/3}$, and $\Delta q/\varepsilon\sim\overline{x}^{2/3}$. At the roughness surface the quantities $(-\overline{p})$, $\Delta\tau/\varepsilon$, and $\Delta q/\varepsilon$ are maximum, while at point $\overline{x}=1$ the viscous stress and heat flux disturbances have sharp minima. With distance from the roughness, as $\overline{x}\to\infty$, all the disturbances decay and the limiting solution of the problem gives $(-\overline{p})\sim\overline{x}^{-2/3}$,

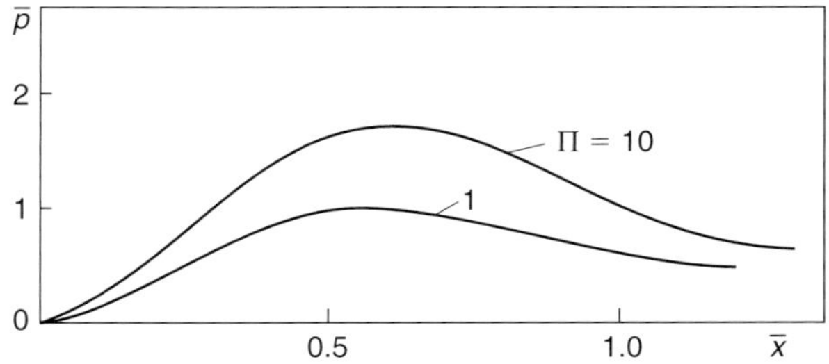


Fig. 8.9.

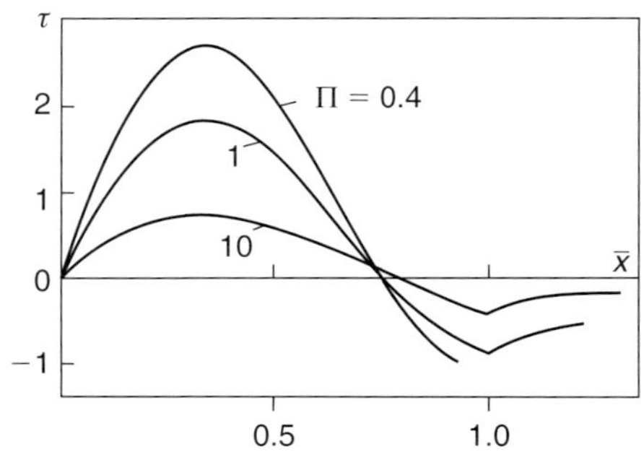


Fig. 8.10.

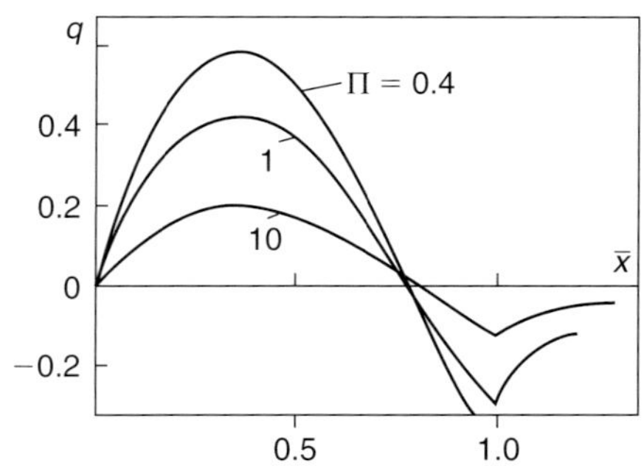


Fig. 8.11.

$\Delta\tau/\varepsilon \sim \overline{x}^{-4/3}$, and $\Delta q/\varepsilon \sim \overline{x}^{-4/3}$ (similar disturbance decay laws were derived in the paper of Hunt, 1971). For large values of the parameter $\Pi$ the flow is separationless; incipient separation occurs for $\Pi \approx 0.6$ at point $\overline{x} = 1$.

If the similarity parameter $\Pi$ is expressed in terms of the characteristic height $a$ of the small roughness and the undisturbed boundary layer thickness $\delta$, then the following relation (Bogolepov, 1977) can be obtained

$$\Pi \sim Re_{\infty}^{-1/2}\left(\frac{\delta}{a}\right)^2\left(\frac{b}{a}\right) \tag{8.94}$$

From this relation it follows that for a laminar $Re_\infty \approx 10^6$ flow and a relative boundary layer thickness $\delta/a \approx 10$ and $b/a \approx 10$ the similarity parameter is $\Pi \approx 1$. This means that the occurrence of separation is actually dependent on the particular shape of the roughness.

### *8.1.8 Classification of the regimes of flow over roughness on a cold surface*

We will now consider the viscous hypersonic $M_\infty \to \infty$ flow over a flat plate in the weak viscous–inviscid interaction regime in which a small pressure disturbance is induced due to the displacing action of the boundary layer (Hayes and Probstein, 1959)

$$\gamma M_\infty^2 \Delta p \sim \delta M_\infty \sim \chi \ll 1 \tag{8.95}$$

where $\chi$ is the interaction parameter. In what follows, for the sake of simplicity, we assume a linear dependence of $\mu$ on $h$.

On the main part of the body, $\Delta x \sim 1$, the pressure disturbance (8.95) induces a small pressure gradient $\partial p/\partial x \sim \chi$ which exhibits itself in the boundary layer equations only in the second approximation. Therefore, in the main part of the boundary layer with the characteristic dimensions $\Delta x \sim 1$ and $\Delta y \sim \delta$ the following asymptotic estimates for the flow functions are valid

$$u \sim h \sim 1, \quad p \sim \rho \sim \frac{1}{M_\infty^2}, \quad \mu \sim M_\infty^2, \quad \delta \sim \frac{M_\infty^2}{Re_\infty^{1/2}}, \quad \tau \sim q \sim \frac{\delta}{M_\infty^2} \tag{8.96}$$

Assuming that the viscous stress $\tau$ and the heat flux $q$ retain their orders throughout almost the entire boundary layer, for its wall region with $\Delta y/\delta \ll 1$ we can obtain

$$u \sim \left(\frac{\Delta y}{\delta} + h_w^2\right)^{1/2} - h_w, \quad h \sim \left(\frac{\Delta y}{\delta} + h_w^2\right)^{1/2} \tag{8.97}$$

See also relations (8.2). It is assumed that the characteristic dimensions of the roughness satisfy the previous constraints (see Section 8.1.1) and are greater than the characteristic free molecular path $\ell_1 \sim \mu/Re_\infty(\rho p)^{1/2}$ (see, e.g., Laytsyanskii, 1962)

$$\ell_1 \ll a \le \delta, \quad a \le b \le 1, \quad \ell_1 \sim \delta^2 \left(\frac{\Delta y}{\delta} + h_w^2\right)^{3/4} \tag{8.98}$$

If the plate surface is not cold and $h_w \le 1$, then in the main part of the boundary layer estimates (8.96) are valid, while in its wall region, $\Delta y/\delta \ll h_w^2 \le 1$, from Eqs. (8.97) and (8.98) we obtain the following distributions and estimates for the flow functions

$$u \sim \frac{\Delta y}{\delta}, \quad h \sim h_w + \frac{\Delta y}{\delta}, \quad \rho \sim \frac{1}{M_\infty^2 h_w}, \quad \mu \sim M_\infty^2 h_w, \quad \ell_1 \sim \delta^2 \tag{8.99}$$

Repeating the analysis of Sections 8.1.2–8.1.6 and using relations (8.99) we can verify that, as $M_\infty \to \infty$, $Re_\infty \to \infty$, $\chi \to 0$, and $h_w \le 1$, the same flow regimes as in the case $(M_\infty^2 - 1) \sim 1$ are realized near small roughnesses with characteristic dimensions given by relations (8.98).

In the limiting case of the flow over a cold surface (Bogolepov and Lipatov, 1991), for $h_w \ll (\Delta y/\delta)^{1/2} \leq 1$, from relations (8.97) and (8.98) the following flow function distributions are obtained

$$u \sim h \sim \left(\frac{\Delta y}{\delta}\right)^{1/2}, \quad \rho \sim \frac{\delta^{1/2}}{M_\infty^2 \Delta y^{1/2}}, \quad \mu \sim M_\infty^2 \left(\frac{\Delta y}{\delta}\right)^{1/2}, \quad \ell_1 \sim \Delta y^{3/4} \delta^{5/4} \quad (8.100)$$

which are obviously valid in both the wall region ($\Delta y/\delta \ll 1$) and the main part ($\Delta y/\delta \sim 1$) of the boundary layer (cf. estimates (8.96)). Relations (8.100) indicate that, since $\min(a, b) \gg \ell_1$, then $\min(a, b) \gg \delta^5$ and in local disturbed flow regions near the roughnesses viscosity is important in a layer with a characteristic thickness $\Delta y \sim \delta b^{2/3}$.

For this reason, near a fairly thick roughness with $a \sim b$ in region 3 with the characteristic dimensions $\delta^5 \ll \Delta x \sim \Delta y^{(3)} \sim b \leq \delta$ we introduce the following variables and asymptotic expansions of the flow functions

$$x = bx_3, \quad \psi = \frac{b}{M_\infty^2}\psi_3, \quad y = by_3 + \cdots \quad (8.101)$$

$$u = \left(\frac{b}{\delta}\right)^{1/2} u_3 + \cdots, \quad v = \left(\frac{b}{\delta}\right)^{1/2} v_3 + \cdots$$

$$p = \frac{1}{\gamma M_\infty^2} + \frac{b^{1/2}}{\delta^{1/2} M_\infty^2} p_3 + \cdots, \quad h = \left(\frac{b}{\delta}\right)^{1/2} h_3 + \cdots$$

$$\rho = \frac{\delta^{1/2}}{b^{1/2} M_\infty^2} \rho_3 + \cdots, \quad \mu = M_\infty^2 \left(\frac{b}{\delta}\right)^{1/2} \mu_3 + \cdots$$

Substituting expansions (8.101) in the Navier–Stokes equations written in the von Mises variables $(x, \psi)$ and passing to the limit $M_\infty \to \infty$, $\delta \to 0$, and $\chi \to 0$ shows that in the most general case $a \sim b \sim \delta^3$ the flow in region 3 is described in the first approximation by the complete compressible Navier–Stokes equations; at the roughness surface $y_3 = f(x_3)$ for $\psi_3 = 0$ the conventional no-slip and impermeability conditions must be fulfilled, while far away from the roughness the solution must be matched with the undisturbed flow in the wall region of the boundary layer (8.100). If $\delta^5 \ll a \sim b \ll \delta^3$, then in region 3 the Stokes equations are valid in the first approximation. Expansions (8.101) show that for $\delta^5 \ll a \sim b \leq \delta^3$ the viscous stress $\tau$ and the heat flux $q$ vary in their leading orders (8.96) along the roughness surface.

For $\delta^3 \ll a \sim b \leq \delta$ the flow in region 3 is described by the Euler equations which can be brought into the following form:

$$\rho_3 \frac{u_3^2 + v_3^2}{2} + p_3 = \rho_3 \frac{u_{30}^2}{2}, \quad \frac{\partial v_3}{\partial x_3} + \frac{\partial p_3}{\partial \psi_3} = 0 \quad (8.102)$$

$$\frac{\partial y_3}{\partial x_3} = \frac{v_3}{u_3}, \quad \frac{\partial y_3}{\partial \psi_3} = \frac{1}{\rho_3 u_3}, \quad h_3 = h_{30}(\psi_3), \quad \rho_3 = \frac{1}{(\gamma - 1) h_{30}(\psi_3)}$$

where $u_{30}(\psi_3)$ and $h_{30}(\psi_3)$ are the longitudinal velocity and enthalpy profiles in the undisturbed flat-plate boundary layer at the point at which the roughness is located. Since, according to Eq. (8.100), for $\delta^3 \ll a \sim b \ll \delta$ or $a \sim b \sim \delta$ and $\psi_3 \to 0$ we have

$$u_{30}(\psi_3) \to C\psi_3^{1/2}, \quad h_{30}(\psi_3) \to D\psi_3^{1/2}, \quad \tau Re_\infty^{1/2} = \frac{C^2}{2(\gamma-1)}, \quad qRe_\infty^{1/2}\sigma = \frac{CD}{2(\gamma-1)}$$

where $C$ and $D$ are some constants, the first equation (8.102) can be rewritten in the form:

$$\frac{u_3^2 + v_3^2}{2} + D(\gamma-1)p_3\psi_3^{1/2} = \frac{C^2\psi_3}{2}, \quad \psi_3 \to 0$$

From the condition that the roughness surface slope is finite, $\mathrm{d}f/\mathrm{d}x_3 \sim 1$, it follows immediately that $u_3 \sim v_3 \sim p_3 \sim \psi_3^{1/2}$ as $\psi_3 \to 0$. This means that near the roughness surface in region 3, even when inviscid nonlinear disturbances are induced ($\Delta p \sim \rho u^2$), the flow functions vary in the same way as in the wall region of the boundary layer on a cold surface (8.100)

$$u \sim h \sim M_\infty\left(\frac{\psi}{\delta}\right)^{1/2}, \quad \rho \sim \frac{\delta^{1/2}}{M_\infty^3\psi^{1/2}}, \quad \Delta y \sim M_\infty^2\psi, \quad \psi \to 0$$

while in the viscous sublayer 4 with the characteristic thickness $\Delta y^{(4)} \sim \delta b^{2/3}$ the viscous stress $\tau$ and the heat flux $q$ vary in their leading orders (8.96). In other words, for $h_w \ll (b/\delta)^{1/2}$ at $\delta^5 \ll b \leq \delta^3$ or for $h_w \ll b^{1/3}$ on all scales from $b \leq \delta$ to $\delta^3 \leq b$ any fairly thick roughnesses ($a \sim b$) can induce only finite disturbances $Re_\infty^{1/2}\Delta\tau$ and $Re_\infty^{1/2}\Delta q$ (of the same order as the quantities $Re_\infty^{1/2}\tau$ and $Re_\infty^{1/2}q$ themselves in the undisturbed boundary layer (8.96)). Therefore, any thin roughnesses ($a \ll b$) can induce only small disturbances $Re_\infty^{1/2}\Delta\tau$ and $Re_\infty^{1/2}\Delta q$ (i.e. they are not able to initiate, e.g., boundary layer separation).

In the case of the flow over thin and short roughnesses ($a \ll b \leq \delta$) for $h_w \ll (\Delta y/\delta)^{1/2} \ll 1$, according to Eq. (8.100), the pressure disturbance $\Delta p \sim \rho^{(3)}u^{(3)}\Delta u^{(3)} \sim \Delta u^{(3)}/M_\infty^2$ is induced in region 3 with the characteristic dimensions $\Delta x \sim b$ and $\Delta y^{(3)} \sim a$; it must decay in region 2 with the characteristic dimensions $\Delta x \sim \Delta y^{(2)} \sim b$. Because of this, the roughness thickness $a$ must be offset by the variation of the thickness $\Delta y^{(2)}$ of region 2:

$$u^{(2)} \sim h^{(2)} \sim \left(\frac{b}{\delta}\right)^{1/2}, \quad \rho^{(2)} \sim \frac{1}{M_\infty^2 h^{(2)}}, \quad \psi^{(2)} \sim \frac{b}{M_\infty^2}, \quad v^{(2)} \sim \frac{\Delta p\Delta x}{\psi^{(2)}} \sim \Delta p M_\infty^2 \tag{8.103}$$

$$\Delta y^{(2)} \sim \frac{v^{(2)}\Delta x}{u^{(2)}} \sim M_\infty^2 b^{1/2}\delta^{1/2}\Delta p \sim a, \quad \Delta p \sim \frac{a}{M_\infty^2 b^{1/2}\delta^{1/2}}$$

Therefore, in region 3 we have $\Delta u^{(3)} \sim M_\infty^2 \Delta p \sim a/(b\delta)^{1/2} \ll u^{(3)} \sim (a/\delta)^{1/2}$, that is to say, only small linear disturbances of the flow functions are induced. In the most general case region 3 is viscous and then we have

$$a \sim \delta b^{2/3}, \quad h_w \ll b^{1/3}, \quad \delta^3 \ll b \leq \delta \tag{8.104}$$

Estimates (8.100), (8.103), and (8.104) make it possible to introduce in regions 2 and 3 the following new variables and asymptotic expansions of the flow functions

$$x = bx_3, \quad \psi = \frac{\delta b^{2/3}}{M_\infty^2}\psi_3, \quad y = \delta b^{2/3} y_3 + \cdots \tag{8.105}$$

$$u = b^{1/3} C \psi_3^{1/2} + \delta^{1/2} b^{1/6} u_3 + \cdots, \quad v = \delta v_3 + \cdots, \quad p = \frac{1}{\gamma M_\infty^2} + \frac{\delta^{1/2} b^{1/6}}{M_\infty^2} p_3 + \cdots$$

$$\rho = \frac{1}{M_\infty^2 b^{1/3} (\gamma - 1) D \psi_3^{1/2}} + \frac{\delta^{1/2}}{M_\infty^2 b^{1/6}} \rho_3 + \cdots, \quad \mu = M_\infty^2 b^{1/3} D \psi_3^{1/2} + \cdots$$

$$h = b^{1/3} D \psi_3^{1/2} + (\delta b)^{1/2} h_3 + \cdots$$

$$x = bx_2, \quad \psi = \frac{b}{M_\infty^2}\psi_2, \quad y = by_{20}(\psi_2) + \delta b^{2/3} y_2 + \cdots \tag{8.106}$$

$$u = \left(\frac{b}{\delta}\right)^{1/2} u_{20}(\psi_2) + \delta^{1/2} b^{1/6} u_2 + \cdots, \quad v = \delta^{1/2} b^{1/6} v_2 + \cdots$$

$$p = \frac{1}{\gamma M_\infty^2} + \frac{\delta^{1/2} b^{1/6}}{M_\infty^2} p_2 + \cdots, \quad \rho = \frac{\delta^{1/2}}{M_\infty^2 b^{1/2}} \rho_{20}(\psi_2) + \frac{\delta}{M_\infty^2 b^{1/3}} \rho_2 + \cdots$$

$$h = \left(\frac{b}{\delta}\right)^{1/2} h_{20}(\psi_2) + b^{2/3} h_2 + \cdots$$

Here, the subscript 20 denotes again the flow function distributions in the undisturbed boundary layer, at the point at which the roughness is located. Substituting expansions (8.105) and (8.106) in the Navier–Stokes equations and passing to the limit $M_\infty \to \infty$, $\delta \to 0$, $\chi \to 0$, and $\delta^3 \ll b \leq \delta$ shows that in the first approximation the flow in region 3 is described by the linearized Prandtl boundary layer equations and the flow in region 2 by the linearized Euler equations

$$\frac{C}{(\gamma - 1) D} \frac{\partial u_3}{\partial x_3} + \frac{dp_3}{dx_3} = \frac{C^2}{(\gamma - 1)^2 D} \frac{\partial}{\partial \psi_3} \left( \psi_3^{1/2} \frac{\partial u_3}{\partial \psi_3} \right), \quad \frac{\partial p_3}{\partial \psi_3} = 0 \tag{8.107}$$

$$\frac{\partial y_3}{\partial x_3} = \frac{v_3}{C\psi_3^{1/2}}, \quad \frac{\partial y_3}{\partial \psi_3} = \frac{(\gamma-1)D}{C}, \quad h_3 = \gamma D\psi_3^{1/2} p_3 - (\gamma-1)D^2\psi_3\rho_3$$

$$\frac{\partial h_3}{\partial x_3} - (\gamma-1)\psi_3^{1/2}\frac{\mathrm{d}p_3}{\mathrm{d}x_3} = \frac{C}{\sigma D(\gamma-1)}\frac{\partial}{\partial \psi_3}\left(\psi_3^{1/2}\frac{\partial h_3}{\partial \psi_3}\right)$$

$$\rho_{20}u_{20}\frac{\partial u_2}{\partial x_2} + \frac{\partial p_2}{\partial x_2} = 0, \quad \frac{\partial v_2}{\partial x_2} + \frac{\partial p_2}{\partial \psi_2} = 0 \tag{8.108}$$

$$\frac{\partial y_2}{\partial x_2} = \frac{v_2}{u_{20}}, \quad \frac{\partial y_{20}}{\partial \psi_2} = \frac{1}{\rho_{20}u_{20}}, \quad \rho_{20}\frac{\partial h_2}{\partial x_2} - \frac{\partial p_2}{\partial x_2} = 0$$

$$\rho_{20}u_{20}\frac{\partial y_2}{\partial \psi_2} = -\frac{u_2}{u_{20}} - \left(\frac{b}{\delta}\right)^{1/2}\frac{\rho_2}{\rho_{20}}, \quad (\gamma-1)\rho_{20}h_{20} = 1, \quad (\gamma-1)\rho_{20}h_2 = \gamma p_2 - \frac{\rho_2}{\rho_{20}}$$

The solution of Eqs. (8.107) must decay, as $x_3 \to -\infty$, and satisfy the conventional inner boundary conditions for $\psi_3 = 0$

$$u_3, v_3, p_3, \rho_3, h_3, y_3 \to 0, \quad x_3 \to -\infty \tag{8.109}$$

$$u_3 = v_3 = h_3 = 0, \quad y_3 = f(x_3), \quad \psi_3 = 0$$

Rewriting Eq. (8.107) in terms of Eq. (8.109) yields

$$y_3 = f(x_3) + \frac{D(\gamma-1)}{C}\psi_3, \quad v_3 = C\psi_3^{1/2}\frac{\mathrm{d}f}{x_3} \tag{8.110}$$

Far away from the roughness the solution of system (8.108) must decay

$$u_2, v_2, p_2, \rho_2, h_2, y_2 \to 0 \quad x_2 \to \pm\infty \text{ or } \psi_2 \to \infty \tag{8.111}$$

Therefore, these equations can be rewritten as follows:

$$\rho_{20}u_{20}u_2 + p_2 = 0, \quad \frac{\partial y_2}{\partial x_2} = \frac{v_2}{u_{20}}, \quad \frac{\partial v_2}{\partial x_2} + \frac{\partial p_2}{\partial \psi_2} = 0 \tag{8.112}$$

$$\frac{\partial y_{20}}{\partial \psi_2} = \frac{1}{\rho_{20}u_{20}}, \quad \rho_{20}h_2 - p_2 = 0, \quad (\gamma-1)\rho_{20}h_{20} = 1, \quad \rho_2 = \rho_{20}p_2$$

$$\frac{\partial y_2}{\partial \psi_2} = p_2\left[\frac{1}{\rho_{20}^2u_{20}^3} - \left(\frac{b}{\delta}\right)^{1/2}\frac{1}{\rho_{20}u_{20}}\right]$$

The deficient boundary conditions for the solutions in region 3 as $\psi_3 \to \infty$ and in region 2 as $\psi_2 \to 0$ are obtained by matching the asymptotic expansions (8.105) and (8.106) with the aid of Eqs. (8.110) and (8.112)

$$x_3 = x_2, \quad p_3(x_3) = p_2(x_2, 0), \quad y_2(x_2, 0) = f(x_3) \tag{8.113}$$

$$u_3 \to -\frac{p_2}{\rho_{20}u_{20}}\bigg|_{\psi_2\to 0} = -\frac{(\gamma-1)D}{C}p_3$$

$$h_3 \to \frac{b^{1/6}}{\delta^{1/2}}\frac{p_2}{\rho_{20}}\bigg|_{\psi_2\to 0} = (\gamma-1)Dp_3\psi_3^{1/2}, \quad \psi_3 \to \infty$$

It can be verified that the outer boundary condition (8.113) for $h_3$ satisfies Eq. (8.107) and the boundary conditions (8.109), that is, it is a solution for region 3. Therefore, here the function $u_3(x_3, \psi_3)$ must be determined from the parabolic equation (8.107) subject to the boundary conditions (8.109) and (8.113). In this boundary value problem the pressure distribution $p_3(x_3)$ is unknown; it must be determined from the solution in region 2 of the elliptic equation

$$u_{20}\frac{\partial^2 y_2}{\partial x_2^2} + \frac{\partial p_2}{\partial \psi_2} = 0, \quad p_2 = \frac{\partial y_2}{\partial \psi_2}\left[\frac{1}{\rho_{20}^2 u_{20}^3} - \left(\frac{b}{\delta}\right)^{1/2}\frac{1}{\rho_{20}u_{20}}\right] \tag{8.114}$$

subject to boundary conditions (8.111) and (8.113). Then the viscous stress $\tau$ and heat flux $q$ distributions are given by the formulas:

$$\tau Re_\infty^{1/2} = \frac{C^2}{2(\gamma-1)}\left(1 + \frac{\delta^{1/2}}{b^{1/6}}\frac{2\psi_3^{1/2}}{C}\frac{\partial u_3}{\partial \psi_3}\right) \tag{8.115}$$

$$qRe_\infty^{1/2}\sigma = \frac{CD}{2(\gamma-1)}[1 + \delta^{1/2}b^{1/6}(\gamma-1)p_3(x_3)]$$

Studying the flow over roughnesses with characteristic lengths greater than the boundary layer thickness, $a \le \delta \ll b \le 1$, can require the consideration of region 1 with the characteristic dimensions $\Delta x \sim b$, $\Delta y^{(1)} \sim b/M_\infty$. In accordance with the work of Neiland (1974), in the main part of the boundary layer (region 2, $\Delta x \sim b$, $\Delta y^{(2)} \sim \delta$) a small pressure disturbance $\Delta p < 1$ induces small disturbances of the flow functions

$$\Delta u^{(2)} \sim v^{(2)} \sim \Delta h^{(2)} \sim \Delta p M_\infty^2, \quad \Delta\rho^{(2)} \sim \Delta p, \quad \Delta y^{(2)} \sim \delta M_\infty^2 \Delta p \tag{8.116}$$

If the roughness does not interact with the uniform outer flow, then the variation of the region 2 thickness must be offset by the thickness of the roughness itself; then from estimates (8.116) there follows:

$$\Delta y^{(2)} \sim \delta M_\infty^2 \Delta p \sim a, \quad \Delta p \sim \frac{a}{\delta M_\infty^2} \tag{8.117}$$

In the case of the interaction between the roughness and the uniform oncoming flow the following pressure disturbance is induced (Hayes and Probstein, 1959)

$$\Delta p \sim \frac{a}{bM_\infty} \tag{8.118}$$

In the transitional regime both estimates (8.117) and (8.118) are valid; then we have $b \sim \delta M_\infty$.

For this reason, for $\delta \ll b \ll \delta M_\infty$, assuming that in region 3 the roughness induces viscous disturbances and, therefore, $a \sim \delta b^{2/3}$, $h_w \ll b^{1/3}$, and using estimates (8.100), (8.116), and (8.117) we can introduce in regions 3 and 2 the following independent variables and the asymptotic expansions of the flow functions

$$x = bx_3, \quad \psi = \frac{\delta b^{2/3}}{M_\infty^2}\psi_3, \quad y = \delta b^{2/3} y_3 + \cdots \tag{8.119}$$

$$u = b^{1/3} C\psi_3^{1/2} + b^{2/3} u_3 + \cdots, \quad v = \delta v_3 + \cdots, \quad p = \frac{1}{\gamma M_\infty^2} + \frac{b^{2/3}}{M_\infty^2} p_3 + \cdots$$

$$\rho = \frac{1}{M_\infty^2 b^{1/3}(\gamma - 1) D\psi_3^{1/2}} + \frac{b^{1/3}}{M_\infty^2}\rho_3 + \cdots, \quad \mu = M_\infty^2 b^{1/3} D\psi_3^{1/2} + \cdots$$

$$h = b^{1/3} D\psi_3^{1/2} + bh_3 + \cdots$$

$$x = bx_2, \quad \psi = \frac{\delta}{M_\infty^2}\psi_2, \quad y = by_{20}(\psi_2) + \delta b^{2/3} y_2 + \cdots \tag{8.120}$$

$$u = u_{20}(\psi_2) + b^{2/3} u_2 + \cdots, \quad v = \frac{\delta}{b^{1/3}} v_2 + \cdots, \quad p = \frac{1}{\gamma M_\infty^2} + \frac{b^{2/3}}{M_\infty^2} p_2 + \cdots$$

$$\rho = \frac{\rho_{20}(\psi_2)}{M_\infty^2} + \frac{b^{2/3}}{M_\infty^2}\rho_2 + \cdots, \quad h = h_{20}(\psi_2) + b^{2/3} h_2 + \cdots$$

Substituting expansions (8.119) and (8.120) in the Navier–Stokes equations and passing to the limit $M_\infty \to \infty$, $\delta \to 0$, $\chi \to 0$, and $\delta \ll b \ll \delta M_\infty$ shows that in the first approximation the flow in region 3 is governed by the linearized Prandtl boundary layer equations (8.107) and the flow in region 2 by the linearized Euler equations; if disturbances decay upstream (as $x_2 \to -\infty$) the latter equations can be brought into the form:

$$\rho_{20} u_{20} u_2 + p_2 = 0, \quad \frac{\partial y_2}{\partial x_2} = \frac{v_2}{u_{20}}, \quad p_2 = p_2(x_2) \tag{8.121}$$

$$\frac{\partial y_{20}}{\partial \psi_2} = \frac{1}{\rho_{20} u_{20}}, \quad \rho_{20} h_2 - p_2 = 0, \quad (\gamma - 1)\rho_{20} h_{20} = 1, \quad \rho_2 = \rho_{20} p_2$$

$$\frac{\partial y_2}{\partial \psi_2} = p_2 \left(\frac{1}{M_{20}^2} - 1\right)$$

where $M_{20}(\psi_2)$ is the Mach number profile in the undisturbed boundary layer. Integrating the equation for the region 2 thickness variation of system (8.121) gives

$$y_2(x_2, \psi_2) - y_2(x_2, 0) = p_2(x_2) \int_0^{y_{20}} \left( \frac{1}{M_{20}^2} - 1 \right) \mathrm{d}y_{20} \tag{8.122}$$

Matching expansions (8.119) and (8.120) shows that conditions (8.113) hold again and, since in the case in question for $\delta \ll b \ll \delta M_\infty$ the roughness thickness must be offset by the variation of the region 2 thickness, condition $y_2(x_2, \infty) \to 0$ applied to Eq. (8.122) makes it possible to determine the pressure distribution $p_3(x_3)$

$$p_3(x_3) = -\frac{f(x_3)}{L}, \quad L = \int_0^{\infty} \left( \frac{1}{M_{20}^2} - 1 \right) \mathrm{d}y_{20} \tag{8.123}$$

where the sign of integral $L$ indicates whether the undisturbed boundary layer is on average subsonic ($L > 0$) or supersonic ($L < 0$). Since for the flat-plate boundary layer the integral $L < 0$, as $M_\infty \to \infty$, $\chi \to 0$, and $h_w \to 0$, a convexity $f(x_3) > 0$ induces a positive pressure disturbance $p_3(x_3) > 0$, as in a supersonic flow.

Formulas similar to (8.115) are valid for the viscous stress $\tau$ and the heat flux $q$

$$\tau Re_\infty^{1/2} = \frac{C^2}{2(\gamma - 1)} \left( 1 + b^{1/3} \frac{2\psi_3^{1/2}}{C} \frac{\partial u_3}{\partial \psi_3} \right)$$

$$q Re_\infty^{1/2} \sigma = \frac{CD}{2(\gamma - 1)} (1 + b^{2/3}(\gamma - 1) p_3(x_3))$$

where $u_3(x_3, \psi_3)$ is the solution of the corresponding Prandtl equation (8.107) at a given pressure distribution (8.123).

If the characteristic roughness length $b \sim \delta M_\infty$, then all formulas (8.116)–(8.123) are valid; only, according to estimates (8.116) and (8.118), the roughness thickness is no longer offset by the variation of the thickness of the main part of the boundary layer (region 2) and the outer edge of the boundary layer is displaced by a quantity equal in the order to the roughness thickness or the viscous wall layer thickness $\sim \delta b^{2/3}$.

For this reason, for the disturbed region of the uniform oncoming flow (region 1) we introduce the following new independent variables and asymptotic expansions of the flow functions

$$x = bx_1, \quad \psi = \frac{b}{M_\infty} \psi_1, \quad y = \frac{b}{M_\infty} \psi_1 + \delta b^{2/3} y_1 + \cdots \tag{8.124}$$

$$u = 1 + \frac{b^{2/3}}{M_\infty^2} u_1 + \cdots, \quad v = \frac{b^{2/3}}{M_\infty} v_1 + \cdots, \quad p = \frac{1}{\gamma M_\infty^2} + \frac{b^{2/3}}{M_\infty^2} p_1 + \cdots$$

$$\rho = 1 + b^{2/3} \rho_1 + \cdots, \quad h = \frac{1}{(\gamma - 1) M_\infty^2} + \frac{b^{2/3}}{M_\infty^2} h_1 + \cdots$$

Substituting expansions (8.124) in the Navier–Stokes equations and passing to the limit $M_\infty \to \infty$, $\delta \to 0$, $\chi \to 0$, and $b \sim \delta M_\infty$ shows that in the first approximation the flow in region 1 is described by the linearized Euler equations

$$\frac{\partial u_1}{\partial x_1} + \frac{\partial p_1}{\partial x_1} = 0, \quad \frac{\partial v_1}{\partial x_1} + \frac{\partial p_1}{\partial \psi_1} = 0, \quad \frac{\partial y_1}{\partial x_1} = v_1 \tag{8.125}$$

$$\frac{\partial y_1}{\partial \psi_1} = -\rho_1, \quad \frac{\partial h_1}{\partial x_1} - \frac{\partial p_1}{\partial x_1} = 0, \quad (\gamma - 1)h_1 + \rho_1 = \gamma p_1$$

If all disturbances decay upstream, the above equations can be reduced to a single hyperbolic equation

$$\frac{\partial^2 y_1}{\partial x_1^2} = \frac{\partial^2 y_1}{\partial \psi_1^2}, \quad y_1 \to 0, \quad x_1 \to -\infty \tag{8.126}$$

The boundary condition for $\psi_1 = 0$ is obtained by matching expansions (8.120) and (8.124) in regions 1 and 2 and using Eq. (8.122)

$$y_1(x_1, 0) = f(x_1) + p_1(x_1, 0)L \tag{8.127}$$

If disturbances propagate from left to right, the boundary value problem (8.126), (8.127) admits the well-known d'Alembert solution $y_1 = \varphi(x_1 - \psi_1)$. Then rewriting Eq. (8.127) in terms of Eq. (8.125) yields the expression for the pressure disturbance distribution

$$L\frac{d\varphi}{dx_1} - \varphi = -f, \quad \varphi \to 0, \quad x_1 \to -\infty, \quad p_1(x_1, 0) = \frac{d\varphi}{dx_1}$$

$$p_3(x_3) = p_2(x_2) = p_1(x_1, 0) = -\frac{e^{x_1/L}}{L^2} \int_{-\infty}^{x_1} f(x_1)e^{-x_1/L}\, dx_1 - \frac{f(x_1)}{L}$$

which closes the solution of the problem of the flow over a roughness for $a \sim \delta^{5/3} M_\infty^{2/3}$ and $b \sim \delta M_\infty$.

For roughnesses with the characteristic length $b \gg \delta M_\infty$ the pressure disturbance is determined by relation (8.118), while the variation of the thickness of the main part of the boundary layer (region 2) is as follows (cf. estimates (8.116)):

$$\Delta y^{(2)} \sim \delta M_\infty^2 \Delta p \sim \frac{a\delta M_\infty}{b} \ll a \tag{8.128}$$

This means that the outer edge of the boundary layer must be displaced by the characteristic roughness thickness $a$. In the most general case the variation of the region 2 thickness is in the order equal to the viscous wall layer 3 thickness $\Delta y^{(3)} \sim \delta b^{2/3}$, so that we have

$$a \sim \frac{b^{5/3}}{M_\infty}, \quad \Delta p \sim \frac{b^{2/3}}{M_\infty^2} \tag{8.129}$$

For this reason, in the case in question it is sufficient to consider only region 1 in which, according to estimates (8.128) and (8.129), the variables and asymptotic expansions of the flow functions (8.124) are valid, with $\delta$ replaced by $b/M_\infty$. Then for $\delta M_\infty \ll b \leq (\delta/M_\infty)^{3/5}$ in region 1 the d'Alembert solution $y_1 = \varphi(x_1 - \psi_1)$ holds again, subject to the obvious boundary condition $y_1 = f(x_1)$ for $\psi_1 = 0$ and, therefore, $p_3(x_3) = p_2(x_2) = p_1(x_1, 0) = \mathrm{d}f/\mathrm{d}x_1$.

Figure 8.12 presents the general diagram of the flow patterns studied in this section. Now the oncoming flow Mach number $M_\infty$ enters additionally (as compared with the case $h_w \leq 1$ and $(M_\infty^2 - 1) \sim 1$, diagram in Fig. 8.2) in the estimates for the characteristic dimensions of the disturbed regions and the orders of the flow functions. However, the dependence on this parameter could hardly be imagined on a two-dimensional diagram; for this reason, the diagram in Fig. 8.12 is formally plotted for $(M_\infty^2 - 1) \sim 1$.

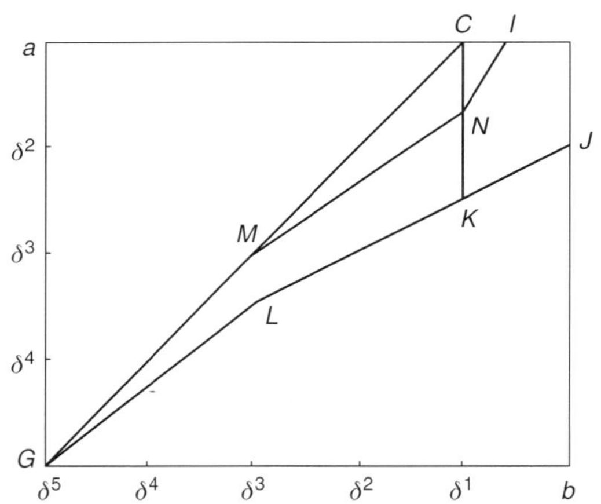


Fig. 8.12.

In the diagram, line $GC$ corresponds to fairly thick ($a \sim b$) roughnesses which induce viscous stress and heat flux disturbances of the same order as their main values near the plate surface, $\Delta\tau \sim \tau \sim \Delta q \sim q \sim \delta/M_\infty^2$. Increasingly thinner roughnesses induce only smaller-in-the-order disturbances $\Delta\tau$ and $\Delta q$; on lines $MN$ ($a \sim \delta b^{2/3}$) and $NI$ ($a \sim b^{5/3}/M_\infty$) the flow in the wall region 3 is viscous. Lines $GL$ ($\ell_1 \sim \delta^{5/4} b^{3/4}$) and $LJ$ ($\ell_1 \sim \delta^2 b^{1/2}$) correspond to variations of the order of the free molecular path. Line $CK$ ($b \sim \delta M_\infty$) separates roughnesses according to whether the disturbed flow over the roughness interacts (to the right of the line) or not (to the left) with the outer uniform flow.

The comparison of the diagrams presented in Figs. 8.2 and 8.12 shows that cooling the plate surface reduces considerably the dimensions of the disturbed regions and the disturbances of the flow functions in these regions.

## 8.2 Regimes of the flow over three-dimensional roughnesses

In the flow over three-dimensional roughnesses it is additionally assumed that their characteristic width $c$ is in the order greater than or equal to the characteristic thickness: $c \geq a$. For $a \gg c$ flows over roughnesses can have the same distinctive features as in the case $a \sim c$, with the only difference that the transverse dimension of the disturbed region is determined by $a$ (see Fig. 8.13).

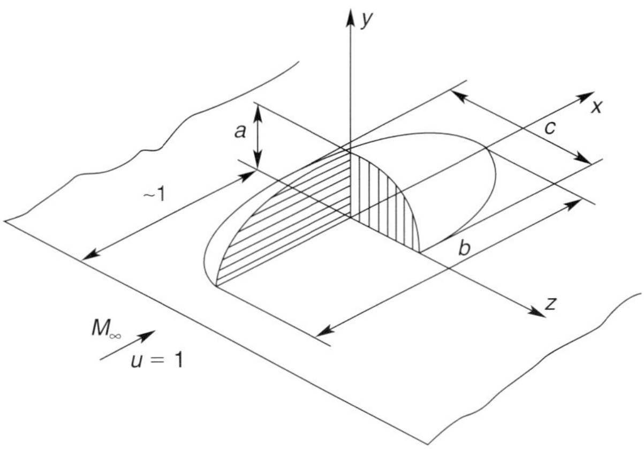


Fig. 8.13.

Among these roughness, we will consider only those which induce high local pressure gradients $\partial p/\partial x \gg 1$ or $\partial p/\partial z \gg 1$ or those for which the convective or diffusive terms of the Navier–Stokes equations are large in the disturbed flow regions (e.g., $\rho u \partial u/\partial x \gg 1$ or $\varepsilon^2 \partial(\mu \partial u/\partial y) \gg 1$). Emphasis is placed on such regimes of the flow over roughnesses in which viscous nonlinear disturbances are induced in wall regions (Bogolepov, 1986). The analysis of the flow over two-dimensional roughnesses drawn in the previous section showed that precisely these cases are most general, while solutions for all other cases can be obtained by the corresponding passages to limit.

### *8.2.1 Flow over fairly narrow roughness of the type of a hole or a hill*

In analyzing the flow over two-dimensional roughnesses it was obtained that in the viscous nonlinear disturbance layer 3 with the characteristic thickness $\Delta y^{(3)} \sim \varepsilon b^{1/3}$ the longitudinal velocity and the pressure disturbance are of the following orders: $u^{(3)} \sim b^{1/3}$ and $\Delta p \sim b^{2/3}$.

Obviously, two-dimensional roughnesses represent the limiting case of wide roughnesses with $c \gg b$. Therefore, from the comparison of the orders of the terms of the Navier–Stokes equations ($\rho u \partial w/\partial x \sim \partial p/\partial z$) we can obtain an estimate for the transverse velocity component $w$

$$w \sim \frac{b^{4/3}}{c}, \quad \frac{\partial w}{\partial z} \sim \frac{b^{4/3}}{c^2} \sim \left(\frac{b}{c}\right)^2 \frac{\partial u}{\partial x} \tag{8.130}$$

These relations show that for wide roughnesses the complete system of equations governing the three-dimensional disturbed flow region breaks up to a system of equations for longitudinal sections of the roughness, which involves $z$ as a parameter, and the equation of transverse momentum conservation linearized with respect to the velocity component $w$, that is, with the term $\rho w \partial w/\partial z$ missing from the convective operator; this equation can be solved separately. Thus, the legitimacy of the above analysis of the flow over two-dimensional roughnesses is supported.

Therefore, with decrease in the characteristic roughness width down to $c \sim b$ and for $h_w \leq 1$ all the estimates and conclusions made in previous Sections 8.1.2–8.1.5 remain valid, only now in the viscous nonlinear disturbance layer 3 the transverse velocity component is equal in the order to the longitudinal velocity

$$w \sim u \sim b^{1/3} \tag{8.131}$$

while the transverse viscous stress is equal in the order to the longitudinal viscous stress $\tau$ and the system of equations governing the three-dimensional disturbed flow region no longer breaks up.

This reasoning makes it possible to construct the solution of the Navier–Stokes equations describing the flow over fairly narrow roughnesses of the type of a round hole or hill for $\varepsilon^2 \ll a \leq b \sim c \ll 1$. For this purpose, in the viscous nonlinear disturbance layer 3 we introduce the following independent variables and asymptotic expansions of the flow functions

$$x = bx_3, \quad y = \varepsilon b^{1/3} y_3, \quad z = bz_3 \tag{8.132}$$

$$u = b^{1/3} u_3 + \cdots, \quad v = \frac{\varepsilon}{b^{1/3}} v_3 + \cdots, \quad w = b^{1/3} w_3 + \cdots$$

$$p = \frac{1}{\gamma M_\infty^2} + b^{2/3} p_3 + \cdots, \quad \rho = \rho_w + \cdots, \quad \mu = \mu_w + \cdots$$

Substituting expansions (8.132) in the Navier–Stokes equations and passing to the limit $\varepsilon \to 0$ shows that in the first approximation region 3 is described by the complete incompressible Navier–Stokes equations for $a \sim b \sim c \sim \varepsilon^{3/2}$

$$\frac{\partial u_3}{\partial x_3} + \frac{\partial v_3}{\partial y_3} + \frac{\partial w_3}{\partial z_3} = 0 \tag{8.133}$$

$$\rho_w\left(u_3\frac{\partial u_3}{\partial x_3}+v_3\frac{\partial u_3}{\partial y_3}+w_3\frac{\partial u_3}{\partial z_3}\right)+\frac{\partial p_3}{\partial x_3}=\mu_w\left(\frac{\partial^2 u_3}{\partial x_3^2}+\frac{\partial^2 u_3}{\partial y_3^2}+\frac{\partial^2 u_3}{\partial z_3^2}\right)$$

$$\rho_w\left(u_3\frac{\partial v_3}{\partial x_3}+v_3\frac{\partial v_3}{\partial y_3}+w_3\frac{\partial v_3}{\partial z_3}\right)+\frac{\partial p_3}{\partial y_3}=\mu_w\left(\frac{\partial^2 v_3}{\partial x_3^2}+\frac{\partial^2 v_3}{\partial y_3^2}+\frac{\partial^2 v_3}{\partial z_3^2}\right)$$

$$\rho_w\left(u_3\frac{\partial w_3}{\partial x_3}+v_3\frac{\partial w_3}{\partial y_3}+w_3\frac{\partial w_3}{\partial z_3}\right)+\frac{\partial p_3}{\partial z_3}=\mu_w\left(\frac{\partial^2 w_3}{\partial x_3^2}+\frac{\partial^2 w_3}{\partial y_3^2}+\frac{\partial^2 w_3}{\partial z_3^2}\right)$$

or the Stokes equations for $\varepsilon^2 \ll a \sim b \sim c \ll \varepsilon^{3/2}$

$$\frac{\partial u_3}{\partial x_3}+\frac{\partial v_3}{\partial y_3}+\frac{\partial w_3}{\partial z_3}=0 \tag{8.134}$$

$$\frac{\partial p_3}{\partial x_3}=\mu_w\left(\frac{\partial^2 u_3}{\partial x_3^2}+\frac{\partial^2 u_3}{\partial y_3^2}+\frac{\partial^2 u_3}{\partial z_3^2}\right)$$

$$\frac{\partial p_3}{\partial y_3}=\mu_w\left(\frac{\partial^2 v_3}{\partial x_3^2}+\frac{\partial^2 v_3}{\partial y_3^2}+\frac{\partial^2 v_3}{\partial z_3^2}\right)$$

$$\frac{\partial p_3}{\partial z_3}=\mu_w\left(\frac{\partial^2 w_3}{\partial x_3^2}+\frac{\partial^2 w_3}{\partial y_3^2}+\frac{\partial^2 w_3}{\partial z_3^2}\right)$$

At the roughness surface $y_3 = f(x_3, z_3)$ the conventional no-slip and impermeability conditions must be fulfilled

$$u_3 = v_3 = w_3 = 0, \quad y_3 = f(x_3, z_3) \tag{8.135}$$

The outer boundary conditions are obtained by matching with the solution for the undisturbed flat-plate boundary layer

$$u_3 \to Ay_3, \quad v_3, w_3, p_3 \to 0, \quad x_3^2 + y_3^2 + z_3^2 \to \infty \tag{8.136}$$

It is well known that systems (8.133), (8.135), and (8.136) or (8.134)–(8.136) represent elliptic boundary value problems.

If $\varepsilon^{3/2} \ll b \ll 1$, then in the first approximation the flow in region 3 is governed by the three-dimensional incompressible boundary layer equations

$$\frac{\partial u_3}{\partial x_3}+\frac{\partial v_3}{\partial y_3}+\frac{\partial w_3}{\partial z_3}=0 \tag{8.137}$$

$$\rho_w\left(u_3\frac{\partial u_3}{\partial x_3}+v_3\frac{\partial u_3}{\partial y_3}+w_3\frac{\partial u_3}{\partial z_3}\right)+\frac{\partial p_3}{\partial x_3}=\mu_w\frac{\partial^2 u_3}{\partial y_3^2},\quad \frac{\partial p_3}{\partial y_3}=0$$

$$\rho_w\left(u_3\frac{\partial w_3}{\partial x_3}+v_3\frac{\partial w_3}{\partial y_3}+w_3\frac{\partial w_3}{\partial z_3}\right)+\frac{\partial p_3}{\partial z_3}=\mu_w\frac{\partial^2 w_3}{\partial y_3^2}$$

At the surfaces of roughnesses whose thicknesses are equal in the order to the region 3 thickness ($a\sim\varepsilon b^{1/3}$) the boundary conditions (8.135) must be fulfilled, while in the flow over roughnesses in which disturbances are produced due to the interaction between the roughness and the oncoming supersonic flow, for $\varepsilon b^{1/3}\ll a\sim b^{5/3}$ we have the boundary conditions of the type

$$u_3=v_3=w_3=0,\quad y_3=0 \tag{8.138}$$

Here, the coordinate $y_3$ is measured normal to the roughness surface. The initial boundary conditions are obtained by matching with the solution for the undisturbed flat-plate boundary layer

$$u_3\to Ay_3,\quad v_3,w_3,p_3\to 0,\quad x_3\to-\infty,\quad z_3\to\pm\infty \tag{8.139}$$

For specifying the outer boundary conditions it is necessary to consider additionally region 2 whose characteristic thickness $\Delta y^{(2)}\sim b$ for $\varepsilon^{3/2}\ll b\ll\varepsilon$ or $\Delta y^{(2)}\sim\varepsilon$ for $\varepsilon\le b\ll 1$. Therefore, in region 2 in the former case the following independent variables and asymptotic expansions of the flow functions are valid

$$x_2=x_3=\frac{x}{b},\quad y_2=\frac{y}{b},\quad z_2=z_3=\frac{z}{b} \tag{8.140}$$

$$u=\frac{b}{\varepsilon}Ay_2+\frac{\varepsilon}{b^{1/3}}u_{22}+\cdots,\quad v=\frac{\varepsilon}{b^{1/3}}v_{22}+\cdots,\quad w=\frac{\varepsilon}{b^{1/3}}w_{22}+\cdots$$

$$p=\frac{1}{\gamma M_\infty^2}+b^{2/3}p_2+\cdots,\quad \rho=\rho_w+\cdots$$

while in the latter case we have the variables and the expansions of the form:

$$x_2=x_3=\frac{x}{b},\quad y_2=\frac{y}{\varepsilon},\quad z_2=z_3=\frac{z}{b} \tag{8.141}$$

$$u=u_{20}(y_2)+b^{1/3}u_{21}+b^{2/3}u_{22}+\cdots,\quad v=b^{2/3}v_{21}+\frac{\varepsilon}{b^{1/3}}v_{22}+\cdots$$

$$w=b^{2/3}w_{22}+\cdots,\quad p=\frac{1}{\gamma M_\infty^2}+b^{2/3}p_2+\cdots$$

$$\rho=\rho_{20}(y_2)+b^{1/3}\rho_{21}+b^{2/3}\rho_{22}+\cdots,\quad h=h_{20}(y_2)+b^{1/3}h_{21}+\cdots$$

Substituting expansions (8.140) or (8.141) in the Navier–Stokes equations and passing to the limit $\varepsilon \to 0$ for $\varepsilon^{3/2} \ll b \ll 1$ shows that in both cases in the first approximation the flow in region 2 is governed by the Euler equations linearized with respect to the oncoming flow ($u = (b/\varepsilon)Ay_2$ or $u = u_{20}(y_2)$) and for $\varepsilon^{3/2} \ll b \ll \varepsilon^{3/4}$ the use of expansions (8.140) or (8.141) leads to the same basic result

$$A\rho_w v_{22} + \frac{\partial p_2}{\partial x_2} \to 0, \quad y_2 \to 0 \tag{8.142}$$

Then matching expansions in regions 2 and 3 and using relations (8.142) makes it possible to derive the outer boundary conditions

$$u_3 \to Ay_3, \quad w_3 \to 0, \quad A\rho_w v_3 + \frac{\partial p_3}{\partial x_3} \to 0, \quad y_3 \to \infty \tag{8.143}$$

The boundary value problem (8.135), (8.137), (8.139), and (8.143) describes the three-dimensional compensation regime of the flow over roughnesses for $a \sim \varepsilon b^{1/3}$, $\varepsilon^{3/2} \ll b \ll \varepsilon^{3/4}$, and $c \sim b$, whose important difference from the corresponding two-dimensional flow regime (8.38)–(8.42) is upstream disturbance propagation, as shown below, in Section 8.2.2.

For $b \sim \varepsilon^{3/4}$ the linearized system of Euler equations governing the flow in region 2 admits partial integration

$$p_2 = p_2(x_2, z_2), \quad u_{21} = D\frac{du_{20}}{dy_2}, \quad v_{21} = -u_{20}\frac{\partial D}{\partial x_2}, \quad D = D(x_2, z_2) \tag{8.144}$$

Matching the expansions for regions 2 and 3 and using relations (8.144) gives the following outer boundary conditions

$$u_3 \to A(y_3 + D), \quad w_3 \to 0, \quad y_3 \to \infty \tag{8.145}$$

When disturbances are produced due to the interaction between roughnesses and the oncoming supersonic flow for $a \sim b^{5/3}$, $\varepsilon^{3/4} \ll b \le \varepsilon^{3/5}$, in region 2 we introduce the following independent variables and asymptotic expansions of the flow functions

$$x_2 = x_3 = \frac{x}{b}, \quad y = \varepsilon y_2 + b^{5/3} f(x_2, z_2) + \cdots, \quad z_2 = z_3 = \frac{z}{b} \tag{8.146}$$

$$u = u_{20}(y_2) + b^{2/3}u_2 + \cdots, \quad v = b^{2/3}v_2 + \cdots, \quad w = b^{2/3}w_2 + \cdots$$

$$p = \frac{1}{\gamma M_\infty^2} + b^{2/3}p_2 + \cdots, \quad \rho = \rho_{20}(y_2) + b^{2/3}\rho_2 + \cdots$$

$$h = h_{20}(y_2) + b^{2/3}h_2 + \cdots$$

Substituting expressions (8.146) in the Navier–Stokes equations and passing to the limit $\varepsilon \to 0$ for $\varepsilon^{3/4} \ll b \leq \varepsilon^{3/5}$ shows that in the first approximation the flow in region 2 is again governed by the linearized Euler equations from which there follows:

$$p_2 = p_2(x_2, z_2), \quad v_2 = u_{20}\frac{\partial f}{\partial x_2} \tag{8.147}$$

while matching the solutions for regions 2 and 3 leads to the outer boundary conditions

$$u_3 \to Ay_3, \quad w_3 \to 0, \quad y_3 \to \infty \tag{8.148}$$

Now for determining the pressure disturbance for $\varepsilon^{3/4} \leq b \leq \varepsilon^{3/5}$ it is necessary to consider the disturbed region 1 of the oncoming uniform flow in which the following new independent variables and asymptotic expansions of the flow functions are valid

$$x_1 = x_2 = x_3 = \frac{x}{b}, \quad y_1 = \frac{y}{b}, \quad z_1 = z_2 = z_3 = \frac{z}{b} \tag{8.149}$$

$$u = 1 + b^{2/3}u_1 + \cdots, \quad v = b^{2/3}v_1 + \cdots, \quad w = b^{2/3}w_1 + \cdots$$

$$p = \frac{1}{\gamma M_\infty^2} + b^{2/3}p_1 + \cdots, \quad \rho = 1 + b^{2/3}\rho_1 + \cdots$$

$$h = \frac{1}{(\gamma - 1)M_\infty^2} + b^{2/3}h_1 + \cdots$$

Substituting expansions (8.149) in the Navier–Stokes equations and passing to the limit $\varepsilon \to 0$ for $\varepsilon^{3/4} \leq b \leq \varepsilon^{3/5}$ shows that in the first approximation the pressure disturbance in region 1 is governed by the solution of the boundary value problem

$$\frac{\partial^2 p_1}{\partial y_1^2} + \frac{\partial^2 p_1}{\partial z_1^2} = (M_\infty^2 - 1)\frac{\partial^2 p_1}{\partial x_1^2}, \quad p_1 \to 0, \quad x_1^2 + y_1^2 + z_1^2 \to \infty \tag{8.150}$$

$$p_1(x_1, 0, z_1) = p_2(x_2, z_2) = p_3(x_3, z_3)$$

subject to the following inner boundary condition for $b \sim \varepsilon^{3/4}$

$$\frac{\partial p_1}{\partial y_1} = \frac{\partial^2 D}{\partial x_1^2}, \quad y_1 = 0 \tag{8.151}$$

or for $\varepsilon^{3/4} \ll b \leq \varepsilon^{3/5}$

$$\frac{\partial p_1}{\partial y_1} = -\frac{\partial^2 f}{\partial x_1^2}, \quad y_1 = 0 \tag{8.152}$$

Therefore, for $a \sim \varepsilon^{5/4}$ and $b \sim c \sim \varepsilon^{3/4}$ the flow over roughnesses is described by the joint solution of the boundary value problems (8.135), (8.137), (8.139), (8.145), and (8.150), (8.151). This pattern of the flow over roughnesses is a three-dimensional counterpart of the two-dimensional free interaction regime (Neiland, 1968, 1969*a*; Stewartson and Williams, 1969); its particular features were studied in the works of Smith et al. (1977), Ryzhov (1980), Kazakov (1983), Manuilovich (1983), Duck and Burggraf (1986), and Bodonyi and Duck (1988).

In the case in which disturbances are produced due to the interaction between a roughness and the oncoming supersonic flow for $a \sim b^{5/3}$ and $\varepsilon^{3/4} \ll b \sim c \leq \varepsilon^{3/5}$ the pressure disturbance distribution is determined by the solution of Eqs. (8.150) and (8.152). Then the three-dimensional incompressible boundary layer equations (8.137)–(8.139) and (8.148) are solved at the given pressure distribution.

The systematic analysis of the flow patterns for the case of fairly narrow roughnesses with $c \sim b$ performed above shows that for these roughnesses the classification diagram of two-dimensional flow regimes presented in Fig. 8.2 remains valid.

### 8.2.2 *Flow over streamwise-elongated narrow roughnesses*

Let now $c \ll b$ which corresponds to the flow over narrow, streamwise-elongated roughnesses with $\varepsilon^2 \ll a \leq c \ll b \leq 1$. Obviously, in this case gas spreading has a considerable effect on the flow function disturbances near the roughnesses. We will also assume that in the transverse direction disturbances decay over a distance equal to the characteristic roughness width.

If the roughness thickness $a$ is so small that the flow function disturbances are produced only due to the interaction between the roughness and the wall region of the flat-plate boundary layer, then in the nonlinear disturbance layer we have

$$u \sim \Delta u \sim \frac{a}{\varepsilon} \tag{8.153}$$

Since the nature of the flow is essentially three-dimensional, from the equations of continuity and transverse momentum conservation we obtain

$$w \sim \Delta w \sim \frac{ac}{\varepsilon b}, \quad \Delta p \sim w^2 \sim \left(\frac{ac}{\varepsilon b}\right)^2 \tag{8.154}$$

For studying the interaction of the roughness with the oncoming uniform flow it is necessary to consider first a disturbed region with the characteristic dimensions $\varepsilon \ll \Delta x \sim \Delta y \sim \Delta z \sim b \leq 1$. However, on the scale of this region the roughness represents a line having neither thickness nor width. Therefore, this region remains undisturbed, so that we should consider a disturbed flow region 1 with the characteristic dimensions $\varepsilon \ll \Delta x^{(1)} \sim b \leq 1$ and $\varepsilon \ll \Delta y^{(1)} \sim \Delta z^{(1)} \sim c \ll b \ll 1$. On interaction between the oncoming uniform flow and the roughness in this region the vertical velocity component

$$v^{(1)} \sim \frac{a}{b} \tag{8.155}$$

is induced.

From the continuity equation we can obtain the following estimates for the disturbances of the velocity components

$$\Delta u^{(1)} \sim \frac{a}{c}, \quad w^{(1)} \sim \frac{a}{b} \tag{8.156}$$

and from the transverse momentum conservation equation (taking into account that in region 1 the disturbed flow must be governed by the equations linearized with respect to the oncoming uniform flow and, therefore, $\partial w/\partial x \sim \partial p/\partial z$) the estimate for the pressure disturbance

$$\Delta p \sim \frac{ac}{b^2} \tag{8.157}$$

In an intermediate case in which the flow function disturbances are produced due to the interaction of the roughness with the flat-plate boundary layer as a whole and estimates (8.154) and (8.157) are simultaneously valid, we have

$$ac \sim \varepsilon^2 \tag{8.158}$$

Then we can derive the estimates for the viscous nonlinear disturbance layer thickness $\Delta y^{(3)}$ and the pressure disturbance $\Delta p$ for all regimes of the flow over narrow roughnesses

$$\Delta y^{(3)} \sim \frac{\varepsilon c^{1/2}}{\Delta p^{1/4}} \sim \varepsilon b^{1/3}, \quad \Delta p \sim \frac{c^2}{b^{4/3}} \tag{8.159}$$

Estimates (8.153)–(8.159) make it possible to construct the solution of the Navier–Stokes equations corresponding to the flow over narrow, streamwise-elongated roughnesses. For this reason, in the layer 3 of viscous nonlinear disturbances we introduce the following independent variables and asymptotic expansions for the flow functions

$$x = bx_3, \quad y = \varepsilon b^{1/3} y_3, \quad z = cz_3 \tag{8.160}$$

$$u = b^{1/3} u_3 + \cdots, \quad v = \frac{\varepsilon}{b^{1/3}} v_3 + \cdots, \quad w = \frac{c}{b^{2/3}} w_3 + \cdots$$

$$p = \frac{1}{\gamma M_\infty^2} + \frac{c^2}{b^{4/3}} p_3 + \cdots, \quad \rho = \rho_w + \cdots, \quad \mu = \mu_w + \cdots$$

Substituting expansions (8.160) in the Navier–Stokes equations and passing to the limit $\varepsilon \to 0$ shows that for $a \sim c \sim \varepsilon b^{1/3}$ and $\varepsilon^{3/2} \ll b \leq 1$ region 3 is described in the first approximation by longitudinally parabolized incompressible Navier–Stokes equations with the term $\partial p/\partial x$ omitted from the equation of longitudinal momentum conservation

$$\frac{\partial u_3}{\partial x_3} + \frac{\partial v_3}{\partial y_3} + \frac{\partial w_3}{\partial z_3} = 0 \tag{8.161}$$

$$\rho_w\left(u_3\frac{\partial u_3}{\partial x_3}+v_3\frac{\partial u_3}{\partial y_3}+w_3\frac{\partial u_3}{\partial z_3}\right)=\mu_w\left(\frac{\partial^2 u_3}{\partial y_3^2}+\frac{\partial^2 u_3}{\partial z_3^2}\right)$$

$$\rho_w\left(u_3\frac{\partial v_3}{\partial x_3}+v_3\frac{\partial v_3}{\partial y_3}+w_3\frac{\partial v_3}{\partial z_3}\right)+\frac{\partial p_3}{\partial y_3}=\mu_w\left(\frac{\partial^2 v_3}{\partial y_3^2}+\frac{\partial^2 v_3}{\partial z_3^2}\right)$$

$$\rho_w\left(u_3\frac{\partial w_3}{\partial x_3}+v_3\frac{\partial w_3}{\partial y_3}+w_3\frac{\partial w_3}{\partial z_3}\right)+\frac{\partial p_3}{\partial z_3}=\mu_w\left(\frac{\partial^2 w_3}{\partial y_3^2}+\frac{\partial^2 w_3}{\partial z_3^2}\right)$$

and for $\varepsilon^2 \ll a \sim c \ll \varepsilon b^{1/3}$, $\varepsilon^2 \ll b \leq 1$ by the Navier–Stokes equations. The solution of the boundary value problem (8.135), (8.136), and (8.161) (with no convective terms) was obtained in the paper of Denisenko (1978) which, obviously, does not describe upstream disturbance propagation.

If $\varepsilon b^{1/3} \ll c \ll b$, then in the first approximation the flow in layer 3 is governed by the three-dimensional incompressible boundary layer equations with the term $\partial p/\partial x$ omitted from the equation of longitudinal momentum conservation

$$\frac{\partial u_3}{\partial x_3}+\frac{\partial v_3}{\partial y_3}+\frac{\partial w_3}{\partial z_3}=0 \tag{8.162}$$

$$\rho_w\left(u_3\frac{\partial u_3}{\partial x_3}+v_3\frac{\partial u_3}{\partial y_3}+w_3\frac{\partial u_3}{\partial z_3}\right)=\mu_w\frac{\partial^2 u_3}{\partial y_3^2},\qquad \frac{\partial p_3}{\partial y_3}=0$$

$$\rho_w\left(u_3\frac{\partial w_3}{\partial x_3}+v_3\frac{\partial w_3}{\partial y_3}+w_3\frac{\partial w_3}{\partial z_3}\right)+\frac{\partial p_3}{\partial z_3}=\mu_w\frac{\partial^2 w_3}{\partial y_3^2}$$

The solution of Eqs. (8.162) is subject to the initial conditions (8.139) and the conditions at the roughness surface (8.135) (for $a \sim \varepsilon b^{1/3}$) or (8.138) (for $\varepsilon b^{1/3} \ll a \sim b^{2/3}c$).

For obtaining the outer boundary conditions it is necessary to consider region 2 with the characteristic thickness $\Delta y^{(2)} \sim c$ for $\varepsilon b^{1/3} \ll c \ll \varepsilon, \varepsilon^{3/2} \ll b \leq 1$ or $\Delta y^{(2)} \sim \varepsilon$ for $\varepsilon \leq c \ll b \leq 1$. Therefore, in the former case in region 2 we introduce the following independent variables and asymptotic expansions of the flow functions

$$x_2 = x_3 = \frac{x}{b},\quad y_2 = \frac{y}{c},\quad z_2 = z_3 = \frac{z}{c} \tag{8.163}$$

$$u = \frac{c}{\varepsilon}Ay_2 + \frac{\varepsilon b^{2/3}}{c}u_{22} + \cdots,\quad v = \frac{\varepsilon}{b^{1/3}}v_{22} + \cdots,\quad w = \frac{\varepsilon}{b^{1/3}}w_{22} + \cdots$$

$$p = \frac{1}{\gamma M_\infty^2} + \frac{c^2}{b^{4/3}}p_2 + \cdots,\quad \rho = \rho_w + \cdots$$

and in the latter case the variables and expansions of the form:

$$x_2 = x_3 = \frac{x}{b},\quad y_2 = \frac{y}{\varepsilon},\quad z_2 = z_3 = \frac{z}{c} \tag{8.164}$$

$$u = u_{20}(y_2) + b^{1/3}u_{21} + b^{2/3}u_{22} + \cdots, \quad v = \frac{c}{b^{1/3}}v_{21} + \frac{\varepsilon}{b^{1/3}}v_{22} + \cdots$$

$$w = \frac{c}{b^{1/3}}w_{22} + \cdots, \quad p = \frac{1}{\gamma M_\infty^2} + \frac{c^2}{b^{4/3}}p_2 + \cdots$$

$$\rho = \rho_{20}(y_2) + b^{1/3}\rho_{21} + b^{2/3}\rho_{22} + \cdots, \quad h = h_{20}(y_2) + b^{1/3}h_{21} + \cdots$$

Substituting expansions (8.163) or (8.164) in the Navier–Stokes equations and passing to the limit $\varepsilon \to 0$ for $\varepsilon^{3/2} \ll b \leq 1$ shows that in both cases the flow in region 2 is described in the first approximation by the Euler equations linearized with respect to the oncoming flow ($u = (c/\varepsilon)Ay_2$ or $u = u_{20}(y_2)$) with the term $\partial p/\partial x$ omitted from the equation of longitudinal momentum conservation, the use of expansions (8.163) or (8.164) for $\varepsilon b^{1/3} \ll c \ll \varepsilon/b^{1/3}$ and $\varepsilon^{2/3} \ll b \leq 1$ leading to the same basic result, namely, $v_{22} \to 0$ as $y_2 \to 0$. The outer boundary conditions take the form:

$$u_3 \to Ay_3, \quad v_3, w_3 \to 0, \quad y_3 \to \infty \tag{8.165}$$

The boundary value problem (8.135), (8.139), (8.162), and (8.165) describes the three-dimensional compensation regime of the flow over narrow roughnesses with the characteristic dimensions $a \sim \varepsilon b^{1/3}$, $\varepsilon^{3/2} \ll b \leq 1$, and $a \ll c \ll \varepsilon/b^{1/3}$. Here, owing to the absence of the term $\partial p/\partial x$ from the equation of longitudinal momentum conservation, as in the compensation flow over two-dimensional roughnesses, disturbances are not transferred upstream. For $c \sim \varepsilon/b^{1/3}$ in region 2 Eqs. (8.144) are valid again and then conditions (8.145) represent the outer boundary conditions.

In the case in which narrow roughnesses for $a \sim b^{2/3}c$, $\varepsilon^{3/4} \ll b \leq 1$, and $\varepsilon/b^{1/3} \ll c \ll \varepsilon/b^{2/3}$ interact only with the outer supersonic flow, in region 2 the following independent variables and asymptotic expansions of the flow functions are introduced

$$x_2 = x_3 = \frac{x}{b}, \quad y = \varepsilon y_2 + b^{2/3}cf(x_2, z_2) + \cdots \quad z_2 = z_3 = \frac{z}{c} \tag{8.166}$$

$$u = u_{20}(y_2) + b^{2/3}u_2 + \cdots, \quad v = \frac{c}{b^{1/3}}v_2 + \cdots, \quad w = \frac{c}{b^{1/3}}w_2 + \cdots$$

$$p = \frac{1}{\gamma M_\infty^2} + \frac{c^2}{b^{4/3}}p_2 + \cdots, \quad \rho = \rho_{20}(y_2) + b^{2/3}\rho_2 + \cdots$$

$$h = h_{20}(y_2) + b^{2/3}h_2 + \cdots$$

Substituting expansions (8.166) in the Navier–Stokes equations and passing to the limit $\varepsilon \to 0$, $\varepsilon^{3/4} \ll b \leq 1$ shows that in the first approximation in region 2 Eq. (8.147) are valid, while in region 3 the solution must satisfy the outer boundary conditions (8.148).

Then we should consider the disturbed region 1 of the oncoming uniform flow with the characteristic dimensions $\Delta x^{(1)} \sim b$, $\Delta y^{(1)} \sim \Delta z^{(1)} \sim c$, $\varepsilon^{3/4} \leq b \leq 1$, and $\varepsilon/b^{1/3} \leq c \leq \varepsilon/b^{2/3}$, in which the following independent variables and asymptotic expansions of the flow functions are valid

$$x_1 = x_2 = x_3 = \frac{x}{b}, \quad y_1 = \frac{y}{c}, \quad z_1 = z_2 = z_3 = \frac{z}{c} \tag{8.167}$$

$$u = 1 + b^{2/3}u_1 + \cdots, \quad v = \frac{c}{b^{1/3}}v_1 + \cdots, \quad w = \frac{c}{b^{1/3}}w_1 + \cdots$$

$$p = \frac{1}{\gamma M_\infty^2} + \frac{c^2}{b^{4/3}}p_1 + \cdots, \quad \rho = 1 + \frac{c^2}{b^{4/3}}\rho_1 + \cdots$$

$$h = \frac{1}{(\gamma - 1)M_\infty^2} + \frac{c^2}{b^{4/3}}h_1 + \cdots$$

Substituting expansions (8.167) in the Navier–Stokes equations and passing to the limit $\varepsilon \to 0$ for $\varepsilon^{3/4} \ll b \le 1$ shows that in the first approximation in region 1 the pressure disturbance is described by the solution of the boundary value problem

$$\frac{\partial^2 p_1}{\partial y_1^2} + \frac{\partial^2 p_1}{\partial z_1^2} = 0 \quad p_1 \to 0, \quad x_1^2 + y_1^2 + z_1^2 \to \infty \tag{8.168}$$

$$p_1(x_1, 0, z_1) = p_2(x_2, z_2) = p_3(x_3, z_3)$$

which must satisfy the inner boundary condition (8.151) for $c \sim \varepsilon/b^{1/3}$ or (8.152) for $\varepsilon/b^{1/3} \ll c \le \varepsilon/b^{2/3}$. Obviously, the solutions of the boundary value problems (8.151), (8.168) or (8.152), (8.168) are independent of $M_\infty$, in spite of the fact that region 1 of the supersonic oncoming uniform flow is disturbed.

The joint solution of the boundary value problems (8.135), (8.139), (8.145), (8.162) and (8.151), (8.168) describes the flow over narrow roughnesses with the characteristic dimensions $a \sim \varepsilon b^{1/3}$, $\varepsilon^{3/4} \ll b \le 1$, and $c \sim \varepsilon/b^{1/3}$ in the free interaction regime. Here, owing to the absence of the term $\partial p/\partial x$ disturbances are no longer transferred upstream.

For narrow roughnesses with the characteristic dimensions $a \sim b^{2/3}c$, $\varepsilon^{3/4} \ll b \le 1$, and $\varepsilon b^{1/3} \ll c \le \varepsilon/b^{2/3}$, when pressure disturbances are produced due to the interaction between the roughness and the oncoming supersonic flow, they are determined from the solution of the boundary value problem (8.152) and (8.168). Then in region 3 the boundary value problem (8.138), (8.139), (8.148), and (8.162) is solved at the given pressure distribution.

In Fig. 8.14 the general classification of the regimes of three-dimensional local flows is presented. Here, the orders of the characteristic dimensions are measured along the axes

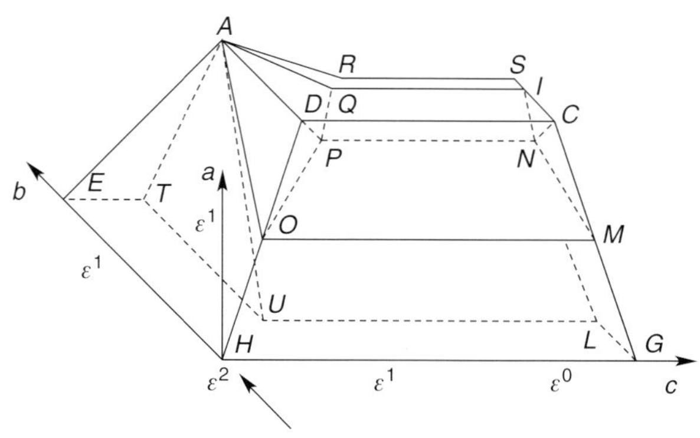


Fig. 8.14.

of a rectilinear coordinate system: the roughness thickness $a$ along the vertical axis, the length $b$ along the axis aligned with the main flow, and the transverse dimension $c$ perpendicular to the main flow direction. Roughnesses induce viscous nonlinear disturbances in planes *APNMO* for $a \sim \varepsilon b^{1/3}$ (compensation regime of the flow over a roughness), *PQIN* for $a \sim b^{5/3}$ (disturbances are produced due to the interaction of fairly wide roughnesses with the oncoming supersonic flow), and *AQP* for $a \sim b^{2/3}c$ (disturbances are produced due to the interaction of narrow roughnesses with the oncoming supersonic flow). Thicker roughnesses induce inviscid nonlinear disturbances, while thinner roughnesses lead to viscous linear disturbances.

The least degenerate regimes of the flow over three-dimensional roughnesses are realized for $b \sim c$. On line *HO* ($\varepsilon^2 \ll a \sim b \sim c \ll \varepsilon^{3/2}$) the flow over roughnesses is governed by the Stokes equations and at point *O* ($a \sim b \sim c \sim \varepsilon^{3/2}$) by the complete three-dimensional incompressible Navier–Stokes equations. On line *OP* the three-dimensional compensation regime of the flow over roughnesses, with upstream disturbance transfer, is realized. For $a \sim \varepsilon^{5/4}$ and $b \sim c \sim \varepsilon^{3/4}$ (point *P*) the three-dimensional free interaction regime of the flow over roughnesses is realized. Finally, on line *PQ* ($a \sim b^{5/3} \sim c^{5/3}$) the flow over roughnesses is described by the Prandtl equations for the three-dimensional boundary layer at a given pressure distribution.

In the case of wide roughnesses ($c \gg b$) the boundary value problems become degenerate in the transverse coordinate $z$: they break up into systems of equations governing the disturbed flow in the $(x, y)$ plane and involving the $z$ coordinate as a parameter and the linearized equations of transverse momentum conservation.

In the case of narrow roughnesses ($c \ll b$) the key role is played by gas spreading, while the boundary value problems become degenerate in the longitudinal coordinate $x$: the terms with the derivatives $\partial p/\partial x$ and $\partial^2 p/\partial x^2$ are absent from all equations. Therefore, in all regimes of the flow over narrow roughnesses, in which disturbances decay in the transverse direction over a distance equal to the characteristic width, disturbances are not transferred upstream. Moreover, in this case the solutions are independent of $M_\infty$ even in the case in which the oncoming uniform flow region is disturbed.

The surfaces *APD* ($a \sim \varepsilon^2/c$) and *PNCD* ($a \sim \varepsilon^2/b$) separate the roughnesses for which the flow function disturbances are produced due to the interaction of a roughness with the wall region of the flat-plate boundary layer and the oncoming supersonic uniform flow.

It can also be shown that the surfaces $a \sim \varepsilon b^{2/3}$, $a \sim c^2/\varepsilon$, $a \sim b^2$, and $a \sim bc$ cut off the regimes of the flow over roughnesses for which the pressure gradients are small, $\partial p/\partial x \ll 1$ or $\partial p/\partial z \ll 1$, and the convective or diffusive terms in the Navier–Stokes equations are also small (e.g., $\rho u \partial u/\partial x \ll 1$ or $\varepsilon^2 \mu \partial^2 u/\partial y^2 \ll 1$) (in Fig. 8.14 these surfaces are not presented).

### *8.2.3 Compensation regime of the flow over roughnesses*

The boundary value problem (8.135), (8.137), (8.139), and (8.143) describes the least degenerate compensation regime of the flow over three-dimensional roughnesses (Bogolepov and Lipatov, 1985; Bogolepov, 1987). Using similarity variables, whose form follows from asymptotic estimates for different disturbed flow regions, makes it possible to formulate boundary value problems for roughnesses of different type.

It is useful to recall that in the flow over fairly wide ($c \geq b$) roughnesses in the viscous nonlinear region 3 the following estimates are valid

$$\Delta x \sim b, \quad \Delta y^{(3)} \sim a, \quad \Delta z^{(3)} \sim c \tag{8.169}$$

$$u^{(3)} \sim \frac{a}{\varepsilon}, \quad v^{(3)} \sim \frac{a^2}{\varepsilon b}, \quad w^{(3)} \sim \frac{ac}{\varepsilon b}, \quad \Delta p \sim \left(\frac{a}{\varepsilon}\right)^2$$

Therefore, the transformation of variables

$$x_3 = b_1\bar{x}, \quad y_3 = a_1\Pi^{1/3}\bar{y}, \quad z_3 = c_1\bar{z} \tag{8.170}$$

$$u_3 = Aa_1\Pi^{1/3}\bar{u}, \quad v_3 = \frac{Aa_1^2}{b_1}\Pi^{2/3}\bar{v}, \quad w_3 = \frac{Aa_1c_1}{b_1}\Pi^{1/3}\bar{w}$$

$$p_3 = \rho_w(Aa_1)^2\Pi^{2/3}\bar{p}, \quad \Pi = \frac{\mu_w b_1}{\rho_w A a_1^3} \sim 1$$

where $a_1$, $b_1$, $c_1 \sim 1$ are the roughness thickness, length, and width on the scales $a$, $b$, and $c$, makes it possible to bring Eqs. (8.135), (8.137), (8.139), and (8.143) into the form:

$$\frac{\partial\bar{u}}{\partial\bar{x}} + \frac{\partial\bar{v}}{\partial\bar{y}} + \frac{\partial\bar{w}}{\partial\bar{z}} = 0, \quad \bar{u}\frac{\partial\bar{u}}{\partial\bar{x}} + \bar{v}\frac{\partial\bar{u}}{\partial\bar{y}} + \bar{w}\frac{\partial\bar{u}}{\partial\bar{z}} + C\frac{\partial\bar{p}}{\partial\bar{x}} = \frac{\partial^2\bar{u}}{\partial\bar{y}^2} \tag{8.171}$$

$$\frac{\partial\bar{p}}{\partial\bar{y}} = 0, \quad \bar{u}\frac{\partial\bar{w}}{\partial\bar{x}} + \bar{v}\frac{\partial\bar{w}}{\partial\bar{y}} + \bar{w}\frac{\partial\bar{w}}{\partial\bar{z}} + D\frac{\partial\bar{p}}{\partial\bar{z}} = \frac{\partial^2\bar{w}}{\partial\bar{y}^2}$$

$$\bar{u} = \bar{v} = \bar{w} = 0, \quad \bar{y} = \frac{1}{\Pi^{1/3}}f(\bar{x}, E\bar{z})$$

$$\bar{u} \to \bar{y}, \quad \bar{v}, \bar{w}, \bar{p} \to 0, \quad \bar{x} \to -\infty, \quad \bar{z} \to \pm\infty$$

$$\bar{u} \to \bar{y}, \quad \bar{w} \to 0, \quad \bar{v} + C\frac{\partial\bar{p}}{\partial\bar{x}} \to 0, \quad \bar{y} \to \infty$$

Here, the similarity parameters $C$, $D$, and $E$ represent the relation between the roughness dimensions

$$C = E = 1, \quad D = \left(\frac{b_1}{c_1}\right)^2, \quad b_1 \leq c_1 \tag{8.172}$$

while the roughness shape $f(\bar{x}, E\bar{z})$ is scaled on unit in height, width, and length.

Obviously, at the values of the parameters satisfying relations (8.172) the boundary value problem (8.171) is applicable for studying flows over different fairly wide ($b_1 \leq c_1$) roughnesses. In the limiting case of wide ($b_1 \ll c_1$) roughnesses, $D \to 0$, $\bar{w} \to 0$, its solution describes the flow past their two-dimensional sections for $\overline{z} = \text{const}$ (see the boundary value problem (8.38)–(8.42)).

In the case of the flow over narrow ($b \gg c$) roughnesses there can occur situations in which disturbances decay in the transverse direction over either a distance equal to the characteristic roughness width $\Delta z \sim c$ or a distance $\Delta z \sim b$. Then in the former case estimates (8.154) are valid, while transformations (8.170), in view of the relation

$$p_3 = \rho_w \left(\frac{Aa_1 c_1}{b_1}\right)^2 \Pi^{2/3}\overline{p} \tag{8.173}$$

make it possible to reduce the boundary value problem under consideration to form (8.171) with the following values of the parameters $C$, $D$, and $E$

$$C = \left(\frac{c_1}{b_1}\right)^2, \quad D = E = 1, \quad c_1 \le b_1 \tag{8.174}$$

Formula (8.173) indicates that in the case of the flow over narrow roughnesses the pressure disturbances induced are by a factor of $(b_1/c_1)^2$ less than those in the flow over fairly wide roughnesses (see transformations (8.170)). In the limiting case $c_1 \ll b_1$, $C \to 0$ the terms that contain $\partial\overline{p}/\partial\overline{x}$ drop out from the boundary value problem under consideration and its solution describes the flow over narrow ($c \ll b$) roughnesses.

If disturbances induced by narrow roughnesses decay in the transverse direction over a width $\Delta z \sim b$, then the same velocity disturbances (8.153) and (8.154) are induced in region 3, while the transverse pressure gradient is as follows:

$$\frac{\partial p}{\partial z} \sim w\frac{\partial w}{\partial z} \sim \frac{a^2 c}{\varepsilon^2 b^2} \tag{8.175}$$

In other words, in region 3 the following pressure disturbances are generated

$$\Delta p(x) \sim b\frac{\partial p}{\partial z} \sim \frac{a^2 c}{\varepsilon^2 b}, \quad \Delta p(x,z) \sim c\frac{\partial p}{\partial z} \sim \left(\frac{ac}{\varepsilon b}\right)^2 \tag{8.176}$$

The first estimate (8.176) determines the leading order of the pressure disturbances in region 3. Therefore, the corresponding boundary value problem can be obtained from the general case (8.135), (8.137), (8.139), and (8.143) using the following transformation of variables

$$x_3 = b_1\overline{x}, \quad y_3 = a_1\Pi^{1/3}\overline{y}, \quad z_3 = b_1\overline{z} \tag{8.177}$$

$$u_3 = Aa_1\Pi^{1/3}\overline{u}, \quad v_3 = \frac{Aa_1^2}{b_1}\Pi^{2/3}\overline{v}, \quad w_3 = Aa_1\Pi^{1/3}\overline{w}, \quad p_3 = \frac{\rho_w A^2 a_1^2 c_1}{b_1}\Pi^{2/3}\overline{p}$$

Thus, we arrive again at the boundary value problem (8.171) for the following values of the parameters $C$, $D$, and $E$

$$C = D = \frac{c_1}{b_1}, \quad E = \frac{b_1}{c_1}, \quad c_1 \le b_1 \tag{8.178}$$

From transformations (8.177) it follows that in this case the pressure disturbances are by a factor of $(b_1/c_1)$ smaller than in the case of fairly wide roughnesses. In the limiting case $c_1 \ll b_1$ we have $C, D \to 0$, and $E \to \infty$; then the transverse variation of the roughness shape is described by a delta function which indicates the necessity of introducing two different scales in the transverse direction: $\Delta z \sim c$ and $\Delta z \sim b$, $c \ll b$.

Obviously, for region 3 with the characteristic dimensions $\Delta x \sim b$, $\Delta y^{(3)} \sim \varepsilon b^{1/3}$, and $\Delta z^{(3)} \sim c$ the previous asymptotic expansions of the flow functions (8.160) are valid with the only difference that the pressure expansion includes one more term

$$p = \frac{1}{\gamma M_\infty^2} + \frac{c}{b^{1/3}} p_{31} + \frac{c^2}{b^{4/3}} p_3 + \cdots \tag{8.179}$$

Substituting expansions (8.160) and (8.179) in the Navier–Stokes equations and passing to the limit $\varepsilon \to 0$ for $\varepsilon b^{1/3} \ll c \ll b$ and $\varepsilon^{3/2} \ll b \leq 1$ shows that in the first approximation the flow in region 3 is described by the same boundary value problem (8.135), (8.139), (8.162), and (8.165) of the compensation regime of the flow over narrow roughnesses for which the boundary conditions for $z_3 \to \pm\infty$ are for a while undetermined. For this reason, we will consider one more disturbed region 4 with the characteristic dimensions $\Delta x \sim \Delta z^{(4)} \sim b$ and $\Delta y^{(4)} \sim \Delta y^{(3)} \sim \varepsilon b^{1/3}$ in which we can introduce the following new variables and construct the asymptotic expansions of the flow functions

$$x_4 = x_3 = \frac{x}{b}, \quad y_4 = y_3 = \frac{y}{\varepsilon b^{1/3}}, \quad z_4 = \frac{z}{b} \tag{8.180}$$

$$u = b^{1/3} A y_4 + \frac{c}{b^{2/3}} u_4 + \cdots, \quad v = \frac{\varepsilon c}{b^{4/3}} v_4 + \cdots, \quad w = \frac{c}{b^{2/3}} w_4 + \cdots$$

$$p = \frac{1}{\gamma M_\infty^2} + \frac{c}{b^{1/3}} p_4 + \cdots, \quad \rho = \rho_w + \cdots, \quad \mu = \mu_w + \cdots$$

Substituting expansions (8.180) in the Navier–Stokes equations and passing to the limit $\varepsilon \to 0$ for $\varepsilon^{3/2} \ll c \ll b \leq 1$ shows that in the first approximation the flow in region 4 is governed by the equations of three-dimensional incompressible flat-plate boundary layer linearized with respect to its wall region

$$\frac{\partial u_4}{\partial x_4} + \frac{\partial v_4}{\partial y_4} + \frac{\partial w_4}{\partial z_4} = 0 \tag{8.181}$$

$$\rho_w \left( A y_4 \frac{\partial u_4}{\partial x_4} + A v_4 \right) + \frac{\partial p_4}{\partial x_4} = \mu_w \frac{\partial^2 u_4}{\partial y_4^2}, \quad \frac{\partial p_4}{\partial y_4} = 0$$

$$\rho_w A y_4 \frac{\partial w_4}{\partial x_4} + \frac{\partial p_4}{\partial z_4} = \mu_w \frac{\partial^2 w_4}{\partial y_4^2}$$

Since region 4 is wider in the order than the roughness, the no-slip and impermeability conditions are imposed now on the plane surface

$$u_4 = v_4 = w_4 = 0, \quad y_4 = 0 \tag{8.182}$$

The initial boundary conditions for $x_4 \to -\infty$ or $z_4 \to \pm\infty$ are obtained again by matching the solution for region 4 with that in the wall region of the flat-plate boundary layer

$$u_4, v_4, w_4, p_4 \to 0, \quad x_4 \to -\infty, \quad z_4 \to \pm\infty \tag{8.183}$$

while the outer boundary conditions by matching with the solution for region 2, as it was done in deriving the "compensation" interaction conditions for $c \sim b$ (see expansions (8.140) or (8.141))

$$u_4, w_4 \to 0, \quad A\rho_w v_4 + \frac{\partial p_4}{\partial x_4} \to 0, \quad y_4 \to \infty \tag{8.184}$$

Matching expansions (8.160), (8.179), and (8.180) makes it possible to determine the missing initial boundary conditions for region 3

$$u_3 \to Ay_3, \quad v_3, \frac{\partial w_3}{\partial z_3} \to 0, \quad z_3 \to \pm\infty \tag{8.185}$$

Therefore, the flow over narrow ($c \ll b$) roughnesses in which the pressure disturbances decay in the transverse direction only over a distance $\Delta z^{(4)} \sim b$, is described by the solution of the boundary value problem (8.135), (8.139), (8.162), (8.165), and (8.185) which for $z_3 \to \pm\infty$ determines the transverse pressure gradient distribution

$$\frac{\partial p_3}{\partial z_3} = \pm G(x_3)\rho_w \frac{A^2 a_1^2 c_1}{b_1^2}, \quad z_3 \to \pm\infty \tag{8.186}$$

where the function $G = G(x_3)$ will be determined below.

For region 4 the matching of expansions (8.160), (8.179), and (8.180) makes it possible, in view of Eq. (8.186), to obtain the missing boundary conditions

$$\frac{\partial p_4}{\partial z_4} = \pm G(x_4)\rho_w \frac{A^2 a_1^2 c_1}{b_1^2}, \quad z_4 \to \pm 0 \tag{8.187}$$

while the solution of the boundary value problem (8.181)–(8.184), (8.187) in region 4 determines

$$p_{31}(x_3) = p_4(x_4, z_4), \quad z_4 \to 0 \tag{8.188}$$

In the new variables (8.170) and (8.173) the boundary value problem (8.135), (8.139), (8.162), (8.165), and (8.185) for region 3 takes form (8.171) with

$$\overline{u} \to \overline{y}, \ \overline{v}, \ \frac{\partial \overline{w}}{\partial \overline{z}} \to 0, \ \overline{z} \to \pm\infty; \quad C = 0, \ D = E = 1 \tag{8.189}$$

For the boundary value problem (8.181)–(8.184), and (8.187) in region 4 we also introduce the new variables

$$x_4 = b_1\bar{x}, \quad y_4 = a_1\Pi^{1/3}\bar{y}, \quad z_4 = b_1\bar{z} \tag{8.190}$$

$$u_4 = \frac{Aa_1c_1}{b_1}\bar{u}, \quad v_4 = \frac{Aa_1^2c_1}{b_1^2}\Pi^{1/3}\bar{v}, \quad w_4 = \frac{Aa_1c_1}{b_1}\bar{w}, \quad p_4 = \frac{\rho_w A^2a_1^2c_1}{b_1}\Pi^{1/3}\bar{p}$$

so that it takes the form:

$$\frac{\partial\bar{u}}{\partial\bar{x}} + \frac{\partial\bar{v}}{\partial\bar{y}} + \frac{\partial\bar{w}}{\partial\bar{z}} = 0, \quad \bar{y}\frac{\partial\bar{u}}{\partial\bar{x}} + \bar{v} + C\frac{\partial\bar{p}}{\partial\bar{x}} = \frac{\partial^2\bar{u}}{\partial\bar{y}^2} \tag{8.191}$$

$$\frac{\partial\bar{p}}{\partial\bar{y}} = 0, \quad \bar{y}\frac{\partial\bar{w}}{\partial\bar{x}} + D\frac{\partial\bar{p}}{\partial\bar{z}} = \frac{\partial^2\bar{w}}{\partial\bar{y}^2}$$

$$\bar{u} = \bar{v} = \bar{w} = 0, \quad \bar{y} = 0$$

$$\bar{u}, \bar{v}, \bar{w}, \bar{p} \to 0, \quad \bar{x} \to -\infty, \quad \bar{z} \to \pm\infty$$

$$\bar{u}, \bar{w} \to 0, \quad \bar{v} + C\frac{\partial\bar{p}}{\partial\bar{x}} \to 0, \quad \bar{y} \to \infty$$

$$\frac{\partial\bar{p}}{\partial\bar{z}} = \pm\frac{G(\bar{x})}{\Pi^{1/3}}, \quad \bar{z} \to \pm 0$$

for

$$C = D = 1 \tag{8.192}$$

Thus, it was shown that various cases of the compensation regime of the flow over roughnesses are described by the boundary value problems of form (8.171) or (8.191) for different values of the similarity parameters $C$, $D$, and $E$. These boundary value problems can be divided into two problems. The first, parabolic problem

$$\frac{\partial\bar{u}}{\partial\bar{x}} + \frac{\partial\bar{v}}{\partial\bar{y}} + \frac{\partial\bar{w}}{\partial\bar{z}} = 0, \quad \frac{\partial}{\partial\bar{y}}\left(\bar{u}\frac{\partial\bar{u}}{\partial\bar{x}} + \bar{v}\frac{\partial\bar{u}}{\partial\bar{y}} + \bar{w}\frac{\partial\bar{u}}{\partial\bar{z}}\right) = \frac{\partial^3\bar{u}}{\partial\bar{y}^3} \tag{8.193}$$

$$\frac{\partial}{\partial\bar{y}}\left(\bar{u}\frac{\partial\bar{w}}{\partial\bar{x}} + \bar{v}\frac{\partial\bar{w}}{\partial\bar{y}} + \bar{w}\frac{\partial\bar{w}}{\partial\bar{z}}\right) = \frac{\partial^3\bar{w}}{\partial\bar{y}^3}$$

makes it possible to determine the velocity components. The second, generally elliptic, problem of the form:

$$C\frac{\partial^2\bar{p}}{\partial\bar{x}^2} + D\frac{\partial^2\bar{p}}{\partial\bar{z}^2} = \frac{\partial^2}{\partial\bar{y}^2}\left(\frac{\partial\bar{u}}{\partial\bar{x}} + \frac{\partial\bar{w}}{\partial\bar{z}}\right) \tag{8.194}$$

(its right side is calculated on the roughness surface $\overline{y} = \Pi^{-1/3} f(\overline{x}, E\overline{z})$), serves for determining the pressure disturbance distribution.

It is useful to recall (Bogolepov and Neiland, 1971) that in the compensation regime of the flow over roughnesses the boundary condition

$$\overline{v} + C\frac{\partial \overline{p}}{\partial \overline{x}} \to 0, \quad \overline{y} \to \infty$$

is equivalent to the condition

$$\overline{u} \to \overline{y}, \quad \overline{y} \to \infty \tag{8.195}$$

This shows the possibility to completely separate the boundary value problems for determining the velocity components and the pressure disturbance. An analogous transformation can be applied to problem (8.191).

For $\Pi \to \infty$ the boundary value problems describing different cases of the compensation regime of the flow over roughnesses on a plane surface admit linearization with respect to the small parameter $\Pi^{-1/3} \ll 1$

$$\begin{aligned} \overline{u} &= \overline{y} + \Pi^{-1/3} U + \cdots, \quad \overline{v} = \Pi^{-1/3} V + \cdots \\ \overline{w} &= \Pi^{-1/3} W + \cdots, \quad \overline{p} = \Pi^{-1/3} P + \cdots \end{aligned} \tag{8.196}$$

In new variables (8.196) from Eqs. (8.193) and (8.194) and the boundary conditions (8.171) and (8.195) it follows:

$$C\frac{\partial^2 P}{\partial \overline{x}^2} + D\frac{\partial^2 P}{\partial \overline{z}^2} = \left.\frac{\partial F}{\partial \overline{y}}\right|_{\overline{y}=0}, \quad \overline{y}\frac{\partial F}{\partial \overline{x}} = \frac{\partial^2 F}{\partial \overline{y}^2}, \quad P = P(\overline{x}, \overline{z}) \tag{8.197}$$

$$F(\overline{x}, \overline{y}, \overline{z}) = \frac{\partial}{\partial \overline{y}}\left(\frac{\partial U}{\partial \overline{x}} + \frac{\partial W}{\partial \overline{z}}\right); \qquad P, F \to 0, \ \overline{x} \to -\infty; \quad P \to 0, \ \overline{x} \to \infty$$

$$F \to 0, \ \int_0^{\overline{y}} F \,\mathrm{d}\overline{y} \to \frac{\partial f(\overline{x}, E\overline{z})}{\partial \overline{x}}, \ \overline{y} \to \infty; \quad P \to 0, \ \overline{z} \to \pm\infty$$

where, if need be, the natural condition of disturbance decay far away from the roughness is also used.

For the similarity parameters $C$, $D$, and $E$ determined by Eq. (8.172) the boundary value problem (8.197) is applicable for studying flows over different fairly wide ($b_1 \le c_1$) roughnesses, the pressure disturbance propagation being determined by an elliptic equation. In the limiting case of wide roughnesses ($b_1 \ll c_1$, $D \to 0$, $W \to 0$) the solution of the boundary value problem (8.197) describes the flow past separate sections of the roughness for $\overline{z} = \text{const}$, while the pressure disturbances are determined from the solution of the equation

$$\frac{\partial P}{\partial \overline{x}} = \left.\frac{\partial^2 U}{\partial \overline{y}^2}\right|_{\overline{y}=0}$$

which does not admit upstream disturbance transfer (Bogolepov and Neiland, 1971; Bogolepov, 1974).

For the similarity parameters $C$, $D$, and $E$ determined by Eq. (8.174) the boundary value problem (8.197) makes it possible to obtain the solution for narrow ($c_1 \ll b_1$) roughnesses for which pressure disturbances decay in the transverse direction over their characteristic width. Let the roughness shape be given by the equation $f(\overline{x}, \overline{z}) = f_1(\overline{x}) f_2(\overline{z})$; then we have $F = f_2(\overline{z}) F_1(\overline{x}, \overline{y})$ and for $c_1 \ll b_1$ the pressure disturbances are determined by the solution of the equation

$$\frac{\partial^2 P}{\partial \overline{z}^2} = f_2(\overline{z}) \left.\frac{\partial F_1}{\partial \overline{y}}\right|_{\overline{y}=0} \tag{8.198}$$

which can satisfy the decay conditions for $\overline{z} \to \pm\infty$ if

$$\int_{-\infty}^{+\infty} f_2(\overline{z})\,\mathrm{d}\overline{z} = 0 \tag{8.199}$$

The form of the function $f_2(\overline{z})$ determines the variation of the pressure disturbances in the transverse direction, while the solution of the parabolic equation for $F_1(\overline{x}, \overline{y})$ that in the longitudinal direction. Thus, in this limiting case there is no upstream disturbance transfer.

The numerical solution of the boundary value problem (8.197) subject to condition (8.199) was obtained for a roughness of the form $f(\overline{x}, \overline{z}) = \exp(-\overline{x}^2 - \overline{z}^2)(1 - 2\overline{z}^2)$. In Fig. 8.15 we have plotted the pressure disturbance distributions $P(\overline{x}, 0)$ along the line of symmetry of the roughness. For $c_1 \gg b_1$ each $\overline{z} = \text{const}$ section of the roughness induces a pressure disturbance independent of other sections and the bump of the form $f = \exp(-\overline{z}^2)$ placed in the subsonic wall region of the flat-plate boundary layer induces expansion (curve 1). For, say, $c_1 = 4b_1$ the roughness induces smaller disturbances than for $c_1 \gg b_1$ and positive pressure disturbances appear for $\overline{x} < 0$ (curve 2) due to upstream disturbance propagation.

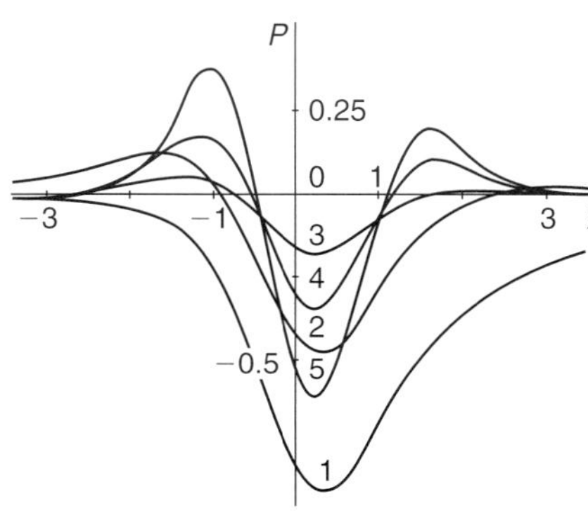


Fig. 8.15.

The pressure disturbance level decreases with the ratio $c_1/b_1$ (for $c_1 \ll b_1$ pressure disturbances are multiplied by $(b_1/c_1)^2$) and the pressure disturbance propagation retains the elliptic nature (curves 3 and 4 for the ratios $c_1/b_1 = 1$ and 0.25, respectively). In the limiting case $c_1 \ll b_1$ the pressure disturbance distribution is described by the equation $P(\bar{x}, \bar{z}) = -0.5\exp(-\bar{z}^2)\partial F_1/\partial\bar{y}|_{\bar{y}=0}$ (curve 5).

The pressure contours plotted in Figs. 8.16–8.20 for the same ratios $b_1/c_1$ as the curves in Fig. 8.15 show the complicated nature of the pressure disturbance propagation in the flow over three-dimensional roughnesses and degeneration of the flows in the longitudinal and transverse directions for $c_1 \gg b_1$ and $c_1 \ll b_1$.

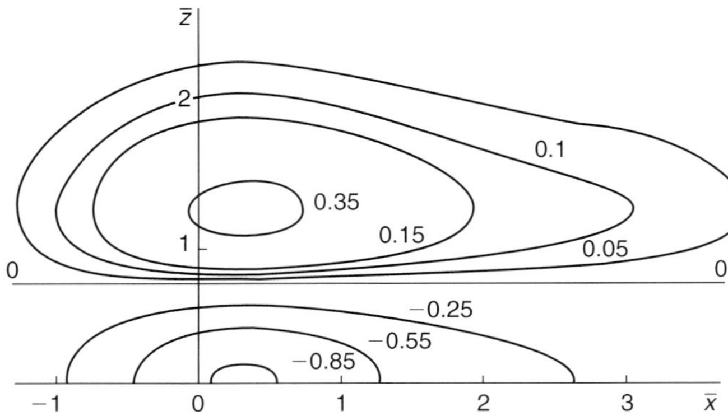


Fig. 8.16.

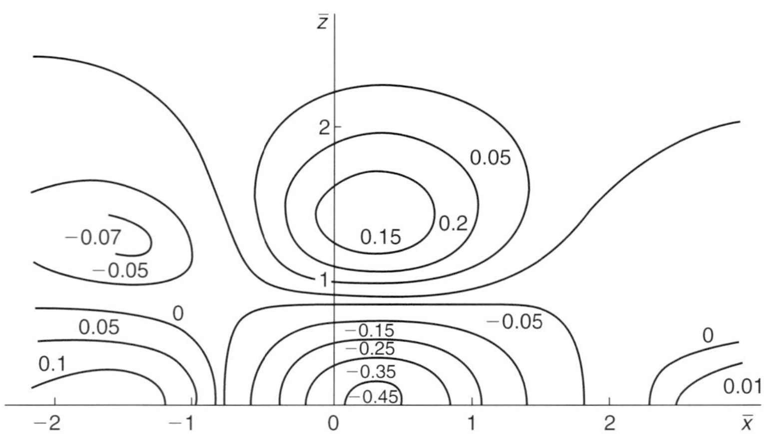


Fig. 8.17.

If condition (8.199) is not fulfilled, then for narrow ($c_1 \ll b_1$) roughnesses pressure disturbances decay in the transverse direction over a distance comparable with the characteristic roughness length; then the solution of the boundary value problem (8.197) is fulfilled for the similarity parameters $C$, $D$, and $E$ determined by Eq. (8.178).

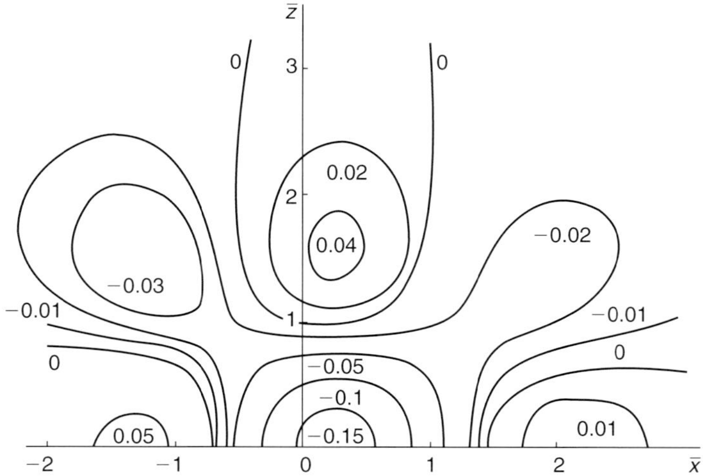


Fig. 8.18.

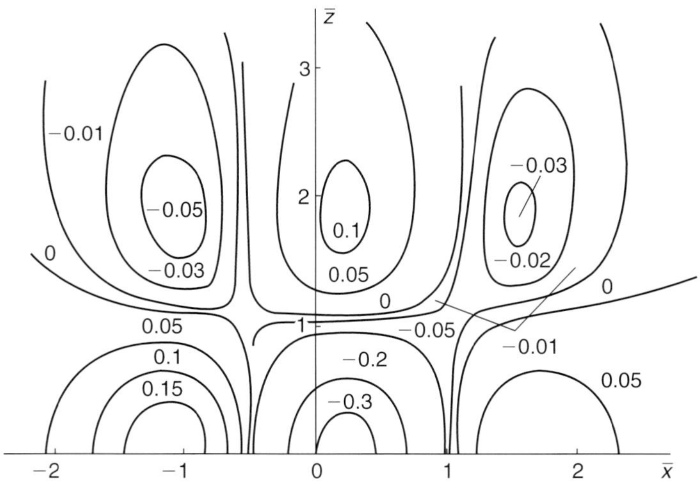


Fig. 8.19.

In the limiting case $c_1 \ll b_1$ in region 3 the pressure disturbance distribution is described by the solution of the boundary value problem (8.197) with the similarity parameters $C$, $D$, and $E$ determined by Eq. (8.189). Obviously, in this case the conditions of the pressure disturbance decay as $\bar{z} \to \pm\infty$ are not satisfied and there is no upstream disturbance transfer. Then for roughnesses possessing a longitudinal symmetry from Eq. (8.197) we obtain

$$G(\bar{x}) = \pm \frac{\partial F_1}{\partial \bar{y}} \int_0^\infty f_2(\bar{z})\, d\bar{z}|_{\bar{y}=0} \tag{8.200}$$

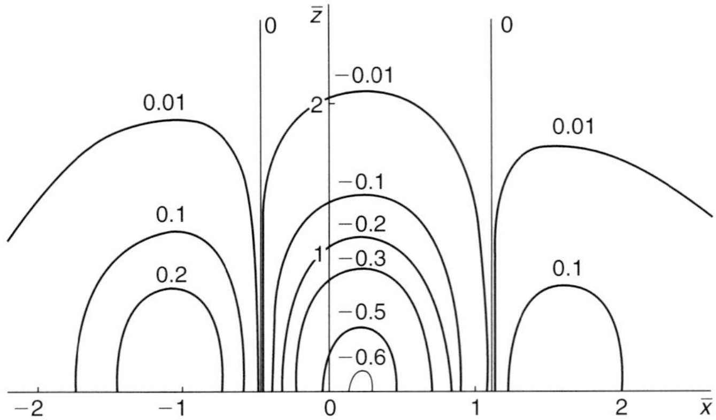


Fig. 8.20.

Pressure disturbances decay only in region 4, for which from Eqs. (8.191) and (8.200) it follows:

$$\frac{\partial^2 \overline{p}}{\partial \overline{x}^2} + \frac{\partial^2 \overline{p}}{\partial \overline{z}^2} = 0 \tag{8.201}$$

$$\overline{p} \to 0, \ \ \overline{x} \to \pm\infty, \ \ \overline{z} \to \pm\infty; \qquad \frac{\partial \overline{p}}{\partial \overline{z}} = \pm G(\overline{x}), \ \ \overline{z} \to \pm 0$$

We note that in the case in question narrow ($b_1 \gg c_1$) roughnesses induce pressure disturbances greater by a factor of $b_1/c_1$ than those induced when condition (8.199) is fulfilled.

Numerical solutions were obtained for a roughness of form $f(\overline{x}, \overline{z}) = \exp(-\overline{x}^2 - \overline{z}^2)$. The pressure disturbance distributions $P(\overline{x}, 0)$ replicate in essence the curves presented in Fig. 8.15. This means that in this case upstream disturbance transfer takes place. In Figs. 8.21–8.25 we have plotted the pressure contours for the previous ratios $b_1/c_1$. Clearly,

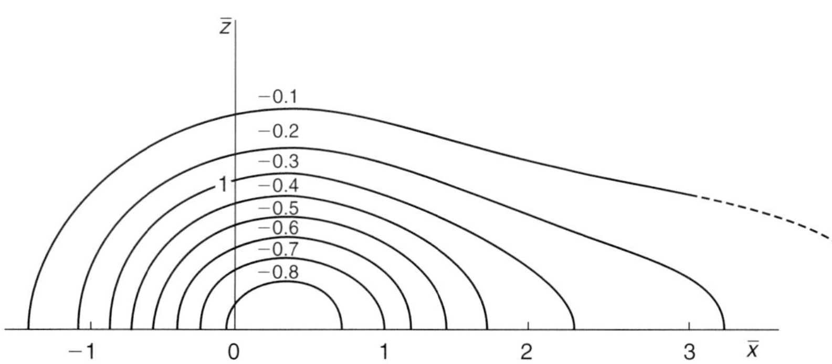


Fig. 8.21.

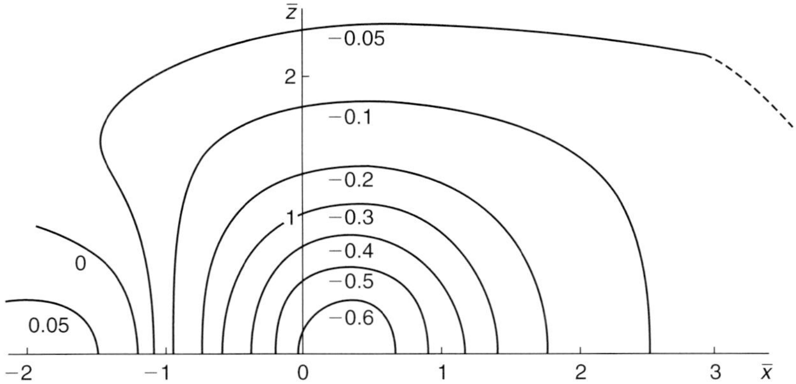


Fig. 8.22.

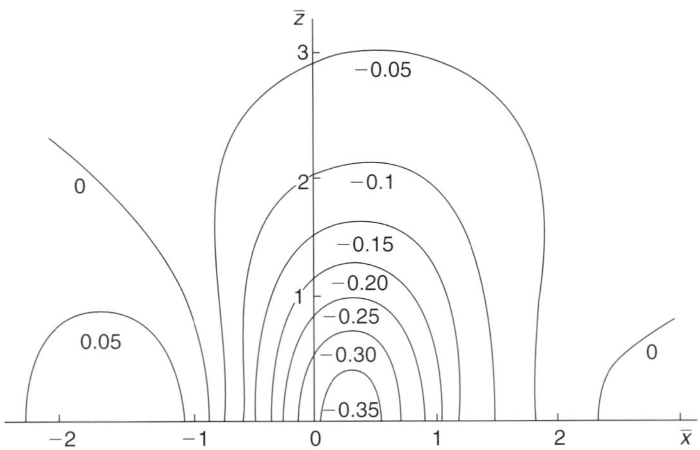


Fig. 8.23.

wide ($c_1 \gg b_1$) bumps in the subsonic wall region of the boundary layer induce only expansion. In all other cases the pressure disturbance distributions are of complicated nature.

## 8.3 Numerical investigation of the three-dimensional flow over roughnesses in the compensation interaction regime

In this section we present the results of the study of the flow past three-dimensional roughnesses in the compensation interaction regime. The corresponding boundary value problem was formulated in the previous sections of this chapter.

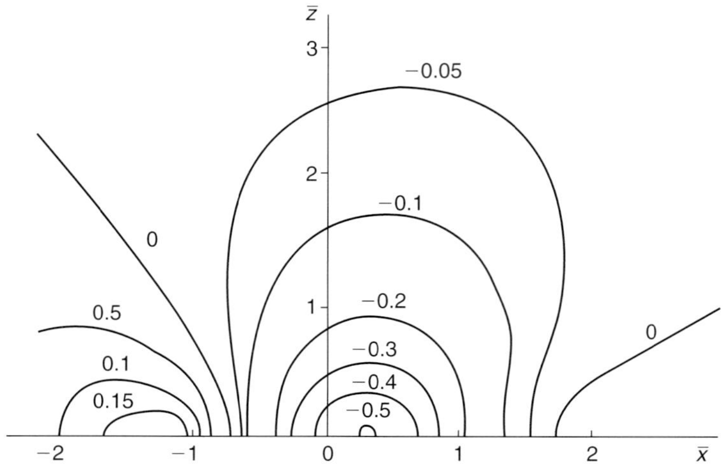


Fig. 8.24.

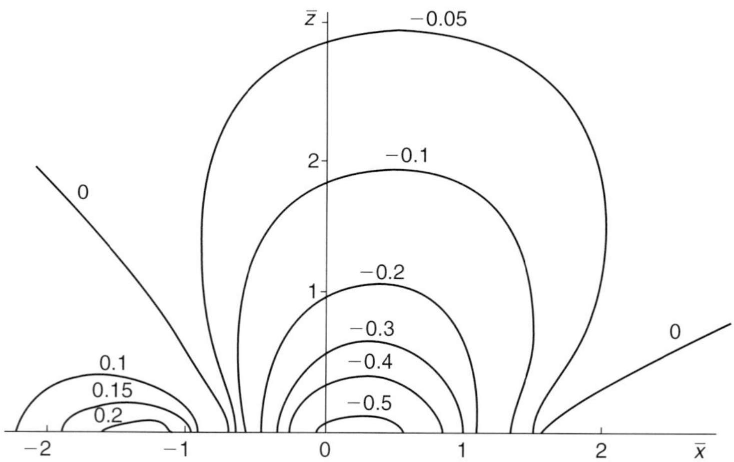


Fig. 8.25.

### *8.3.1 Formulation of the problem and estimates for the scales*

Let on the flat plate surface there be a small three-dimensional roughness at a distance $\ell$ from the leading edge of the plate. The plate is in uniform viscous flow, either subsonic or supersonic, with $(M_\infty^2 - 1) \sim 1$ and a high but subcritical Reynolds number $Re_\infty = \rho_\infty u_\infty \ell / \mu_\infty = \varepsilon^{-2}$, where $\rho$, $u$, and $\mu$ are the density, the velocity, and the dynamic viscosity, respectively, and the subscript $\infty$ refers to the oncoming flow. We will introduce a Cartesian coordinate system with the $x$ axis aligned with the undisturbed flow, the $y$ axis normal to the plate, and the $z$ axis perpendicular to the $x$ and $z$ axes. In what follows, the linear dimensions are scaled on $\ell$, the velocity components $u$, $v$, and $w$ (along the $x$, $y$, and $z$ axes) on $u_\infty$, the pressure on $\rho_\infty u_\infty^2$, the enthalpy on $u_\infty^2$, and $\rho$ and $\mu$ on $\rho_\infty$ and $\mu_\infty$, respectively; only dimensionless variables are used. The characteristic dimensions of the roughness are its thickness $a$, length $b$, and width $c$.

We will consider the flow over roughnesses with the length and the width of the same order, $b \sim c$. Obviously, the limiting problems for asymptotically different values characterizing the length and the width can be obtained by corresponding passages to limit from the boundary value problem under consideration.

We will also assume that the characteristic length is greater in the order than the boundary layer thickness and the characteristic height is smaller in the order than the boundary layer thickness. In accordance with the matched asymptotic expansion method, it is necessary to bring into consideration region 1 with asymptotically equal dimensions in all directions. It can be shown that in this region for a small relative thicknesses of the roughness the flow is characterized by the velocity component, pressure, and density disturbances of the same order which is determined by the vertical velocity at the outer edge of the boundary layer. This quantity can be estimated as the characteristic height-to-length ratio of the roughness

$$\Delta p \sim v \sim \frac{a}{b}$$

Having the estimate for the pressure disturbance we can determine the variation of the boundary layer displacement thickness generated by this disturbance. In the undisturbed boundary layer, at a height comparable with the roughness height, the longitudinal velocity is as follows:

$$u \sim \frac{a}{\varepsilon}$$

Then for nonlinear variation of the velocity this quantity and the variation of the wall region thickness can be expressed in terms of the pressure disturbance as follows:

$$u \sim \Delta u \sim \Delta p^{1/2}, \quad y \sim \Delta y \sim \varepsilon \Delta p^{1/2}$$

Therefore, at the outer edge of the boundary layer the vertical velocity is induced due to both the roughness itself and the variation of the boundary layer displacement thickness. It can be shown that the variation of the displacement thickness in the leading term is determined by the variation of the wall region thickness. Then the estimate for the pressure disturbance takes the form:

$$\Delta p \sim \left(\frac{a}{b}\right) + \frac{\varepsilon \Delta p^{1/2}}{b}$$

It can easily be verified that this estimate has a meaning only provided that there is a certain relation between the roughness dimensions

$$\frac{b}{a} \sim \frac{b^2}{\varepsilon^2}$$

If

$$\frac{b^2}{\varepsilon^2} \ll \frac{b}{a}$$

we arrive at a contradiction: the displacement thickness variation generates a pressure disturbance greater in the order than the original pressure disturbance which has led to the displacement thickness variation.

The solution of the problem is presented above; it is shown that if the last inequality is fulfilled, disturbances in the outer flow are absent from the leading approximation. In practice, this means that the estimate for the pressure disturbance follows from the compensation condition (integral zero variation of the boundary layer displacement thickness)

$$\frac{a}{b} \sim \frac{\varepsilon \Delta p^{1/2}}{b}, \quad \Delta p \sim \frac{a^2}{\varepsilon^2}$$

It should be noted that in this case an analysis of the solutions in regions 1 and 2 (region 2 includes streamtubes in the boundary layer above the roughness) is required only for formulating the boundary condition at the outer boundary of region 3. This analysis is drawn at the beginning of Section 8.2. The relations obtained should be supplemented by the condition that the effects of the viscosity and inertia forces are of the same order, which is assumed in this study. The condition follows from the consideration of the equation of the longitudinal momentum

$$a \sim \varepsilon b^{1/3}$$

Then the condition of existence of the compensation interaction regime takes the form:

$$b \ll \varepsilon^{3/4}$$

In this case, a counterpart of the interaction condition is represented by the condition of the absence of disturbances from regions 1 and 2 which determines the longitudinal velocity at the outer boundary of region 3

$$u = Ay\varepsilon^{-1} + o(1)$$

where $A$ is the dimensionless viscous stress at the surface in the undisturbed boundary layer upstream of the roughness.

This condition is obtained by matching the solutions in regions 2 and 3. Since in region 2 the solution is as follows:

$$u = u_0(y_2) + B(x, z)\frac{du_0}{dy}, \quad v_2 = -u_0\frac{dB}{dx}$$

the absence of disturbances at the outer edge of the boundary layer means that $B = 0$.

From this condition we can derive a relation between the pressure disturbance and the vertical velocity at the outer boundary of region 3 which is used below.

### 8.3.2 Boundary value problem

In region 3 with the characteristic dimensions $x \sim b$, $y \sim \varepsilon^{1/3}$, and $z \sim c$ we introduce the following asymptotic expansions

$$
\begin{aligned}
&x = bx_3, \quad y = \varepsilon b^{1/3}\mu_w^{1/3}\rho_w^{-1/3}A^{-1/3}y_3, \quad z = cz_3\\
&u = A^{1/3}\mu_w^{1/3}\rho_w^{-1/3}b^{1/3}u_3 + \cdots, \quad v = A^{1/3}\mu_w^{2/3}\rho_w^{-2/3}\varepsilon b^{-1/3}v_3 + \cdots\\
&w = A^{2/3}\mu_w^{1/3}\rho_w^{-1/3}b^{-2/3}cw_3 + \cdots \qquad (8.202)\\
&p = \frac{1}{\gamma M_\infty^2} + A^{4/3}\mu_w^{2/3}\rho_w^{1/3}b^{2/3}p_3 + \cdots, \quad \rho = \rho_w + \cdots, \quad \mu = \mu_w + \cdots
\end{aligned}
$$

Provided that the limiting relations $a \sim \varepsilon b^{1/3}$, $\varepsilon^{3/2} < b < \varepsilon^{3/4}$, and $c \sim b$ are fulfilled, substituting expansions (8.202) in the Navier–Stokes equations yields that in the first approximation the flow in region 3 is described by the three-dimensional incompressible boundary layer equations

$$
\begin{aligned}
&\frac{\partial u_3}{\partial x_3} + \frac{\partial v_3}{\partial y_3} + \frac{\partial w_3}{\partial z_3} = 0\\
&u_3\frac{\partial u_3}{\partial x_3} + v_3\frac{\partial u_3}{\partial y_3} + w_3\frac{\partial u_3}{\partial z_3} + \frac{\partial p_3}{\partial x_3} = \frac{\partial^2 u_3}{\partial y_3^2}, \quad \frac{\partial p_3}{\partial y_3} = 0 \qquad (8.203)\\
&u_3\frac{\partial w_3}{\partial x_3} + v_3\frac{\partial w_3}{\partial y_3} + w_3\frac{\partial w_3}{\partial z_3} + D\frac{\partial p_3}{\partial z_3} = \frac{\partial^2 w_3}{\partial y_3^2}\\
&D = \frac{b^2}{c^2}
\end{aligned}
$$

The boundary conditions are as follows:

$$
\begin{aligned}
&u_3 = v_3 = w_3 = 0 \quad \text{for } y_3 = hF(x_3, z_3) \qquad (8.204)\\
&u_3 \to y_3, \quad v_3,\ w_3,\ p_3 \to 0 \quad \text{for } x_3 \to -\infty,\ z_3 \to \pm\infty\\
&u_3 \to y_3, \quad v_3 + \frac{\partial p_3}{\partial x_3} \to 0, \quad w_3 \to 0, \quad y_3 \to \infty\\
&h = A^{1/3}\mu_w^{-1/3}\rho_w^{1/3}ab^{-1/3}\varepsilon^{-1}
\end{aligned}
$$

Below subscripts of the functions and variables are omitted.

The boundary value problem (8.203), (8.204) incorporates two similarity parameters, $D$ and $h$. The former is determined by the characteristic length-to-width ratio of the roughness. It can be shown that as this ratio vanishes, the boundary value problem (8.203), (8.204) reduces to the problem governing two-dimensional flow.

The latter similarity parameter is proportional to the ratio of the orders of the inertia and viscosity forces in region 3. For large values of this parameter a local inviscid flow occurs near the roughness, whereas for its small values it is viscosity forces that are predominant in the flow in the vicinity of the roughness.

It can be shown that the problem formulated represents a lubrication problem for the case in which the gap width is asymptotically greater than the roughness height.

We will begin our analysis with the study of the problem solutions for small values of the parameter $h$; this solution can be sought in the form:

$$u = y + hU + \cdots, \quad v = hV + \cdots, \quad p = hP + \cdots, \quad w = hW + \cdots$$

The corresponding linear system of equations for $D = 1$ takes the form:

$$y\frac{\partial U}{\partial x} + V + \frac{\partial P}{\partial x} = \frac{\partial^2 U}{\partial y^2}$$

$$y\frac{\partial W}{\partial x} + \frac{\partial P}{\partial z} = \frac{\partial^2 W}{\partial y^2}$$

$$\frac{\partial U}{\partial x} + \frac{\partial V}{\partial y} + \frac{\partial W}{\partial z} = 0, \quad \frac{\partial P}{\partial y} = 0$$

Differentiating the first equation of the system with respect to $x$ and $y$ and the second equation with respect to $z$ and $y$ and summing up the results we arrive at the following system:

$$y\frac{\partial S}{\partial x} = \frac{\partial^2 S}{\partial y^2}, \quad S = \frac{\partial^2 V}{\partial y^2}$$

$$\frac{\partial S(x, 0, z)}{\partial y} = \frac{\partial^2 P}{\partial x^2} + \frac{\partial^2 P}{\partial z^2}, \quad \int_0^\infty S\,\mathrm{d}y = 0$$

From an analysis of this system it follows directly that a nontrivial solution for function $S$ exists if $F \neq 0$ or if the convective derivative of the function $S$ (wake downstream of the roughness) is nonzero even for nonzero values of the derivative of this function at the surface. Then at any point on the surface, outside the roughness and the wake behind it, the solution of this equation is as follows:

$$S(x, y, z) = 0, \quad V(x, y, z) = 0$$

In this case, for the velocity components $U$ and $W$ we obtain the following problem that describes a quasi-two-dimensional flow in a region outside the roughness and the wake behind it

$$y\frac{\partial U}{\partial x} + \frac{\partial P}{\partial x} = \frac{\partial^2 U}{\partial y^2}$$

$$y\frac{\partial W}{\partial x} + \frac{\partial P}{\partial z} = \frac{\partial^2 W}{\partial y^2}$$

$$\frac{\partial U}{\partial x} + \frac{\partial W}{\partial z} = 0, \quad \frac{\partial P}{\partial y} = 0$$

In the case of nonlinear flow regime the situation is more complicated; however, even in this case a relation between the pressure Laplacian and the normal derivative of the function $S$ can be derived by analyzing the local flow near a certain point on the surface. In the nonlinear problem a domain of influence is determined by subcharacteristics represented in the case in question by streamlines (Wang, 1975). Then we can formulate the assumption that the equality of the vertical velocity to zero is fulfilled for the flow outside the roughness and for those points in the $xz$ plane which do not lie on the projections of streamlines passing above the roughness. Generally, the vertical velocity is nonzero in the wake behind the roughness and in the separation zone in the vicinity of the roughness. Below this inference is confirmed by the results of numerical investigation.

Let us dwell upon the question of upstream disturbance propagation. In accordance with the inferences of many studies, the compensation flow regime does not include the disturbance transfer in the two-dimensional case (in three-dimensional flows the disturbance propagation takes place). It can be shown that in the disturbed two-dimensional flow, apart from the compensation interaction region, there is a longer free interaction region. This region also appears in considering three-dimensional flows in which, however, its effect manifests itself only in the higher approximations, since the disturbances decay directly in the compensation interaction region. However this region is necessarily present in two-dimensional flow and pressure variations in it turn out to be comparable with those in the compensation interaction region. The pressure directly ahead of the roughness is generally nonzero and is determined from the condition that downstream of the compensation interaction region disturbances must decay. This follows from, for example, the consideration of the linear solution for the free interaction regime as the roughness length vanishes (Smith, 1973).

### *8.3.3 Numerical solution*

The spectral method proposed by Duck and Burggraf (1986) was applied for solving problem (8.203) and (8.204). Later this method was used for calculating longitudinal-transverse interaction in the work of Kravtsova (1993). The method made it possible to obtain for the first time solutions of nonlinear problems describing the compensation regime for roughnesses of general shape (Duck, 1986; 1988). Let us apply the Prandtl transposition transformation

$$\widetilde{x} = x, \quad \widetilde{y} = y - F(x, z), \quad \widetilde{z} = z, \quad \widetilde{u} = u$$

$$\widetilde{v} = v - u\frac{\partial F}{\partial x} - w\frac{\partial F}{\partial z}, \quad \widetilde{w} = w, \quad \widetilde{p} = p$$

(in what follows the tilde is omitted). Then the equations do not change, while the boundary conditions take the form:

$$u = v = w = 0 \quad y = 0$$

$$u \to y, \quad v,\ w,\ p \to 0 \quad x \to -\infty,\ z \to \pm\infty$$

$$u \to y + F(x, z), \quad w \to 0, \quad y \to \infty$$

Let us apply to the system thus obtained the Fourier transformation of the form:

$$u^{**}(k, m, y) = \frac{1}{(2\pi)^2} \int_{-\infty}^{+\infty}\int_{-\infty}^{+\infty} u(x, y, z) \exp(-ikx - imz)\,\mathrm{d}x\,\mathrm{d}z$$

and introduce a variable $\widetilde{u} = u_1 - y$. As a result, the system takes the form:

$$iku^{**} + v^{**\prime} + imw^{**} = 0$$

$$u^{**\prime\prime} - ikyu^{**} - v^{**} - ikP = \left(u\frac{\partial u}{\partial x} + v\frac{\partial u}{\partial y} + w\frac{\partial u}{\partial z}\right)^{**} = R_1^{**}$$

$$w^{**\prime\prime} - ikyw^{**} - imP^{**} = \left(u\frac{\partial u}{\partial x} + v\frac{\partial u}{\partial y} + w\frac{\partial u}{\partial z}\right)^{**} = R_2^{**}$$

$$u^{**} = v^{**} = w^{**} = 0, \quad y = 0$$

$$u^{**} \to F^{**}, \quad w^{**} \to 0, \quad y \to \infty$$

The solution of this problem can be so organized that its right-hand side is taken from a previous iteration stage, while all other relations entering in the formulation of the interaction problem are considered in the current stage. Let us multiply the equation for the longitudinal momentum by $k$ and that for the transverse momentum by $m$, add the results, and bring into the consideration the function $f = ku^{**} + mw^{**}$ (in what follows the symbol $^{**}$ is omitted). Then we arrive at a single equation which, after having been differentiated with respect to $y$, takes the form:

$$f''' - ikyf' = (ikR_1 + imR_2)' \tag{8.205}$$

$$f(0) = 0, \quad f = i(k^2 + m^2)P, \quad f(\infty) = kF$$

Let us introduce the function $g = df/dy$ in Eq. (8.205) and represent the solution in the form of the sum of the solution $g_1$ of a homogeneous equation and $g_2$ of an inhomogeneous equation: $g = Pg_1 + g_2$. The functions $g_1$ and $g_2$ are determined from the solution of the following problems:

$$g_1'' - ikyg_1 = 0$$

$$g_1'(0) = i(k^2 + m^2), \quad g_1(\infty) = 0$$

$$g_2'' - ikyg_2 = (kR_1 + mR_2)'$$

$$g_2'(0) = 0, \quad g_2(\infty) = \left(-\frac{1}{iky}(kR_1 + mR_2)'\right)_{y=\infty}$$

Now, integrating the function $g$ we determine the value of $f$ in the current iteration stage. Then $P = (kF - f_2)/f_1$, where $f_1 = \int_0^\infty g_1\,dy$ and $f_2 = \int_0^\infty g_2\,dy$. Then from the second momentum equation there follows the boundary value problem for the Fourier image of the transverse velocity

$$w'' - ikyw = imP + R_2$$

$$w(0) = 0, \quad w(\infty) = 0$$

The function $f$ determines the Fourier image of the longitudinal velocity $u = (f - mw)/k$, while $v$ is determined by the expression $v = -i\int_0^y f\,dy$. Then we go over from the spectral to the physical variables and obtain the nonlinear terms $R_1$ and $R_2$. For this purpose the fast Fourier transformation algorithm (Cooley and Tukey, 1965) was applied. Finally, 6–90 iteration stages were required for obtaining convergence with an error of $10^{-5}$, depending on the cavity depth. The relaxation parameter was varied from $r = 1$ for $h = -1$ to $r = 0.5$ for $h = -3$. The roughness shape was given by the function $F(x, z) = h \times \exp(-(x^2 + z^2))$. The computation grid consisted of $NX \times NY \times NZ = 64 \times 26 \times 32$ gridpoints with the steps in the corresponding variables $\Delta x = \Delta z = 0.3$, $\Delta y = 0.4$, and $D = 1$.

In Fig. 8.26 the pressure disturbance distributions in region 3 are plotted in the plane of symmetry for different values of the parameter $h$. Negative values of this parameter correspond to the flow over a cavity in the surface. Clearly that an increase in its depth leads to an increase of the pressure disturbance peak. The distributions are also characterized by the presence of two minima, upstream and downstream of the origin at which the roughness center is placed.

It should be noted that, when a return flow region is formed, a tendency to the formation of the region of near-constant values (plateau) in the pressure distribution is observable. Similar results for subsonic flows were obtained by Duck and Burggraf (1986).

In Fig. 8.27 the longitudinal friction distribution over the surface is presented. Clearly that an increase in the cavity depth is accompanied by a decrease in the longitudinal friction minimum. There exists a limiting depth for which the longitudinal friction vanishes.

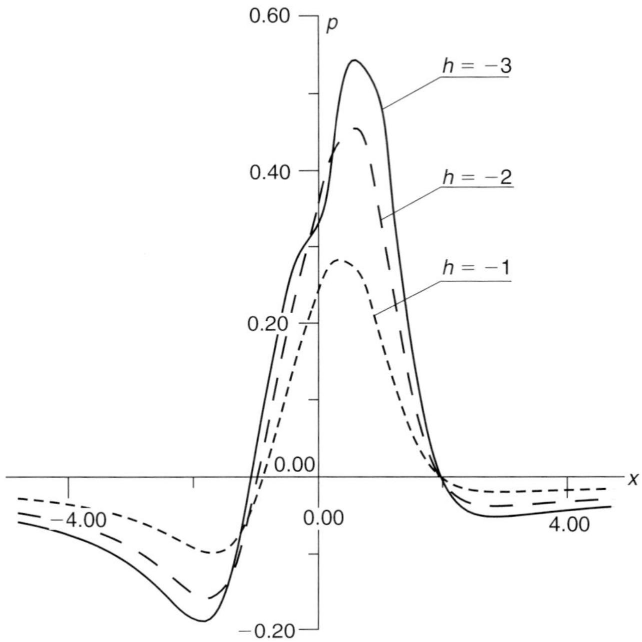


Fig. 8.26.

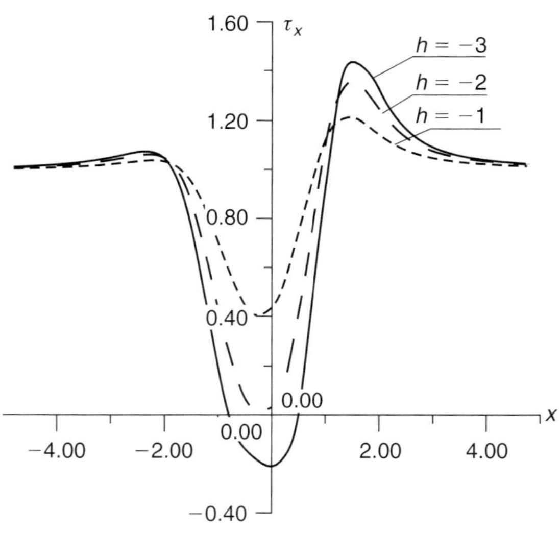


Fig. 8.27.

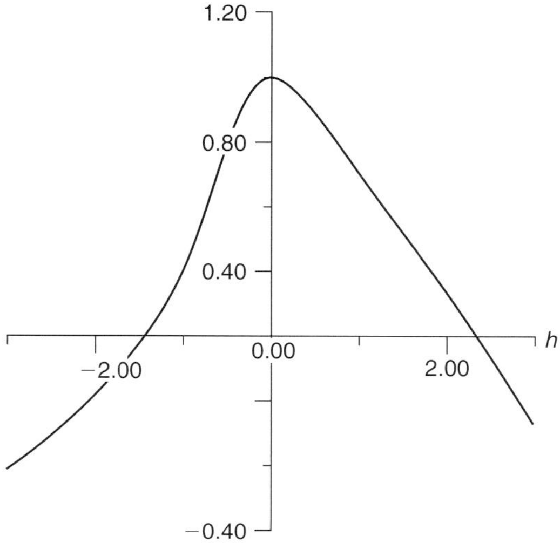


Fig. 8.28.

Figure 8.28 presents the dependence of the longitudinal friction minimum at the surface on the parameter $h$. It can be suggested that for deeper cavities there appears a limiting state in which the longitudinal friction minimum tends to a finite value. Opposite tendency is observable with increase in the roughness height. At the same time, an additional study is required to make final conclusions.

We also investigated the disturbed flow near finite roughnesses of the form:

$$F(x,z)=0,\quad \text{if } R\geq 1,\qquad \text{and}\qquad F(x,z)=h\cos^2\left(\frac{\pi R}{2}\right)\quad \text{for } R<1$$

where $R=x^2+z^2$.

In Fig. 8.29 we have plotted the $z$-dependence of the pressure disturbance for $h=1.5$ in both linear and nonlinear cases (the solid curve corresponds to the nonlinear case) for different values of the longitudinal coordinate. Clearly that, in accordance with the results obtained above, the roughness effect manifests itself throughout the entire flowfield due to the elliptic nature of the equation for the pressure disturbance.

In Fig. 8.30 the vertical velocity distribution is plotted in the plane parallel to the surface and located at a distance $y=0.8$ from it. Clearly, the vertical velocity is actually zero outside the roughness and the zone of its convective influence, so that the disturbed flow is quasi-two-dimensional.

The fact of the formation of layered flows near three-dimensional roughnesses seems very important and can find an application in controlling boundary layer flows.

It should be noted that the model of compensation interaction regime considered above is applicable to a wide class of flows. The results obtained on its basis can be applied for estimating the means for controlling laminar boundary layer flows.

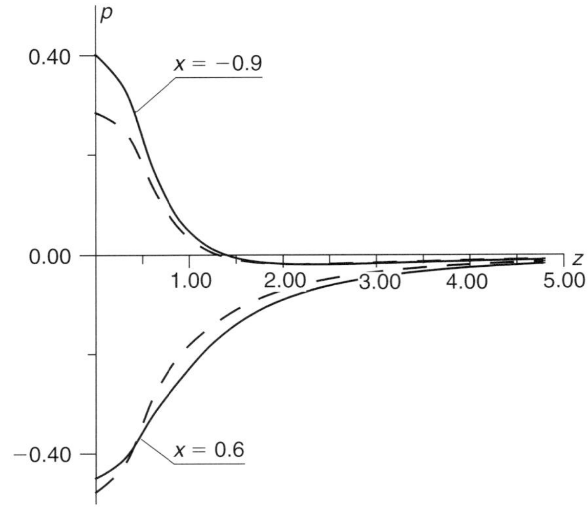


Fig. 8.29.

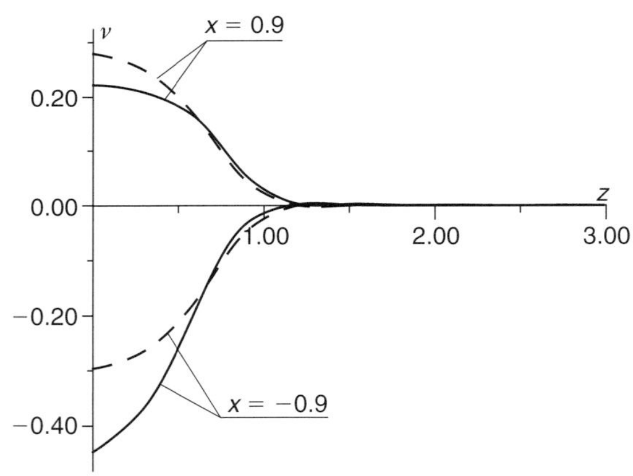


Fig. 8.30.

# Bibliography*


1. Abott, D. E., M. Holt, and J. N. Nielsen (1963), "Studies of separated laminar boundary layers at hypersonic speed with some low Reynolds number data," AIAA Paper No. 172.
2. Abramowitz, M. and I. A. Stegun (1964), *Handbook of Mathematical Functions with Formulas, Graphs and Mathematical Tables*, Dover, New York.
3. Ai, D. K. (1967), "On the hypersonic laminar near wake critical point on the Crocco–Lees mixing theory," AIAA Paper No. 60.
4. Akatnov, N. N. (1953), "Spreading of a two-dimensional incompressible laminar jet over a rigid wall," *Tr. Leningrad. Politekhn. Inst.*, 5(24).
5. Ball, H. O. W. and R. H. Korkegi (1968), "An investigation of the effect of suction on hypersonic laminar boundary layer separation," *AIAA J.*, 6(2), 239.
6. Barry, F. W., A. H. Shapiro, and E. P. Neumann (1951), "The interaction of shock waves with boundary layers on a flat plate," *J. Aeronaut. Sci.*, 18(4), 229.
7. Bashkin, V. A. (1984), *Delta Wings in Hypersonic Flow* [in Russian], Mashinostroenie, Moscow.
8. Bashkin, V. A. and G. N. Dudin (2000), *Three-Dimensional Hypersonic Viscous Flows* [in Russian], Nauka, Moscow.
9. Batchelor, G. K. (1956), "On steady laminar flow with closed streamlines at large Reynolds number," *J. Fluid Mech.*, 1(2), 177.
10. Batchelor, G. K. (1967), *Introduction to Fluid Dynamics*, Cambridge University Press, Cambridge.
11. Baum, E. (1968), "An interaction model of a supersonic laminar boundary layer on sharp and rounded backward facing steps," *AIAA J.*, 6(3), 440.
12. Baum, E., H. H. King, and M. R. Denison (1964), "Recent studies of the laminar base-flow region," *AIAA J.*, 2(9), 1527.
13. Becker, R. (1921/1922), "Stosswelle und Detonation," *Zeitschr. Phys.*, 8(321).
14. Bespalov, V. L. and V. G. Voronkov (1980), "Theory of a catalytic calorimeter," *Fluid Dynam.*, 15(1), 162.
15. Bodonyi, R. J. and P. W. Duck (1988), "A numerical method for treating strongly interactive three-dimensional viscous-inviscid flows," *J. Computers Fluids*, 16(3), 279.


* References to the books by Russian and other non-English-speaking authors are given in accordance with their English translations, if there are any and they are known to the authors or the translator. As for the papers from Russian scientific journals cited in this book, most of them were published in the following three journals: Prikladnaya Matematika i Mekhanika, Izvestiya Akademii Nauk, ser. Mekhanika Zhidkosti i Gaza, and Zhurnal Prikladnoi Meknaniki i Tekhnicheskoi Fiziki. All the three journals are translated into English – the first by Elsevier under the name Applied Mathematics and Mechanics and the other two by Springer as Fluid Dynamics and Journal of Applied Mechanics and Technical Physics. References to the papers published in these journals are given in accordance with the above translations, while references to the papers from other Russian journals in accordance with the original publications.

16. Bogdonoff, S. M. and C. E. Kepler (1955), "Separation of a supersonic turbulent boundary layer," *J. Aeronaut. Sci.*, 22(6), 414.
17. Bogolepov, V. V. (1974), "Calculation of the interaction of a supersonic boundary layer with a thin obstacle," *Uch. Zap. TsAGI*, 5(6).
18. Bogolepov, V. V. (1975), "Shearing flow of a viscous liquid around a semicylindrical protuberance on the surface of a plate," *Fluid Dynam.*, 10(6), 896.
19. Bogolepov, V. V. (1977), "Analysis of the limiting solutions for the case of supersonic viscous flow over small roughnesses on the body surface," *Tr. TsAGI*, No. 1812.
20. Bogolepov, V. V. (1980), "Aerodynamic heating of the walls of a small rectangular trench in the surface of a plate," *Uch. Zap. TsAGI*, 11(3).
21. Bogolepov, V. V. (1983), "Flow past forward-facing small step," *J. Appl. Mech. Techn. Phys.*, 24(2), 166.
22. Bogolepov, V. V. (1984), "Local viscous, chemically nonequilibrium flows," *Fluid Dynam.*, 19(3), 459.
23. Bogolepov, V. V. (1985a), "Investigation of local disturbances of a laminar boundary layer in the flow past elements of thermal protection of flight vehicles," *Tr. TsAGI*, No. 2286.
24. Bogolepov, V. V. (1985b), "Flow structure near the trailing edge of a plate," *J. Appl. Mech. Techn. Phys.*, 26(3), 391.
25. Bogolepov, V. V. (1986), "General arrangement of regimes for spatial local flows," *J. Appl. Mech. Techn. Phys.*, 27(6), 860.
26. Bogolepov, V. V. (1987), "Small perturbations of a laminar boundary layer," *J. Appl. Mech. Techn. Phys.*, 28(5), 706.
27. Bogolepov, V. V. and I. I. Lipatov (1982), "Similarity solutions for boundary layer equations with interaction," *J. Appl. Mech. Techn. Phys.*, 23(4), 518.
28. Bogolepov, V. V. and I. I. Lipatov (1985), "Locally three-dimensional laminar flows," *J. Appl. Mech. Techn. Phys.*, 26(1), 24.
29. Bogolepov, V. V. and I. I. Lipatov (1988a), "Laminar flow past a step," *J. Appl. Mech. Techn. Phys.*, 29(1), 73.
30. Bogolepov, V. V. and I. I. Lipatov (1988b), "Tangential injection effect on laminar boundary layer flow," *Uch. Zap. TsAGI*, 19(1).
31. Bogolepov, V. V. and I. I. Lipatov (1991), "Structure of the disturbed boundary layer near a cold surface," *Fluid Dynam.*, 26(1), 15.
32. Bogolepov, V. V. and V. Ya. Neiland (1971), "Supersonic viscous flow over small roughnesses at the surface of a body," *Tr. TsAGI*, No. 1363.
33. Bogolepov, V. V. and V. Ya. Neiland (1976), "Study of local disturbances in viscous supersonic flows," in: *Aeromechanics* [in Russian], Nauka, Moscow, p. 104.
34. Bogolepov, V. V., I. I. Lipatov, and L. A. Sokolov (1990), "Structure of chemically nonequilibrium flows with a sudden change in the temperature and the catalytic properties of the surface," *J. Appl. Mech. Techn. Phys.*, 31(3), 367.
35. Bondarev, E. N. (1966), "Approximate calculation of the interaction of supersonic flow with a laminar layer in the flow separation region," *Fluid Dynam.*, 1(5), 81.
36. Bondarev, E. N. and G. I. Petrov (1960), "Experimental investigation of the interaction of a turbulent boundary layer with shocks," in: *Abstracts of the All-Union Congress on Theoretical and Applied Mechanics* [in Russian], Moscow, p. 50.

37. Bondarev, E. N. and M. Ya. Yudelovich (1960), "On the possibility of an increase of the base pressure behind a wedge in hypersonic flight," *Izv. Akad. Nauk SSSR. Otd. Tekhn. Nauk. Mekh. Mash.*, No. 5.
38. Brailovskaya, I. Yu. (1967) "Calculation of viscous compressible gas flow past a corner," *Fluid Dynam.*, 2(3), 49.
39. Brailovskaya, I. Yu. and L. A. Chudov (1967), "Two methods for constructing the higher-order approximations in a small viscosity parameter," in: *Computational Methods and Programming*, Issue 8 [in Russian], Moscow University Press, Moscow, p. 173.
40. Bray, K. (1957), "The effect of heat transfer on interactions involving laminar boundary layers," *ARC CP*, No. 339.
41. Bray, K., G. Gadd, and M. Woodger (1961), "Some calculations by the Crocco-Lees and other methods of interaction between shock waves and laminar boundary layers including effects of heat transfer and suction," *ARC CP*, No. 556.
42. Brower, W. (1961), "Leading-edge separation of laminar boundary layer in supersonic flow," *J. Aerospace Sci.*, 28(12), 957.
43. Brown, S. N. and K. Stewartson (1969), "Laminar separation," *Annu. Rev. Fluid Mech.*, 1(45).
44. Brown, S. N., H. K. Cheng, and C. J. Lee (1990), "Inviscid-viscous interaction on triple deck scales in a hypersonic flow with strong wall cooling," *J. Fluid Mech.*, 220(309).
45. Brykina, I. G. "An analytical solution of the problem of convective diffusion in the neighbourhood of a discontinuity of the catalytic properties of a surface," *Appl. Math. Mech.*, 52(2), 191.
46. Burgers, D. M. (1948), "A mathematical model illustrating the theory of turbulence," in: R. von Mises and Th. von von Kàrmàn (eds.), *Advances in Applied Mechanics*, Vol. 1, New York.
47. Burggraf, O. R. (1966), "Analytical and numerical studies of the structure of steady separated flows," *J. Fluid Mech.*, 24(1), 113.
48. Bush, W. B. (1966), "Hypersonic strong-interaction similarity solutions for flow past a flat plate," *J. Fluid Mech.*, 25(1), 51.
49. Bushnell, D. M. and L. M. Weinstein (1968), "Correlation of peak heating for reattachment of separation flows," *J. Spacecraft Rockets*, 5(9), 1111.
50. Catherall, D. and K. W. Mangler (1966), "The integration of the two-dimensional laminar boundary layer equations past the point of vanishing skin friction," *J. Fluid Mech.*, 26(1), 163.
51. Catherall, D., K. Stewartson, and P. G. Williams (1965), "Viscous flow past a flat plate with uniform injection," *Proc. Roy. Soc., London. Ser. A*, 284(1398), 370.
52. Chang, P. K. (1970), *Separation of Flow*, Pergamon, Oxford.
53. Chapman, D. R. (1950), "Laminar mixing of a compressible fluid," *NACA Rep.*, No. 958.
54. Chapman, D. R. (1951), "An analysis of base pressure at supersonic velocities and comparison with experiment," *NACA Rep.*, No. 1051.
55. Chapman, D. R., D. Kuehn, and H. Larson (1958), "Investigation of separated flows in supersonic and subsonic streams with emphasis on the effect of transition," *NACA Rep.*, No. 1356.

56. Charwat, A. F. (1970), "Supersonic flows with imbedded separation regions," in: *Advances in Heat Transfer*, Vol. 6, Academic Press, New York and London.
57. Chernyi G. G. (1952), "Influence of subsonic part of boundary layer on position of shock wave," *Theoretical Hydromeh*. No. 9, 2, Moscow, p. 63–96.
58. Chernyi, G. G. (1955), "Boundary layer with a discontinuity surface. Flow over a plate with fluid permeation through the surface," *Dokl. Akad Nauk SSSR*, 100(5), 867.
59. Chernyi, G. G. (1966), *Introduction to Hypersonic Flow Theory*, Academic Press, New York.
60. Chudov, L. A. (1963), "Certain shortcomings of the classical boundary layer theory," in: *Numerical Methods in Gasdynamics*, Issue 2, Moscow University Press, Moscow, p. 98.
61. Chung, P. M., S. W. Liu, and H. Mirels (1963), "Effect of discontinuity of surface catalycity on boundary layer flow of dissociated gas," *Int. J. Heat Mass Transfer*, 6(3).
62. Cohen, C. B. and E. Reshotko (1956), "Similar solutions for the compressible laminar boundary layer with heat transfer and pressure gradient," *NACA Rep.*, No. 1293.
63. Cole, J. D. (1968), *Perturbation Methods in Applied Mathematics*, Blaisdel Publishing Company, Waltham, MA.
64. Cole, J. D. and J. Aroesty (1968), "The blow hard problem inviscid flow with surface injection," *Int. J. Heat Mass Transfer*, 11(7).
65. Cooley, J. W. and J. W. Tukey (1965), "An algorithm for machine calculation of complex Fourier Series," *Math. Comput.*, 19(297).
66. Crawford, D. R. and M. Holt (1968), "Method of integral relation as applied to the problem of laminar free mixing," *AIAA J.*, 6(2), 372.
67. Crocco, L. (1955), "Consideration on the shock-boundary layer interaction," in: *Proc. Conf. High-Speed Aeron.*, p. 75.
68. Crocco, L. and L. Lees (1952), "A mixing theory for the interaction between dissipative flows and nearly isentropic streams," *J. Aeron. Sci.*, 19(10), 649.
69. Cross Jr., E. J. and W. L. Hankey (1968), "Investigation of the lee-ward side of a delta wing at hypersonic speeds," AIAA Paper No. 675.
70. Curle, N. (1960), "The effects of heat transfer of laminar boundary layer separation in supersonic flow," ARC Report No. 21986.
71. Curle, N. (1961), "The effects of heat transfer of laminar boundary layer separation in supersonic flow," *Aeronaut. Quart.*, 12.
72. Daniels, P. G. (1974), "Numerical and asymptotic solutions for the supersonic flow near the trailing edge of a flat plate," *Quart. J. Mech. Appl. Math.*, 27(2), 175.
73. Dem'yanov, Yu. A. (1969), "A method for constructing the solution of the equations of the type Prandtl in the vicinity of points of violation of analyticity of boundary conditions," *Zh. Vychisl. Mat. Mat. Fiz.*, 9(4).
74. Dem'yanov, Yu. A. and Yu. N. Shmanenkov (1960), "Investigation of reverse flows in the region of separation of the turbulent boundary layer," *Appl. Math. Mech.*, 24(2), 340.
75. Denisenko, O. V. (1978a), "Flow past the trailing edges of planar delta and diamond-shaped wings in the strong viscous interaction regime," *Tr. TsAGI*, No. 1939, 3.
76. Denisenko, O. V. (1978b), "Viscous gas flow over longitudinal grooves," *Uch. Zap. TsAGI*, 9(4).

77. Denison, M. R. and E. Baum (1961), "Compressible free shear layer with finite initial thickness," IAS Paper No. 62.
78. Dewey Jr., C. F. (1963), "Use of local similarity concepts in hypersonic interaction problems," *AIAA J.*, 1(1), 20.
79. Dorodnitsyn, A. A. (1958), "A method for numerically solving certain nonlinear problems of aerodynamics," in: *Proceedings of the 3rd All-Union Congress on Mathematics*, Vol. III [in Russian], USSR Academy of Sciences, Moscow.
80. Dorodnitsyn, A. A. (1962), "General method of integral relations and its application to boundary layer theory," *Advan. Aeron. Sci.*, 3.
81. Duck, P. W. (1988), "The effect of small surface perturbation on the pulsatile boundary layer on a semi-infinite flat plate," *J. Fluid Mech.*, 197(259).
82. Duck, P. W. and O. R. Burggraf (1986), "Special solutions for three-dimensional triple-deck flow over surface topography," *J. Fluid Mech.*, 162(1).
83. Dudin, G. N. (1978a), "Interaction between a hypersonic flow and the boundary layer on a thin delta wing," *Tr. TsAGI*, No. 1912, 3.
84. Dudin, G. N. (1978b), "Hypersonic viscous flow past a thin yawed delta wing," *Uch. Zap. TsAGI*, 9(3), 30.
85. Dudin, G. N. (1978c), "Calculation of the boundary layer on a triangular plate in the strong viscous interaction regime," *Uch. Zap. TsAGI*, 9(5), 65.
86. Dudin, G. N. (1979), "Hypersonic boundary-layer flow with mass transfer on power-law bodies," *J. Appl. Mech. Techn. Phys.*, 20(5), 583.
87. Dudin, G. N. (1980), "Calculation of the laminar boundary layer equations at the line of symmetry of a delta wing," *Tr. TsAGI*, No. 2046, 58.
88. Dudin, G. N. (1982a), "Calculation of three-dimensional boundary layer on a triangular plate of finite length in a hypersonic stream," *Fluid Dynam.*, 17(5), 693.
89. Dudin, G. N. (1982b), "Calculation of three-dimensional boundary layer on a delta wing of finite length in the strong viscous interaction regime," in: *Mathematical Methods of Liquid and Gas Mechanics* [in Russian], Dnepropetrovsk University Press, Dnepropetrovsk, p. 90.
90. Dudin, G. N. (1983a), "Flow in the hypersonic boundary layer on a finite triangular wing in the presence of an angle of attack," *J. Appl. Mech. Techn. Phys.*, 24(3), 383.
91. Dudin, G. N. (1983b), "Finite-difference method for solving three-dimensional boundary layer equations in the strong viscous interaction regime," *Tr. TsAGI*, No. 2190, 3.
92. Dudin, G. N. (1985), "Numerical method for solving three-dimensional boundary layer equations on a delta wing," in: *Problems of Viscous Fluid Dynamics* [in Russian], Novosibirsk, p. 135.
93. Dudin G. N. (1985) "Problems of dynamic of viscous fluid," abstracts of. X – Allunion seminar, Novosibirsk, p. 135.
94. Dudin, G. N. (1988a), "Calculation of supersonic viscous flow over a delta wing," *Tr. TsAGI*, No. 2376, 30.
95. Dudin, G. N. (1988b), "Heat transfer at the surface of a triangular body in a hypersonic viscous gas flow," *Fluid Dynam.*, 23(1), 62.
96. Dudin, G. N. (1991a), "Three-dimensional boundary layer on a plane delta wing in the intermediate hypersonic interaction regime," *Fluid Dynam.*, 26(3), 409.

97. Dudin, G. N. (1991b), "Investigation of the three-dimensional boundary layer on the leeward side of a delta wing in the regime of viscous interaction with a hypersonic flow," in: *Abstracts of the 7th All-Union Congress on Theoretical and Applied Mechanics* [in Russian], Moscow, p. 142.
98. Dudin G. N. (1991b) Russian National Comitet on Theoretical and Appl. Mechan. Institute Problem Mechan., Moscow 15–21, VIII, 1991.
99. Dudin, G. N. (1992), "Calculation of the three-dimensional boundary layer on the leeward side of a delta wing of finite length in the hypersonic flow in viscous interaction regime," *Fluid Dynam.*, 27(2), 251.
100. Dudin, G. N. (1995), "Supercritical regimes of hypersonic flow past a planar yawed delta wing," *Fluid Dynam.*, 30(4), 615.
101. Dudin, G. N. (1997), "On the existence of self-similar solutions in the supercritical region on a delta wing in a hypersonic flow," *Fluid Dynam.*, 32(2), 282.
102. Dudin, G. N. (1998), "Influence of intense surface cooling on hypersonic viscous flow past a delta wing," *Fluid Dynam.*, 33(4), 512.
103. Dudin, G. N. (2000), "Gas injection from the surface of a triangular plate in a hypersonic flow," *Fluid Dynam.*, 35(1), 101.
104. Dudin G. N. (2001) "About Masstransfer on Triangular Wing with Region of Supercritical Interaction in Boundary Layer," Uch. Zap. TsAGI 32(37), pp. 22–33.
105. Dudin G. N. (2002) "Masstransfer on Triangular Wing in Hypersonic Flow," *Fluid Dynam* 37(2), p. 178.
106. Dudin, G. N. and I. I. Lipatov (1985), "Supersonic flow over a triangular wing," *J. Appl. Mech. Techn. Phys.*, 26(3), 391.
107. Dudin, G. N. and D. O. Lyzhin (1983), "A method of calculating a regime of strong viscous interaction on a delta wing," *Fluid Dynam.*, 18(4), 596.
108. Dudin, G. N. and V. Ya. Neiland (1976), "Law of transverse sections for the three-dimensional boundary layer on a thin wing in a hypersonic stream," *Fluid Dynam.*, 11(2), 232.
109. Dudin G. N. and V. Ya. Neiland (2002) "About one limiting solution for flow near triangular wing with supercritical region of flow," *Fluid Dynam.* 37(6).
110. Dunawant, J. C., K. Y. Narayan, and G. D. Walberg (1976), "A survey of leeside flow and heat transfer on delta planform configurations," AIAA Paper No. 118.
111. Einstein A. (1905) "Zur Electrodinamik Bewegter Körper," Ann. d. Physik, 17, 891.
112. Eliseev, V. I. (1972), "Boundary layer theory with suddenly changing boundary conditions," *Fluid Dynam.*, 7(4), 587.
113. El'kin, Yu. G. and V. Ya. Neiland (1965), "Calculation of the parameters of laminar separation zones," *Inzh. Zh.*, 5(5).
114. El'kin, Yu. G., V. Ya. Neiland, and L. A. Sokolov (1963), "Base pressure behind a wedge in a supersonic flow," *Inzh. Zh.*, 3(2).
115. Ellinwood, J. W. and H. Mirels (1968), "Hypersonic viscous interaction theory for slender axisymmetric," AIAA Paper No. 1.
116. Emmons, H. W. and D. C. Leigh (1954), "Tabulation of the Blasius function with blowing and suction," *Aeronaut. Res. Council*, Current Paper No. 157.
117. Erdos, J. and A. Pallone (1962), "Shock-boundary layer interaction and flow separation," *Proc. Heat Transfer Fluid. Mech. Inst.*

118. Ferri, A. 1940, "Experiments at supersonic speed on a biplane of the Busemann type," British R.T.P. Trans. No. 1407, Ministry Aircraft Production (from *Atti di Guidonia*, No. 37–38, p. 517).
119. Fershbein and Kruppa (1986) "About longitudal diffusion near a point of discontinuity in the catalytic activity of a surface," *Fluid Dynam.* 6(4).
120. Fletcher, L. S., D. G. Briggs, and R. H. Page (1970), "A review of heat transfer in separated and reattached flows," AIAA Paper No. 767.
121. Frankl, F. L. (1947), "Supersonic jet outflow from a vessel with plane walls," *Dokl. Akad Nauk SSSR*, 58(3).
122. Friedrichs, K. O. (1953), *Special Topics in Fluid Dynamics*, New York University Press, New York.
123. Friedrichs, K. O. (1955), "Asymptotic phenomena in mathematical physics," *Bull. Am. Soc.*, 61(6), 485.
124. Gadd, G. E. (1953), "Interaction between wholly laminar or wholly turbulent boundary layers and shock waves strong enough to cause separation," *J. Aeron. Sci.*, 20(11), 729.
125. Gadd, G. E. (1957a), "An experimental investigation of heat transfer effects on boundary layer separation in supersonic flow," *J. Fluid Mech.*, 2(2), 105.
126. Gadd, G. E. (1957b), "A theoretical investigation of laminar separation in supersonic flow," *J. Aeronaut. Sci.*, 24(10), 759.
127. Gadd, G. E. (1960), "Boundary layer separation in the presence of heat transfer," AGARD Report No. 280.
128. Gadd, G. E. and D. W. Holder (1959), "The behavior of supersonic boundary layers in the presence of shock waves," IAS Paper No. 138.
129. Gadd, G. E., D. W. Holder, and J. D. Regan (1954), "An experimental investigation of the interaction between shock waves and boundary layers," *Proc. Roy. Soc., London. Ser. A*, 226.
130. Gershbein, É. A. and V. Yu. Kazakov (1988), "Determining the rate constants of heterogeneous catalytic reactions in experimental and theoretical investigations of gas flows over plane surfaces," *Teplofiz. Vys. Temp.*, 26(1).
131. Gershbein, É. A., V. Yu. Kazakov, and V. S. Shchelin (1985), "Hypersonic viscous shock layer on the surface with jumpwise variation of the catalytic activity," *Teplofiz. Vys. Temp.*, 23(5).
132. Gershbein, É. A., V. Yu. Kazakov, and G. A. Tirskii (1986), "Development of the laminar boundary layer behind a point of discontinuity in the catalytic activity of a surface," *Teplofiz. Vys. Temp.*, 24(6).
133. Gladkov, A. A., O. Yu. Polyanskii, V. P. Agafonov, and V. K. Vertushkin (1972), *Nonequilibrium Physical and Chemical Processes in Aerodynamics* [in Russian], Mashinostroenie, Moscow.
134. Glick, H. S. (1962), "Modified Crocco–Lees mixing theory for supersonic separated and reattaching flows," *J. Aeron Sci.*, 29(10), 1238.
135. Gogish, L. V. and G. Yu. Stepanov (1966), "Calculating base pressure in two-dimensional supersonic flows," *Fluid Dynam.*, 1(3), 72.
136. Gogish, L. V., V. Ya. Neiland, and G. Yu. Stepanov (1975), "Theory of two-dimensional separated flows," in: *Advances in Science and Engineering. All-Union*

*Institute of Science and Technical Information. Fluid Mech. Series*, Vol. 8 [in Russian], Moscow.

137. Goldstein, S. (1930), "Concerning some solutions of boundary layer equations in hydrodynamics," *Proc. Camb. Phyl. Soc.*, 26(1), 1.
138. Goldstein, S. (1931), "Concerning some solutions of the boundary layer equations," *Philos. Mag.*, 26(1).
139. Goldstein, S. (1948), "On laminar boundary-layer flow near a position of separation," *Quart. J. Mech. Appl. Math.*, 1(43).
140. Goldstein, S. (1960), "Flow of an incompressible viscous fluid along a semiinfinite flat plate," in: *Lectures on Fluid Mechanics*, Wiley, New York.
141. Golubinskii, A. I., G. I. Maikapar, and V. Ya. Neiland (1973), "New results of the study of separated flows," in: P. K. Chang, *Separation of Flows*, Vol. III [Russian translation], Mir, Moscow.
142. Gorenbukh, P. I. and I. G. Kozlova (1973), "Experimental investigation of upstream disturbance propagation in the strong viscous interaction regime," *Uch. Zap. TsAGI*, 4(2).
143. Gorislavskii, V. S. and Z. A. Stepchenkova (1971), "Experimental investigation of separation zones on a flat plate in a hypersonic flow," *Uch. Zap. TsAGI*, 2(5).
144. Gosman, A. D., W. M. Pun, A. K. Runchal, D. B. Spalding, and M. Wolfstein (1969), *Heat and Mass Transfer in Recirculating Flows*, Academic Press, New York.
145. Govorukhin, V. N. and V. G. Tsibulin (1997), *Introduction to Maple. Mathematical Package for Everybody* [in Russian], Mir, Moscow.
146. Grange, J. M., J. M. Klineberg, and L. Lees (1967), "Laminar boundary-layer separation and near wake-flow for a smooth blunt body at supersonic and hypersonic speeds," *AIAA J.*, 5(6), 1089.
147. Greber, J., R. I. Hakkinen, and L. Trilling (1957), "Some problem of laminar boundary layer shock wave interaction," *Heat Transfer Fluid Mech. Inst. Preprint*, Stanford.
148. Guiraud, J. P. (1966), *Topics in Hypersonic Flow Theory*, Academic Press, New York.
149. Hakkinen, R. J. and N. Rott (1965), "Similar solutions for merging shear flows," *AIAA J.*, 3(8), 1553.
150. Hakkinen, R. J., I. Greber, L. Trilling, and S. S. Abarbanel (1959), "The interaction of an oblique shock wave with a laminar boundary layer," *NASA Memo.* No. 2-19-59 W.
151. Hama, F. R. (1966), "Experimental investigations of wedge base pressure and lip-shock," *Jet Propulsion Laboratory, California Institute of Technology*, TN No. 32-1033.
152. Hammit, A. G. (1958), "The interaction of shock waves and turbulent boundary layers," *J. Aeronaut. Sci.*, 25(6), 345.
153. Hayes, W. D. and R. F. Probstein (1959), "Viscous hypersonic similitude," *J. Aeron. Sci.*, 26(12), 815.
154. Hayes, W. D. and R. F. Probstein (1966), *Hypersonic Flow Theory*, Academic Press, New York.
155. Holt, M. (1966), "Separation of laminar boundary-layer flow past a concave corner," *AGARD CP*, 1(4).
156. Holt, B. and J. C. S. Meng (1968), "The calculation of base flow and near wake properties by the method of integral relation," IAS Paper No. RE 65.

157. Howarth, L. (1948), "The propagation of steady disturbances in a supersonic stream bounded on one side by a parallel subsonic stream," *Proc. Cambridge Philos. Soc.*, 44(3).
158. Hunt, J. C. R. (1971), "A theory for the laminar wake of a two-dimensional body in a boundary layer," *J. Fluid Mech.*, 49(1), 159.
159. Imai, I. (1957), "Second approximation to the laminar boundary layer flow over a flat plate," *J. Aeron. Sci.*, 24.
160. Inger, G. R. and G. A. Gaitatzes (1971), "Strong blowing into supersonic laminar flows around two-dimensional and axisymmetric bodies," *AIAA J.*, 9(3), 436.
161. Jaffe, N. A. (1970), "The laminar boundary layer with uniform injection of a foreign gas," *Proc. Roy. Soc., London. Ser. A*, 317(1530).
162. Kaplun, S. (1954), "The role of coordinate systems in boundary-layer theory," *Z. Angew. Math. Phys.*, 5(2), 111.
163. Karabalaev, A. H., Lipatov I. I (1996) "Nonlinear process in laminar boundary layers initiated by discontinuos boundary conditions," AIAA Paper. No. 2020.
164. Kassoy, D. R. (1974), "A resolution of the blow-off singularity for similarity flow on a flat plate," *J. Fluid Mech.*, 62(1), 145.
165. Kazakov, A. V. (1983), "Interaction of an unsteady three-dimensional boundary layer with a hypersonic flow near a rapidly heated section of the surface," *Fluid Dynam.*, 18(5), 712.
166. Kirk, F. N. (1959), "An approximate theory of base pressure in two-dimensional flow at supersonic speeds," *RAE TN Aero.*, No. 2377 (AD-233-651).
167. Kiya, M. and M. Arie (1973), "A perturbation analysis of the laminar far wake behind a symmetrical two-dimensional body in a uniform shear flow," *J. Fluid Mech.*, 61(2), 305.
168. Klineberg, J. M. and L. Lees (1969), "Theory of laminar viscous–inviscid interaction in supersonic flow," AIAA Paper No. 7.
169. Klineberg, J. M. and J. L. Steger (1974), "On laminar boundary-layer separation," AIAA Paper, No. 94.
170. Kogan, M. N., V. S. Galkin, and O. G. Fridleneder (1976), "Stresses in gases due to temperature and concentration in homogeneities. New types of free convection," *Usp. Fiz. Nauk*, 119(1).
171. Korst H. H., Page R. H., Childs M. E. (1955) "A theory for base pressure in transonic and supersonic flow," Univ. of Illinois, ME-CN-392-2. OSRTN 55–89.
172. Korst H. H., Page R. H., Childs M. E. (1956) "A Theory for base pressures in transonic and supersonic flow," *J. Appl. Mech.* V.23, No. 4.
173. Kovalenko A. A. (1974) "Investigation of separation of boundary layer for strong interaction with hypersonic flow of gase," *Uch. Zap. TsAGI* 5(6), p. 39.
174. Kozlova, I. G. and V. V. Mikhailov (1970), "Strong viscous interaction on triangular and slipping wings," *Fluid Dynam.*, 5(6), 982.
175. Kozlova, I. G. and V. V. Mikhailov (1971), "Effect of boundary-layer perturbations on hypersonic flows with a viscous interaction," *Fluid Dynam.*, 6(4), 593.
176. Kravtsova, M. A. (1993), "Numerical solution for s criss-cross interaction problem," in: *Proceedings of the International Workshop on Advances in Analytical Methods in Aerodynamics, Poland*, 1993, p. 42.

177. Krupa, V. G. and G. A. Tirskii (1988), "Asymptotic analysis of convective diffusion in problems with a discontinuity in the catalytic properties of the surface around which the flow takes place," *Appl. Math. Mekh.*, 52(3), 348.
178. Kuo, Y. H. (1952), "On the flow of an incompressible viscous fluid past a flat plate at moderate Reynolds numbers," *J. Math. Phys.*, 32(2–3), 83.
179. Lachmann, G. V. (1961), *Boundary Layer and Flow Control*, Vols. I and II, Pergamon, London.
180. Ladyzhenskii, M. D. (1964), "On three-dimensional hypersonic flow past slender airfoils," *Apll. Math. Mech.*, 28(5), 1017.
181. Ladyzhenskii, M. D. (1965), "Strong interaction between the boundary layer and the inviscid flow past a triangular wing," *Apll. Math. Mech.*, 29(4), 760.
182. Lagerstrom, P. A. (1957), "Note on the preceding two papers," *J. Math. Mech.*, 6(5), 605.
183. Lagerstrom, P. A. and J. D. Cole (1955), "Examples illustrating expansion procedures for Navier–Stokes equations," *J. Rat. Mech. Anal.*, 4(6), 817.
184. Landau, L. D. and E. M. Lifshitz, (1944), Moscow-Leningrad, p. 624.
185. Landau, L. D. and E. M. Lifshitz (1959), *Fluid Mechanics*, Pergamon, London.
186. Lapin, Yu. V., L. G. Loytsianskii, Yu. P. Lun'kin, V. Ya. Neiland, V. V. Sychev, and G. A. Tirskii (1970), *Mechanics of Viscous Liquids and Gases. Theory of Laminar and Turbulent Boundary Layers. Mechanics in USSR for Fifty Years*, Vol. II [in Russian], Nauka, Moscow.
187. Laytsyanskii, L. G. (1962), *Laminar Boundary Layer* [in Russian], Fizmatfiz, Moscow.
188. Laytsyanskii, L. G. (1966), *Mechanics of Liquids and Gases*, Pergamon, Oxford.
189. Lee, R. S. and H. K. Cheng (1969), "On the outer-edge problem of a hypersonic boundary layer," *J. Fluid Mech.*, 38(1), 161.
190. Lees, L. (1949), "Interaction between the laminar boundary layer over a plane surface and an incident oblique shock wave," Princeton University, Engineering Laboratory, Paper 143, No. 24.
191. Lees, L. J. (1953), "On the boundary-layer equations and their approximate solutions," *J. Aeron. Sci.*, 20(2).
192. Lees, L. J. (1955), *Proceedings of the 5th International Aeronautic Conference, Los Angeles*, Institute of Aeronautic Science, New York.
193. Lees, L. and B. L. Reeves (1964), "Supersonic separated and reattaching laminar flows. I. General theory and application to adiabatic boundary-layer/shock-wave interaction," *AIAA J.*, 2(11), 1907.
194. Lees, L. and R. L. Chapkis (1971), "Surface mass injection at supersonic and hypersonic speeds as a problem in turbulent mixing. Part II: Axially symmetric flow," *AIAA J.*, 9(6), 1067.
195. Levin, V. A. (1973), "Strong blowing on the surface of a body immersed in a supersonic gas stream," *Fluid Dynam.*, 8(5), 760.
196. Levin, V. A., I. I. Lipatov, and V. Ya. Neiland (1980), *Asymptotic Methods for Studying the Problems of Flow Past Bodies with Intense Mass Transfer* [in Russian], Znanie, Kiev.

197. Lew, H. G. (1949), "On the compressible boundary layer over a flat plate with uniform suction," in: J. W. Edwards, *Reissner Annivera*, Vol. 1, Ann Arbor.
198. Lewis, J. E., T. Kubota, and L. Lees (1968), "Experimental investigation of supersonic laminar, two-dimensional boundary-layer separation in a compression corner with and without cooling," *AIAA J.*, 6(1), 7.
199. Li, T. Y. and H. T. Namagatsu (1953), "Shock-wave effect on the laminar skin friction of an insulated flat plate at hypersonic speeds," *J. Aeronaut. Sci.*, 20(5), 345.
200. Li, T. Y. and H. T. Namagatsu (1955a), "Similar solutions of compressible boundary-layer equations," *J. Aeronaut. Sci.*, 22(9), 607.
201. Li, T. Y. and H. T. Namagatsu (1955b), *Proceedings 4th Midwestern Conference on Fluid Mechanics*, Lafayette, IN.
202. Liepmann, H. W. (1946), "The interaction between boundary layer and shock waves in transonic flows," *J. Aeronaut. Sci.*, 13(12), 623.
203. Lighthill, M. J. (1949), "A technique for rendering approximate solutions to physical problems uniformly valid," *Phil. Mag.*, 7(40).
204. Lighthill, M. J. (1950), "Reflection at a laminar boundary layer of a weak steady disturbance to a supersonic stream, neglecting viscosity and heat conduction," *Quart. J. Mech. Appl. Math.*, 3(3), 303.
205. Lighthill, M. J. (1953), "On boundary layers and upstream influence. 2. Supersonic flows without separation," *Proc. Roy. Soc. London. Ser. A*, 217(1131), 478.
206. Lighthill, M. J. (1995), "Chapter 10," in: L. M. Brown, A. Pais, and B. Pippard (eds.), *Twentieth Century Physics*, Vol. 2, IOP Publishing, Bristol & Philadelphia, and American Institute of Physics, New York, p. 795.
207. Linn, J. and A. P. Rothmayer (1993), "Viscous flows past short-scaled humps and surface blowing," *Eur. J. Mech. B/Fluids*, 12(3).
208. Lipatov, I. I. (1976), "Flow in the vicinity of the point of beginning of intense suction of a laminar boundary layer in a supersonic flow," *Uch. Zap. TsAGI*, 7(2), 37.
209. Lipatov, I. I. (1977), "Boundary layer detachment on uniform gas injection into supersonic flow," *Tr. TsAGI*, No. 1864, 8.
210. Lipatov, I. I. (1979), "Effect of intense injection in a small region of a surface on the laminar boundary layer flow in a supersonic stream," *Tr. TsAGI*, No. 2080.
211. Lipatov, I. I. (1980a), "Three-dimensional flow over a small roughness in the weak hypersonic interaction regime," *Uch. Zap. TsAGI*, 11(2).
212. Lipatov, I. I. (1980b), "Three-dimensional laminar boundary-layer flow over a small roughness," *Uch. Zap. TsAGI*, 11(2).
213. Lipatov, I. I. (1983), "Interaction between the laminar boundary layer flow and the outer supersonic flow in the presence of tangential slot injection," *Tr. TsAGI*, No. 2190, 38.
214. Lipatov, I. I. (1987), "Distributed injection of a gas into a hypersonic flow," *J. Appl. Mech. Techn. Phys.*, 28(6), 855.
215. Lipatov, I. I. (1995), "Disturbance propagation in supersonic boundary layers," in: *Proceedings of the IUTAM Symposium*, Kluwer.
216. Lipatov I. I. and V. Ya. Neiland (1974) "Influence of strong suction on flow near separation point of laminar boundary layer in supersonic flow," *Fluid Dynam.* 9(7), p. 17.

217. Lipatov, I. I. and V. Ya. Neiland (1982), "Effect of sudden change of the plate surface velocity on the laminar boundary layer flow in a supersonic stream," *Uch. Zap. TsAGI*, 13(5).
218. Lipatov, I. I. and V. Ya. Neiland (1987), "Theory of unsteady separation and interaction of a laminar boundary layer," *Uch. Zap. TsAGI*, 18(1), 36.
219. Lo, A. K.-F. and P. A. Sullivan (1970), "A complete solution to the viscous–inviscid equations for the interaction of a laminar boundary layer with a corner expansion wave," AIAA Paper No. 807.
220. Lunev, V. V. (1959), "On the similarity of hypersonic viscous gas flows around slender bodies," *Appl. Math. Mech.*, 23(1), 273.
221. Lunev, V. V. (1975), *Hypersonic Aerodynamics* [in Russian], Masinostroenie, Moscow.
222. Lun'kin, V. V., F. D. Popov, T. Ya. Timofeeva, and Yu. P. Lipnitskii (1956), "Passage through singular points in numerically solving the problem of supersonic flow past a body," in: *Proceedings of Leningrad Polytechnical Institute*, No. 248 [in Russian], Mashinostroenie, Moscow.
223. Mahony, J. J. (1962), "An expansion method for singular perturbation problems," *J. Austral. Math. Soc.*, 2(4), 440.
224. Makofski, R. A. (1963), "A two-parameter method for shock wave laminar boundary layer interaction and flow separation," in: *Proceedings of the Heat Transfer and Fluid Mechanics Institute*, Stanford.
225. Mann, W. M. and R. G. Bradley (1963), *J. Astronaut. Sci.*, 10(1).
226. Manuilovich, S. V. (1983), "Free interaction of a three-dimensional boundary layer with an exterior irrotational flow," *Fluid Dynam.*, 18(5), 695.
227. Matveeva, N. S. and V. Ya. Neiland (1967), "The laminar boundary layer near a body corner point," *Fluid Dynam.*, 2(4), 42.
228. Matveeva, N. S. and V. Ya. Neiland (1970), "Strong injection from a body of finite length in a supersonic flow," *Uch. Zap. TsAGI*, 1(5), 13.
229. Matveeva, N. S. and V. V. Sychev (1965), "On the theory of strong interaction of the boundary layer with an inviscid hypersonic flow," *Appl. Math. Mech.*, 29(4), 770.
230. Meredith, F. W. and A. A. Griffith (1938), "Control with auxiliary power. Diffusers," in: *Modern Development in Fluid Dynamics*, Vol. 2, Oxford University Press, Oxford.
231. Messiter, A. F. (1970), "Boundary layer near the trailing edge of a flat plate," *SIAM J. Appl. Math.*, 18(1), 241.
232. Mikhailov, V. V., V. Ya. Neiland, and V. V. Sychev (1971a), "Disturbance propagation in viscous hypersonic flows," in: *Problems of Applied Mathematics and Mechanics* [in Russian], Nauka, Moscow.
233. Mikhailov, V. V., V. Ya. Neiland, and V. V. Sychev (1971b), "The theory of viscous hypersonic flow," *Annu. Rev. Fluid Mech.*, 3(371).
234. Minyatov, A. V. (1961), "Calculation of the base pressure in a supersonic flow past a body of revolution," *Izv. Akad Nauk SSSR. Otd. Tekhn. Nauk. Mekh. Mash.*, No. 3.
235. Nagakura, T. and H. Naruse (1957), "An approximate solution of the hypersonic laminar boundary-layer equations and its application," *J. Phys. Soc. Japan*, 12(11), 1298.
236. Nash, J. F. (1963), "An analysis of two-dimensional base flow including the effect of approaching boundary layer," *ARC Rep.*, No. 33.

237. Neiland, V. Ya. (1963), "Hypersonic viscous flow over a flat plate at an angle of attack," *Inzh. Zh.*, 3(3).
238. Neiland, V. Ya. (1964), "Effect of heat transfer and turbulent flow in the mixing region on separation zone parameters," *Inzh. Zh.*, 4(1).
239. Neiland, V. Ya. (1965a), "Calculation of the separation zone parameters and the base pressure in supersonic flow past bodies," *Inzh. Zh.*, 5(1).
240. Neiland, V. Ya. (1965b), "Solving the equations of a laminar boundary layer under arbitrary initial conditions," *Appl. Math. Mech.*, 30(4), 807.
241. Neiland V. Ya. (1966) "About solution boundary layer equation with arbitrary initial condition," *Appl. Math. Mech.* 30(4).
242. Neiland, V. Ya. (1968a), "Supersonic viscous flow near a separation point," in: *Abstracts of the 3rd All-Union Congress on Theoretical and Applied Mathematics* [in Russian], Nauka, Moscow, p. 224.
243. Neiland, V. Ya. (1968b), "Laminar boundary layer near a body corner," in: *Proceedings of XVIII Astronaut. Congress*, Belgrad, 1967, Vol. 3, Pergamon, Oxford.
244. Neiland, V. Ya. (1969a), "Theory of laminar boundary layer separation in supersonic flow," *Fluid Dynam.*, 4(4), 33.
245. Neiland, V. Ya. (1969b), "Asymptotic theory for calculating heat transfer near the corner of a body," *Fluid Dynam.*, 4(5), 33.
246. Neiland, V. Ya. (1970a), "Asymptotic theory of plane steady supersonic flows with separation zones," *Fluid Dynam.*, 5(3), 372.
247. Neiland, V. Ya. (1970b), "Propagation of perturbations upstream with interaction between a hypersonic flow and a boundary layer," *Fluid Dynam.*, 5(4), 559.
248. Neiland, V. Ya. (1971a), "Flow behind the boundary layer separation point in a supersonic stream," *Fluid Dynam.*, 6(3), 378.
249. Neiland, V. Ya. (1971b), "The asymptotic theory of the interaction of a supersonic flow with a boundary layer," *Fluid Dynam.*, 6(4), 587.
250. Neiland, V. Ya. (1972), "Gas injection into a hypersonic flow," *Uch. Zap. TsAGI*, 3(6).
251. Neiland, V. Ya. (1973), "Boundary layer separation on a cooled body and interaction with a hypersonic flow," *Fluid Dynam.*, 8(6), 931.
252. Neiland, V. Ya. (1974a), "Asymptotic problems of theory of supersonic viscous flows," *Tr. TsAGI*, No. 1529.
253. Neiland, V. Ya. (1974b), "On the theory of interaction with boundary layer for separated two-dimensional and three-dimensional flows. Part 1. Three-dimensional flows," *Uch. Zap. TsAGI*, 5(2), 70.
254. Neiland, V. Ya. (1974c), "On the theory of interaction with the boundary layer for separated two-dimensional and three-dimensional flows. Part 2. Two-dimensional flows and delta wing," *Uch. Zap. TsAGI*, 5(3), 28.
255. Neiland, V. Ya. (1981), "Asymptotic theory of boundary layer separation and interaction with a supersonic gas flow," *Usp. Mekh.*, 4(2), 3.
256. Neiland, V. Ya. (1987), "Transcritical boundary layer interaction and separation," *Uch. Zap. TsAGI*, 18(2), 30.
257. Neiland, V. Ya. and N. I. Kukanova (1965), "Investigation of flows with separation zones," *Rev. TsAGI*, No. 129.

258. Neiland, V. Ya. and L. A. Sokolov (1964), "Base pressure behind a wedge at an angle of attack in a supersonic flow," *Inzh. Zh.*, 4(2).
259. Neiland, V. Ya. and L. A. Sokolov (1967), "Calculation of turbulent separation zones," *Tr. TsAGI*, No. 1057.
260. Neiland, V. Ya. and L. A. Sokolov (1977), "Laminar boundary layer on a cone at an angle of attack in a supersonic flow," *Tr. TsAGI*, No. 1812.
261. Neiland, V. Ya. and V. V. Sychev (1966), "Asymptotic solutions of the Navier–Stokes equations in regions with large local perturbations," *Fluid Dynam.*, 1(4), 29.
262. Neiland, V. Ya. and G. I. Taganov (1961), "On the nature of return flows in separation zones," *Inzh. Zh.*, 1(2).
263. Neiland, V. Ya. and G. I. Taganov (1963a), "On the configuration of forward separation zones in symmetric supersonic flow past bodies," *Inzh. Zh.*, 3(2).
264. Neiland, V. Ya. and G. I. Taganov (1963b), "Forward separation zone in nonsymmetric supersonic flow past a spiked body," *Inzh. Zh.*, 3(3).
265. Nielsen, J. N. (1965), "Calculation of laminar separation with free interaction by the method of integral relation," AIAA Paper No. 65.
266. Nielsen, J. N., L. L. Lynes, and F. K. Goodwin (1968), "Theory of laminar separated flows on flared surfaces including supersonic flow with heating and cooling," *AGARD CP*, No. 84.
267. Nikolaev, V. S. (1967), "Maximum L/D profile in viscous hypersonic flow," *Fluid Dynam.*, 2(6), 88.
268. Olsson, G. R. and A. F. Messiter (1969), "Hypersonic laminar boundary layer approaching the base of a slender body," *AIAA J.*, 7(7), 1261.
269. Oseen, C. W. (1910), "Über die stokes'sche Formel und über eine verwandte Aufgabe in der Hydrodynamik," *Ark. Math. Atronom. Fys.*, 6(29).
270. Pearson, H., J. B. Holliday, and S. F. Smith (1958), "Theory of the cylindrical ejector propelling nozzle," *J. Roy. Aeron. Soc.*, 62(574), 746.
271. Petrova, L. I. and O. S. Ryzhov (1969), "Hypersonic viscous gas flow past half-bodies," *Fluid Dynam.*, 4(1), 43.
272. Petukhov, I. V. (1964), "Numerical calculation of two-dimensional boundary layer flows," in: *Numerical Methods for Solving Differential and Integral Equations and Quadrature Formulas* [in Russian], Nauka, Moscow.
273. Poincaré, H. (1892), *Les méthodes nouvelles de la mécanique céleste*, Vol. 1, Dover, New York.
274. Popov, D. A. (1975), "Problem with discontinuous boundary conditions and the diffusion boundary layer approximation," *Appl. Math. Mech.*, 39(1), 98.
275. Prandtl, L. (1904), "Über Flüssigkeitsbewegung bei sehr kleiner Reibung," in: *Int. Math. Kongr.*, Heidelberg, S. 484.
276. Prandtl, L. and O. Tietjens (1929, 1931), *Hydro- und Aeromechanik, Bd. 1, 2*, Springer, Berlin.
277. Reeves, B. L. and L. Lees (1965), "Theory of laminar near wake of blunt bodies in hypersonic flow," *AIAA J.*, 3(11), 2061.
278. Rom, J. (1962), "Theory for supersonic two-dimensional laminar base-type flows using Crocco–Lees mixing concepts," *J. Aeronaut. Sci.*, 29(8), 963.

279. Ruban, A. I. (1982), "Asymptotic theory of short separation regions on the leading edge of a slender airfoil," *Fluid Dynam.*, 17(1), 33.
280. Ruban, A. I. and V. V. Sychev (1973), "Hypersonic viscous gas flow past a low aspect ratio wing," *Uch. Zap. TsAGI*, 4(5).
281. Ryzhov, O. S. (1977), "Equations of the time-dependent boundary layer with self-induced pressure," *Dokl. Akad. Nauk SSSR*, 234(4).
282. Ryzhov, O. S. (1980), "On the unsteady three-dimensional boundary layer freely interacting with the external stream," *Appl. Math. Mech.*, 44(6), 739.
283. Ryzhov, O. S. and E. D. Terent'ev (1977), "On unsteady boundary layer with self-induced pressure," *Appl. Math. Mech.*, 41(6), 1024.
284. Schlichting, H. (1942), "Grenzschicht mit Absaugung und Ausblasen," *Luftfahrt Forschung*, 19(5).
285. Schlichting, H. (1968), *Boundary Layer Theory*, McGraw-Hill, New York.
286. Schneider, W. I. (1974), "Upstream propagation of unsteady disturbances in supersonic boundary layers," *J. Fluid Mech.*, 63(3), 465.
287. Sears, W. R. (ed.) (1955), *General Theory of High Speed Aerodynamics*, London.
288. Shen, S. F. (1952), "An estimate of viscosity effect on the hypersonic flow over an insulated wedge," *J. Math. Phys.*, 31(3), 192.
289. Shmanenkov, V. N. (1966), "Study of the laminar boundary layer behind a point of abrupt change of boundary conditions," *Fluid Dynam.*, 1(3), 76.
290. Siriex, M., J. Mirande, and J. Delery, "Recherches expérimentales fondamentales sur les écoulements séparés et applications," ONERA-(092).
291. Siriex, M., J. Mirande, and J. Delery (1966), "Expériences fondamentales sur le recollement turbulent d'un jet supersonique," *ONERA TD*, No. 326.
292. Smirnov, V. I. (1958), *Handbook of Higher Mathematics*, Vol. 2 [in Russian], Gostekhizdat, Moscow.
293. Smith, F. T. (1973), "Laminar flow over a small hump on a flat plate," *J. Fluid Mech.*, 57(4), 803.
294. Smith, F. T. (1976), "Pipe flows distorted by non-symmetric indentation or branching," *Mathematika (London)*, 23(1), 45.
295. Smith, F. T. (1977), "The laminar separation of an incompressible fluid streaming past a smooth surface," *Proc. Roy. Soc. London. Ser. A*, 356(1687), 443.
296. Smith, F. T. (1982), "On the high Reynolds number theory of laminar flows," *IMA J. Appl. Math.*, 28(3), 207.
297. Smith, F. T. and K. Stewartson (1973), "Plate injection into a separated supersonic boundary layer," *J. Fluid Mech.*, 58(1), 143.
298. Smith, F. T. and A. G. Walton (1998), "Flow past a two- or three-dimensional steep-edges roughness," *Proc. Roy. Soc., London. Ser. A*, 454.
299. Smith, F. T., R. I. Sykes, and P. W. M. Brighton (1977), "A two-dimensional boundary layer encountering a three-dimensional hump," *J. Fluid Mech.*, 83(1), 163.
300. Sokolov, L. A. (1975), "On the asymptotic theory of two-dimensional laminar boundary layer flows with a temperature discontinuity at the body surface," *Tr. TsAGI*, No. 1650, 18.
301. Sokolov, L. A. (1980), "Boundary layers with self-induced pressure on a surface in motion," *Uch. Zap. TsAGI*, 11(5).

302. Springer, S. G. and T. J. Pedley, (1973), "The solution of heat-transfer problems by Wiener–Hopf technique. 1. Leading edge of hot film," *Proc. Roy. Soc., London. Ser. A*, 333(1594).
303. Stewart, D. A., J. V. Rakich, and M. J. Lafranco (1981), "Catalytic surface effects experiment on Space Shuttle," AIAA Paper No. 1143.
304. Stewartson, K. (1954), "Further solution of the Falkner–Skan equation," *Proc. Cambridge Phil. Soc.*, 50(3), 454.
305. Stewartson, K. (1955a), "On the motion of a flat plate at high speed in a viscous compressible fluid. I. Impulsive motion," *Proc. Cambridge Phil. Soc.*, 51(1), 202.
306. Stewartson, K. (1955b), "On the motion of a flat plate at high speed in a viscous compressible fluid. I. Steady motion," *J. Aeronaut. Sci.*, 22(5), 303.
307. Stewartson, K. (1958), "On the motion of a sphere along the axis of a rotating fluid," *Quart. J. Mech. Appl. Math.*, 11(1), 39.
308. Stewartson, K. (1968), "On the flow near the trailing edge of a flat plate," *Proc. Roy. Soc. London. Ser. A*, 306(1486).
309. Stewartson, K. (1969), "On the flow near the trailing edge of a flat plate," *Mathematika (London)*, 16(1), 106.
310. Stewartson, K. (1970a), "On laminar boundary layers near corners," *Quart. J. Mech. Appl. Math.*, 23(2), 137.
311. Stewartson, K. (1970b), "Is the singularity at separation removable?" *J. Fluid Mech.*, 44(2), 347.
312. Stewartson, K. and P. G. Williams (1969), "Self-induced separation," *Proc. Roy. Soc. London. Ser. A*, 312(1509), 181.
313. Stokes, G. G. (1851), "On the effect of the internal friction of fluid on the motion of pendulums," *Trans. Camb. Phil. Soc.*, 9.
314. Stollery, J. L. (1972), "Laminar and turbulent boundary layer studies at hypersonic speeds," ICAS Paper No. 09.
315. Stratford, B. S. (1954), *ARC Pep. Memo.,* No. 3002, London.
316. Sychev, V. V. (1972), "Laminar separation," *Fluid Dynam.*, 7(3), 407.
317. Sychev, V. V., A. I. Ruban, Vik. V. Sychev, and G. L. Korolev (1987), *Asymptotic Theory of Separated Flows* [in Russian], Nauka, Moscow.
318. Sykes, R. I. (1980), "On three-dimensional boundary layer flow over surface irregularities," *Proc. Roy. Soc., London. Ser. A*, 373.
319. Taganov, G. I. (1968), "Contribution to the theory of stationary separation zones," *Fluid Dynam.*, 3(5), 1.
320. Tagirov, R. K. (1961), "Determining the base pressure and temperature at sudden expansion of sonic and supersonic flows," *Izv. Akad. Nauk SSSR. Otd. Tekhn. Nauk. Mekh. Mash.*, No. 5.
321. Tagirov, R. K. (1966), "Influence of the initial boundary layer on base pressure," *Fluid Dyman.*, 1(2), 99.
322. Terrill, R. M. (1960), *Phil. Trans. Roy. Soc. A*, 253(55).
323. Thomas, P. D. (1969), "Flow over a finite plate with massive injection," *AIAA J.*, 7(4), 681.
324. Tom, A. (1928), "An investigation of fluid flow in two dimensions," *Aeron. Res. Council Repts. Memo.*, No. 1194.

325. Tsien, H. and H. Finston (1949), "Interaction between parallel streams of subsonic and supersonic velocities," *J. Aeronaut. Sci.*, 16(9), 515.
326. Tyler, R. D. and A. H. Shapiro (1953), "Pressure rise required for separation in interaction between turbulent boundary layer and shock wave," *J. Aeronaut. Sci.*, 20(12).
327. Van Dyke, M. (1962), "Second-order compressible boundary-layer theory with application to blunt bodies in hypersonic flow," in: F. D. Riddell (ed.), *Hypersonic Flow Research*, Academic Press, New York, p. 37.
328. Van Dyke, M. (1964). *Perturbation Methods in Fluid Dynamics*, Academic Press, New York.
329. Vasiliu, S. (1962), "Pressure distribution in regions of step-induced turbulent separation," *J. Aerospace Sci.*, 29(5), 596.
330. Veldman, A. E. P. (1976), *Boundary-Layer Flow past a Finite Flat Plate*, Rijksuniv., Groningen.
331. Vishik, M. I. and L. A. Lusternik (1957), "Regular degeneration and boundary layer for linear differential equations with a small parameter," *Usp. Mat. Nauk*, 5(59), 12.
332. Vishik, M. I. and L. A. Lusternik (1958), "Asymptotics of the solutions of boundary value problems for quasilinear differential equations," *Dokl. Akad. Nauk SSSR*, 121.
333. Voitkova, G. V. and V. V. Lunev (1991), "Discontinuous solution of boundary layer equations with a positive pressure gradient," *Fluid Dynam.*, 26(5), 680.
334. Wang, K. C. (1971), "On the determination of the zones of influence and dependence for three-dimensional boundary-layer equations," *J. Fluid Mech.*, 48(2), 397.
335. Wang, K. C. (1975), "Aspects of 'multitime initial-value problem' originating from boundary layer equations," *Phys. Fluids*, 18(8), 951.
336. Webb, W. H., R. J. Golik, F. W. Vogenitz, and L. Lees (1965), "A multimoment integral theory for the laminar supersonic near wake," *Proc. Heat Transfer Fluid Mech. Inst.*
337. Weinbaum, S. (1966), "Rapid expansion of a supersonic boundary layer and its application to a near wake," *AIAA J.*, 4(2), 217.
338. Weiss, R. F. and W. Nelson (1968), "Upstream influence of the base pressure," *AIAA J.*, 6(3), 466.
339. Werle, M. T., D. L. Dwoyer, and W. L. Hankey (1973), "Initial conditions for the hypersonic-shock/boundary-layer interaction problem," *AIAA J.*, 11(4), 525.
340. Whitehead, A. H., J. N. Hefner, and D. M. Rao (1972), "Lee-surface vortex effects over configurations in hypersonic flow," AIAA Paper No. 77.
341. Yalamanchili, R. V. S. and D. R. Jeng (1968), *19th Congress of the International Asronautic Federation*, New York.
342. Yasuhara, M. J. (1962), "Axisymmetric viscous flow past very slender bodies of revolution," *J. Aerospace. Sci.*, 29(6), 667.
343. Young, A. D. (1948), "Note on the velocity and temperature distributions attained with suction on a flat plate of infinite extent in compressible flow," *Quart. J. Mech. Appl. Math.*, 1(1), 70.
344. Zakkay, V. and T. Tani (1962), "Theoretical and experimental investigation of the laminar heat transfer downstream of a sharp corner," in: *Proceedings of the Fourth Congress of Applied Mechanics,* Vol. 4.
345. Zakkay, V., K. Toba, and T.-J. Kuo (1964), "Laminar, transitional, and turbulent heat transfer after a sharp convex corner," *AIAA J.*, 2(8), 1389.

346. Zavadskii, A. Yu. and G. I. Taganov (1968), "Calculation of asymmetric supersonic gas flow past an ellipsoid with forward separation zone," *Fluid Dynam.*, 3(6), 72.
347. Zhuk, V. I. (1980), "Formation of recirculation zones in the boundary layer on a moving surface," *Fluid Dynam.*, 15(5), 637.
348. Zhuk, V. I. and O. S. Ryzhov (1979a), "Boundary layer with self-induced pressure on a surface in motion," *Dokl. Akad. Nauk SSSR*, 248(2).
349. Zhuk, V. I. and O. S. Ryzhov (1979b), "Locally inviscid disturbances in a boundary layer with self-induced pressure," *Dokl. Akad. Nauk SSSR*, 263(1), 56.
350. Zubtsov, A. V. (1971), "Effect of a single roughness on the boundary layer flow," *Uch. Zap. TsAGI*, 2(1).
351. Zukoski, E. E. (1967), "Turbulent boundary layer separation in front of a forward-facing step," *AIAA J.*, 5(10), 1746.

# Index

## H